WELINGTON DE MELO -
Selected Works

Welington de Melo, the sailor
Courtesy of Maria José Pacífico

WELINGTON DE MELO - Selected Works

Edited by
Paolo Piccione

SBM
SOCIEDADE BRASILEIRA DE MATEMÁTICA

Springer

impa
Instituto de Matemática Pura e Aplicada

Welington de Melo
IMPA
Rio de Janeiro, Brazil

Editor
Paolo Piccione
Departamento de Matemática
Universidade de São Paulo
São Paulo, Brazil

ISBN 978-3-031-15830-8

Mathematics Subject Classification (2020): 37-06, 37E05, 37E10, 37C05, 37E20, 37F25, 37F31, 37F34, 37F44, 37F20, 37E45, 37C15, 37C20

This Springer imprint is published by the registered company Springer Nature Switzerland AG
The registered company address is: Gewerbestrasse 11, 6330 Cham, Switzerland

Selected Works of Outstanding Brazilian Mathematicians

Mathematics in Brazil is a very young endeavor. Regular research activities and graduate studies go back to the 1950s, when the biannual Brazilian Mathematical Colloquium was launched and Brazil joined the International Mathematical Union (IMU). The Instituto de Matemática Pura e Aplicada (IMPA) was founded in 1952. Creation of the SBM, Sociedade Brasileira de Matemática, would follow, in 1969.

Half a century later, Brazil has reached Group IV in the IMU's ranking and is responsible for about 1.8% of the world's mathematical production (according to MathSciNet). The Brazilian mathematical community is growing steadily, and has earned international reputation in several fields of research. While many great challenges remain ahead, one can certainly be proud of what has been achieved in such a short time.

This progress was largely due to a number of leading mathematicians who, in addition to their own fundamental contributions to Mathematics, created genuine "schools" around them that raised entire research fields in Brazil to the best international standards. Through this collection, the SBM and IMPA aim to highlight their scientific trajectories and to pay tribute to their pioneering efforts.

Editors
Hilário Alencar
César Camacho
André Nachbin
Marcelo Viana

Preface

Welington de Melo (Guapé, 1946 – Rio de Janeiro, 2016) was a gifted mathematician who had a huge impact on the development of research in the field of Dynamical Systems in Brazil. He was one of Jacob Palis' first students in the 1970s, and he is indelibly attached to the success of IMPA and its dynamical systems group.

Most of Welington's research was focused on one-dimensional dynamics, an area that has developed at an explosive rate since the 1970s, and which remains an interesting field with multiple connections with Hyperbolic Geometry, Teichmüller Theory, Ergodic Theory and Complex Dynamics. He followed this progress closely, and played a very important role in the growth of the field, as is shown in this volume. His book in collaboration with Sebastian van Strien, *One-dimensional dynamics*, remains an undisputed essential reference in the field.

The selected works here constitute a fairly representative sample of his contributions, starting with the work on structural stability of dynamical systems that resulted from his Ph.D. thesis, to the study of typical behavior in families of unimodal maps with Artur Avila (one of his students, who obtained the Fields Medal in 2014) and Misha Lyubich, and the deep studies on rigidity and renormalization theory for circle maps with Edson de Faria, Pablo Guarino (his last student) and Marco Martens.

The list of talented younger researchers he attracted under his umbrella over the years, such as Sebastian van Strien, Marco Martens, Edson de Faria and Alberto Pinto, is quite remarkable. Several of the beautiful papers in this volume are co-authored with them, which witnesses the importance of his legacy in the Brazilian and the international mathematical community. His collaboration with Sebastian van Strien was particularly long and productive, and four of their works can be appreciated in this volume.

The Brazilian Mathematical Society is proud to celebrate one of its most distinguished authors with this volume. My most sincere thanks go to all the people whose work and dedication have made the publication of this volume come true, and among them I want to mention Marcelo Viana, Hilário Alencar, Daniel Smania, and Martin Peters.

IME-USP, São Paulo, and SBM, Rio de Janeiro — *Paolo Piccione*
May, 2022.

Contents

Welington and his Odyssey

Daniel Smania

"Gray-eyed Athena sent them a favorable breeze, a fresh west wind, singing over the wine-dark sea."

- Homer, *Odyssey*

Welington de Melo (1946–2016) was a Brazilian mathematician. Brazilians often treat each other on a first-name basis, and curiously even more so with someone we admire and respect, so we will refer to him just as Welington.

Welington was born in Guapé, a small town (today it has a little more than 14000 souls) in the state of Minas Gerais, Brazil. The town lies on the beautiful margins of the Rio Grande river. Indeed the old town where Welington was born does not exist anymore: it was replaced by another one in higher ground in the 1950s since a huge dam was constructed in the river.

Due to the nature of his father's profession, a tax collector, his family moved quite often, living in cities in three Brazilian states until he finished high school [1]. As for many of his generation, his attraction to the sciences started during his (electrical) engineering undergraduate studies at the Federal University of Minas Gerais at Belo Horizonte, Minas Gerais.

At the end of his undergraduate years, he was settled to do graduate studies in physics, but it seems that destiny had other plans. He attended the Brazilian Colloquium of Mathematics in July 1969, at Poços de Caldas, Minas Gerais. This event, in its seventh edition, was quite modest by today's standards. There were minicourses by Brazilian mathematicians and by a few foreign participants, such as Paul Halmos and Jack Hale.

Daniel Smania
Departamento de Matemática, Instituto de Ciências Matemáticas e de Computação (ICMC), Universidade de São Paulo (USP), Avenida Trabalhador São-carlense, 400, São Carlos-SP, CEP 13566-590, Brazil.
e-mail: smania@icmc.usp.br
https://sites.icmc.usp.br/smania/

Left: Welington and Pedro Mendes at the social event of the Dynamical Systems Congress in Salvador, 1971. Courtesy of Alberto Verjovsky. Right: Antonio Gaspar Ruas, Welington and Maurício Peixoto, 1970s. Courtesy of Maria José Pacífico.

However, this event set the start of a new chapter for mathematics in Brazil, with the foundation of the Brazilian Mathematical Society (SBM – Sociedade Brasileira de Matemática) on July 24th, 1969. Indeed some states even extinguished their own mathematical societies on the very same day to strengthen SBM's influence from the beginning. Welington was in the right place, at the right time. Just an undergraduate student, he signed with other participants the founding act of SBM.

Another (more mature) participant was Elon Lages Lima (1929–2017). Elon had entered the ranks of the Institute of Pure and Applied Mathematics (IMPA – Instituto de Matemática Pura e Aplicada) in Rio de Janeiro in 1968, a few years after his participation in the University of Brasília had been cut short by the terrible political events in Brazil of the 1960s.

Welington attended a mini-course by Elon, and he was told that Jacob Palis (1940–) was very excited about working on the organization of a major event on dynamical systems and geometry to be held in Salvador, Bahia, in 1971. Some of the foreign researchers would visit IMPA before the event and there would be very intense research activities at IMPA in the coming years. Elon invited Welington to do his graduate studies there.

Shortly after his marriage to Gilza, Welington became a student at IMPA in January 1970, initially as a master's student. At the time, mathematics in Brazil was far from having the prominence it has today. Indeed higher study education in Brazil had a very late start when compared with many Latin American countries. Welington and his fellow students shared serious doubts about the worth of doing a Ph.D. in Brazil since they believed that most of the best mathematics by Brazilians had been done on visits abroad [20].

In another stroke of luck for Welington, in the 1960s a few Brazilians went to the United States to do graduate studies with major names in the mathematical landscape. Perhaps more surprisingly, given the precarious conditions, political atmosphere in Brazil and tempting offers to some of them in the US, many were willing to come back to work in Brazil.

Among them was Jacob Palis (1940–), who finished his Ph.D. under the supervision of Stephen Smale (1930–) in 1968. The connection with Smale, who had a very productive visit to IMPA in 1960 and got his Fields Medal in 1966 for his contributions to differential topology, was nurtured by Elon, a topologist, and Maurício Peixoto (1921–2019), the pioneer of dynamical systems research in Brazil [4].

Due to Welington's insistence on studying dynamical systems and his wish to participate in Palis' working seminar, Palis asked him to present a topic there [21]. He was successful and Welington became a Ph.D. student of Palis. Two weeks later, Welington was already pointing out typos in his papers [2]. Among the participants of this seminar were students such as Alberto Verjovsky (Peixoto's student) and Pedro Mendes (Palis's student), and researchers such as Sheldon Newhouse, Clark Robinson and Jorge Sotomayor. Ricardo Mañé (1948–1995) became one of Palis' students shortly after. It was a remarkable beginning for Palis' long career as an advisor, and Palis himself was gladly surprised by the success of his first students.

According to Palis [20], a fortuitous encounter after a bus ride in Rio secured a (by then) sizable sum of 150,000 dollars to organize the Salvador congress and visits of researchers to IMPA. Those resources and the well-known persistence of the organizers were a winning combination that allowed the event to get the likely unparalleled dimension it achieved at that time. Over the years, the influence of this event in the development of the school of dynamical systems in Brazil reached almost mythical proportions. J. Mather, S. Smale, R. Thom, E.C. Zeeman took part in the event.

His work

Welington gave a talk on his thesis in the event, and it was published shortly after in the prestigious *Inventiones Mathematicae* [9]. The result was at the center of essential questions at the time: to understand when dynamical systems are structurally stable. He shows that Axiom A C^1-diffeomorphisms with the strong transversality condition are C^1-structurally stable, generalizing the result by J.W. Robbins for C^2-diffeomorphisms.

In 1972, after finishing his Ph.D., Welington visited the University of California at Berkeley, where he got in touch with S. Smale, C. Pugh (1940–), and R. Bowen (1947–1978). He also visited the University of Warwick (England) and IHES (France) [1]. He got a position at IMPA in 1974 and became a full professor in 1980.

One of Welington's earliest and most lasting contributions was his book *Introdução aos sistemas dinâmicos* written with Jacob Palis in 1978. It was translated into many languages.

But one can fairly say that his most important contributions were to one-dimensional dynamics, which studies maps acting on either an interval or the circle. Those seem to be at first glance quite simple, but in one of the most surprising developments in dynamics, early on it became clear that one-dimensional maps can have a very complicated behavior. On the other hand, during the 1970s, 1980s and 1990s,

Maria José Pacífico, Yves Lequain, Welington, and Aron Simis. Circa 1975. Courtesy of Maria José Pacífico.

an impressively diverse and rich set of tools were developed to study them, using ergodic theory, complex dynamics, quasi-conformal maps, and hyperbolic geometry.

It seems that Smale and Rene Thom (1923–2002) were the first to direct Welington's attention to the field, however his frequent visits to IHES in the 1970s and 1980s were likely the determinant factor. France was the center of effervescent activity in one-dimensional dynamics. Michael Herman was working on the linearization of circle diffeomorphisms [22], Dennis Sullivan (1941–), one of the Welington's heroes and a main influence, held a position in IHES from 1974 to 1997, and he induced a rebirth of complex dynamics by introducing quasi-conformal methods. The rise of personal computers allowed the first images of Julia sets and related sets, and, in the early 1980s, in nearby Orsay, Douady and Hubbard were working on quadratic polynomials and the Mandelbrot set [15].

On the other side of the Atlantic, Milnor and Thurston were studying interval maps, and Ricardo Mañé and Paulo Sad (IMPA) were studying the structural stability of rational maps, that in the end gave origin to an influential work by Ricardo Mañé, Paulo Sad and Sullivan.

Feigenbaum and Coullet-Tresser observed the phenomenon of *universality* on families of unimodal maps (maps with just one turning point) and they proposed a series of intriguing conjectures to explain these observations. The key concept was *renormalization*, an idea borrowed from statistical mechanics. The renormalization operator is an infinite-dimensional dynamical system acting *on* one-dimensional dynamical systems. We note that Oscar E. Lanford III was also a professor at IHES from 1982 to 1989. Lanford made some of the earliest contributions in the

Left: Welington and Sheldon Newhouse. 1970s. Courtesy of Maria José Pacífico. Right: Sebastian van Strien and Welington. 1980s. Courtesy of Sebastian van Strien.

field, including a partial computer-assisted proof of the Feigenbaum–Coullet–Tresser conjectures.

Welington visited IHES and CUNY numerous times from 1977 to 1995 (Sullivan stayed part of each year on each side of the Atlantic), which was part of constant exchanges between young and senior researchers between IHES-Orsay, New York, and Rio. These exciting times had a huge impact at IMPA, that led to far more links of the dynamical system group, with early strong connections in the US, and with the French mathematicians that we see today.

His fascination for the subject played a crucial role in his future work. Firstly he just want to understand and clarify Sullivan contributions on renormalisaton theory, which was already a quite demanding task given the multiplicity of new ideas involved. In his paper, Sullivan acknowledges that Welington "has studied the proof relentlessly and provided crucial assists at various difficult points" [23].

Indeed one can argue that for Welington one of the motivations for the daunting task of writing the 1993 book *One-dimensional dynamics* [10] with Sebastian van Strien was to give a full account of Sullivan's results on renormalization and other recent developments. The book has more than 600 pages, and its last chapter, on Sullivan's results, takes one-fifth of its length and it is far more detailed than Sullivan's article. It is a testament to Welington's character and his deep admiration for D. Sullivan that he spent uncountable hours working in this chapter. The book

Welington, Sebastian van Strien and Marco Martens, in a conference celebrating the 60th birthday of S. van Strien in Leiden, 2016. Courtesy of Edson Vargas.

contains an up-to-date account of the progress in the field achieved by many other authors. It had a huge impact.

This author did his undergraduate degree in the small city of Rio Claro, São Paulo, in the middle of the 1990s. Chaos, fractals, and universality were a constant presence in scientific vulgarization (even in Brazil) and James Gleick's *Chaos* was popular in the US. I found the ambitious, encyclopedic and thick yellow-covered book by de Melo and van Strien and tried to read something of it during my summer break, without any success. I was glad that not too far away, in Rio de Janeiro, there was someone that could help me to unveil those mysteries, and soon I applied to the graduate program of IMPA.

This book was also part of Welington's most prolific collaboration. Sebastian van Strien (1956–) was a student of Floris Takens (1940–2010), a long-time friend and collaborator of Jacob Palis, when he met Welington and Palis at a conference in Warwick in 1980. He managed to read Palis and Welington book, even though it was only available in Portuguese at that time. At the end of the 1970s Palis and Welington had been working on the moduli of stability of two-dimensional diffeomorphisms, a natural topic arising from Welington's Ph.D. thesis. They were both very interested in van Strien's work on the subject. This became a collaboration with Palis and, later on, with Welington [11]. Besides the book, they published six papers together.

Most of their collaborations were on one-dimensional dynamics, and that subject would remain his main topic of research for the rest of his life. Indeed, he considered his work with Marco Martens and van Strien [18] one of his greatest achievements. Their results bring to real one-dimensional maps (maps on the interval and the circle) a description of the dynamics on their Fatou set similar to that obtained by Julia, Fatou, and Dennis Sullivan for rational maps acting on the Riemann sphere. Here, on the other hand, they make very weak assumptions about the regularity of the maps, and they use completely different methods, such as new cross-ratio distortion arguments. There was a lot of interest in this study, with previous works by J. Guckenheimer, J.C. Yoccoz, and Blokh & Lyubich.

Left: Maria José Pacífico, Welington, Floris Takens and Mitsuhiro Shishikura, in a conference in Japan. Courtesy of Sheldon Newhouse. Right: Jacob Palis, Gilza de Melo and Welington, in the same conference. Courtesy of Maria José Pacífico.

Later on, at the end of 1990, Welington started another long collaboration with Edson de Faria (1962–), a former student of D. Sullivan. The subject of their collaboration was the very renormalization theory that was his object of fascination in Sullivan's work. It turns out that renormalization slowly became one of the main tools to study one-dimensional maps, from unimodal maps to multimodal maps, from diffeomorphisms on the circle to homeomorphism with either critical points or breakpoints, and generalized interval exchange transformations.

A reason for the success of renormalization theory in the study of unimodal maps is that Sullivan proved that (under certain combinatorial conditions) it is possible to *complexifiy* the map, obtaining a *quadratic-like* extension as defined by Douady and Hubbard. This sort of result is known as *complex bounds*. This allowed Sullivan (and later McMullen) to use methods of complex dynamics (such as quasiconformal maps and hyperbolic geometry) to study infinitely renormalizable maps. In his Ph.D. thesis Edson de Faria introduced a way to complexify unicritical analytic homeomorphisms of the circle, which allowed him to use complex methods in the study of renormalization and universality for such maps [8].

Among other results, Welington and de Faria [5] [6] proved that unicritical analytic homeomorphisms of the circle whose rotation number satisfies certain diophantine condition are rigid, that is, the conjugacy between two such maps is always $C^{1+\alpha}$, with $\alpha > 0$. Welington and de Faria also published two books.

At the end of the 1990s, Welington had two students, Daniel Smania (1974–) and Artur Avila (1979–). Welington started a seminar on one-dimensional dynamics, and Federico Rodriguez Hertz also took part in it before becoming Palis' student. In the beginning, there was not too much focus, but soon Welington shared with them a couple of preprints by Misha Lyubich (1959–) (some were later published in the *Annals of Mathematics* [16][17]). It was quite intense, with Welington leading the way with his unparalleled experience. This was an important factor in the choice of problems for the students' theses.

In particular, Welington asked Artur Avila to talk with M. Lyubich during a conference at IMPA. Part of his Ph.D. thesis developed into an article with Lyubich and de Melo [3].

It is difficult not to see this work as a synthesis of all his influences. Smale and later Palis and his school had for a long time tried to understand, if not all, at least *most* of finite-dimensional smooth dynamical systems. Lyubich [17] proved that, in the logistic family, for almost every parameter the map is either hyperbolic (has a periodic attractor) or stochastic (has an absolutely continuous invariant probability). In the spirit of Palis' school, Avila, Lyubich and de Melo were able to generalize this result for *all* non-degenerate real-analytic families of quadratic unimodal maps. This result uses sophisticated techniques of complex dynamics, parameter exclusion, renormalization theory, and dynamics in infinite dimensions since the renormalization operator can be seen as a highly non-linear operator acting on an infinite-dimensional space of real analytic unimodal maps. The regularity of the holonomies along the hybrid classes plays an important role in this work.

Soon after, in another work[7] in collaboration de Edson de Faria and Alberto Pinto (1964–), Lyubich's results on the hyperbolicity of the renormalization operator acting on quadratic-like maps were extended to unimodal maps with finite smoothness. This involves heavy analysis since composition operators, and consequently, the renormalization operator itself, are not really smooth in this setting, which generates all sorts of difficulties.

Welington's first student was Maria José Pacífico, who completed her Ph.D. in 1980, now a well-known researcher on singular hyperbolic flows. He had just six more students: Antonio Gaspar Ruas (1982), Edson Vargas (1989), Daniel Smania (2001), Artur Avila (2001), Alejandro Kocsard (2007) and Pablo Guarino (2012). He was quite proud of their achievements.

Welington was unbearably happy when Avila got the Fields Medal on the occasion of the International Congress of Mathematics in Seoul (2014). The concept of renormalization and the method of parameter exclusion remain quite important in Avila's work [13].

His last publication [14] was with Pablo Guarino, one of his students, and Marco Martens, a former student of Sebastian van Strien. The theme of renormalization again plays a central role. They proved the rigidity of certain circle homeomorphisms of the circle with a critical point of odd order and finite smoothness.

Welington gained many distinctions during his career. I remember the occasion of one, in a satellite conference in Vadodara, India, after the International Congress of Mathematics (2010). The large auditorium was packed with people, and Welington was on the stage receiving a condecoration. He was red with emotion, quite happy, with a broad smile. He did not move his eyes from the place in the audience where his wife, Gilza, was. He was an invited lecturer at the International Congress of Mathematicians in Berlin (1998) and recived the Third World Academy of Science Award in Mathematics in 2003.

The man

Welington had a small stature and sometimes suffered from a little stuttering, but, after a few moments of talking with him, his imposing voice and strong-minded, carefree opinions left little doubt of who was in charge in the conversation. This

Thesis defense of Pablo Guarino, Welington's last student. Daniel Smania, Edson de Faria, Welington, Pablo Guarino and Artur Avila, IMPA, 2012. Courtesy of Pablo Guarino.

sometimes drove students away, but interesting enough, his confidence also attracted young researchers under his wing and worked in his favor for his teaching.

His lectures, especially his advanced courses, paid little attention to the details, and instead offered a wide view of the subject that allowed him to go farther away. They were grandiose epic poems. His passion for the subject was never hidden from the students. He also often taught advanced courses on subjects he was not an expert on but just an appreciative outsider. There was always a buzz among IMPA students when he taught something new.

Welington's magnetic teaching skills are well illustrated by an anecdote told by Pablo Guarino, his last student. His first encounter with Wellington was disheartening. We can only imagine Welington seated in the very comfy but deranged and old lounge chair he kept in his office, while Guarino presented himself at the door for half-hour, just to be received with Welington's muted glaze, followed by a short "Good luck!" in the end! Gilza met Pablo just leaving the office and asked him who his advisor was. Pablo said it was Welington and Gilza half-joking replied "Poor guy!" [2]. Pablo considered changing advisors. But after a lecture on hyperbolic dynamics by Welington he was undeniably captivated, and commented to a friend that the only thing missing was "three guys playing violins following the professor as he lectured".

It was a running joke among his students how often Welington said "it is trivial!" when someone asked for more details during a lecture. This author, one of his Ph.D. students, got an explanation for this behavior from Welington's own mouth many years later, at a massive party at Federal University of Bahia in Salvador commemorating his 60th birthday, in the same place the famous 1971 congress took place, now with young and prolific professors, some of them former students of IMPA. We met by chance in the huge crowd. Welington, a little bit worrisome, asked us if we also thought he used "trivial" too often. After taking a taste of our *umbu*'s cocktail (umbu is one of the many native fruits one can find in the northeastern region of Brazil) we reluctantly agreed, to which Welington replied "but you need to realize that trivial did not mean that it was simple, but just that *it was not important*".

Left: Welington, Misha Lyubich and Alberto Verjovsky in a Conference in Cancún, México, 2016. Courtesy of Alberto Verjovsky. Right: Welington and Steve Smale sailing in Angra dos Reis with *Doisdu*. Courtesy of Maria José Pacífico.

Welington had a soft spot for two former students of Palis from Bahia, Vilton Pinheiro and Paulo Brandão, two of his greatest admirers, and he and Gilza always had a good time visiting them. They helped to organize the event.

Later in his life, his willingness to give a frank point of view on many topics and his rigorous standards of mathematics gained an aura of folklore. Artur Avila and a collaborator played with the idea of a journal called "Non-triviality", that would only publish contributions that passed the strictly rigorous standards set by Welington, however, they decided that there would not be enough accepted articles to sustain the enterprise [2]. In another episode, some universities in Rio de Janeiro and Niterói (a nearby city), many of them with former students of IMPA as faculty, decided to create an inter-institutional seminar on dynamical systems (that is alive to this day). An apocryphal story tells that one of the founders decided to tell Welington about it, and his prompt response was "E daí?", something like "so what?" in Portuguese, so they decided to baptize the seminar "EDAÍ".

There were a few occasions when his guard lifted. Many of the events organized by IMPA took place in hotels in small touristic towns in the state of Rio de Janeiro, since empty off-season hotels are cheaper than hotels in the always busy city of Rio de Janeiro. It was not uncommon that on a hot evening, after a long day of talks, Welington would, by chance, pass a group of students in the bar of the hotel, sit with them, order the classic Brazilian drink *caipirinha* and chat, sharing entertaining anecdotes having his mathematical heroes as protagonists, such as S. Smale and Dennis Sullivan. The group of students would slowly increase, fascinated by his stories. Those events were also a good opportunity to appreciate Welington and Gilza dancing *forró*, a rhythm from the northeastern region of Brazil.

A writer famously said that Brazilians are "*cordiais*", meaning that they often make decisions under the influence of passion, or strong emotions. One can certainly disagree with this, but it fits well with Welington's relation with mathematics, a very passionate one, that shows in every pore of his being.

One place he could be at peace was Angra dos Reis, a small town with a landscape formed by hills covered with pristine tropical forests. And these hills are surrounded

Front: Steve Smale, Edson Vargas, Jacob Palis, Artur Avila, Welington, Maurício Peixoto, Maria José Pacífico, Antonio Gaspar Ruas. Back: Daniel Smania and Pablo Guarino. IMPA, in the conference to honor the 70th birthday of Welington de Melo, 2016. Credits: Courtesy of Maria José Pacífico.

by the Atlantic ocean that bifurcates in a myriad of small channels. It is a breathtaking combination. Welington and Gilza had a second house there, and he sailed on these channels weekly in his little boat *Doisdu*, and often invited visiting researchers and students to sail with him. The last ones had to help clean the boat at the end of the journey, which was a small cost to pay for their wonderful stay at Angra. Once Welington told us how a great but irascible mathematician became quite pleasant during a stay with him and Gilza at Angra. We think that Angra had the same effect on him.

He often sailed alone and he did so up to a few weeks before passing away. Welington did not seem to be a spiritual man, but we wonder how these frequent, contemplative, solitary, and silent sailings in this amazing setting did not have a similar role in his life. Welington passed away in December of 2016, just after his 70th birthday. His journey here with us ended, but what an amazing journey it was, and we cherished every moment.

It did not slip our minds that we are writing this after two years of the COVID pandemic, which has deprived us of the presence of our loved ones, some (too many) forever. The memory of Welington reminds us how much those shared moments with family, friends, advisors, colleagues, and students are important to us, and how much we miss them.

ICMC-USP, São Carlos, Brazil *Daniel Smania*
April 2022

References

1. Acadêmia Brasileira de Ciências. Biografia de Welington de Melo. https://www.abc.org.br/membro/welington-celso-de-melo/
2. A. Avila. Palestra Especial em Homenagem ao Prof. Welington de Melo. 31° Colóquio Brasileiro de Matemática. https://impa.br/en_US/noticias/artur-avila-homenageia-welington-de-melo-no-31o-coloquio/
3. A. Avila, M. Lyubich, and W. de Melo. Regular or stochastic dynamics in real analytic families of unimodal maps. *Invent. Math.* **154**(3), 451–550 (2003).
4. S. Batterson. *Stephen Smale: the mathematician who broke the dimension barrier*. American Mathematical Society, Providence, RI (2000).
5. E. de Faria and W. de Melo. Rigidity of critical circle mappings. I. *J. Eur. Math. Soc. (JEMS)* **1**(4), 339–392 (1999).
6. E. de Faria and W. de Melo. Rigidity of critical circle mappings. II. *J. Amer. Math. Soc.* **13**(2), 343–370 (2000).
7. E. de Faria, W. de Melo and A. Pinto. Global hyperbolicity of renormalization for C^r unimodal mappings. *Ann. of Math. (2)* **164**(3), 731–824 (2006).
8. E. de Faria and S. van Strien. Welington de Melo (1946–2016). In: *New trends in one-dimensional dynamics*, volume **285** of Springer Proc. Math. Stat., pp. 7–20. Springer, Cham (2019).
9. W. de Melo. Structural stability of diffeomorphisms on two-manifolds. *Invent. Math.* **21**, 233–246 (1973).
10. W. de Melo and S. van Strien. *One-dimensional dynamics*. Volume 25 of Ergebnisse der Mathematik und ihrer Grenzgebiete (3) [Results in Mathematics and Related Areas (3)]. Springer-Verlag, Berlin (1993).
11. W. de Melo and S.J. van Strien. Diffeomorphisms on surfaces with a finite number of moduli. *Ergodic Theory Dynam. Systems* **7**(3), 415–462 (1987).
12. B. Esteves. Velejador regrado. *Revista Piauí*. August. 2014.
13. E. Ghys. The work of Artur Avila. In: *Proceedings of the International Congress of Mathematicians – Seoul 2014*. Vol. 1, pp. 47–54. Kyung Moon Sa, Seoul (2014).
14. P. Guarino, M. Martens, and W. de Melo. Rigidity of critical circle maps. *Duke Math. J.* **167**(11), 2125–2188 (2018).
15. J. Hubbard. Preface. In: *The Mandelbrot set, theme and variations*, volume **274** of London Math. Soc. Lecture Note Ser., pp. xiii–xx, Cambridge Univ. Press, Cambridge (2000).
16. M. Lyubich. Feigenbaum–Coullet–Tresser universality and Milnor's hairiness conjecture. *Ann. of Math. (2)* **149**(2), 319–420 (1999).
17. M. Lyubich. Almost every real quadratic map is either regular or stochastic. *Ann. of Math. (2)* **156**(1), 1–78 (2002).
18. M. Martens, W. de Melo and S. van Strien. Julia–Fatou–Sullivan theory for real one-dimensional dynamics. *Acta Math.* **168**(3-4), 273–318 (1992).
19. M.J. Pacifico and P. Guarino, editors. *New trends in one-dimensional dynamics*. In honour of Welington de Melo on the occasion of his 70th birthday, IMPA 2016, Rio da Janeiro, Brazil, November 14–17, volume **285** of Springer Proceedings in Mathematics & Statistics, Springer, Cham (2019).
20. J. Palis. Entrevistas com Eméritos IV – Marcelo Viana e Welington de Melo entrevistam Jacob Palis. *Recorded interview*. IMPA (2011).
21. J. Palis and F. Lenarduzzi. Welington de Melo and Jacob Palis: their first meeting, some of their work on structural stability and a lifetime of friendship. In: *New trends in one-dimensional dynamics*, volume **285** of Springer Proc. Math. Stat., pp. 1–5, Springer, Cham (2019).
22. D. Sullivan. Reminiscences of Michel Herman's first great theorem. https://www.math.stonybrook.edu/~dennis/publications/PDF/DS-pub-0112.pdf
23. D. Sullivan. Bounds, quadratic differentials, and renormalization conjectures. In: *American Mathematical Society centennial publications*, Vol. II (Providence, RI, 1988), pp. 417–466. Amer. Math. Soc., Providence, RI (1992).
24. S. van Strien. For Jacob Palis. https://impa.br/wp-content/uploads/2020/03/Sebastian-van-strien.pdf

Structural Stability of Diffeomorphisms on Two-Manifolds

W. de Melo

Inventiones math. **21**, 233–246 (1973)

Introduction

In [8] Robbin proved that the C^2-diffeomorphisms on a compact manifold which satisfy Axiom A and the Transversality Condition are structurally stable in the C^1 sense. However, some basic results in the theory of Dynamical Systems, such as the Closing-lemma [7], are only known in the C^1-topology. Therefore, as Robbin observed, it would be of interest to extend his result for C^1 diffeomorphisms.

We construct in this paper the analogue of Palis-Smale tubular families [6], for such diffeomorphisms on two dimensional manifolds. These families provide a rather fine geometric structure of the diffeomorphism near the basic sets and the relation between them. From this fact it follows the structural stability of the diffeomorphisms. We will also prove, the stability of diffeomorphisms satisfying Axiom A in a neighborhood of a basic set for two dimensional manifolds and some other cases.

We will now set some notation, basic definitions, and the statements of our results.

Let $\mathrm{Diff}^1(M)$ be the set of C^1-diffeomorphisms of M with the C^1-topology, where M is a compact n-dimensional manifold without boundary. If $f \in \mathrm{Diff}^1(M)$, $\Omega(f)$ will denote the non-wandering set of f and $\mathrm{Per}(f)$ the set of periodic points of f. A diffeomorphism f is said to satisfy Axiom A if

(i) $\Omega(f)$ is hyperbolic, namely, the tangent bundle of M over Ω splits in a Df-invariant Whitney sum, $T_\Omega M = E^s \oplus E^u$, and there exists a riemannian metric $\|\cdot\|$ on M such that

$$\begin{aligned} &\|Df/E^s\| < \lambda \\ &\|Df^{-1}/E^u\| < \lambda \end{aligned} \qquad \text{for some } \lambda \in (0,1).$$

(ii) $\mathrm{Cl}(\mathrm{Per}(f)) = \Omega(f)$, that is, the periodic points are dense in $\Omega(f)$.

If $f \in \mathrm{Diff}^1(M)$ satisfies Axiom A then $\Omega(f) = \Lambda_1 \cup \ldots \cup \Lambda_k$ where the Λ_i's are disjoint, closed, f-invariant and topologically transitive, i.e. f/Λ_i has a dense orbit. Such Λ_i's are called basic sets of f. We refer the

16*

Originally published in *Inventiones math.* **21**, 233–246 (1973).

reader to [9] or [5] for the proof of the above statement. If $f \in \mathrm{Diff}^1(M)$ satisfies Axiom A we define the stable and unstable manifolds of a point $x \in \Omega$ by

$$W^s(x) = \{y \in M;\, d(f^k(x), f^k(y)) \to 0 \text{ as } k \to +\infty\}$$
$$W^u(x) = \{y \in M;\, d(f^{-k}(x), f^{-k}(y)) \to 0 \text{ as } k \to +\infty\}$$

where d stands for a riemannian metric on m. If Λ is a basic set of f we define the stable set of Λ by $W^s(\Lambda) = \bigcup_{x \in \Lambda} W^s(x)$, and this coincides with the set of points whose positive orbits converge to Λ [2]. In the same way we define the unstable set of Λ. If β is a positive real number we define the stable manifold of x of size β by

$$W^s_\beta(x) = \{y \in M;\, d(f^k(x), f^k(y)) < \beta \text{ for } k \geqq 0\}.$$

An attractor of f is a basic set Λ such that for any $\beta > 0$, $W^s_\beta(\Lambda)$ is a neighborhood of Λ. In [1] the reader will find the basic material concerning hyperbolic structure and stable manifold theory.

Let $f \in \mathrm{Diff}^1(M)$ be such that the Axiom A holds. We say that f satisfies the Transversality Condition if, for any $x, y \in \Omega(f)$, $W^s(x)$ is transversal to $W^u(y)$.

We are going to prove the following theorems:

Theorem A. *If M is two-dimensional and $f \in \mathrm{Diff}^1(M)$ satisfies Axiom A and the Transversality Condition then f is structurally stable.*

Let $f \in \mathrm{Diff}^1(M)$ satisfy Axiom A, and Λ be a basic set of f. We say that f is locally stable with respect to Λ if there are neighborhoods N of f in $\mathrm{Diff}^1(M)$ and V of Λ in M with the property: each $g \in N$ has a unique basic set Λ_g in V and there exists a homeomorphism h_g of a neighborhood of Λ onto a neighborhood of Λ_g such that $h_g f = g h_g$.

Theorem B. *Let M be an n-dimensional manifold, $f \in \mathrm{Diff}^1(M)$ satisfy Axiom A and Λ be an attractor of f whose stable manifolds have codimension 1. Then f is locally stable with respect to Λ.*

Theorem C. *If M is a two-dimensional manifold, $f \in \mathrm{Diff}^1(M)$ satisfies Axiom A, and Λ is a basic set of f, then f is locally stable with respect to Λ.*

This paper is based in my doctoral thesis at IMPA under the guidance of J. Palis. We are grateful to E. Lima, P. Mendes, S. Newhouse, G.L. dos Reis, C. Robinson, R. Williams, and especially to J. Palis for many helpful conversations.

Section 1

In this section M will be an n-dimensional manifold, $f \in \mathrm{Diff}^1(M)$ will satisfy Axiom A and Λ will be a basic set of f.

Definition 1.1. An unstable tubular family for Λ is a continuous foliation $\mathscr{F}^u$ in a neighborhood V of Λ with the properties:

(1) the leaves are C^1 manifolds immersed 1-1 in M;

(2) the leaf through a point $x \in \Lambda$ is the connected component of $W^u(x) \cap U$ containing x;

(3) the tangent spaces to the leaves vary continuously;

(4) the leaves are transversal to the stable manifolds of points in Λ;

(5) the foliation is f invariant, i.e. $f^{-1}(F^u(x)) \subset F^u(f^{-1}(x))$ where $F^u(x)$ stands for the leaf containing the point x.

In this section we will construct an unstable tubular family for an attractor whose stable manifolds have codimension 1. Using this tubular family we will prove Theorem B.

Lemma 1.1. *Let $f \in \mathrm{Diff}^1(M)$ satisfy Axiom A and Λ be an attractor of f whose stable manifolds have codimension* 1. *If $K \subset W^s(\Lambda)$ is a compact set then there exists a C^∞ 1-dimensional foliation in a neighborhood of K whose leaves are transversal to the stable manifolds of Λ.*

Proof. Since Λ is an attractor the stable manifolds of points in Λ constitute a C^0-foliation on $W^s(\Lambda)$ which is a neighborhood of Λ. This foliation satisfies conditions (1), (3) and (4) of Definition 1.1, [1].

If $x \in K$ we can choose small neighborhoods A_x and B_x with $\mathrm{Cl}(A_x) \subset B_x$ such that there is a C^∞ foliation in B_x transversal to the stable manifolds of points in Λ. Let $A_1, \ldots, A_r$ be a cover of K. Suppose we have a C^∞ foliation F_j^u in $U_j = A_1 \cup \ldots \cup A_j$ satisfying the above property. We are going to extend it to U_{j+1}. Let $V = U_j \cup B_{j+1}$. Since there is a trivial foliation in B_{j+1} we can pick a C^∞ vector field X in V which is unitary and tangent to the leaves of this foliation. Take a unitary vector field Y in V with the property: for each $x \in V$, $Y(x)$ is tangent to $F_j^u(x)$ and its direction is such that $Y(x)$ induces the same orientation as $X(x)$ on the stable manifold that contains x. Since the foliation F_j^u is C^∞, so is Y. Let $\varphi\colon M \to R$ be a C^∞ function such that $\varphi(M) \subset [0, 1]$, $\varphi(x) = 0$ if $x \in A_{j+1} - \mathrm{Cl}(U_j)$ and $\varphi(x) = 1$ if $x \in \mathrm{Cl}(U_j) - B_{j+1}$. Consider the vector field $Z = (1-\varphi)\, X + \varphi\, Y$, which is an average of X and Y. Clearly Z is C^∞ and does not vanish in V. The trajectories of Z give us a foliation in V which extends the foliation in $U_j - V$ and the foliation in $A_{j+1} - V$. Therefore we get a C^∞ foliation in U_{j+1} which is transversal to the stable manifolds of Λ. By induction, the lemma is proved.

Lemma 1.2. *Let $U \subset M$ be an open set and $\mathscr{F}$ a C^1-foliation in U whose leaves have dimension* 1. *Let $B \subset U$ be a ball such that:*

(i) *B is contained in a neighborhood where the foliation is trivial;*

(ii) *the frontier of* $B = C \cup D \cup D^*$, *where* D *and* D^* *are* C^∞ *discs transversal to the foliation and* C *is a cylinder invariant by the foliation, i.e.* C *is a union of plaques.*

Then, given $\delta > 0$, *there exists a* C^1 *foliation* $\mathscr{F}_1$ *in* U *with the properties:*

(1) $\mathscr{F}_1$ *is equal to* $\mathscr{F}$ *outside* B;

(2) *the projections along the leaves* $p_1: B \to D$ *and* $p_1^*: B \to D^*$ *are* C^∞ *(this implies the foliation is* C^∞ *in* B*)*

(3) *if* $p: B \to D$, $p^*: B \to D^*$ *are the projections along the leaves of* $\mathscr{F}$, *then* $\|p - p_1\|_{C^1} < \delta$ *and* $\|p^* - p_1^*\|_{C^1} < \delta$.

Proof. Let $p: B \to D$, $p^*: B \to D^*$ be the projections along the leaves of $\mathscr{F}$. Clearly p and p^* are C^1 submersions (at each point the derivative is surjective). Let $\varepsilon: M \to R$ be a continuous function such that $\varepsilon(M - B) = 0$ and ε is positive in B. By well known lemma in differential topology, [3], there exist C^∞ functions $\pi: B \to D$, $\pi^*: B \to D^*$ such that $\|p - \pi\|_{C^1} < \varepsilon$ and $\|\pi^* - p^*\|_{C^1} < \varepsilon$. If we take ε small enough, π and π^* will be submersions and will give us foliations in B which extend the foliation $\mathscr{F}/(M - B)$ in the C^1-sense. Let $\mathscr{F}$ and $\mathscr{F}^*$ be the corresponding foliations in U. Let $S \subset B$ be a C^∞ disc which is a transversal section to both foliations in B. Let $p_s: B \to S$ and $p_s^*: B \to S$ be the corresponding projections along leaves. Clearly p_s and p_s^* are C^∞ submersions and are C^1 close to each other. Averaging p_s and p_s^* we get a C^∞ submersion $\pi_s: B \to S$ which coincides with p_s in a neighborhood of D and coincides with p_s^* in a neighborhood of D^*. The foliation induced by π_s in B extends $\mathscr{F}/(M - B)$ and gives us a new foliation $\mathscr{F}_1$ in U which satisfies the requirements of the lemma.

Lemma 1.3. *Let* $U \subset M$ *be an open set and* $\mathscr{F}$ *a* C^1*-foliation in* U *whose leaves have dimension* 1. *Let* $B_1 \subset B_2 \subset U$ *be balls such that:*

(i) *the frontier of* $B_i = C_i \cup D_i \cup D_i^*$, *where* C_i *is a* C^1 *cylinder invariant by* $\mathscr{F}$, *and* D_i *and* D_i^* *are* C^∞ *discs transversal to the foliation;*

(ii) $D_1 \subset D_2$, $D_1^* \subset D_2^*$ *and* $\mathrm{Cl}(B_1) - D_1 \cup D_1^* \subset B_2$.

Then, given $\delta > 0$, *there exists a* C^1 *foliation* $\mathscr{F}_1$ *in* U *with the properties:* (1) $\mathscr{F}_1$ *is equal to* $\mathscr{F}$ *outside* B_2; (2) *the projections along the leaves of* $\mathscr{F}_1$, $p_1: B_2 \to D_2$, $p_1^*: B_2 \to D_2^*$ *are* C^∞ *in* B_1; (3) *if* $p: B_2 \to D_1$, $p^*: B_2 \to D_2^*$ *are the projections along the leaves of* $\mathscr{F}$ *then* $\|p - p_1\|_{C^1} < \delta$, $\|p^* - p_1^*\|_{C^1} < \delta$ *and* p_1 (p_1^*) *is at least as differentiable as* $p(p^*)$ *in* B_2.

Proof. Analogous to the proof of Lemma 1.2.

Lemma 1.4. *Let* $U \subset M$ *be an open set and* $\mathscr{F}$ *a* C^1*-foliation in* U *whose leaves have dimension* 1. *Let* $K \subset U$ *be compact. Given* $\delta > 0$ *there exists a* C^1*-foliation* $\tilde{\mathscr{F}}$ *in* U *such that:* (1) $\tilde{\mathscr{F}}$ *is* C^∞ *in a neighborhood* W *of* K *and is equal to* $\mathscr{F}$ *outside a neighborhood* V *of* $\mathrm{Cl}(W)$ *with* $\mathrm{Cl}(V) \subset U$: (2) *the* *tangent spaces to the leaves of $\mathscr{F}$ are δ-close to the tangent space of the leaves of $\tilde{\mathscr{F}}$.*

Proof. Let $A_1, \ldots, A_s$, $B_1, \ldots, B_s$ be balls such that: (a) $\bigcup_{i=1}^{s} A_i \supset K$ $\mathrm{Cl}\, A_i \subset B_i$; (b) B_i is contained in a neighborhood where the foliation $\mathscr{F}$ is trivial; (c) frontier of $B_i = C_i \cup D_i \cup D_i^*$ where C_i is a C^1 cylinder which is union of plaques of the foliation, D_i and D_i^* are C^∞ discs transversal to the foliation and are pairwise disjoint.

Suppose there exists a C^1-foliation $\mathscr{F}^k$ in U and balls $B_i^k, \ldots, B_s^k$ satisfying the properties: (i) $\mathrm{Cl}\, A_i \subset B_i^k$, $\mathrm{Cl}\, B_i^k \subset B_i$; (ii) the frontier of $B_i^k = C_i^k \cup D_i^k \cup D_i^{*k}$ satisfies the property c above; (iii) the foliation $\mathscr{F}^k$ is C^∞ in $\bigcup_{i=1}^{k} B_i^k$, is trivial in $B_j, j=1, \ldots, s$ and coincides with $\mathscr{F}$ in $M - \bigcup_{i=1}^{s} B_i$.

Using Lemma 1.2 in $B_{k+1}^k - \bigcup_{j=1}^{k} B_j^k$ we get a C^1 foliation $\mathscr{F}_1^k$ in U which is C^∞ in $B_{k+1}^k - \bigcup_{j=1}^{k} B_j^k$ and equal to $\mathscr{F}^k$ otherwise. The new foliation $\mathscr{F}_1^k$ is, therefore, C^∞ in $\bigcup_{j=1}^{k+1} B_j^k$ except possibly in the intersections of D_j^k, D_j^{*k} and C_j^k with $B_{k+1}^k (j=1, \ldots, k)$. Decompose B_{k+1}^k in a union of balls $E_1, \ldots, E_\gamma$ such that frontier of $E_i = S_{i-1} \cup S_i \cup C_{k+1,i}^k$ where $C_{k+1,i}^k = C_{k+1}^k \cap \mathrm{Cl}\, E_i$, S_i are pairwise disjoint transversal sections to the foliation in B_{k+1}^k with $S_0 = D_{k+1}^k$ and $S_\gamma = D_{k+1}^{*k}$ are such that if $D_j^k \cap B_{k+1}^k \neq \emptyset (D_j^{*k} \cap B_{k+1}^k \neq \emptyset)$ then $D_j^k \cap B_{k+1}^k (D_j^{*k} \cap B_{k+1}^k)$ is contained in some S_i.

Let $\pi_i \colon E_i \to S_i$ be the projections along the leaves of $\mathscr{F}_1^k$. It is clear that π_i is C^∞ except possibly in points of $C_j^k \cap E_i$, $j=1, \ldots, k$, that is, π_i has the same differentiability as the foliation. Let $\underset{\sim}{B}_1$ and $\underset{\sim}{B}_2$ be balls satisfying the frontier condition c above, $\mathrm{Cl}\, A_{k+1} \subset \underset{\sim}{B}_1 \subset \underset{\sim}{B}_2 \subset B_{k+1}^k$, $\mathrm{Cl}\, \underset{\sim}{B}_1 - \underset{\sim}{D}_1 \cup \underset{\sim}{D}_1^* \subset \underset{\sim}{B}_2$, $\mathrm{Cl}\, \underset{\sim}{B}_2 - \underset{\sim}{D}_2 \cup \underset{\sim}{D}_2^* \subset B_{k+1}^k$ and $\underset{\sim}{D}_1 \subset \underset{\sim}{D}_2 \subset D_{k+1}^k$, $\underset{\sim}{D}_1^* \subset \underset{\sim}{D}_2^* \subset D_{k+1}^{*k}$. Using Lemma 1.2 with $B_1 = E_1 \cap \underset{\sim}{B}_1$, $B_2 = E_1 \cap \underset{\sim}{B}_2$, we get a C^1 foliation $\mathscr{F}_{1,1}^k$ in U which is C^∞ in $E_1 \cap \underset{\sim}{B}_1$, coincides with $\mathscr{F}_1^k$ outside E_1 and is at least as differentiable as $\mathscr{F}_1^k$ in all points except possibly in points of $S_0 \cup S_1$. Making this approximation for each E_j we get a C^1-foliation $\mathscr{F}_1^k$, in U which is equal $\mathscr{F}$ outside $B_1 \cup \ldots \cup B_s$ and is C^∞ in $B_1^k \cup \ldots \cup B_k^k \cup \underset{\sim}{B}_1$ except possibly in points of $S_1 \cup \ldots \cup S_s$. By the property (2) of Lemmas 1.2 and 1.3, each point $x \in S_i$, $x \in B_1^k \cup \ldots \cup B_k^k \cup \underset{\sim}{B}_1$, has a neighborhood N_x such that the projection along the leaves p: $N_x \to N_x \cap S_i$ is C^∞ except in points of $S_i \cap N$. Let $N_1 \subset \mathrm{Cl}\, N_1 \subset N_2 \subset \mathrm{Cl}\, N_2 \subset N_x$ be neighborhoods of x. Using a well known lemma in differential topology we can approximate p by a submersion $\tilde{p}$ which is C^∞ in N_1, equal p outside N_2 and is at least as differentiable as p in all points. Hence the foliation in N given by p extends the foliation outside N. Therefore we get a C^1 foliation in U which is C^∞ in N_1 and is at least as differentiable as the previous one in all points. Using this argument we can get a C^1 foliation $\mathscr{F}^{k+1}$ in U which is C^∞ in $B_1^{k+1} \cup \ldots \cup B_{k+1}^{k+1}$ where B_j^{k+1} satisfies the properties above. It is clear that if we take all those approximations small enough, the tangent spaces to the leaves of $\mathscr{F}^{k+1}$ will be C^0 close to the tangent spaces to the leaves of $\mathscr{F}$.

By induction the lemma is proved.

Let Λ be a basic set for $f \in \mathrm{Diff}^1(M)$. A fundamental domain for $W^s(\Lambda)$ is a compact set $D^s \subset W^s(\Lambda)$ such that $W^s(\Lambda) - \Lambda = \bigcup_{n \in Z} f^n(D^s)$ and each orbit in $W^s(\Lambda) - \Lambda$ has at most two points in D^s. If $D^s \cap \Lambda = \emptyset$ we say that D^s is a proper fundamental domain. It is proved in [2] that if β is a small positive number then $D^s = \mathrm{Cl}(W_\beta^s(\Lambda) - f(W_\beta^s(\Lambda)))$ is a proper fundamental domain for $W^s(\Lambda)$.

Lemma 1.5. *Let $f\colon M \to M$ be a diffeomorphism satisfying Axiom A and Λ an attractor of f whose stable manifolds have codimension* 1. *Then there exists a C^1 foliation $\mathscr{F}^u$ in a proper fundamental neighborhood of $W^s(\Lambda)$ which is f-invariant and transversal to the stable manifolds of points in Λ.*

Proof. Let β be a small positive number and $D^s = \mathrm{Cl}(W_\beta^s(\Lambda) - f(W_\beta^s(\Lambda)))$ a fundamental domain for $W^s(\Lambda)$. Let $B^E = \mathrm{Cl}(W_\beta^s(\Lambda)) - (W_\beta^s(\Lambda))$ and $B^I = \mathrm{Cl}(f(W_\beta^s(\Lambda))) - f(W_\beta^s(\Lambda))$. Then $B^E \cap B^I = \emptyset$, $f(B^E) = B^I$, and the orbit of each points in $D^s - B^E \cup B^I$ has a unique point in D^s. By Lemma 1.1 there exists a C^∞ foliation $\mathscr{F}_E$ in a neighborhood V^E of B^E which is transversal to the stable manifolds of points in Λ. Let V^E be so small that $f(V^E) \cap V^E = \emptyset$ and $V^E \cap \Lambda = \emptyset$. Let $\mathscr{F}_I$ be the C^1 foliation in $f(V^E)$ induced by $\mathscr{F}_E$ and f; that is, the leaves of $\mathscr{F}_I$ are the f-images of the leaves of $\mathscr{F}_E$. Using Lemma 1.4 we can pick neighborhoods U^I, V^I and W^I of B^I such that there exists a C^1 foliation $\tilde{\mathscr{F}}_I$ in W^I which coincides $\mathscr{F}_I$ in U^I and is C^∞ in $W^I - \mathrm{Cl}(V^I)$, where $\mathrm{Cl}(U^I) \subset V^I \subset \mathrm{Cl}\, V^I \subset W^I \subset \mathrm{Cl}\, W^I \subset f(V^E)$. Let $\tilde{W}^I \subset W^I$ be a neighborhood of $D^s \cap (\mathrm{Cl}\, V^I - U^I)$ such that $f^{-1}(\tilde{W}^I) \cap D^s = \emptyset$ and $\tilde{U}^I = U^I \cup \tilde{W}^I$. Let $\tilde{W}^E$ be aneighborhood of $D^s \cap f^{-1}(\mathrm{Cl}\, V^I - U^I)$ such that $f(\tilde{W}^E) \cap (\tilde{W}^I \cup D^s) = \emptyset$. Let $\tilde{U}^E = f^{-1}(U^I) \cup \tilde{W}^E$. Hence we have a C^1-foliation in $\tilde{U}^I \cup \tilde{U}^E$ which is f-invariant and is C^∞ outside U^I. Using the method of averaging local vector fields as in Lemma 1.1 we can extend the C^∞ foliation in $\tilde{U}^I \cup \tilde{U}^E - (U^I \cup U^E)$ to a C^∞ foliation in a neighborhood V of $D^s - \tilde{U}^I \cup \tilde{U}^E$. Take V so small that $f^{-1}(V) \cap U^E = \emptyset$ and $f(V) \cap U^I = \emptyset$. Therefore we get a C^1-foliation $\mathscr{F}^u$ in $N = U^E \cup U^I \cup V$ which is f-invariant and whose leaves are transversal to the stable manifolds of points in Λ.

Let Λ be a basic set of $f \in \mathrm{Diff}^1(M)$ and $D^s = \mathrm{Cl}\, W^s(\Lambda) - f(W^s(\Lambda))$ a proper fundamental domain. Let N^s be a proper fundamental neighborhood of D^s. Let $x \in \Lambda$, and A a small ball around x such that if $y \in A \cap \Lambda$, then the connected component of $W^u(y) \cap A$ which contains y is a disc.

Lemma 1.6 (Generalized λ-lemma). *Let $S \subset N^s$ be a disc transversal to the stable manifolds of points in Λ and whose dimension is equal to the codimension of the stable manifolds of Λ. Given $\varepsilon > 0$ there exists a positive integer n_0 such that if $n > n_0$ then $f^n(S) \cap A = \emptyset$ or $f^n(S) \cap A = S_1 \cup \ldots \cup S_r$ where S_i are connected, and for each i there exists $x \in A \cap \Lambda$ such that S_i is ε C^1-close to the connected component of $W^s(x) \cap A$ containing x.*

The proof of this lemma follows from computations similar to those used in the proof of λ-lemma, [4].

Theorem 1.1. *Let f: $M \to M$ be a diffeomorphism satisfying Axiom A and Λ a basic set of f which is an attractor whose stable manifolds have codimension 1. Then there exists a stable tubular family for Λ.*

Proof. Let N^s be a fundamental neighborhood of $W^s(\Lambda)$ and $\mathscr{F}^u$ a foliation in N^s as in Lemma 1.5. Iterating by f we get the tubular family, by Lemma 1.6.

Theorem 1.2. *Let $f \in \mathrm{Diff}^1(M)$ be a diffeomorphism satisfying Axiom A and Λ a basic set of f which is an attractor whose stable manifolds have codimension 1. Then f is locally stable with respect to Λ.*

Proof. Let N^s be a proper fundamental neighborhood for $W^s_f(\Lambda)$ and $\mathscr{F}^u$ a C^1 foliation in N^s which is f-invariant and transversal to the stable manifolds of points in Λ, just like in Lemma 1.5.

Let $\varepsilon_i > 0$ and $B_i'' \subset B_i' \subset B_i \subset N^s$, $i = 1, \ldots, r$ be balls such that: (1) B_i has diameter less than ε_i and the foliation is trivial in B_i; (2) if $y \in B_i$ and $\tilde{y} \in \Lambda$ is such that $y \in W^s_\beta(\tilde{y})$ then $W^s_{\beta+\varepsilon_i}(\tilde{y}) \cap B_i$ is connected; (3) if $B_i \cap B_j \neq \emptyset$ then $B_i \cup B_j$ is contained in some neighborhood where the foliation $\mathscr{F}^u$ is trivial; (4) $\bigcup_{i=1}^{r} B_i'' \supset D^s$ and $\mathrm{Cl}\, B_i'' \subset B_i'$, $\mathrm{Cl}\, B_i' \subset B_i$. Let

$$0 < \delta < \min\{d(B_i'', M - B_i'); d(B_i', M - B_i), \varepsilon_i\}.$$

Let N be a neighborhood of f in $\mathrm{Diff}^1(M)$, V a neighborhood of Λ in M with $\mathrm{Cl}\, V \cap \mathrm{Cl}\, N = \emptyset$, and U a neighborhood of the inclusion in $C^0(\Lambda, M)$ with the following property: if $g \in N$ then g has unique maximal hyperbolic set in V (which is an attractor) and there exists a continuous map $N \to U$ such that each $g \in N$ assigns a homeomorphis h_g: $\Lambda \to \Lambda_g$ conjugating f and g ($h_g f = g h_g$). Furthermore, $W^s_{\beta+\varepsilon}(\tilde{x})$ is δ C^1-close to $W^s_{\beta+\varepsilon}(h_g(\tilde{x}))$ where $\varepsilon = \max \varepsilon_i$ [1].

Let $g \in N$. We are going to extend the homeomorphis h_g: $\Lambda \to \Lambda_g$ to a neighborhood of Λ. Clearly we can suppose $g = f$ outside V. Hence, $\mathscr{F}^u$ is also a g-invariant foliation transversal to the stable manifolds of Λ_g. First we are going to define h_g in D^s. If $x \in D^s$, then $x \in B_i''$ for some i and $x \in W^s(\tilde{x})$ for some $\tilde{x} \in \Lambda$. It follows from the construction above that

$W^s_{\beta+\varepsilon}(\tilde{x}, f)$ is a transversal section of the foliation $\mathscr{F}^u$ in B_i, i.e. it intersects all plaques. Since $W^s_{\beta+\varepsilon}(h_g(\tilde{x}), g)$ is $\delta - C^1$-close to $W^s_{\beta+\varepsilon}(\tilde{x})$ it follows that $W^s_{\beta+\varepsilon}(h_g(\tilde{x}), g)$ is a transversal section to the foliation $\mathscr{F}^u$ in B'_i. Therefore, the plaque $P^u_i(x)$ intersects $W^s_{\beta+\varepsilon}(h_g(\tilde{x}))$ transversally in a unique point. Then we define $h_g(x) = P^u_i(x) \cap W^s_{\beta+\varepsilon}(h_g(\tilde{x}))$. Using properties (2) and (3) above, it follows easily that $h_g(x)$ does not depend on the choice of $\tilde{x}$ and B_i. Since the foliation $\mathscr{F}^u$ is C^1, and the stable manifolds vary continuously, h_g is continuous. Now we extend h_g to $W^s(\Lambda)$ by conjugacy, that is, if $x \in W^s(\Lambda) - \Lambda$ there exists an integer n such that $f^{-n}(x) \in D^s$, and we define $h_g(x) = g^n h_g f^{-n}(x)$. Since h_g/D^s is a conugacy between f and g, h_g is well defined. The continuity of h_g in $W^s(\Lambda) - \Lambda$ is obvious, and the continuity of h_g in Λ follows from Lemma 1.6.

Section 2

In this section M will be a compact two-dimensional manifold. Let $f \in \mathrm{Diff}^1(M)$ satisfy Axiom A and the Transversality Condition. Let $\{\Lambda_1, \ldots, \Lambda_m\}$ be the basic sets of f.

Definition 2.1. A system of unstable tubular families for f is a collection $\mathscr{F}^u_1, \ldots, \mathscr{F}^u_m$ of unstable tubular families, one for each basic set, satisfying the following compatibility condition: if $F^u_i(x) \cap F^u_j(y) \neq \emptyset$ then

$$F^u_i(x) \subset F^u_j(y) \quad \text{or} \quad F^u_j(y) \subset F^u_i(x).$$

Here $F^u_i(x)$ is the leaf of $\mathscr{F}^u_i$ containing the point x.

In the same way we define a system of stable tubular families.

In this section we are going to construct a system of tubular families for a diffeomorphism f as above.

Lemma 2.1. *Let $f \in \mathrm{Diff}^1(M)$ satisfy Axiom A, and Λ be a basic set of f. Let β be a small positive number such that $\{W^s_\beta(x); x \in \Lambda\}$ is a continuous family of C^1 submanifolds of M. Let $K \subset W^s_\beta(\Lambda)$ be a compact. Then there exists a C^∞ foliation in a neighborhood of K which is transversal to the submanifolds $W^s_\beta(x)$, $x \in \Lambda$.*

Proof. Since M is two-dimensional, we can suppose that the stable manifolds of points in Λ have dimension 1 otherwise, the lemma would be trivial. The proof of this lemma will be more involved than that of Lemma 1.1 because here the stable manifolds of points in Λ do not fill a neighborhood of Λ. Let A_i, B_i, $i = 1, \ldots, r$ be balls such that $\bigcup_{i=1}^r A_i \supset K$, $\mathrm{Cl}\, A_i \subset B_i$, and in B_i there exists a trivial C^∞ foliation which is transversal to the stable manifolds of points in Λ. Let $C_i = A_1 \cup \ldots \cup A_i$. Suppose, by induction, that we have found open sets D_i and E_i with the properties: (1) in E_i there exists a C^∞ foliation transversal to the stable manifolds of

Λ; (2) $\mathrm{Cl}\,D_i \subset E_i$; (3) $D_i \supset K \cap C_i$. Let E_i' be an open set such that $\mathrm{Cl}\,D_i \subset E_i' \subset \mathrm{Cl}\,E_i' \subset E_i$. Take balls A_{i+1}', B_{i+1}' such that $\mathrm{Cl}\,A_{i+1} \subset A_{i+1}' \subset \mathrm{Cl}\,A_{i+1}' \subset B_{i+1}' \subset \mathrm{Cl}\,B_{i+1}' \subset B_i$. Let $\varphi: M \to R$ be a C^∞ function such that $\varphi(M) \subset [0,1]$, $\varphi(x)=0$ if $x \in \mathrm{Cl}(A_{i+1}' - E_i')$, $\varphi(x)=1$ if $x \in \mathrm{Cl}(E_1' - B_{i+1}')$. Let X be a unitary C^∞ vector field tangent to the leaves of B_{i+1}. Consider the compact set $C = K \cap E_i' \cap B_{i+1}''$ where B_{i+1}'' is a ball such that $\mathrm{Cl}\,A_{i+1}' \subset B_{i+1}'' \subset \mathrm{Cl}\,B_{i+1}'' \subset B_{i+1}'$. Suppose there exist neighborhoods V and W of C, with $\mathrm{Cl}\,V \subset W \subset E_i$, such that in W we have a unitary C^∞ vector field Y tangent to the leaves of the foliation in E_i and inducing the same orientation as X on the stable manifolds of Λ. Then we can define the vector field $Z = \varphi X + (1-\varphi) Y$ which is C^∞ and does not vanish on W. The trajectories of Z give us a foliation in W which extends the foliations in $E_i' - \mathrm{Cl}\,B_{i+1}''$ and in $A_{i+1}'' - \mathrm{Cl}\,D_i$ where A_{i+1}'' is a ball such that $\mathrm{Cl}\,A_{i+1} \subset A_{i+1}'' \subset \mathrm{Cl}\,A_{i+1}'' \subset A_{i+1}'$. Then we get a C^∞ foliation in the open set $E_{i+1} = (E_i' - \mathrm{Cl}\,B_{i+1}') \cup W \cup (A_{i+1}'' - D_i)$ which is transversal to the stable manifolds of Λ. If we take $D_{i+1} = (D_i - \mathrm{Cl}\,B_{i+1}') \cup V \cup (A_{i+1} - \mathrm{Cl}\,E_i')$ we have $\mathrm{Cl}\,D_{i+1} \subset E_{i+1}$, $D_{i+1} \supset K \cap C_{i+1}$ and the lemma follows by induction. It remains to show the existence of neighborhoods V and W and of the vector field Y with the above properties.

Let $F_1, \ldots, F_t$ be an open cover of C such that $\mathrm{Cl}\,F_j \subset G_j$, and $G_j \subset E_i$ is so small that there exists a unitary C^∞ vector field $\tilde{Y}_{j+1}$ in G_j which is tangent to the leaves of the foliation in E_i and induces the same orientation as X on the stable manifolds of Λ. Suppose, by induction, that we have found open sets V_j and W_j such that: (i) $\mathrm{Cl}\,V_j \subset W_j$; (ii) in W_j there exists a C^∞ vector field Y_j which is unitary, tangent to the leaves of the foliation in E_i, and if $x \in W^s(\tilde{x})$, $\tilde{x} \in \Lambda$, then $Y_j(x)$ induces the same orientation as $X(x)$ on $W^s(\tilde{x})$; (iii) $V_j \supset C \cap \left(\bigcup_{k=1}^{j} F_k \right)$. Let W_j' be an open set such that $\mathrm{Cl}\,V_j \subset W_j' \subset \mathrm{Cl}\,W_j' \subset W_j$. Let F_{j+1}' be an open set such that $\mathrm{Cl}\,F_{j+1} \subset F_{j+1}' \subset \mathrm{Cl}\,F_{j+1}' \subset G_{j+1}$. Let $K_j = \{x \in \mathrm{Cl}\,W_{j+1}' \cap \mathrm{Cl}\,F_{j+1}';\ Y_j(x) = -\tilde{Y}_{j+1}(x)\}$. Clearly K_j is a compact set. Since each point of C belongs to a stable manifold of Λ we have that $Y_j(x)$ and $\tilde{Y}_{j+1}(x)$ induces the same orientation on the stable manifold containing x. Hence $C \cap K_j = \emptyset$. Let $V(K_j)$ and $V'(K_j)$ be compact neighborhoods of K_j such that $V'(K_j) \subset$ interior of $V(K_j)$ and $V(K_j) \cap C = \emptyset$. Let $W_{j+1} = (W_j' \cup F_{j+1}') - V'(K_j)$ and $V_{j+1} = (V_j \cup F_{j+1}) - V(K_j)$. Then $\mathrm{Cl}\,V_{j+1} \subset W_{j+1}$. Define a vector field Y_{j+1} in W_{j+1} by $Y_{j+1}(x) = Y_j(x)$ if $x \in W_j'$ and $Y_{j+1}(x) = \tilde{Y}_{j+1}(x)$ if $x \in F_{j+1}'$. Clearly Y_{j+1} is well defined and satisfies the induction hypothesis.

Let β be a small positive real number and $D^s = \mathrm{Cl}(W_\beta^s(\Lambda)) - f(W_\beta^s(\Lambda))$ a proper fundamental domain for $W^s(\Lambda)$.

Lemma 2.2. *There exist a proper fundamental neighborhood N^s of D^s and a C^1 foliation $\mathscr{F}^u$ in N^s which is f-invariant and transversal to the* *stable manifolds of size β of Λ.*

Proof. Analogous to the proof of Lemma 1.5.

Iterating this foliation by f and using Lemma 1.6 we get

Theorem 2.1. *If $f \in \mathrm{Diff}^1(M)$ satisfies Axiom A and Λ is a basic set of f then there exists an unstable tubular family for Λ.*

Similarly we prove the existence of a stable tubular family for Λ.

Remark. The foliation obtained in Theorem 2.1 is C^1 except possibly in points of the unstable manifolds of Λ.

Theorem 2.2. *If $f \in \mathrm{Diff}^1(M)$ satisfies Axiom A and the Transversality Condition then there exists a system of unstable tubular families for f.*

Proof. Let $\Lambda_1, \ldots, \Lambda_m$ be an ordering of the basic sets of f such that $W^s(\Lambda_i) \cap W^u(\Lambda_j) = \emptyset$ if $i<j$. Let $\mathscr{F}_1^u$ be a tubular family for Λ_1 as in Theorem 2.1. Suppose we have constructed tubular families $\mathscr{F}_1^u, \ldots, \mathscr{F}_{k-1}^u$ satisfying the compatibility condition of definition 2.1. Let D_k^s be a proper fundamental domain for $W^s(\Lambda_k)$ and B^I, B^E be disjoint compact sets as before. Pick a neighborhood N of D^s so small that $N \cap W^u(\Lambda_k) = \emptyset$ and $W^u(\Lambda_j) \cap W^s(\Lambda_k) \neq \emptyset$ if $W^u(\Lambda_j) \cap N \neq \emptyset$. We are going to construct a tubular family $\mathscr{F}_k^u$ for Λ_k compatible with the previous ones. We can suppose that the stable manifolds of Λ_k have dimension 1, otherwise the construction would be trivial.

Recall that the behavior of a basic set Λ with respect to Λ_k is the smallest integer i with the following property: if $\Lambda_{i_0} = \Lambda, \ldots, \Lambda_{i_n} = \Lambda_k$ is a sequence of basic sets such that $W^u(\Lambda_{i_j}) \cap W^s(\Lambda_{i_{j+1}}) \neq \emptyset$ then $n \leq i$. Let Γ_i be the collection of basic sets Λ with behavior i with respect to Λ_k and such that the stable manifolds of Λ have dimension 1. If $\Lambda \in \Gamma_1$, it is easy to see that the unstable manifolds of Λ do not accumulate in N. Therefore $\mathrm{Cl}\, N \cap (\bigcup_{\Lambda \in \Gamma_1} W^u(\Lambda))$ is compact.

Then by transversality condition we can pick open sets $U_1 \subset V_i \subset W_1 \subset N$ such that (i) there is a 1-dimensional foliation in W_1 transversal to the stable manifolds of Λ_k and compatible with $\mathscr{F}^u(\Lambda)$, for $\Lambda \in \Gamma_1$; (ii) $W^u(\Lambda) \cap N \subset U_1$ for all $\Lambda \in \Gamma_1$; $\mathrm{Cl}(U_1 \cap D^s) \subset V_1$, $\mathrm{Cl}(V_1 \cap D^s) \subset W_1$. If $\Lambda \in \Gamma_2$ then the points of N where the unstable manifolds of Λ accumulate belong to U_1. Therefore $\mathrm{Cl}\, N_1 \cap (\bigcup_{\Lambda \in \Gamma_2} W^u(\Lambda)) - U_1$ is compact. Then we can pick open sets $U_2 \subset V_2 \subset W_2 \subset N$ satisfying the above properties for all $\Lambda \in \Gamma_1 \cup \Gamma_2$. By induction on the behavior we get open sets $U \subset V \subset W \subset N$ satisfying the above properties for all basic sets whose unstable manifolds have dimension 1. Hence the foliation in W is compatible with all the previous tubular families. This foliation is C^1 outside U (see remark bellow Theorem 2.1). Using Lemma 1.4 and the method of averaging vector fields we can extend the foliation in V to a foliation in a neighborhood V^E of B^E. Iterating this foliation by f we get an f-invariant foliation in $V^E \cup f(V^E)$ which is C^1 outside V. Let $U^E \subset V^E$ be a small neighborhood of B^E and $U^I = f(U^E)$. Using again Lemma 1.4 and the method of averaging vector fields we extend the f-invariant foliation in $U \cup U^E \cup U^I$ to a neighborhood N^s of D^s. Then we have obtained an f-invariant foliation in a fundamental neighborhood N^s which is compatible with the tubular families $\mathscr{F}_1^u, \ldots, \mathscr{F}_{k-1}^u$. Iterating by f we get the tubular family.

Remark. If $g \in \mathrm{Diff}^1(M)$ is close to f, then g, also satisfies Axiom A [9]. Looking at the constructions above we can see that if g is close enough to f then g also has a system of unstable tubular families. Furthermore the tubular families of g depend continuously on g (that is, the tangent spaces to the leaves depend continuously on g). In particular, g also satisfies the Transversality Condition. Similarly we can construct systems of stable tubular families.

Section 3

Here M will be a compact two-dimensional manifold and $f\colon M \to M$ a diffeomorphism satisfying Axiom A and the Transversality Condition. In this section we are going to construct a conjugacy between f and a diffeomorphism C^1-close to f.

Let $\Lambda_1(f), \ldots, \Lambda_m(f)$ be the basic sets of f. Let N be a neighborhood of f in $\mathrm{Diff}^1(M)$ such that each $g \in N$ has systems of tubular families $\mathscr{F}^s(g) = \{\mathscr{F}_1^s(g), \ldots, \mathscr{F}_m^s(g)\}$, $\mathscr{F}^u(g) = \{\mathscr{F}_1^u(g), \ldots, \mathscr{F}_m^u(g)\}$. Consider an open cover of Λ_k, $\{A_k^1, \ldots, A_k^r\}$ where $\mathrm{Cl}\, A_k^i \subset B_k^i \subset \mathrm{Cl}\, B_k^i \subset C_k^i$ and $C_k^i \cup C_k^j$ is contained in a neighborhood where the foliations $\mathscr{F}_k^u(f)$, $\mathscr{F}_k^s(f)$ are trivial whenever $C_k^i \cap C_k^j \neq \emptyset$. Since the foliations $\mathscr{F}_k^s(g)$ and $\mathscr{F}_k^u(g)$ depend continuously on g, the same property holds for $g \in N$ if we take N small enough. Furthermore, we can choose those neighborhoods such that if $x, y \in B_k^i$, and $P_{k,i}^s(x, g)$, $P_{k,i}^u(y, g)$ are the plaques of the foliations in C_k^i, then $P_{k,i}^s(x, g)$ intersects transversally $P_{k,i}^u(y, g)$ in a unique point. Let $D_k = \bigcup_{i=1}^r A_k^i$, $E_k = \bigcup_{i=1}^r B_k^i$, $F_k = \bigcup_{i=1}^r C_k^i$. Let U_k, V_k and W_k be neighborhoods of Λ_k such that $\mathrm{Cl}\, U_k \subset V_k \subset \mathrm{Cl}\, V_k \subset W_k \subset \mathrm{Cl}\, W_k \subset D_k$. Take N small enough so that $\Lambda_k(g) \subset U_k$ for each $g \in N$.

Lemma 3.1. *There exists a neighborhood $N_1 \subset N$ of f such that if $g \in N_1$, $g = f$ and $g^{-1} = f^{-1}$ outside W_k then there is a homeomorphism $h_g\colon M \to M$ such that $g h_g = h_g f$ and h_g depends continuously on g.*

Proof. Let $N_1 \subset N$ be a neighborhood of f such that if $g \in N_1$ there exists a homeomorphism $h_g\colon \Lambda_k(f) \to \Lambda_k(g)$ conjugating g and f. This homeomorphism h_g depends continuously on g [1]. With the same procedure as in Theorem 1.2 we get an extention of h_g to $W^s_\beta(\Lambda_k) \cup W^u_\beta(\Lambda_k)$ which preserves the foliations $\mathscr{F}^s_k$, $\mathscr{F}^u_k$ and conjugates f and g.

1. Extention of h_g to a neighborhood D_k of Λ_k

Let $z \in D_k$. Hence $z \in A^i_k$ for some i. Pick $\tilde{z} \in A^i_k \cap \Lambda_k$. If we take N_1 small enough we can suppose h_g is ε C^0-close to the inclusion $W^s_\beta(\Lambda_k) \cup W^u_\beta(\Lambda_k) \to M$, where $\varepsilon = \min\{d(A^i_k, M - B^i_k)\}$. Let $x = P^u_{k,i}(z, f) \cap P^s_{k,i}(\tilde{z}, f)$ and $y = P^s_{k,i}(z, f) \cap P^u_{k,i}(\tilde{z}, f)$, that is, x and y are the coordinates of z in the coordinate system given by the foliations $\mathscr{F}^s_k(f)$ and $\mathscr{F}^u_k(f)$ in A^i_k. Since $x, y \in A^i_k$, h_g is C^0-close to the inclusion, $x \in W^s_\beta(\Lambda_k)$ and $y \in W^u_\beta(\Lambda_k)$, it follows that $h_g(x)$ and $h_g(y)$ are defined and belong to B^i_k. Hence $P^u_{k,i}(h_g(x), g)$ intersects $P^s_{k,i}(h_g(y), g)$ in a unique point. Then we define $h_g(z) = P^s_{k,i}(h_g(y), g) \cap P^u_{k,i}(h_g(x), g)$. It is easy to verify that h_g is well defined, continuous, and conjugates f and g. Obviously h_g has a continuous inverse.

2. Extention of h_g to M

If $x \in \bigcup_{n \in Z} f^n(D_k)$ there exists an integer n_0 such that $f^{n_0}(x) \in D_k$. We define $h_g(x) = g^{-n_0} h_g f^{n_0}(x)$. If $x \in M - \bigcup_{n \in Z} f^n(D_k)$ we define $h_g(x) = x$. Since h_g/D_k is a conjugacy between f and g it follows that h_g is well defined on M and obviously conjugates f and g. It remains to show that h_g is continuous. Observe that h_g preserves the foliations $\mathscr{F}^s_k$ and $\mathscr{F}^u_k$. By the compactibility of the systems of tubular families we have that h_g preserves the stable and unstable manifolds of basic sets. Let Λ_j be a basic set of f such that Λ_k has behavior 1 with respect to Λ_j. Therefore $W^u(\Lambda_k) \cap W^s(\Lambda_j) \neq \emptyset$ and the unstable manifolds of Λ_k do not accumulate in points of the fundamental domain D^s_j. Now let us prove that h_g is continuous in a neighborhood of Λ_j. We can take the fundamental domain D^s_j disjoint from W_k. By the compatibility of the tubular families we have that h_g/D^s_j is a map from D^s_j to $W^s(\Lambda_j)$ whose image is a fundamental domain for $W^s(\Lambda_j)$ because h_g is a conjugacy. Consider the set $D^s_j \cap (\bigcup_{n \in Z} f^n(D_k))$ where h_g is obviously continuous. Looking at the construction in (1) we conclude that there exists a neighborhood D'_k of Λ_k such that $\mathrm{Cl}\, D'_k \subset D_k$ and $h_g = \text{identity}$ in $D^s_j \cap (\bigcup_n f^n(D_k) - \bigcup_n f^n(D'_k))$. Since $h_g = \text{identity}$ in $D^s_j - \bigcup_n f^n(D^k)$ we have that h_g/D^s_j is continuous. Since $W^u(\Lambda_j) \cap (\bigcup_n f^n(D_k)) = \emptyset$, $h_g = \text{identity}$ in $W^u(\Lambda_j)$ and, by the compatibility of the systems of tubular families, h_g preserves the foliation $\mathscr{F}^s_j$. Clearly h_g also preserves the foliation $\mathscr{F}^u_j$. Using the "local coordinate systems" in points of Λ_j given by the foliations $\mathscr{F}^s_j$ and $\mathscr{F}^u_j$ we have that h_g is continuous in a neighborhood of Λ_j, because $h_g/W^s(\Lambda_j)$ and $h_g/W^u(\Lambda_j)$ are continuous. Continuing the argument we have by induction that h_g is continuous in neighborhoods of basic sets greater than Λ_k. Similarly we show the continuity in basic sets smaller than Λ_k, and it follows that h_g is continuous in M. Clearly h_g has a continuous inverse, and so we are done.

Remark. In the proof above we used the transversality condition only in (2). Therefore, as a consequence of the first part of the proof we have the following.

Theorem 3.1. *If $f\colon M^2 \to M^2$ is a diffeomorphism satisfying Axiom A and Λ is a basic set of f then f is locally stable with respect to Λ.*

Let Λ_k, $k=1,\dots,m$ be the basic sets of f and $U_k \subset \mathrm{Cl}\, U_k \subset V_k \subset \mathrm{Cl}\, V_k \subset W_k \subset \mathrm{Cl}\, W_k \subset D_k$ neighborhoods of Λ_k where W_k and D_k are as in lemma 3.1. Consider the compact set $K = M - \bigcup_{k=1}^{m} V_k$. Since K is disjoint from $\Omega(f)$ we can take for each $x \in K$ a neighborhood W_x disjoint from $\bigcup_k U_k$ and a positive integer n_x such that for all integer n, with $|n| > n_x$, $f^n(W_x) \cap f^\alpha(W_x) = \emptyset$ with $\alpha = 0, 1, -1$. Let $W_{m+1}, \dots, W_{m+t}$ be a finite cover of K and $n_0 = \max n_i$. Hence, if n is an integer with $|n| > n_0$ then

$$f^n(W_{n+j}) \cap f^\alpha(W_{n+j}) = \emptyset$$

for $\alpha = 0, 1, -1$ and $j = 1, \dots, t$. If we take the neighborhood N_1, small enough, the same property holds for all $g \in N_1$. We are going to use the following lemma to localize the pertubations.

Lemma 3.2. *Let W_k, $k=1,\dots,n+t$ be an open cover of M and N a neighborhood of the identity in* $\mathrm{Diff}^1(M)$. *There exists a neighborhood $\tilde{N}_1 \subset N$ of the identity such that if $d \in \tilde{N}_1$, then $d = d_{m+t} \circ \cdots \circ d_1$ where $d_i \in N$ and* $\mathrm{supp}\, d_i \subset W_i$. *Here* $\mathrm{supp}\, d_i = \mathrm{Cl}\{p \in M;\, d_i(p) \neq p\}$.

Proof. See [6].

Theorem 3.2. *Let M be a compact, C^∞ two-dimensional manifold. If $f\colon M \to M$ is a diffeomorphism satisfying Axiom A and the Transversality Condition then f is structurally stable.*

Proof. Consider the open cover of M, $\{W_1, \dots, W_{n+t}\}$ as above. Take a neighborhood N of f such that if $g \in N$ then g satisfies Axiom A and Transversality Condition. If N is small enough we have, by Lemma 3.2, $gf^{-1} = f_{n+t} \circ \cdots \circ f_1$ where f_i is close to the identity and $\mathrm{supp}\, f_i \subset W_i$. Hence each $\bar{f}_i = f_i \circ \cdots \circ f_1 \circ f$ is C^1 close to f and therefore satisfies Axiom A and the Transversality Condition. So we will be done if we show that $\bar{f}_i = f_i \circ \cdots \circ f_1 \circ f$ and $\bar{f}_{i-1} = f_{i-1} \circ \cdots \circ f_1 \circ f$ are conjugated. Therefore we can suppose $g = f_i \circ f$ where f_i is C^1 close to the identity and $\mathrm{supp}(f_i) \subset W_i$.

If $i \leqq n$, Lemma 3.1 gives a conjugacy between f and g. If $i > n$ there exists a positive integer n_0 such that if $|k| > n_0$ then $f^k(W_i) \cap f^\alpha(W_i) = \emptyset$ and $g^k(W_i) \cap g^\alpha(W_i) = \emptyset$ for $\alpha = 0, 1, -1$. We define

$$h(x) \lim_{n \to \infty} g^n f^{-n}(x)$$

If $x \in \bigcup_{n \in Z} f^n(W_i)$, the sequence $(g^n f^{-n}(x))$ is eventually constant and the limit exists. If $x \notin \bigcup_{n \in Z} f^{-n}(W_i)$ then $h(x) = x$. Therefore h is well defined. Clearly h is continuous and is a conjugacy between f and g. It is easy to see that $k(x) = \lim f^n g^{-n}(x)$ is a continuous inverse for f.

References

1. Hirsch, M., Pugh, C.: Stable manifolds and hyperbolic sets. Global Analysis. Proc. Symp. in Pure Math. **XIV**, AMS, 1970
2. Hirsch, M., Palis, J., Pugh, C., Shub, M.: Neighborhoods of hyperbolic sets. Inventiones math. **9**, 121–134 (1970)
3. Munkers, J. R.: Elementary differential topology. Princeton, N.J.: Princeton University Press 1961
4. Palis, J.: On Morse-Smale dynamical systems. Topology **8**, 385–405 (1969)
5. Palis, J.: Seminario de Sistemas Dinamicos. IMPA, 1971
6. Palis, J., Smale, S.: Structural stability theorems, global analysis. Proc. Symp. in Pure Math. **XIV**, AMS, 1970
7. Pugh, C.: An improved closing lemma and a general density theorem. American Journal of Math. **89**, 1010–1021 (1967)
8. Robbin, J.: A structural stability theorem. Annals of Math. 94 (1971)
9. Smale, S.: Differentiable dynamical systems. Bull. AMS. 73 (1967)

W. de Melo
Instituto de Matematica Pura
e Aplicada (IMPA)
Rio de Janeiro, Brasil

(Received March 23, 1973)

Moduli of Stability of Two-Dimensional Diffeomorphisms

W. de Melo

Topology **19**, 9–21 (1980)

INTRODUCTION

IN THIS paper we consider diffeomorphisms on a compact two-dimensional manifold having a pair of hyperbolic periodic points such that the unstable manifold of one of them is tangent to the stable manifold of the other along an orbit. They occur naturally in Bifurcation Theory: there is an open set of arcs, starting at a Morse–Smale diffeomorphism, whose first bifurcation point is such a diffeomorphism[3].

It is well known that a diffeomorphism f which exhibits an orbit of tangency between stable and unstable manifolds of hyperbolic periodic points cannot be structurally stable[7]. In fact, this situation gives rise to interesting invariants of topological equivalence, as pointed out in [4], which implies the existence of an uncountable number of different topological equivalence classes in any small neighborhood of f. However, it might be possible to parametrize all these equivalence classes with finitely many real parameters which would provide a pretty good description of the diffeomorphisms near f. In this case we say that the modulus of stability of f is finite and equal to the minimum number of parameters. Here we will prove that in some cases the modulus of stability is finite and in other cases it is infinite and that both situations occur for diffeomorphisms which are first bifurcation points for an open set of arcs starting at Morse–Smale diffeomorphisms.

The existence of those tangencies is not a persistent phenomenum in the space of all diffeomorphisms. However, if we restrict our perturbations to a specific subspace it may become persistent. This is the case when we consider the set of G-equivariant diffeomorphisms, where G is a finite group acting on the manifold. The action of G induces a partition of the manifold in submanifolds whose points have the same orbit type and this partition is let invariant by any equivariant diffeomorphism f. If f has two hyperbolic periodic points p_1 and p_2 such that the stable manifold of p_1 intersects the unstable manifold of p_2 and both are contained in the same one-dimensional submanifold of the partition then the same happens for any equivariant diffeomorphism near f. We prove here that, given any integer k, there is an open set of equivariant diffeomorphisms whose modulus of stability is exactly k.

§1. STATEMENT OF RESULTS

Let $\mathrm{Diff}^r(M)$ be the set of C^r, $2 \le r \le \infty$, diffeomorphisms on a compact C^∞ manifold M, endowed with C^r topology. The diffeomorphisms f and g are topologically equivalent if there is a homeomorphism h, called a conjugacy between f and g, such that $hf = gh$. This is clearly an equivalence relation and the equivalence classes are called conjugacy classes. If there is a neighborhood of f contained in its conjugacy class we say that f is structurally stable. If a small neighborhood of f intersects only finitely many conjugacy classes then the modulus of stability of f is equal to 0. In particular, a structurally stable diffeomorphism has zero modulus of stability.

Originally published in *Topology* **19**, 9–21 (1980).

Let p be a periodic point of f of period π. Then p is hyperbolic if the eigenvalues of $df^{\pi}(p)$ have absolute value different from one. If an eigenvalue has absolute value less than one and the other bigger than one we say that p is a saddle. The stable manifold of p, $W^s(p)$, is the set of points $x \in M$ such that $f^{n\pi}(x)$ converges to p as $n \to \infty$. The stable manifold of the orbit of p, $W^s(0(p))$, is the union of $W^s(q)$ for q in the orbit of p. Similarly, the unstable manifold of p, $W^u(p)$, is the set of points $x \in M$ such that $f^{-n\pi}(x)$ converges to p as $n \to \infty$. $W^s(p)$ and $W^u(p)$ are C^r immersed submanifolds of M which are transversal to each other at the point p.

Let p and q be hyperbolic periodic points of f. We say that $z \in W^s(p) \cap W^u(q)$ is a point of quasi-transversal intersection if $W^s(p)$ is tangent to $W^u(q)$ at z and the contact between them is parabolic. The same happens for any point in the orbit of z and it is called an orbit of quasi-transversal intersection.

In §1 we prove the following theorem.

THEOREM B. *Let $f \in \mathrm{Diff}^r(M)$. If there are k orbits of tangency between stable and that $W^u(p_1)$ has a tangency with $W^s(p_2)$. If there are infinitely many orbits in $W^s(0(p_1))$ belonging to unstable manifolds of periodic saddle points of f then the modulus of stability of f is infinite.*

In the same section we define a new conjugacy invariant which gives rise to the following theorem.

THEOREM B. *Let $f \in \mathrm{Diff}^r(M)$. If there are k orbitals of tangency between stable and unstable manifolds of hyperbolic saddle orbits of f then the modulus of stability of f is at least k.*

Recall that a cycle of periodic points of f is a sequence $p_1, p_2, \ldots, p_n = p_1$, of hyperbolic periodic points such that $W^u(p_i) \cap W^s(p_{i+1}) \neq \phi$ for $i = 1, \ldots, n-1$.

Let $U_k \subset \mathrm{Diff}^\infty(M)$ be the set of diffeomorphisms f satisfying the following properties:

(1) The set of non-wandering points of f is finite and hyperbolic;

(2) f has no cycles;

(3) There are only k orbits of tangency between stable and unstable manifolds of periodic points of f and these are orbits of quasi-transversal intersection;

(4) If p_1 and p_2 are periodic points of f such that $W^u(0(p_1))$ is not transversal to $W^s(0(p_2))$ then there is no saddle point of f whose unstable manifold intersects $W^s(0(p_1))$ or whose stable manifold intersects $W^u(0(p_2))$;

(5) If p is a periodic point of f of period π then if g is any diffeomorphism near f, g^{π} is C^2-linearizable in a neighborhood of p.

We observe that from [8], the set of diffeomorphisms satisfying condition (5) is open and dense in the set of diffeomorphisms satisfying conditions (1)–(4).

Section 3 is dedicated to the proof of the theorem below.

THEOREM C. *If $f \in U_k$ then the modulus of stability of f is equal to k.*

We notice that the set U_1 defined above is open in the boundary of the set of Morse–Smale diffeomorphisms. On the other hand, let us consider the set υ of diffeomosphisms satisfying conditions (1)–(3) above, for $k = 1$, and also the following property: if $W^u(0(p_1))$ is not transversal to $W^s(0(p_2))$ then there are saddle points q_1, q_2 such that $W^u(0(q_1)) \cap W^s(0(p_1)) \neq \phi$ and $W^u(0(q_2)) \cap W^s(0(q_1)) \neq \phi$. Then υ is also open in the boundary of the set of Morse–Smale diffeomorphisms and, by

Theorem A, any $f \in \upsilon$ has modulus of stability infinite. After the completion of this paper I realized upon remarks of S. Newhouse, C. Pugh and the referee that Theorem C can be extend to C^2 diffeomorphisms satisfying properties (1)–(4).

In §4 we deal with a situation which has codimension infinite in the set of all diffeomorphisms. We consider the set $\mathfrak{m}$ of C^∞ diffeomorphisms on M satisfying properties (1), (2), (4) and also the following condition (3'): if p_1 and p_2 are periodic points such that $W^u(p_1)$ is not transversal to $W^s(p_2)$ then one of the connected components of $W^s(p_2) - \{p_2\}$ is contained in $W^u(p_1)$.

The modulus of stability of any $f \in \mathfrak{m}$ is infinite. However, it may become finite if we restrict our perturbations to $\mathfrak{m}$. In fact we have the following theorem.

THEOREM D. *For each integer k there is a subset $\mathfrak{m}_k \subset \mathfrak{m}$ such that $\mathfrak{m}_k$ is open in $\mathfrak{m}$ and the modulus of stability in $\mathfrak{m}$ of any $f \subset \mathfrak{m}_k$ is equal to k. Furthermore, $\bigcup_{k=1}^{\infty} \mathfrak{m}_k$ is dense in $\mathfrak{m}$.*

Using Theorem D we can describe open sets in the space of equivariant diffeomorphisms having modulus of stability equal to k for any integer k. This occurs in the case where G is a finite group and the partition of M in orbit types has at least one submanifold of dimension one.

We finish this section by stating some open questions.

Let $f \in \mathrm{Diff}^\infty(M)$ satisfy properties (1), (2), (3) and (5) above. Suppose also that if p_1 and p_2 are periodic points of f such that $W^u(0(p_1))$ is not transversal to $W^s(0(p_2))$, then there are finitely many orbits in $W^s(0(p_1))$ of transversal intersection with unstable manifolds of saddle points and there is no orbit in $W^u(0(p_2))$ belonging to stable manifold of saddle point. Is the modulus of stability of f finite?

Suppose now that there are finitely many orbits in $W^s(0(p_1))$ and in $W^u(0(p_2))$ belonging to unstable and stable manifolds of saddle points. Is the modulus of stability of f infinite? We suspect the answer is yes.

Problem. Let $f \in \mathrm{Diff}^\infty(M)$ satisfy properties (1), (2), (4) and also the following condition: there are finitely many orbits of tangency between stable and unstable manifolds of periodic points and the contact between these manifolds is of finite order. Find the modulus of stability of f.

We are grateful to A. Lins, J. Palis and F. Takens for many stimulating discussions.

§2. PROOF OF THEOREMS A AND B

Let f be a C^2 diffeomorphism of a compact two dimensional manifold M having a pair of hyperbolic periodic points p_1 and p_2 such that $W^u(p_1)$ meets $W^s(p_2)$ non-transversally. Let π_i be the periods of p_i, $i = 1, 2$, and let λ_i and μ_i be the eigenvalues of $df^{\pi_i}(p_i)$ with $0 < |\lambda_i| < 1$ and $|\mu_i| > 1$. We say that $q \in W^u(p_1) \cap W^s(p_2)$ is a one side tangency point if there is a neighborhood V of q such that $W^u(p_1)$ intersects at most one connected component of $V - W^s(p_2)$. If q is a point of quasi-transversal (parabolic) intersection of $W^u(p_1)$ and $W^s(p_2)$ then q is a one side tangency point.

It follows from [4] that if there is a one side tangency point of $W^u(p_1)$ and $W^s(p_2)$ then $T_1(f, p_1, p_2) = [(\log|\mu_2|)/(\log|\lambda_1|)]$ is a conjugacy invariant. This means that if h: $M \to M$ is a conjugacy between f and $\bar{f}$, with $h(p_1) = \bar{p}_1$ and $h(p_2) = \bar{p}_2$, where $\bar{p}_i$ are hyperbolic periodic points of $\bar{f}$ such that $W^u(\bar{p}_1)$ meets $W^s(\bar{p}_2)$ non-transversally, then $T_1(f, p_1, p_2) = T_1(\bar{f}, \bar{p}_1, \bar{p}_2)$. From this it follows that the modulus of stability of f is at least one. Here we will prove that the restriction of f to $W^s(p_1)$ and $W^u(p_2)$ is determined once we know the image of a single point in $W^s(p_1)$ or in $W^u(p_2)$. This is the main step in the proof of Theorem A. Furthermore we will associate to each pair of one side tangency points q_1 and q_2 a new conjugacy invariant $T_2(f, p_1, p_2, q_1, q_2)$ which will be used in the proof of Theorem B and C.

To simplify the exposition we will assume that p_1 and p_2 are fixed points and the eigenvalues λ_i and μ_i are positive. This can be done by considering, instead of f, the diffeomorphism f^n for a suitable integer n.

Let q be a one side tangency point of $W^u(p_1)$ and $W^s(p_2)$. Since f is of class C^2, there are neighborhoods U_i of p_i, with $q \in U_1$, and C^1 coordinate systems φ_i: $U_i \to R^2$ linearizing f[1]. Hence, $\varphi_i \circ f \circ \varphi_i^{-1}(x^1, x^2) = (\lambda_i x^1, \mu_i x^2)$ for $i = 1, 2$. For $p \in U_i$ we denote by (p^1, p^2) the coordinates of $\varphi_i(p)$. Let k be a positive integer such that $f^k(q) \in U_2$. Consider the mapping g defined on a neighborhood of $\varphi_1(q) = (0, q^2)$ by $g(x^1, x^2) = \varphi_2 \circ f^k \circ (\varphi_1)^{-1}(x^1, x^2)$. Let $\alpha(q, k) = [(\partial g^2)/(\partial x^1)](0, q^2)$ where g^2 is the second component of g. The following lemma is easy to prove.

Lemma 2.1. (1) *$\alpha(q, k)$ is different from zero*; (2) *if $q_n \to q$, $z_n \doteq f^k(q_n)$ and $[(q_n^2 - q^2)/q_n^1]$ is bounded then $[z_n^2/q_n^1]$ converges to $\alpha(q, k)$*; (3) *if $q_n \to q$ and $q_n^1 \neq 0$ for all n then $[(g^2(q_n^1, q_n^2) - g^2(0, q_n^2))/q_n^1]$ converges to $\alpha(q, k)$*; (4) *if m is a positive integer then $\alpha(q, k + m) = (\mu_2)^m \cdot \alpha(q, k)$ and $\alpha(f^{-m}(q), k + m) = (\lambda_1)^m \alpha(q, k)$.*

Lemma 2.2. *Let $q_1, q_2 \in U_1$ be one side tangency points of $W^u(p_1)$ and $W^s(p_2)$. If $\beta(f, p_1, p_2, q_1, q_2) = [\alpha(q_2, k)/\alpha(q_1, k)]$, where k is a positive integer such that $f^k(q_1)$ and $f^k(q_2)$ belong to U_2, then β doen not depend on the choices of the integer k and of the coordinate systems φ_i. Furthermore, if m_1 and m_2 are integers then*

$$\beta(f, p_1, p_2, f^{m_1}(q_1), f^{m_2}(q_2)) = \left(\frac{\mu_2}{\lambda_1}\right)^{m_2 - m_1} \beta(f, p_1, p_2, q_1, q_2).$$

Proof. Using (4) of Lemma 2.1 it is easy to see that β does not dependent on k. Let $\tilde{\varphi}_i$: $U_i \to R^2$ be others C^1 coordinate systems linearizing f. Then $\Phi_1 = \varphi_1 \circ \tilde{\varphi}_1^{-1}$ and $\Phi_2 = \tilde{\varphi}_2 \circ \varphi_2^{-1}$ are diffeomorphisms, defined on a neighborhood of the origin of R^2, which commute with the linear isomorphisms L_i: $R^2 \to R^2$, $L_i(x^1, x^2) = (\lambda_i x^1, \mu_i x^2)$, for $i = 1, 2$ respectively. Thus,

$$\frac{\partial \phi_i}{\partial x^j}(x^1, 0) = \frac{\partial \phi_i}{\partial x^j}(0, x^2) = \frac{\partial \phi_i}{\partial x^j}(0, 0)$$

for $i, j = 1, 2$ and

$$\frac{\partial \phi_i^1}{\partial x^2}(0, 0) = \frac{\partial \phi_i^2}{\partial x^1}(0, 0) = 0$$

for $i = 1, 2$. From these formulas it follows easily that

$$\alpha(f, p_1, p_2, q_i, \tilde{\varphi}_1, \tilde{\varphi}_2) = \frac{\partial \phi^2}{\partial x^2}(0, 0) \cdot \alpha(f, p_1, p_2, q_i, \varphi_i, \varphi_2) \cdot \frac{\partial \phi^1}{\partial x^1}(0, 0).$$

Therefore β does not depend on the choice of the coordinate systems. The equality in Lemma 2.2 follows easily from (4) of Lemma 2.1.

We recall that a fundamental domain for the stable manifold of a hyperbolic fixed point p of f is a compact manifold with boundary $D^s \subset W^s(p)$ such that $\bigcup_{n \in Z} f^n(D^s) = W^s(p) - \{p\}$ and, for each x in the interior of D^s, $f^n(x)$ does not belong to

D^s if $n \neq 0$. If p is a saddle point in dimension 2 then D^s is either a closed interval or a pair of closed intervals.

LEMMA 2.3. *Let $f \in \mathrm{Diff}^2(M)$ be as above. Suppose that $[(\log \mu_2)/(\log \lambda_1)]$ is irrational. Let $q \in U_1$ be a one side tangency point of $W^u(p_1)$ and $W^s(p_2)$. Let $w \in W^u(p_2)$. Then there exist a set G_w, contained in a fundamental domain D^s of $W^s(p_1)$, with the following properties: (a) G_w is dense in one of the connected components of D^s; (b) for each $x \in G_w$ there are sequences $x_j \to x$, $m_j \to \infty$ and $n_j \to \infty$ such that $q_j = f^{m_j}(x_j)$ converges to q, $[(q_j^2 - q^2)/q_j^1]$ is bounded and $f^{n_j+k}(q_j)$ converges to w where k is an integer such that $f^k(q_j) \in U_2$.*

Proof. Let $z = f^k(q)$ and $S \subset U_2$ be a 1-dimensional disc transversal to $W^s(p_2)$ at z. It is clear that there exist a sequence z_n in S such that $f^n(z_n)$ converges to $w = (0, w^2)$. Hence $z_n^2 = (\mu_2)^{-n} w^2$. If $q_n = f^{-k}(z_n)$ then q_n converges to q and $z_n^2 = \alpha_n q_n^1$. Since $q_n \in f^{-k}(S)$ which is transversal to $W^u(p_1)$ it follows that $[(q_n^2 - q^2)/q_n^1]$ converges. In particular $[(q_n^2 - q^2)/q_n^1]$ is bounded and α_n converges to $\alpha = \alpha(q, k)$. We may assume that z_n^2 and q_n^1 are positive and $\alpha(q, k) > 0$. It suffices to prove that the set of limit points of the double sequence $f^{-j}(q_n)$ is dense in $D_+^s = \{(x^1, 0) \in U_1;\ \lambda_1 \le x^1 \le 1\}$. Let $s_n = [(\log q_n^1)/(\log \lambda_1)]$. Then

$$s_n = \frac{\log \dfrac{1}{\alpha} w^2}{\log \lambda_1} + \frac{\log \dfrac{\alpha}{\alpha_n}}{\log \lambda_1} - n \frac{\log \mu_2}{\log \lambda_1}.$$

Hence

$$s_n = a + b_n - n \frac{\log \mu_2}{\log \lambda_1}$$

with b_n converging to zero. For each n we may write $s_n = \bar{s}_n + m_n$ where $\bar{s}_n \in [0, 1]$ and m_n is an integer. Hence $m_n \to \infty$ and since $[(\log \mu_2)/(\log \lambda_1)]$ is irrational, it follows that $\{\bar{s}_n, n \ge 0\}$ is dense in $[0, 1]$. Hence the set $\bar{G}$ of limit points of the sequence $\bar{s}_n$ is dense in $[0, 1]$. Consequently the set $G_w = \{(x^1, 0);\ x^1 = (\lambda_1)^\delta$ for some $\delta \in \bar{G}\}$ is dense in D_+^s. We claim that any point in G_w is a limit point of the double sequence $f^{-j}(q_n)$. In fact, if $x \in G_w$ then $x = (x^1, 0)$ with $x^1 = (\lambda_1)^\delta$ and $\delta = \lim_{j\to\infty} \bar{s}_{n_j}$. Hence

$$x^1 = \lim_{j\to\infty} (\lambda_1)^{\bar{s}_{n_j}} = \lim_{j\to\infty} (\lambda_1)^{s_{n_j}} (\lambda_1)^{-m_{n_j}} = \lim_{j\to\infty} (\lambda_1)^{-m_{n_j}} \times \exp\left(\frac{\log q_{n_j}^1}{\log \lambda_1} \cdot \log \lambda_1\right) = \lim_{j\to\infty} (\lambda_1)^{-m_{n_j}} q_{n_j}^1.$$

Therefore $x_{n_j} = f^{-m_{n_j}}(q_{n_j})$ converges to x as $j \to \infty$. This proves the claim and the lemma.

Remarks. If μ_2 is positive, λ_1 negative and $[(\log \mu_2)/(\log |\lambda_1|)]$ is irrational then G_w is dense in the fundamental domain D^s. If both eigenvalues are negative and $[(\log |\mu_2|)/(\log |\lambda_1|)]$ is irrational the lemma remains true.

Let $\bar{f}$ be another C^2 diffeomorphism having a one side tangency point $\bar{q}$ of $W^u(\bar{p}_1)$ and $W^s(\bar{p}_2)$. Let $0 < \bar{\lambda}_i < 1$ and $\bar{\mu}_i > 1$ be the eigenvalues of $d\bar{f}(\bar{p}_i)$, $i = 1, 2$. As before we consider C^1 coordinate systems in neighborhoods $\bar{U}_i$ of $\bar{p}_i$, $\bar{U}_1 \supset \bar{q}$, linearizing $\bar{f}$. We denote by $(\bar{x}^1, \bar{x}^2)$ the coordinates of a point $x \in \bar{U}_i$.

Lemma 2.4. *Let $h: M \to M$ be a conjugacy between f and $\bar{f}$ such that $h(p_i) = \bar{p}_i$ and $h(q) = \bar{q}$. Then there exist constants a_-, a_+, b_-, b_+ such that:*

$$h(x^1, 0) = (a_+(x^1)^c, 0) \quad \text{if} \quad (x^1, 0) \in U_1 \cap W^s(p_1) \quad \text{and} \quad x^1 \geq 0;$$

$$h(x^1, 0) = (a_-|x^1|^c, 0) \quad \text{if} \quad (x^1, 0) \in U_1 \cap W^s(p_1) \quad \text{and} \quad x^1 \leq 0;$$

$$h(0, y^2) = (0, b_+(y^2)^c) \quad \text{if} \quad y \in U_2 \cap W^u(p_2) \quad \text{and} \quad y^2 \geq 0;$$

$$h(0, y^2) = (0, b_-|y^2|^c) \quad \text{if} \quad y \in U_2 \cap W^u(p_2) \quad \text{and} \quad y^2 \leq 0.$$

Here

$$c = \frac{\log \bar{\mu}_2}{\log \mu_2}.$$

Proof. Let $V_q \in U_1$ and $V_z \subset U_2$ be small neighborhoods of q and $z = f^k(q)$ respectively. We may choose the coordinates in U_1 and U_2 in such a way that $V_q \cap W^s(p_2) \subset \{x \in V_q; x^1 \geq 0\}$ and $V_z \cap W^u(p_1) \subset \{x \in V_z; x^2 \leq 0\}$. Similarly we can choose the coordinates in $\bar{U}_i$ in such a way that $V_{\bar{q}} \cap W^s(\bar{p}_2) \subset \{x \in V_{\bar{q}}; x^1 \geq 0\}$ and $V_{\bar{z}} \cap W^u(\bar{p}_1) = \{x \in V_{\bar{z}}; x^2 \leq 0\}$, where $V_{\bar{q}}$ and $V_{\bar{z}}$ are small neighborhoods of $\bar{q}$ and $\bar{z}$ respectively. Let h be a conjugacy between f and $\bar{f}$ sending p_i on $\bar{p}_i$ and q on $\bar{q}$. For $p \in M$ we denote by $\bar{p}$ the image of p under h. Since $h(W^s(p_i)) = W^s(\bar{p}_i)$, $h(W^u(p_i)) = W^u(\bar{p}_i)$ it follows that if p is near q and p^1 is positive then $\bar{p}^1$ is positive; if p is near z and p^2 is positive then $\bar{p}^2$ is positive. Let w be a point in $W^u(p_2)$ with $w^2 > 0$. Let G_w be as in Lemma 2.3. Let $x \in G_w$, $x_j \to x$, $m_j \to \infty$ and $n_j \to \infty$ be such that $q_j = f^{m_j}(x_j)$ converges to q and $w_j = f^{k+n_j}(q_j) = f^{n_j}(z_j)$ converges to w. Hence q_j^1 and z_j^2 are positive and $z_j^2 = \alpha_j q_j^1$ where α_j converges to $\alpha = \alpha(q, k)$. Since $w_j^2 = \alpha_j \mu_2^{n_j} \lambda_1^{m_j} x_j^1$ converges to w^2 and x_j^1 converges to x^1, it follows that $\mu_2^{n_j}\lambda_1^{m_j}$ converges to $(w^2/\alpha x^1)$. Using the fact that h is a conjugacy we conclude that $\bar{x}_j$ converges to $\bar{x} = (\bar{x}^1, 0)$, $\bar{q}_j = \bar{f}^{m_j}(\bar{x}_j)$ converges to $\bar{q}$, $\bar{z}_j = \bar{f}^k(\bar{q}_j)$ converges to $\bar{z}$ and $\bar{w}_j = \bar{f}^{n_j}(\bar{z}_j)$ converges to $\bar{w}$. Furthermore $\bar{z}_j^2$ and $\bar{q}_j^1$ are positive. Hence $\bar{w}_j^2 = d_j \bar{\mu}_2^{n_j} \bar{\lambda}_1^{m_j} \bar{x}_j^1$ where $d_j = (\bar{z}_j^2/\bar{q}_j^1)$ is positive. Since $[(\log \bar{\mu}_2)/(\log \mu_2)] = [(\log \bar{\lambda}_1)/(\log \lambda_1)] = c$ it follows $\bar{\mu}_2^{n_j}\bar{\lambda}_1^{m_j} = (\mu_2^{n_j}\lambda_1^{m_j})^c$ converges to $(w^2/\alpha x^1)^c$. Thus d_j converges to $(\bar{w}^2/\bar{x}_1) \cdot [\alpha^c (x^1)^c/(w^2)^c]$. We claim that $\lim_{j\to\infty} d_j$ is less or equal $\bar{\alpha}(\bar{f}, \bar{p}_1, \bar{p}_2, \bar{q}, k)$. In fact, if $\bar{g}^2$ is the second component of the mapping $\bar{g} = \bar{\varphi}_2 \circ \bar{f}^k \circ \bar{\varphi}_1^{-1}$ then

$$\frac{\bar{z}_j^2}{\bar{q}_j^1} = \frac{\bar{g}^2(\bar{q}_j^1, \bar{q}_j^2)}{\bar{q}_j^1} \leq \frac{\bar{g}^2(\bar{q}_j^1, \bar{q}_j^2) - \bar{g}^2(0, \bar{q}_j^2)}{\bar{q}_j^1}$$

because $g^2(0, q_j^2)$ is negative. By Lemma 2.1, the right member converges to $\bar{\alpha}$. Hence

$$\lim_{j\to\infty} d_j = \frac{\bar{w}^2}{\bar{x}^1} \frac{\alpha^c (x^1)^c}{(w^2)^c} \leq \bar{\alpha}.$$

Therefore $\bar{x}^1 \geq a_+(x^1)^c$ for every $x \in G_w$ where $a_+ = [\bar{w}^2 \cdot \alpha^c/(w^2)^c \cdot \bar{\alpha}]$. Since G_w is dense in one of the connected components of a fundamental domain, the inequality remains true for every $x \in W^s(p_1)$ with $x^1 \geq 0$. By reasoning with h^{-1} instead of h we prove that $(x^1)^c \geq (1/a_+)\bar{x}_1$ for every $\bar{x} \in W^s(\bar{p}_1)$ with $\bar{x}^1 \geq 0$. Therefore $\bar{x}^1 = a_+(x^1)^c$ for every $x \in W^s(p_1)$ with $x^1 \geq 0$. Now, if $y \in W^u(p_2)$ is such that $y^2 > 0$ then we conclude, using the same arguments as above, that $a_+ = [\bar{y}^2 \cdot \alpha^c/(y^2)^c\bar{\alpha}]$. Hence $\bar{y}^2 = b_+(y^2)^c$ where $b_+ = [a_+\bar{\alpha}/(\alpha)^c]$. To prove the existence of the constants a_- and b_- we use the same reasoning with f^{-1} and $\bar{f}^{-1}$. Thus the lemma is proved.

LEMMA 2.5. *Let q_1, q_2 be one side tangency points of $W^s(p_2)$ and $W^u(p_1)$. Let h: $M \to M$ be a conjugacy between f and $\bar{f}$ such that $h(p_i) = \bar{p}_i$ and $h(q_i) = \bar{q}_i$, $i = 1, 2$, where q_i are one side tangency points of $W^s(\bar{p}_2)$ and $W^u(\bar{p}_1)$. If $[(\log \mu_2)/(\log \lambda_1)]$ is irrational them $|\beta(\bar{f}, \bar{p}_1, \bar{p}_2, \bar{q}_1, \bar{q}_2)| = |\beta(f, p_1, p_2, q_1, q_2)|^c$ where $c = [(\log \bar{\mu}_2)/(\log \mu_2)]$ and β is as in Lemma 2.2.*

Proof. Let $w \in W^u(p_2)$ and $x \in W^s(p_1)$ be such that $w^2 > 0$ and $x^1 > 0$. Since q^1 is a one side tangency point, it follows from the proof of Lemma 2.4 that

$$\frac{\bar{x}^1}{(x^1)^c} = a_+ = \frac{\bar{w}^2 \cdot |\alpha(q_1, k)|^c}{(w^2)^c|\bar{\alpha}(\bar{q}_1, k)|}.$$

On the other hand, q_2 is also a one side tangency point and, therefore,

$$a_+ = \frac{\bar{w}^2|\alpha(q_2, k)|^c}{(w^2)^c|\bar{\alpha}(\bar{q}_2, k)|}.$$

Thus

$$|\beta(\bar{f}, \bar{p}_1, \bar{p}_2, \bar{q}_1, \bar{q}_2)| = \frac{|\bar{\alpha}(\bar{q}_2, k)|}{|\bar{\alpha}(\bar{q}_1, k)|} = \left|\frac{\alpha(q_2, k)}{\alpha(q_1, k)}\right|^c$$

$$= |\beta(f, p_1, p_2, q_1, q_2)|^c.$$

Since the sign of β is also preserved by a conjugacy we have the following corollary.

COROLLARY. *Let q_1, q_2 be one side tangency points of $W^u(p_1)$ and $W^s(p_2)$. Let $T_2(f, p_1, p_2, q_1, q_2)$ be equal to $\pm|\beta(f, p_1, p_2, q_1, q_2)|^{1/(\log|\lambda_1|)}$ and having the same sign as β. If $[(\log|\mu_2|)/(\log|\lambda_1|)]$ is irrational then T_2 is a conjugacy invariant.*

Proof of Theorem B. Let f be a C^2 diffeomorphism having two hyperbolic periodic points p_1, p_2 such that $W^u(p_1) \cap W^s(p_2)$ contains k orbits of tangency. We can approximate f by a C^2 diffeomorphism $\bar{f}$ which has two hyperbolic periodic points $\bar{p}_1$, $\bar{p}_2$, near p_1 and p_2 resp., such that $W^u(\bar{p}_1) \cap W^s(\bar{p}_2)$ contains k orbits of quasi-transversal intersection and $[(\log|\bar{\mu}_2|)/(\log|\bar{\lambda}_1|)]$ is irrational, where $\bar{\lambda}_i$ and $\bar{\mu}_i$ are the eigenvalues of the periodic points $\bar{p}_i$ with $|\bar{\lambda}_i| < 1$ and $|\bar{\mu}_i| > 1$, $i = 1, 2$. Since a point of quasi-transversal intersection is a one side tangency point, it follows from Lemma 2.5 that the modulus of stability of $\bar{f}$ is at least k. Hence the modulus of stability of f is at least k and the theorem is proved.

Proof of Theorem A. Let f be a C^2 diffeomorphism having two periodic points p_1, p_2 such that $W^u(p_1) \cap W^s(p_2)$ contains an orbit of quasi-transversal intersection. Suppose that $[(\log|\mu_2|)/(\log|\lambda_1|)]$ is irrational where λ_i and μ_i are the eigenvalues of $df^{n_i}(p_i)$, $i = 1, 2$, with $|\lambda_i| < 1$ and $|\mu_i| > 1$. If $W^s(p_1)$ contains k orbits whose α-limit are hyperbolic periodic points of f of saddle type then it follows easily from Lemma 2.4 that the modulus of stability of f is at least k. It is easy to see that the same holds if $k = \infty$.

§3. PROOF OF THEOREM C

Let U_k be the set of C^∞ diffeomorphisms satisfying properties (1)–(4) of §1. Let us prove that the modulus of stability of $f \in U_k$ is equal to k. To simplify the exposition we assume that the k orbits of tangency are in $W^u(p_1) \cap W^s(p_2)$ where p_1 and p_2 are fixed points of f. We assume also that the eigenvalues λ_i and μ_i of $df(p_i)$ are positive, $\lambda_i < 1$ and $\mu_i > 1$. The proof when there are orbits of tangency between stable and unstable manifolds of several pairs of periodic orbits is similar.

Let $D^s \subset W^s(p_2)$ be a fundamental domain such that no unstable manifold of saddle point of f meets the boundary of D^s. Let $z_1, \ldots, z_k \in D^s$ be the tangency points of $W^u(p_1) \cap W^s(p_2)$ in D^s. Let $V_1, \ldots, V_k$ be disjoint neighborhoods of $z_1, \ldots, z_k$ such that z_i is the only point in V_i belonging to unstable manifold of saddle point of f.

If $\mathscr{n}$ is a small neighborhood of f then each $g \in \mathscr{n}$ has fixed points $p_i(g)$ near $p_i = p_i(f)$ and the eigenvalues $\lambda_i(g)$ and $\mu_i(g)$ of $dg(p_i(g))$ are near λ_i and μ_i respectively. Furthermore, for $i = 1, \ldots, k$, $W^u(p_1(g)) \cap W^s(p_2(g)) \cap V_i$ is either empty or a pair of points of transversal intersection or a point $z_i(g)$ of quasi-transversal intersection. It is easy to see that if $\mathscr{n}$ is small enough then any $g \in \mathscr{n}$ is either a Morse–Smale or an element of U_j for some $j \leq k$.

For each $g \in \mathscr{n}$ we choose a fundamental domain D_g^s of $W^s(p_2(g))$ such that D_g^s varies continuously with g. For any sequence of integers $I = (I_1, \ldots, I_j)$ with $1 \leq I_1 < I_2 < \cdots < I_j \leq k$ we denote by $\mathscr{n}_I$ the set of diffeomorphisms $g \in \mathscr{n}$ such that the points of tangency of $W^u(p_1(g))$ and $W^s(p_2(g))$ in D_g^s are $z_{I_1}(g), \ldots, z_{I_j}(g)$. Let $T_1(g)$ denote the number $[(\log \mu_2(g))/(\log \lambda_1(g))]$ and let $T_2(z_{I_1}(g), z_{I_j}(g)), \ldots, T_2(z_{I_{j-1}}(g), z_{I_j}(g))$ be the conjugacy invariants defined in §2.

It is easy to see that Theorem C follows from the theorem below.

THEOREM 3.1. *Let $g, \bar{g} \in \mathscr{n}_I$ be such that for each $i = 1, \ldots, j$, the number of points in $V_i \cap W^s(p_2(g)) \cap W^u(p_1(g))$ is equal to the number of points in $V_i \cap W^s(p_2(\bar{g})) \cap W^u(p_1(\bar{g}))$. If $T_1(g) = T_1(\bar{g})$ and $T_2(z_{I_m}(g), z_{I_j}(g)) = T_2(z_{I_m}(\bar{g}), z_{I_j}(\bar{g}))$ for $m = 1, \ldots, j-1$ then g is topologically equivalent to $\bar{g}$.*

To prove Theorem 3.1 we need several geometric constructions.

Recall that an unstable tubular family for a periodic point p of g is a continuous retraction π^u: $V^u \to W^s(O(p))$, where V^u is a neighborhood of $W^u(O(p))$, such that $(\pi^u)^{-1}(p) = W^u(p)$; $\pi^u(g(x)) = g\pi^u(x)$ whenever x and $g(x) \in V^u$; the fibers of π^u are C^1 submanifolds transversal to $W^s(O(p))$ and the mapping which assigns to each $x \in V^u$ the tangent space of the fiber through x is continuous. Similarly we define a stable tubular family π^s: $V^s \to W^u(O(p))$. Let $p_1, p_2, \ldots, p_s$ be the periodic points of g, of saddle type, with $p_1 = p_1(g)$ and $p_2 = p_2(g)$. For each $l > 2$ we construct a tubular family π_l^u: $V_l^u \to W^s(O(p_l))$ satisfying the following compatibility condition: if $W^u(O(p_l))$ intersects $W^s(O(p_i))$ then any fiber of π_i^u which intersects a fiber of π_l^u is contained in this fiber. This can be done by induction in the phase diagram of g as in [5] or [6]. For p_1 and p_2 we will consider tubular families π_i^u: $V_i^u \to W^s(p_i)$ and π_i^s: $V_i^s \to W^u(p_i)$, $i = 1, 2$, which are of class C^2. For this we take C^2 coordinate systems in a neighborhood U_i of p_i linearizing g. If $x \in U_i$ we denote by (x^1, x^2) the coordinates of x. Hence $g(x) = (\lambda_i(g)x^1, \mu_i(g)x^2)$. In these coordinates we define $\pi_i^u(x) = x^1$ and $\pi_i^s(x) = x^2$. It is easy to extend π_i^u to a neighborhood V_i^u of $W^u(p_i)$.

We may assume that $D_g^s \subset U_2$. Since for each $m = 1, \ldots, j$, $z_{I_m}(g)$ is a quasi-transversal point of intersection of $W^u(p_1)$ and $W^s(p_2)$ and the tubular families π_1^u

and π_2^s are C^2, it follows that, in a neighborhood of $z_{l_m}(g)$, the fibers of π_1^u are transversal to the fibers of π_2^s except along a C^1 curve Σ_m which is transversal to $W^s(p_2)$. *Let* $J_m \subset D_g^s$ be an interval containing z_{l_m}, so small that z_{l_m} is the only point in J_m whose α-limit set is a saddle point. Shrinking V_1^u if necessary we can assume that $V_1^u \cap D_g^s$ is contained in $\bigcup_{m=1}^{j} J_m$. Through the points of the boundary of J_m we take a pair of discs S_m^1, S_m^2 contained in the fibers of π_1^u. In the strip bounded by S_m^1 and S_m^2 we consider a foliation, having these discs as leaves, satisfying the following conditions: all leaves are transversal to the fibers of π_2^s except along Σ_m and the foliation coincides with the fibration π_1^u in V_1^u. Over $D_g^s - \bigcup_{m=1}^{j} J_m$ we raise a continuous fibration, whose fibers are discs transversal to the fibers of π_2^s, satisfying the properties: the fibers over the points in the boundary of D_g^s is contained in the fibers of π_2^u; the fiber over a point which belongs to V_i^u, for some $i \neq 2$, is contained in the fiber of π_i^u through this point. Now we iterate this foliation by g. Let us denote by $\mathcal{F}^u$ the foliation constructed above. We notice that the leaves of $\mathcal{F}^u$ in the strips bounded by S_m^2 and S_{m+1}^1 are discs transversal to $W^s(p_2)$.

We perform all the constructions above for $\bar{g}$. Hence $\bar{\Sigma}_m$ are the curves of tangency of the unstable tubular family of $\bar{p}_1 = p_1(\bar{g})$ with the stable tubular family of $\bar{p}_2 = p_2(\bar{g})$, $\bar{z}_{l_m} = z_{l_m}(\bar{g}) = \bar{\Sigma}_m \cap D_{\bar{g}}^s$ and $\bar{\mathcal{F}}^u$ denote the foliation whose leaf through any point in $\bar{V}_i^u$, $i \neq 2$, is contained in the fiber of $\bar{\pi}_i^u$ through this point.

Let us define the conjugacy h between g and $\bar{g}$. For each $i > 2$ we construct a homeomorphism h_i: $W^s(0(p_i)) \to W^s(0(\bar{p}_i))$, where $\bar{p}_i$ is the periodic point of $\bar{g}$ near p_i, such that $h_i g = \bar{g} h_i$ and h_i preserves the unstable tubular families of the periodic points. This means that if $x \in W^s(p_i) \cap V_n^u$ for $i, n > 2$ then $\bar{\pi}_n^u(h_i(x)) = h_n(\pi_n^u(x))$. This can be done by induction on the phase diagrams of g and $\bar{g}$ as in the proof of the stability of Morse–Smale diffeomorphisms [5]. Now we define a homeomorphism h_1: $W^s(p_1) \to W^s(\bar{p}_1)$ conjugating g and $\bar{g}$. Let $x \in W^s(p_1)$ be in a small neighborhood of p_1 contained in U_1. Then $x = (x^1, 0)$ in the coordinate system in U_1 which linearizes g. We set $h_1(x) = \bar{x}$ whose coordinates $(\bar{x}^1, 0)$, in the coordinate system linearizing $\bar{g}$, are given by $\bar{x}^1 = (x^1)^c$ if $x^1 \geq 0$ and $\bar{x}^1 = -|x^1|^c$ if $x^1 < 0$, where $c = [(\log \bar{\lambda}_1)/(\log \lambda_1)]$. Clearly h_1 is a homeomorphism of a neighborhood N of p_1 in $W^s(p_1)$ onto a neighborhood of $\bar{p}_1$ in $W^s(\bar{p}_1)$ and $h_1 g = \bar{g}^{h_1}$. Now we can extend h_1 to $W^s(p_1)$: if $x \in W^s(p_1)$ then there exists an integer n such that $g^n(x) \in N$; we then set $h_1(x) = \bar{g}^{-n} h g^n(x)$

For each $m = 1, \dots, j$, we denote by φ_m: $\Sigma_m \to W^s(p_1)$ and by $\bar{\varphi}_m$: $\bar{\Sigma}_m \to W^s(\bar{p}_1)$ the restrictions of π_1^u and $\bar{\pi}_1^u$ respectively. Then φ_m and $\bar{\varphi}_m$ are C^1 diffeomorphisms onto neighborhoods of p_1 and $\bar{p}_1$ respectively. We define homoemorphisms h_{Σ_m}: $\Sigma_m \to \bar{\Sigma}_m$ by $\bar{z} = h_{\Sigma_m}(z) = \bar{\varphi}_m^{-1} h_1 \varphi_m(z)$. Now we extend h_{Σ_m} to $\bigcup_{n \in Z} g^n(\Sigma_m)$ using the equation $h_{\Sigma_m} g = \bar{g} h_{\Sigma_m}$. We claim that there is a homeomorphism h^u: $W^u(p_2) \to W^u(\bar{p}_2)$ such that: (i) $h^u g = \bar{g} h^u$; (ii) if $w_i \to w \in W^u(p_2)$, $w_i \in g^{n_i}(\Sigma_m)$ when $w_i = h_{\Sigma_m}^{(w_i)}$ converges to $\bar{w} = h^u(\omega)$, for all $m = 1, \dots, j$. In fact, let $w = (0, w^2)$ and $l > 0$ be an integer such that $g^{-l}(\Sigma_m) \subset U_1$ and $\bar{g}^{-l}(\bar{\Sigma}_m) \subset U_1$ for $m = 1, \dots, j$. If $y_i = g^{-(l+n_i)}(w_i)$ and $\bar{y}_i = h_{\Sigma_m}(y_i) = \bar{g}^{-(l+n_i)}(w_i)$ then $\bar{y}_i^1 = \pm |y_i^1|^c$. Hence $\bar{w}_i$ converges to $\bar{w} = (0, \bar{w}^2)$ where $\bar{w}^2 = \pm b|w^2|^c$ and $b = [|\alpha(\bar{g}^{-1}(\bar{z}_{l_m}), l|/|\alpha(g^{-1}(z_{l_m}), l|^c]$. Since $T_2(z_{l_m}(g), z_{lj}(g)) = T_2(z_{l_m}(\bar{g})$ for $m = 1, \dots, j-1$, it follows that b does not depend on m. Hence we may define $h^u(w) = \bar{w}$ by $\bar{w}^2 = b|w^2|^c$ if $w^2 \geq 0$ and $\bar{w}^2 = -b|w^2|^c$ if $w^2 \leq 0$. Since $[(\log \bar{\mu}_2)/(\log \mu_2)] = c$, h^u is a homeomorphism of a neighborhood of p_2 in $W^u(p_2)$ onto a neighborhood of $\bar{p}_2$ in $W^u(\bar{p}_2)$, conjugating g and $\bar{g}$. Now we can extend h^u to $W^u(p_2)$ using the equation $h^u g = \bar{g} h^u$. The claim is proved.

Next we define a homeomorphism h_2: $W^s(p_2) \to W^s(\bar{p}_2)$ such that $h_2 g = \bar{g} h_2$ and h_2 preserves the unstable tubular families π_i^u and $\bar{\pi}_i^u$ for $i \neq 2$. Let $A_m \subset D_g^s$ be a neighborhood of z_{l_m} such that for any $z \in A_m$ the leaf of π_1^u through z intersects Σ_m in a point $\gamma_m(z)$. Let $A_m^+ \subset A_m$ be the set of points $z \in A_m$ such that $z^1 > z_{l_m}^1$ where $(z^1, 0)$ and $(z_{l_m}^1, 0)$ are the coordinates of z and z_{l_m} respectively. The restriction γ_m^+ of γ_m to A_m^+ is a homeomorphism onto an open subset of Σ_m. Similarly we consider a neighborhood $\bar{A}_m$ of $\bar{z}_{l_m}$ in $W^s(\bar{p}_2)$ and a mapping $\bar{\gamma}_m$: $\bar{A}_m \to \bar{\Sigma}_m$. For $z \in A_m^+$ we set $h_2(z) = (\bar{\gamma}_m^+)^{-1} h_{\Sigma_m} \gamma_m(z)$. If $z \in A_m - A_m^+$ we define $h_2(z) = (\bar{\gamma}_m^-)^{-1} h_{\Sigma_m} \gamma_m(z)$, where $\bar{\gamma}_m^-$ is the restriction of $\bar{\gamma}_m$ to $\bar{A}_m - \bar{A}_m^+$. Hence h_2: $\bigcup_{m=1}^{j} A_m \to \bigcup_{m=1}^{j} \bar{A}_m$ is a homeomorphism and preserves the unstable tubular families π_1^u and $\bar{\pi}_1^u$. Let p_i be a periodic point of g with behavior 1 with respect to p_2 (see [5], p. 389). Then $W^u(p_i)$ intersects D_g^s in finitely many points $q_1, \ldots, q_\delta$ which are contained in $D_g^s - \bigcup_{m=1}^{j} A_m$. The restriction of π_i^u to neighborhoods $B_{i,l} \subset D_g^s$ of q_l are homeomorphisms $\varphi_{i,l}$ of $B_{i,l}$ onto a neighborhood of p_i in $W^s(p_i)$. For $z \in B_{i,l}$ we define $h_2(z) = (\bar{\varphi}_{i,l})^{-1} h_i \varphi_{i,l}(z)$ where $\bar{\varphi}_{i,l}$ is the restriction of $\bar{\pi}_i^u$ to a neighborhood $\bar{B}_{i,l}$ of $\bar{q}_l$ in $W^s(\bar{p}_2)$. Here $\bar{q}_1, \ldots, \bar{q}_\delta$ are the intersection of $W^u(\bar{p}_i)$ with $W^s(\bar{p}_2)$, $\bar{p}_i$ is the periodic point of $\bar{g}$ near p_i and $\bar{q}_l$ is near q_l for $l = 1, \ldots, \delta$, since $\bar{g}$ is near g. Repeating the construction for every periodic point of g whose behavior with respect to p_2 is 1 we extend h_2 to a homeomorphism h_2: $A^1 \to \bar{A}^1$ preserving the unstable tubular families, where $A^1 \subset D_g^s$ is an open set which contains the intersection with D_g^s of all the unstable manifolds of periodic points with behavior one with respect to p_2. If p is a periodic point of g of behavior 2 with respect to p_2 then $W^u(p)$ intersects $D_g^s - A^1$ in finitely many points. So we can, using the same arguments, extend h_2 to an open set A^2 which contains A^1 and the intersection of D_g^s with the unstable manifolds of all periodic points with behavior 2 with respect to p_2. By induction, we extend h_2 to an open set A in D_g^s which contains the intersections of D_g^s with the unstable manifolds of all saddle points of g, in such way that h_2 preserves the unstable tubular families π_i^u and $\bar{\pi}_i^u$ for every $i \neq 2$. It is clear from the construction above that A is the union of finitely many intervals. Furthermore, since g and $\bar{g}$ are close to each other and we can construct tubular families depending continuously on the diffeomorphism, it follows that $h_2(x)$ is close to x for every $x \in A$. Hence we can extend h_2 to a homeomorphism h_2: $D_g^s \to D_{\bar{g}}^s$. Next we extend h_2 to homeomorphism of $W^s(p_2)$ onto $W^s(\bar{p}_2)$ using the equation $h_2 g = \bar{g} h_2$.

Up to now we have defined a homeomorphism h on the union of the stable manifolds of the saddle points of g, conjugating g and $\bar{g}$ and preserving the tubular families π_i^u and $\bar{\pi}_i^u$ for $i \neq 2$. This homeomorphism is also defined on $\bigcup_{m=1}^{j} \bigcup_{n \in Z} g^n(\Sigma_m)$, on $W^u(p_2)$ and preserves the foliations $\mathscr{F}^u$ and $\bar{\mathscr{F}}^u$. To extend h to a neighborhood of p_2 we are going to construct a continuous retraction π: $\bar{N}_2^u \to W^u(\bar{p}_2)$, where $\bar{N}_2^u$ is a neighborhood of $W^u(\bar{p}_2)$ containing $D_{\bar{g}}^s$, having the following properties:

(i) $\pi^{-1}(\bar{p}_2) = W^s(\bar{p}_2) \cap \bar{N}_2^u$;

(ii) π is $\bar{g}$ invariant, namely, $\pi(\bar{g}(p)) = \bar{g}(\pi(p))$;

(iii) If $q \in \Sigma_m$ then $\pi(h(q)) = h(\pi_2^s(q))$ for $m = 1, \ldots, j$;

(iv) The fibers of π in $U_2 \cap \bar{N}_2^u$ coincide with the fibers of $\bar{\pi}_2^s$ except on the iterates of the strip bounded by $\bar{S}_m^2$ and $\bar{S}_{m+1}^1$, $m = 1, \ldots, j-1$.

(v) In the interior of the strips bounded by $\bar{S}_m^2$ and $\bar{S}_{m+1}^1$ the fibers of π are differentiable and transversal to the leaves of the foliation $\bar{\mathscr{F}}^u$.

Before proving the existence of π we extend h to M.

It is easy to see that there is a unique extension of $h_2 = h|W^s(p_2)$ to a neighborhood N^s of D_g^s satisfying the following conditions: (a) $\pi(h(x)) = h(\pi_2^s(x))$; (b) $x, y \in N^s$ belong to the same leaf of $\mathcal{F}^u$ iff $h(x)$ and $h(y)$ belong to the same leaf of $\bar{\mathcal{F}}^u$; (c) if x and $g(x)$ are in N^s then $\bar{g}h(x) = hg(x)$. Now let $N_2^u = \bigcup_{n\geq 0} g^n(N^s) \cup W^u(p_2)$. Then N_2^u is a neighborhood of $W^u(p_2)$[5], and we can extend h to N_2^u by $h(x) = h^u(x)$ if $x \in W^u(p_2)$ and $h(x) = \bar{g}^n h g^{-n}(x)$ if $g^{-n}(x) \in N^s$. We claim that h is a homeomorphism of N_2^u onto a neighborhood $\bar{N}_2^u$ of $W^u(\bar{p}_2)$. Clearly, the restriction of h to $N_2^u - W^u(p_2)$ is a homeomorphism onto $\bar{N}_2^u - W^u(\bar{p}_2)$. So it remains to prove the continuity of h at $W^u(p_2)$. Let $x_i \in N_2^u - W^u(p_2)$ be a sequence converging to $x \in W^u(p_2)$. Then there exists a sequence $n_i \to \infty$ such that $g^{-n_i}(x_i) \in N^s$. Since $\pi_2^s(x_i) \to x$ and $h|W^u(p_2) = h^u$ is continuous it follows that $\pi h(x_i) = h\pi_2^s(x_i) = h^u(\pi_2^s(x_i))$ converges to $h(x)$. We have also that $\bar{\pi}_2^u(h(x_i))$ converges to zero because $n_i \to \infty$. Hence $h(x_i)$ converges to $h(x)$. Thus h is continuous. It is easy to see that h has a continuous inverse. Therefore h: $N_2^u \to \bar{N}_2^u$ is a homeomorphism. We notice that if $x \in N_2^u \cap V_i^u$ for some $i \neq 2$ then $\bar{\pi}_i^u h(x) = h_i \pi_i^u(x)$.

To extend h to M we proceed as in the proof of the stability of Morse–Smale diffeomorphisms. Let $Q_1, \ldots, Q_t$ be the sinks of g and let $\bar{Q}_1, \ldots, \bar{Q}_t$ be the corresponding sinks of $\bar{g}$. For each $i = 1, \ldots, t$ we choose a fundamental domain $G_g^s(Q_i)$ of Q_i whose boundary is transversal to the unstable manifolds of the saddle points of g. We also choose a fundamental domain $G_{\bar{g}}^s(\bar{Q}_i)$ near $G_g^s(Q_i)$ such that $G_{\bar{g}}^s(\bar{Q}_i)$ contains $h(N_2^u \cap G_g^s(Q_i))$ whenever $G_{\bar{g}}^s(\bar{Q}_i)$ intersects $\bar{N}_2^u$. Using the methods of [5] we extend h to a homeomorphism h: $G_g^s(Q_i) \to G_{\bar{g}}^s(\bar{Q}_i)$, for $i = 1, \ldots, t$, such that: (i) if $x, g(x) \in G_g^s(Q_i)$ then $hg(x) = \bar{g}h(x)$; (ii) if $x \in V_n^u \cap G_g^s(Q_i)$ for some n then $\bar{\pi}_n^u(h(x)) = h_n(\pi_n^u(x))$. Now we can extend h to M. If $x \in W^s(p_l)$ for some l we define $h(x) = h_l(x)$; if $g^n(x) \in G_g^s(Q_i)$ for some n and some i then we set $h(x) = \bar{g}^{-n} h g^n(x)$; if x is a source of g, $h(x)$ is defined as the source of $\bar{g}$ near x. It is easy to see that h is $1-1$ and onto. It remains to prove that h is continuous. It is clear that h is continuous on the stable manifolds of the sinks. The continuity of h on $W^s(p_2)$ follows from the fact that the restriction of h to N_2^u is continuous. To prove the continuity of h at the stable manifolds of the others saddle points and at the sources we proceed by induction on the phase diagram of g as in [5].

To finish the proof of Theorem 3.1 we have to prove the existence of the retraction π. Let $\bar{N}^s$ be a small neighborhood of $D_{\bar{g}}^s$. Denote by $\bar{N}_+^s$ the connected component of $\bar{N}^s$ containing $\bar{\Sigma}_j$. It is easy to construct a continuous fibration ρ: $\bar{N}_+^s \to \bar{\Sigma}_j$ satisfying the following properties: (a) in the coordinate system on U_2 linearizing $\bar{g}$, the fibers of ρ are graphs of piecewise linear functions; (b) if $x \in \bar{\Sigma}_j$ and $x_m = \rho^{-1}(x) \cap \bar{\Sigma}_m$ then $\pi_2^s(h^{-1}(x)) = \pi_s^2(h^{-1}(x_m))$; (c) if $y \in \rho^{-1}(x)$ is in the strip bounded by S_m^1 and S_m^2, $m = 1, \ldots, j$, then $y = (0, y^2)$, i.e. the fibers of ρ coincide with the fibers of $\bar{\pi}_2^s$ in these strips.

Now let $\bar{N}_2^u = \bigcup_{n\geq 0} \bar{g}^n(\bar{N}^s)$. Let π: $\bar{N}_2^u \to W^u(\bar{p}_2)$ be defined by $\pi(x) = x$ if $x \in W^u(\bar{p}_2)$, $\pi(x) = \bar{\pi}_2^s(x)$, $\bar{g}^{-n}(x) \in \bar{N}^s - \bar{N}_+^s$ for some n and $\pi(x) = \bar{g}^n h \pi_2^s h^{-1} \rho \bar{g}^{-n}(x)$ if $\bar{g}^{-n}(x) \in \bar{N}_+^s$ for some n. It is easy to see that π satisfy the required properties. This finish the proof of Theorem 3.1.

§4. PROOF OF THEOREM D

Let m be the subset of $\mathrm{Diff}^\infty(M)$ defined in §1. We recall that if $f \in m$ then there is a unique pair of periodic saddle points of f, $p_1(f)$ and $p_2(f)$, such that one of the connected components of $W^u(p_1(f)) - \{p_1(f)\}$ is contained in $W^s(p_2(f))$. To simplify the notation we assume that $p_1(f)$ and $p_2(f)$ are fixed points and that the eigenvalues $\lambda_i(f)$ and $\mu_i(f)$ of $df(p_i(f))$ are positive with $\lambda_i(f) < 1$ and $\mu_i(f) > 1$ for $i = 1, 2$, and $f \in m$. The proof without this assumption is similar.

From p. 256 of [2], it follows that if the eigenvalues $\lambda_i(f)$ and $\mu_i(f)$ satisfy a certain number of non-resonance conditions then f is C^3-linearizable in neighborhoods of $p_i(f)$, $i = 1, 2$. Hence the set $\mathfrak{m}^1 \subset \mathfrak{m}$ of diffeomorphisms satisfying these conditions is open and dense in $\mathfrak{m}$.

To each $f \in \mathfrak{m}^1$ and $q \in W^u(p_1(f)) \cap W^s(p_2(f))$ we assign a C^2 function $T_{f,q}$: $[0, 1] \to \mathbb{R}$ which will be used in describing the conjugacy class of f. First we choose C^3 coordinate systems φ_i: $U_i \to \mathbb{R}^2$ linearizing f, where U_i is a neighborhood of $p_i(f)$, $i = 1, 2$, and U_1 contains q and $f(q)$. Let θ: $[0, 1] \to M$ be defined by $\varphi_1\theta(t) = \varphi_1(q) + t(\varphi_1(f(q)) - \varphi_1(q))$. Then θ is a C^3 embedding, $\theta(0) = q$, $\theta(1) = f(q)$ and $\theta([0, 1])$ is contained in $W^u(p_1(f)) \cap W^s(p_2(f))$. Now, let $T_{f,q}$: $[0, 1] \to \mathbb{R}$ be given by $T_{f,q}(t) = T_2(f, p_1, p_2, \theta(t), q)$ where T_2 is the conjugacy invariant defined in §2. We notice that $T_{f,q}$ is C^2 and the definition does not depend on the choice of the coordinate systems linearizing f.

LEMMA 4.1. *Let* $\mathfrak{n} \subset \mathfrak{m}^1$ *be a small neighborhood of* $f \in \mathfrak{m}^1$. *Suppose* $f' \in \mathfrak{n}$ *satisfy the conditions*: (i) $[(\log \mu_2(f'))/(\log \lambda_1(f'))] = [(\log \mu_2(f))/(\log \lambda_1(f))]$; (ii) *there is a homeomorphism* ψ: $[0, 1] \to [0, 1]$ *such that* $\psi(0) = 0$, $\psi(1) = 1$ *and* $T_{f',q'}(\psi(t)) = T_{f,q}(t)$. *Then* f' *is topologically equivalent to* f. *Furthermore, if* $[(\log \mu_2(f))/(\log \lambda_1(f))]$ *is irrational then the conditions* (*i*) *and* (*ii*) *are also necessary for the existence of a conjugacy* h *between* f *and* f' *such that* $h(q) = q'$.

Proof. To prove the necessity of the conditions (i) and (ii) we can use the same arguments of Lemma 2.5. The methods of §3 can be used to prove that the conditions (i) and (ii) are also sufficient for the existence of a conjugacy.

LEMMA 4.2. *Given* $\epsilon > 0$ *there are neighborhoods* $\mathfrak{n}$ *of* f *and* N *of* q *such that if* $f' \in \mathfrak{n} \cap \mathfrak{m}^1$ *and* $q' \in W^u(p_1(f')) \cap W^s(p_2(f')) \cap N$ *then* $T_{f',q'}$ *is* ϵC^2 *near* $T_{f,q}$.

Proof: From the proof of Sternberg's theorem [8], we can choose neighborhoods U_i of $p_i(f)$ and, for each f' near f, C^3 coordinate systems $\varphi_{i,f'}$ on U_i linearizing f', such that $\varphi_{i,f'}$ depend continuously on f' in the C^3 topology. Therefore the lemma follows from the expression of T_2 in these coordinate systems (see §2).

Let $\mathfrak{m}^2 \subset \mathfrak{m}^1$ be the set of diffeomorphisms such that $f \in \mathfrak{m}^2$ iff the critical points of $T_{f,q}$ are non-degenerate. By Lemma 4.2, $\mathfrak{m}^2$ is open in $\mathfrak{m}^1$. The lemma below is not difficult to prove.

LEMMA 4.3. $\mathfrak{m}^2$ *is open and dense in* $\mathfrak{m}^1$.

Let $\mathfrak{m}_k \subset \mathfrak{m}^2$ be such that $f \in \mathfrak{m}_k$ if and only if the number of critical points of $T_{f,q}$ is equal to k-1, where $q \in W^u(p_1(f)) \cap W^s(p_2(f))$ is such that no critical point of $T_{f,g}$ occurs in the boundary of [0, 1]. By Lemma 4.2, $\mathfrak{m}_k$ is open in $\mathfrak{m}$. Furthermore, by Lemma 4.1, a diffeomorphism f' near f is topologically equivalent to f if $[(\log \mu_2(f'))/(\log \lambda_1(f'))] = [(\log \mu_2(f))/(\log \lambda_1(f))]$ and the critical values of $T_{f',q'}$ are equal to the corresponding critical values of $T_{f,q}$. Thus the modulus of stability of f is equal to k. This proves Theorem D.

REFERENCES

1. P. HARTMAN: On local homeomorphisms of Euclidean spaces, *Bol. Soc. Math. Mexicana* **5** (1960), (2).
2. P. HARTMAN: *Ordinary Differential Equations*. Wiley, New York (1964).
3. S. NEWHOUSE and J. PALIS: Bifurcations of Morse-Smale dynamical systems, *Dynamical Systems* (Edited by M. Peixoto). Academic Press, New York (1973).
4. J. PALIS: A differentiable invariant of topological conjugacies and moduli of stability, *Astérisque* **51** (1978).

5. J. Palis: On Morse–Smale dynamical systems, *Topology* **8** (1969).
6. J. Palis and S. Smale: Structural stability theorems, global analysis, *Proc. Symp. in Pure Math. XIV*, *AMS* (1970).
7. C. Robinson: C^r structural stability implies Kupka–Smale, *Dynamical Systems* (Edited by M. Peixoto). Academic Press, New York (1973).
8. S. Sternberg: On local structure of local homeomorphisms of Euclidean n-spaces II, *Am. J. Math.* **80** (1958).

Instituto de Matematica Pura e Aplicada
Rio de Janeiro, Brazil

Diffeomorphisms on Surfaces with a Finite Number of Moduli

W. de Melo and S.J. van Strien

Ergod. Th. & Dynam. Sys. **7**, 415–462 (1987)

Abstract. This paper generalises the well known structural stability theorem which says that a diffeomorphism is structurally stable if it is Axiom A and if all invariant manifolds are transversal to each other.

If these transversality conditions are not satisfied then the diffeomorphism not only fails to be stable, but also this gives rise to the appearance of moduli. That is, one needs several real parameters to parameterise all conjugacy classes of nearby diffeomorphisms. (The minimum number of parameters needed is called the number of moduli).

Here we deal with diffeomorphisms on two dimensional manifolds, whose asymptotic dynamics are well understood (the class of Axiom A diffeomorphisms). The main result characterises those Axiom A diffeomorphisms which have a finite number of moduli. This result can be regarded as a generalisation of the structural stability theorem. From the proofs it follows that the dynamics of these diffeomorphisms can also be well understood.

In the proof of our main theorem we need certain invariant foliations to be quite smooth. In an appendix we prove a differentiable version of the Lambda Lemma.

One of the purposes of the theory of Dynamical Systems is to understand the orbit structure of diffeomorphisms. Here we say that two diffeomorphisms f and g have the same orbit structure if they are conjugate that is, if there is a homeomorphism h of the ambient manifold such that $hf = gh$. Clearly this defines an equivalence relation $\sim$ on the space of diffeomorphisms. Since the dynamics of many diffeomorphisms is very complicated and sensitive to perturbations (of the diffeomorphisms) we cannot hope to understand the space of conjugacy classes in general. However this space is very neat in neighbourhoods of many conjugacy classes. This paper is in the direction of characterising all diffeomorphisms representing these conjugacy classes. One may measure the degree of complexity of this local structure by looking at the local dimensions of this space. This dimension we call the number of moduli. This number is defined as follows. If a diffeomorphism f is structurally stable or if there are at most a countable number of different conjugacy classes in a neighbourhood U of f then we say that the number of moduli is zero. If a sufficiently small neighbourhood U contains a countable number of k-parameter C^1 families of diffeomorphisms such that each diffeomorphism in U is conjugate

Originally published in *Ergod. Th. & Dynam. Sys.* **7**, 415–462 (1987).

to at least one diffeomorphism from these families, then f has a finite number of moduli. The minimum number k of parameters needed is the number of moduli (or the modality of f).

Here we deal with diffeomorphisms on two dimensional manifolds whose asymptotic dynamics is well understood. In fact we consider the set $\mathscr{A}$ of diffeomorphisms satisfying Axiom A and the no cycle condition. Such diffeomorphisms are Ω-stable, namely, the dynamics of the non-wandering set does not change with small perturbations [**Sm1**], [**Sm2**]. However the intersection pattern of invariant manifolds is the same for conjugate diffeomorphisms. Transversality conditions of stable and unstable manifolds are needed to have structural stability [**R**]. If these transversality conditions are not satisfied then the diffeomorphism not only fails to be stable but this also gives rise to the appearance of moduli [**P2**], [**NPT**], [**MPS**], [**S1**], [**S2**]. However even in this case the diffeomorphism may still have finite modality [**Me2**], [**MP**], [**S2**]. The purpose of this paper is to characterise the class $\mathscr{M}$ of diffeomorphisms in $\mathscr{A}$ having finite modality.

Our results provide a much more precise description of the conjugacy classes in a sufficiently small neighbourhood $\mathscr{N}$ of a diffeomorphism f in $\mathscr{M}$. One has that $\mathscr{N}$ is a countable union of submanifolds $\mathscr{F}_i$. Each $\mathscr{F}_i$ has finite codimension and contains a finite dimensional submanifold S_i. Moreover there exists a differentiable fibration $\pi_i : \mathscr{F}_i \to S_i$ so that all diffeomorphisms in the same fiber are conjugate. Furthermore each S_i contains a dense subset T_i such that no two distinct diffeomorphisms in T_i are conjugate to each other. The codimension of $\mathscr{F}_i$ is uniformly bounded and the maximum of all the dimensions of S_i is the modality of f.

1. *Statement of results*

From now on let M be a compact, C^∞, two-dimensional manifold without boundary and $\mathrm{Diff}^\infty(M)$ be the set of C^∞ diffeomorphisms on M with the C^∞ topology.

We denote by $\mathscr{A} \subset \mathrm{Diff}^\infty(M)$ the set of diffeomorphisms satisfying Axiom A and the no-cycle condition. Recall that $f \in \mathrm{Diff}^\infty(M)$ satisfies Axiom A if the non-wandering set $\Omega(f)$ is hyperbolic and the periodic orbits are dense in $\Omega(f)$, [**Sm1**]. If f satisfies Axiom A then $\Omega(f)$ is a union of finitely many closed invariant sets Ω_i's, called basic sets, such that f has an orbit dense in each Ω_i. Finally we say that f satisfies the no-cycle condition if there are no basic sets $\Omega_1, \ldots, \Omega_n$, $\Omega_{n+1} = \Omega_1$ with $W^u(\Omega_i) \cap W^s(\Omega_{i+1}) \neq \emptyset$. Here $W^u(\Omega_i)$ (resp. $W^s(\Omega_i)$) is the set of points whose α-limit set (resp. ω-limit set) is contained in Ω_i. We recall that if $f \in \mathscr{A}$ then f is Ω-stable [**Sm2**]. The main result of this paper is the characterisation of the set $\mathscr{M} \subset \mathscr{A}$ of diffeomorphisms having finite moduli. Here the notion of modulus is defined as in the introduction. Remark that if f has k moduli then f has a neighbourhood U such that each g in U has modality at most k. In particular the set of diffeomorphisms in $\mathscr{A}$ with finite modality is open.

Let $\mathscr{M} \subset \mathscr{A}$ be the set of diffeomorphisms satisfying the conditions below:

(1) if $x, y \in \Omega(f)$ are such that $W^u(x)$ is not transversal to $W^s(y)$ then the basic sets containing x and y are trivial (i.e. consist of periodic orbits);

(2) there is only a finite number of orbits of non-transversal intersections between stable and unstable manifolds and the contact between these manifolds along each of these orbits is of finite order;

(3) if $p, q \in \text{Per}(f)$ are such that $W^u(p)$ has an orbit of non-transversal intersection with $W^s(q)$ then the number of orbits in $W^s(p)$ (resp. in $W^u(q)$) belonging to some unstable (resp. stable) manifolds of periodic saddle points of f is finite;

(4) if x is a point of non-transversal intersection of $W^u(p)$ and $W^s(q)$ then there exists an arc Σ transversal to $W^u(p)$ at x such that no connected component of $\Sigma - \{x\}$ contains points of both stable and unstable manifolds of saddles;

(5) if $W^u(p)$ has a point of non-transversal intersection with $W^s(q)$, and $W^u(q)$ has a point of non-transversal intersection with $W^s(r)$, then there is no saddle point of f whose unstable manifold (resp. stable manifold) intersects $W^s(p)$ (resp. $W^u(r)$).

Remark. It is well known that every diffeomorphism in $\mathcal{A}$ which satisfies the transversality conditions on invariant manifolds is structurally stable [**Rb1**]. The class $\mathcal{M}$ relaxes these conditions. The following theorem generalizes this structural stability result.

MAIN THEOREM. *If $f \in \text{Diff}^\infty(M^2)$ is in $\mathcal{A}$ then f has finite modality if and only if $f \in \mathcal{M}$.*

Remark. As in [**P1**] it is not hard to show that if $f \in \text{Diff}^\infty(M^2)$ satisfies the Axiom A condition and f has cycles then f has infinite modality.

2. *Existence of moduli*

Let M be a compact C^∞ manifold of dimension two and $f: M \to M$ be a C^∞ diffeomorphism having a pair of periodic points p and q such that the unstable manifold of p has an orbit of non-transversal intersection with the stable manifold of q. In this section we will construct all conjugacy invariants generated by the existence of these tangencies. We will use these conjugacy invariants to prove the following.

THEOREM. *If $f \in \mathcal{A}$ has finite modality then $f \in \mathcal{M}$.*

Later in §§ 3, 4 and 5 we will prove that two diffeomorphisms in $\mathcal{M}$, having an equivalent intersection pattern of stable and unstable manifolds, are conjugate if they have the same conjugacy invariants.

2(a). *Some technical lemmas.* In order to show the existence of moduli we will have to compare metrics induced on M by two C^1 coordinate systems. So we start by describing some properties of C^r metrics. By this we mean metrics which are induced by C^r Riemannian structures. We say that $d: M \times M \to \mathbb{R}$ is a C^r metric, $0 \le r \le \infty$, on M if there exists a C^r Riemannian structure g on M such that $d(x, y)$ is given by the infimum of the lengths of all paths that connect x and y. In formulae:

$$d(x, y) = \inf\{l_g(\gamma);\ \gamma: [0, 1] \to M \text{ is a piecewise } C^1 \text{ curve with } \gamma(0) = x \text{ and } \gamma(1) = y\}.$$

Here

$$l_g(\gamma)=\int_0^1 \sqrt{g(\gamma'(t),\gamma'(t))}\,dt.$$

The distance from a point x to a set S will be denoted by $d(x, S)=\inf\{d(x,y);\, y\in S\}$. It will be convenient to use the following notation. For sequences α_i, β_i of real numbers we will write

$$\alpha_i\sim\beta_i \quad \text{if} \quad \left|\frac{\alpha_i}{\beta_i}\right| \text{ is bounded and bounded away from zero,}$$

$$\alpha_i\simeq\beta_i \quad \text{if} \quad \frac{\alpha_i}{\beta_i} \text{ converges to 1.}$$

LEMMA 2.1. *Let $d:\mathbb{R}^n\times\mathbb{R}^n\to\mathbb{R}$ be a C^0 metric on $\mathbb{R}^n$ induced by the Riemannian structure g. Let d_0 be the metric induced by the constant Riemannian structure g_0 which coincides with g at 0. If $S\subset\mathbb{R}^n$ contains 0 and $x_i\in\mathbb{R}^n-S$ converges to zero, then $d(x_i, S)\simeq d_0(x_i, S)$.*

Proof. Since g is continuous, for every $\rho>0$ there are positive constants c_ρ, c'_ρ such that

$$c_\rho\sqrt{g((0,v),(0,v))}\le\sqrt{g((x,v),(x,v))}\le c'_\rho\sqrt{g((0,v),(0,v))} \tag{2.1}$$

for every $v\in\mathbb{R}^n$ and every $x\in\mathbb{R}^n$ with $\|x\|<\rho$. Clearly we can take c_ρ and c'_ρ arbitrarily near 1 by taking ρ sufficiently small. From (2.1) we have that if γ is a piecewise C^1 curve contained in the ball with centre 0 and radius ρ then

$$c_\rho l_{g_0}(\gamma)\le l_g(\gamma)\le c'_\rho l_{g_0}(\gamma),$$

where g_0 is the constant Riemannian structure, i.e. $g_0((x,v),(x,w))=g((0,v),(0,w))$ for all $x, v, w\in\mathbb{R}^n$. Hence, if $\bar\rho>0$ is sufficiently small then

$$c_\rho d_0(x,y)\le d(x,y)\le c'_\rho d_0(x,y)$$

for all $x, y\in\mathbb{R}^n$ with $\|x\|, \|y\|\le\bar\rho$. Therefore for $\bar{\bar\rho}>0$ sufficiently small we have that

$$c_\rho d_0(x,S)\le d(x,S)\le c'_\rho d_0(x,S)$$

for all $x\in\mathbb{R}^n$ with $\|x\|\le\bar{\bar\rho}$. This proves the lemma. □

LEMMA 2.2. *Let $S\subset\mathbb{R}^n$ be a codimension one C^1 submanifold containing 0 and d_j, $j=1,2$ be C^0 metrics on $\mathbb{R}^n$. Then there exists a positive real number λ such that $d_1(x_i,S)\simeq\lambda d_2(x_i,S)$ for any sequence $x_i\in\mathbb{R}^n-S$ converging to $0\in\mathbb{R}^n$.*

Proof. By taking C^1 coordinates on $\mathbb{R}^n$ we may assume that S is a hyperplane. By lemma 2.1 we can assume that d_1 and d_2 are induced by Riemannian structures which are constant with respect to this coordinate system. From these observations the lemma follows easily. □

Remark. If $\tilde S$ is another C^1 codimension one submanifold tangent to S at 0 and $x_i\in\mathbb{R}^n-\tilde S$ converges to 0 then $d_1(x_i,\tilde S)\simeq\lambda d_2(x_i,\tilde S)$, where λ is the same constant as for S. This follows from the proof of the lemma and the fact that we can find a pair of C^1 diffeomorphisms $\varphi_1, \varphi_2:(\mathbb{R}^n,0)\circlearrowleft$ such that $\varphi_1(S)=\varphi_2(\tilde S)$ is a hyperplane and the derivative of $\varphi_1\circ\varphi_2^{-1}$ at 0 is the identity.

Using C^0 metrics we can introduce the notion of contact of C^1 submanifolds.

Definition. Let x be a point of tangency of two C^1 submanifolds $S_1, S_2 \subset M$. We say that S_1 *has a contact of order n* with S_2 at x if for some metric d on M the limit:

$$\lim_{\substack{w \to x \\ w \in S_1}} \frac{d(w, S_2)}{[d(w, x)]^n}$$

exists and is positive. If this limit is infinite we say that the contact is less than n. If the limit is zero for all n we say that the contact is infinite.

Remark. From lemmas 2.1 and 2.2 it follows that the above definition is independent of the metric. Notice also that the contact may not exist. However if there is a C^1 coordinate system φ on a neighbourhood of x such that $\varphi(S_1)$ and $\varphi(S_2)$ are both C^∞ submanifolds of $\mathbb{R}^2$ then the contact is either defined or it is infinite.

Definition. Let p be a hyperbolic fixed point of a C^∞ diffeomorphism $f: M \to M$. By a *linearising metric at p* we mean a C^0 metric d on a neighbourhood U of $W^s(p) \cup W^u(p)$ such that in a C^1 coordinate system in U linearising f, d coincides with the Euclidean metric.

Remark. If $\overline{W^s(p)} \cap \overline{W^u(p)} = \{p\}$, such a linearising metric always exists. In fact by a theorem of Hartman [**Ha**] (see also the appendix) there is a C^1 coordinate system in a neighbourhood of p linearising f. If $W^s(p) \cap W^u(p) = \{p\}$, this coordinate system can be extended to a full neighbourhood of $W^s(p) \cup W^u(p)$. Thus we obtain a linearising metric at p. These metrics are not unique. However, since any C^1 diffeomorphism commuting with a linear contraction $l: \mathbb{R} \to \mathbb{R}$ is linear, it is easy to see that if $\tilde{d}$ is another linearising metric then the restrictions of d and $\tilde{d}$ to each connected component of $W^s(p) \cup W^u(p) - \{p\}$ differ only by a multiplicative constant.

The following two lemmas will be used repeatedly.

LEMMA 2.3. *Let p be a hyperbolic fixed point of saddle type of a C^2 diffeomorphism $f: M \to M$. Let $x \in W^u(p) - \{p\}$, d be a C^0 metric on M and α be the contracting eigenvalue of df_p. For any sequence $x_i \to x$ we have:*

(i) *if there exists a sequence $n(i) \to \infty$ such that $f^{-n(i)}(x_i)$ converges to a point $z \in W^s(p)$, then $d(x_i, W^u(p)) \simeq cd(z, p)|\alpha|^{n(i)}$ for some constant c which depends on x, z and d but not on the sequence; if d is a linearising metric then c is independent of x and z;*

(ii) *if $d(x_i, W^u(p)) \simeq c|\alpha|^{n(i)}$ for some constant $c > 0$ and some sequence $n(i) \to \infty$ then the sequence $f^{-n(i)}(x_i)$ has precisely one limit point if $\alpha > 0$ and two limit points if $\alpha < 0$. These limit points are contained in $W^s(p)$.*

Proof. It is clear that both statements hold for a linearising metric. Therefore the lemma follows from lemma 2.2. □

Definition. Let x be a one sided tangency (for example a quadratic tangency or a tangency of even order) between $W^u(p)$ and $W^s(q)$, where p and q are hyperbolic fixed points of f. Let $y \in W^u(q)$. We say that a sequence $x_i \to x$ *is nice with respect to the pair* (x, y) if (i) $f^i(x_i)$ converges to y; (ii) $d(x_i, W^u(p)) \simeq d(x_i, W^s(q))$.

We will now prove that the property of being a nice sequence is preserved under a topological conjugacy.

LEMMA 2.4. *Let h be a conjugacy between f and $\bar{f}$. Then $x_i \to x$ is a nice sequence with respect to (x, y) if and only if $h(x_i)$ is a nice sequence with respect to $(h(x), h(y))$.*

Proof. We have two cases to consider: (a) y is accumulated by the iterates of any small interval in $W^u(p)$ which contains x; (b) case (a) does not occur. In both cases it is easy to see that we have the following characterisation of nice sequences. The sequence $x_i \to x$ is nice with respect to (x, y) if (i) $f^i(x_i) \to y$; (ii) given sequences of integers $n(j) \to \infty$ and $i(j) \to \infty$ such that $f^{-n(j)}(x_{i(j)})$ converges to $z \in W^s(p)$ then for any other sequence $x_i' \to x$ such that $f^i(x_i') \to y$ we have:

$$\left.\begin{array}{ll}\limsup d(f^{-n(j)}(x'_{i(j)}), p) \leq d(z, p) & \text{in case (a),} \\ \liminf d(f^{-n(j)}(x'_{i(j)}), p) \geq d(z, p) & \text{in case (b).}\end{array}\right\} \qquad (*)$$

In fact let x_i be nice with respect to (x, y). Then $d(x_i, W^u(p)) \simeq d(x_i, W^s(q))$ and $f^i(x_i) \to y$. If for some other sequence x_i' we also have $f^i(x_i') \to y$, then by lemma 2.3 one has $d(x_i, W^s(q)) \simeq d(x_i'. W^s(q))$. Moreover since $W^u(p)$ and $W^s(q)$ have one-sided tangencies

$$\begin{array}{ll} d(x_i', W^u(p)) \leq d(x_i', W^s(q)) & \text{in case (a),} \\ d(x_i', W^u(p)) \geq d(x_i', W^s(q)) & \text{in case (b).}\end{array}$$

Hence the limit points of $\{d(x_i', W^u(p))/d(x_i, W^u(p))\}$ are at most 1 in case (a), and at least 1 in case (b). From this and lemma 2.3 one deduces (*). Similarly (*) implies that x_i is a nice sequence. Since $W^s(p)$ has dimension one, it follows easily that these characterisations are preserved under conjugacy. □

2(b). *Construction of moduli.* Now we will construct all conjugacy invariants arising from non-transversal intersection of invariant manifolds of periodic points. To simplify the exposition we will assume that all periodic points involved are in fact fixed points. The general case is treated in the same way by looking at appropriate iterates of the diffeomorphism. The next lemma introduces the first conjugacy invariant found in [**P2**] and shows the rigidity of the conjugacy if this invariant is irrational as in [**Me2**].

LEMMA 2.5. *Let x (resp. $\bar{x}$) be a tangency of even order between $W^u(p)$ and $W^s(q)$ (resp. $W^u(\bar{p})$ and $W^s(\bar{q})$) where p and q (resp. $\bar{p}$ and $\bar{q}$) are hyperbolic fixed points of the C^2 diffeomorphism f (resp. $\bar{f}$). Let α (resp. $\bar{\alpha}$) be the contracting eigenvalue of df_p (resp. $d\bar{f}_{\bar{p}}$) and β (resp. $\bar{\beta}$) be the expanding eigenvalue of df_q (resp. $d\bar{f}_{\bar{q}}$). Let h be a conjugacy between f and $\bar{f}$ with $h(p) = \bar{p}$, $h(q) = \bar{q}$ and $h(x) = \bar{x}$. Then we have the following properties:*

(1) $$\frac{\log|\alpha|}{\log|\beta|} = \frac{\log|\bar{\alpha}|}{\log|\bar{\beta}|}.$$

(2) *Let d_p be a linearising metric at p and d_q be a linearising metric at q. If* $\log|\alpha|/\log|\beta|$ *is irrational then*

$$\frac{d_{\bar{p}}(h(z), \bar{p})}{[d_p(z, p)]^{\log|\bar{\alpha}|/\log|\alpha|}}$$

is constant in each connected component of $W^s(p)-\{p\}$, *and*

$$\frac{d_{\bar{q}}(h(w),\bar{q})}{[d_q(w,q)]^{\log|\bar{\beta}|/\log\beta|}}.$$

is constant in each connected component of $W^u(q)-\{q\}$.

Proof. (1) Let $x_i\to x$ be a nice sequence with respect to (x,y) where $y\in W^u(q)$. Hence by lemma 2.4 $\bar{x}_i=h(x_i)$ is a nice sequence with respect to $(h(x),h(y))$. Choose subsequence $i(j)$ and $n(j)$ such that $f^{-n(j)}(x_{i(j)})$ converges to a point $z\in W^s(p)$. From lemma 2.3 it follows that

$$cd_p(z,p)|\alpha|^{n(j)}\simeq d(x_{i(j)},W^u(p))\simeq d(x_{i(j)},W^s(q)\simeq c'd_q(y,q)|\beta|^{i(j)}.$$

Hence $i(j)/n(j)\simeq\log|\alpha|/\log|\beta|$. Since h is a conjugacy it follows also that $i(j)/n(j)\simeq\log|\bar{\alpha}|/\log|\bar{\beta}|$. Thus

$$\frac{\log|\alpha|}{\log|\beta|}=\frac{\log|\bar{\alpha}|}{\log|\bar{\beta}|}.$$

(2) Let x_i be nice with respect to (x,y). From the first part of the proof it follows that for any $i(j)$, $n(j)\to\infty$,

$$d_p(f^{-n(j)}(x_{i(j)}),p)\simeq\Delta_x\cdot d_q(y,q)\cdot|\beta|^{i(j)}\cdot|\alpha|^{-n(j)},$$

where $\Delta_x=c'/c$. It follows that for any z in one component of $W^s(p)-\{p\}$ there exists $i(j)$, $n(j)$ such that $f^{-n(j)}(x_{i(j)})\to z$, since $\log|\alpha|/\log|\beta|$ is irrational. Because the same holds for $\bar{f}$ one has

$$\frac{d_{\bar{p}}(h(z),\bar{p})}{(d_p(z,p))^\delta}=\frac{\Delta_{\bar{x}}}{(\Delta_x)^\delta}\cdot\frac{d_{\bar{q}}(\bar{y},\bar{q})}{(d_q(y,q))^\delta},$$

where $\delta=\log|\bar{\alpha}|/\log|\alpha|=\log|\bar{\beta}|/\log|\beta|$. Since the right hand side of this equation is independent of z, the result follows. □

Remarks. (1) From the second part of lemma 2.5 it follows that the restriction of h to $W^s(p)-\{p\}$ and to $W^u(q)-\{q\}$ is a C^1 diffeomorphism.

(2) The restriction of h to each component of $W^s(p)-\{p\}$ and $W^u(q)-\{q\}$ is determined by the image of one point. This is the rigidity of the conjugacy we have mentioned before.

COROLLARY 1. *Each orbit of tangency between stable and unstable manifolds of periodic orbits gives rise to at least one modulus condition.*

Proof. If f has k orbits of tangency, then arbitrarily near f there is a diffeomorphism $\bar{f}$ with at least k orbits of tangency of even order. Let $W^u(p)$ and $W^s(q)$ have tangencies of even order along r orbits $O(x_1),\ldots,O(x_r)$. Then from the proof of lemma 2.5 (using the same notation) one obtains the following r conditions:

$$\frac{\log|\alpha|}{\log|\beta|}=\frac{\log|\bar{\alpha}|}{\log|\bar{\beta}|}\quad\text{and}\quad\frac{(\Delta x_i)^\delta}{\Delta_{\bar{x}_i}}=\frac{(\Delta_{x_j})^\delta}{(\Delta_{\bar{x}_j})}.$$

From this corollary 1 follows. □

COROLLARY 2. *If* $f\in\mathcal{A}$ *has finite moduli then satisfies conditions* (1), (2), (3) *and* (5) *of* § 1.

Proof. If $f \in \mathcal{A}$ does not satisfy (1), (2), (3) or (5) then for any $k \in \mathbb{N}$ there exists a diffeomorphism f arbitrarily near f satisfying one of the following properties:

(a) $\bar{f}$ has at least k orbits of tangency between stable and unstable manifolds of periodic orbits;

(b) there exist saddles $\bar{p}, \bar{q}$ so that $W^u(\bar{p})$ and $W^s(\bar{q})$ are tangent and so that there exist an infinite number of orbits of intersections of invariant manifolds of saddles with $W^s(\bar{p})$ or $W^u(\bar{q})$. In the latter case $\bar{f}$ has modality at least k because of remark 2 above. In the former case $\bar{f}$ has modality ∞ because of corollary 1. Since this holds for any k the modality of f is ∞. □

LEMMA 2.6. *Let $\bar{f}$ and f be as in lemma 2.5 with $\log|\alpha|/\log|\beta|$ irrational. Let r (resp. $\bar{r}$) be a hyperbolic fixed point of f (resp. $\bar{f}$) of saddle type whose stable manifold intersects transversally the unstable manifold of q (resp. $\bar{q}$). If h is a conjugacy between f and $\bar{f}$ then the expanding eigenvalue of $df(r)$ is equal to the expanding eigenvalue of $d\bar{f}(\bar{r})$.*

Proof. Let d (resp, $\bar{d}$) be C^0 metrics as before. Let $y_i \in W^u(q)$ be a sequence converging to $y \in W^u(q) \cap W^s(r)$ and such that $f^i(y_i)$ converges to a point $w \in W^u(r)$. Since $h|W^u(q)-\{q\}$ is C^1, there exists a constant $c_1 > 0$ such that $\bar{d}(h(y_i), h(y)) \simeq c_1 d(y_i, y)$. If γ is the expanding eigenvalue of $df(r)$ we have, by lemma 2.3,

$$d(y_i, W^s(r)) \simeq c_2 d(w, r)|\gamma|^{-i}$$

for some constant $c_2 > 0$. Similarly

$$\bar{d}(h(y_i), W^s(\bar{r})) \simeq \bar{c}_2 \bar{d}(\bar{w}, \bar{r})|\bar{\gamma}|^{-i}.$$

From these equations we have $|\gamma| = |\bar{\gamma}|$. Clearly this implies $\gamma = \bar{\gamma}$. □

Remark. Similarly to the above situation we have that the contracting eigenvalue at each saddle point whose unstable manifold intersects $W^s(p)$ transversally is a conjugacy invariant.

Now we will introduce some other conjugacy invariants which arise in the following situation: there are three fixed points p, q, r such that $W^u(p)$ has a point x of tangency with $W^s(q)$ of order $2n$ and $W^u(q)$ has a point y of tangency with $W^s(r)$ of order $2m$. Take C^1 curves Σ_x, Σ_y at x, y, transversal to $W^s(q)$ and $W^u(q)$ respectively. Take a neighbourhood R of q which contains x and y, and on which f has a linearising coordinate system. Let $\tilde{\Sigma}_{n,x}$ be the component of $f^n(\Sigma_x) \cap R$ containing $f^n(x)$, and similarly $\tilde{\Sigma}_{n,y}$ the component of $f^{-n}(\Sigma_y) \cap R$ containing $f^{-n}(y)$. There are four cases to consider: (see figure 2.1).

Case A. $\Sigma_{n,x} \cap W^s(r)$ and $\Sigma_{n,y} \cap W^u(p)$ do not accumulate to y respectively x as $n \to \infty$;

Case B. $\Sigma_{n,x} \cap W^s(r)$ does accumulate to y, but $\Sigma_{n,y} \cap W^u(p)$ does not accumulate to x;

Case C. $\Sigma_{n,x} \cap W^s(r)$ does not accumulate to y, but $\Sigma_{n,y} \cap W^u(p)$ does accumulate to x;

Case D. $\Sigma_{n,x} \cap W^s(r)$ and $\Sigma_{n,y} \cap W^u(p)$ do accumulate to y respectively x.

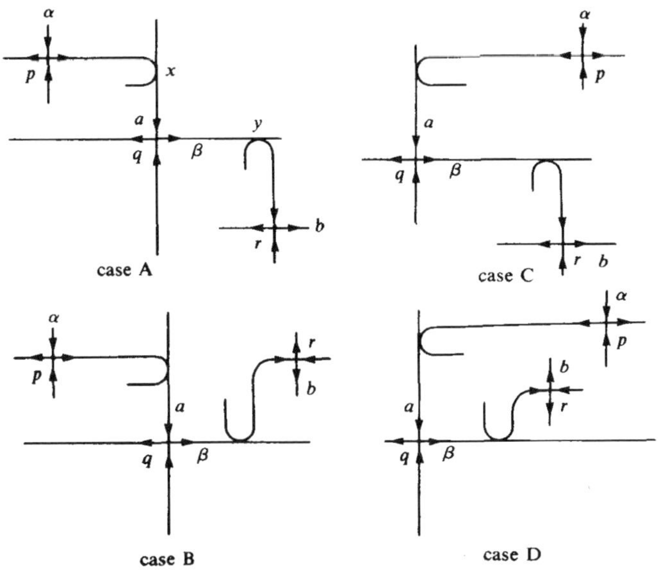

FIGURE 2.1

LEMMA 2.7. *Let $\bar{f}: M \to M$ be a diffeomorphism having the same intersection pattern of stable and unstable manifolds as f. Consider the following equations*:

(M1) $\alpha = \bar{\alpha}$

(M2) $\beta = \bar{\beta}$

(M3) $a = \bar{a}$

(M4) $b = \bar{b}$

(M5) $$\frac{Q_x[d(x,q)]^{2n}}{d(y,q)} = \frac{Q_{\bar{x}}[\bar{d}(\bar{x},\bar{q})]^{2n}}{\bar{d}(\bar{y},\bar{q})}$$

(M6) $$\frac{Q_y[d(y,q)]^{2m}}{d(x,q)} = \frac{Q_{\bar{y}}[\bar{d}(\bar{y},\bar{q})]^{2m}}{\bar{d}(\bar{x},\bar{q})}$$

where α, a are contracting eigenvalues of df_p and df_q; β, b are expanding eigenvalues of df_q, df_r; d is a linearising metric at q;

$$Q_x = \lim_{\substack{w \in W^s(q) \\ w \to x}} \frac{d(w, W^u(p))}{[d(w,x)]^{2n}}; \qquad Q_y = \lim_{\substack{w \in W^u(q) \\ w \to y}} \frac{d(w, W^s(r))}{[d(w,y)]^{2m}}.$$

If $\log|\alpha|/\log|\beta|$ and $\log|a|/\log|b|$ are irrational and h is a conjugacy between f and $\bar{f}$ such that $h(p)=\bar{p}$, $h(q)=\bar{q}$, $h(r)=\bar{r}$, $h(x)=\bar{x}$ and $h(y)=\bar{y}$ then the following conditions are satisfied:

Case A. *Conditions* (M1) *to* (M6).

Case B. *Conditions* (M1) *to* (M4) *plus condition* (M5).

Case C. *Conditions* (M1) *to* (M4) *plus condition* (M6).

Case D. *Conditions* (M1) *to* (M4).

Proof. From lemma 2.5 we have that

$$\frac{\log|\alpha|}{\log|\beta|}=\frac{\log|\bar{\alpha}|}{\log|\bar{\beta}|} \tag{1}$$

and

$$\frac{\log|a|}{\log|b|}=\frac{\log|\bar{a}|}{\log|\bar{b}|}. \tag{2}$$

Furthermore if these numbers are irrational we have that $h|W^s(q)-\{q\}$ and $h|W^u(q)-\{q\}$ are C^1 maps. We will show now that in all cases equalities (M1) to (M4) are satisfied. Take a sequence $x_i \in W^s(q)$ converging to the point of tangency x and such that $f^{-n(i)}(x_i)$ converges to some point $z \in W^s(p)$. By lemma 2.3 we have that

$$d(x_i, W^u(p)) \simeq cd(z,p)|\alpha|^{n(i)}, \tag{3}$$

where c is a positive constant independent of the sequence. Now $W^u(p)$ and $W^s(q)$ have a tangency of order $2n$ at x. Therefore if we consider a C^∞ metric d' induced by a C^∞ coordinate system in which $W^u(p)$ is a straight line and $W^s(q)$ is the graph of a homogeneous polynomial of degree $2n$ then

$$\frac{d'(x_i, W^u(p))}{[d'(x_i,x)]^{2n}} \to Q'_x, \tag{4}$$

where Q'_x is a positive number. But since d is a C^0 metric, lemma 2.2 tells us that $d(x_i, W^u(p))/d'(x_i, W^u(p))$ converges to a positive constant which does not depend on the metric. Clearly the sequence $d(x_i,x)/d'(x_i,x)$ also converges to a positive constant because $x_i \in W^s(q)$. This and (4) imply that

$$\frac{d(x_i, W^u(p))}{[d(x_i,x)]^{2n}} \to Q_x \tag{5}$$

and $Q_x>0$. Now $h|W^s(q)-\{p\}$ is C^1 so that

$$d(x_i,x) \sim \bar{d}(h(x_i),\bar{x}). \tag{6}$$

Equations (3), (5) and (6) imply that $|\alpha|=|\bar{\alpha}|$. Hence $\alpha=\bar{\alpha}$. Similarly we prove that $b=\bar{b}$ and from (1) and (2) it follows that $\beta=\bar{\beta}$ and $a=\bar{a}$. If we are in case D then we are finished. So let us show that there are additional moduli in cases A, B, C. Let us prove that (M5) is satisfied in cases A and B. For that, take a sequence $x_i \to x$ nice with respect to (x,y). Choose subsequences $i(j)$ and $n(j)$ such that $f^{-n(j)}(x_{i(j)})$ converges to a point $z \in W^s(p)$. Since we are in case A or in case B we can take a sequence $z_j \in W^s(q)$ such that $f^{-n(j)}(z_j) \to z$. Since $\{x_i\}$ is nice with respect to (x,y) and d is a linearizing metric we have

$$d(x_{i(j)}, W^s(q)) \simeq d(y,q)|\beta|^{-i(j)},$$
$$d(x_{i(j)}, W^u(p)) \simeq d(x_{i(j)}, W^s(q)).$$

So we have

$$d(x_{i(j)}, W^u(p)) \simeq d(y,q)|\beta|^{-i(j)}. \tag{7}$$

Since $f^{-n(j)}(x_{i(j)})$ and $f^{-n(j)}(z_j)$ both converge to z we have from lemma 2.3 that $d(x_{i(j)}, W^u(p)) \simeq d(z_j, W^u(p))$. Finally we have equation (5): $d(z_j, W^u(p)) \simeq$

$Q_x\{d(z_j, x)\}^{2n}$. This and (7) imply:

$$d(y, q) \cdot |\beta|^{-i(j)} \simeq Q_x\{d(z_j, x)\}^{2n} \tag{8}$$

i.e.

$$\frac{Q_x\{d(x, q)\}^{2n}}{d(y, q)} \simeq \left\{\frac{d(x, q)}{d(z_j, x)}\right\}^{2n} \cdot |\beta|^{-i(j)}. \tag{9}$$

Since $h \mid W^s(q) - \{q\}$ is linear and d is a linearising metric we have

$$\frac{d(z_j, x)}{d(h(z_j), h(x))} = \frac{d(x, q)}{d(h(x), h(q))}.$$

From this, $\beta = \bar{\beta}$, and (9) it follows that

$$\frac{Q_x \cdot \{d(x, q)\}^{2n}}{d(y, q)}$$

is a topological invariant.

Similarly, we prove that in cases A and C condition (M6) is satisfied. □

Remark. It is not obvious that the numbers which appear in the modulus conditions (M5) and (M6) are independent of the choices made for the metric d. That these numbers are independent of the metrics is proved in § 3(b). (In case A this also follows from the conclusion of lemma (2.6)).

In § 5 we shall show that the equations from lemma (2.6) are the only obstructions for constructing conjugacies between two diffeomorphisms as above.

Proof of the theorem. It remains to prove that if $f \in \mathscr{A}$ has finite modality then conditions 4 and 5 of § 1 are satisfied. Notice that if f has finite modality then the modality of any g sufficiently near f is at most equal to the modality of f. So we will prove that if $f \in \mathscr{A}$ does not satisfy condition 4 of § 1 then f can be approximated by diffeomorphisms with arbitrarily high modality. So let $f \in \mathscr{A}$ be a diffeomorphism having an orbit of non transversal intersection between $W^s(q)$ and $W^u(p)$. We may assume that all periodic points are in fact fixed points by considering a sufficiently high iterate of f. By taking a small perturbation of f we may assume that $W^u(p)$ has a point x of quasi-transversal (i.e. quadratic contact) intersection with $W^s(q)$. Now assume that condition 4 of § 1 does not hold. Then any arbitrarily small arc Σ at x transversal to $W^u(p)$ contains unstable (stable) manifolds of saddles. Then, by taking a small perturbation of f one can find a saddle P (resp. Q) such that $W^u(P)$ (resp. $W^s(Q)$) has an intersection with $W^s(p)$ (resp. $W^u(q)$). We also may assume that $W^u(P)$ (respectively $W^s(Q)$) contains a small interval I (resp. J) which intersects transversally $W^s(p)$ (resp. $W^u(q)$) in a unique point (see figure 2.2). Now assume that condition 4 of § 1 does not hold. Then $f^k(I)$ has an intersection with J for any k sufficiently big.

From the appendix it follows that we can construct an invariant C^1 unstable (resp. stable) foliation $\mathscr{F}_p^u$ (resp. $\mathscr{F}_q^s$), with a C^1 tangent line field, having I (resp. J) as a leaf.

Claim. If $\log|\alpha|/\log|\beta|$ is irrational then the points of tangency of $f^{-k}(J)$ with $\mathscr{F}_p^u$ accumulate at $I \cap W^s(p)$.

The theorem then follows easily from this claim. In fact, if the claim is correct then by a C^∞ small perturbation of f with support in a small neighbourhood of $I \cap W^s(p)$

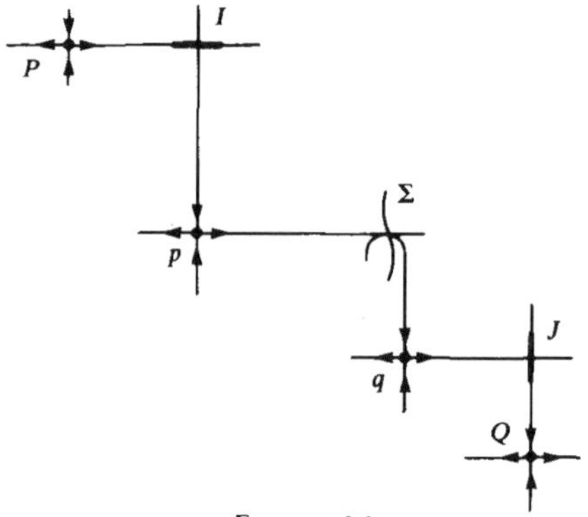

FIGURE 2.2

we can create an arbitrary number of orbits of tangency between $W^u(P)$ and $W^s(Q)$. Since by corollary 1 each orbit of tangency gives rise to a conjugacy invariant we have that f has infinite modality. So it remains to prove the claim. Since x is a point of quasi-transversal intersection of $W^u(p)$ with $W^s(q)$ and the tangent line fields to the foliations $\mathcal{F}_p^u$ and $\mathcal{F}_q^s$ are C^1 it follows, from the implicit function theorem, that in a neighbourhood of x, $\mathcal{F}_p^u$ is transversal to $\mathcal{F}_q^s$ except along a C^1 curve Σ which is transversal to $W^u(p)$ at x (see [**Me2**]). Let

$$a_n = f^n(I) \cap \Sigma \quad \text{and} \quad b_n = f^{-n}(J) \cap \Sigma.$$

If d is a C^0 metric on M then we have

$$\begin{aligned} d(a_n, x) &\simeq c_1 |\alpha|^n, \\ d(b_n, x) &\simeq c_2 |\beta|^{-n}, \end{aligned} \tag{$*$}$$

where c_1, c_2 are positive constants. This follows from lemma 2.2, the transversality of Σ with $W^u(p)$ and the fact that $(*)$ is obviously true for appropriate linearising metrics. Now since $\log|\alpha|/\log|\beta|$ is irrational we can find sequences $n(k), m(k) \to \infty$ such that $|\beta|^{-n(k)}/|\alpha|^{m(k)} \to c_1/c_2$. From $(*)$ therefore follows

$$\frac{d(a_{m(k)}, b_{n(k)})}{|\alpha|^{m(k)}} \to 0,$$

and therefore:

$$d(f^{-m(k)}(b_{n(k)}), f^{-m(k)}(a_{m(k)})) \to 0.$$

Hence $f^{-m(k)}(b_{n(k)})$ converges to $W^s(p) \cap I$ as $k \to \infty$. This proves the claim. □

3. *A few techniques for constructing conjugacies*

In this section we will develop a few techniques for constructing conjugacies. In §§ 4 and 5 we will employ these techniques in order to construct conjugacies between two nearby maps in $\mathcal{A}$ which satisfy the relevant moduli conditions.

Let us start with making a general comment. It is well known that every diffeomorphism in $\mathcal{A}$ which satisfies certain transversality conditions on invariant manifolds is structurally stable, see [**Rb1**]. In order to prove these results one has to construct conjugacies between two nearby diffeomorphisms. Basically there are two ways of doing this. One method is functional analytic. Here one defines an operator on an appropriate space of homeomorphisms and shows this operator has a hyperbolic fixed point, see [**Rb1**]. The other method is a geometric proof using invariant foliations, see [**P1**], [**PS**], [**Me1**], [**Me2**]. (For Axiom A diffeomorphisms this geometric method has only been developed in dimension two, see [**Me1**]). In any case it is probably difficult to use the functional analytic methods since all the operators will be non-hyperbolic, see [**Rb2**]. Therefore we will give geometric constructions using invariant foliations. These methods also enable us to keep track of all the freedom one has in constructing the conjugacies. Let us now define these invariant foliations and explain how they are used.

Take a hyperbolic periodic point p (of saddle type) of a diffeomorphism f. An unstable foliation $\mathscr{F}_p^u$ (tubular family) for p is a continuous retraction $\pi_u: V^u \to W^s(\mathcal{O}(p))$, where $\mathcal{O}(p)$ is the orbit of p and V^u is a neighbourhood of $W^u(\mathcal{O}(p))$ with: (i) $(\pi_u)^{-1}(p) = W^u(p)$; (ii) the foliation whose leaves are fibers of π_u is f-invariant. Similarly we define the stable foliation $\mathscr{F}_p^s$ for p. Using the λ-Lemma it is easy to construct invariant unstable foliations, see [**P1**]. The leaves in fact can be chosen differentiably (C^k if f is C^k) but the field of tangent lines to the leaves is only continuous in general (however see the appendix where we prove this foliation is sometimes more smooth). The use of this foliation is explained by the following lemma whose proof is straightforward.

LEMMA. *Let $p, \bar{p}$ be hyperbolic fixed points of f and $\bar{f}$ with unstable foliations $\mathscr{F}_p^u$, $\mathscr{F}_{\bar{p}}^u$. Let h be a map from a neighbourhood $\mathcal{N}_p$ of $W^u(p)$ to a neighbourhood of $W^u(\bar{p})$ with the following properties*: (i) $hf = \bar{f}h$; (ii) $(h \mid W^s(p)): W^s(p) \to W^s(\bar{p})$ *is continuous*; (iii) $h \mid (\mathcal{N}_p - W^s(p))$ *is continuous*; (iv) *h preserves the unstable foliations. Then h is continuous.*

In order to construct global conjugacies it is necessary that all unstable foliations fit together nicely. So let p, q be hyperbolic periodic points of f with $W^u(p) \cap W^s(q) \neq \emptyset$. We say that the unstable foliations $\mathscr{F}_p^u$ and $\mathscr{F}_q^u$ are compatible if each leaf of $\mathscr{F}_p^u$ contains a leaf of $\mathscr{F}_q^u$. In order to construct a conjugacy h between two nearby Morse-Smale diffeomorphisms one first constructs conjugacies on the stable manifolds of periodic orbits of saddle type preserving the unstable foliations. Here one proceeds inductively using the natural dynamical ordering of periodic orbits starting at sources and ending up at sinks. Then we extend h to fundamental domains of the sinks again preserving these foliations. The conjugacy equation ($hf = \bar{f}h$) gives a unique extension of h to the whole manifold. From the lemma above it follows that h is continuous. In this approach the compatibility of the foliations is essential.

In our case we have tangencies of invariant manifolds. Therefore some leaves of unstable foliations do not have a unique intersection with stable manifolds. Hence it is impossible to construct a compatible system of unstable foliations. So we will have to use both the unstable and the stable foliations. In the Morse-Smale case these foliations are transversal but here they can be tangent. The geometry of their intersection is extremely complicated (especially when higher order tangencies are involved). So we are forced to make careful adjustments of these foliations in order to get a good control of their intersection. These modifications near the tangencies must be globalised in order to keep the foliations invariant. For that we need some careful estimates which we will make in § 3(a) using a generalised λ-Lemma proved in the appendix.

There is another difference with the Morse-Smale case. Here we will have to define the conjugacy in a whole neighbourhood of the tangency points first. Then we have the problem of extending h continuously to the closure of the orbit of this neighbourhood. This is done in § 3(b).

3(a). *Construction of foliations.* Let p be a hyperbolic fixed point (of saddle type) of a C^∞ diffeomorphism f. As usual we say that D is a fundamental domain of $W^s(p)$ if each orbit in $W^s(p)$ has exactly one intersection with D. Similarly N is called a fundamental neighbourhood of $W^s(p)$ if each orbit sufficiently near $W^s(p)$ has precisely one intersection with N. Take a point $x \in W^u(p)-\{p\}$ and a small neighbourhood V_x of x. Let $C_{x,n}$ be a cone-like region which has a tangency of order n with $W^u(p)$ at x:

$$C_x = C_{x,n} = \{w \in V_x;\ d(w, W^u(p)) \geq c(d(w,x))^n\}.$$

Here d is a C^0 metric on M and c is a positive constant. Let $\mathcal{F}$ be a C^1 invariant foliation on V_x whose leaf through x is $W^u(p)$. For each $a \in V_x$ let L_a denote the leaf of $\mathcal{F}$ through a.

LEMMA 3.1 (Extension of foliations defined on cones). *If*

$$\lim_{a\to x} \frac{\sup\{d(w, W^u(p));\ w \in L_a \cap C_x\}}{\inf\{d(w, W^u(p));\ w \in L_a \cap C_x\}} = 1 \tag{3.1}$$

then there exists an invariant unstable C^0 foliation $\mathcal{F}_p^u$ whose restriction to $C_{x,n}$ coincides with $\mathcal{F}$.

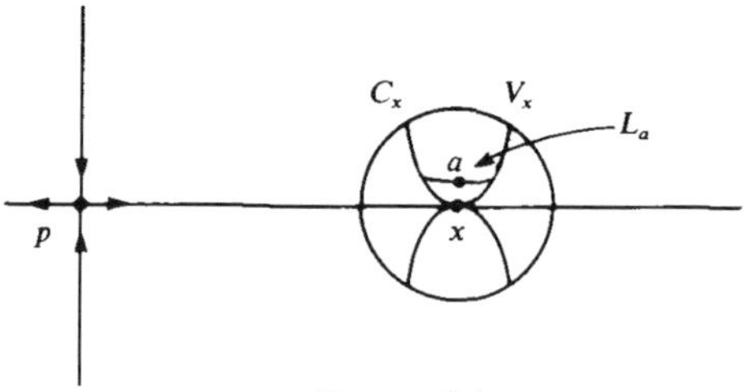

FIGURE 3.1

Remark. If d is a linearizing metric then lemma 3.1 remains true if we take for C_x a whole neighbourhood of x, provided V_x is contained in a fundamental domain.

Proof. Take any invariant unstable foliation $\hat{\mathcal{F}}_p^u$ of p. We will modify $\hat{\mathcal{F}}_p^u$ so that restricted to C_x it will coincide with $\mathcal{F}$. Let N_x be the connected component containing x of a fundamental neighbourhood of $W^u(p)$ such that $N_x - C_x$ consists of two components as in figure 3.2. For N_x sufficiently small $\hat{\mathcal{F}}_p^u \cap N_x \cap \partial C_x$ has precisely two intersections.

Define the new foliation $\mathcal{F}_p^u$ in N_x by taking this foliation identical to $\hat{\mathcal{F}}_p^u$ on $N_x - C_x$ and identical to $\mathcal{F}$ on C_x. Clearly $\mathcal{F}_p^u$ extends uniquely into an invariant foliation

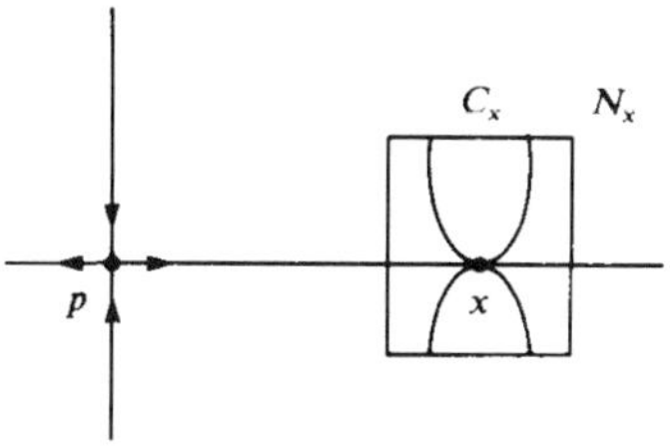

FIGURE 3.2

on a neighbourhood of $W^u(p)-\{p\}$. We must show that the closure of each leaf of $\mathcal{F}_p^u$ has exactly one intersection with $W^s(p)$. More precisely, we must check that there is a continuous projection π_p on $W^s(p)$ along the leaves of $\mathcal{F}_p^u$. In order to show this it suffices to prove that for any sequence $a(i)\in N_x$ such that $f^{-i}(a(i))\to y\in W^s(p)$ as $i\to\infty$ one has $f^{-i}(L_{a(i)})\to y$. Here $L_{a(i)}$ denotes the leaf of $\mathcal{F}_p^u\cap N_x$ through $a(i)$. Now if $a(i)$ and $x(i)\in L_{a(i)}$ are both contained in the same component of N_x-C_x it follows from the fact that $\mathcal{F}_p^u$ is identical to the unstable foliation $\hat{\mathcal{F}}_p^u$ (on N_x-C_x) that $f^{-i}(x(i))\to y$. So it suffices to consider sequences $a(i)\in N_x\cap C_x$ and show that for $x(i)\in L_{a(i)}\cap C_x$ one has $f^{-i}(x(i))\to y$. But from equation (3.1) it follows that

$$d(a(i), W^u(p))\simeq d(x(i), W^u(p)). \tag{3.2}$$

From equation (3.2) and lemma 2.3 it follows that $f^{-i}(a(i))\to y$ if and only if $f^{-i}(x(i))\to y$. This finishes the proof of the lemma. □

It will be convenient to give a general way of constructing foliations $\mathcal{F}$ which satisfy equation (3.1), in terms of some C^1-coordinate systems near x. In fact take a C^1-coordinate system (u, v) near x so that

$$\{v=0\}=W^u(p), \qquad \{(u, v)=(0,0)\}=\{x\}.$$

In these coordinates the leaves of $\mathcal{F}$ are in the form

$$L_a=\{v=g(u, a)\},$$

with $g(0, a)=a$. Write $g_a(u)=g(u, a)$.

LEMMA 3.2 (Construction of foliations in cases which can be extended to tubular families). *Let n be the order of tangency of the cone $C_{x,n}$ with $W^u(p)$ and let $g:\mathbb{R}^2\to\mathbb{R}$ be so that*:

(1) *the n-jet $j^n g_a(u)$ varies C^1 or, in other words, $a\mapsto j^n g_a(u)$ is C^1.*
(2) *$u\mapsto g_a(u)$ is C^∞.*
(3) *$j^n g_0(u)=0$.*

Then the conclusion of lemma 3.1 holds, i.e. the foliation $\mathcal{F}$ on $C_{x,n}$ (corresponding to g as above) can be extended to an unstable foliation.

Proof. According to the theorem of Taylor one has

$$g(u, a)=a+\varphi_2(a)\cdot u+\cdots+\varphi_n(a)\cdot u^n+R(u, a)\cdot u^n, \tag{3.3}$$

with $R(u, a)\to 0$ as $(u, a)\to(0,0)$, and with φ_i C^1 so that $\varphi_i(0)=0$. Now for $(i, v)\in$

$C_{x,n}$ one has for some constant c_1,

$$|u|^n \leq c_1 \cdot |v|. \tag{3.4}$$

Now let $(u, v) = (u, g(u, a))$. Since φ_i is C^1 and $\varphi_i(0) = 0$ one has from (3.3) and (3.4) that:

$$|v| = |g(u, a)| \leq c_2 \cdot |a| + c_1 \cdot |v| \cdot |R(u, a)|$$

for some constant c_2. From this it follows that

$$|v| \leq c_3 |a| \quad \text{and} \quad |u|^n \leq c_4 \cdot |a| \tag{3.5}$$

for some constants c_3 and c_4. Therefore

$$\begin{aligned} \left|\frac{v}{a} - 1\right| &= \left|\frac{g(u, a)}{a} - 1\right| \\ &\leq \left|\frac{\varphi_2(a)}{a}\right| \cdot |u| + \cdots + \left|\frac{\varphi_n(a)}{a}\right| \cdot |u|^n + \left|\frac{u^n}{a}\right| \cdot |R(u, a)| \\ &\leq c_5 \cdot |u| + c_4 \cdot |R(u, a)|. \end{aligned}$$

From this it follows that

$$\left|\frac{v}{a} - 1\right| = \left|\frac{g(u, a)}{a} - 1\right| \to 0 \tag{3.6}$$

as $(x, a) \to (0, 0)$. Obviously equation (3.6) implies equation (3.1) in lemma 3.1.

□

We will use lemma 3.2 in order to modify foliations near certain parts of the manifold. Let f be a diffeomorphism with a non-transversal saddle-connection as in figures (3.3) and (3.4).

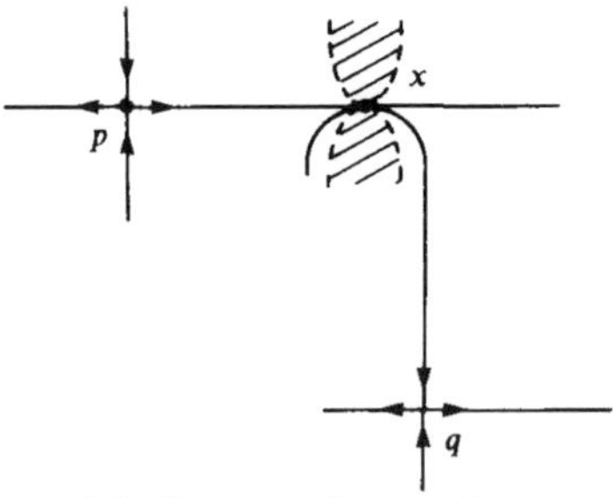

FIGURE 3.3. Tangency of even order

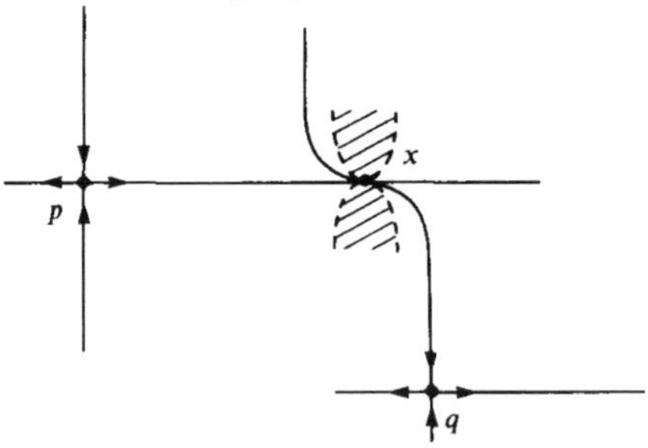

FIGURE 3.4. Tangency of odd order

Assume that $W^u(p)$ and $W^s(q)$ have a tangency of order n at x. Let $\mathscr{F}_p^u$ and $\mathscr{F}_q^s$ be respectively the unstable foliation at p and the stable foliation at q. According to the theorem in the appendix one can choose these foliations C^1 so that the fields of tangent lines are C^1 and so that the leaves of these foliations are C^∞. We first show that $\mathscr{F}_p^u$ and $\mathscr{F}_q^s$ can only be tangent in a certain cone $C_{x,n}$.

LEMMA 3.3. *Let $\mathscr{F}$, $\mathscr{F}'$ be C^1 foliations on a neighbourhood of the origin of $\mathbb{R}^2$ such that the leaves are C^{r+1}, $r \geq 1$, the fields of tangent lines are C^1 and the leaf $\mathscr{F}'(0)$ of $\mathscr{F}'$ through 0 has contact $r+1$ with $\mathscr{F}(0)$. Let T be the set of points where $\mathscr{F}$ is tangent to $\mathscr{F}'$. Then the contact of T with $\mathscr{F}(0)$ at 0 is less than $r+1$, that is more precisely, either $T-\{0\}=\varnothing$ or*

$$\lim_{\substack{z \in T-\{0\} \\ z \to 0}} \frac{d(z, \mathscr{F}(0))}{(d(z,0))^{r+1}} = \infty,$$

where d is some C^0 metric.

Proof. Choose a C^{r+1} coordinate system (u, v) in a neighbourhood of 0 such that $\mathscr{F}(0,0)=\{(u,v) \mid v=0\}$ and $\mathscr{F}'(0,0)=\{(u,v) \mid v=u^{r+1}\}$. Denote by $\varphi(u,v)$ the tangent of the angle of $\mathscr{F}'(u,v)$ with the horizontal line through (u,v) and by $\psi(u,v)$ the tangent of the angle between $\mathscr{F}(u,v)$ and $\mathscr{F}'(u,v)$. Hence φ and ψ are C^1, $\psi(u,0)=\varphi(u,0)$, $\varphi(u,u^{r+1})=(r+1)\cdot u^r$ and $T=\{(u,v) \mid \psi(u,v)=0\}$. By the mean value theorem we have

$$\psi(u,v)=\psi(u,0)+\frac{\partial \psi}{\partial v}(u,\bar{v})\cdot v,$$

and

$$\psi(u,0)=\varphi(u,0)=\varphi(u,u^{r+1})-\frac{\partial \varphi}{\partial v}(u,\tilde{v})\cdot u^{r+1},$$

for some $\bar{v}$ between 0 and v and some $\tilde{v}$ between 0 and u^{r+1}. Thus

$$\psi(u,v)=(r+1)\cdot u^r-\frac{\partial \varphi}{\partial v}(u,\tilde{v})\cdot u^{r+1}+\frac{\partial \psi}{\partial v}(u,\bar{v})\cdot v,$$

and

$$\frac{\psi(u,v)}{u^{r+1}}=\frac{r+1}{u}-\frac{\partial \varphi}{\partial v}(u,\tilde{v})+\frac{\partial \psi}{\partial v}(u,\bar{v})\cdot\frac{v}{u^{r+1}}.$$

If $(u,v)\in T$ we have $\psi(u,v)=0$ and since $\partial\psi/\partial v$ and $\partial\varphi/\partial v$ are bounded, we have

$$\lim_{\substack{(u,v)\in T \\ u \to 0}} \frac{|v|}{|u|^{r+1}}=+\infty.$$

This clearly proves the lemma. □

Now we will put lemma 3.2 and lemma 3.3 together and show how to modify the foliations $\mathscr{F}_p^u$ and $\mathscr{F}_q^s$ so that they are tangent only along a curve.

Consider any C^0 metric d. Since $W^u(p)$ and $W^s(q)$ have a tangency of order n one has (as in lemma (2.6)) that

$$\lim_{\substack{w \in W^s(q) \\ w \to x}} \frac{d(w, W^u(p))}{(d(w,x))^n} = Q_x$$

for some positive number Q_x. Now let T be the set where $\mathscr{F}_p^u$ and $\mathscr{F}_q^s$ are tangent, and let as before

$$C_{x,n}=\{w \mid d(w, W^u(p)) \geq c\cdot(d(w,x))^n\}.$$

According to lemma 3.3, for any $c>0$ there is a neighbourhood V_x of x so that

$$T \cap V_x \subset C_{x,n}.$$

So take $c>Q_x$, and V_x so small that $C_{x,n} \cap W^s(q)=\{x\}=C_{x,n} \cap W^u(p)$. In other words, outside the set $C_{x,n}$ the foliations $\mathscr{F}_p^u$ and $\mathscr{F}_q^s$ have no tangencies (in V_x), so we will modify these foliations in $C_{x,n}$. We can modify either $\mathscr{F}_p^u$ or $\mathscr{F}_q^s$ according to whichever is more convenient. Let us change $\mathscr{F}_q^s$. In order to do this take C^1-coordinates (u, v) near x so that $\{v=0\}=W^u(p)$, and so that the leaves of $\mathscr{F}_p^u$ are of the form $\{v=c\}$, i.e. parallel to the $\{v=0\}$ line. From the appendix it follows that this is possible. Clearly we can write $W^s(q)$ in the form $W^s(q)=\{v=q(u)\}$ where q is some C^1 function and $q(0)=0$. From the remark above lemma 2.3 it follows that $q(u)/u^n$ converges to a number $Q \neq 0$ as $u \to 0$. We will treat the case that n is even and the case that n is odd separately.

Case 1: *n is even.* Consider the foliation $\{v=g(u, a)\}$, $a \in \mathbb{R}$, where $g(u, a)=Qu^n+a$. According to lemma 3.2 we can change the foliation $\mathscr{F}_q^s$ so that inside $C_{x,n}$ the leaves of this foliation are of the form $\{v=g(u, a)\}$. From this and lemma 3.3 it follows that then $\mathscr{F}_p^u$ and $\mathscr{F}_q^s$ are tangent exactly along the line $\{u=0\}$.

Case 2: *n is odd.* In this case change the foliation $\mathscr{F}_q^s$ so that inside $C_{x,n}$ the leaves of this foliation are of the form $\{v=g(u, a)\}$, where $g(u, a)=Qu^n+a^2 \cdot u+a$. From this it follows that then $\mathscr{F}_p^u$ and $\mathscr{F}_q^s$ are tangent only at the point x.

Remark. The set $C_{x,n} \backslash \{x\}$ consists of two components. Sometimes it will be necessary to modify $\mathscr{F}_p^u$ in one component and $\mathscr{F}_q^s$ in the other component.

3(b). *Construction of conjugacies.* Now we give a lemma which will show us how to use these foliations in the construction of conjugacies. As before let V_x be a neighbourhood of a point $x \in W^u(p)-\{p\}$ and let,

$$C_{x,n}=\{w \in V_x;\ d(w, W^u(p)) \geq c \cdot \{d(w, x)\}^n\}.$$

(Here d is a C^0 metric as before).

We consider the following situation. Let $p, \bar{p}$ be hyperbolic fixed points of C^2 diffeomorphisms $f, \bar{f}$. Let the contracting eigenvalue of f (resp. $\bar{f}$) be α (resp. $\bar{\alpha}$), and let

$$\delta=\frac{\log|\bar{\alpha}|}{\log|\alpha|}, \qquad \text{i.e.} \quad |\alpha|^\delta=|\bar{\alpha}|.$$

Let $\mathscr{F}_p^u$ (resp. $\mathscr{F}_{\bar{p}}^u$) be the unstable foliation for f (resp. $\bar{f}$). Let $d, \bar{d}$ be two C^0 metrics. We have the following.

LEMMA 3.4. *Let $h: V_x \to V_{\bar{x}}$ be a homeomorphism satisfying the properties*:

(i) $\displaystyle\lim_{\substack{w \in C_{x,n} \\ w \to x}} \frac{\bar{d}(h(w), W^u(\bar{p}))}{(d(w, W^u(p)))^\delta}$ *exists and is positive.*

(ii) *The restriction of h to each connected component of $V_x - C_x$ maps leaves of the restriction of the foliation $\mathscr{F}_p^u$ to this connected component into leaves of $\mathscr{F}_{\bar{p}}^u$.*
Then h extends to a homeomorphism of a neighbourhood of p, conjugating f and $\bar{f}$.

Proof. Let d' and $\bar{d}'$ be metrics induced by C^1-linearising coordinates for f (resp. $\bar{f}$). From lemma 2.2 in § 2 it follows that

$$\lim_{\substack{w\to x\\ w\in C_{x,n}}} \frac{\bar{d}'(h(w), W^u(\bar{p}))}{\{d'(w, W^u(p))\}^{\delta}} = l \neq 0. \tag{3.6}$$

Furthermore since d' and $\bar{d}'$ are induced by linearising coordinates one has for sequences $x_i \to x$ that $f^{-n(i)}(x_i)$ converges to some point $z \in W^s(p)$ if and only if

$$d'(x_i, W^u(p)) \approx d'(p, z)\cdot|\alpha|^{n(i)},$$

and a similar statement holds for $\bar{f}$. From this, equation (3.6), and the fact that $|\alpha|^{\delta} = |\bar{\alpha}|$, it follows that h extends to a homeomorphism of the closure of $\bigcup_{k=0}^{\infty} f^{-k}(C_{x,n})$ conjugating f and $\bar{f}$.

Now assume V_x is sufficiently small so that $f(V_x)\cap V_x = \emptyset$. We can extend h to V_x by taking any extension of $h|C_{x,n}$ which maps leaves of $\mathscr{F}_p^u$ to leaves of $\mathscr{F}_{\bar{p}}^u$. It follows that h extends to the closure of $\bigcup_{k=0}^{\infty} f^{-k}(V_x)$.

Let us now extend h to a neighbourhood of $W^u(p)$. Take a leaf F of $\mathscr{F}_p^u$. For simplicity we can assume that V_x is chosen so that if F intersects V_x, then $F\backslash(V_x \cup f^{-1}(V_x))$ consists of three components, see figure 3.5. Consider the component of $F\backslash(V_x\cup f^{-1}(V_x))$ with boundary points $x_1 \in V_x$ and $x_2 \in f^{-1}(V_x)$.

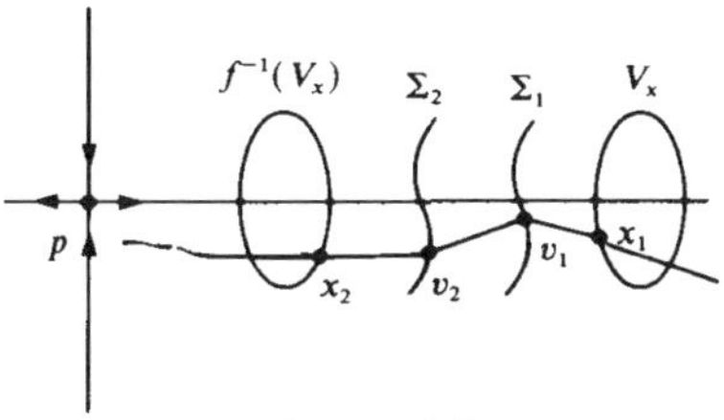

FIGURE 3.5

Consider $h(x_1)$ and $h(x_2)$. In general $h(x_1)$ and $h(x_2)$ need not lie on one leaf of $\mathscr{F}_{\bar{p}}^u$. Therefore we will modify $\mathscr{F}_{\bar{p}}^u$. In fact take two curves $\bar{\Sigma}_1, \bar{\Sigma}_2$ transversal to $W^u(\bar{p})$ between $h(V_x)$ and $\bar{f}^{-1}(h(V_x))$ as in figure 3.5. The leaf $\bar{F}_i$ of $\mathscr{F}_{\bar{p}}^u$ through $h(x_i)$ has an intersection v_i with $\bar{\Sigma}_i$. Take a leaf L_{v_1} in the strip between Σ_1 and Σ_2 which is piecewise linear with respect to the linearising coordinates for $\bar{f}$ such that $L_{v_1}\cap\bar{\Sigma}_1 = v_1$, $L_{v_1}\cap\bar{\Sigma}_2 = v_2$. Do this for every $v_1 \in \bar{\Sigma}_1$. Now define a new foliation '$\mathscr{F}_{\bar{p}}^{u}$' which is identical to $\mathscr{F}_{\bar{p}}^u$ outside the strip between Σ_1 and Σ_2 and identical to the piecewise linear one inside this strip. It is easy to see that '$\mathscr{F}_{\bar{p}}^{u}$' is an honest invariant unstable foliation for $\bar{p}$, using lemma 3.1.

Now define a conjugacy h on a neighbourhood of $W^u(p)$ by taking any extension of $h|V_x$ which maps leaves of $\mathscr{F}_p^u$ onto leaves of '$\mathscr{F}_{\bar{p}}^{u}$'. From the construction above it follows that this is possible. □

Now we consider the situation that a homeomorphism h is only defined on a cone-like set C_x. More specifically consider the situation that f has hyperbolic saddle-points p, q such that $W^u(p)$ and $W^s(q)$, have a tangency of (even) order n at a point x.

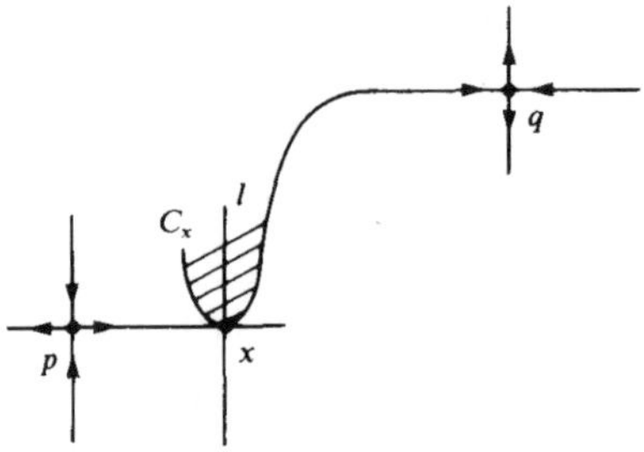

FIGURE 3.6

Let C_x be the cone-like set as in figure 3.6, and let $\bar{f}$ be a diffeomorphism so that $W^u(\bar{p})$ and $W^s(\bar{q})$ also have a tangency of order n at $\bar{x}$. Assume $h: C_x \to \bar{C}_x$ is a homeomorphism. When can we extend h to a conjugacy on a neighbourhood of p? In order to investigate this question take a C^0-metric d near x and define

$$Q_x = \lim_{\substack{w \in W^s(q) \\ w \to x}} \frac{d(w, W^u(p))}{\{d(w, x)\}^n}.$$

As we have shown before, Q_x is well defined and $Q_x \neq 0$. Define $Q_{\bar{x}}$ similarly. Let l be a C^∞ curve through x transversal to $W^u(p)$ and take a similar curve $\bar{l}$ for $\bar{f}$. We assume that $h(l) = \bar{l}$ and that the following limits exist and are positive:

$$C_1 = \lim_{\substack{w \to x \\ w \in C_x - l}} \frac{\bar{d}(h(w), \bar{l})}{d(w, l)} \quad \text{and} \quad C_2 = \lim_{\substack{w \to x \\ w \in C_x - W^s(q)}} \frac{\bar{d}(h(w), W^s(\bar{q}))}{d(w, W^s(q))}.$$

Also assume the contracting eigenvalue of f at p is equal to the contracting eigenvalue of $\bar{f}$ at $\bar{p}$. The numbers $Q_{\bar{x}}$, C_1, C_2 do depend on the choice made for the metrics d, $\bar{d}$. However we have the following result.

LEMMA 3.5. *Let h be a homeomorphism as above. Then*:

(1) *the equality $Q_x = (C_1^n / C_2) Q_{\bar{x}}$ does not depend on the metric $d, \bar{d}$; (i.e. if the equality holds for some C^0 metrics d, $\bar{d}$ then it holds for all C^0 metrics)*;

(2) *if $Q_x = (C_1^n / C_2) Q_{\bar{x}}$ then the homeomorphism h can be extended to a conjugacy on a neighbourhood of p.*

Proof. (1) Let d' be another C^0 metric instead of d. From lemma 2.2 and the remark below this lemma we have that there exists $\lambda > 0$ so that

$$\lim_{\substack{w \to x \\ w \in W^s(q)}} \frac{d(w, W^u(p))}{d'(w, W^u(p))} = \lambda = \lim_{\substack{w \to x \\ w \in C_x - W^s(p)}} \frac{d(w, W^s(q))}{d'(w, W^s(q))}.$$

Furthermore if $w \in W^s(q)$ converges to x then from lemma 2.1 it follows that $d(w, x)/d'(w, x)$ converges to a positive constant μ. On the other hand one has $d(w, l) \simeq d(w, x)$, $d'(w, l) \simeq d'(w, x)$ for $w \in W^s(q)$ converging to x since $W^s(q)$ and $W^u(p)$ are tangent to each other and transversal to l. Hence for some $\mu > 0$

$$\lim_{\substack{w \in W^s(q) \\ w \to x}} \frac{d(w, x)}{d'(w, x)} = \mu = \lim_{\substack{w \in W^s(q) \\ w \to x}} \frac{d(w, l)}{d'(w, l)}.$$

It follows immediately that the equality $Q_x = (C_1^n/C_2)Q_{\bar{x}}$ does not depend on the choice of d. Similarly it does not depend on the choice of $\bar{d}$.

(2) Take C^∞ coordinates (u, v) at x so that $W^u(p) = \{v = 0\}$, $W^s(q) = \{v = Q_x u^n\}$ and $l = \{u = 0\}$. From (1) we can take for d the usual Euclidean metric in this coordinate system. Take $w \in C_x$ and let w have coordinates (u, v). Then

$$d(w, W^u(p)) = v = Q_x(d(w, l))^n + v - Q_x(d(w, l))^n.$$

Furthermore

$$v - Q_x(d(w, l))^n \simeq d(w, W^s(q))$$

as $w \to x$ (see the arguments used in the remark below lemma 2.2). Now we use the following elementary fact. Take sequences a_i, b_i, c_i, d_i of positive numbers such that $a_i \simeq b_i + c_i$ and c_i/d_i converges to 1. Then $a_i \simeq b_i + d_i$. It follows that

$$d(w, W^u(p)) \simeq Q_x(d(w, l))^n + d(w, W^s(q))$$

and similarly

$$\bar{d}(h(w), W^u(\bar{p})) \simeq Q_{\bar{x}}(\bar{d}(h(w), \bar{l}))^n + \bar{d}(h(w), W^s(\bar{q})).$$

Using a similar elementary computation as above one concludes that

$$\frac{\bar{d}(h(w), W^u(\bar{p}))}{d(w, W^u(p))}$$

converges for $w \in C_x$ converging to x if and only if

$$Q_x = \frac{C_1^n}{C_2} Q_{\bar{x}}.$$

(In this case the limit is C_2). Using lemma 3.4 it follows that we can extend h to a conjugacy near p. □

4. *Construction of conjugacies when all invariant manifolds except $W^u(p)$ and $W^s(q)$ intersect each other transversally*

Suppose that p and q are hyperbolic saddle-points and that $W^u(p)$ and $W^s(q)$ have k-orbits of non-transversal intersection.

In order to state the next theorem we need some notation. Let α (resp. β) be the contracting (resp. expanding) eigenvalue of f at p (resp. q). Choose a fundamental domain D_p^u in $W^u(p)$ and let $z_1, \ldots, z_k$ be the points in D_p^u where $W^u(p)$ and $W^s(q)$ are tangent. Order these points so that z_i is between p and z_{i+1} on the curve $W^u(p)$. Then take a C^1 curve Σ_i at z_i transversal to $W^u(p)$ and $W^s(q)$ and take linearising metrics d_p, d_q at p and q respectively. Define

$$\Delta_{z_i} = \lim_{\substack{w \to z_i \\ w \in \Sigma_i}} \frac{d_p(w, z_i)}{d_q(w, z_i)}.$$

Let $p_1, \ldots, p_{n(p)}$ (resp. $q_1, \ldots, q_{n(q)}$) be the saddles whose unstable (resp. stable) manifolds intersect $W^s(p)$ (resp. $W^u(q)$). Denote the contracting (resp. expanding) eigenvalue of f at p_i (resp q_i) by A_i (resp. B_i), and the linearising metrics by respectively d_{p_i} and d_{q_i}. Choose fundamental domains D_p^s in $W^s(p)$ and D_q^u in $W^u(q)$. Let $x_1, \ldots, x_{l(p)}$ (resp. $y_1, \ldots, y_{l(q)}$) be the intersections of the unstable (resp. stable) manifolds of the saddles p_i's (resp. q_i's) with D_p^s (resp. D_q^u) ordered as above. Let

$$I = \{i \mid W^u(p) \text{ and } W^s(q) \text{ have a tangency of even order at } z_i\},$$

$$I_{p_i} = \{j;\ x_j \in W^u(p_i) \cap D_p^s\}$$

and

$$I_{q_i} = \{j;\ y_j \in W^s(q_i) \cap D_q^u\}.$$

Define

$$\Delta_{p_i, x_j} = \lim_{\substack{w \to x_j \\ w \in W^s(p)}} \frac{d_{p_i}(w, x_j)}{d_p(w, x_j)} \qquad \text{if } x_j \in W^u(p_i) \cap D_p^s$$

$$\Delta_{q_i, y_j} = \lim_{\substack{w \to y_j \\ w \in W^u(q)}} \frac{d_{q_i}(w, y_j)}{d_q(w, y_j)} \qquad \text{if } y_j \in W^s(q_i) \cap D_q^u.$$

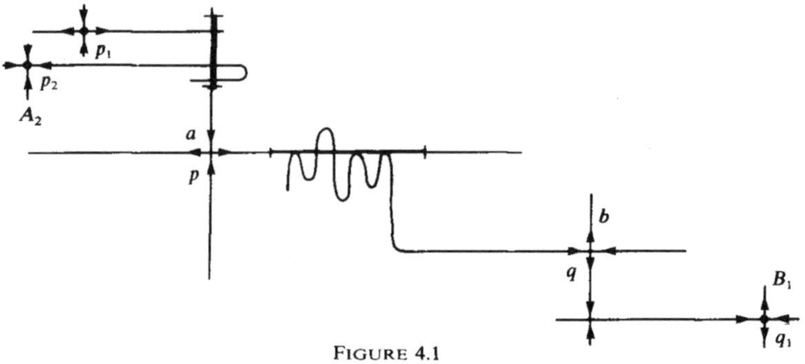

FIGURE 4.1

Let $f, \bar{f} \in \mathcal{M}$ be as above and C^1 near each other. Assume that all invariant manifolds (except $W^u(p)$ and $W^s(q)$) are transversal to each other. Furthermore suppose that $\bar{f}$ has points $\bar{z}_i$, $\bar{x}_i$, $\bar{y}_i$, $\bar{p}_i$, $\bar{q}_i$ corresponding to z_i, x_i, y_i, p_i, q_i, which are ordered in the same way, such that $W^u(p)$, $W^s(q)$ have a tangency of even order at z_i if and only if $W^u(\bar{p})$, $W^s(\bar{q})$ have a tangency of even order at $\bar{z}_i$.

THEOREM 4.1. *The diffeomorphisms $f, \bar{f} \in \mathcal{M}$ as above are conjugate if conditions* (M1)-(M5) *hold. Here* (Mi) *is given by*:

(M1)
$$\frac{\log|\alpha|}{\log|\beta|} = \frac{\log|\bar{\alpha}|}{\log|\bar{\beta}|},$$

i.e. there exists $\delta > 0$ such that $|\alpha|^\delta = |\bar{\alpha}|$ and $|\beta|^\delta = |\bar{\beta}|$;

(M2)
$$\frac{(\Delta_{z_i})^\delta}{\Delta_{\bar{z}_i}} = \frac{(\Delta_{z_j})^\delta}{\Delta_{\bar{z}_j}} \quad \textit{for all } i, j \in I;$$

(M3)
$$A_i = \bar{A}_i \quad i = 1, \ldots, n(p),$$
$$B_j = \bar{B}_j \quad j = 1, \ldots, n(q);$$

(M4)
$$\frac{d_p(x_i, p)}{d_p(x_j, p)} = \left[\frac{d_{\bar{p}}(\bar{x}_i, \bar{p})}{d_{\bar{p}}(\bar{x}_j, \bar{p})}\right]^\delta \quad i = 1, \ldots, l(p),$$
$$\frac{d_q(y_i, q)}{d_q(y_j, q)} = \left[\frac{d_{\bar{q}}(\bar{y}_i, \bar{q})}{d_{\bar{q}}(y_j, \bar{q})}\right]^\delta \quad j = 1, \ldots, l(q);$$

(M5) $$\frac{\Delta_{\bar{p}_i,\bar{x}_j}}{|\Delta_{p_i,x_j}|^{\delta-1}}\cdot[d_p(x_j,p)]^{\delta-1}=\frac{\Delta_{\bar{p}_i,\bar{x}_k}}{|\Delta_{p_i,x_k}|^{\delta-1}}\cdot[d_p(x_k,p)]^{\delta-1}\quad \textit{for all } j,k\in I_{p_i},$$
$$\frac{\Delta_{\bar{q}_i,\bar{y}_j}}{|\Delta_{q_i,y_j}|^{\delta-1}}\cdot[d_q(y_j,q)]^{\delta-1}=\frac{\Delta_{\bar{q}_i,y_k}}{|\Delta_{q_i,y_k}|^{\delta-1}}\cdot[d_q(y_k,q)]^{\delta-1}\quad \textit{for all } j,k\in I_{q_i};$$

If $\log|\alpha|/\log|\beta|$ *is irrational these conditions are also necessary for the existence of a conjugacy.*

Remark. Only tangencies of even order lead to conjugacy invariants. However the presence of tangencies of odd order causes the modality to increase because they generate, by perturbations of the diffeomorphism, new tangencies of even order.

Proof. The necessity of the moduli conditions follows from lemma 2.5 and its corollaries. In order to show the sufficiency of these conditions we have to construct conjugacies. We have subdivided the proof in a few cases.

Case 1. Suppose that there is no periodic saddle point P with $W^u(P)\cap W^s(p)\neq\emptyset$ and similarly no periodic saddle-point Q with $W^u(q)\cap W^s(Q)\neq\emptyset$. As before let $W^u(p)$ and $W^s(q)$ be tangent at k orbits $\mathcal{O}(z_1),\ldots,\mathcal{O}(z_k)$. It will turn out that the number of conjugacy invariants is equal to the cardinality of I. In particular if $\#I=0$ then there are no conjugacy invariants. Let us now construct a conjugacy between f and $\bar{f}$. First we define $h: W^s(p)\to W^s(\bar{p})$ so that

$$d_{\bar{p}}(h(v),\bar{p})=c[d_p(v,p)]^{\delta},\tag{4.1}$$

where c is a positive constant and

$$\delta=\frac{\log|\bar{\alpha}|}{\log|\alpha|}.\tag{4.2}$$

From (4.1), (4.2) and the fact that d_p is a linearising metric it follows that $h\circ f=\bar{f}\circ h$ restricted to $W^s(p)$.

Case 1(a): *All tangencies are of odd order* (i.e. $\#I=0$). Take an unstable family $\mathcal{F}^u_p$ for p and a stable family $\mathcal{F}^s_q$ for q. Using the construction at the end of § 3(a) one can modify $\mathcal{F}^u_p$ and $\mathcal{F}^s_q$ so that, near z_i, $\mathcal{F}^u_p$ and $\mathcal{F}^s_q$ are tangent only in z_i. Define $h: W^u(q)\to W^u(\bar{q})$ to be any homeomorphism so that $h\circ f=\bar{f}\circ h$. The homeomorphisms $h: W^s(p)\to W^s(\bar{p})$ and $h: W^u(q)\to W^u(\bar{q})$ induce maps on the space of leaves of $\mathcal{F}^u_p$ and $\mathcal{F}^s_q$. This defines $h: V_{z_i}\to V_{\bar{z}_i}$ uniquely, where V_{z_i} is a neighbourhood of z_i. Now we show how to extend h globally. Take a system of invariant foliations $\mathcal{F}^s_{\Omega_i}$, $\mathcal{F}^u_{\Omega_i}$ for all basic sets Ω_i in $\Omega(f)$, so that $\mathcal{F}^s_{\Omega_i}$ and $\mathcal{F}^s_{\Omega_j}$ (and similarly $\mathcal{F}^u_{\Omega_i}$ and $\mathcal{F}^u_{\Omega_j}$) are compatible for all basic sets Ω_i,Ω_j except where $\Omega_i=\{p\}$ and $\Omega_j=\{q\}$. In **[P1]**, **[MP]** and **[Me2]** it is shown how to do this when $\Omega(f)$ is finite. If $\Omega(f)$ is infinite then one has to use the methods of **[Me1]**. Remark that $f\in\mathcal{M}$ implies that there is no point x which is part of a non-trivial basic set (i.e. which is not a periodic orbit) such that $W^u(x)$ or $W^s(x)$ has intersection with $W^s(p)$ or $W^u(q)$. As in **[Pa1]** and **[Me1]** we proceed by defining $h: W^u(\Omega_i)\to W^u(\bar{\Omega}_i)$ for any basic set Ω_i such that $W^s(\Omega_i)\cap W^u(p)\neq\emptyset$. This induces a map on the space of leaves of $\mathcal{F}^s_{\Omega_i}$ and from the transversality of $W^s(x)$, $x\in\Omega_i$, with $W^u(p)$ this gives a map $h: W^u(p)\cap V\to W^u(\bar{p})\cap\bar{V}$, where V is a neighbourhood of $W^u(p)\cap W^s(\Omega_i)$. Extend these maps to a conjugacy $h: W^u(p)\to W^u(\bar p)$. Since $\mathcal{F}^u_p$ and $\mathcal{F}^s_q$ intersect transversally, except at z_i, the conjugacies $h: W^s(p)\to W^s(\bar p)$ and $h: W^u(p)\to W^u(\bar p)$ can be extended to a unique conjugacy on a neighbourhood of p which respects all invariant foliations. As in **[P1]**, **[MP]**, **[Me1]** and **[Me2]** one extends h globally.

Case 1(b): *One even tangency, i.e.* $\# I=1$. Let $z=z_i$ be a tangency of $W^u(p)$ and $W^s(q)$ of even order. Using the construction of the end of § 3(a) one can modify $\mathcal{F}^u_p$ and $\mathcal{F}^s_q$ so that, near z, $\mathcal{F}^u_p$ and $\mathcal{F}^s_q$ are tangent along a C^1-curve Σ. The homeomorphism $h: W^s(p)\to W^s(\bar p)$ induces a map on the fibers of $\mathcal{F}^u_p$ and $\mathcal{F}^u_{\bar p}$, and we get a homeomorphism $h:\Sigma\to\bar\Sigma$. From equation (4.1) we get

$$d_{\bar p}(h(v),\bar z)\simeq c[d_p(v,z)]^\delta \qquad \text{for } v\in\Sigma. \tag{4.3}$$

Hence

$$d_{\bar q}(h(v),\bar z)\simeq c\frac{(\Delta_z)^\delta}{\Delta_{\bar z}}[d_q(v,z)]^\delta. \tag{4.4}$$

Using the conjugacy equation $h\circ f=\bar f\circ h$ we can extend h to $\bigcup_{j\in\mathbf{Z}} f^j(\Sigma)$. This set accumulates on $W^u(q)$. From (4.4) it follows that h extends continuously to a homeomorphism $h: W^u(q)\to W^u(\bar q)$ defined by

$$d_{\bar q}(h(v),\bar q)=c\frac{(\Delta_z)^\delta}{\Delta_{\bar z}}[d_q(v,q)]^\delta. \tag{4.5}$$

From (M1) we have that $h\,|\,W^u(q)$ is a conjugacy. Of course the map $h:\Sigma\cup W^u(q)\to\bar\Sigma\cup W^u(\bar q)$ may not map leaves of $\mathcal{F}^s_q$ into leaves of $\mathcal{F}^s_{\bar q}$, but as in the proof of lemma 3.4 one can modify $\mathcal{F}^s_{\bar q}$ in the complement of a neighbourhood of the orbit of $\bar z$ so this is the case. As before we can use these foliations to extend h.

Case 1(c): *More tangencies of even order*; *i.e.* $\# I>1$. As before we can induce homeomorphisms $h:\Sigma_i\to\bar\Sigma_i$ for $i\in I$ and we can define h on $\bigcup_j f^j(\Sigma_i)$ using $h\circ f=\bar f\circ h$. If (M1) holds, then each map $h:\Sigma_i\to\bar\Sigma_i$ induces a conjugacy $h: W^u(q)\to W^u(\bar q)$ satisfying:

$$d_{\bar q}(h(v),\bar q)=c\cdot\frac{(\Delta_{z_i})^\delta}{\Delta_{\bar z_i}}\cdot[d_{\bar p}(v,q)]^\delta.$$

These conjugacies only coincide if (M2) holds. As before one needs to modify $\mathcal{F}^s_{\bar q}$ so that h maps leaves of $\mathcal{F}^s_q$ into leaves of $\mathcal{F}^s_{\bar q}$. Extend h as before.

Case 2. Suppose we are in the same situation as before except that there is exactly one saddle-point P with $W^u(P)\cap W^s(p)\neq\emptyset$ and exactly one saddle point Q with $W^s(Q)\cap W^u(q)\neq\emptyset$. Assume also that $W^u(P)\cap W^s(p)$ (resp. $W^s(Q)\cap W^u(q)$) is a unique orbit $\mathcal{O}(x)$ (resp. $\mathcal{O}(y)$). Suppose that all intersections of invariant manifolds (except of $W^u(p)$ and $W^s(q)$) are transversal, see figure 4.2.

We will now show how to construct conjugacies in this case. Take C^1 foliations $\mathcal{F}^u_P$ and $\mathcal{F}^s_Q$. According to the first theorem in the appendix this is possible. Now in the constructions we made in cases 1(b) and 1(c) we had to modify the foliation $\mathcal{F}^s_q$. But since we do not wish to modify the foliation $\mathcal{F}^s_Q$ we will have to be careful. In fact here we need condition 4 in the definition in the class $\mathcal{M}$, see § 1. Consider the

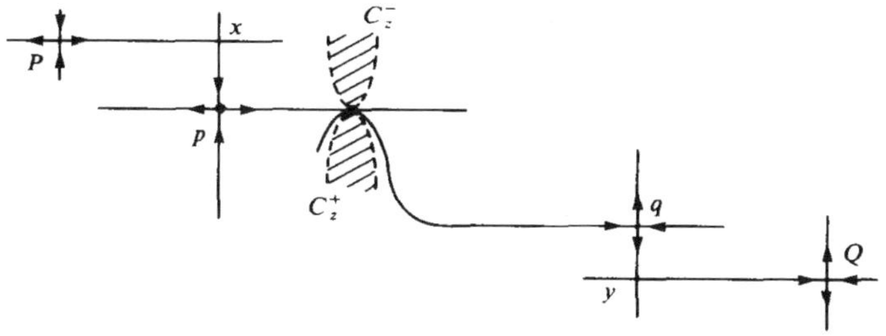

FIGURE 4.2

cone-like sets C_{z_i} as in § 3(a). Let $C^+_{z_i}$ be the component of $C_{z_i}\backslash\{z_i\}$ such that $C^+_{z_i}$ is disjoint from $W^u(P)$, and $C^-_{z_i}$ be the other component. From condition 4 in § 1 it follows that $C^+_{z_i}$ is non-empty and $C^-_{z_i}$ is disjoint from $W^s(Q)$. Let $W^s_{z_i^+}(p)$ be the component of $W^s(p)-\{p\}$ on to which $f^n(C^+_{z_i})$ accumulates and $W^s_{z_i^-}(p)$ the other component of $W^s(p)-\{p\}$. Condition 4 implies that $W^s_{z_i^+}(p)$ does not depend on the tangency point z_i. We also denote this set by $W^s_+(p)$. Similarly define $W^u_{z^+}(q)$ and $W^s_{z^-}(q)$. Clearly the foliation $\mathscr{F}^u_P$ is disjoint from $C^+_{z_i}$ and the foliation $\mathscr{F}^s_Q$ is disjoint from $C^-_{z_i}$. So we can modify $\mathscr{F}^u_p$ in $C^+_{z_i}$ and $\mathscr{F}^s_q$ in $C^-_{z_i}$ as in § 3(a), without changing $\mathscr{F}^u_P$ or $\mathscr{F}^s_Q$. In this way we get foliations $\mathscr{F}^u_p$ and $\mathscr{F}^s_q$ which either are tangent along a curve Σ_i (if the tangency at z_i is of even order) or only at z_i (if the tangency of $W^u(p)$ and $W^s(q)$ is of odd order). Let us now construct conjugacies.

Case 2(a): $\# I=0$. The construction is basically the same as in case 1(a). We start now by defining conjugacies $h: W^s(P)\to W^s(\bar{P})$ and $h: W^u(Q)\to W^u(\bar{Q})$. Via the leaves of $\mathscr{F}^u_P$ and $\mathscr{F}^s_Q$ this induces homeomorphisms on subsets of $W^s(p)$ into $W^s(\bar{p})$ and on subsets of $W^u(q)$ into $W^u(\bar{q})$. Extend these homeomorphisms to conjugacies $h: W^s(p)\to W^s(\bar{p})$ and $h: W^u(q)\to W^u(\bar{q})$. Now proceed as before.

Case 2(b): $\# I=1$. Let Σ be the curve of tangency of $\mathscr{F}^u_p$ and $\mathscr{F}^s_q$, and let Σ_- be the component of $\Sigma-\{z\}$ such that $f^j(\Sigma_-)$ accumulates on $W^s_-(p)\,(=W^s_{z^-}(p))$. Now define $h: W^s_-(p)\to W^s_-(\bar{p})$ as in equation (4.1) where we choose $c>0$ such that $h(x)=\bar{x}$, where $x\in W^s(p)\cap W^u(P)$ and $\bar{x}\in W^s(\bar{p})\cap W^u(\bar{P})$. That is:

$$c=\frac{d_{\bar{p}}(\bar{x},\bar{p})}{(d_p(x,p))^{\delta}}. \tag{4.6}$$

The map h induces a map on the space of leaves of $\mathscr{F}^u_p$ and therefore a map $h:\Sigma_-\to\bar{\Sigma}_-$. As in case 1(b) this induces a map $h: W^u_-(q)\to W^u_-(\bar{q})$, which is a conjugacy if and only if condition (M1) is satisfied. So up to now we have defined h on one side only: starting in $W^s_-(p)$ and ending in $W^u_-(q)$. Similarly one can start in $W^u_+(q)$ and end in $W^s_+(p)$. It follows now from the construction and equation (4.1) that $h: W^s(p)\to W^s(\bar{p})$ and $h: W^u(q)\to W^u(\bar{q})$ are C^1 outside p and q. From this it follows that h extends to $h: W^s(P)\to W^s(\bar{P})$ and $h: W^u(Q)\to W^u(\bar{Q})$ if and only if

(M3) $$A=\bar{A},\quad B=\bar{B},$$

see lemma 3.4. As in the proof of lemma 3.4 one has to change the foliations $\mathcal{F}^u_{\bar{P}}$ (resp. $\mathcal{F}^s_{\bar{Q}}$) in order to make sure that h maps leaves of $\mathcal{F}^u_P$ (resp. $\mathcal{F}^s_Q$) into leaves of $\mathcal{F}^u_{\bar{P}}$ (resp. $\mathcal{F}^s_{\bar{Q}}$). As before one extends h.

Case 2(c): $\# I > 1$. One can construct conjugacies in this case in the same way as in cases 2(b) and 1(c). As in case 1(c) one needs the additional modulus conditions (M2).

Proof of theorem 4.1: *Remaining cases.* Suppose now that f and $\bar{f}$ are again as in case 2, but that there is more than one saddle-point P_i with $W^u(P_i) \cap W^s(p) \neq \emptyset$. Let us first treat the case that there is no saddle-point Q with $W^s(Q) \cap W^u(q) \neq \emptyset$. Then one can modify $\mathcal{F}^s_q$ so that $\mathcal{F}^u_p$ and $\mathcal{F}^s_q$ are tangent only in curves Σ_i or at the points z_i. Now construct h as in case 2. There is one problem left. There are several orbits of intersections $\mathcal{O}(x_i) \subset W^u(P_i) \cap W^s(p)$. But in the construction of $h: W^s(p) \to W^s(\bar{p})$ one has to choose a constant $c > 0$ as in equation (4.5) so that $h(x_i) = \bar{x}_i$ for all i. This is possible if and only if

$$\text{(M4)} \qquad \frac{d(x_i, p)}{d(x_1, p)} = \left\{ \frac{d(\bar{x}_i, \bar{p})}{d(\bar{x}_1, \bar{p})} \right\}^{\delta}, \quad \text{for all } i.$$

Since $h \mid W^s(p)$ is differentiable we need (M3) in order to extend h to $W^s(P_i)$. As before $h \mid W^s(P_i)$ must be a linear map (because $A_i = \bar{A}_i$). If $W^u(P_i) \cap W^s(p)$ contains more than one orbit we will need (M5) because each orbit will induce a linear map $W^s(P_i) \to W^s(\bar{P}_i)$ and these maps must coincide. If all these modulus conditions are satisfied one can extend h as before.

The general case goes similarly. If there are also saddle-points Q_j with

$$W^s(Q_j) \cap W^u(q) \neq \emptyset$$

then it follows from $f \in \mathcal{M}$ that $W^u_+(q) \neq W^u(q)$, $W^s_+(p) \neq W^s(p)$ and either

(i) $W^u(P_i) \cap W^s_+(p) \neq \emptyset$ and $W^s(Q_j) \cap W^u_+(q) \neq \emptyset$; or

(ii) $W^u(P_i) \cap W^s_-(p) \neq \emptyset$ and $W^s(Q_j) \cap W^u_-(q) \neq \emptyset$.

In this case one modifies $\mathcal{F}^u_p$ on one side of z_i and $\mathcal{F}^s_q$ on the other side, see case 2(b). The construction of a conjugacy goes exactly as before. This finishes the proof of theorem (4.1). □

5. *Construction of conjugacies: general case*

First we consider the case that:

5(a). $W^u(p)$, $W^s(q)$ *and* $W^u(q)$, $W^s(r)$ *have non-transversal intersections. All other invariant manifolds intersect transversally.* Assume that there are three hyperbolic periodic points p, q, r so that both $W^u(p)$ and $W^s(q)$ as well as $W^u(q)$ and $W^s(r)$ have exactly one orbit of non-transversal intersections. As in lemma 2.7 there are exactly four cases A, B, C and D to consider. Let (M1)-(M6) be the modulus conditions from lemma 2.7. We will prove:

THEOREM 5.1. *Let $f, \bar{f} \in \mathcal{M}$ be as above and C^∞ near each other. Assume that no other invariant manifold of saddle point intersects $W^s(q) \cup W^u(q)$. The diffeomorphisms f and $\bar{f}$ are conjugate if and only if the modulus conditions* (Mi) *from lemma* 2.7. (*i.e.* 6, 5 *or* 4 *conditions depending on the case we are looking at from lemma* 2.7) *are satisfied.*

Proof of theorem 5.1. In figure 5.1 we have drawn the situations A, B, C and D. The necessity of the modulus conditions (Mi) are proved in lemma 2.6. As before the numbers α, β, a, b denote the eigenvalues at p, q, r as indicated in this figure. We will assume that α, β a, b are positive. The general case can be dealt with similarly.

Also we can assume that the tangency of $W^u(p)$ and $W^s(q)$ and the tangency of $W^u(q)$ and $W^s(r)$ are of even order, because if one of these tangencies is of odd order, then one can construct the conjugacy as in the proof of theorem 4.1. Finally we will assume that $W^u(p) \cap W^s(q)$ and $W^u(q) \cap W^s(r)$ both consist of only one orbit. The general case is treated similarly.

Case A: Denote by R the upper right hand quadrant near q, bounded by $W^u(q)$ and $W^s(q)$. That is, let U_q be a rectangular region on which f is linearisable. Then R is the connected component of $U_q - W^s(q) \cup W^u(q)$ having the orbits of tangencies in its boundary. First choose C^1 invariant foliations $\mathscr{F}^s_q$ and $\mathscr{F}^u_q$ having C^1 tangent line fields. These foliations $\mathscr{F}^s_q$ and $\mathscr{F}^u_q$ induce a C^1-linearising system for f. Let d be the metric induced by this coordinate system.

(a) Let $W^u(p)$, $W^s(q)$ (resp. $W^u(q)$ and $W^s(r)$) have a tangency of order n (resp. m), with n and m even numbers. Define

$$Q_x = \lim_{\substack{z \in W^s(q) \\ z \to x}} \frac{d(z, W^u(p))}{\{d(z, x)\}^n},$$

and define Q_y similarly. In lemma 2.6 it is shown that the numbers

$$\frac{Q_x \cdot [d(x, q)]^n}{d(y, q)}, \frac{Q_y \cdot \{d(y, q)\}^m}{d(x, q)}, \alpha, \beta, a, b \tag{5.1}$$

are topological invariants. Let us show that these invariants are sufficient.

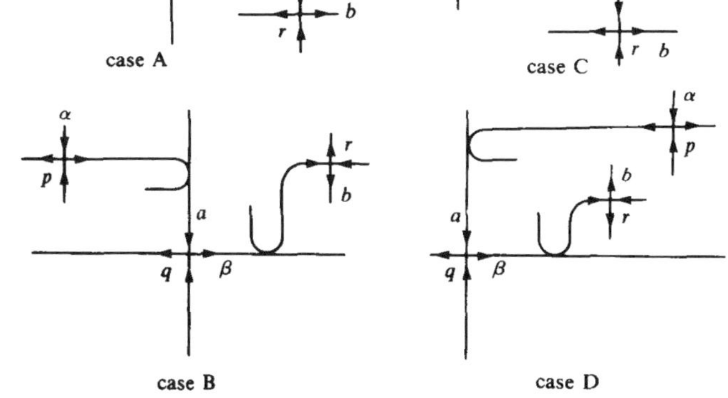

FIGURE 5.1

(b) So we need to define a conjugacy between f and $\bar{f}$. First we define a conjugacy h on the quadrant R. We do this so that h is linear with respect to the linearising coordinates and so that $h(x)=\bar{x}$, $h(y)=\bar{y}$. Since

$$a=\bar{a}, \qquad \beta=\bar{\beta} \tag{5.2}$$

then h indeed defines a conjugacy on R near q.

(c) Let us now show that we can extend h to a neighbourhood of p, provided $\alpha=\bar{\alpha}$ and the modulus condition on Q_x and Q_y is satisfied.

Let l be the line through x which is horizontal with respect to the linearising coordinates, and let $\bar{l}$ be similar. Clearly $h(l)=\bar{l}$ and for $x_i \to x$,

$$\frac{d(h(x_i), W^s(\bar{q}))}{d(x_i, W^s(q))} \to \frac{\bar{d}(\bar{y}, \bar{q})}{d(y, q)}, \tag{5.3}$$

$$\frac{\bar{d}(h(x_i), l)}{d(x_i, l)} \to \frac{\bar{d}(\bar{x}, \bar{q})}{d(x, q)}, \tag{5.4}$$

where d, $\bar{d}$ are induced by linearising coordinates at q and $\bar{q}$. Take any unstable invariant foliations $\mathscr{F}_p^u$ and $\mathscr{F}_{\bar{p}}^u$. We claim that we can modify $\mathscr{F}_{\bar{p}}^u$ so that $\mathscr{F}_{\bar{p}}^u$ agrees with $h(\mathscr{F}_p^u)$ inside R. In fact from lemma 3.5, § 3(b), it follows that h extends to a conjugacy near p if and only if

$$\alpha=\bar{\alpha}, \qquad Q_x=\frac{C_1^n}{C_2}\cdot Q_{\bar{x}},$$

where C_1, C_2 are respectively the limits in equations (5.4) and (5.3). Clearly this corresponds to two equations from (5.1). From lemma 3.5 it follows that one can extend h to a neighbourhood of p, and after a slight modification of $\mathscr{F}_{\bar{p}}^u$, the conjugacy h maps leaves of $\mathscr{F}_p^u$ onto $\mathscr{F}_{\bar{p}}^u$.

(d) In the same way one can extend h to a conjugacy near r, provided $b=\bar{b}$ and the conjugacy condition on Q_y and $Q_{\bar{y}}$ is satisfied.

(e) Now one can extend h as in § 4.

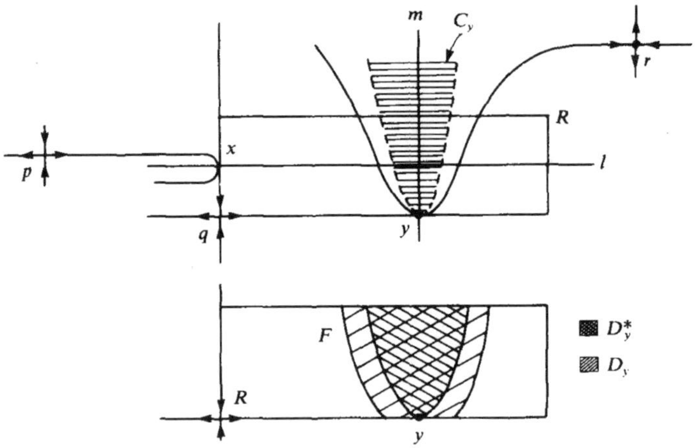

FIGURE 5.2.

Remark. If there are invariant manifolds intersecting with $W^s(q)$ and $W^u(q)$ then one has new necessary moduli conditions, see § 2 and theorem 4.1. As in § 4 one can extend h to M provided all these additional moduli conditions are satisfied.

Cases B *and* C. Cases B and C are dual since one obtains case B out of case C by taking f^{-1}. Therefore we will just treat case B. In lemma 2.6 it is shown that the numbers

$$Q_x \cdot \frac{\{d(x, q)\}^n}{d(y, q)}, \alpha, \beta, a, b$$

are topological invariants. Since we do not have any moduli conditions on Q_y and $Q_{\bar{y}}$ the construction of a conjugacy is more complicated than in the previous case.

(a) As in the previous case choose C^1-invariant foliations $\mathscr{F}_q^s$ and $\mathscr{F}_q^u$ having C^1 tangent line fields. Let l be the leaf of $\mathscr{F}_q^u$ through x and m the leaf of $\mathscr{F}_q^s$ through y, see figure 5.2.

(b) We want to change $\mathscr{F}_r^s$ so that $\mathscr{F}_r^s$ and $\mathscr{F}_q^u$ have only tangencies at m. So consider a cone-like set C_y near y as in lemma 3.3, § 3(a). From this lemma and the first theorem in the appendix it follows that $\mathscr{F}_r^s$ and $\mathscr{F}_q^u$ can only have tangencies inside the sets $f^i(C_y)$. Now modify $\mathscr{F}_r^s$ inside C_y so that $\mathscr{F}_r^s$ restricted to C_y is given by homogeneous polynomials (+constants) in terms of the coordinates induced by $\mathscr{F}_q^s$ and $\mathscr{F}_q^u$. From lemma 3.2 this is possible. It follows that $\mathscr{F}_r^s$ and $\mathscr{F}_q^u$ have only tangencies along $f^i(m)$.

(c) Let R be as before. Take a leaf F of $\mathscr{F}_r^s$ as in figure 5.2 and let D_y (resp. D_y^*) be the region bounded by F and ∂R (resp. $W^s(r)$ and ∂R). Take F and R so that the sets $f^i(D_y)$ are mutually disjoint. Now let $\mathscr{F}^s$ be an invariant foliation on R which is identical to $\mathscr{F}_r^s$ in $\bigcup f^i(D_y)$ and identical to $\mathscr{F}_q^s$ away from $\bigcup f^i(D_y)$. By lemma 3.3 one can choose the rectangle R and $\mathscr{F}^s$ so that all its leaves are transversal to $\mathscr{F}_q^u$ outside the cones $f^i(C_y)$. From the construction of $\mathscr{F}_r^s$ and since $\mathscr{F}^s$ coincides with $\mathscr{F}_r^s$ inside D_y, $\mathscr{F}^s$ and $\mathscr{F}_q^u$ are tangent precisely along iterates of m.

(d) Now we define the conjugacy on R. First define linear conjugacies $h: W^s(q) \to W^s(\bar{q})$ and $h: W^u(q) \to W^u(\bar{q})$ so that $h(x) = \bar{x}$ and $h(y) = \bar{y}$, see equation (4.2). Since $\beta = \bar{\beta}$ and $a = \bar{a}$ this is possible. These maps induce maps on the space of leaves of $\mathscr{F}_q^u$ and of $\mathscr{F}^s$ outside iterates of D_y^*. Therefore these maps induce homeomorphisms $h: m \to \bar{m}$ and $h: l \backslash D^+ \to \bar{l} \backslash \bar{D}^+$, where $D^+ = \bigcup f^i(D_y^*)$ and $\bar{D}^+ = \bigcup f^i(\bar{D}_y^*)$. Since $\mathscr{F}_q^u$ and $\mathscr{F}^s$ are precisely tangent along (iterates of) m, the homeomorphism $h: m \to \bar{m}$ induces also a homeomorphism $h: l \cap D^+ \to l \cap \bar{D}^+$, via the leaves of $\mathscr{F}^s$. Hence there is a unique homeomorphism h on int(R) which preserves $\mathscr{F}_q^u$ and $\mathscr{F}^s$. Let us check that h is a continuous conjugacy. In order to do this we want to apply lemma 3.5 as in the previous case, A. We have to make some estimates similar to (5.3) and (5.4).

LEMMA 5.2. *Take metrics d, $\bar{d}$ corresponding to the linearising coordinates for f and $\bar{f}$*

near q and $\bar{q}$. Then one has for a sequence $x_i \to x$,

$$\left.\begin{aligned}\frac{\bar{d}(h(x_i), \bar{l})}{d(x_i, l)} &\to \frac{\bar{d}(\bar{x}, \bar{q})}{d(x, q)},\\ \frac{\bar{d}(h(x_i), W^s(\bar{q}))}{\bar{d}(x_i, W^s(q))} &\to \frac{\bar{d}(\bar{y}, \bar{q})}{d(y, q)}.\end{aligned}\right\} \tag{5.5}$$

Proof. The first limit follows from the fact that h preserves $\mathscr{F}^u_q$ and from the fact that $h \mid W^s(q)$ is linear with $h(x) = \bar{x}$. The second limit is proved as follows. If $x_i \to x$ and $f^{j(i)}(x_i) \to w \in W^u(q)$, then

$$\begin{aligned} d(x_i, W^s(q)) &\simeq |\beta|^{j(i)} \cdot d(w, q),\\ \bar{d}(h(x_i), W^s(\bar{q})) &\simeq |\bar{\beta}|^{j(i)} \cdot \bar{d}(h(w), \bar{q}).\end{aligned} \tag{5.6}$$

Since $h \mid W^u(q)$ is linear and $h(y) = \bar{y}$ the second equation in (5.5) follows by taking the ratio of the two equations in (5.6). □

Now it follows from equations (5.5) and the modulus conditions on α, Q_x, that h can be extended to a conjugacy near p. This is done exactly as in case A.

(e) As before one can extend h to M. In fact since h maps $\mathscr{F}^s$ into $\bar{\mathscr{F}}^s$ it follows from lemma 3.4 in § 3(b), and from $b = \bar{b}$ that h extends to conjugacy near r. The extension to M goes as before.

Case D. In lemma 2.6 it is shown that in this case the numbers

$$\alpha, \beta, a, b$$

are topological invariants. Now we show that there are no modulus conditions on Q_x and Q_y in this case. It follows that any conjugacy h between f and $\bar{f}$ will be highly non-linear in this case. We construct the conjugacy in a number of steps:

(a) As before take C^1-unstable and stable foliations $\mathscr{F}^u_q$ and $\mathscr{F}^s_q$.

(b) Take a region D_y as in the previous case. Here the boundary of D_y is a leaf of $\mathscr{F}^s_r$ (or is a piece of ∂R). Let N_y be a neighbourhood of D_y. By lemma 3.3 all the leaves of $\mathscr{F}^s_r$ are transversal to the leaves of $\mathscr{F}^s_q$ and $\mathscr{F}^u_q$ inside $N_y \setminus C_y$. Hence, using the implicit function theorem we can take a new invariant foliation $\mathscr{F}^{s'}_q$ for q (which is C^1 except possibly at iterates of y and identical to $\mathscr{F}^s_q$ outside iterates of $N_y \setminus C_y$) so that each leaf of $\mathscr{F}^s_r$ in $D_y \setminus C_y$ is the graph of a homogeneous polynomial function (+constants) with respect to the coordinates induced by $\mathscr{F}^{s'}_q$ and $\mathscr{F}^u_q$. Take a similar set D_x and a similar foliation $\mathscr{F}^{u'}_q$ related to $\mathscr{F}^u_p$.

(c) Modify the foliations $\mathscr{F}^u_p$ and $\mathscr{F}^s_r$ as before. That is, modify $\mathscr{F}^s_r$ inside C_y so that the leaves are given by homogeneous polynomials with respect to the C^1 linearising coordinates induced by $\mathscr{F}^{s'}_q$ and $\mathscr{F}^{u'}_q$. Do the corresponding modification for $\mathscr{F}^s_r$.

(d) Now take an invariant foliation $\mathscr{F}^s$ as before which is identical to $\mathscr{F}^s_r$ inside D_y and identical to $\mathscr{F}^s_q$ away from iterates of D_y. Take a similar foliation $\mathscr{F}^u$. According to lemma 3.3 and the last theorem in the appendix all tangencies of $\mathscr{F}^s$ and $\mathscr{F}^u$ are contained in $f^i(C_x) \cup f^j(C_y)$. Let us study these tangencies.

(e) First we study the tangencies of $\mathscr{F}^s$ and $\mathscr{F}^u$ in $f^i(C_x) \cap f^j(C_y)$. Since $\mathscr{F}^s$ and $\mathscr{F}^u$ are given by polynomial functions it suffices to prove the following lemma.

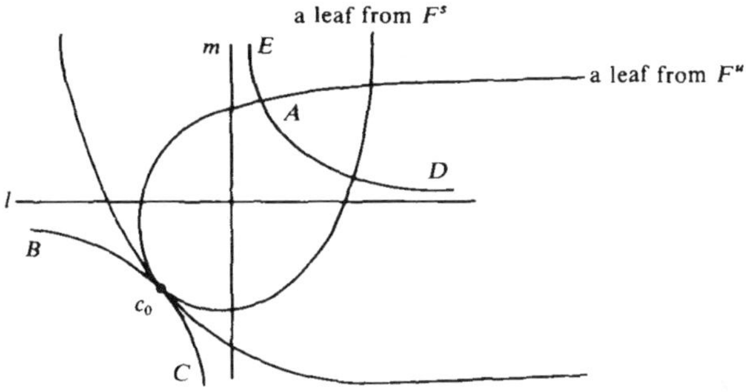

FIGURE 5.3

LEMMA 5.3. *Consider the following two foliations in* $\mathbb{R}^2$:

$$F^u = \{(u, v) \mid u = Q_1 \cdot v^n + a, \quad a \in \mathbb{R}\}$$
$$F^s = \{(u, v) \mid v = Q_2 \cdot u^m + b, \quad b \in \mathbb{R}\},$$

where n, m are even. Then:

(i) F^s *and* F^u *are tangent along a curve c consisting of two components*;

(ii) *there is a unique point* c_0 *where* F^s *and* F^u *have a tangency of odd order.* F^s *and* F^u *are transversal to c except at* c_0.

Proof. If F^s and F^u are tangent at (u, v) then the tangency vector $(1, mQ_2u^{m-1})$ to F^s at (u, v) and the tangency vector $(nQ_1v^{n-1}, 1)$ to F^u are proportional. Hence F^s and F^u are tangent along the curve:

$$n \cdot m \cdot Q_1 \cdot Q_2 \cdot u^{m-1} \cdot v^{n-1} = 1.$$

Statement (ii) follows similarly. □

We have drawn the curve c as a union of the curves A, B, C, D, E in figure (5.3). We can also make the following:

Observations. (1) Each leaf of F^s and each leaf of F^u intersects the curve c at least once and at most three times.

(2) There is a curve A in c, as drawn in figure 5.3, so that if a leaf in F^s or a leaf in F^u intersects A then this leaf has no other intersections with c.

Let A, B, C, D, E be the curves in c as is shown in figure 5.3.

From the lemma it follows that all the tangencies of $\mathcal{F}^u$, $\mathcal{F}^s$ inside $f^i(C_x) \cap f^{-j}(C_y)$ are subsets of (scaled down) copies c_{ij} of the curve c as above. The last theorem in the appendix implies that the leaves of $\mathcal{F}^u$ (respectively $\mathcal{F}^s$) outside C_y (C_x) accumulate in a C^1 sense to $W^s(q)$ $(W^u(q))$. Hence the set c_{ij} consists of two components, and contains the point c_0 and the arc A.

(f) Since $\mathcal{F}_q^{u'}$ is equal to $\mathcal{F}_q^u$ in C_x it follows that inside $f^i(C_x) \cap f^{-j}(D_y)$ the foliations $\mathcal{F}^s$, $\mathcal{F}^u$ are polynomial with respect to $\mathcal{F}_q^u$ and $\mathcal{F}_q^s$. Hence inside these sets the tangencies of $\mathcal{F}^u$ and $\mathcal{F}^s$ are as in (e). Furthermore $\mathcal{F}^s$ is identical to $\mathcal{F}_q^s$ outside the iterates of a neighbourhood N_y of D_y. Hence all tangencies of $\mathcal{F}^s$ with $\mathcal{F}^u$ outside $f^j(N_y)$ are contained in the lines $f^i(l)$. Now choose $\mathcal{F}^s$ on $N_y \backslash D_y$ so

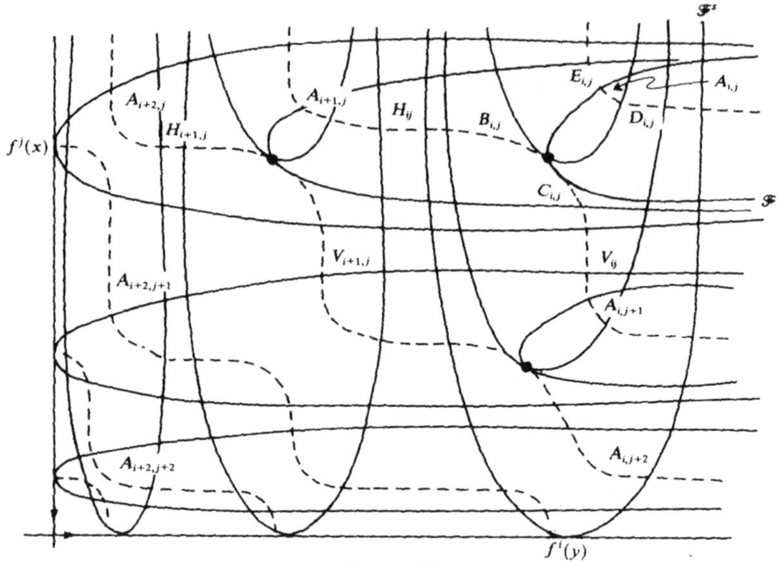

FIGURE 5.4

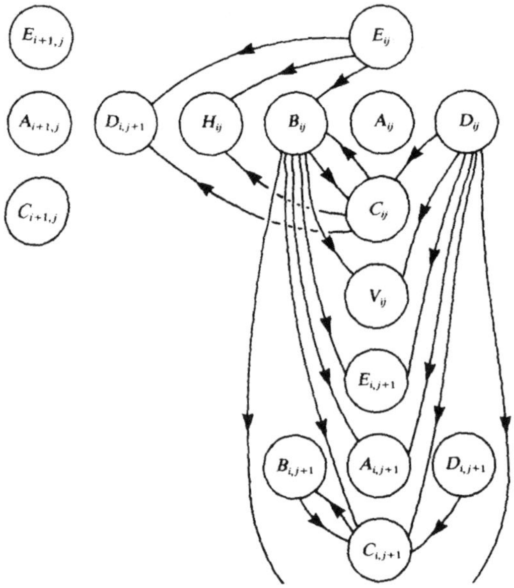

FIGURE 5.5

that $\mathcal{F}^s$ has only tangencies with $\mathcal{F}^u$ in $(N_y \backslash D_y) \cap f^i(C_x)$ along line segments. Similarly inside $f^{-j}(C_y) \cap f^i(D_x)$ the foliations $\mathcal{F}^s$, $\mathcal{F}^u$ are polynomial w.r.t. $\mathcal{F}_q^{u'}$ and $\mathcal{F}_q^s$. Now choose $\mathcal{F}^u$ or $N_x \backslash D_x$ similarly as above. Since all tangencies of $\mathcal{F}^s$ and $\mathcal{F}^u$ are contained in $f^i(C_x) \cup f^{-j}(D_y)$ it follows that $\mathcal{F}^s$ and $\mathcal{F}^u$ are tangent along a curve c (with a countable number of components), as drawn in figure 5.4.

(g) Now one can decompose the tangency curve c in curves A_{ij}, B_{ij}, C_{ij}, D_{ij}, F_{ij}, H_{ij}, V_{ij} as is shown in figure 5.4. Here A_{ij} etc. are contained in $f^i(C_x) \cap f^{-j}(C_y)$. The curves H_{ij}, V_{ij} are bounded by $W^s(r)$ and $W^u(p)$, and the curves A_{ij}, B_{ij}, C_{ij}, D_{ij}, E_{ij}, are as in figure 5.3. Let $l_i = f^i(l)$ and $m_j = f^{-j}(m)$. Using the leaves of these foliations we can define a diagram of maps between those intervals as is shown in figure 5.5. More precisely: to each point z on the tangency curve c we associate a new point $\psi(z) \in c$. We do this as follows.

(i) For $z \in A_{ij}$, let $\psi(z) = z$.

(ii) For $z \in B_{ij}$ take the leaf F of $\mathcal{F}^s$ through z. Follow F downward and let $\psi(z)$ be the first intersection of F with the curve c in one of the components $C_{i,j'}$, $V_{i,j'}$, $E_{i,j'}$, or $A_{ij'}$, see the diagram in figure 5.5.

(iii) For $z \in C_{ij}$ follow a leaf F of F^u to the left and let $\psi(z)$ be the first intersection of F with c in a component $B_{i',j}$, $H_{i',j}$, $D_{i',j}$ or $A_{i',j}$.

(iv) $z \in D_{ij}$. Then follow F downwards as in case (ii).

(v) $z \in E_{ij}$ as (iii).

(vi) $z \in V_{ij}$, then let $\psi(z)$ be the intersection of the leaf of $\mathcal{F}^u$ through z with $W^s(q)$.

(vii) $z \in H_{ij}$, then $\psi(z)$ is the intersection of the leaf of $\mathcal{F}^s$ through z with $W^u(q)$.

Clearly if h is a conjugacy between two diffeomorphisms $f, \bar{f}$ as above and if h preserves the foliations $\mathcal{F}^s$, $\mathcal{F}^u$, then one must have $h(c) = \bar{c}$ and $h \circ \psi = \bar{\psi} \circ h$. From this it follows that we have very little freedom in choosing a conjugacy h.

For example ψ^{-1} maps $B_{ij} \cup C_{ij}$ into $C_{ij} \cup B_{ij}$. We will show shortly that this map in fact is a contraction. Clearly for the point $c_{ij} \in C_{ij} \cap B_{ij}$ where $\mathcal{F}^u$ and $\mathcal{F}^s$ are tangent with c, one has $\psi(c_{ij}) = c_{ij}$. Since h must respect all foliations, it must map the end-points of $B_{ij} \cup C_{ij}$ onto end-points of $\overline{B_{ij}} \cup \overline{C_{ij}}$ since these points belong to stable and unstable manifolds. Hence h is completely fixed in a sequence of points in $B_{ij} \cup C_{ij}$ converging to a fixed point c_{ij} of ψ.

Clearly $\psi^k(z)$ is either contained in $W^s(q)$, $W^u(q)$ or A_{ij} for some sufficiently big k or the sequence $\{\psi^k(z)\}$ is infinite. If this sequence is infinite assign to it a symbol-sequence $\{S_k\}$, where S_k is the component B_{ij}, C_{ij}, D_{ij} or E_{ij} of c which contains $\psi^k(z)$.

LEMMA 5.4(i). *If the sequence $\{\psi^k(z)\}$ is infinite then no other point $z' \in c$ has the same symbol sequence as z.*

(ii) *If this sequence is finite it must end with an interval A_{ij}, $W^s(q)$ or $W^u(q)$. Moreover in this case there is a small interval (in the curve c) of points having the same sequence.*

Proof. The second statement is obvious from the definitions and by continuity. So suppose the sequence $\{\psi^k(z)\}$ is infinite. Then for each $k \in \mathbb{N}$, $\psi^k(z)$ is contained in intervals B_{ij}, C_{ij}, D_{ij}, E_{ij} for some i, j. Let π_l be the projection from $B_{ij} \cup D_{ij}$ on $f^i(l)$

defined as follows. Take $z \in B_{ij} \cup D_{ij}$, and the leaf F of $\mathscr{F}^s$ through z. Let $\pi_l(z)$ be the intersection of F with $f^i(l)$ (near z). Similarly define a projection π_m: $C_{ij} \cup E_{ij} \to f^{-j}(m)$. Then define

$$\theta(z) = \begin{cases} \pi_m \circ \psi \circ \pi_l^{-1}(z) & \text{for } z \in \pi_l(B_{ij} \cup D_{ij}), \\ \pi_l \circ \psi \circ \pi_m^{-1}(z) & \text{for } z \in \pi_m(C_{ij} \cup E_{ij}). \end{cases}$$

Now $f^i(l)$ and $f^{-j}(m)$ are all copies of $\mathbb{R}^+$ so we can consider θ as a map from $\theta: \mathbb{R}^+ \to \mathbb{R}^+$ (which is not defined everywhere). We claim that θ is an expansion:

$$|\theta'(z)| > 1. \tag{5.7}$$

From this claim the lemma follows. In fact take a point $z' \in c$ near z. If $\psi^k(z'')$ is an infinite sequence for every $z'' \in c$ between z and z', then it follows that the length of ψ^k (segment on c between z and z') has finite length. Since θ is an expansion, this is impossible.

So let us prove (5.7). From the construction above it follows that A_{ij}, B_{ij}, C_{ij}, D_{ij}, E_{ij} are contained in the set

$$\{f^i(C_x) \cap f^{-j}(D_y)\} \cup \{f^i(D_x) \cap f^{-j}(C_y)\}. \tag{5.8}$$

Inside $f^{-j}(D_y) \cap f^i(C_x)$ the foliations $\mathscr{F}^s$, $\mathscr{F}^u$ are given by polynomials (+constants) w.r.t. $\mathscr{F}_q^u$ and $\mathscr{F}_q^{s'}$. Similarly inside $f^i(D_x) \cap f^{-j}(C_y)$ the foliations $\mathscr{F}^s$, $\mathscr{F}^u$ are polynomial w.r.t. $\mathscr{F}_q^{u'}$ and $\mathscr{F}_q^s$. It follows that it is sufficient to prove the following lemma.

LEMMA 5.5. *Consider the following two foliations*:

$$F^u = \{(u, v);\, u = Q_1 \cdot v^n + a, \quad a \in \mathbb{R}\}$$
$$F^s = \{(u, v);\, v = Q_2 \cdot u^m + b, \quad b \in \mathbb{R}\},$$

where n, m are even. Let F_a^u (F_b^s) be the leaf of F^u (F^s) through $(a, 0)$ (resp. $(0, b)$). Let $\theta(b)$ be the number $a \in \mathbb{R}$ so that F_a^u is tangent to F_b^s in the intervals $C \cup E$, see figure 5.3. Then $\theta: \mathbb{R} \to \mathbb{R}$ is an expansion, i.e. $|\theta'| > 1$. A similar result holds in $B \cup D$, if we change the role of F^u and F^s.

Proof. The foliations F^u and F^s are tangent in the curve

$$v = c(u) = \frac{1}{(n \cdot m \cdot Q_1 \cdot Q_2)^{1/n-1} \cdot u^{m-1/n-1}}.$$

So

$$b = c(u) - Q_2 \cdot u^m = B(u),$$
$$a = u - Q_1 \cdot (c(u))^n = A(u).$$

Now a is related to b by:

$$a = A \circ B^{-1}(b) = \theta(b).$$

Hence since $nQ_1 \cdot (c(u))^{n-1} \cdot m \cdot Q_2 u^{m-1} = 1$ one has

$$\theta'(b) = \frac{A'(u)}{B'(u)} = \frac{1 - Q_1 \cdot n \cdot (c(u))^{n-1} \cdot c'(u)}{c'(u) - Q_2 \cdot m \cdot u^{m-1}} = \frac{1}{m \cdot Q_2 \cdot u^{m-1}}.$$

Here the last equality follows by using the definition of $c(u)$ explicitly. For the point where the curve c is tangent to F^u and F^s one has $m \cdot Q_2 \cdot u^{m-1} = 1$. Since by assumption F_a^u and F_b^s are tangent in $C \cup E$ it follows that $|\theta'(b)| > 1$. □

(h) Up to now we have investigated the set of tangencies of $\mathcal{F}^u$ and $\mathcal{F}^s$, and a diagram related to these tangencies. Now we can define the conjugacy h as follows.

Take conjugacies $h: W^s(q) \to W^s(q)$, $h: W^u(q) \to W^u(q)$ which are linear and so that $h(x) = x$. Since

$$a = \bar{a}, \qquad \beta = \bar{\beta}$$

this is possible. Then take a homeomorphism

$$h: A_{i0} \to \bar{A}_{i0}.$$

Here we have freedom. Extend h to $\bigcup_{ij} A_{ij}$ by forcing the conjugacy $h \circ f = \bar{f} \circ h$. From lemma 5.4 it follows that there is a unique extension of h to the tangency curve c so that $\bar{\psi} \circ h = h \circ \psi$, i.e. so that h respects the diagram in figure 5.5. It follows that there is a unique extension of h to R which respects the foliations $\mathcal{F}^s$ and $\mathcal{F}^u$. From the way h is constructed it follows that h is monotone. By interchanging the role of f and $\bar{f}$ it follows that h has a monotone inverse. Hence $h: c \to \bar{c}$ is a homeomorphism. Since $\mathcal{F}_q^{s'}$ is a C^1 foliation, except possibly in y, it follows as in lemma 5.2 that

$$\lim_{\substack{z \to x \\ z \in l}} \frac{\bar{d}(h(z), x)}{d(z, x)}$$

converges. Here $d, \bar{d}$ are C^0-metrics. It follows from lemma 3.4 that h extends to a conjugacy near p, since

$$\alpha = \bar{\alpha}.$$

Similarly h can be extended to a conjugacy near r, since

$$b = \bar{b}.$$

5(b). *Construction of conjugacies: remaining cases.* Let $f, \bar{f} \in \mathcal{M}$ be as in theorem 5.1 except that there are finitely many saddle-points whose invariant manifolds intersect $W^s(q) \cup W^u(q)$ transversally in a finite number of orbits. To define a conjugacy h in this case we start by constructing stable and unstable foliations for these saddle points and then we construct the foliations $\mathcal{F}^s$ and $\mathcal{F}^u$ compatible with these foliations. We then perform the construction of the proof of theorem 5.1. In order to extend this conjugacy to a neighbourhood of the saddle points whose invariants manifolds intersect $W^s(q) \cup W^u(p)$ we need more moduli conditions as in § 4. If these extra moduli conditions are satisfied we extend the conjugacy using the same arguments of the proof of theorem 4.1.

Another situation that may occur for a diffeomorphism $f \in \mathcal{M}$ having a cascade of tangencies is a combination of the four cases treated in § 5(a). In fact we may have a finite number of saddle points $p_1, \ldots, p_q$ such that $W^u(p_i)$ has a finite number of orbits of non-transversal intersection with $W^s(q)$ and also a finite number of saddles $q_1, \ldots, q_s$ whose stable manifolds have orbits of non-transversal intersection with $W^u(q)$. If $\bar{f}$ is a nearby diffeomorphism having the same intersection pattern of stable and unstable manifolds we can construct a conjugacy between f and $\bar{f}$ by putting the previous techniques together provided the appropriate moduli conditions are satisfied.

6. *A bound for the number of tangencies*

In § 4 and § 5 we proved that two nearby diffeomorphisms in $\mathcal{M}$, having the same intersection pattern of stable and unstable manifolds, are conjugate provided a finite number of moduli conditions are satisfied. In order to conclude the proof of the main theorem it remains to show that every diffeomorphism $f \in \mathcal{M}$ has a neighbourhood $\mathcal{N}$ which contains a countable number of k-parameter families of diffeomorphisms such that any diffeomorphism in $\mathcal{N}$ has the same intersection pattern of stable and unstable manifolds as some diffeomorphism in one of these families. In this section we will achieve this by proving the following result.

THEOREM 6.1. *If $f \in \mathcal{M}$ then there exists a neighbourhood $\mathcal{N}$ of f and a number $K > 0$ such that $\mathcal{N} \subset \mathcal{M}$ and the number of orbits of non-transversal intersection of stable and unstable manifolds of every $g \in \mathcal{N}$ is at most K.*

Proof. Since $\mathcal{A}$ is an open set **[Sm1]**, it is easy to see that every $f \in \mathcal{M}$ has a neighbourhood $\mathcal{N} \subset \mathcal{A}$ such that every $g \in \mathcal{N}$ satisfies conditions (1), (3), (4) and (5) of the definition of $\mathcal{M}$ in § 1. Now we prove that if $\mathcal{N}$ is small enough then condition (2) is also satisfied and we get a bound for the number of orbits of tangencies. Let $V \subset M$ be a small open set such that each orbit of tangency of f has a unique point in V. So, in order to prove the theorem it is enough to show that for each $g \in \mathcal{N}$, the number of tangencies of g in V is at most K and these tangencies are of finite order. Let $x \in V$ be a point of tangency of order r of $W^u(p)$ and $W^s(q)$ where p and q are saddle points of f. For each $g \in \mathcal{N}$ let $U(g) = W^u(p(g)) \cap V$ and $S(g) = W^s(q(g)) \cap V$ where $p(g)$ (resp. $q(g)$) is the periodic point of g near p (resp. q). If $P(g)$ is a saddle point of g whose unstable manifold intersect $W^s(p(g))$ then, by the λ-Lemma **[P1]**, $W^u(P(g)) \cap V$ is a sequence of submanifolds $U_n(g)$ which converges to $U(g)$ in the C^∞ topology. Similarly if $Q(g)$ is a saddle point whose stable manifold intersects $W^u(q(g))$ then $W^s(Q(g)) \cap V$ is a sequence of submanifolds $S_m(g)$ converging to $S(g)$ in the C^∞ topology. Since $f \in \mathcal{M}$ we have that $S_m(f)$ is transversal to $U_n(f)$ for all $m, n \in \mathbb{N}$ and $U(f) \cap S(f) = \{x\}$. Hence if $\mathcal{N}$ and V are small enough and $z \in U_n(g) \cap S_m(g)$ then the contact between $U_n(g)$ and $S_m(g)$ at z is at most r. So we need to prove that

$$\#\{m \in \mathbb{N};\ U_m(g) \text{ is not transversal to } S_n(g) \text{ for some } n \in \mathbb{N}\}$$

is uniformly bounded.

For each $g \in \mathcal{N}$ let $\mathcal{F}^u_{p(g)}$ be a C^1 unstable foliation at $p(g)$ such that each $U_n(g)$ is contained in a leaf of $\mathcal{F}^u_{p(g)}$ and the r-jet of $\mathcal{F}^u_{p(g)}$ is C^1 and varies continuously with g (see the appendix). Similarly we consider a stable foliation $\mathcal{F}^s_{q(g)}$ compatible with $W^s(Q(g))$. Let (u, v) be a C^∞ coordinate system in a neighbourhood V_x of x such that $U(f) = \{(u, 0);\ u \in (-a, a)\}$ and $S(f) = \{(u, v);\ v = u^r, u \in (-a, a)\}$. We may take V and $\mathcal{N}$ so that for each $g \in \mathcal{N}$

$$U_n(g) = \{(u, v);\ v = \psi^g_n(u)\}$$

$$U(g) = \{(u, v);\ v = \psi^g_\infty(u)\}.$$

$$S_m(g) = \{(u, v);\ v = \varphi^g_m(u)\}$$

$$S(g) = \{(u, v);\ v = \varphi^g_\infty(u)\}$$

where $\varphi_m^g, \varphi_\infty^g, \psi_n^g, \psi_\infty^g : [-a, a] \to \mathbb{R}$ are C^∞ functions satisfying the following properties:

(a) The maps $g \mapsto D^j\varphi_m^g, D^j\varphi_\infty^g, D^j\psi_n^g, D^j\psi_\infty^g$, from $\mathcal{N}$ to $C^1([-a, a], \mathbb{R})$ are continuous for $j = 0, 1, \ldots, r$.

(b) $\varphi_m^g(u) - \varphi_\infty^g(u) > 0$, $\psi_n^g(u) - \psi_\infty^g(u) < 0$ for all $u \in [-a, a]$.

(c) There exists $1 \le C < \infty$ such that

(i) $$\frac{1}{C}\beta_g^{-m}|\varphi_n^g(u) - \varphi_\infty^g(u)| \le |\varphi_{n+m}^g(u) - \varphi_\infty^g(u)| \le C\beta_g^{-m}|\varphi_n^g(u) - \varphi_\infty^g(u)|$$

$$\frac{1}{C}\alpha_g^m|\psi_n^g(u) - \psi_\infty^g(u)| \le |\psi_{n+m}^g(u) - \psi_\infty^g(u)| \le C\alpha_g^m|\psi_n^g(u) - \psi_\infty^g(u)|;$$

(ii) $$|D(\varphi_m^g - \varphi_\infty^g)(u)| \le C|\varphi_m^g(u) - \varphi_\infty^g(u)|$$
$$|D(\psi_n^g - \psi_\infty^g)(u)| \le C|\psi_n^g(u) - \psi_\infty^g(u)|,$$

where α_g is the contracting eigenvalue of $Dg(p(g))$ and β_g is the expanding eigenvalue of $Dg(q(g))$.

(d) Each $U_n(g)$ (resp. $S_m(g)$) has at most $r-1$ points of tangencies with leaves of the foliation $\mathcal{F}_{q(g)}^s$ (resp. $\mathcal{F}_{p(g)}^u$).

Conditions (a) and (d) follow from the continuity of the r-jet of the foliations $\mathcal{F}_{p(g)}^u$ and $\mathcal{F}_{q(g)}^s$; (b) follows from condition (4) of the definition of $\mathcal{M}$ in § 1; (c) follows from the differentiability of the foliations $\mathcal{F}_{p(g)}^u$, $\mathcal{F}_{q(g)}^s$ and its 1-jet. We need some lemmas.

Lemma 6.2. *If $D\varphi_n^g(u) = D\psi_m^g(u)$ and $\varphi_n^g(u) = \psi_m^g(u)$ for some $u \in (-a, a)$ and $m, n \in \mathbb{N}$ then*

$$|D\psi_\infty^g(u) - D\varphi_\infty^g(u)| \le 2C|\psi_\infty^g(u) - \varphi_\infty^g(u)|.$$

Proof. From (b) it follows that

$$0 < \varphi_n^g(u) - \varphi_\infty^g(u) = \psi_m^g(u) - \varphi_\infty^g(u) < \psi_\infty^g(u) - \varphi_\infty^g(u),$$
$$\varphi_\infty^g(u) - \psi_\infty^g(u) < \varphi_n^g(u) - \psi_\infty^g(u) = \psi_m^g(u) - \psi_\infty^g(u) < 0.$$

Hence (using (c(ii))):

$$|D\varphi_n^g(u) - D\varphi_\infty^g(u)| \le C|\varphi_n^g(u) - \varphi_\infty^g(u)| < C|\varphi_\infty^g(u) - \psi_\infty^g(u)|.$$

Therefore using $D\varphi_n^g(u) = D\psi_m^g(u)$ and the inequalities above:

$$\begin{aligned}|D\varphi_\infty^g(u) - D\psi_\infty^g(u)| &< C|\varphi_\infty^g(u) - \psi_\infty^g(u)| + |D\psi_m^g(u) - D\psi_\infty^g(u)|\\ &\le C|\varphi_\infty^g(u) - \psi_\infty^g(u)| + C|\psi_m^g(u) - \psi_\infty^g(u)|\\ &\le 2C|\varphi_\infty^g(u) - \psi_\infty^g(u)|.\end{aligned}$$

□

Lemma 6.3. *Let $J_g = \{u \in (-a, a); \varphi_\infty^g(u) \le \psi_\infty^g(u)$ and $|D\psi_\infty^g(u) - D\varphi_\infty^g(u)| \le 2C|\psi_\infty^g(u) - \varphi_\infty^g(u)|\}$. If $\mathcal{N}$ is small enough then J_g has at most $2r$ connected components.*

Proof. Consider the functions $\theta_\pm^g : [-a, a] \to \mathbb{R}$,

$$\theta_\pm^g(u) = (D\psi_\infty^g(u) - D\varphi_\infty^g(u)) \pm 2C(\psi_\infty^g(u) - \varphi_\infty^g(u)).$$

Since the map $g \to \theta_\pm^g$ from $\mathcal{N}$ to $C^r([-a, a], \mathbb{R})$ is continuous and $\theta_\pm^f(u) = ru^{r-1} \pm 2Cu^r$ it follows that for $\mathcal{N}$ small enough the set

$$\{u \in [-a, a]; |D\psi_\infty^g(u) - D\varphi_\infty^g(u)| = 2C|\psi_\infty^g(u) - \varphi_\infty^g(u)|\}$$

has at most $2r$ points. This proves the lemma because the boundary of J_g is contained in this set. □

LEMMA 6.4. *If $\mathcal{N}$ is small enough there exists an integer m such that for each $g \in \mathcal{N}$ and each connected component J_g^i of J_g there are integers $m_0 = m_0(g, i)$ and $n_0 = n_0(g, i)$ such that*

$$\psi_{n_0-m}^g(u) < \varphi_\infty^g(u) < \psi_{n_0+m}^g(u)$$

and

$$\varphi_{m_0+m}^g(u) < \psi_\infty^g(u) < \varphi_{m_0-m}^g(u)$$

for every $u \in J_g^i$.

Proof. Let $\tilde{\varphi}^g, \tilde{\psi}_n^g : J_g \to \mathbb{R}$ be the maps $\tilde{\varphi}^g(u) = -\mathrm{Log}\,(\psi_\infty^g(u) - \varphi_\infty^g(u))$ and $\tilde{\psi}_n^g(u) = -\mathrm{Log}\,(\psi_\infty^g(u) - \psi_n^g(u))$. From the definition of J_g and from property (c(ii)) one has $|D\tilde{\psi}_n^g(u)| \le C$, and $|D\tilde{\varphi}^g(u)| < 2C$ for every $u \in J_g$. If we take V and $\mathcal{N}$ small enough so that $a < 1/(2C)$ we have that $|\tilde{\varphi}^g(u_1) - \tilde{\varphi}^g(u_2)| < 1$ and $|\psi_n^g(u_1) - \psi_n^g(u_2)| < \frac{1}{2}$, if $u_1, u_2 \in J_g^i$. Moreover, from (c(i)), $|\tilde{\psi}_{n+m}(u) - \tilde{\psi}_n(u)| > 2$ for every $n \in \mathbb{N}$ and $u \in J_g$ if m is such that $-m\,\mathrm{Log}\,\alpha_g - \mathrm{Log}\,C > 2$. Hence, if there exist $n_0 \in \mathbb{N}$ and $u_0 \in J_g^i$ such that $\tilde{\psi}_{n_0}^g(u_0) = \tilde{\varphi}^g(u_0)$ then

$$\begin{aligned}\tilde{\psi}_{n_0-m}(u) &< -1 + \tilde{\psi}_{n_0}^g(u_0) = -1 + \tilde{\varphi}^g(u_0)\\ &\le \tilde{\varphi}^g(u) \le 1 + \tilde{\varphi}_g(u_0)\\ &= 1 + \tilde{\psi}_{n_0}^g(u_0) < \tilde{\psi}_{n_0+m}(u)\end{aligned}$$

for every $u \in J_g^i$ If this is not the case then there is an integer n_0 such that

$$\tilde{\psi}_{n_0-m}^g(u) < \tilde{\psi}_{n_0}^g(u) < \tilde{\varphi}^g(u) < \tilde{\psi}_{n_0+1}^g(u) < \tilde{\psi}_{n_0+m}^g(u) \qquad \text{for } u \in J_g^i.$$

This proves the first inequality of the lemma. The proof of the second one is similar. □

LEMMA 6.5. *Let J_g^i be a connected component of J_g and $n_0 = n_0(g, i)$ be an integer such that $\psi_{n_0-m}^g(u) < \varphi_\infty^g(u) < \psi_{n_0+m}^g(u)$ for every $u \in J_g^i$. If $1 - C^2\alpha_g^m > 0$ then*

$$\#\{n \in \mathbb{N};\ \exists u \in J_g^i,\ \varphi_n^g(u) \in [\psi_{n_0+2m}^g(u), \psi_\infty^g(u)]\} < \frac{-\mathrm{Log}\left(\frac{1}{C^3}\alpha_g^{2m}(1 - C^2\alpha_g^m)\right)}{\mathrm{Log}\,\beta_g}.$$

Proof. Since $-\psi_{n_0-m}^g(u) > -\varphi_\infty^g(u) > -\psi_{n_0+m}^g(u)$ we have that

$$\begin{aligned}&\mathrm{Log}\,(\psi_\infty^g(u) - \varphi_\infty^g(u)) - \mathrm{Log}\,(\psi_{n_0+2m}^g(u) - \varphi_\infty^g(u))\\ &\quad \le \mathrm{Log}\,(\psi_\infty^g(u) - \psi_{n_0-m}^g(u)) - \mathrm{Log}\,(\psi_{n_0+2m}^g(u) - \psi_{n_0+m}^g(u))\\ &\quad \le -\mathrm{Log}\left(\frac{1}{C^2}\alpha_g^{2m}(1 - C^2\alpha_g^m)\right).\end{aligned}$$

Let $m_1 \in \mathbb{N}$ be such that $\varphi_{m_1}^g(u) < \psi_\infty^g(u) \le \varphi_{m_1-1}^g(u)$. If $k \in \mathbb{N}$ is such that $\psi_\infty^g(u) >$

$\varphi^g_{m_1+k}(u) > \psi^g_{n_0+2m}(u)$ then

$$-\text{Log}\left(\frac{1}{C^2}\alpha_g^{2m}(1-C^2\alpha_g^m)\right) \geq \text{Log}\,(\psi^g_\infty(u)-\varphi^g_\infty(u)) - \text{Log}\,(\psi^g_{n_0+2m}(u)-\varphi^g_\infty(u))$$
$$\geq \text{Log}\,(\varphi^g_{m_1}(u)-\varphi^g_\infty(u)) - \text{Log}\,(\varphi^g_{m_1+k}(u)-\varphi^g_\infty(u))$$
$$\geq -\text{Log}\,C + k\,\text{Log}\,\beta_g.$$

Hence

$$k \leq \frac{-\text{Log}\left(\frac{1}{C^3}\alpha_g^{2m}(1-C^2\alpha_g^m)\right)}{\text{Log}\,\beta_g}.$$

This proves the lemma. □

LEMMA 6.6. *Let J^i_g be a connected component of J_g. The cardinality of the set $\{n \in \mathbb{N};$ the graph of $\varphi_n | J^i_g$ has a tangency with the graph of $\psi_k | J^i_g$ for some $k\}$ is at most*

$$(r-1)\left(5m + \frac{-\text{Log}\left(\frac{1}{C^3}\alpha_g^{2m}(1-C^2\alpha_g^m)\right)}{\text{Log}\,\beta_g}\right).$$

Proof. Let $u^k_j \in J^i_g$, $j = 1, \ldots, l$, be such that $(u^k_j, \psi^g_k(u^k_j))$ are all the points where the graph of $\psi^g_k | J^g_i$ is tangent to the foliation $\mathscr{F}^s_{q(g)}$. Clearly $l \leq r-1$. If $n > m_0 + m$ is such that the graph of $\varphi^g_n | J^i_g$ is tangent to the graph of ψ^g_k for some $k \geq n_0 + 2m$ then

$$\varphi^g_n(u^k_j) \in [\psi^g_{n_0+2m}(u^k_j), \varphi^g_{m_0+m}(u^k_j)] \subset [\psi^g_{n_0+2m}(u^k_j), \psi^g_\infty(u^k_j)]$$

for some $j = 1, \ldots, l$. By lemma 6.5 there are at most

$$l\,\frac{-\text{Log}\left(\frac{1}{C^3}\alpha_g^{2m}(1-C^2\alpha_g^m)\right)}{\text{Log}\,\beta_g}$$

such integers. To finish the proof we notice that for each k there are at most $l \leq r-1$ integers n such that the graph of φ^g_n has a tangency with the graph of ψ^g_k and that $\varphi^g_n(u) > \psi^g_k(u)$ for every $u \in J^i_g$ if $k \leq n_0 - m$ and $\varphi^g_n(u) > \psi^g_l(u)$ for every $u \in J^i_g$ if $n \leq m_0 - m$. A simple combinatorial argument finishes the proof of the Lemma. □

End of the proof of theorem 6.1. If the graph of φ^g_n has a tangency with the graph of ψ^g_k then, by lemma 6.2, there exists an integer i such that the graph of $\varphi^g_n | J^i_g$ has a tangency with the graph of ψ^g_k. Since J_g has at most $2r$ connected components we have that the number of integers n such that the graph of φ^g_n is tangent to the graph of ψ^g_k for some $k \in \mathbb{N}$ is at most

$$2r \cdot (r-1) \cdot \left(5m + \frac{-\text{Log}\left(\frac{1}{C^3}\alpha_g^{2m}(1-C^2\alpha_g^m)\right)}{\text{Log}\,\beta_g}\right).$$

The graph of φ^g_n has at most $r-1$ points of tangency with the foliation $\mathscr{F}^u_{p(g)}$. Hence the total number of tangencies is at most

$$(r-1) \cdot 2r \cdot (r-1) \cdot \left(5m + \frac{-\text{Log}\left(\frac{1}{C^3}\alpha_g^{2m}(1-C^2\alpha_g^m)\right)}{\text{Log}\,\beta_g}\right)$$

and this is clearly bounded by some $K > 0$ in a small neighbourhood of f. □

7. *Appendix: Smoothness of Invariant Foliations*

In this appendix we will prove that certain invariant foliations are highly differentiable in some sense. More specifically, let f be a C^∞ diffeomorphism on a neighbourhood of p in $\mathbb{R}^n$, with $f(p)=p$ and p a hyperbolic saddle-point. Take a point $x \in W^s(p)$, a C^∞ disc Σ at x transversal to $W^s(p)$ (with $\dim \Sigma = \dim W^u(p)$) and let V be a compact neighbourhood of p. According to the λ-Lemma the discs $f^k(\Sigma) \cap V$ converge in the C^r sense to $W^u(p) \cap V$ as $k \to \infty$. Here $r \in \mathbb{N}$. For a proof of the λ-Lemma see **[P1]**, **[MP]**.

In this way one can construct invariant foliations. Fill a strip N between Σ and $f(\Sigma)$ with C^r-discs which are all transversal to $W^s(p)$. Iterating this foliation on N, one obtains an invariant unstable foliation $\mathcal{F}$ on V. According to the λ-Lemma the r-jet along leaves varies continuously, (we shall make this statement more precise below). But in many applications this is not sufficient. One needs to have that the r-jet along leaves varies 'in a C^1 sense'.

In this appendix we want to extend the λ-Lemma in two ways.

(a) Let $\mu_1, \ldots, \mu_u$ $(\lambda_1, \ldots, \lambda_s)$ be the expanding (contracting) eigenvalues of $Df(p)$ and order these eigenvalues as follows:

$$|\mu_u| \geq \cdots \geq |\mu_1| > 1 > |\lambda_1| \geq \cdots \geq |\lambda_s|.$$

Assume that

$$|\lambda_1| < |\mu_1| \cdot |\lambda_s|.$$

Then the r-jet varies C^1 along leaves of $\mathcal{F}$, see theorem 7.3. Here $r \in \mathbb{N}$ is arbitrary. For $r=1$ this result is already contained in [**HP**, theorem (6.3)].

(b) Even if the disc Σ has a tangency with $W^s(p)$ of polynomial type the result from above still holds under appropriate conditions, see theorem 7.4 below.

Clearly if $|\lambda_1| = |\lambda_s|$ (i.e. all contracting eigenvalues of $Df(p)$ have the same norm) then the condition $|\lambda_1| < |\mu_1| \cdot |\lambda_s|$ is automatically satisfied. In particular if $W^u(p)$ has codimension 1 then this condition can be dropped. From this one easily deduces that one can find C^1 linearising coordinates near a hyperbolic saddle-point p if we are in the two-dimensional case. This result is not new, see **[Ha]**. The additional smoothness we obtain here is new and turns out to be essential in the estimates in this paper.

7.1. *A fiber-contraction on a jet-bundle.* Take a neighbourhood V of p. Using local coordinates we can assume that V is of the form $V = E_1(r) \times E_2(r)$, where $E_1(r)$

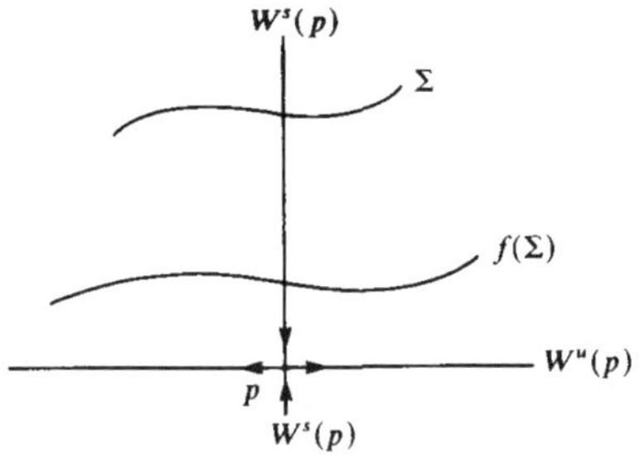

FIGURE 7.1

$(E_2(r))$ is a $u(s)$-dimensional ball with radius r. Assume $W^u(p)\cap V = E_1(r)$ and $W^s(p)\cap V = E_2(r)$.

Now let $\mathscr{E}^k(x_0, y_0)$ be the space of germs of C^k functions $(E_1, x_0)\to(E_2, y_0)$. We say that $g_1, g_2\in\mathscr{E}^k(x_0, y_0)$ are k-equivalent, i.e. $g_1\sim_k g_2$ if the Taylor jet of g_1 and g_2 at x_0 agree up to k-th order (i.e. if $j^k g_1(x_0) = j^k g_2(x_0)$). Now define

$$J^k(x_0, y_0) = \mathscr{E}^k(x_0, y_0)/\sim$$

and

$$J^k(V) = \bigcup_{(x_0,y_0)\in V} J^k(x_0, y_0).$$

$J^k(V)$ is a smooth manifold. In fact one has the natural identification $D_k: J^k(V)\to V\times\mathbb{R}^{l(1)}\times\cdots\times\mathbb{R}^{l(k)}$ which identifies the k-jet with the coefficients in the Taylor expansion. Here $l(j)$ is the dimension of the space of homogeneous polynomials from $\mathbb{R}^u$ to $\mathbb{R}^s$ of degree j. Similarly one has the map $\pi_{k,k-1}: J^k(V)\to J^{k-1}(V)$ which maps the k-jet of a function to its $(k-1)$-jet. In this way one gets the commutative diagram

$$\begin{array}{ccc}
J^k(V) & \xrightarrow[\simeq]{D_k} & V\times\mathbb{R}^{l(1)}\times\cdots\times\mathbb{R}^{l(k)} \\
\downarrow \pi_{k,k-1} & & \downarrow \tilde{\pi}_{k,k-1} \\
\vdots & & \vdots \\
J^2(V) & \xrightarrow[\simeq]{D_2} & V\times\mathbb{R}^{l(1)}\times\mathbb{R}^{l(2)} \\
\downarrow \pi_{2,1} & & \downarrow \tilde{\pi}_{2,1} \\
J^1(V) & \xrightarrow[\simeq]{D_1} & V\times\mathbb{R}^{l(1)} \\
\downarrow \pi_1 & & \downarrow \tilde{\pi}_1 \\
V & \xrightarrow{\text{id}} & V.
\end{array}$$

Here $\tilde{\pi}_{j,j-1}: V\times\mathbb{R}^{l(1)}\times\cdots\times\mathbb{R}^{l(j)}\to V\times\mathbb{R}^{l(1)}\times\cdots\times\mathbb{R}^{l(j-1)}$ is the projection $(x,(v_1,\ldots,v_j))\to(x,(v_1,\ldots,v_{j-1}))$. It will also be useful to work with jet spaces of functions with bounded derivatives. So let $B^j = \{v\in\mathbb{R}^{l(1)}\times\cdots\times\mathbb{R}^{l(j)};\ v = (v_1,\ldots,v_j),\ v_i\in\mathbb{R}^{l(i)}$ and $|v_i|\le 1, i = 1, 2,\ldots, j\}$. Define $X^j(V) = D_j^{-1}(V\times B^j)$. As before we have a commutative diagram.

Now assume that $f: V\to\mathbb{R}^m$ is a diffeomorphism with 0 as a hyperbolic fixed point. Let $\lambda_1,\ldots,\lambda_s$ $(\mu_1,\ldots,\mu_u)$ be the contracting (expanding) eigenvalues of $Df(0)$. Assume that they are ordered as follows:

$$|\mu_u|\ge\cdots\ge|\mu_1|>1>|\lambda_1|\ge\cdots\ge|\lambda_s|.$$

For $x\in V = E_1(r)\times E_2(r)$ one can write $Df(x), Df^{-1}(x): E_1\times E_2\to E_1\times E_2$ as follows:

$$Df(x) = \begin{pmatrix} A(x) & B(x) \\ C(x) & D(x)\end{pmatrix},\qquad Df^{-1}(x) = \begin{pmatrix}\tilde{A}(x) & \tilde{B}(x)\\ \tilde{C}(x) & \tilde{D}(x)\end{pmatrix}.$$

Since $W^u(p)\cap V = E_1(r)$ and $W^s(p)\cap V = E_2(r)$ the matrices $B(x)$, $\tilde{B}(x)$, $C(x)$ and $\tilde{C}(x)$ vanish for $x = 0$. Moreover for each $\delta>0$ one can choose $r>0$ so small that for $x \in V$,

$$\|A(x)\| \le |\mu_u| + \delta, \qquad \|B(x)\| \le \delta,$$
$$\|C(x)\| \le \delta, \qquad \|D(x)\| \le |\lambda_1| + \delta. \tag{7.1}$$

Similarly

$$\|\tilde{A}(x)\| \le (|\mu_1| - \delta)^{-1}, \qquad \|\tilde{B}(x)\| \le \delta,$$
$$\|\tilde{C}(x)\| \le \delta, \qquad \|\tilde{D}(x)\| \le (|\lambda_s| - \delta)^{-1}. \tag{7.2}$$

Furthermore take a neighbourhood U of 0, $U \subset V$, such that $f(U) \subset V$. In other words if $(x_0, y_0) \in U$, then $f(x_0, y_0) = (f^1(x_0, y_0), f^2(x_0, y_0)) = (x_1, y_1) \in V$. Furthermore take $\mathscr{E}_1^k(x_0, y_0) = \{g \in \mathscr{E}^k(x_0, y_0) : |Dg(x_0)| < 1\}$. Suppose f is C^n. Then define for $k \le n$,

$$\Gamma_{(x_0,y_0)} : \mathscr{E}^k(x_0, y_0) \to \mathscr{E}^k(x_1, y_1)$$

by defining $\Gamma_{(x_0,y_0)}(g) : (E_1, x_1) \to (E_2, y_1)$ to be the germ of the function

$$t \to (f^2(\mathrm{id}, g)) \circ (f^1(\mathrm{id}, g))^{-1}(t),$$

see figure 7.2.

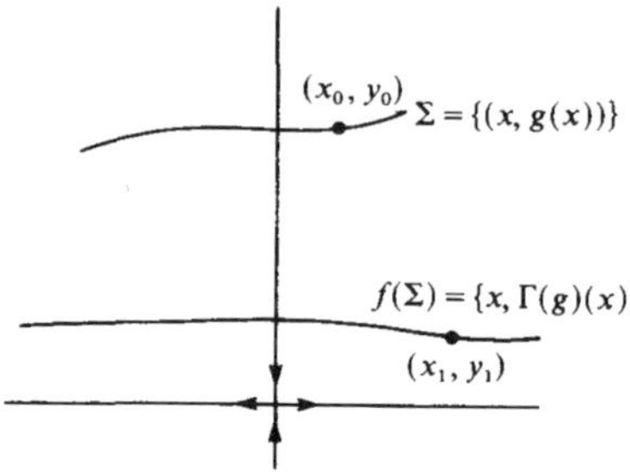

FIGURE 7.2

Let V be so small that $\|A(x)\| > \|B(x)\|$, for all x in V. Since for $g \in \mathscr{E}^k(x_0, y_0)$, $|Dg(x_0)| < 1$, the map $t \to f^1 \circ (\mathrm{id}, g)(t)$ is invertible at $t = x_0$, for (x_0, y_0) in V. Hence $\Gamma_{(x_0,y_0)}$ is a well defined map for (x_0, y_0) in V. (This also follows from the λ-Lemma). Clearly Γ induces a map on the k-jet level: This map $\Gamma_k : X^k(U) \to J^k(V)$ is defined by

$$\Gamma_k(j^k g(x_0)) = J^k(\Gamma_{(x_0,y_0)} g)(x_1).$$

Presently we will show that the map Γ_k contracts the fibers of $\pi_{k,k-1} : J^k \to J^{k-1}$. So define for $g \in \mathscr{E}^k(x_0, y_0)$,

$$|g|_k = \max_{1 \le i \le k} |Dg^i(x_0)|.$$

LEMMA 7.1. *Let $k \le n$ and $\varepsilon > 0$ be given. Then we can choose V so small that for any $g_1, g_2 \in J^k(V)$ with $\pi_{k,k-1}(g_1) = \pi_{k,k-1}(g_2)$ one has*

(a) $|\Gamma_k(g_1) - \Gamma_k(g_2)| \le ((|\lambda_1|/|\mu_1|^k) + \varepsilon) \cdot |g_1 - g_2|_k$;
(b) $D(f^{-1}|V)) \le (1/|\lambda_s|) + \varepsilon$.

Proof. Statement (b) is obvious from equations (7.2) so let us prove (a). Here we are going to use the following formulae: if ϕ, ψ_1, ψ_2 are C^k and $|\psi_1-\psi_2|_{k-1}=0$ then

$$|\phi\circ\psi_1-\phi\circ\psi_2|_k=|\phi|_1\cdot|\psi_1-\psi_2|_k,$$
$$|\psi_1\circ\phi-\psi_2\circ\phi|_k=|\psi_1-\psi_2|_k\cdot(|\phi|_1)^k.$$
$$|(\psi_1)^{-1}-(\psi_2)^{-1}|_k\le|\psi_1-\psi_2|_k\cdot(|\psi_1^{-1}|_1)^{k+1}.$$

Now by assumption the $(k-1)$-jets of $\Gamma^k g_1$ and $\Gamma^k g_2$ are equal. Therefore $|\Gamma^k g_1-\Gamma^k g_2|_k$ can be estimated (using the equalities from above) by:

$$|f^2\circ(\mathrm{id},g)|_1\cdot|(f^1\circ(\mathrm{id},g_1))^{-1}-(f^1\circ(\mathrm{id},g_2))^{-1}|_k$$
$$+|f^2\circ(\mathrm{id},g_1)-f^2\circ(\mathrm{id},g_2)|_k\cdot\{|(f^1\circ(\mathrm{id},g))^{-1}|_1\}^k.$$

Here we can take for g either g_1 or g_2. For V sufficiently small this is majorised by (see equations (7.1), (7.2)).

$$\frac{(|\lambda_1|+\delta)}{(|\mu_1|-\delta)^{k+1}}\cdot\delta\cdot|g_1-g_2|_k+\frac{(|\lambda_1|+\delta)}{(|\mu_1|-\delta)^k}\cdot|g_1-g_2|_k.$$

So for V sufficiently small, inequality (a) holds. □

COROLLARY. $\Gamma_k(X^k(U))\subset X^k(V)$.

7.2. *Differentiability of invariant foliations; a differentiable version of the λ-Lemma.* As before let $V=E_1(r)\times E_2(r)$ and f be a C^n diffeomorphism with a saddle-point at 0. Let U be a neighbourhood of 0 so that $\Gamma_k(X^k(U))\subset X^k(V)$ (and in particular $f(U)\subset V$). Such a neighbourhood exists, see the corollary above.

Let $\mathscr{F}$ be some foliation on U, not necessarily invariant, whose leaves are graphs of C^n functions $E_1(r)\to E_2(r)$. For $k\le n$, this defines a section $\sigma_k: U\to J^k(U)$ where $\sigma_k(x_1,y_1)$ is the k-jet at $x_1\in E_1(r)$ of the function whose graph is the leaf of $\mathscr{F}$ through the point $(x_1,y_1)\in U$. Suppose that the foliation $\mathscr{F}$ is chosen so that $\sigma_k(U)\subset X^k(U)$. In particular this implies that the leaves of $\mathscr{F}$ are transversal to $W^s(p)$. This foliation $\mathscr{F}$ is invariant precisely when the following diagram commutes:

$$\begin{array}{ccc} X^k(f^{-1}(U)\cap U) & \xrightarrow{\Gamma_k} & X^k(U) \\ \uparrow\sigma_k & & \uparrow\sigma_k \\ f^{-1}(U)\cap U & \xrightarrow{f} & U. \end{array}$$

Let us now show how to construct such invariant foliations. In order to do this take a foliation $\mathscr{F}$ so that for each leaf Σ in $U-f(U)$ the image $f(\Sigma)\cap U$ is also a leaf of $\mathscr{F}$. We claim that we can find a new invariant foliation $\mathscr{F}^*$ which coincides with $\mathscr{F}$ on $U-f(U)$. In fact:

LEMMA 7.2. *Let $\mathscr{F}$ be as above and such that for the corresponding section σ_k one has $\sigma_k(U)\subset X^k(U)$, $k\le n$. Then there exists an invariant foliation $\mathscr{F}^*$ which coincides with $\mathscr{F}$ on $U-f(U)$. The sections $\sigma_k^*: U\to X^k(U)$ corresponding to $\mathscr{F}^*$ are C^0 for $k\le n$.*

Proof. Follows from the λ-Lemma [**P1**].

Remark that if we assume that $\sigma_{n-1}: U\to X^{n-1}(U)$ is C^1 on $U-f(U)$, then the restriction of $\sigma_{n-1}^*: U\to X^{n-1}(U)$ is also C^1 on $U-W^u(0)$. Let us show that in fact $\sigma_{n-1}^*: U\to X^{n-1}(U)$ is C^1 on U.

THEOREM 7.3 (C^1-version of the λ-Lemma). *Let $\mathcal{F}^*$ be an invariant foliation inducing a continuous section $\sigma_n^*: U \to X^n(U)$ so that the restriction of $\sigma_{n-1}^*: U \to X^{n-1}(U)$ to $U - W^u(0)$ is C^1. Then $\sigma_{n-1}^*: U \to X^{n-1}(U)$ is C^1, provided $|\lambda_1| < |\mu_1| \cdot |\lambda_s|$.*

Proof. Let $\mathcal{G}^k$ be the space of sections $\sigma_k: U \to J^k(U)$ such that σ^k restricted to $U - f(U)$ coincides with σ_k^* (the section corresponding to $\mathcal{F}^*$). Endow this space with the supremum norm $\|\sigma\|_k = \sup\{|\sigma(x)|_k, x \in U\}$. $\mathcal{G}^k$ is a complete metric space. Let $\mathcal{G}^k(1)$ be the unit ball in $\mathcal{G}^k$.

Using the graph transformation Γ_k we define $\phi_k: \mathcal{G}^k(1) \to \mathcal{G}^k(1)$ as follows. Let $\sigma \in \mathcal{G}^k(1)$. Then

$$\phi_k(\sigma)(x) = \begin{cases} \sigma_k^*(x) & \text{if } x \in U - f(U) \\ \Gamma_k(\sigma(f^{-1}(x))) & \text{if } x \in U \cap f(U). \end{cases}$$

Since we had chosen U so that $\Gamma_k(X^k(U)) \subset X^k(V)$ it follows that ϕ_k is well defined. Remark that $\sigma \in \mathcal{G}^k(1)$ is an invariant section precisely when the diagram above commutes, i.e. when $\sigma = \Gamma_k(\sigma \circ f^{-1})$. This is equivalent to $\phi_k(\sigma) = \sigma$. From the λ-Lemma, see lemma 7.2, it follows that ϕ_k has a unique attracting fixed point σ (for $k \leq n$). We will prove that σ is C^1 (for $k \leq n-1$) by induction on k.

The idea will be the following. Suppose we can show that for some $\sigma \in \mathcal{G}^k(1)$ one has that $D((\phi_k)^m(\sigma))$ converges as $m \to \infty$. (Here $(\phi_k)^m$ is the mth iterate of ϕ_k.) Since $(\phi_k)^m(\sigma)$ converges to the fixed point σ_k^* it follows that σ_k^* is C^1. So it suffices to show that $D((\phi_k)^m(\sigma))$ converges as $m \to \infty$. We will do this using the fibre contraction theorem.

We start by noting that $J^k(U)$ is homeomorphic to $U \times \mathrm{Fib}^1 \times \cdots \times \mathrm{Fib}^k$, where $U \simeq \mathbb{R}^n$, $\mathrm{Fib}^{(i)} \simeq \mathbb{R}^{l(i)}$. Let $p_i: J^k(U) \to \mathrm{Fib}^{(i)}$ be the projection on the ith fiber. Now define $\mathcal{H}^k$ to be the space of continuous maps $H: U \times \mathbb{R}^m \to \mathrm{Fib}^k$ with $H(x, v) = H_x(v)$, where $H_x: \mathbb{R}^m \to \mathrm{Fib}^k$ is linear and H_x coincides with the derivative of $p_k \circ \sigma_k^*$ at x for x in $U - f(U)$. Endow $\mathcal{H}^k$ with the norm $\|H\| = \sup_{x \in U} \|H_x\|$.

Let us now prove the first induction step that σ_1^* is C^1. In order to do this fix $\sigma \in \mathcal{G}^1(1)$. For this σ we define a transformation $\psi_{1,\sigma}: \mathcal{H}^1 \to \mathcal{H}^1$ by

$$(\psi_{1,\sigma} H)_x(v) = \begin{cases} D(p_1 \circ \sigma_1^*)_x(v) & \text{if } x \in U - f(U) \\ D(p_1 \circ \Gamma_1)_{\sigma(y)}(w, H_y(w)) & \text{if } x \in U \cap f(U). \end{cases}$$

Here $w = (df_x)^{-1}(v)$ and $y = f^{-1}(x)$. Remark that for $x \in U \cap f(U)$ one has $y = f^{-1}(x) \in U \cap f^{-1}(U)$ and hence $(w, H_y(w)) \in T_{\sigma(y)} X^1(U \cap f^{-1}(U))$. Therefore for $x \in U \cap f(U)$, $(\psi_{1,\sigma} H)_x(v)$ is the derivative of

$$p_1 \circ \Gamma_1: X^1(U \cap f^{-1}(U)) \to \mathrm{Fib}^1$$

at $\sigma(y)$ in the direction $(w, H_y(w))$, and therefore $\psi_{1,\sigma}(D\sigma) = D(\phi_1 \circ \sigma)$ if σ is C^1. Let us show that $\psi_{1,\sigma}$ is a contraction. Indeed take $H, K \in \mathcal{H}^1$. Then

$$(\psi_{1,\sigma} H - \psi_{1,\sigma} K)_x(v) = \begin{cases} 0 & \text{for } x \in U - f(U) \\ D(p_1 \circ \Gamma_1)_{\sigma(y)}(0, (H_y - K_y)(w)) & \text{otherwise,} \end{cases}$$

where $w \doteq (Df_x)^{-1}(v)$. From lemma 7.1 it follows that

$$\begin{aligned}\|D(p_1\circ\Gamma_1)_{\sigma(y)}(0,(H_y-K_y)(w))\| &\le \left(\frac{|\lambda_1|}{|\mu_1|}+\varepsilon\right)\cdot\|(H_y-K_y)(w)\|\\ &\le \left(\frac{|\lambda_1|}{|\mu_1|}+\varepsilon\right)\cdot\|H_y-K_y\|\cdot\|Df_x^{-1}(v)\|\\ &\le \left(\frac{|\lambda_1|}{|\mu_1|}+\varepsilon\right)\cdot\left(\frac{1}{|\lambda_s|}+\varepsilon\right)\cdot\|H_y-K_y\|\cdot\|v\|.\end{aligned}$$

Hence

$$\|\psi_{1,\sigma}H-\psi_{1,\sigma}K\|\le\lambda\cdot\|H-K\|,$$

where $\lambda=((|\lambda_1|/|\mu_1|)+\varepsilon)\cdot(1/|\lambda_s|)+\varepsilon)$ can be chosen smaller than one, by taking $\varepsilon>0$ sufficiently small. Hence

$$\|\psi_{1,\sigma}H-\psi_{1,\sigma}K\|\le\lambda\cdot\|H-K\|,$$

i.e. $\psi_{1,\sigma}$ is a contraction. Now we are in the position to prove that σ_1^* is C^1. In fact let $\theta_1:\mathscr{G}^1\times\mathscr{H}^1\to\mathscr{G}^1\times\mathscr{H}^1$ be defined by

$$\theta_1(\sigma,H)=(\phi_1(\sigma),\psi_{1,\sigma}(H)).$$

Remark that σ_1^* is the unique attracting fixed point of ϕ_1 and since $\psi_{1,\sigma}$ is a contraction for every σ, it follows from the Fiber Contraction Theorem (see [**HP**]) that θ_1 has a unique attracting fixed point (σ_1^*, H_1^*). So let σ be C^1. Then

$$\theta_1(\sigma,D\sigma)=(\phi_1(\sigma),\psi_{1,\sigma}(D\sigma))=(\phi_1(\sigma),D\phi_1\sigma)$$

and therefore $(\theta_1)^l(\sigma,D\sigma)=(\phi_1^l(\sigma),D\phi_1^l(\sigma))$. Hence $D(\phi_1^l)(\sigma)$ converges to H_1^* and $\phi_1^l(\sigma)$ to σ_1^* as $l\to\infty$. It follows that σ_1^* is C^1 and $D\sigma_1^*=H_1^*$. This finishes the first induction step.

Let us now prove that $\sigma_2^*:U\to X^2(U)$ is C^1. Let $\sigma\in\mathscr{G}^2$ and $H_1\in\mathscr{H}^1$. Define $\psi_{2,\sigma,H_1}:\mathscr{H}^2\to\mathscr{H}^2$ by

$$\psi_{2,\sigma,H_1}(H_2)=\begin{cases}D(p_2\circ\sigma_2^*)_x(v) & \text{if } x\in U-f(U),\\ D(p_2\circ\Gamma_2)_{\sigma(y)}(w,H_{1,y}(w),H_{2,y}(w)) & \text{if } x\in U\cap f(U).\end{cases}$$

Here $w=(Df_x)^{-1}(v)$ and $y=f^{-1}(x)$. As before take $H_2, K_2\in\mathscr{H}^2$. Then

$$(\psi_{2,\sigma,H_1}(H_2)-\psi_{2,\sigma,H_1}(K_2))_x(v)=\begin{cases}0 & \text{if } x\in U-f(U)\\ D(p_2\circ\Gamma_2)_{\sigma(y)}(0,0,(H_{2,y}-K_{2,y})(w)) & \text{otherwise,}\end{cases}$$

where $y=f^{-1}(x)$ and $w=(Df_x)^{-1}(v)$. As before one deduces that ψ_{2,σ,H_1} is a contraction, using lemma 7.1. Now define a map $\theta_2:(\mathscr{G}^2\times\mathscr{H}^1)\times\mathscr{H}^2 \circlearrowleft$ by

$$((\sigma,H_1),H_2)\mapsto((\phi_2(\sigma),\psi_{1,\sigma}(H_1)),\psi_{2,\sigma,H_1}(H_2)).$$

As before the Fiber Contraction Theorem implies that θ_2 has a unique attracting fixed point $(\sigma_2^*, H_1^*, H_2^*)$. If $\sigma\in\mathscr{G}^2$ is C^1, then

$$\theta_2(\sigma,D(p_1\circ\sigma),D(p_2\circ\sigma))=(\phi_2(\sigma),D(p_1\circ\phi_2\circ\sigma),D(p_2\circ\phi_2\circ\sigma)).$$

As before it follows that $D(p_2\circ\sigma_2^*)=H_2^*$. From the previous induction step $D(p_1\circ\sigma_2^*)=D(\sigma_1^*)=H_1^*$. Hence σ_2^* is C^1. Similarly one proves by induction that the fixed point $\sigma_k^1\in\mathscr{G}^k$ of ϕ_k is C^1 for $k\le n-1$. □

COROLLARY 1. *Let $f: M \to M$ be a diffeomorphism with a saddle-point p and let* codim $(W^u(p)) = 1$. *Then there exists an invariant unstable foliation $\mathcal{F}^*$ for p, such that the k-jet along leaves is a C^1 function.*

Proof. This follows from theorem 7.3 because in this case the condition $|\lambda_1| < |\mu_1| \cdot |\lambda_s|$ is trivially satisfied. □

Remark. Let $\mathcal{F}$ be an invariant foliation on $V - W^u(p)$, where V is a neighbourhood of $W^u(p)$. If $\mathcal{F}$ satisfies the conclusion of corollary 1, then the foliation $\mathcal{F}^* = \mathcal{F} \cup W^u(p)$ on V also satisfies the conclusion of corollary 1. In other words the foliation $\mathcal{F}^*$ can be found as the extension of a given one on $V - W^u(p)$.

COROLLARY 2. *If the diffeomorphism f and the foliation $\mathcal{F}$ depend C^s on parameters, then the corresponding invariant foliation $\mathcal{F}^*(f, \mathcal{F})$ also depends C^s on this parameter.*

Proof. This can be proved with the same methods as used in [**HP**] for showing that the unstable manifolds depend continuously on *f*. □

7.3. *C^1-linearisability near saddle-points in the two-dimensional case.* Let $f: M \to M$ be a diffeomorphism with a hyperbolic saddle-point p. Assume dim $(M) = 2$. From corollary 1 of theorem 7.3 one can obtain invariant stable and unstable foliations for p, $\mathcal{F}^s$ and $\mathcal{F}^u$, which are C^1. From this one obtains a C^1 invariant projection π_s (π_u) from a neighbourhood V of p onto $W^s(p)$ $(W^u(p))$ by projecting along the leaves of $\mathcal{F}^u$ $(\mathcal{F}^s)$. By construction one has

$$\pi_s \circ f = f \circ \pi_s, \qquad \pi_u \circ f = f \circ \pi_u.$$

Then take C^1 coordinates on $W^s(p)$ and on $W^u(p)$ so that $f| W^s(p)$ and $f| W^u(p)$ are linear with respect to these coordinates. (This is not hard to do, since $W^s(p)$ and $W^u(p)$ are 1-dimensional). Using these coordinates and π_s, π_u one obtains a C^1-linearising coordinate system for f near p. This result was already known, see [**Ha**] and [**HP**, theorem (6.1)]. The fact that there is additional smoothness i.e. that $\sigma^*_{n-1}: V \to X^{n-1}(V)$ is C^1, is new.

7.4. *Differentiability of invariant foliations with polynomial tangencies.* In § 7.2 we have extended foliations which were smooth and transversal to $W^s(p)$. In this section we will consider foliations with leaves which have a tangency of finite order with $W^s(p)$, see figure 7.3.

More precisely take a neighbourhood V of p. Using local coordinates we can assume that $V = E_1(r) \times E_2(r)$ where $E_1(r)$ (resp. $E_2(r)$) is a u(resp. s)-dimensional ball of

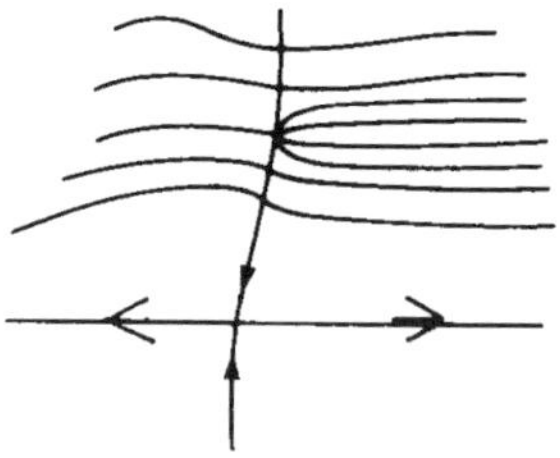

FIGURE 7.3

radius r. Assume $W^u(p)\cap V=E_1(r)$, $W^s(p)\cap V=E_2(r)$. As before define the space of germs $\mathscr{E}^k(x_0,y_0)$. For $g\in\mathscr{E}^k(x_0,y_0)$ and $0\le\delta<1$ define a new norm

$$|g|_{k,\delta}=\max_{i=1,2,\ldots,k}\{|D^ig(x_0)|\cdot|x_0|^{i-\delta}\}.$$

Furthermore let

$$X^{k,\delta}(V)=\{g\in J^k(V);\ |g|_{k,\delta}\le 1\}.$$

(We could also introduce extra coefficients $a_i>0$ and consider $|D^ig(x_0)|\cdot|x_0|^{i-\delta}\cdot a_i$, but this would have no essential effect on the sequel.)

Now consider some foliation $\mathscr{F}^u$ on V whose leaves are graphs of functions $E_1(r)\to E_2(r)$ which are C^k on $E_1(r)-0$. This defines a section $\sigma_k: V-W^s(p)\to J^k(V-W^s(p))$ as before. We assume now that

$$\sigma_k(V-W^s(p))\subset X^{k,\delta}(V-W^s(p)),$$

for some $\delta>0$. In other words we allow the foliation $\mathscr{F}^u$ to have tangencies with $W^s(p)$ of order at most $1/(1-\delta)$. We have now the following generalization of the λ-Lemma and of theorem 7.3.

THEOREM 7.4. *Assume that σ_k is as above and that for the corresponding δ one has,*

$$\left|\frac{\lambda_1}{\lambda_s}\right|\cdot\frac{|\mu_u|^{k-\delta}}{|\mu_1|^k}<1.$$

(a) *If $k\le n$, where n is the degree of differentiability of f, then $\mathscr{F}$ can be extended to an invariant foliation $\mathscr{F}^*$ on V such that $\sigma: V-W^s(p)\to X^{k,\delta}(V)$ is continuous.*

(b) *If $\sigma: V-(W^s(p)\cup W^u(p))\to X^{k,\delta}(V)$ is C^1, then in fact $\sigma: V-W^s(p)\to X^{k,\delta}(V)$ is C^1.*

Proof. One cannot deduce (a) from the λ-Lemma because $\mathscr{F}^u$ has tangencies with $W^s(p)$. So define as in theorem 7.3:

$$\mathscr{G}^{k,\delta}=\{\sigma: V\backslash W^s(p)\to X^{k,\delta}(V),\ \sigma=\sigma^*\text{ on } U-f(U)\}.$$

For $\sigma\in\mathscr{G}^k$ define $\phi_k(\sigma)$ exactly as before. Let us show that ϕ_k maps $\mathscr{G}^{k,\delta}$ into $\mathscr{G}^{k,\delta}$. Let $(x_0,y_0)\in V\backslash W^s(p)$, i.e. assume $x_0\neq 0$. Then take $g_i\in\mathscr{E}^k(x_0,y_0)$ with $\pi_{k,k-1}(g_1)=\pi_{k,k-1}(g_2)$ and let $(x_1,y_1)=f(x_0,y_0)$. From lemma 7.1(a) for any $\varepsilon>0$ we can choose V so small that:

$$\begin{aligned}
|\Gamma_k(g_1)-\Gamma_k(g_2)|_{k,\delta}&=|\Gamma_k(g_1)-\Gamma_k(g_2)|_k\cdot\|x_1\|^{k-\delta}\\
&\le\left(\frac{|\lambda_1|}{|\mu_1|^k}+\varepsilon\right)|g_1-g_2|_k\cdot\|x_1\|^{k-\delta}\\
&=\left(\frac{|\lambda_1|}{|\mu_1|^k}+\varepsilon\right)|g_1-g_2|_{k,\delta}\cdot\left(\frac{\|x_1\|}{\|x_0\|}\right)^{k-\delta}\\
&\le\left(\frac{|\lambda_1|}{|\mu_1|^k}+\varepsilon\right)\cdot(\mu_u+\varepsilon)^{k-\delta}\cdot|g_1-g_2|_{k,\delta}.
\end{aligned}$$

Furthermore $\|Df_x^{-1}\|\le((1/|\lambda_s|)+\varepsilon)$. Now one can estimate the Lipschitz expansion:

$$L(\phi_k)\le L(\Gamma_k)\cdot\|Df_x^{-1}\|\le\left(\frac{|\lambda_1|}{|\mu_1|^k}+\varepsilon\right)\cdot(|\mu_u|+\varepsilon)^{k-\delta}\left(\frac{1}{|\lambda_s|}+\varepsilon\right).$$

Since $\delta>0$ this last number is smaller than one, provided $(|\lambda_1|/|\lambda_s|)\cdot(|\mu_u|^{k-\delta}/|\mu_1|^k)<1$ and provided we take $\varepsilon>0$ small enough. It follows that ϕ_k maps $\mathcal{G}^{k,\delta}$ into $\mathcal{G}^{k,\delta}$ and moreover that this map is a contraction. Hence there is a unique fixed point σ_k^* which extends σ_k. Statement (a) follows.

The proof of statement (b) goes exactly as the proof of theorem 7.3 if we replace $|\cdot|_k$ by $|\cdot|_{k,\delta}$ and use the estimates from above. □

COROLLARY. *Let $f: M\to M$ be a diffeomorphism with a saddle-point p and let* dim $M=2$. *Then one can extend a foliation $\mathcal{F}$ with tangencies along $W^s(p)$ (as above) to an invariant unstable foliation $\mathcal{F}^*$ for p, such that the k-jet along leaves is a C^1 function away from $W^s(p)$.*

Proof. For the two-dimensional case the condition $(|\lambda_1|/|\lambda_s|)\cdot(|\mu_u|^{k-\delta}/|\mu_1|^k)<1$ is trivially satisfied. Therefore apply theorem 7.4. □

Remark. As before $\mathcal{F}^*$ depends continuously on f and $\mathcal{F}$.

Finally we wish to thank Freddy Dumortier and Anthony Manning for making many detailed suggestions after reading an earlier version of this paper.

REFERENCES

[Ha] P. Hartman. On local homeomorphisms of Euclidean spaces. *Bol. Soc. Math. Mexicana* **5** (1960).

[HP] M. Hirsch & C. Pugh. Stable manifolds and hyperbolic sets. In *Global Analysis. Proc. Symp. in Pure Math.* **XIV**. Amer. Math. Soc. (1970), 133-164.

[Me1] W. de Melo. Structural stability of diffeomorphisms on two manifolds. *Inv. Math.* **21** (1973), 233-246.

[Me2] W de Melo. Moduli of stability of two-dimensional diffeomorphisms. *Topology* **19** (1980), 9-21.

[MP] W. de Melo & J. Palis. Moduli of stability for diffeomorphisms. *Lecture Notes in Math.* **819**. Springer-Verlag (1980), 315-339.

[MPS] W. de Melo, J. Palis & S. J. van Strien. Characterizing diffeomorphisms with modulus of stability one. In *Dynamical Systems and Turbulence, Warwick* 1980. *Springer Lecture Notes in Math.* **898** (1981), 266-285.

[NPT] S. Newhouse, J. Palis & F. Takens. Bifurcations and stability of families of diffeomorphisms. *Publ. Math. Inst. Hautes Études Scientifiques* **57** (1983).

[P1] J. Palis. On Morse-Smale dynamical systems. *Topology* **8** (1969), 385-404.

[P2] J. Palis. Moduli of stability and bifurcation theory. *Proc. Int. Congress of Mathematicians Helsinki* (1978).

[PS] J. Palis & S. Smale. Structural stability theorems. In. *Global Analysis. Proc. Symp. in Pure Math.* **XIV**. Amer. Math. Soc. (1970), 223-232.

[R] C. Robinson. C^r structural stability implies Kupka-Smale. In *Dynamical Systems* (ed. M. Peixoto) Academic Press (1973), 443-449.

[R1] J. Robbin. A structural stability theorem. *Ann. of Math.* **94** (1971), 447-493.

[R2] J. Robbin. Unfoldings of discrete dynamical systems. *Ergod. Th. & Dynam. Sys.* **4** (1984), 421-486.

[Sm1] S. Smale. Differentiable dynamical systems. *Bull. Amer. Math. Soc.* **73** (1967), 747-817.

[Sm2] S. Smale. The Ω-stability theorem. In *Global Analysis. Proc. Symp. in Pure Math.* **XIV**. Amer. Math. Soc. (1970), 289-297.

[S1] S.J. van Strien. Saddle connections of arcs of diffeomorphisms, moduli of stability. In *Dynamical Systems and Turbulence, Warwick* 1980. *Springer Lecture Notes in Math.* **898** (1981), 352-366.

[S2] S. J. van Strien. One parameter families of vectorfields. Bifurcations near saddle-connections. Ph.D. Thesis, Utrecht (1982). (An extended version will appear as a monograph.)

A Structure Theorem in One Dimensional Dynamics

W. de Melo and S. van Strien

Annals of Mathematics **129**, 519–546 (1989)

Introduction

The study of the dynamics of one dimensional maps commenced with Poincaré in the 1880s. We shall briefly describe his contribution. For each homeomorphism f of the circle S^1 without periodic points there exists a unique rigid rotation R_μ such that for each integer n and each $x \in S^1$, the points $\{x, f^{-1}(x), \ldots, f^{-n}(x)\}$ and $\{x, R_\mu^{-1}(x), \ldots, R_\mu^{-n}(x)\}$ are ordered in the same way in S^1. From this, one can easily show that there exists a continuous surjective and monotone map $h\colon S^1 \to S^1$ which sends orbits of f into orbits of R_μ, i.e., $hf = R_\mu h$. In his celebrated paper of 1932, Denjoy observed that even for C^1 diffeomorphisms this semiconjugacy h need not be a conjugacy. He constructed a C^1 diffeomorphism f of S^1, without periodic points, having a wandering interval, i.e., an interval $L \subset S^1$ such that $L, f(L), f^2(L), \ldots$ are disjoint. Since such an interval cannot exist for a rigid rotation, h must collapse L to a point. However, he showed that for C^2 diffeomorphisms such wandering intervals cannot exist. Therefore, any C^2 diffeomorphism of S^1 either has a periodic point or is conjugate to a rigid rotation. In this way Denjoy completed the topological classification of C^2 diffeomorphisms of S^1.

The next step in the analysis of one dimensional dynamics is to consider smooth non-invertible maps of the interval. Consider the class $\mathscr{A}$ of C^∞ maps $f\colon [0,1] \to [0,1]$ such that $f(0) = f(1) = 0$ and f has a unique critical point $c \in (0,1)$. In the 1910s, Fatou and Julia, [F1], [F2], [J], showed that these maps can have very rich dynamics. In particular, they may have infinitely many periodic points. However, Fatou showed that if f is a polynomial map then only a finite number of these periodic points can be attractors. Here we say that a periodic point is an attractor if its basin of attraction, i.e., the set of points forward asymptotic to its orbit, has a nonempty interior. In particular, each map of the family $f_\mu\colon [0,1] \to [0,1]$, $f_\mu(x) = \mu x(1-x)$, $\mu \in (0,4)$, has at most one attractor.

Following the programme of Poincaré, W. Parry [Pa] and more completely J. Milnor and W. Thurston, [M-T], have shown that the one parameter family

Originally published in *Annals of Mathematics* **129**, 519–546 (2020).

$f_\mu(x) = \mu x(1-x)$, $0 < \mu \leq 4$, plays the same role in the class $\mathscr{A}$ as the rigid rotation R_μ inside the class of diffeomorphisms of the circle. More precisely, given a map $f \in \mathscr{A}$ there exists a parameter value $\mu \in (0,4]$ such that for each n the points

$$\{c\} \cup \{f^{-1}(c)\}, \ldots, \{f^{-n}(c)\}$$

and

$$\{\tfrac{1}{2}\} \cup \{f_\mu^{-1}(\tfrac{1}{2})\} \cup \cdots \cup \{f_\mu^{-n}(\tfrac{1}{2})\},$$

are in the same order in $[0,1]$. From this we can construct an order-preserving map $h\colon \bigcup_{k=0}^\infty f^{-k}(c) \to \bigcup_{k=0}^\infty f_\mu^{-k}(\frac{1}{2})$ such that $hf = f_\mu h$. J. Guckenheimer showed in [G] that $\bigcup_{k=0}^\infty f_\mu^{-k}(\frac{1}{2})$ is either dense in $[0,1]$ or in the complement of the basin of the attractor of f_μ. Hence h extends to a semiconjugacy, i.e., to a monotone surjective map $h\colon [0,1] \to [0,1]$ with $hf = f_\mu h$. This shows that the analogue of Poincaré's theory also holds for maps in $\mathscr{A}$. In particular if $L = h^{-1}(x)$ is a non-trivial interval then $\operatorname{int} L \cap \bigcup_{k=0}^\infty f^{-k}(c) = \varnothing$ and we have two possibilities: (i) L is a *periodic interval*, or (ii) L is a *wandering interval*. Here we say that L is a *periodic interval* if there exists a positive integer n such that $f^{k+n}(L) = f^n(L)$ for some positive integer k. L is said to be a *wandering interval* if $f^k(L) \cap f^{k+l}(L) = \varnothing$ for all positive integers k, l, and if moreover no point of L is contained in the basin of a periodic attractor.

Our main result shows that the analogue of Denjoy's theory also holds for a map f in $\mathscr{A}$ if the critical point c is non-flat; i.e., some derivative of f at c is non-zero.

MAIN THEOREM. *If the critical point of $f \in \mathscr{A}$ is non-flat then f has no wandering interval.*

More precisely, we show:

THEOREM. *Let f: $[0,1] \to [0,1]$ be C^3 with $f(0) = f(1) = 0$ and with a unique critical point $c \in (0,1)$ such that the Schwarzian derivative $Sf = D^3f/Df - \frac{3}{2}\{D^2f/Df\}^2$ is negative in some small neighbourhood of c. (This last statement is true if f is C^{k+1} near c and $D^kf(c) \neq 0$, for some $k \geq 2$.) Then f has no wandering intervals.*

Using an example of Hall, [Ha], one can show that the Main Theorem is false if one drops the non-flatness condition. There are C^∞ maps f: $[0,1] \to [0,1]$ with a unique (flat) critical point $c \in (0,1)$ which have wandering intervals; see [S-I], [Me].

In a forthcoming paper we intend to show that if the critical point of f is non-flat then the (minimal) period of periodic intervals of f is uniformly bounded.

Added in proof: This statement and the multinodal generalisation of the main theorem have recently been proved in [M-M-S].

It follows from our main theorem together with Milnor-Thurston's theory that if $f \in \mathcal{A}$ has no attractor then f is conjugate to a quadratic map f_μ for some value of the parameter μ. If f has periodic attractors then one has to be more specific. A periodic point p of f is said to be an attractor if the component $B_f(p)$, which contains p, of the set $\{x; f^n(x) \to \{p, f(p), \ldots,\} \text{ as } n \to \infty\}$ has a non-empty interior. Let A^l_f, A^r_f, A^t_f be the set of periodic points p of f such that $B_f(p)$ is respectively a left-sided, right-sided or a two-sided neighbourhood of p. Furthermore let A^z_f be the set of periodic points p of f such that for some $n > 0$, $f^n(p) = p$ and for some sequence $x_i \to p$, $f^n(x_i) = x_i$. Finally let $A_f = A^l_f \cup A^r_f \cup A^t_f \cup A^z_f$ and $O(c_f) = \bigcup_{n \geq 0} f^n(c_f)$ where c_f is the critical point of f. Using the kneading coordinates introduced in [M-T] our main theorem easily implies that, up to conjugacy, a map $f \in \mathcal{A}$ is determined by the position of the periodic attractors and the orbit of the critical point. In the special case that $f \in \mathcal{A}$ and $Sf = D^3f/Df - \frac{3}{2}\{D^2f/Df\}^2 < 0$, this corollary was proved before by J. Guckenheimer; see [G].

COROLLARY. *Let $f, g \in \mathcal{A}$. Furthermore, assume that there exists a bijection h: $A_f \cup O(c_f) \to A_g \cup O(c_g)$ which is order preserving (with respect to the induced interval ordering) and such that*

$$h \circ f(x) = g \circ h(x) \quad \textit{for all} \quad x \in A_f \cup O(c_f),\ h(A^i_f) = A^i_g,$$
$$\textit{for} \quad i = l, r, t, z \quad \textit{and} \quad h(c_f) = c_g.$$

Then f and g are conjugate; one can extend h to a homeomorphism h: $[0,1] \to [0,1]$ such that

$$h \circ f = g \circ h.$$

Proof. Let B_f be the union of the collection of open intervals I such that for some $n > 0$, $f^n|I$ is a diffeomorphism, $f^n(I) \subset I$. Similarly define B_g. Since h maps attracting periodic points of f to corresponding attracting periodic points in a strictly order-preserving way, one can extend h to a homeomorphism k: $B_f \to B_g$ such that $k \circ f = g \circ k$ for all $x \in B_f$.

Furthermore, h: $O(c_f) \to O(c_g)$, $h \circ f = g \circ h$ and h is order-preserving. Therefore the kneading invariants of f and g are the same. Let $\theta_f(x), \theta_g(x) \in \{-1, 0, 1\}^Z$ be the kneading coordinates of x; see [M-T] and also [C-E1], [Str2]. Note that $x \to \theta_f(x)$, $x \to \theta_g(x)$ are monotone (with respect to the lexicographical ordering) and $\theta_f([0,1]) = \theta_g([0,1])$.

Now the Main Theorem implies that f has no wandering intervals. Therefore, if θ_f is constant on some interval I then $f^n(I) \subset B_f$ for some $n \geq 0$. It follows that θ_f is strictly monotone on $[0,1] \setminus \bigcup_{n \geq 0} f^{-n}(B_f)$. Similarly for g. Hence there exists a unique homeomorphism h: $[0,1] \setminus \bigcup_{n \geq 0} f^{-n}(B_f) \to [0,1] \setminus \bigcup_{n \geq 0} f^{-n}(B_g)$ so that $\theta_f(x) = \theta_g(h(x))$ (this implies $h \circ f = g \circ h$). On components of $f^{-n}(B_f)$ extend h continuously so that $g^n \circ h = k \circ f^n$. Then h maps components of $\bigcup_{n \geq 0} f^{-n}(B_f)$ homeomorphically to components of $\bigcup_{n \geq 0} f^{-n}(B_g)$. In this way one extends h to a homeomorphism h: $[0,1] \to [0,1]$ such that $h \circ f = g \circ h$.

Our main theorem is the analogue in the real case of a well known result about iterations of holomorphic maps of the Riemann sphere: Sullivan proved in [Su] that there are no wandering components in the complement of the Julia set. The main ingredient in Sullivan's proof is the theory of quasiconformal deformations.

The principal new tool introduced in this paper can be regarded as an extension of a technique of A. Schwartz; see [Sc]. Let $N = [0,1]$ or S^1. This result of Schwartz states that if f: $N \to N$ is a C^2 map without critical points and J an interval such that $J, \ldots, f^{n-1}(J)$ are disjoint then f^n has bounded non-linearity on an interval slightly larger than J. More precisely, there exist $\delta > 0$ and $C < \infty$ with the following properties. For any interval J such that $\sum_{i=0,\ldots,n-1} |f^i(J)| \leq 1$ and any interval $T \supset J$ with $|T| \leq (1+\delta)|J|$ one has $|Df^n(x)|/|Df^n(y)| \leq C$, for all $x, y \in T$. In particular this means that the ratio of the length of two intervals in T is not distorted too much by f^n.

Our tools also apply in situations where one does have critical points, at least provided they are all non-flat. But instead of considering how f^n distorts the ratio of the length of certain intervals, we estimate the distortion of the *cross-ratio* of intervals. We think that the technique proposed in this paper will be useful in many problems related to interval maps with critical points. Indeed, R. Mañé, [Ma], used the techniques of Schwartz extensively and proved an extremely general hyperbolicity result for C^2 maps without critical points. Combining these ideas of Mañé with our tools (rather than those of Schwartz) one can extend Mañé's results considerably. It turns out to be possible to prove the results on invariant measures from [Mi] and [C-E2] without the assumption that $Sf < 0$; see [Strl] and [N-S].

Let us make a few remarks on the history of the main theorem and its proof which has two aspects: a topological one and an analytical one.

Denjoy's analytical tool was that if $\log|Df|$ has bounded variation, then the derivatives of iterates of f, Df^n, have bounded distortion on an interval J whose iterates, $J, f(J), \ldots, f^n(J), \ldots$ are disjoint. His topological tool was the semiconjugacy with the rigid rotation and a detailed understanding of the dynamics of rotations of the circle; see [D].

Schwartz [Sc] gives another proof of this theorem which is almost completely analytic. His assumption is that $\log|Df|$ is Lipschitz. Then he gets bounded distortion for the derivatives of iterates on an interval L if the sum of the lengths of its iterates is finite. This theorem is the main tool of Mañé who considers C^2 maps of an interval (or a circle); see [Ma]. He considers points whose forward orbit stays away from the set of critical points $C(f)$. Using disjointness properties and a C^2 theorem of Schwartz he proves that any compact set K not containing any critical points or non-hyperbolic periodic points is hyperbolic.

The main problem in the study of orbits which pass close to critical points is that one gets a lot of non-linearity: the bounded non-linearity tools of Schwartz completely break down. In [Si], Singer introduced a completely different analytic tool: the Schwarzian derivative, $Sf = D^3f/Df - \frac{3}{2}\{D^2f/Df\}^2$. He pointed out that if Sf is negative then Sf^n is also negative for every n. Using the observation that if $Sf < 0$ then $|Df|$ cannot have a positive local minimum he proved that, for maps with negative Schwarzian derivative, the basin of any periodic attractor must contain a critical point.

Misiurewicz and Guckenheimer, [Mi], [G], showed that the Schwarzian derivative was an important analytic tool for the study of several dynamical properties. In particular, Guckenheimer proved in [G] the non-existence of wandering intervals for maps which everywhere have negative Schwarzian derivative. However this additional condition is extremely strong (it implies that $|Df|$ cannot have any positive local minima). Also this condition is not very natural (for example, it is not coordinate invariant).

Yoccoz in [Y] proved the non-existence of wandering intervals for C^∞ homeomorphisms of the circle having only non-flat critical points. He combines techniques of Denjoy away from the critical points with some analytical estimates near the critical points related to the Schwarzian derivative.

Our topological tools are the same as Guckenheimer's. The analytical aspect of our proof is related to the distortion of the cross-ratios under iterates of f. This part of the proof has two ingredients. The first one is the use of the Schwarzian derivative near the critical point to estimate the distortion of the cross-ratio. The other is the use of Lipschitz conditions on D^2f away from the critical points. The way we combine these two ingredients resembles the proof of Schwartz.

In Section 5 we prove also the non-existence of wandering intervals for a C^∞ map f with a finite number of critical points, all of them non-flat, provided the forward orbit of the critical points does not accumulate at critical points. (This last condition is usually called the Misiurewicz condition.)

During the preparation of this paper, de Melo visited TH-Delft and IHES; van Strien visited IMPA. We thank these institutions for their hospitality.

We are grateful to D. Sullivan for many stimulating conversations.

1. How does a map distort the cross-ratio ?

Let N be either the circle S^1 or the interval $[0, 1]$ and $f\colon N \to N$ be a smooth map with non-flat critical points. Later in this paper we need to compare the size of intervals $f^n(J)$ and $f^n(T)$ for intervals J and T such that $J \subset T$ and $f^n|T$ is a diffeomorphism. However, if f has critical points then by varying the intervals J and T the ratio of $|f^n(T)|/|f^n(J)|$ and $|T|/|J|$ can be made arbitrarily large (here $|W|$ denotes the length of an interval W). So instead of considering ratios of intervals we will consider *cross-ratios* of pairs of intervals. In this section we will study how a map f distorts the cross-ratio of a pair of intervals. In the next section we will obtain estimates of the distortion of this cross-ratio under high iterates of f.

More precisely let $T \subset N$ be an open interval, $\mathrm{Clos}(T) = [a, b]$, and $g\colon N \to N$ be a C^3 map with $g|\mathrm{Clos}(T)$ a diffeomorphism onto its image.

1.1. *Definition*. Let $J \subset T$ be an open and bounded interval in R such that $T - J$ has two connected components L and R. Define two cross-ratios of intervals as

$$C(T, J) = \frac{|J|\,|T|}{|L \cup J|\,|J \cup R|}, \tag{1.1}$$

$$D(T, J) = \frac{|J|\,|T|}{|L|\,|R|}, \tag{1.2}$$

where $|J|$ denotes the length of J. If g is monotone on T:

$$A(g, T, J) = \frac{C(g(T), g(J))}{C(T, J)}, \tag{1.3}$$

$$B(g, T, J) = \frac{D(g(T), g(J))}{D(T, J)}. \tag{1.4}$$

The next result is the main result in this section and shows that a smooth map $f\colon N \to N$ with non-flat critical points does not contract these cross-ratios too much. Here we say that f is non-flat at a critical point c if there exists $k \geq 2$ such that f is C^{k+1} near c and $D^k f(c) \neq 0$.

1.2 THEOREM. *Let $f\colon N \to N$ be a C^3 map whose critical points are non-flat. Then there exists a constant $C_0 = C_0(f)$ such that if $T \supset J$ are intervals (such that $T - J$ consists of two connected components L and R) and $Df(x) \neq 0$ for all $x \in T$ then*

$$(1.5) \quad \log A(f, T, J) \geq -C_0|L|\,|R|, \qquad A(f, T, J) \geq 1 - C_0|L|\,|R|,$$

$$(1.6) \quad \log B(f, T, J) \geq -C_0|T|^2, \qquad B(f, T, J) \geq 1 - C_0|T|^2.$$

Remark. It is essential in the rest of the paper that one have estimates as in (1.5) and (1.6) rather than estimates of the type $\log A(f, T, J) \geq -C_0|T|$, and $\log B(f, T, J) \geq -C_0|T|$ (which were obtained in [Y]).

In the remainder of this section we will prove this theorem. First we will show that the above operators A and B are connected with the Schwarzian derivative.

1.3. *Definition.* Let $g\colon R \to R$ be a C^3 map. If $Dg(x) \neq 0$ the Schwarzian derivative of g at x is

$$Sg(x) = \frac{D^3g(x)}{Dg(x)} - \frac{3}{2}\left(\frac{D^2g(x)}{Dg(x)}\right)^2.$$

Remark. If $N = S^1$ we fix once and for all a covering map $\pi\colon R \to S^1$. If $T \subset S^1$ is an interval and $g\colon S^1 \to S^1$ is a C^3 map which is monotone on T we can lift T to an interval $T' \subset R$ and g to a map $g'\colon R \to R$ and define $A(g, T, J) = A(g', T', J')$, $B(g, T, J) = B(g', T', J')$ and $Sg(x) = S(g'(x'))$, $\pi(x') = x$. This is clearly independent of the lift.

1.4. *Elementary facts about Schwarzian derivatives and cross-ratios.*

1) If p is a critical point of g which is not flat then there is a neighbourhood W of p such that $Sg(x) < 0$ for $x \in W - \{p\}$.

2) If g is monotone on an interval $T \subset R$ and $Sg < 0$ on T then $|Dg|$ does not have a positive local minimum on T. In particular, if $Dg \neq 0$ on T and $x < y < z$ are in T then

$$|Dg(y)| > \min\{|Dg(x)|, |Dg(z)|\}.$$

3) If $g_1, g_2\colon N \to N$ are C^∞ functions then

$$S(g_1 \circ g_2) = (S(g_1) \circ g_2)(Dg_2)^2 + Sg_2.$$

In particular if $Sg < 0$ then $Sg^n < 0$ for all $n > 0$.

4) Let M be the group of real Moebius transformations; i.e., $\phi \in M$ if and only if $\phi(t) = (\alpha t + \beta)/(\gamma t + \delta)$ with $\alpha, \beta, \gamma, \delta$ real numbers. If $\phi \in M$ then $S\phi \equiv 0$ (in fact $S\phi \equiv 0$ iff $\phi \in M$).

5) There is a similar relationship between Moebius transformations and the above operators. In fact, if $\phi \in M$ is monotone on $T \subset R$ and $J \subset T$ then, $A(\phi, T, J) = 1$ and $B(\phi, T, J) = 1$.

6) From (3) and (4) it follows that if Sg is negative on W and $\phi, \psi \in M$ then $S(\phi \circ g \circ \psi)$ is negative on $\psi^{-1}(W)$.

7) The operators A and B are multiplicative:

$$A(g \circ f, T, J) = A(g, f(T), f(J)) \times A(f, T, J)$$

and

$$B(g \circ f, T, J) = B(g, f(T), f(J)) \times B(f, T, J)$$

if f is monotone on T and g is monotone on $f(T)$.

1.5 Lemma. *Let f: $[a, b] \to R$ be a C^3 function such that $Df(x) \neq 0$ and $Sf(x) < 0$ for all $x \in [a, b]$. Then for* $\mathrm{Clos}(J) \subset \mathrm{Int}(T) \subset [a, b]$:

$$A(f, T, J) > 1, \tag{1.7}$$

$$B(f, T, J) > 1. \tag{1.8}$$

Proof. Given three points $x < y < z$ in $[-\infty, \infty]$ there is a unique Moebius transformation ϕ such that $\phi(x) = 0$, $\phi(y) = 1$ and $\phi(z) = \infty$. Hence we can find $\phi \in M$ mapping T diffeomorphically onto $[0, 1]$. For every $\lambda < 1$, the Moebius transformation $\phi_\lambda(x) = x/(\lambda x + 1 - \lambda)$ maps the interval $[0, 1]$ diffeomorphically onto itself. If $z \in (0, 1)$, $\phi_\lambda(z) \to 1$ as $\lambda \to 1$ and $\phi_\lambda(z) \to 0$ as $\lambda \to -\infty$. Hence, given $\alpha, \alpha' \in (0, 1)$ there exists λ such that $\phi_\lambda(\alpha') = \alpha$. Since composition with Moebius transformations preserves cross-ratio and the sign of the Schwarzian derivative, we may assume that $T = [0, 1]$, f maps T onto T preserving orientation and maps L onto L. So we let $L = [0, \alpha]$ and $J = [\alpha, \beta]$. We have that $f(0) = 0$, $f(\alpha) = \alpha$ and $f(1) = 1$. Since f has negative Schwarzian derivative on $[0, 1]$, f has no other fixed point. We claim that $f(x) > x$ for all $x \in (\alpha, 1)$. Suppose, by contradiction that this is not the case. Since f has no fixed point in $(\alpha, 1)$ we have then that $f(x) < x$ for all $x \in (\alpha, 1)$. This implies $Df(1) \geq 1$ and $Df(\alpha) \leq 1$. If $f(x) > x$ for all $x \in (0, \alpha)$ then $Df(0) \geq 1$ and $Df(\alpha)$ is not strictly bigger than $\min(Df(0), Df(1))$. This contradicts $Sf < 0$. If $f(x) < x$ for all $x \in (0, \alpha)$ then $Df(\alpha) = 1$ and there is an $x \in (\alpha, 1)$ such that $Df(x) \leq 1$. So $Df(x)$ is not strictly bigger than min($Df(\alpha)$, $Df(1)$) again contradicting $Sf < 0$. Since f has no fixed point in $(0, \alpha)$ these are the only two possibilities, proving the claim by contradiction.

Let us show that $A(f, T, J) > 1$. From the claim, it follows that $f(\beta) > \beta$. Now,

$$A(g, T, J) = \frac{\dfrac{f(\beta) - \alpha}{\beta - \alpha} \dfrac{1}{1}}{\dfrac{f(\beta)}{\beta} \dfrac{1 - \alpha}{1 - \alpha}} > 1$$

since $f(\beta) > \beta > \alpha$. Similarly, $B(f, T, J) > 1$; see Figure 1.

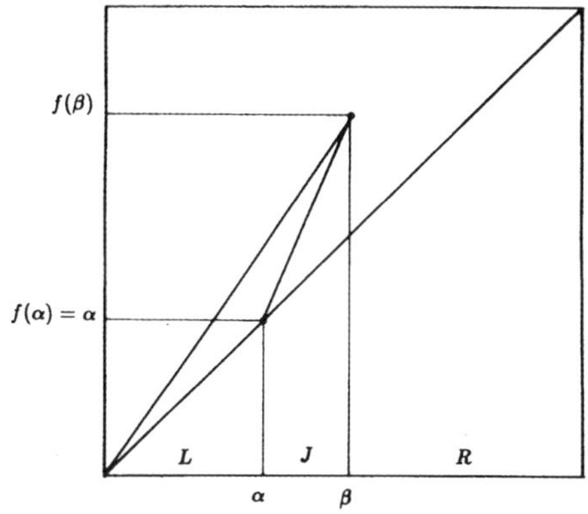

FIGURE 1

1.6. LEMMA. *Let f: $N \to N$ be a C^3 map and V be a neighbourhood of the critical points $C(f)$. There exists a constant $C_0' = C_0'(f, V)$ such that if $T \subset N - V$ then*

$$A(f, T, J) \geq 1 - C_0'|L|\,|R|,$$

$$B(f, T, J) \geq 1 - C_0'|T|^2.$$

Proof. Let us prove only the first inequality. The second one is similar and, in fact, easier. Suppose, by contradiction that the first inequality is not satisfied.

Then there is a sequence of intervals $J_n \subset T_n \subset N - V$ such that

$$A_n = \frac{A(f, T_n, J_n) - 1}{|L_n|\,|R_n|} \to -\infty$$

as $n \to \infty$. By taking subsequences we may assume that $a_n \to a$, $b_n \to b$, $c_n \to c$, $d_n \to d$ where $T_n = [a_n, d_n]$, $J_n = (b_n, c_n)$. There are two possibilities:

$$\sup \frac{b_n - a_n}{d_n - c_n} < \infty \quad \text{or} \quad \sup \frac{d_n - c_n}{b_n - a_n} < \infty.$$

Let us assume we are in the first case. The other is similar (interchange the roles of L_n and R_n). Since f is C^3 and $a_n \in N - V$ we have

$$f(a_n + x) = f(a_n) + \lambda_n x + \mu_n(x) x^2 \tag{1.9}$$

where $|\lambda_n| \geq \min\{|Df(x)|, x \in N - V\} > 0$, μ_n is differentiable and $\max\{|\mu_n(x)|, |D\mu_n(x)|; x \in N - V\} < \infty$. Let us calculate A_n:

$$A_n = \frac{\dfrac{f(c_n) - f(b_n)}{c_n - b_n}\,\dfrac{f(d_n) - f(a_n)}{d_n - a_n} - \dfrac{f(c_n) - f(a_n)}{c_n - a_n}\,\dfrac{f(d_n) - f(b_n)}{d_n - b_n}}{\dfrac{f(c_n) - f(a_n)}{c_n - a_n}\,\dfrac{f(d_n) - f(b_n)}{d_n - b_n}(b_n - a_n)(d_n - c_n)}. \tag{1.10}$$

Since we are in the case $\sup(b_n - a_n)/(d_n - c_n) < \infty$ and $T_n \subset N - V$, then $A_n \to -\infty$ implies that $b_n - a_n \to 0$ (i.e. $a = b$) and

$$B_n = \frac{\dfrac{f(c_n) - f(b_n)}{c_n - b_n}\,\dfrac{f(d_n) - f(a_n)}{d_n - a_n} - \dfrac{f(c_n) - f(a_n)}{c_n - a_n}\,\dfrac{f(d_n) - f(b_n)}{d_n - b_n}}{(b_n - a_n)(d_n - c_n)} \to -\infty. \tag{1.11}$$

Let us calculate B_n using (1.9). We get, using the notation $b_n' = b_n - a_n$, $c_n' = c_n - a_n$, $d_n' = d_n - a_n$,

$$B_n = \frac{1}{b_n'(d_n' - c_n')}\left\{\left[\left(\lambda_n + \frac{\mu_n(c_n')c_n'^2 - \mu_n(b_n')b_n'^2}{c_n' - b_n'}\right)\cdot(\lambda_n + \mu_n(d_n')d_n')\right]\right.$$
$$\left. - \left[(\lambda_n + \mu_n(c_n')c_n')\cdot\left(\lambda_n + \frac{\mu_n(d_n')d_n'^2 - \mu_n(b_n')b_n'^2}{d_n' - b_n'}\right)\right]\right\}.$$

Since

$$\frac{\mu_n(c_n')c_n'^2 - \mu_n(b_n')b_n'^2}{c_n' - b_n'} = \mu_n(c_n')(c_n' + b_n') + \frac{\mu_n(c_n') - \mu_n(b_n')}{c_n' - b_n'} \times b_n'^2$$

and

$$\frac{\mu_n(d_n')d_n'^2 - \mu_n(b_n')b_n'^2}{d_n' - b_n'} = \mu_n(d_n')(d_n' + b_n') + \frac{\mu_n(d_n') - \mu_n(b_n')}{d_n' - b_n'} \times b_n'^2$$

we have

$$B_n = \frac{\lambda_n(\mu_n(c_n') - \mu_n(d_n'))b_n' + \mu_n(c_n')\mu_n(d_n')b_n'(d_n' - c_n') + O(b_n'^2)}{b_n'(d_n' - c_n')}.$$

Since we have assumed $\sup b_n'/(d_n' - c_n') < \infty$, $\sup [O(b_n'^2)]/[b_n'(d_n' - c_n')] < \infty$. Moreover, from $\sup |\mu_n(c_n')| < \infty$, $\sup |\mu_n(d_n')| < \infty$ and

$$\sup [\mu_n(c_n') - \mu_n(d_n')]/(c_n' - d_n') < \infty,$$

we have that $\sup |B_n| < \infty$. This is a contradiction to $A_n \to -\infty$.

Remark. From the proof we get that if $\sup (b_n - a_n)/(d_n - c_n) < \infty$ and if a_n and b_n are not in V, then A_n cannot go to $-\infty$, independently whether or not c_n and d_n are in V. This is because the only thing we need in (1.9) is that $\lambda = Df(a)$ is non-zero and μ has bounded derivatives.

1.7. Proposition. *Let $f\colon N \to N$ be a C^3 map whose critical points are non-flat. Then there exists a constant $C_0 = C_0(f)$ such that if $T \supset J$ are intervals and $Df(x) \neq 0$ for all $x \in T$ then*

$$A(f, T, J) \geq 1 - C_0|L|\,|R|, \tag{1.12}$$

$$B(f, T, J) \geq 1 - C_0|T|^2, \tag{1.13}$$

where L and R are the connected components of $T - J$.

Proof. Since the critical points of f are non-flat there exists a neighbourhood W of the critical points so that $Sf(x) < 0$ for all $x \in W$. Suppose, by contradiction, that (1.12) is not satisfied. Then there exists a sequence of intervals $T_n = [a_n, d_n]$, $J_n = [b_n, c_n] \subset (a_n, d_n)$ such that $A_n = [A(f, T_n, J_n) - 1]/(|L_n|\,|R_n|)$ goes to $-\infty$. Taking subsequences we may assume that $a_n \to a$, $c_n \to c$, $d_n \to d$. Let us consider the case

$$\sup (b_n - a_n)/(d_n - c_n) < \infty;$$

the other case is similar. From this and the expression (1.10) for A_n we can deduce that $b_n - a_n$ goes to zero because otherwise $\inf(d_n - c_n) > 0$ since $\sup (d_n - a_n)/(d_n - c_n) < \infty$. Moreover, from the remark below Lemma (1.6) we have that a_n, b_n both converge to a critical point $a \in C(f)$. Also, we may assume that T_n is not contained in W because otherwise we could use Lemma (1.5). In particular $d_n \to d \neq a$. We consider two cases.

Case 1. $c_n \to c \neq a$. In this case, $A_n \to -\infty$ implies $B_n \to -\infty$ (compare equations (1.10) and (1.11)). But now, from $c \neq a$, $d \neq a$, $b = a$,

$$\frac{f(c_n) - f(b_n)}{c_n - b_n} = \frac{f(c_n) - f(a_n)}{c_n - a_n} + \Theta_1, \tag{1.14}$$

$$\begin{aligned}\frac{f(d_n) - f(b_n)}{d_n - b_n} &= \frac{f(d_n) - f(a_n)}{d_n - a_n} + \Theta_2 \\ &= \frac{f(c_n) - f(a_n)}{c_n - a_n} + \Theta_2 + \Theta_3,\end{aligned} \tag{1.15}$$

where $\Theta_1 = O(b_n - a_n)$, $\Theta_2 = O(b_n - a_n)$, $\Theta_3 = O(d_n - c_n)$. More precisely,

$$\begin{aligned}\Theta_1 &= \frac{1}{(c_n - b_n)(c_n - a_n)} \\ &\quad \times \left\{ f(c_n) - (c_n - b_n)\frac{f(b_n) - f(a_n)}{b_n - a_n} - f(b_n) \right\}(b_n - a_n),\end{aligned}$$

$$\begin{aligned}\Theta_2 &= \frac{1}{(d_n - b_n)(d_n - a_n)} \\ &\quad \times \left\{ f(d_n) - (d_n - b_n)\frac{f(b_n) - f(a_n)}{b_n - a_n} - f(b_n) \right\}(b_n - a_n).\end{aligned}$$

Substituting (1.14) and (1.15) in the expression (1.11) we get

$$\begin{aligned} B_n &= \frac{1}{(b_n - a_n)(d_n - c_n)} \left\{ \left(\frac{f(c_n) - f(a_n)}{c_n - a_n} + \Theta_1 \right) \left(\frac{f(c_n) - f(a_n)}{c_n - a_n} + \Theta_3 \right) \right. \\ &\qquad \left. - \left(\frac{f(c_n) - f(a_n)}{c_n - a_n} \right) \left(\frac{f(c_n) - f(a_n)}{c_n - a_n} + \Theta_2 + \Theta_3 \right) \right\} \\ &= \frac{1}{(b_n - a_n)(d_n - c_n)} \left\{ \frac{f(c_n) - f(a_n)}{c_n - a_n} (\Theta_1 - \Theta_2) + \Theta_1 \Theta_3 \right\}. \end{aligned}$$

Since $1/[(d_n - b_n)(d_n - a_n)] = [1 + O(d_n - c_n)]/[(c_n - b_n)(c_n - a_n)]$ we have that $(\Theta_1 - \Theta_2)/(b_n - a_n) = O(d_n - c_n)$. Thus B_n does not tend to $-\infty$.

Case 2. $c_n \to c = a$. Let W_0 be the component of W containing a. Choose $d'_n \in W_0$ and a constant $\alpha > 0$ such that $[f(d'_n) - f(a_n)]/(d'_n - a_n) = \alpha \cdot [f(d_n) - f(a_n)]/(d_n - a_n)$. Then

$$[f(d'_n) - f(b_n)]/(d'_n - b_n) = \alpha \cdot [f(d_n) - f(b_n)]/(d_n - b_n) + o(b_n - a_n).$$

Substituting this into (1.10) one gets:

$$A_n = \frac{d'_n - c_n}{d_n - c_n} \times \left\{1 + \frac{o(b_n - a_n)(d'_n - b_n)}{f(d'_n) - f(b_n)}\right\}^{-1}$$
$$\times \{\text{expression (1.10) with } d_n \text{ replaced by } d'_n\}$$
$$+ \frac{o(b_n - a_n)\dfrac{f(c_n) - f(a_n)}{c_n - a_n}}{\dfrac{f(c_n) - f(a_n)}{c_n - a_n}\dfrac{f(d_n) - f(b_n)}{d_n - b_n}(b_n - a_n)(d_n - c_n)}.$$

The first term is ≥ 0 because the interval $[a_n, d'_n] \subset W$ (see Lemma (1.5)). The last term is bounded because $d_n \to d \neq a$, $a_n, c_n, b_n \to a$.

Proof of Theorem (1.2). Clearly Theorem (1.2) follows from Proposition (1.7).

2. Distortion of the cross-ratio C under iteration

Let f: $N \to N$ be a C^3 map. If f has negative Schwarzian derivative, then $S(f^m)$ is negative for all m. Therefore, in this case, f^m expands the cross-ratio. In general this is not true. However, Theorem (2.1) below tells us that if $\sum_{k=0}^{m-1}|f^i(J)|$ is not too big then $A(f^m, T, J)$ is bounded away from 0. This is the analogue of a result of A. Schwartz, [Sc], for C^2 diffeomorphisms $\tilde{f}$: $N \to N$ *without* critical points which says that there exist $C > 0$ and $\delta > 0$ with the following property: For each interval $T \supset J$, and each integer m such that $\sum_{k=0}^{m-1}|\tilde{f}^i(J)| \leq 1$ and $|T| \leq (1 + \delta)|J|$ one has

$$\frac{|D\tilde{f}^m(x)|}{|D\tilde{f}^m(y)|} \leq C$$

for each $x, y \in T$. We will prove and use this result of A. Schwartz in Lemma (4.1).

2.1. Theorem. *Let f: $N \to N$ be a C^3 map whose critical points are non-flat. There exist constants $\delta > 0$ and $\frac{1}{18} > \varepsilon > 0$ such that if $T \supset J$ are intervals satisfying:*

i) *f^m is a diffeomorphism on* Clos(T);

ii) $\sum_{k=0}^{m-1}|f^k(J)| < \delta$;

iii) $|L|\,|R| < \varepsilon|J|^2$, *where L and R are the components of $T - J$,*

then

$$A(f^m, T, J) \geq 1 - 8\frac{|L|\,|R|}{|J|^2}.$$

To prove this theorem we first need a lemma.

2.2. Lemma. *If g is monotone on an interval $T \supset J$ and $C_1 > 0$ is such that*

$$A(g, T, J) \geq 1 - C_1|L|\,|R| \tag{2.1}$$

and $|L|\,|R|(C_1 + 1/(|J \cup L|\,|J \cup R|)) < 1$ *then*

$$\frac{|g(L)|}{|L|}\frac{|g(R)|}{|R|} \leq C_2|g(L \cup J)|\,|g(J \cup R)| \tag{2.2}$$

and

$$\frac{|g(L)|}{|L|}\frac{|g(R)|}{|R|} \leq C_2'|g(J)|\,|g(T)| \tag{2.3}$$

where $C_2 = 1/(|J \cup L|\,|J \cup R|) + C_1$ *and* $C_2' = C_2/(1 - C_2|L|\,|R|) > 0$.

Proof. Notice that

$$C(T, J) = 1 - \frac{|L|\,|R|}{|L \cup J|\,|J \cup R|}. \tag{2.4}$$

Equation (2.1) implies:

$$\begin{aligned}
&- \frac{|g(L)|\,|g(R)|}{|g(L) \cup g(J)|\,|g(J) \cup g(R)|} \\
&\quad = \frac{|g(J)|\,|g(T)|}{|g(L) \cup g(J)|\,|g(J) \cup g(R)|} - 1 \\
&\quad = C(g(T), g(J)) - 1 \\
&\quad \geq (1 - C_1|L|\,|R|) \cdot C(T, J) - 1 \\
&\quad = (1 - C_1|L|\,|R|)\left(1 - \frac{|L|\,|R|}{|J \cup L|\,|J \cup R|}\right) - 1 \\
&\quad \geq - \left(\frac{1}{|J \cup L|\,|J \cup R|} + C_1\right)|L|R| > 0.
\end{aligned}$$

This implies (2.2).

Let us now prove (2.3). We have:

$$|g(L \cup J)|\,|g(J \cup R)| = |g(J)|\,|g(T)| + |g(L)|\,|g(R)|.$$

Hence, from (2.2),

$$\frac{|g(L)|}{|L|}\frac{|g(R)|}{|R|} \leq C_2|g(J)|\,|g(T)| + C_2|g(L)|\,|g(R)|.$$

Taking the last term to the left-hand side, we get

$$(1 - C_2|L|\,|R|)\frac{|g(L)|}{|L|}\frac{|g(R)|}{|R|} \leq C_2|g(L)|\,|g(T)|.$$

Since $1 - C_2|L|\,|R| > 0$, this proves the lemma.

Proof of Theorem 2.1. Choose $1 > \eta > 0$ so that

$$|x| < \eta \Rightarrow |\log(1-x)| < 2|x|, \tag{2.5(a)}$$

$$|\log(y)| < \eta \Rightarrow |y-1| < 2|\log y|. \tag{2.5(b)}$$

Choose $\delta > 0$ so that

$$36C_0\delta < 1, \quad \delta^2 < \frac{1}{C_0}, \quad \delta C_0 < \eta < 1, \tag{2.6}$$

where $C_0 = C_0(f)$ is as in Theorem 1.2. Choose $\varepsilon > 0$ so that $\varepsilon < \eta$, $\varepsilon < 1/18C_0$, $\varepsilon < \frac{1}{18}$. Let $L_n = A(f^n, T, J)$. Write $L_n = 1 - K_n|L|\,|R|$. Let us prove by induction that $K_n < 8/|J|^2$. From Theorem (1.2),

$$L_1 = A(f, T, J) \geq 1 - C_0|L|\,|R|. \tag{2.7}$$

From assumption (ii) of the theorem and inequalities (2.6),

$$|J|^2 \leq \left(\sum_{k=0}^{\infty} |f^k(J)|\right)^2 \leq \delta^2 < \frac{1}{C_0} < \frac{8}{C_0}.$$

Using this in (2.7) we get:

$$K_1 \leq C_0 < \frac{8}{|J|^2}.$$

Suppose, by induction, that

$$K_j < \frac{8}{|J|^2} \qquad \text{for} \qquad j = 1, \dots, n < m.$$

Notice that

$$L_j = 1 - K_j|L|\,|R| > 1 - \frac{8}{|J|^2}|L|\,|R| > 1 - 8\varepsilon > 0.$$

Taking in Lemma (2.2) $C_1 = 8/|J|^2$ we get $C_2 = 1/|J|^2 + 8/|J|^2 = 9/|J|^2$. Hence $C_2|L|\,|R| < 9/|J|^2|L|\,|R| < 9\varepsilon < \frac{1}{2} < 1$. Then, by Lemma (2.2),

$$\frac{|f^j(L)|}{|L|}\frac{|f^j(R)|}{|R|} \leq C_2'|f^j(J)|\,|f^j(T)| \leq C_2'|f^j(J)|.$$

Since $C_2' = C_2/(1 - C_2|L|\,|R|)$ and $C_2|L|\,|R| < \frac{1}{2}$ as above, we have $C_2' < 2C_2$. Hence

$$\frac{|f^j(L)|}{|L|}\frac{|f^j(R)|}{|R|} \le 2C_2|f^j(J)|\,|f^j(T)| \le 2C_2|f^j(J)| = \frac{18}{|J|^2}|f^j(J)|.$$

Thus,

$$C_0|f^j(L)|\,|f^j(R)| \le \frac{18C_0}{|J|^2}|f^j(J)|\,|L|\,|R|. \tag{2.8}$$

In particular, $C_0|f^j(L)|\,|f^j(R)| \le 18C_0\varepsilon|f^j(J)| < 18C_0\varepsilon < 1$. From Theorem 1.2 we have:

$$L_{j+1} = L_j \cdot A\big(f, f^j(T), f^j(J)\big) \ge L_j \cdot \big(1 - C_0|f^j(L)|\,|f^j(R)|\big).$$

So, using equation (2.8) we get:

$$L_{j+1} \ge L_j \cdot \left(1 - \frac{18C_0}{|J|^2}|f^j(J)|\,|L|\,|R|\right) > 0. \tag{2.9}$$

From (2.9),

$$L_{n+1} \ge L_1 \cdot \prod_{j=1}^{n}\left(1 - \frac{18C_0}{|J|^2}|f^j(J)|\,|L|\,|R|\right). \tag{2.10}$$

If $L_{n+1} > L_1$ then $L_{n+1} > 1 - 8(|L|\,|R|)/|J|^2$. If $L_{n+1} < L_1$ then

$$\left|\log\frac{L_{n+1}}{L_1}\right| \le \sum_{j=1}^{n}\left|\log\left(1 - \frac{18C_0}{|J|^2}|f^j(J)|\,|L|\,|R|\right)\right|. \tag{2.11}$$

Since $18C_0(|f^j(J)|\,|L|\,|R|)/|J|^2 < 18C_0\varepsilon\delta < C_0\delta < \eta$, we have

$$\left|\log\frac{L_{n+1}}{L_1}\right| \le 2\sum_{j=1}^{n} 18C_0\frac{|f^j(J)|\,|L|\,|R|}{|J|^2} \le 36C_0\delta\frac{|L|\,|R|}{|J|^2} < \frac{|L|\,|R|}{|J|^2} < \eta.$$

Therefore, $|L_{n+1}/L_1 - 1| \le 2(|L|\,|R|)/|J|^2$. Hence,

$$L_{n+1} \ge L_1 \cdot \left(1 - 2\frac{|L|\,|R|}{|J|^2}\right) \ge (1 - C_0|L|\,|R|)\left(1 - 2\frac{|L|\,|R|}{|J|^2}\right),$$

and

$$L_{n+1} \ge 1 - \big(C_0|J|^2 + 2\big)\frac{|L|\,|R|}{|J|^2} \ge 1 - \big(C_0\delta^2 + 2\big)\frac{|L|\,|R|}{|J|^2}$$
$$\ge 1 - 3\frac{|L|\,|R|}{|J|^2} \ge 1 - 8\frac{|L|\,|R|}{|J|^2}.$$

2.4. Corollary. *Under the conditions of Theorem* (2.1):

$$\frac{|f^m(L)|}{|L|}\frac{|f^m(R)|}{|R|} \le \frac{18}{|J|^2}|f^m(J)|\,|f^m(T)|. \tag{2.12}$$

Proof. From Theorem (2.1) we have

$$A(f^m, T, J) \ge 1 - \frac{8}{|J|^2}|L|\,|R|.$$

Now, taking $C_1 = 8/|J|^2$, $C_1 + 1/|J|^2 = 9/|J|^2$, we have that

$$\left(C_1 + 1/|J|^2\right)|L|\,|R| < 9\varepsilon < 1.$$

So we can use Lemma (2.2) and we get

$$\frac{|f^m(L)|}{|L|}\frac{|f^m(R)|}{|R|} \le C_2'|f^m(J)|\,|f^m(T)|,$$

where

$$C_2' = \frac{C_2}{1 - C_2|L|\,|R|} < 2C_2 = \frac{18}{|J|^2}$$

because $C_2 = C_1 + 1/|J|^2 = 9/|J|^2$.

3. Distortion of the cross-ratio D under iteration

In the last section we have shown that f^n cannot contract the cross-ratio C too much on $T \supset J$ provided $\Sigma_{k=0}^{\infty}|f^k(J)|$ and $(|L|\,|R|)/|J|^2$ are small. In this section we will show that f^n cannot contract the cross-ratio D too much either, but now we need a stronger condition: $\Sigma_{k=0}^{n}|f^k(T)|$ must be not too big. However, from this cross-ratio D we can get some infinitesimal information whereas C describes more global properties of f on T.

3.1. Theorem. *Let f: $N \to N$ be a C^3 map whose critical points are non-flat. If $T \supset J$ are intervals satisfying:*

i) *f^m is a diffeomorphism on* Clos(T);

ii) *$\Sigma_{i=0}^{m-1}|f^i(T)| = S < 3$; then:*

$$\log B(f^m, T, J) \ge -C_0\left(\sum_{i=0}^{m}|f^i(T)|^2\right) \ge -3C_0 \cdot S, \tag{3.1}$$

where C_0 is the constant from Theorem 1.2.

Proof. From the multiplicative property of the operator B and Theorem 1.2 we get:

$$\log B(f^m, T, J) = \log \prod_{i=0}^{m-1} B(f, f^i(T), f^i(J)) = \sum_{i=0}^{m-1} \log B(f, f^i(T), f^i(J))$$
$$\geq \sum_{i=0}^{m-1} \left(-C_0 |f^i(T)|^2\right).$$

The lemma below describes the infinitesimal information which can be obtained by the operator B.

3.2. Lemma. *Let $g\colon T \to R$ be a C^3 diffeomorphism with $T = [a, b]$. Let $x \in (a, b)$. If for every $J \subset T^* \subset T$*

$$B(g, T^*, J) \geq C_2 > 0, \tag{3.2}$$

then one has either

$$|Dg(x)| \geq C_2^3 |Dg(a)| \quad \textit{or} \quad |Dg(x)| \geq C_2^3 |Dg(b)| \tag{3.3}$$

or both.

Proof. Let us consider the following two operators:

$$B_0(g, T^*) = \frac{|g(T^*)|^2}{|T^*|^2} \frac{1}{|Dg(a^*)|\,|Dg(b^*)|},$$

$$B_1(g, T, x) = \frac{|Dg(x)| \dfrac{|g(T)|}{|T|}}{\dfrac{|g(L)|}{|L|} \dfrac{|g(R)|}{|R|}},$$

where $T^* = [a^*, b^*] \subset T$ and L and R are the connected components of $T - \{x\}$. Observe that

$$B_0(g, T^*) = \lim_{J \to T^*} B(g, T^*, J),\ B_1(g, T, x) = \lim_{J \to x} B(g, T, J).$$

Hence $B_0(g, L)$, $B_0(g, R)$, $B_1(g, T, x) \geq C_2 > 0$. Let

$$\lambda_1 = \frac{|g(L)|}{|L|}, \qquad \lambda_2 = \frac{|g(R)|}{|R|} \qquad \text{and} \qquad \lambda = \frac{|g(T)|}{|T|}.$$

Since $B_0(g, L)$, $B_0(g, R) \geq C_2$ we have:

$$\lambda_1^2 \geq C_2 |Dg(a)|\,|Dg(x)|, \tag{3.4}$$

$$\lambda_2^2 \geq C_2 |Dg(x)|\,|Dg(b)|. \tag{3.5}$$

Since $B_1(g, T, x) \geq C_2 > 0$ we have:

$$|Dg(x)|\lambda \geq C_2\lambda_1\lambda_2. \tag{3.6}$$

It is clear that if $\lambda_1 \leq \lambda$ then $\lambda_2 \geq \lambda$ and if $\lambda_2 \leq \lambda$ then $\lambda_1 \geq \lambda$. So, suppose $\lambda_1 \leq \lambda$. Then:

$$|Dg(x)|^2 \overset{(3.6)}{\geq} C_2^2\frac{\lambda_1^2\lambda_2^2}{\lambda^2} \geq C_2^2\lambda_1^2 \overset{(3.4)}{\geq} C_2^3|Dg(a)|\,|Dg(x)|.$$

Hence $|Dg(x)| \geq C_2^3|Dg(a)|$ and similarly for $\lambda_2 \leq \lambda$. Since it is impossible that both λ_1 and λ_2 are bigger than λ, the lemma follows.

Remark. For a diffeomorphism $g\colon T \to R$ satisfying $Sg < 0$, one has $|Dg(x)| > \min\{|Dg(a)|, |Dg(b)|\}$. So the above lemma gives a slightly weaker version of this inequality.

3.3. THEOREM. *Let $f\colon N \to N$ be a C^3 map which is non-flat at all of its critical points, and let $C_0 = C_0(f)$ be the constant from Theorem* (1.2). *If $T = [a, b] \subset N$ is such that*:

i) *f^m is a diffeomorphism on* Clos(T),

ii) $\sum_{i=0}^{m-1}|f^i(T)|^2 = S < 3$,

then either

$$|Df^m(x)| \geq \exp(-gC_0S)|Df^m(a)|, \tag{3.7}$$

or

$$|Df^m(x)| \geq \exp(-gC_0S)|Df^m(b)|, \tag{3.8}$$

or both.

Proof. This follows immediately from Theorem (3.1) and Lemma (3.2).

4. Non-existence of wandering intervals for unimodal maps

Assume $N = [0, 1]$ and $f\colon N \to N$ is a C^3 unimodal map, i.e., a map with just one critical point. In this section we show that if the critical point is non-flat then f cannot have a wandering interval. We start with a more general result which follows from the proof of Schwartz's theorem (see for example [C-E1, pp. 111] or [Ma]). Versions of this result are proved in several papers but in order to be complete we include the proof here.

4.1. LEMMA. *Let N be either* $[0, 1]$ *or S^1 and $f\colon N \to N$ be a C^2 map. If J is a wandering interval of f then there exists a sequence $n(i) \to \infty$ such that $f^{n(i)}(J) \to C(f)$ where $C(f)$ is the critical set of f.*

Proof. For simplicity assume that $\operatorname{diam}(S^1) \leq 1$. Let J be a wandering interval of f. We may assume that J is a maximal wandering interval, i.e., that J is not strictly contained in some bigger wandering interval. Suppose, by contradiction, that the forward iterates of J do not accumulate at the critical set of f. Then, there is a neighbourhood V of $C(f)$ such that $f^n(J) \cap V = \varnothing$ for all $n \geq 0$. Let W be a neighbourhood of $C(f)$ such that $\operatorname{Clos}(W) \subset V$. Since f is C^2, there exists a constant $K > 0$ such that $|D\log|Df(x)|\,| < K$ for every $x \in N - W$. Let T be an interval such that $f^i(T) \cap W = \varnothing$ for every $i \leq n$. Then, for $x, y \in T$ we have, from the chain-rule and the mean value theorem, that

$$\left|\log\frac{|Df^n(x)|}{|Df^n(y)|}\right| = \left|\sum_{i=0}^{n-1}\left(\log|Df(f^i(x))| - \log|Df(f^i(y))|\right)\right| \leq \sum_{i=0}^{n-1} K|f^i(T)|.$$

Hence,

$$(*) \qquad \exp\left(-K\sum_{i=0}^{n-1}|f^i(T)|\right) \leq \frac{|Df^n(x)|}{|Df^n(y)|} \leq \exp\left(K\sum_{i=0}^{n-1}|f^i(T)|\right),$$

for all $x, y \in T$. From this and the mean value theorem we get

$$(^{**}\text{a}) \qquad |Df^n(x)| \leq \exp\left(K\sum_{i=0}^{n-1}|f^i(T)|\right) \times \frac{|f^n(T)|}{|T|}$$

and

$$(^{**}\text{b}) \qquad |Df^n(x)| \geq \exp\left(-K\sum_{i=0}^{n-1}|f^i(T)|\right) \times \frac{|f^n(T)|}{|T|},$$

for every $x \in T$. In particular, since the intervals $f^i(J)$ are disjoint, we get from (**a) that

$$(^{***}) \qquad |Df^n(x)| < \exp(K)\frac{|f^n(J)|}{|J|}.$$

Let $T \supset J$ be an interval such that $L \neq \varnothing$ and $R \neq \varnothing$ are the connected components of $T - J$. Choose $\delta > 0$ so small that $2 \cdot \delta \cdot \exp(2K)$ is less than the length of each of the connected components of $V - W$. We claim that if $|L|, |R| < \delta \cdot |J|$, then: i) $\sum_{i=0}^n |f^i(T)| < 2$, ii) $\exp(-2K)|f^{n+1}(J)|/|J| \leq |Df^{n+1}(x)| \leq \exp(2K)|f^{n+1}(J)|/|J|$ and iii) $f^{n+1}(T) \cap W = \varnothing$ for every $n > 0$ and every $x \in T$. We prove the claim by induction. Suppose i), ii) and iii) hold for every $n = 1, \ldots, m-1$. From ii) we have $|Df^{i+1}(x)| \leq \exp(2K)1/|J|$ for $i < m$. Hence $|f^{i+1}(T)| \leq |f^{i+1}(J)| + 2\delta \cdot |J| \cdot \exp(2K)|f^{i+1}(J)|/|J|$ for $i < m$ and $\sum_{i=0}^{m}|f^i(T)| \leq (1 + 2\delta \cdot \exp(2K)) \cdot (\sum_{i=0}^{m}|f^i(J)|) < 2$, since $\sum|f^i(J)| < 1$ and $\delta < \frac{1}{2}\exp(-2K)$. This shows that i) holds for m. Using this, (**a) and (**b) we get that ii) holds for m. From this we have

$$|f^{m+1}(T)| - |f^{m+1}(J)| < 2\delta \cdot \exp(2K)|f^{m+1}(J)| \leq 2\delta \cdot \exp(2K).$$

By the choice of δ this is less than the minimum of the lengths of the connected components of $V - W$. Hence, since $f^{m+1}(J) \cap V = \emptyset$, we conclude that iii) also holds for m. This proves the claim.

Let $T \supsetneq J$ be an interval such that i), ii) and iii) are satisfied. From ii) it follows that $f^n|T$ is a diffeomorphism for every $n > 0$. Since $T \supset J$ and $T \neq J$ it follows from the maximality of J that T is not a wandering interval. So either a) T contains points which are contained in the basin of some periodic attractor or, b) there exist integers n and k such that $f^{n+k}(T) \cap f^n(T) \neq \emptyset$. First assume we are in case a). Since J does not contain points which are in the basin of some periodic attractor, T contains a point in the boundary of the (immediate) basin of the periodic attractor. Hence $|f^k(T)|$ does not tend to zero, contradicting $|f^n(T)| \leq (1 + 2\delta\exp(2K)) \cdot |f^n(J)| \to 0$. In case b) it follows that, for every integer j, $f^{n+(j+1)k}(T) \cap f^{n+jk}(T) \neq \emptyset$. Thus, $I = \bigcup_{j=0}^{\infty} f^{n+jk}(T)$ is an interval and f^n maps I homeomorphically into itself. Hence all points in I are asymptotic to periodic points. This is a contradiction because I contains $f^n(J)$ and J does not contain points which are in the basin of periodic attractors. So in both cases we get a contradiction, and it follows that f cannot have a wandering interval.

4.2. COROLLARY. *If f is a C^2 unimodal map and the critical point of f is in the basin of an attractor then f has no wandering interval.*

From now on $f: N \to N$ will be a C^3 unimodal map with a non-flat critical point c. Assume by contradiction that f has a wandering interval J. By replacing J with another wandering interval containing some iterate of J we may assume that:

$$f^n(\mathrm{Clos}(J)) \cap C(f) = \emptyset, \tag{4.1}$$

for all $n \geq 0$ and

$$\sum_{k=0}^{\infty} |f^k(J)| \leq \delta_0, \tag{4.2}$$

where δ_0 is the constant from Theorem 2.1. In fact (4.1) holds for any wandering interval of unimodal maps because of Lemma (4.1). Let K_m be the maximal interval containing J such that $f^m|K_m$ is a homeomorphism. From (4.1) $K_m - \text{Clos}(J)$ is a disjoint union of two non-empty intervals L_m and R_m. To be definite choose R_m so that $d(f^m(R_m), c) < d(f^m(J), c)$.

4.3. *Remark*. It follows that f^{m+1} is a homeomorphism on $L_m \cup J$ and $K_{m+1} \subset K_m$.

4.4. LEMMA. *The lengths of the intervals L_m and R_m go to zero as $m \to \infty$.*

Proof. If this were not the case then $K = \bigcap_{m=0}^{\infty} K_m$ would be an interval that strictly contains J because $K_{m+1} \subset K_m$. Let us show that K is a wandering interval. Since $f^n|K$ is monotone for all $n \geq 0$ and no point of $J \subset K$ is contained in the basin of a periodic attractor, no point of K can be in the basin of a periodic attractor. Hence it suffices to prove that $f^i(K)$, $i \geq 0$, are disjoint.

So assume by contradiction that for some $n > 0$ and some $k \geq 0$,

$$f^k(K) \cap f^{k+n}(K) \neq \varnothing .$$

Then $f^{k+jn}(K) \cap f^{k+(j+1)n}(K) \neq \varnothing$ for every $j \geq 0$. Therefore $K^* = \bigcup_{j=0}^{\infty} f^{k+jn}(K)$ is an interval and f^n maps K^* into itself monotonically. Hence every non-periodic point in K^* is asymptotic to a periodic attractor. This is a contradiction because $K \supset J$. Hence K is a wandering interval. But this contradicts the maximality of J (J was chosen so that it is not contained in a strictly bigger wandering interval).

Let us introduce some notation. For x near the critical point c we denote by x' the "symmetric" of x; i.e., $f(x') = f(x)$ and $x' \neq x$. The symbol $x^{\#}$ will denote either x' or x. If $U_1, U_2 \subset N$ we define $\langle U_1, U_2 \rangle$ as the smallest interval containing $U_1 \cup U_2$. In particular $\langle a, b \rangle$ is the (closed) segment connecting a and b. Let us define, by induction, a sequence of integers $k(n)$ that describe the closest approach of iterates of J to the critical point. More precisely, $k(0) = 0$ and, for $n \geq 1$,

$$k(n) = \min\left\{k;\, f^k(J) \subset \left\langle f^{k(n-1)}(J), \left(f^{k(n-1)}(J)\right)' \right\rangle\right\}.$$

By Lemma 4.1 there exists a sequence $n(i) \to \infty$ such that $f^{n(i)}(J) \to c$. Hence the sequence $k(n)$ is well-defined and $f^{k(n)}(J) \to c$.

4.5. LEMMA. *$f^{k(n)}(K_{k(n)})$ contains either*

$$\langle f^{k(n-1)}(J), c \rangle \text{ or } \langle \left(f^{k(n-1)}(J)\right)', c \rangle .$$

Proof. Since $K_{k(n)}$ is maximal, there exist integers $i, j < k(n)$ such that $f^i(\overline{R_{k(n)}})$ and $f^j(\overline{L_{k(n)}}) \not\ni c$. Since $i, j < k(n)$,

$$f^i(J) \cap \langle f^{k(n-1)}(J), \left(f^{k(n-1)}(J)\right)' \rangle = \varnothing .$$

Therefore $f^i(\overline{J \cup R_{k(n)}})$ and $f^j(\overline{L_{k(n)} \cup J})$ contain either $\langle f^{k(n-1)}(J), c\rangle$ or $\langle (f^{k(n-1)}(J))', c\rangle$ (the ith and the jth iterates may contain different segments). Suppose, by contradiction that $f^{k(n)}(L_{k(n)})$ does not contain $f^{k(n-1)}(J)^{\#}$. From the choice of $L_{k(n)}$ and $R_{k(n)}$ one has $d(f^{k(n)}(L_{k(n)}), c) > d(f^{k(n)}(J), c)$ and hence $f^{k(n)}(L_{k(n)} \cup J)$ is contained in $f^j(L_{k(n)} \cup J)^{\#} = \langle (f^{k(n-1)}(J))', c\rangle$ and $f^{k(n)-j}$ maps $f^j(L_{k(n)} \cup J)^{\#}$ onto $f^{k(n)}(L_{k(n)} \cup J)$ homeomorphically. This implies that $f^j(J)$ is contained in the basin of an attractor and, therefore, is not a wandering interval. This contradiction shows that $f^{k(n)}(L_{k(n)})$ contains $f^{k(n-1)}(J)^{\#}$.

Let us now prove by contradiction that $f^{k(n)}(R_{k(n)})$ contains c. If this is not the case, then $f^{k(n)-i}$ maps $f^i(J \cup R_{k(n)})^{\#}$ homeomorphically into itself. As before this is a contradiction. So $f^{k(n)}(K_{k(n)})$ contains both $(f^{k(n-1)}(J))^{\#}$ and c. This finishes the proof of Lemma (4.5).

Define

$$V_n = \{x;\, f^n(x) \in \langle x, x'\rangle \text{ and } f^i(x) \notin \langle x, x'\rangle \text{ for } i < n\}.$$

4.6. Lemma. *Let (a, b) be a connected component of V_n. Then $f^n|(a, b)$ is a diffeomorphism, $f^n(a) = a^{\#}$ and $f^n(b) = b^{\#}$.*

Proof. Since b is in the boundary of V_n there is a smallest $0 < i \le n$ such that $f^i(b) = b^{\#}$. Let us show that $f^n(b) = b^{\#}$. Write $n = ki + l$ with $0 \le l < i$. Then

$$f^n(b) = f^l(f^{ki}(b)) = f^l(b^{\#}) = f^l(b) \notin \langle b, b'\rangle$$

if $l > 0$, since $l < i$. Hence $l = 0$ and $f^n(b) = f^{ki}(b) = b^{\#}$. Similarly $f^n(a) = a^{\#}$. Now, $f^n|(a, b)$ is a diffeomorphism because, otherwise, there would exist $x \in (a, b)$ and $i < n$ such that $f^i(x) = c \in (x, x')$.

4.7. *Remark*. Let $A = a$ or b. Assume $f^i(A) = A^{\#}$ for some $0 < i < n$. If $Df^i(A^{\#}) > 0$ then $i = n$ and if $Df^i(A^{\#}) < 0$ then either $i = n/2$ or $i = n$.

Proof. Let $Df^i(A^{\#}) > 0$. Then, since $n = ki$ we have, for $x \in (a, b)$ near $A^{\#}$, $f^i(x^{\#}) \in \langle x^{\#}, f^n(x^{\#})\rangle \subset \langle x', x\rangle$. Hence $i = n$ because $(a^{\#}, b^{\#})$ is also a component of V_n. Let $Df^i(A^{\#}) < 0$. We distinguish between two cases. If A is the inner boundary of (a, b), i.e., $\langle A, A'\rangle \cap (a, b) = \varnothing$, then $i = n$ because $f^i(x^{\#}) \in \langle A, A'\rangle \subset \langle x, x'\rangle$ for x near A. If A is the outer boundary then for $x \in (a, b)$ near A, $x^{\#}$ and $f^n(x^{\#})$ are on the same side of $A^{\#}$. Hence n/i is even and $i < n$. Since $f^{2i}(x) \in \langle x, f^n(x)\rangle \subset \langle x, x'\rangle$ we have $2i = n$.

4.8. Lemma. *Let (a, b) be a connected component of V_n. If $f^n(a) = a$ and $f^n(b) = b$ or $f^n(a) = a'$ and $f^n(b) = b'$ then (a, b) is contained in the basin of an attractor.*

Proof. This follows because in this case f^n maps either (a, b) or (a', b') homeomorphically onto itself.

In order to use the estimates from Section 3 we need some disjointness of iterates of intervals.

4.9. LEMMA. *Let (a, b) be a connected component of V_n. Let I be the interval $\langle (a,b),(a,b)' \rangle$ (this is $\mathrm{int}(\langle a,a'\rangle \cup \langle b,b'\rangle)$). For $x \in (a,b)$ let $n(x) = \inf\{i > 0, f^i(x) \in I\}$. Then $n(x) = n$ for all $x \in (a, b)$.*

Proof. To be definite let $I = (b, b')$. For x near b, $n(x) = n$. In fact assume that $f^i(x) \in \langle b, b'\rangle$ for some $0 \le i < n$. If $f^i(b) \notin \langle b, b'\rangle$ then this is impossible. But if $f^i(b) \in \langle b, b'\rangle$ then $i = n/2$ and $Df^i(b) < 0$ (see Remark above Proposition 1.7). But then $f^i(x) \notin \langle b, b'\rangle$. Hence, by contradiction, $n(x) = n$. We claim that $n(a) = n$. Suppose by contradiction that $f^i(a) \in \langle b, b'\rangle$ for some $i < n$. Hence $f^i((a, b)) \ni b^\#$ because $f^i(b) \notin$ interior $\langle b, b'\rangle$. So let $y \in (a, b)$ such that $f^i(y) = b^\#$. Thus $f^n(y) = f^{n-i}f^i(y) = f^{n-i}(b^\#) = f^{n-i}(b) \notin \langle y, y'\rangle$. On the other hand $f^n(y) \in \langle y, y'\rangle$. This shows that $n(a) = a$. Summarizing, we have that if $i < n$ then $f^i(a) \notin \langle b, b'\rangle$ and $f^i(b) \notin$ interior $\langle b, b'\rangle$. Therefore, if $f^i(a, b) \cap \langle b, b'\rangle \neq 0$ then $f^i((a, b)) \supset \langle b, b'\rangle$ because it is a homeomorphism. Then f^{i+1} is not a homeomorphism. This contradiction shows that $n(x) = n$ for every $x \in (a, b)$.

4.10. LEMMA. *Let (a, b) be a connected component of V_n. Then the intervals $\{f^i(a, b);\ i = 0, \dots, n-1\}$ are pairwise disjoint.*

Proof. Suppose not. Then there exists $i < j < n - 1$ with

$$f^i((a, b)) \cap f^j((a, b)) \neq 0.$$

Let $z = f^i(x) = f^j(y), (x, y) \in (a, b)$. Then $f^{n-(j-i)}(x) = f^{n-j}f^i(x) = f^{n-j}(f^j(y)) = f^n(y) \in (b, b')$. Hence $n(x) \le n - (j - i) < n$ and we have a contradiction to Lemma 4.9.

4.11. PROPOSITION. *Let $k(n)$ be the sequence of closest approach as before. Then there exists a constant $e > 0$ such that $|f^{k(n+1)}(J)| > e|f^{k(n)}(J)|$.*

Proof. By the definition of the sequence $k(n)$, $f^{k(n)}(J) \subset V_{k(n+1)-k(n)}$. Let (a, b) be the component of $V_{k(n+1)-k(n)}$ that contains $f^{k(n)}(J)$. Since we assume the existence of a wandering interval, a point in this interval cannot be in the basin of an attractor (see Corollary 4.2). Hence if x is in the boundary of the immediate basin of an attractor of period m then x is a periodic point of period either m or $2m$ and $|Df^{2m}(x)| \ge 1$. Let $(\tilde{a}, \tilde{b}) \supset (a, b)$ be such that $\tilde{a} = a$ if $\tilde{a}^\#$ is not a hyperbolic attractor; otherwise take $\tilde{a}^\#$ to be equal to the periodic point in the boundary of the immediate basin of $a^{\#}$, and similarly for $\tilde{b}$. Hence $Df^{2m}(\tilde{a}^{\#}) \geq 1$ and $Df^{2m}(\tilde{b}^{\#}) \geq 1$. Since $Df^{2m}(\tilde{a}^{\#}) = Df^{m}(f^{m}(\tilde{a}^{\#}))$, $Df^{m}(\tilde{a}^{\#}) \geq 1$ and $\mathrm{int}\langle \tilde{a}^{\#}, f^{m}(\tilde{a}^{\#})\rangle$ is in the basin of $\tilde{a}^{\#}$, we can choose $\bar{a}, \bar{b} \in (\tilde{a}, \tilde{b})$ so that $|Df^{m}(\bar{a}^{\#})| \geq 1$, $|Df^{m}(\bar{b}^{\#})| \geq 1$ and $(\tilde{a}, \tilde{b}) - (\bar{a}, \bar{b})$ is in the basin of attractors. So $f^{m}(J) \subset (\bar{a}, \bar{b})$. Let $\alpha = \inf\{|Df(x)|/|Df(x')|,\ x$ near $c\}$. Notice that α is positive because f is non-flat at c. Then $|Df^{m}(\bar{a})| \geq \alpha$ and $|Df^{m}(\bar{b})| \geq \alpha$.

Claim: Each point $x \in N$ belongs to at most three of the intervals $f^{i}(\bar{a}, \bar{b})$ for $i = 0, \ldots, m-1$. In fact, $[\bar{a}, \bar{b}] = \langle \bar{a}, a\rangle \cup (a, b) \cup \langle \bar{b}, b\rangle$, $\langle \bar{a}, a\rangle$ is in the basin of $a^{\#}$ and $\mathrm{int}\langle \bar{b}, b\rangle$ is in the basin of $b^{\#}$. Therefore, each point x belongs to at most one of the intervals $f^{i}(a, b)$, $i = 1, \ldots, m$ (because they are disjoint), to at most two of the sets $f^{i}(\langle \bar{a}, a\rangle \cup \langle \bar{b}, b\rangle)$ because the basins of two different attractors are disjoint and the period of $a^{\#}$ and $b^{\#}$ are either n or $n/2$. Hence each point belongs to at most three of the intervals $f^{i}(\bar{a}, \bar{b})$ and the claim is proved.

Let us take, in Theorem (3.3), $T = (\bar{a}, \bar{b})$. Then we have

$$|Df^{m}(x)| > C_2 \min\{|Df^{m}(\bar{a})|, |Df^{m}(\bar{b})|\} > C_2\alpha,$$

for all $x \in f^{k(n)}(J)$. In particular, $|f^{k(n+1)}(J)| > C_2\alpha |f^{k(n)}(J)|$. This finishes the proof of Proposition (4.11).

Let us now show that there are no wandering intervals for unimodal maps as above. Choose a subsequence $n_i \to \infty$ such that

$$|f^{k(n_i-1)}(J)| > |f^{k(n_i)}(J)|.$$

From Lemma (4.5) we have that $f^{k(n_i)}(L_{k(n_i)}) \supset (f^{k(n_i-1)}(J))^{\#}$ and $f^{k(n_i)}(R_{k(n_i)}) \supset (f^{k(n_i+1)}(J))^{\#}$. Hence

$$\left|f^{k(n_i)}\left(L_{k(n_i)}\right)\right| > \alpha \left|f^{k(n_i-1)}(J)\right|$$

and

$$\left|f^{k(n_i)}\left(R_{k(n_i)}\right)\right| > \alpha\left|f^{k(n_i+1)}(J)\right| \overset{\text{Prop 4.11}}{>} e\alpha\left|f^{k(n_i)}(J)\right|,$$

where, as above, $\alpha = \inf\{|Df(x)|/|Df(x')|\}$. From Corollary (2.4) we have

$$\frac{\left|f^{k(n_i)}\left(L_{k(n_i)}\right)\right|}{|L_{k(n_i)}|} \cdot \frac{\left|f^{k(n_i)}\left(R_{k(n_i)}\right)\right|}{|R_{k(n_i)}|}$$

$$\leq \frac{18}{|J|^2}\left|f^{k(n_i)}(J)\right|\left(\left|f^{k(n_i)}\left(L_{k(n_i)}\right)\right| + \left|f^{k(n_i)}(J)\right| + \left|f^{k(n_i)}\left(R_{k(n_i)}\right)\right|\right).$$

By shrinking K_{n_i} we get

$$\frac{|f^{k(n_i)}(\tilde{L}_i)|}{|L_{k(n_i)}|}\frac{|f^{k(n_i)}(\tilde{R}_i)|}{|R_{k(n_i)}|} \le \frac{|f^{k(n_i)}(\tilde{L}_i)|}{|\tilde{L}_i|}\frac{|f^{k(n_i)}(\tilde{R}_i)|}{|\tilde{R}_i|}$$

$$\le \frac{18}{|J|^2}|f^{k(n_i)}(J)|$$

$$\times\left(|f^{k(n_i)}(\tilde{L}_i)| + |f^{k(n_i)}(J)| + |f^{k(n_i)}(\tilde{R}_i)|\right),$$

for every interval $\tilde{K}_i = \tilde{L}_i \cup J \cup \tilde{R}_i \subset K_{n_i}$. Choose $\tilde{L}_i$ and $\tilde{R}_i$ so that

$$|f^{k(n_i)}(\tilde{L}_i)| = \min\left\{|f^{k(n_i)}(J)|, \alpha|f^{k(n_i-1)}(J)|\right\},$$

$$|f^{k(n_i)}(\tilde{R}_i)| = \min\left\{|f^{k(n_i)}(J)|, e\alpha|f^{k(n_i)}(J)|\right\}.$$

Then

$$\frac{|f^{k(n_i)}(\tilde{L}_i)|}{|L_{k(n_i)}|}\frac{|f^{k(n_i)}(\tilde{R}_i)|}{|R_{k(n_i)}|} \le 3 \times \frac{18}{|J|^2}|f^{k(n_i)}(J)|^2,$$

$$\frac{|f^{k(n_i)}(J)|}{|f^{k(n_i)}(\tilde{R}_i)|} = \max\left(1, \frac{1}{e\alpha}\right), \frac{|f^{k(n_i)}(J)|}{|f^{k(n_i)}(\tilde{L}_i)|} \le \max\left(1, \frac{1}{\alpha}\right),$$

because $|f^{k(n_i-1)}(J)| > |f^{k(n_i)}(J)|$. Hence

$$\frac{1}{|L_{k(n_i)}|\,|R_{k(n_i)}|} \le 3 \times \frac{18}{|J|^2}\max\left(1, \frac{1}{e\alpha}\right)\max\left(1, \frac{1}{\alpha}\right).$$

This contradicts Lemma (4.4) and proves the theorem.

5. Non-existence of wandering intervals for some other maps

In this section we consider maps, with an arbitrary number of critical points, which satisfy the so-called Misiurewicz condition.

5.1. THEOREM. *Let $f\colon N \to N$ be a C^3 endomorphism, where N is either the interval* $[0, 1]$ *or the circle S^1. Suppose f satisfies the following properties:* a) *every critical point is non-flat;* b) *the forward orbit of each of the critical points does not accumulate to the critical set $C(f)$. Then f has no wandering intervals.*

Proof. Suppose, by contradiction, that f has a wandering interval J. We may assume that $\sum_{k=0}^{\infty}|f^k(J)| < \delta$ where δ is the constant in Theorem 2.1. For each integer n, let T_n be the maximal interval containing J so that f^n is monotone on T_n. Let L_n and R_n be the connected components of $T_n - J$. As in Lemma 4.4 we have that the lengths of L_n and R_n tend to zero as $n \to \infty$. In particular, if n is big enough, we have $|L_n|\,|R_n| < \varepsilon|J|^2$, where ε is as in Theorem 2.1. Therefore, from this theorem and Lemma 2.2 we get:

$$|L_n|\,|R_n| \geq \frac{1}{C_2}\frac{|f^n(L_n)|\,|f^n(R_n)|}{|f^n(L_n \cup J)|\,|f^n(J \cup R)|}, \tag{5.1}$$

where $C_2 = |J|^2/(|J|^4 + 8)$. On the other hand, by maximality of T_n,

$$\mathrm{Clos}(T_n) - T_n \subset P = \bigcup_{k=1}^{\infty} f^k(C(f)). \tag{5.2}$$

By Lemma 4.1, there exists a sequence $n(i) \to \infty$ such that $f^{n(i)}(J)$ converges to $C(f)$. Since, by hypothesis, $\mathrm{Clos}(P) \cap C(f) = \varnothing$ we get that $|f^{n(i)}(L_{n(i)})|$ and $|f^{n(i)}(R_{n(i)})|$ do not tend to zero. This and (5.1) imply that $|L_{n(i)}|\,|R_{n(i)}|$ does not tend to zero. This contradiction proves the theorem.

IMPA, RIO DE JANEIRO, BRAZIL
TECHNICAL UNIVERSITY, DELFT, THE NETHERLANDS

REFERENCES

[C-E1] P. COLLET and J. ECKMANN, *Iterated Maps of the Interval as Dynamical Systems*, Birkhauser (1980).

[C-E2] ———, Positive Liapounov exponents and absolute continuity for maps of the interval, Ergod. Th. and Dynam. Sys. 3 (1983), 13–46.

[D] A. DENJOY, Sur les courbes definie par les equations différentielles à la surface du tore, J. Math. Pure et Appl. 11, ser. 9 (1932), 333–375.

[F1] P. FATOU, Sur les equations fonctionnelles, Bull. Soc. Math. France, **47** (1919), 161–271.

[F2] ———, Sur les equations fonctionnelles, Bull. Soc. Math. France, **48** (1920), 33–94; 208–314.

[G] J. GUCKENHEIMER, Sensitive dependence on initial conditions for one-dimensional maps, Comm. Math. Phys. **70** (1979), 133–160.

[Ha] C. R. HALL, A C^∞ Denjoy counterexample, Ergod. Th. and Dynam. Syst. **1** (1981), 261–272.

[J] G. JULIA. Mémoire sur l'iteration des fonctions rationnelles, J. Math. **8** (1918), 47–245.

[Ma] R. MAÑÉ, Hyperbolicity, sinks, and measure in one dimensional dynamics, Commun. Math. Phys. **100** (1985), 495–524.

[Me] W. DE MELO, A finiteness problem for one dimensional maps, Proc. A.M.S. **101** (1987), 721–727.

[M-M-S] M. MARTENS, W. DE MELO and S. VAN STRIEN, Julia-Fatou-Sullivan theory for real one-dimensional dynamics, preprint, Delft University.

[M-T] J. MILNOR and W. THURSTON, On iterated maps of the interval: I, II, preprint, Princeton University (1977).

[Mi] M. MISIUREWICZ, Absolutely continuous measures for certain maps of an interval, Publ. Math. IHES **53** (1981), 17–51.

[N-S] T. NOWICKI and S. VAN STRIEN, Absolutely continuous measures for Collet-Eckmann maps without Schwarzian derivative conditions, Invent. Math. **93** (1988), 619–635.

[Pa] W. PARRY, Symbolic dynamics and transformations of the unit interval, Trans. A.M.S. **122** (1966), 368–378.

[Si] D. SINGER, Stable orbits and bifurcations of maps of the interval, SIAM J. Appl. Math. **35** (1978), 260–267.

[Sc] A. SCHWARTZ, A generalization of a Poincaré-Bendixon theorem to closed two dimensional manifolds, Amer. J. Math. **85** (1963), 453–458.

[S-I] A. N. SHARKOVSKII and A. F. IVANOV, C^∞-mappings of an interval with attracting cycles of arbitrary large periods, Ukrainskii Mathematicheskii Zhurnal, **4** (1983), 537–539.

[Str1] S. VAN STRIEN, Hyperbolicity and invariant measures for general C^3 interval maps satisfying the Misiurewicz condition, to appear in Comm. Math. Phys.

[Str2] ______, Smooth dynamics on the interval, in *New Directions in Chaos*, Cambridge University Press, England (1987), 57–119.

[Su] D. SULLIVAN, Quasiconformal homeomorphisms and dynamics I: Solution of the Fatou-Julia problem on wandering domains, Ann. of Math. **122** (1985), 401–418, and II: Structural stability implies hyperbolicity for Kleinian groups, Acta Math. **155** (1985), 243–260.

[Y] J. C. YOCCOZ, Il n'y a pas de contre exemple de Denjoy analytique, C. R. Acad. Sc. Paris, **298**, Serie I, no. 7 (1984), 141–144.

(Received September 18, 1986)

On Cherry Flows

Marco Martens[†], Sebastian van Strien[‡], Welington de Melo and Pedro Mendes[*]

Ergod. Th. & Dynam. Sys. **10**, 531–554 (1990)

Abstract. The purpose of this research is to describe all smooth vector fields on the torus T^2 with a finite number of singularities, no periodic orbits and no saddle-connections. In this paper we are able to complete the description within the class of vector fields which are area contracting near all singularities. In particular we give a large class of analytic vector fields on the torus T^2 which have non-trivial recurrence and also sinks.

This result for vector fields follows from a result dealing with continuous monotone circle mappings which are possibly constant on a finite number of intervals.

1. *Statement of results for vector fields on T^2*

In this paper we are interested in recurrence of vector fields on surfaces. From the Poincaré–Bendixson theorem the dynamics of vector fields on the sphere is clear: the ω-limit of every point is equal to a point or to a closed orbit. On the torus T^2 the situation is more interesting. If a vector field on T^2 has periodic orbits then there are again no non-trivial recurrent orbits, because then the situation reduces to a planar one.

In his celebrated 1932 paper, Denjoy studied vector fields on the torus T^2 *without singularities* and *without periodic orbits*. He showed that if a vector field X on the torus is C^∞, then such a vector field either has a periodic orbit or every orbit of X is dense in T^2. In 1963, A. J. Schwartz extended Denjoy's result and showed that any minimal set of a C^∞ vector field on a compact connected, two dimensional manifold M is either a singleton (consisting of a singularity), or a single closed orbit, or all of M in which case $M = T^2$. (Here a set L is called *minimal* if it is a

† Supported by Netherlands Organisation for the Advancement of Pure Research (N.W.O.).
‡ Since most of the results in this paper were proved independently by M.M. & S.v.S and by W. de M. & P.M. it was decided to publish this paper jointly.
* Partially supported by the Brazilian CNPq, grants 300376/79/0 and 150013/87/MA/Fv.

Originally published in *Ergod. Th. & Dynam. Sys.* **10**, 531–554 (1990).

closed, non-empty set which is invariant under the vector field such that L contains no smaller set with these properties.) From this it follows that the ω-limit of any point is equal to a periodic orbit, to $M = T^2$ or contains a singularity. But because no minimal set can strictly contain a singularity, the result of Schwartz gives a rather incomplete description of the dynamics of vector fields with singularities.

The next simplest vector fields on the torus have

(i) a finite number of singularities all of which are hyperbolic;
(ii) no periodic orbits;
(iii) no saddle-connections;
(iv) no sources.

We shall call a vector field which satisfies properties (i)-(iv) a Cherry vector field, because Cherry gave in 1938 an example of an analytic vector field on T^2 satisfying these properties with a sink and also a recurrent orbit. His example is quite specific, but shows that Cherry vector fields can have recurrent behaviour.

In this paper we will generalise Denjoy's result to all Cherry vector fields which have negative divergence at each saddle-point. This last condition is unnecessary if X is a Cherry vector field with at most one saddle-point. In this case, index arguments give that X can have only one other singularity. We may assume that the divergence in these singularities is of opposite signs, because otherwise either X or $-X$ would satisfy the conditions from before. So in this case we can assume that X has precisely one sink, no sources and one saddle-point. At this saddle-point the divergence of X is positive. This situation can be treated using the result of C. Gutierrez, A. Lins and W. de Melo described in § 6.

Theorem A implies that there exists a large class of vector fields which display the same phenomenon as Cherry's example.

Let $\mathcal{B}$ be the class of C^∞ Cherry vector fields X on T^2 such that at each singularity p the divergence of X is smaller or equal to 0. Furthermore, let $\mathcal{B}'$ be the class of C^∞ Cherry vector fields X on T^2 with precisely one hyperbolic sink and one saddle-point; moreover assume that X has positive divergence at this saddle-point. (As we noted above, for each vector field X with only hyperbolic singularities and one saddle-point either $X \in \mathcal{B} \cup \mathcal{B}'$ or $-X \in \mathcal{B} \cup \mathcal{B}'$.)

Denote the flow of X through x by $t \to X_t(x)$. The α and ω limit set of x are defined as $\alpha(x) = \{y; \exists t_n \to \infty$ with $X_{-t_n}(x) \to y\}$, respectively $\omega(x) = \{y; \exists t_n \to \infty$ with $X_{t_n}(x) \to y\}$. We say that x is *recurrent* if $x \in \alpha(x) \cup \omega(x)$.

THEOREM A. *Let $X \in \mathcal{B} \cup \mathcal{B}'$. Then there exists an infinite set $L \subset T^2$ such that every point $x \in L$ is recurrent, $\omega(x) = \alpha(x) = L$. Moreover, for every $x \in T^2$ which is not contained in the unstable manifold of a saddle point one has $\alpha(x) = L$ and for every point $x \in T^2$ which is not contained in the stable manifold of a sink or of a saddle, $x \in L$.*

Remark 1. The condition that $X \in \mathcal{B} \cup \mathcal{B}'$ implies that X cannot have any sources.

Remark 2. It is not very difficult to construct many vector fields which satisfy the assumptions of Theorem A.

Remark 3. Of course Denjoy's result also holds for C^2 vector fields. Our result also holds for C^2 vector fields, but in this case we have to require that there exist C^2

linearising coordinates near every saddle-point of X, or alternatively that X is C^6 near saddle-points (see the Appendix).

The next result says that vector fields $X \in \mathscr{B}$ can be described by circle maps.

THEOREM B. *Let $X \in \mathscr{B} \cup \mathscr{B}'$. Then there exists a neighbourhood U of X such that for every $Y \in U$ one can associate a continuous order-preserving return-map $f_Y : S^1 \to S^1$ with the following properties*:
(a) *f_Y depends continuously on $Y \in U$;*
(b) *if Y, $Z \in U$ have no periodic orbits and no saddle-connections then they are equivalent if f_Y and f_Z are topologically conjugate circle maps.*

In fact, these circle maps may be constant on some intervals. Also they will satisfy some smoothness conditions and some non-flatness conditions at the boundary of the intervals where they are constant. In the next section we will study a special class $\mathscr{A}$ of circle mappings and show that if these maps have no periodic points then they are determined by a finite number of parameters, see Theorem D. In § 6 we will show that the maps f_X are contained in this class $\mathscr{A}$, provided some transition map is sufficiently smooth and non-flat. In the Appendix we will show that these conditions hold for all smooth vector fields. (These conditions would follow immediately if we assumed that X was C^2 linearizable at singularities. However, it is rather pretty to see how natural the conditions that one obtains without assuming linearizability correspond to the conditions needed in the class $\mathscr{A}$.)

2. *Statements of results for circle maps*

Let $S^1 = \mathbb{R}(\text{mod } 1)$ and $f: S^1 \to S^1$ be a continuous, order-preserving, degree one map which is possibly constant on a finite number of intervals. Since f is order preserving it has a well-defined rotation number $\rho(f) \in [0, 1)$. $\rho(f)$ is irrational if and only if f has no periodic points. It is well known that if $\rho(f)$ is irrational then f is semi-conjugate to the rotation $R_{\rho(f)}: x \to x + \rho(f) \bmod 1$. The semi-conjugacy is continuous, monotone, has degree one and maps orbits of f to orbits of $R_{\rho(f)}$.

Assume that $f: S^1 \to S^1$ is continuous, order-preserving, degree one and everywhere C^1 except possibly in a finite number of points. Let K_f be the closure of the set of points x with either
- f is not C^1 in x, or
- f is constant on some neighbourhood of x, or
- $Df(x) = 0$.

Let C_f be the closure of the set of points where f is locally constant.

We say that such a map f is in $\mathscr{A}$ if f satisfies the conditions from above, K_f has finitely many components and also the following smoothness and non-flatness conditions. For each point $x_0 \in S^1 \backslash K_f$ there exists a neighbourhood U of x_0 such that f is a local C^1 diffeomorphism on U and Df has bounded variation on U. f is constant on each component of K_f. (In particular f is everywhere C^1 except possibly in points $x \in \partial K_f$.) Moreover we require that for every point $x_0 \in \partial K_f$, f satisfies the following non-flatness condition. There exist neighbourhoods U, V of

x_0 respectively $f(x_0)$, a constant $\alpha \geq 1$, and two C^1 diffeomorphisms $\phi_r:((-1,1),0)\to(U,x)$ and $\phi_l:(f(U),f(x))\to((-1,1),0)$ such that $D\phi_l$ and $D\phi_r$ have bounded variation and $\phi_l \circ f \circ \phi_r(x)$ is equal to 0, or $\pm|x|^\alpha$.

Similarly we say that $f \in \mathscr{A}'$ if f has degree one, is everywhere strictly increasing except on one interval K_f where it is constant and $f|(S^1 - K_f)$ is C^1. Moreover at the boundary points of $K_f = [a, b]$ the (one-sided) derivative of f is infinite and there exists $\varepsilon > 0$ such that $\log Df|[a-\varepsilon, a)$ and $\log Df|(b, b+\varepsilon]$ are monotone.

Remark. If f is C^2 on $S^1\backslash K_f$, and the limits $\lim_{x\to x_0, x\notin K_f} D^2 f(x)$ are non-zero for each point $x_0 \in \partial K_f$, then $f \in \mathscr{A}$. However, we will not restrict ourselves to such maps, because, in applying our results to vector fields on the torus, we get that singularities of the vector field of saddle-type give rise to critical points of a corresponding circle map of the form $|x|^\delta$, where $\delta > 0$ and is the absolute value of the ratio of the eigenvalues at the saddle-point and need not be an integer.

We can also consider the slightly more general class of maps such that there exists neighbourhoods $V \subset \mathrm{Clos}\,(V) \subset U$ of K_f such that the maps $f|(S^1\backslash V)$, $\phi_l|U$ and $\phi_r|U$ are absolutely continuous and the maps $\log Df|(S^1\backslash V)$, $\log D\phi_l|U$ and $\log D\phi_r|U$ (which therefore exist almost everywhere) are almost everywhere equal to maps with bounded variation. The proofs of our results go through without much change; we indicate the changes needed in the remark following Proposition 3.4.

We say that I is a *wandering interval* of $f: S^1 \to S^1$ if

- $f^n(I) \cap f^m(I) = \emptyset$, $\forall n, m \geq 0$ with $n \neq m$;
- there exists no $n \geq 0$ such that $f^n(I) \subset C_f$.

THEOREM C. *Let f be in $\mathscr{A} \cup \mathscr{A}'$ with irrational rotation number. Then f has no wandering intervals.*

COROLLARY. *For every $x \in S^1\backslash\bigcup_{n\geq 0} f^{-n}(C_f)$ both the forward and the backward orbit of x is dense in $S^1\backslash\bigcup_{n\geq 0} f^{-n}(C_f)$. In particular there exists a non-finite recurrent orbit of f.*

Proof of Corollary. Since f has no wandering intervals, the semi-conjugacy h_f between f and $R_{\rho(f)}$ is constant only on preimages of C_f. Since $S^1\backslash\bigcup_{n\geq 0} f^{-n}(C_f)$ cannot be empty the Corollary follows. □

Remark 1. J. C. Yoccoz has proved Theorem C for maps f which are strictly monotone (isolated critical points are still allowed). However, if a vector field on T^2 has sinks then the corresponding return-map is constant on some interval. So the result of Yoccoz cannot be applied to vector fields. Unfortunately the proof of his result cannot be directly generalized to maps which are allowed to be constant on some intervals, see the remarks below: Definition 3.1, Lemma 3.7 and in § 5.

Remark 2. Consider order-preserving maps $f: S^1 \to S^1$ of degree one which are constant on at least one interval and which are discontinuous in at least one point. Let K_f be the closure of the set of points where f is locally constant or is discontinuous. Assume that f satisfies the same smoothness conditions as the maps from $\mathscr{A}$ (except that f is not continuous). Then Gutierrez has shown that either $\rho(f)$ is rational or some iterate of one of the components of K_f is mapped into some other component of K_f.

So the main thing which is missing in the general description of these maps is the case where $\alpha > 1$ in some points in ∂K_f and $\alpha < 1$ in other points of ∂K_f.

From this result one can easily show that the topological conjugacy class of f is determined by a finite number of invariants. Indeed, if $\rho(f)$ is irrational, then there exists an order-preserving map $h_f : S^1 \to S^1$ such that $h_f \circ f = R_\rho \circ h_f$, where $R_\rho(x) = x + \rho \pmod 1$ and $\rho = \rho(f)$, see [**He**]. For simplicity write $R = R_\rho$. Clearly h_f is constant on the components of C_f.

Take $f \in \mathcal{A} \cup \mathcal{A}'$. If $\rho(f)$ is irrational and h_f is constant on some interval I then $R^n(h_f(I))$ is disjoint for all $n \geq 0$. Hence the intervals $f^n(I)$ are mutually disjoint for all n and from Theorem C there exists $n \geq 0$ with $f^n(I) \subset C_f$. It follows that $h_f^{-1}(h_f(x))$ is a non-trivial interval if and only if $x \in \bigcup_{n \geq 0} f^{-n}(C_f)$. Let d be the metric on $S^1 = \mathbb{R} \pmod 1$ induced by $\mathbb{R}$.

THEOREM D. *Let $f, g \in \mathcal{A} \cup \mathcal{A}'$ with irrational rotation number such that the number of components of C_f is the same as that of C_g. Call this number k. Assume that*

$$\rho(f) = \rho(g) \tag{2.1}$$

and that we can order the components C_f^i and C_g^i of C_f respectively C_g cyclically on S^1 such that

$$d(c_f^i, c_f^{(i+1)}) = d(c_g^i, c_g^{(i+1)}), \quad i = 1, 2, \dots, k-1. \tag{2.2}$$

Here $c_f^i = h_f(C_f^i)$ and $c_g^i = h_g(C_g^i)$. Then f and g are conjugate, i.e. there exists a homeomorphism $h : S^1 \to S^1$ such that $h \circ f = g \circ h$.

Proof of Theorem D. We begin by taking an arbitrary order-preserving homeomorphism $k : C_f \to C_g$ such that $k(C_f^i) = C_g^i$. Then, let h_f and h_g be the order-preserving maps such that $h_f \circ f = R_\rho \circ h_f$ and $h_g \circ g = R_\rho \circ h_g$. Choosing θ so that R_θ maps $h_f(C_f^1)$ onto $h_g(C_g^1)$, one gets from (2.2) that $R_\theta(h_f(C_f)) = h_g(C_g)$. Moreover taking $\tilde{h}_f = R_\theta \circ h_f$, the following diagram commutes

$$\begin{array}{ccc} S^1 & \xrightarrow{\ f\ } & S^1 \\ {\scriptstyle \tilde{h}_f}\big\downarrow & & \big\downarrow{\scriptstyle \tilde{h}_f} \\ S^1 & \xrightarrow{\ R_\rho\ } & S^1 \\ {\scriptstyle h_g}\big\uparrow & & \big\uparrow{\scriptstyle h_g} \\ S^1 & \xrightarrow{\ g\ } & S^1 \end{array}$$

and that $\tilde{h}_f(C_f^i) = C_g^i$. So

$$h(x) = \begin{cases} h_g^{-1} \circ \tilde{h}_f(x), & \text{for } x \notin \bigcup_{n \geq 0} f^{-n}(C_f) \\ k(x), & \text{for } x \in C_f, \end{cases}$$

defines a strictly order-preserving homeomorphism from $S^1\backslash\bigcup_{n\geq 1} f^{-n}(C_f)$ to $S^1\backslash\bigcup_{n\geq 1} g^{-n}(C_g)$ such that $h\circ f = g\circ h$. Finally, for $x\in\bigcup_{n\geq 1} f^{-n}(C_f)$, let $n\geq 1$ be the minimal number such that $f^n(x)\in C_f$ and define $h(x)=g^{-n}\circ h\circ f^n$. In this way we have defined a homeomorphism h such that $h\circ f = g\circ h$. □

In the Proof of Theorem C, we will need a simple Proposition which produces collections of disjoint intervals. Since it does not fit logically in the next two sections we state and prove it here.

PROPOSITION 2.1. *Let* $\{W_i | i=1,\ldots,m\}$ *be a collection of intervals in* S^1 *with the following two properties.*

(i) *every point in* S^1 *is in at most three elements of the collection;*

(ii) $W_i\subset W_k$ *implies* $i=k$.

Then there exists a partition of the collection

$$\{W_i;\ i=0,1,\ldots,m\} = A_1\cup A_2\cup\cdots\cup A_5,$$

such that A_k consists of mutually disjoint intervals for each $k=1,2,\ldots,5$.

Proof. We prove this Proposition inductively for each $m\geq 1$. The Proposition is trivially true for $m=1$. Suppose that it is proved inductively for $m-1$. Then we can write

$$\{W_i;\ i=1,2,\ldots,m-1\} = \bigcup_{j=1,2,\ldots,5} A_j'$$

where A_j' consists of mutually disjoint intervals. From property (ii) it follows that $\partial W_m\cap W_i\neq\emptyset$ when $W_m\cap W_i\neq\emptyset$. So assumption (i) implies that there exists at most four is with $1\leq i<m$ such that W_m has a non-empty intersection with W_i. Hence there exists $j'\in\{1,2,\ldots,5\}$ such that W_m does not intersect any interval from $A_{j'}'$. So by letting $A_j=A_j'$ for $j\in\{1,2,\ldots,5\}$, $j\neq j'$ and $A_{j'}=A_{j'}'\cup\{W_m\}$ we get the required partition of $\{W_i;\ i=1,2,\ldots,m\}$. This proves the Proposition by induction. □

The organization of the paper is as follows. In the next three sections we will prove Theorem C. In § 6 we will prove Theorems A and B, except a technical Lemma which is proved in an Appendix.

It will be convenient to introduce the following notation. For an interval I let $|I|$ denote the length of I. Moreover let d be the metric on $S^1=\mathbb{R}\ (\mathrm{mod}\ 1)$ induced by $\mathbb{R}$.

3. *Some analytic tools*

In this section we will develop some analytic tools which give estimates on the shape of non-linearity of f^n on intervals T such that $T, f(T),\ldots,f^n(T)$ are disjoint.

Definition 3.1. Let $g: S^1\to S^1$ be a continuous order-preserving map and I, T intervals in S^1 with $\mathrm{Clos}(I)\subset\mathrm{int}(T)$. Then

$$D(T,I)=\frac{|T|\cdot|I|}{|L|\cdot|R|},\ \text{where } L \text{ and } R \text{ are the components of } T\backslash I;$$

$$B(g,T,I)=\frac{D(g(T),g(I))}{D(T,I)}.$$

Remark. J. C. Yoccoz uses the operator $B_0(g, T) = \lim_{I \to T} B(g, T, I)$.

Definition 3.2. Let $g: T \to S^1$ be a C^3-map on the interval $T \subset S^1$. Then the operator

$$Sg(x) = \frac{g'''(x)}{g'(x)} - \frac{3}{2}\left(\frac{g''(x)}{g'(x)}\right)^2$$

is called the Schwarzian derivative.

It is easy to see that the operator B is multiplicative:

$$B(g \circ h, T, I) = B(g, h(T), h(I)) \times B(h, T, I). \tag{3.1}$$

The proof of the following proposition can be found in [**MS**].

PROPOSITION 3.3. *Let $g: T \to T$ be a C^3 diffeomorphism on the open interval T such that for every $x \in T$, $Sg(x) < 0$. Let $\mathrm{Clos}(I) \subset \mathrm{int}(T)$. Then*

$$B(g, T, I) > 1. \tag{3.2}$$

Let $\mathcal{V} \subset \mathrm{Clos}(\mathcal{V}) \subset \mathrm{int}\,(\mathcal{U})$ be interval neighbourhoods of K_f such that the number of components of $\mathcal{V}$ and $\mathcal{U}$ are both the same as $\# K_f$.

PROPOSITION 3.4. *Let $f \in \mathcal{A}$. Then there exists $V < \infty$ with the following property. Let $\Xi = \{T_1, T_2, \ldots, T_n\}$ be a collection consisting of n intervals in S^1 with the following properties. Every point of S^1 is contained in at most three intervals T_i, $T_i \subset T_j$ implies $i = j$ and none of the intervals T_i contains a component of $\mathcal{U} \backslash \mathcal{V}$ or contains points of K_f. Then*

$$\sum_{i=1}^{n} \log B(f, T_i, I_i) \geq -5 \cdot V. \tag{3.3}$$

Proof. Let $V_f = \mathrm{Var}\,(\log Df \,|\, (S^1 \backslash \mathcal{V}))$. Since on $S^1 \backslash \mathcal{V}$ the map f is a C^1 diffeomorphism and Df has bounded variation, $V_f < \infty$. Moreover on each component $\mathcal{U}_i$ of $\mathcal{U}$, f has the form $f(x) = \phi_{l,i} \circ \phi_\alpha \circ \phi_{r,i}$, where $\phi_\alpha(x) = \pm |x|^\alpha$ as above. Here $\phi_{l,i}$ and $\phi_{r,i}$ are C^1 diffeomorphisms such that $\log D\phi_{l,i}$ and $\log D\phi_{r,i}$ have bounded variations on $\mathcal{U}_i$. Hence $V_l = \sum_i \mathrm{Var}\,(\log(D\phi_{l,i}))$ and $V_r = \sum_i \mathrm{Var}\,(\log(D\phi_{r,i}))$ are both finite. Let $V = V_f + V_l + V_r$.

Let $I_1 = \{i;\ T_i \cap \mathcal{V} = \emptyset\}$ and $I_2 = \{i;\ T_i \subset \mathcal{U}\}$. Since the intervals T_i never contain a component of $\mathcal{U} \backslash \mathcal{V}$, $I_1 \cup I_2 = \{1, 2, \ldots, n\}$.

First assume that $i \in I_1$. Let L_i and R_i be the components of $T_i \backslash I_i$. For $u_i, v_i \in T_i$ let (u_i, v_i) be the open interval connecting u_i and v_i. Because f is C^1 on $S^1 \backslash \mathcal{V}$ the mean-value theorem is valid on T_i. Using this in the definition of $B(f, T_i, I_i)$ we find that there exist $l_i \in L_i$, $r_i \in R_i$, $m_i \in I_i$ and $\tau_i \in T_i$ such that

$$\log B(f, T_i, I_i) = \log\left(\frac{Df(m_i)\, Df(\tau_i)}{Df(l_i)\, Df(r_i)}\right), \tag{3.4}$$

and

$$m_i \in (l_i, r_i). \tag{3.5}$$

From (3.4) one has

$$|\log B(f, T_i, I_i)| \le |\log Df(m_i) - \log Df(l_i)| + |\log Df(\tau_i) - \log Df(r_i)| \qquad (3.6)$$

and also

$$|\log B(f, T_i, I_i)| \le |\log Df(m_i) - \log Df(r_i)| + |\log Df(\tau_i) - \log Df(l_i)|. \qquad (3.7)$$

Rename the points l_i, m_i, r_i, τ_i in increasing order a_i, b_i, c_i, d_i. From (3.5) one gets that either $(l_i, m_i) \cap (\tau_i, r_i) = \emptyset$ or $(\tau_i, l_i) \cap (m_i, r_i) = \emptyset$, and so we can use either (3.6) or (3.7) and get

$$|\log B(f, T_i, I_i)| \le |\log Df(b_i) - \log Df(a_i)| + |\log Df(d_i) - \log Df(c_i)|$$
$$\le \mathrm{Var}\,(\log Df \,|\, T_i)$$

and therefore

$$\log B(f, T_i, I_i) \ge -\mathrm{Var}\,(\log Df \,|\, T_i). \qquad (3.8)$$

Now consider $i \in I_2$. Then T_i is contained in some component $\mathcal{U}_i$ of $\mathcal{U}$ (and does not intersect K_f) and so f has the form $f(x) = \phi_{l,i} \circ \phi_\alpha \circ \phi_{r,i}$. Hence

$$B(f, T_i, I_i) = B(\phi_{l,i}, T_i'', I_i'') \times B(\phi_\alpha, T_i', I_i') \times B(\phi_{r,i}, T_i, I_i).$$

Here $T_i'' = \phi_\alpha \circ \phi_{r,i}(T_i)$, $T_i' = \phi_{r,i}(T_i)$, $I_i'' = \phi_\alpha \circ \phi_{r,i}(I_i)$ and $I_i' = \phi_{r,i}(I_i)$. Since the Schwarzian derivative of ϕ_α is less or equal to 0 (because $\alpha \ge 1$) one gets $B(\phi_\alpha, T_i', I_i') \ge 1$. Hence, as above,

$$\log B(f, T_i, I_i) \ge \log B(\phi_{l,i}, T_i'', I_i'') + 0 + \log B(\phi_{r,i}, T_i, I_i)$$
$$\ge -\{\mathrm{Var}\,(\log D\phi_{l,i} | T_i'') + \mathrm{Var}\,(\log D\phi_{r,i} | T_i)\}. \qquad (3.9)$$

Since every point of S^1 is contained in at most three intervals, using Proposition 2.1, one can write $\{T_i ; i = 1, 2, \dots, n\} = A_1 \cup A_2 \cdots \cup A_5$ where A_j consists of a collection of mutually disjoint intervals. Therefore also the collection of intervals T_i'' corresponding to $T_i \in A_j$ consists of disjoint intervals. Hence from (3.8) and (3.9) one gets

$$\sum_{i=1}^{n} \log B(f, T_i, I_i) \ge -5 \cdot (V_f + V_l + V_r).$$

The Proposition follows. □

Remark. In eq. (3.4) in this Proposition we have used that Df exists and has bounded variation on $S^1 \backslash \mathcal{V}$. Of course it would have been sufficient to assume that f is absolutely continuous, $\log Df | (S^1 \backslash \mathcal{V})$ is almost everywhere equal to a map with bounded variation. In this case we could have still found $m_i \in I_i$, $l_i \in L_i$, $r_i \in R_i$ and $\tau_i \in T_i$ such that Df exists in these points and such that

$$\frac{|f(I_i)|}{|I_i|} \le Df(m_i), \quad \frac{|f(T_i)|}{|T_i|} \le Df(\tau_i), \quad \frac{|f(L_i)|}{|L_i|} \ge Df(l_i) \quad \text{and} \quad \frac{|f(R_i)|}{|R_i|} \ge Df(r_i).$$

Hence (3.6), (3.7) and (3.8) would still hold. Similarly it suffices to assume that ϕ_l, ϕ_r are absolutely continuous and $\log D\phi_l$ and $\log D\phi_r$ are almost everywhere equal to maps with bounded variation.

LEMMA 3.5. *Let $f \in \mathcal{A}$. Then there exists $A_0 > 0$ such that if $I \subset W$ are intervals in S^1 such that*

(a) $W \setminus I$ *consists of one component* H;

(b) $|H| \leq |I|$;

(c) $I \cap K_f = \emptyset$,

then

$$\frac{|f(I)|}{|f(H)|} \geq A_0 \cdot \frac{|I|}{|H|}. \tag{3.10}$$

Proof. Let us prove this Proposition by contradiction. So suppose that there exists a sequence of intervals I_n, W_n, $H_n = W_n \setminus I_n$ satisfying (a), (b) and (c) above such that

$$\lim_{n \to \infty} \frac{|f(I_n)|/|I_n|}{|f(H_n)|/|H_n|} \to 0. \tag{3.11}$$

Since $|H_n| \leq |I_n|$ this implies that $|W_n| \to 0$ and $W_n \to K_f$. By taking a subsequence we may assume that $W_n \to p \in K_f$.

Since $f \in \mathcal{A}$ we may assume that p is a critical point and near p, f is of the form $f = \phi_l \circ \phi \circ \phi_r$ where ϕ_α is equal to x^α or 0 on $x \geq 0$ and to 0 or $-|x|^\alpha$ on $x \leq 0$ as before. Hence

$$\frac{|f(I_n)|/|I_n|}{|f(H_n)|/|H_n|} \geq \frac{\min D\phi_l}{\sup D\phi_l} \times \frac{\min D\phi_r}{\sup D\phi_r} \times \frac{|\phi_\alpha(I'_n)|/|I'_n|}{|\phi_\alpha(H'_n)|/|H'_n|}. \tag{3.12}$$

Here $H'_n = \phi_r(H_n)$ and $I'_n = \phi_r(I_n)$. Since ϕ_r is a C^1 diffeomorphism, writing $K_n = |H'_n|/|I'_n|$ we get $\limsup K_n < \infty$ and therefore there exists $K \in (0, \infty)$ such that $K_n \leq K$, $\forall n \geq 0$. Let us distinguish between the following three possibilities.

Case 1. I_n is between p and H_n.

Writing $I'_n = [a_n, b_n]$, $H'_n = [b_n, c_n] \subset (0, 1)$ one has

$$\frac{|\phi_\alpha(I'_n)|/|I'_n|}{|\phi_\alpha(H'_n)|/|H'_n|} = \frac{\left[\dfrac{b_n^\alpha - a_n^\alpha}{b_n - a_n}\right]}{\left[\dfrac{c_n^\alpha - b_n^\alpha}{c_n - b_n}\right]} \geq \frac{\left[\dfrac{b_n^\alpha}{b_n}\right]}{\left[\dfrac{c_n^\alpha - b_n^\alpha}{c_n - b_n}\right]} \geq \frac{\left[\dfrac{b_n^\alpha}{b_n}\right]}{\left[\dfrac{((1+K)b_n)^\alpha - (b_n)^\alpha}{(1+K)b_n - b_n}\right]}$$

$$= \frac{K}{(1+K)^\alpha - 1} > 0. \tag{3.13}$$

Here, the second inequality follows because $\alpha \geq 1$ implies that the function $x \to Dx^\alpha = \alpha \cdot x^{\alpha-1}$ is increasing. The third inequality follows from this and because $|H'_n| \leq K'_n \cdot |I_n|$ gives

$$c_n \leq b_n + K_n \cdot (b_n - a_n) \leq b_n + K_n \cdot b_n \leq (1+K) \cdot b_n.$$

Using (3.13) in (3.12) gives a contradiction in (3.11).

Case 2. H_n is between p and I_n, but $H_n \cap K_f = \emptyset$.

Since the derivative of ϕ_α is increasing we have in this case

$$\frac{|\phi_\alpha(I'_n)/|I'_n|}{|\phi_\alpha(H'_n)|/|H'_n|} \geq 1. \tag{3.14}$$

Using (3.14) in (3.12) we again get a contradiction with (3.11).

Case 3. H_n is between p and I_n, but $H_n \cap K_f \neq \emptyset$.

If H_n intersects an interval component of K_f, then we can choose the maximal interval $\tilde{W}_n$ with $I_n \subset \tilde{W}_n \subset W_n$ such that $\tilde{W}_n \cap K_f = \emptyset$. Let $\tilde{H}_n = \tilde{W}_n \backslash I_n$ and $\tilde{H}'_n = \phi_r(H'_n)$. By construction $|\tilde{H}'_n| \leq |H'_n|$ and $\phi_\alpha(\tilde{H}'_n) = \phi_\alpha(H'_n)$, and therefore

$$\frac{|\phi_\alpha(I'_n)|/|I'_n|}{|\phi_\alpha(H'_n)|/|H'_n|} \geq \frac{|\phi_\alpha(I'_n)|/|I'_n|}{|\phi_\alpha(\tilde{H}'_n)|/|\tilde{H}'_n|}.$$

As in case 2 this last term is bounded away from zero and we get a contradiction as before.

So we may assume that $H_n \cap K_f$ is an isolated point. So writing $H'_n = [-a_n, b_n]$, $I'_n = [b_n, c_n]$, with $a_n, b_n, c_n > 0$ one has

$$\frac{|\phi_\alpha(I'_n)|/|I'_n|}{|\phi_\alpha(H'_n)/|H'_n|} = \frac{\left[\dfrac{c_n^\alpha - b_n^\alpha}{c_n - b_n}\right]}{\left[\dfrac{b_n^\alpha + a_n^\alpha}{b_n + a_n}\right]} \geq \frac{\left[\dfrac{c_n^\alpha - b_n^\alpha}{c_n - b_n}\right]}{\left[\dfrac{b_n^\alpha}{b_n} + \dfrac{a_n^\alpha}{a_n}\right]} \geq \frac{\left[\dfrac{c_n^\alpha}{c_n}\right]}{\left[\dfrac{c_n^\alpha}{c_n} + \dfrac{a_n^\alpha}{a_n}\right]} \geq \frac{1}{(1+K)^\alpha}. \tag{3.15}$$

Here the first inequality follows from

$$\frac{x+u}{y+v} \leq \frac{x}{y} + \frac{u}{v} \quad \text{for all } x, y, u, v > 0.$$

The last three inequalities follow from the concavity of x^α and because $b_n, a_n \leq |H'_n| \leq K \cdot |I'_n| \leq K \cdot c_n$. So using (3.15) in (3.12) we get again a contradiction with (3.11).

Since cases 1, 2 and 3 cover all possibilities we have proved the Proposition by contradiction. □

PROPOSITION 3.6. *Let $f \in \mathscr{A}$ and I, T be intervals in S^1 with* Clos $(I) \subset$ int (T) *and L and R be the components of $T \backslash I$. Let $\gamma \in (0, 1)$. If*

$$\frac{|f(I)|}{|f(R)|} \geq \gamma \cdot \frac{|I|}{|R|} \quad \textit{and} \quad \frac{|f(I)|}{|f(L)|} \geq \gamma \cdot \frac{|I|}{|L|},$$

then

$$B(f, T, I) \geq \gamma^2.$$

Proof.

$$B(f, T, I) = \frac{\dfrac{|f(I)|}{|f(R)|} + \dfrac{|f(I)|}{|f(R)|}\dfrac{|f(I)|}{|f(L)|} + \dfrac{|f(I)|}{|f(L)|}}{\dfrac{|I|}{|R|} + \dfrac{|I|}{|R|}\dfrac{|I|}{|L|} + \dfrac{|I|}{|L|}} \geq \frac{\gamma + \gamma^2 \cdot d}{1+d} = \gamma \cdot \frac{1 + \gamma \cdot d}{1+d} \geq \gamma^2.$$

Here

$$d = \frac{\dfrac{|I||I|}{|R||L|}}{\dfrac{|I|}{|R|} + \dfrac{|I|}{|L|}}$$

and the last inequality follows since $\gamma \in (0,1)$ and therefore $(0,\infty)\ni x \to (1+\gamma x)/(1+x)$ is always greater than γ. The result follows. □

PROPOSITION 3.7. *Let $f \in \mathcal{A}$. Then there exists $A_1 > 0$ such that if I, T are intervals in S^1 with* $\mathrm{Clos}(I) \subset \mathrm{int}(T)$ *and L and R the components of $T \backslash I$ such that*
(a) $|L| \le |I|$ *or* $|R| \le |I|$;
(b) $I \cap K_f = \emptyset$;
then

$$B(f, T, I) \ge A_1. \tag{3.16}$$

Proof. Let A_0 be the number from Proposition 3.5. We may assume that $A_0 \in (0,1)$. We will prove the Proposition for $A_1 = \frac{1}{3} \cdot (A_0)^2$. By possibly renaming L and R, we may consider the case that $|R| \le |I|$. Then from Proposition 3.5 we get

$$\frac{|f(I)|/|I|}{|f(R)|/|R|} \ge A_0, \tag{3.17}$$

and hence

$$B(f, T, I) \ge \frac{|f(T)|}{|f(L)|} \cdot \frac{|L|}{|T|} \cdot A_0 \ge A_0 \cdot \frac{|L|}{|T|}.$$

If $|L| \ge |I|$ then it follows from this and $|R| \le |I|$ that

$$B(f, T, I) \ge A_0 \cdot \frac{|L|}{|L|+|I|+|R|} \ge A_0 \cdot \frac{|I|}{2|I|+|R|} \ge A_0 \cdot \tfrac{1}{3} \ge A_1$$

and the Proposition is proved. So assume that $|L| \le |I|$. Then applying Proposition 3.5 again we get

$$\frac{|f(I)|/|I|}{|f(L)|/|L|} \ge A_0, \tag{3.18}$$

and it follows from (3.17), (3.18) and Proposition 3.6 that $B(f, T, I) \ge A_0 \cdot A_0 = A_1$. □

Remark. In general the interval T we need to consider may intersect or even overlap a component of K_f. It is easy to check that there exists no universal lower bound for $B_0(f, T)$ (as in (3.16)) in this case. Here B_0 is the operator of Yoccoz mentioned in the remark below Definition 3.1. Therefore we have to replace the operator B_0 by B.

Let $\mathcal{V} \subset \mathcal{U}$ be the neighbourhoods of K_f from above.

THEOREM 3.8. *Let $f \in \mathcal{A}$. Then there exists $\varepsilon > 0$ and B_0 such that for any intervals* $I \subset \mathrm{Clos}(I) \subset \mathrm{int}(T)$ *in S^1, and any $n \ge 0$, satisfying the following conditions*
(a) $|L| < \varepsilon \cdot |I|$ *and* $|R| < \varepsilon \cdot |I|$; *(Here L and R are the component of $T \backslash I$.)*
(b) *every point in S^1 is contained in at most three members of the sequence T, $f(T), \ldots, f^{n-1}(T)$;*
(c) $f^i(I) \cap K_f = \emptyset$, $i \in \{0, 1, \ldots, n-1\}$ *then*

$$B(f^n, T, I) \ge B_0. \tag{3.19}$$

Proof. Let A_1 be the number from Proposition 3.7 and assume that $A_1 < 1$ and $e^{-5V} < 1$. Let V be the number from Proposition 3.3. We may assume that $e^{-5V} < 1$.

Let $B_0=\{A_1\, e^{-5V}\}^{12\#K_f+1}$ and $\varepsilon=\sqrt{B_0/3}$. Let $0\le t_1<t_2<\cdots<t_s\le n-1$ be integers such that either $f^i(T)$ contains a component of $\mathcal{U}\setminus\mathcal{V}$, or $f^i(T)\cap K_f\neq\varnothing$. From the disjointness property of the orbit of T we get $s\le 12\#K_f$. From the definition of t_1 one has that $f^i(T)$ does not contain a component of $\mathcal{U}\setminus\mathcal{V}$ for $i<t_1$ and because f does not have periodic points we have that $f^i(T)\subset f^j(T)$ implies $i=j$. Using this and the disjointness property of $f^i(T)$, Proposition 3.4 gives

$$B(f^{t_1},T,I)\ge e^{-5V}.$$

So

$$\frac{|f^{t_1}(T)||f^{t_1}(I)|}{|f^{t_1}(L)||f^{t_1}(R)|}\ge e^{-5V}\cdot\frac{|T||I|}{|L||R|}\ge B_0\cdot\frac{|T||I|}{|L||R|}.$$

This implies

$$\frac{|f^{t_1}(I)|}{|f^{t_1}(R)|}+\frac{|f^{t_1}(I)|}{|f^{t_1}(R)|}\frac{|f^{t_1}(I)|}{|f^{t_1}(L)|}+\frac{|f^{t_1}(I)}{|f^{t_1}(L)|}\ge B_0\cdot\frac{1}{\varepsilon^2}=3.$$

So $|f^{t_1}(R)|\le|f^{t_1}(I)|$ or $|f^{t_1}(L)|\le|f^{t_1}(I)|$. Hence we can apply Proposition 3.7 and get $B(f,f^{t_1}(T),f^{t_1}(I))\ge A_1$ and hence

$$B(f^{t_1+1},T,I)\ge B(f,f^{t_1}(T),f^{t_1}(I))\times B(f^{t_1},T,I)\ge e^{-5V}\cdot A_1. \tag{3.20}$$

In the same way we get

$$B(f^{t_2-(t_1+1)},f^{t_1+1}(T),f^{t_1+1}(I))\ge e^{-5V},$$

and hence, using (3.20),

$$B(f^{t_2},T,I)\ge[e^{-5V}]^2\cdot A_1\ge B_0.$$

From this we get as before $|f^{t_2}(R)|\le|f^{t_2}(I)|$ or $|f^{t_2}(L)|\le|f^{t_2}(I)|$. Hence we can apply Proposition 3.7 again and get

$$B(f^{t_2+1},T,I)\ge[e^{-5V}\cdot A_1]^2\ge B_0.$$

This procedure has to be repeated at most $12\cdot\#K_f$ times, i.e., $s\le 12\cdot\#K_f$. Since $[e^{-5V}\cdot A_1]^{12\cdot\#K_f}\ge B_0$ we get by induction that for each $k\le s$, $|f^{t_k}(R)|\le|f^{t_k}(I)|$ or $|f^{t_k}(L)|\le|f^{t_k}(I)|$ and hence

$$B(f^{t_s+1},T,I)\ge[e^{-5V}\cdot A_1]^{12\cdot\#K_f}.$$

Using Proposition 3.4 we can handle the last piece from t_s+1 to n and get

$$B(f^{t_s+1},T,I)\ge[e^{-5V}\cdot A_1]^{12\cdot\#K_f}\cdot e^{-5V}.$$ □

4. *The topological situation*

Throughout this section we will assume that f has a wandering interval I. We may assume that I is a maximal wandering interval, i.e. not contained in a bigger wandering interval. Since the forward orbit of I is disjoint and K_f has at most a finite number of components, by possibly replacing I by some iterate of I we can assume that $f^i(I)\cap K_f=\varnothing$. We want to show that there exist iterates q_n and neighbourhoods T_n of I such that f^{q_n} is contracting on I but strongly expanding on $T_n\setminus I$. In order to obtain a contradiction from this we need to show we can control the non-linearity of $f^{q_n}|T_n$. From the previous section it follows that this can be done if the intervals $T_n, f(T_n), \ldots, f^{q_n}(T_n)$ are more or less disjoint. This disjointness will follow from the fact that f preserves ordering on S^1. The proofs of the results in this section can be found in section V of **[He]**.

In fact, because f is order-preserving and has no periodic points, there exists an order-preserving and continuous map $h: S^1 \to S^1$ such that $h \circ f = R_\rho \circ h$, where $\rho = \rho(f)$ and R_ρ is the rigid rotation over ρ (see **[He]**). In particular, the order of points in an orbit of f is the same as the order of points in an orbit of R_ρ. Since we keep ρ fixed in this section, we write $R = R_\rho$. Take some $x \in S^1$ and inductively define $q(n)$ to be the time of the nth closest approach:

$$q(0) = 1;$$

$$q(n+1) = \min\{i \in \mathbb{N};\ i > q(n),\ d(x, R^i(x)) < d(x, R^{q(n)}(x))\}.$$

Remark. These numbers $q(n)$ are independent of x (they only depend on $\rho(f)$).

PROPOSITION 4.1. $d(I, f^{q(n)}(I)) \to 0$ as $|n| \to \infty$;

In the next two Propositions we will describe the way the sequence $f^{q(n)}(I)$ is ordered in S^1.

PROPOSITION 4.2. *The intervals* $\{f^{q(2n)}(I);\ n \geq 0\}$ *approach* I *from the right and* $\{f^{q(2n+1)}(I);\ n \geq 0\}$ *approach* I *from the left (or vice versa).*

From Propositions 4.1 and 4.2 we know that the sequence $\{f^{q(2n)}(I)\}$, $n \geq 0$, accumulates arbitrarily close to one side of I. In the next Proposition the order of the intervals $\{f^i(I);\ q(2n) \leq i \leq q(2n+2)\}$ is described.

Let $Q(2n)$ be the set of integers $t \in \mathbb{N}$, such that $q(2n) \leq t \leq q(2n+2)$ and such that $f^t(I)$ is between $f^{q(2n+2)}(I)$ and $f^{q(2n)}(I)$.

PROPOSITION 4.3. $t \in Q(2n)$ *if and only if there exists* $i \in \mathbb{N}$ *such that* t *is of the form* $t = q(2n) + iq(2n+1)$ *and* $q(2n) \leq t \leq q(2n+2)$. *In particular, there exists* $a(2n+2) \in \mathbb{N}$ *such that* $q(2n+2) = q(2n) + a(2n+2) \cdot q(2n+1)$.

PROPOSITION 4.4. *Let* T_n *be the smallest interval containing* $f^{-q(n)}(I)$, I *and* $f^{q(n)}(I)$. *Then every point in* S^1 *is an at most three intervals of the sequence*

$$T_n, f(T_n), \ldots, f^{q(n+1)-1}(T_n).$$

5. *Conclusion of the proof of Theorem C for maps* $f \in \mathcal{A}$

PROPOSITION. 5.1. *Let* $f \in \mathcal{A}$ *and assume that* I *is a wandering interval of* f. *There exists* $n_0 < \infty$ *such that if* $n > n_0$ *then*

$$|f^{q(2n)+a(2n+2)\cdot q(2n+1)}(I)| = |f^{q(2n+2)}(I)| > |f^{q(2n+2)-q(2n+1)}(I)|. \tag{5.1}$$

Proof. Let T_n be the smallest interval containing $f^{-q(2n+2)}(I)$, I and $f^{-q(2n+1)}(I)$. Since $T_n \subset [f^{q(2n+1)}(I), f^{-q(2n+1)}(I)]$ one has from Proposition 4.4 that each point of S^1 is contained in at most three of the intervals T_n, $f(T_n)$, $f^2(T_n), \ldots, f^{q(2n+2)-1}(T_n)$. Moreover from Proposition 4.1 the length of the two components L_n and R_n of $T_n \backslash I$ tends to zero as $n \to \infty$. Hence we can apply Theorem

3.8 and get that there exist $B_0>0$ and $n_0<\infty$ such that for all $n\ge n_0$, $B(f^{q(2n+2)}, T_n, I)\ge B_0$. Since $f^{q(2n+2)}(R_n)\supset f^{q(2n+2)-q(2n+1)}(I)$ one gets

$$\frac{|f^{q(2n+2)-q(2n+1)}(I)|}{|f^{q(2n+2)}(I)|}\le\frac{|f^{q(2n+2)}(R_n)|}{|f^{q(2n+2)}(I)|}\le\frac{1}{B_0\cdot|I|^2}|L_n||R_n|\frac{|f^{q(2n+2)}(T_n)|}{|f^{q(2n+2)}(L_n)|}\le\frac{|L_n||R_n|}{B_0\cdot|I|^3}.$$

For large n the last term tends to 0 and since $q(2n+2)=q(2n)+a(2n+2)\cdot q(2n+1)$, the Proposition follows. □

PROPOSITION 5.2. *Let $f\in\mathscr{A}$ and I be a wandering interval of f. Then there exists sequences $\{n_k\}$ and $\{i_{n_k}\}$ with $0<i_{n_k}<a(2n_k+2)$ such that for $i=i_{n_k}$,*

$$|f^{q(2n_k)+i\cdot q(2n_k+1)}(I)|<|f^{q(2n_k)+(i-1)\cdot q(2n_k+1)}(I)|, |f^{q(2n_k)+(i+1)\cdot q(2n_k+1)}(I)|. \tag{5.2}$$

(In particular, if $f\in\mathscr{A}$ has a wandering interval then there are infinitely many integers $n\ge 0$ such that $a(2n)>1$.)

Proof. For $n\ge n_0$ (see Proposition 5.1) we have the following. Suppose that for some $n>n_0$ we cannot find an integer $0<i<a(2n+2)$ such that (5.2) is satisfied. Then eq. (5.1) implies that for all $0<i\le a(2n+2)$ one has

$$|f^{q(2n)+i\cdot q(2n+1)}(I)|>|f^{q(2n)+(i-1)\cdot q(2n+1)}(I)|.$$

Hence, because $q(2n+2)=q(2n)+a(2n+2)\cdot q(2n+1)$,

$$|f^{q(2n+2)}(I)|>|f^{q(2n)}(I)|. \tag{5.3}$$

But since $f^i(I)$ are mutually disjoint, $\sum_i |f^i(I)|\le 1$. So there must exist a sequence $n_k\to\infty$ such that (5.3) does not hold and for each of these n_k there exists $0<i<a(2n_k+2)$ such that (5.2) holds. □

Now we can finish the proof of Theorem C.

The proof of Theorem C for maps $f\in\mathscr{A}$. Let n_k and i_{n_k} be as in Proposition 5.2, $q(n)=q(2n_k)+i_{n_k}\cdot q(2n_k+1)$ and K_n the smallest interval containing $f^{q(2n_k+1)}(I)$, I and $f^{-q(2n_k+1)}(I)$. As before let L_n, R_n be the components of $K_n\setminus I$. Proposition 4.4 guarantees the disjointness properties needed in order to apply Theorem 3.8 for $f^{q(n)}|K_n$. From eq. (5.2) and since $B(f^{q(n)}, K_n, I)\ge B_0$,

$$3\cdot\frac{|L_n||R_n|}{|I|^2}\ge\frac{|L_n||R_n|}{|I|^2}\cdot\left[\frac{|f^{q(n)}(I)|}{|f^{q(n)}(R_n)|}+\frac{|f^{q(n)}(I)|}{|f^{q(n)}(R_n)|}\frac{|f^{q(n)}(I)|}{|f^{q(n)}(L_n)|}+\frac{|f^{q(n)}(I)|}{|f^{q(n)}(L_n)|}\right]$$
$$\ge B(f^{q(n)}, K_n, I)\ge B_0.$$

Since $|L_n|$, $|R_n|\to 0$ as $k\to\infty$ we get a contradiction. □

Remark. The difference with the case Yoccoz considers (where there are no intervals where f is constant) is that he can use the operator B_0. This allows him to choose L_n and R_n to be intervals in the backward orbit of I. In this way he can compare $|f^{q(n)}(L_n)|$ and $|R_n|$. This makes his proof easier.

6. *The proof of Theorem C for maps in $\mathscr{A}'$*

The result in this section is a version of a result from unpublished work of C. Gutierrez, A. Lins and W. de Melo.

Assume that $f\in\mathscr{A}'$. This means that f has degree one, is everywhere increasing except on one interval K_f where it is constant and $f|(S^1-K_f)$ is C^1. Moreover outside every neighbourhood of K_f is $\log Df$ of bounded variation, at the boundary points of $K_f = [a, b]$ the (one-sided) derivative of f is infinite and there exists $\varepsilon > 0$ such that $\log Df|[a-\varepsilon, a)$ and $\log Df|(b, b+\varepsilon]$ are monotone.

Let $\Lambda = S^1 \setminus \bigcup_{i \geq 0} f^{-i}(K_f)$.

THEOREM 6.1. *Let $f \in \mathscr{A}'$. Assume that the rotation number of f is irrational. Then there exists $N \in \mathbb{N}$ such that for all $x \in \Lambda$,*

$$Df^N(x) > 1.$$

This Theorem implies Theorem C for maps in $\mathscr{A}'$:

COROLLARY. *$f \in \mathscr{A}'$ (with irrational rotation number) cannot have wandering intervals.*

Proof. This immediately follows from the expansion of $f^N|\Lambda$ and the disjointness of wandering intervals. □

The most important step in proving Theorem 6.1 is the following Proposition.

PROPOSITION 6.2. *Let $f \in \mathscr{A}'$. Assume that the rotation number of f is irrational. Then for each $x \in \Lambda$ there exists $N \in \mathbb{N}$ such that*

$$Df^N(x) > 1.$$

Proof. Choose $x \in \Lambda$, i.e., $f^i(x) \notin K_f$ for all $i \geq 0$. Let $\alpha = -\rho(f)$. We first consider the rotation $R_\alpha : S^1 \to S^1$. From **[He]** (V.8.4) we get that for each $y \in S^1$ and each $n \in \mathbb{N}$,

$$\bigcup_{j=0}^{q_{n+1}-1} R_\alpha^j([y, R_\alpha^{q_n}(y)]) \cup \bigcup_{j=0}^{q_n - 1} R_\alpha^j([y, R_\alpha^{q_{n+1}}(y)]) = S^1, \tag{6.1}$$

where the interiors of the above intervals are pairwise disjoint. Using the semi-conjugacy between f^{-1} and R_α and applying (6.1) to f^{-1} we get that x is contained in one of the two unions corresponding to those in (6.1). In particular for all $k \in \mathbb{N}$, there exist $n \geq k$ and $j \in \mathbb{N}$ such that

$$x \in f^{-j}(U), \quad \{\text{int}(U), \text{int}(f(U)), \ldots, \text{int}(f^{-j}(U))\} \text{ are pairwise disjoint}, \tag{6.2}$$

where U is the smallest interval containing K_f and $f^{-q_n}(K_f)$. Because $x \notin \bigcup_{i \geq 0} f^{-i}(K_f)$ one has (for fixed x) that $j \to \infty$ as $k \to \infty$. Let $\varepsilon > 0$ be the number from the definition of the class of maps $\mathscr{A}'$ and let U be a ε neighbourhood of K_f. Let

$$V = \text{Var}(\log Df|U^c), \quad A = \max(\log Df|U^c), \quad B = \min(\log Df|U^c).$$

We may assume that $Df|(U \setminus K_f) > 1$ (and therefore $\log Df|(U \setminus K_f) > 0$). Furthermore since $K_f, f^{-1}(K_f), f^{-2}(K_f), \ldots$ is disjoint we may assume that $k \in \mathbb{N}$ is so big that

$$\log \frac{|K_f|}{|f^{-j}(K_f)|} > V + 2A + 3|B|.$$

Now choose $y \in f^{-j}(K_f)$ such that $Df^j(y) = |K_f|/|f^{-j}(K_f)|$. Let T be the smallest interval containing x and y. Notice that $x \in f^{-j}(U)$ and since $U \supset K_f$ and $y \in f^{-j}(K_f)$ one has $T \subset f^{-j}(U)$. Moreover, since $x \notin \bigcup_{i \geq 0} f^{-i}(K_f)$ the interval T is a proper subset of $f^{-j}(U)$. From (6.2) this gives that

$$\{T, f(T), \ldots, f^j(T)\} \text{ is pairwise disjoint}. \tag{6.3}$$

This disjointness and the fact that f is orientation-preserving then implies that the points $\{x, f(x), \ldots, f^j(x)\}$ and $\{y, f(y), \ldots, f^j(y)\}$ are alternatively distributed over S^1 (more precisely, between each two points of the first set there is a point of the second set and vice versa). Choose the usual covering $\mathbb{R} \to S^1 = \mathbb{R} \bmod 1$, $t \to t \bmod 1$, and for a point $z \in S^1$ let $\bar{z}$ be one of the points in $\mathbb{R}$ covering $z \in S^1$. Write the sets

$$\{f^i(x); i=0,\ldots,j\} = \{x_i; i=0,\ldots,j\},$$
$$\{f^i(y); i=0,\ldots,j\} = \{y_i; i=0,\ldots,j\},$$

in such a way that there are points $\bar{x}_i$, $\bar{y}_i$, $\bar{a}$, $\bar{b}$ covering x_i, y_i, a, b such that

$$\bar{a} < y_0 = f^j(y) < \bar{b} < \bar{x}_0 < \bar{y}_1 < \bar{x}_1 < \bar{y}_2 < \cdots < \bar{y}_j < \bar{x}_j < \bar{a}+1.$$

Choose $0 \le s \le r \le j$ such that

$$\bar{y}_s < \bar{b} + \varepsilon < \bar{y}_{s+1} < \cdots \bar{y}_r \le \bar{a} + 1 - \varepsilon < \bar{y}_{r+1}.$$

Let $N = j+1$.

$$\begin{aligned}
\log Df^N(x) &= \sum_{i=0}^{j} \log Df(x_i) \\
&> \sum_{i=0}^{j} \log Df(x_i) - \log Df^j(y) + V + 2A + 3|B| \\
&= \sum_{i=0}^{j} \log Df(x_i) - \sum_{i=1}^{j} \log Df(y_i) + V + 2A + 3|B|.
\end{aligned}$$

Note that there are $j+1$ of the form $\log Df(x_i)$ and only j terms of the form $\log Df(y_i)$. Now we will rewrite (6.6) by pairing the term $\log Df(x_i)$ either with $\log Df(y_i)$ or with $\log Df(y_{i+1})$. Indeed, notice that

$$\begin{aligned}
\bar{a} < \bar{y}_0 = f^j(\bar{y}) < \bar{b} < \bar{x}_0 < \bar{y}_1 < \bar{x}_1 < \bar{y}_2 < \cdots < \bar{x}_{s-1} < \bar{y}_s < b + \varepsilon \\
< \bar{y}_{s+1} < \bar{x}_{s+1} < \bar{y}_{s+2} < \cdots \\
< \bar{x}_{r-2} < \bar{y}_{r-1} < \bar{x}_{r-1} < \bar{y}_r \le a + 1 - \varepsilon \\
< \bar{x}_r < \bar{y}_{r+1} < \bar{x}_{r+1} < \cdots < \bar{y}_j < \bar{x}_j < \bar{a} + 1. \qquad (6.5)
\end{aligned}$$

Using that $\log Df$ is decreasing on $(b, b+\varepsilon]$ and increasing on $[a-\varepsilon, a)$ and using (6.4), one gets

$$\begin{aligned}
\log Df^N(x) &> \sum_{i=0}^{j} \log Df(x_i) - \sum_{i=1}^{j} \log Df(y_i) + V + 2A + 3|B| \\
&= \sum_{i=0}^{s-1} [\log Df(x_i) - \log Df(y_{i+1})] + \log Df(x_s) - \log Df(y_{s+1}) \\
&\quad + \sum_{i=s+1}^{r-2} [\log Df(x_i) - \log Df(y_{i+1})] + \log Df(x_{r-1}) - \log Df(y_r) \\
&\quad + \log Df(x_r) + \sum_{i=r+1}^{j} [\log Df(x_i) - \log Df(y_i)] + V + 2A + 3|B| \qquad (6.6) \\
&\ge 0 + \log Df(x_s) - \log Df(y_{s+1}) + \sum_{i=s+1}^{r-2} [\log Df(x_i) - \log Df(y_{i+1})] \\
&\quad \times \log Df(x_{r-1}) - \log Df(y_r) + \log Df(x_r) + 0 + V + 2A + 3|B|.
\end{aligned}$$

Now, distinguishing between $\bar{x}_s < \bar{b}+\varepsilon$ or $\bar{x}_s \geq \bar{b}+\varepsilon$, one has

$$\log Df(x_s) - \log Df(y_{s+1}) \geq \min(0-A, B-A) \geq -|B| - A \tag{6.7a}$$

and similarly

$$\log Df(x_{r-1}) - \log Df(y_r) + \log Df(x_r) \geq \min(B-A+B, B-A+0) \geq -2|B| - A. \tag{6.7b}$$

Using (6.7) in (6.6) and the definition of V gives

$$\log Df^N(x) \geq 0 + (-|B| - A) - V + (-2|B| - A) + V + 2A + 3|B| > 0.$$

This finishes the proof of the Proposition.

Proof of Theorem 6.1. Let $\Lambda' = S^1 \backslash \bigcup_{i\geq 0} f^{-1}\,\mathrm{int}(K_f)$. We claim that for all $x \in \Lambda'$ there exists an interval U of x and a number $N \in \mathbb{N}$ such that $Df^N(y) > 1$ for all $y \in U \cap \Lambda$. Indeed if $x \in \Lambda$ this follows from Proposition 6.2, since f^N is C^1 in a neighbourhood of x. If $x \in \Lambda' \backslash \Lambda$ then there exists $n \geq 0$ such that $f^n(x) \in \partial K_f$. Let $N = n+1$. Since $Df^N(y) \to \infty$ as $y \in \Lambda$ and $y \to x$ the claim also holds for $x \in \Lambda' \backslash \Lambda$. Since Λ' is compact one can cover this set with a finite number of these neighbourhoods U_i as above such that for each i there exists N_i such that $Df^{N_i}|(U_i \cap \Lambda) > 1$. Let

$$N' = \max_i \{N_i\},$$

$$\rho_1 = \min_i \{\min\{\log Df^{N_i}|(U_i \cap \Lambda)\}\} > 0,$$

$$\rho_2 = \min_{0 \leq l \leq N'} \{\min\{\log Df^l|\Lambda\}\}.$$

Since $\rho_1 > 0$ we can choose N so that

$$\rho_1 \cdot \frac{N-N'}{N'} + \rho_2 > 0.$$

Now we will show that $Df^N(x) > 1$ for all $x \in \Lambda$. So choose $x \in \Lambda$. Then there exists a sequence i_l such that $x \in U_{i_1}$, $f^{N_{i_1}}(x) \in U_{i_2}$, $f^{N_{i_1}+N_{i_2}}(x) \in U_{i_3}$, $f^{N_{i_1}+N_{i_2}+N_{i_3}}(x) \in U_{i_4}, \ldots$. Let $k_i(x, N)$ be the number of times U_i appears in the sequence $x, f^{N_{i_1}}(x)$, $f^{N_{i_1}+N_{i_2}}(x), \ldots, f^{N_{i_1}+N_{i_2}+\cdots+N_{i_l}}(x)$ where l is the largest number such that $N_{i_1} + N_{i_2} + \cdots + N_{i_l} \leq N-1$. By definition of N',

$$N - (N_{i_1} + N_{i_2} + \cdots + N_{i_l}) \leq N'. \tag{6.7}$$

Now write

$$\begin{aligned}\log Df^N(x) &= \log Df^{N_{i_1}}(x) + \cdots + \log Df^{N_{i_l}}(f^{N_{i_1}+N_{i_2}+\cdots+N_{i_{l-1}}}(x)) \\ &\quad + Df^{N-(N_{i_1}+N_{i_2}+\cdots+N_{i_l})}(f^{N_{i_1}+N_{i_2}+\cdots+N_{i_l}}(x)) \\ &\geq \rho_1 \sum_i k_i(x, N) + \log Df^{N-(N_{i_1}+N_{i_2}+\cdots+N_{i_l})}(f^{N-(N_{i_1}+N_{i_2}+\cdots+N_{i_l})}(x) \\ &\geq \rho_1 \sum_i k_i(x, N) + \rho_2 \geq \rho \cdot \frac{N-N'}{N'} + \rho_2 > 1.\end{aligned}$$

Here the second inequality holds because of (6.7) and because of the definition of ρ_2. This finishes the proof of Theorem 6.1. □

7. *The proof of Theorems A and B*

Consider the class $\mathcal{B}$ of C^∞ Cherry vector fields on T^2 such that at each singularity p, div $(X)(p) \leq 0$. Furthermore, let $\mathcal{B}'$ be the class of C^∞ Cherry vector fields X on T^2 with precisely one hyperbolic sink and one saddle-point and such that X has positive divergence at this saddle-point. We will denote the flow through a point x by $t \to X_t(x)$. Let Sing (X) be the set of singularities of X.

PROPOSITION 7.1. *Let* $X \in \mathcal{B} \cup \mathcal{B}'$. *Then there exists a closed* C^∞ *curve* Σ *on* $T^2 \backslash \text{Sing}(X)$ *without self-intersections and with the following properties*

(a) Σ *is everywhere transversal to* X;

(b) Σ *is not retractable to a point.*

Proof. It is enough to show that there exists a recurrent orbit γ which is non-trivial (i.e. not equal to a point or a closed curve), see for example page 144 of [**PaMe**]. Because by assumption f has no periodic orbits, the theorem of Denjoy implies that if X has no singularites then every orbit of X is recurrent and we are done. So assume that X has at least one singularity. Since the Euler characteric of T^2 is zero, this implies that X has at least one saddle-point. Let us call a stable separatrix a component of $W^s(p) \backslash \{p\}$ where p is some saddle-point. We claim that there exists a stable separatrix γ such that the α-limit set of γ, $\alpha(\gamma)$, contains γ. Indeed, take a stable separatrix γ_0. Because of the Denjoy-Schwartz theorem $\alpha(\gamma_0)$ must contain at least one singularity p. Since all singularities of X are sinks or saddles, p is a saddle. Let γ_1 be the stable separatrix of p. Because X has no saddle-connections, $\alpha(\gamma_0) \supset \gamma_1$. By induction we define in this way a sequence of stable separatrices such that $\alpha(\gamma_n) \supset \gamma_{n+1}$ for all $n \geq 0$. Since there are only finitely many separatrices there exist $n \geq 0$, $m > 0$ such that $\gamma_n = \gamma_{n+m}$. One has that

$$\alpha(\gamma_n) \supset \alpha(\gamma_{n+1}) \supset \cdots \supset \alpha(\gamma_{n+m-1}) \supset \gamma_{n+m} = \gamma_n.$$

Hence γ_n is a non-trivial recurrent orbit. As we remarked above this implies the existence of a transversal circle. □

Let $X \in \mathcal{B} \cup \mathcal{B}'$ and let Σ be the closed transversal to X on T^2 from the previous Proposition. Notice that $T^2 \backslash \Sigma$ is an annulus $\Sigma \times (0,1)$ and we can write $T^2 \cong \Sigma \times [0,1]/\sim$, where $(s,0) \sim (s,1)$. Consider X as a flow on $T^2 \cong \Sigma \times [0,1]$ where we identify $\Sigma \times \{0\}$ and $\Sigma \times \{1\}$. Since X has no sources it follows that

(c) for every $x \in \Sigma \times \{0\}$ which is not contained in the stable manifold of one of the saddles or one of the sinks, there exists a $t > 0$ such that $X_t(x) \in \Sigma \times \{1\}$;

(d) there exists at least one $x \in \Sigma \times \{0\}$ such that $X_t(x) \in \Sigma \times \{1\}$ for some $t > 0$.

Now let Σ and Σ' be two closed curves transversal to X. Denote the points $x \in \Sigma$ such that there exists $t > 0$ with $X_t(x) \in \Sigma'$ for some $t > 0$ by Σ_0. For $x \in \Sigma_0$, let $t(x)$ be the minimal $t > 0$ such that $X_t(x) \in \Sigma'$ and define the map $f: \Sigma_0 \to \Sigma'$ by $f(x) = X_{t(x)}(x)$. This map is called the *transition map* between Σ and Σ'.

Now let $\Sigma' = \Sigma$ be the section from Proposition 7.1 and take the transition f from Σ to Σ'. This map is called the *return-map* to Σ.

Since orbits of X cannot intersect, f is order preserving. Let I be a component of $(\Sigma \backslash \Sigma_0) \times \{0\}$. Take $x \in \partial I$. Since the basin of a sink is open, x must be contained in the stable manifold of some saddle-point $p \in \Sigma \times (0,1)$. Since $p \in \partial I$ and X has no saddle-connections, $W^u(p)$ intersects $\Sigma \times \{1\}$ in some point v and for every point in $u \in (\Sigma \backslash I) \times \{0\}$, near ∂I there exists $t > 0$ such that $X_t(u)$ intersects $\Sigma \times \{1\}$ near u. (This can be seen by considering the backward orbits of $X|(\Sigma \times [0,1])$ of points in $\Sigma \times \{1\}$ near v intersecting $\Sigma \times \{0\}$. This set consists of a neighbourhood of ∂I in Σ.) In particular $\lim_{x \to \partial I, x \notin I} f(x)$ consists of one single point and we can define f on components of $\Sigma \backslash \Sigma_0$ to be constant. From the smooth dependence on initial condition, f is then everywhere continuous, and as smooth as the vector field outside boundary points of $\Sigma \backslash \Sigma_0$. From Theorem 7.1 in the Appendix, f also satisfies the non-flatness conditions from the introduction. It follows that $f \in \mathcal{A}$.

Remark. Let Σ be the section from Proposition 7.1, and $f: \Sigma_0 \to \Sigma$ its transition map. Now $T^2 \backslash \Sigma$ is an annulus A, see figure 7.1. It is easy to see that the orbits through boundary points of Σ_0 go directly to a saddle without leaving the annulus. From this one gets that $\Sigma \backslash \Sigma_0$ consits of finitely many components. Let $I \subset \Sigma$ be one of the components.

(i) I is not equal to a point. Indeed if I consists of one point then this point goes to some saddle-point and both unstable separatrices of this saddle-point intersect Σ. Let $D \subset A$ be the region bounded by Σ and these two unstable separatrices, see figure 7.1. But since X has no periodic orbits and no sources, this contradicts the Poincaré–Bendixon theorem.

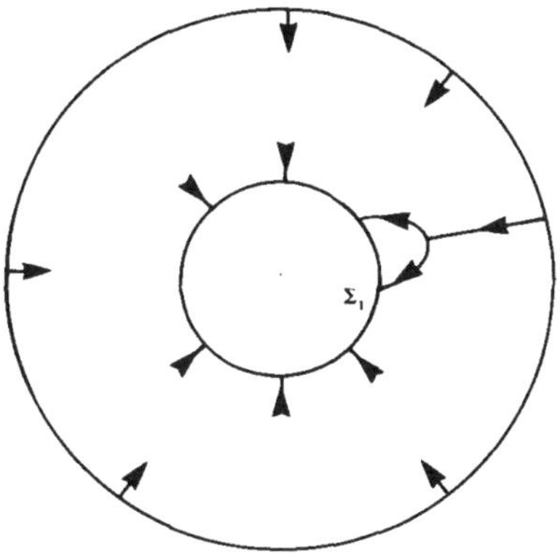

FIGURE 7.1. This situation cannot occur for $X \in \mathcal{B} \cup \mathcal{B}'$.

(ii) Let $I = [a, b]$. If p is the saddle-point such that $a \in W^s(p)$ then also $b \in W^s(p)$. Indeed, let γ be the stable separatrix of p which does not contain a. From the assumption on X and Poincaré–Bendixon theorem it follows again that γ cannot be contained in the annulus. So let c be the intersection of the closure of $\gamma \cap A$ and Σ. It is easy to see that $c = b$. From all this it follows that the situation is as drawn in figure 7.2.

Now we are in the position to prove Theorems A and B.

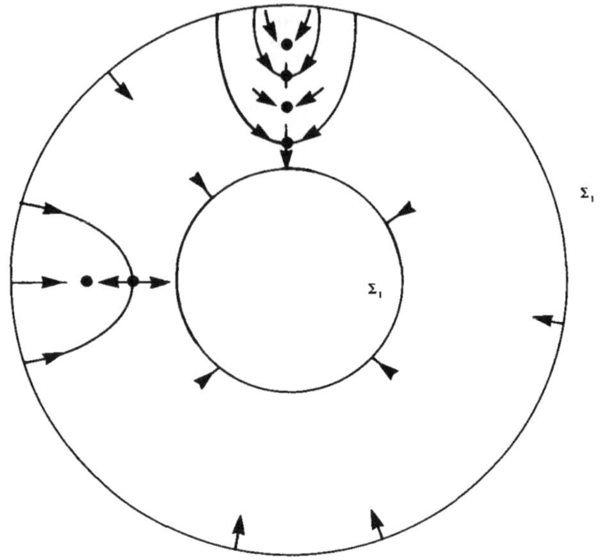

FIGURE 7.2. A typical situation for $x \in \mathcal{B} \cup \mathcal{B}'$.

Proof of Theorem A. Since to a vector field X we associated a map $f \in \mathcal{A}$, and since X has no periodic orbits and has no saddle-connections, $\rho(f)$ is irrational.

But if $\rho(f)$ is irrational, we can apply the Corollary to Theorem C to $f: S^1 \cong \Sigma \to \Sigma \cong S^1$ and the orbit of every point in $\{x; f^n(x) \in \Sigma_0, \forall n \geq 0\}$ is dense in this set. But now notice that every point in $\Sigma \backslash \Sigma_0$ is contained in the basin of a periodic attractor or in the stable manifold of a saddle-point. The statement follows. □

Proof of Theorem B. Take X as above and let Σ be the section from Proposition 7.1. There exists a neighbourhood U of X such that for each $Y \in U$, Σ is also a section of Y. Associate $f_Y: \Sigma \to \Sigma$ to Y as above. From the construction f_Y depends continuously on $Y \in U$. If there exists a conjugacy between $f_X, f_Y: \Sigma \to \Sigma$ then one can construct an equivalence between X and Y in the same way as in the proof of the structural stability of Morse–Smale flows on surfaces, see for example, [**PaMe**].

8. *Appendix: The transition map near singularities*

Let X be a C^∞ vector field on $\mathbb{R}^2$ which has a hyperbolic singularity at 0 of saddle-type with eigenvalues $\lambda > 0 > \mu$. Let p_1 and p_2 be points in $W^s(0)$ respectively $W^u(0)$. Furthermore let Σ_i be a C^2 curve through p_i which is transversal to X. If we choose Σ_1 sufficiently small, then for every x in one of the components of $\Sigma_1 - \{p_1\}$ there exists $t \geq 0$ such that $X_t(x) \in \Sigma_2$. Call this component Σ_1' and let $t(x) \in \mathbb{R}$ be the smallest number so that $X_{t(x)}(x) \in \Sigma_2$ and define $T: \Sigma_1' \to \Sigma_2$ by

$$T(x) = X_{t(x)}(x).$$

We call this map the *transition map.*

Since Σ_i are C^2 curves, it is a well defined notion to say that maps $\phi:\Sigma_1\to\mathbb{R}$ and $\psi:\mathbb{R}\to\Sigma_2$ have a derivative which has bounded variation.

In this section we want to show that $T:\Sigma_1\to\Sigma_2$ is equal to a map ϕ_α, up to maps whose derivatives have bounded variations. Here $\alpha=|\mu|/\lambda>0$, and ϕ_α is defined by $\phi_\alpha=\pm|z|^\alpha$.

THEOREM 8.1. *Let X be a C^∞ vector field on $\mathbb{R}^2$ which has a hyperbolic singularity at 0 of saddle-type with eigenvalues $\lambda>0>\mu$. Let $\alpha=|\mu|/\lambda$. There exist maps ϕ, $\psi:\mathbb{R}^+\to\mathbb{R}^+$ which are C^1, such that $D\phi$ and $D\psi$ have bounded variation and such that the map T from above is of the form*

$$T(x)=\phi\circ\phi_\alpha\circ\psi(x). \tag{8.1}$$

Proof.

Step 1. Let us first show that (8.1) is true if X is linear near 0.

If Σ_i are the straight lines $\{x=1\}$, $\{y=1\}$ and $T_{\{x=1\},\{y=1\}}$ the transition map from $\{x=1\}$ and $\{y=1\}$, then (8.1) follows from explicit integration (and choosing the natural parametrisations of $\{x=1\}$, $\{y=1\}$. If Σ_i are different C^2 curves (8.1) still holds. In fact consider the transition map T_1 from Σ_1 to $\{x=1\}$, resp. $T_2:\{y=1\}\to\Sigma_2$. Then $T_{\Sigma_1,\Sigma_2}=T_2\circ T_{\{x=1\},\{y=1\}}\circ T_1$. Since X is linear (and in particular the flow map is C^2) it follows from the implicit function theorem that the maps T_i are C^2. This last argument also shows that if (8.1) is true for one choice of C^2 curves Σ_i, then it is also true for any other choice of C^2 curves $\tilde{\Sigma}_i$ as above.

Step 2. X is C^2 linearisable if $|\lambda|\notin\{|\mu|,2|\mu|,\frac{1}{2}|\mu|\}$.

According to the Linearisation Theorem of Sternberg (see **[Ster]**, **[Bel]**, **[BD]** or for example Theorem 4 in **[Sto]**) there exists a C^2 coordinate system ϕ near 0 such that ϕ_*X is linear if

$$|\lambda|\neq|\mu|,2|\mu|,\tfrac{1}{2}|\mu|. \tag{8.2}$$

That is, in this case

$$\phi_*X(x,y)=\lambda\cdot x\frac{\partial}{\partial x}+\mu\cdot y\frac{\partial}{\partial y}.$$

Using step 1 we are finished if (8.2) holds.

Step 3. A normal form for X when $|\mu|=2|\lambda|$ or $|\mu|=|\lambda|$.

Let us now see what happens if (8.2) fails. We deal with the cases that $|\mu|=2|\lambda|$ and that $|\mu|=|\lambda|$. The case that $|\lambda|=2|\mu|$ is similar. By considering a multiple of the vector field we may assume that X is of the form

$$X(x,y)=x\cdot\frac{\partial}{\partial x}-2y\cdot\frac{\partial}{\partial y}+o|\underline{x}|^2. \tag{8.3a}$$

$$X(x,y)=x\cdot\frac{\partial}{\partial x}-y\cdot\frac{\partial}{\partial y}+o|\underline{x}|^2. \tag{8.3b}$$

Here $\underline{x}=(x,y)$.

LEMMA 8.2. *Let X be a C^∞ vector field.*

(a) *If X is of the form* (8.3a), *then there exists a C^∞ change of coordinates ϕ such that*

$$\phi_*(X) = x(1+ax^2y)\frac{\partial}{\partial x} - 2y(1+bx^2y)\frac{\partial}{\partial y} + o|\underline{x}|^6. \tag{8.4a}$$

(b) *If X is of the form* (8.3b), *then there exists a C^∞ change of coordinates ϕ such that*

$$\phi_*(X) = x(1+axy)\frac{\partial}{\partial x} - y(1+bxy)\frac{\partial}{\partial y} + o|\underline{x}|^4. \tag{8.4b}$$

Proof. Let X_1 be the vector field on $\mathbb{R}^2$ which has the same 1-jet in 0 as X and whose coefficients are linear. Let $[X_1, -]_h : H^h \to H^h$ be the linear map which assigns to each homogeneous vector field $Y \in H^h$ of degree h the Lie-product $[X_1, Y]$ (which is again in H^h). For each $h \in \mathbb{N}$ choose G^h so that $H^h = \mathrm{Im}([X_1, -]_h) + G^h$. According to Theorem 2.1 of Takens [**Ta**] there exists for each $l < \infty$, a C^∞ diffeomorphism Φ such that $\Phi_*(X)$ is of the form

$$\Phi_*(X) = X_1 + g_2 + \cdots + g_l + o|\underline{x}|^l)$$

where $g_i \in G^i$, $i = 2, 3, \ldots, l$.

Let us determine $\mathrm{Im}([X_1, -]_h)$. Taking $\beta \in \{1, 2\}$,

$$\left[x\frac{\partial}{\partial x} - \beta y\frac{\partial}{\partial y}, x^n y^m \frac{\partial}{\partial x}\right] = (n - \beta m - 1)x^n y^m \frac{\partial}{\partial x},$$

$$\left[x\frac{\partial}{\partial x} - \beta y\frac{\partial}{\partial y}, x^n y^m \frac{\partial}{\partial y}\right] = (n - \beta m + \beta)x^n y^m \frac{\partial}{\partial y}.$$

For $\beta = 2$, we can choose $G^h = 0$ for $h \le 6$, $h \ne 4$, and

$$G^4 = \left\langle x^3 \cdot y \cdot \frac{\partial}{\partial x}, x^2 \cdot y^2 \cdot \frac{\partial}{\partial y}\right\rangle.$$

Case (a) follows.

For $\beta = 1$, we can choose $G^h = 0$ for $h \le 4$, $h \ne 3$, and

$$G^3 = \left\langle x^2 \cdot y \cdot \frac{\partial}{\partial x}, x \cdot y^2 \cdot \frac{\partial}{\partial y}\right\rangle.$$

Case (b) follows and the proof of the Lemma is finished. □

Step 4. A C^1+'bounded variation' linearisation of X.

From step 1, we may assume that X is of the form

$$X = x(1+a\cdot x^2y)\frac{\partial}{\partial x} - 2y(1+b\cdot x^2y)\frac{\partial}{\partial y} + o|\underline{x}|^6; \tag{8.5a}$$

or

$$X = x(1+a\cdot x\cdot y)\frac{\partial}{\partial x} - y(1+b\cdot x\cdot y)\frac{\partial}{\partial y} + o|\underline{x}|^4. \tag{8.5b}$$

Step 4*A*. Let us first consider the case of (8.5a). Consider the C^1 coordinate transformation $\Psi(x, y) = (\tilde{x}, \tilde{y})$ defined by:

$$\tilde{x} = x - ax^3 y \log|x|,$$
$$\tilde{y} = y - bx^2 y^2 \log|y|.$$

Then $\{x=1\} = \{\tilde{x}=1\}$ and $\{y=1\} = \{\tilde{y}=1\}$. Moreover, the transformation between y and $\tilde{y}$ coordinates on the line $\{x=1\} = \{\tilde{x}=1\}$, $\{(x, y); x=1\} \to \{(\tilde{x}, \tilde{y}); \tilde{x}=1\}$, is equal to $y \mapsto \tilde{y} = y - by^2 \log|y|$, which has a derivative of bounded variation on $[-1, 1]$. Similarly $\tilde{y} \mapsto y$ is also C^1 and has a derivative of bounded variation. The same holds for the coordinate transformation $\{(x, y); y=1\} \to \{(\tilde{x}, \tilde{y}); \tilde{y}=1\}$. Since $T_{\{x=1\},\{y=1\}}$ is equal to $T_{\{\tilde{x}=1\},\{\tilde{y}=1\}}$ composed with $\tilde{y} \mapsto y$ on the left and with $x \mapsto \tilde{x}$ on the right, it suffices therefore to prove the result for $\Psi_* X$. Let us calculate $\Psi_* X$.

$$\begin{aligned}\left(X, \frac{\partial}{\partial \tilde{x}}\right) &= \left(X, \frac{\partial}{\partial x}\right) \cdot \frac{\partial \tilde{x}}{\partial x} + \left(X, \frac{\partial}{\partial y}\right) \cdot \frac{\partial \tilde{x}}{\partial y}\\ &= x(1+ax^2y)(1 - a3x^2 y \log|x| - ax^2 y) - 2y(1+bx^2y)(-ax^3 \log|x|) + o|\underline{x}|^6\\ &= x - ax^3 y \log|x| + o|\underline{x}|^6 + o|\underline{\tilde{x}}|^6 = \tilde{x} + o|\underline{\tilde{x}}|^6,\end{aligned}$$

and

$$\begin{aligned}\left(X, \frac{\partial}{\partial \tilde{y}}\right) &= \left(X, \frac{\partial}{\partial x}\right) \cdot \frac{\partial \tilde{y}}{\partial x} + \left(X, \frac{\partial}{\partial_y}\right) \cdot \frac{\partial \tilde{y}}{\partial y}\\ &= x(1+ax^2y)(-2bxy^2 \log|y|) - 2y(1+bx^2y)(1 - 2bx^2 y \log|y| - bx^2 y)\\ &\quad + o|\underline{x}|^6\\ &= -2y + 2bx^2y^2 \log|y| + o(|\underline{x}|^6) + o|\underline{\tilde{x}}|^6 = -2\tilde{y} + o|\underline{\tilde{x}}|^6.\end{aligned}$$

Here $\underline{\tilde{x}} = (\tilde{x}, \tilde{y})$. Hence

$$\Psi_* X = \tilde{x} \cdot \frac{\partial}{\partial \tilde{x}} - 2 \cdot \tilde{y} \cdot \frac{\partial}{\partial \tilde{y}} + o|\underline{\tilde{x}}|^6.$$

Using the notation of Theorem 2 of **[Sto]** we take $\sigma = 2$, $r = 6$, and it follows that there exists a transformation of Hölder class C^s, where $s = \sigma(r-1)/1+\sigma = 10/3$ linearising $\Psi_* X$ at 0. (We use here the vector field analogue of these linearisation results for diffeomorphisms, see the end of the introduction in **[Sto]**.) Using step 1, the result follows for (8.5a).

Step 4*B*. Now we consider the case of (8.5b). Consider the C^1 coordinate transformation $\Psi(x, y) = (\tilde{x}, \tilde{y})$ defined by:

$$\tilde{x} = x - ax^2 y \log|x|,$$
$$\tilde{y} = y - bxy^2 \log|y|.$$

Then as before $\{(x, y); x=1\} = \{(\tilde{x}, \tilde{y}); \tilde{x}=1\}$ and $\{(x, y); y=1\} = \{(\tilde{x}, \tilde{y}); \tilde{y}=1\}$ and the coordinate transformation $\tilde{y} \mapsto y$ and $x \mapsto \tilde{x}$ on these curves are C^1 and have derivatives which have bounded variation. So as before it suffices to prove the result for $\Psi_* X$. Let us calculate $\Psi_* X$.

$$\left(X, \frac{\partial}{\partial \tilde{x}}\right) = \left(X, \frac{\partial}{\partial x}\right) \cdot \frac{\partial \tilde{x}}{\partial x} + \left(X, \frac{\partial}{\partial y}\right) \cdot \frac{\partial \tilde{x}}{\partial y}$$
$$= x(1+axy)(1-a2xy\log|x|-axy) - y(1+bxy)(-ax^2\log|x|) + o|\underline{x}|^4$$
$$= x - ax^3y\log|x| + o|\underline{x}|^4 + o|\underline{\tilde{x}}|^4 = \tilde{x} + o|\underline{\tilde{x}}|^4,$$

and

$$\left(X, \frac{\partial}{\partial \tilde{y}}\right) = \left(X, \frac{\partial}{\partial x}\right) \cdot \frac{\partial \tilde{y}}{\partial x} + \left(X, \frac{\partial}{\partial y}\right) \cdot \frac{\partial \tilde{y}}{\partial y}$$
$$= x(1+axy)(-by^2\log|y|) - y(1+bxy)(1-2bxy\log|y|-bxy) + o|\underline{x}|^4$$
$$= y + bxy^2\log|y| + o|\underline{x}|^4) + o|\underline{\tilde{x}}|^4 = \tilde{y} + o|\underline{\tilde{x}}|^4.$$

Hence

$$\Psi_* X = \tilde{x} \cdot \frac{\partial}{\partial \tilde{x}} - \tilde{y} \cdot \frac{\partial}{\partial \tilde{y}} + o|\underline{\tilde{x}}|^4.$$

Using the notation of Theorem 2 of **[Sto]** we take $\sigma = 1$, $r = 4$ and it follows that there exists a transformation of Hölder class C^s, where $s = (r+\sigma)/(1+\sigma) = \frac{5}{2}$ linearising $\Psi_* X$ at 0. Using step 1, the result follows.

Since we had reduced everything to these two cases we are finished with the proof of Theorem 8.1. □

Remark. All the results in this section are valid if X is C^6.

REFERENCES

[Bel] G. R. Belitskii. Equivalence and normal forms of germs of smooth mappings. *Russian Math. Surveys* **33** (1978), 107–177.

[BD] P. Bonckaert and F. Dumoriter. On a linearization theorem of Sternberg for germs of diffeomorphisms. *Math. Z.* **185** (1984), 115–135.

[Ch] T. M. Cherry. Analytic quasi-periodic curves of discontinuous type on the torus. *Proc. London Math. Soc.* **44** (1938), 175–215.

[Den] A. Denjoy. Sur les courbes définies par les équations différentielles à la surface du tore. *J. de Math. Pures et Appl.* **11** (1932), 333–375.

[Ha] G. Hall. A C^∞ Denjoy counter example. *Ergod. Th. & Dynam. Sys.* **1** (1981), 261–272.

[He] M. R. Herman. Sur la conjugaison différentiable des difféomorphismes du cercle à des rotations. *Publ. Math. I.H.E.S.* **49** (1979), 5–233.

[MS1] W. de Melo and S. van Strien. A structure theorem in one dimensional dynamics. *Ann. Math.* **189** (1989), 519–546.

[MS2] W. de Melo and S. van Strien. One-dimensional dynamics: The Schwarzian derivative and beyond. *Bull. Amer. Math. Soc.* **18** (1988), 159–162.

[PaMe] J. Palis and W. de Melo. *Geometric Theory of Dynamical Systems, an Introduction.* Springer Verlag: New York, 1982.

[Sch] A. J. Schwartz. A generalization of a Poincaré–Bendixson Theorem to closed two dimensional manifolds. *Amer. J. Math.* **85** (1963), 453–458.

[Ster] S. Sternberg. On the structure of local homeomorphisms in Euclidean *n*-space, II. *Amer. J. Math.* **80** (1958), 623–631.

[Sto] D. Stowe. Linearisation in two dimensions. *J. Diff. Eq.* **63** (1986) 183–226.

[Ta] F. Takens. Singularities of vector fields. *Publ. Math. I.H.E.S.* **43** (1974), 47–100.

[Y] J. C. Yoccoz. Il n'y a pas de contre-exemples de Denjoy analytique. *C.R. Acad. Sci. Paris* **298**, ser. 1, no. 7 (1984), 141–144.

Julia–Fatou–Sullivan Theory for Real One-Dimensional Dynamics

M. Martens[1], W. de Melo and S. van Strien

Acta Math. **168**, 273–318 (1992)

Introduction

Our aim is to show that the Julia–Fatou–Sullivan structure theory for the dynamics of rational maps is also valid for smooth endomorphisms of the circle (and of the interval) under extremely mild smoothness and non-flatness conditions.

In order to stress the similarity between real and complex one-dimensional dynamics let us recall the main results from the Fatou–Julia–Sullivan theory. If f is a rational map then there is a dynamical decomposition of the Riemann sphere into the disjoint union of two totally invariant (i.e., both forward and backward invariant) sets $J(f), F(f)$. Here, $F(f)$ is the domain of normality of the family of iterates of f, and is called the Fatou set. Its complement, which is called the Julia set of f, is a compact set, which contains all the complications of the dynamics of f. The connected components of the open set $F(f)$ are mapped onto each other by f. Hence the orbit of a component of $F(f)$ is the union of some components of $F(f)$. Julia proved in the beginning of the century that if a component of $F(f)$ is periodic and contains an attracting periodic point then the orbit of this component must contain a critical point. Sullivan, in the remarkable paper [Su], showed via quasi-conformal deformations that the components are eventually periodic and fall into finitely many orbits.

Let N be either the circle S^1 or a compact interval of the real line and $f: N \to N$ be a smooth endomorphism. A *critical point* of f is a point where the derivative vanishes. A critical point is *non-flat* if some (higher) derivative is non-zero. A critical point is an *inflection point* if it has a neighbourhood where f is monotone. Otherwise it is called a *turning point*. Assume f is not a homeomorphism (if f is a homeomorphism then these maps correspond to degree ± 1 rational maps of the Riemann sphere and the situation is

---

(1) Supported by N.W.O.

Originally published in *Acta Math.* **168**, 273–318 (1992).

simple). If f has turning points then we define the singular set of f, Sing(f), as the union of the set of turning points of f and the boundary points of N. If f has no turning points then f is a circle map of degree ≥ 2 (or ≤ -2) and we define Sing(f) to be the set of fixed points of f. Finally we define the *Julia set* of f to be the α-limit set of Sing(f), i.e., $x \in \alpha(\text{Sing}(f))$ iff there exists $y_i \to x$ and a sequence $n_i \to \infty$ such that $f^{n_i}(y_i) \in \text{Sing}(f)$. Denote this set by $J(f)$. It is forward invariant. Thus its complement, called the *Fatou set* $F(f)$, is backward invariant: $f^{-1}(F(f)) \subset F(f)$. In general the Fatou set is not forward invariant but if U is a connected component of $F(f)$ which does not contain a turning point then $f(U)$ is also a component of $F(f)$.

MAIN THEOREM. *Let $f: N \to N$ be a smooth map such that all its critical points are non-flat. Then:*

(1) *All the connected components of $F(f)$ are eventually periodic* (*i.e., eventually mapped into a periodic component of $F(f)$*);

(2) *The number of periodic components of $F(f)$ is finite.*

Part (1) of the above theorem can be reformulated by stating that there are no *wandering intervals* for such maps. By a wandering interval we mean an interval J such that all forward iterates of J are disjoint and such that the ω-limit set of J is not a periodic orbit. The first result in this direction was obtained by Denjoy in 1932, [D]. He proved that a C^2 diffeomorphism of the circle does not have wandering intervals (more precisely, his proof applies to all C^1 diffeomorphisms $f: S^1 \to S$ such that $\log Df$ has bounded variation). His proof relies on a detailed understanding of the dynamics of rotations and on the control of the distortion of iterates of the map on intervals whose iterates are all disjoint. Later, in 1963, A. Schwartz [Sc] gave a different proof of Denjoy's result. His proof does not rely on precise dynamical properties but requires that $\log Df$ is Lipschitz. For maps with critical points the techniques of Denjoy and Schwartz cannot be used. In 1979, J. Guckenheimer was able to deal with critical points in some special cases. He proved in [Gu1] the non-existence of wandering intervals for unimodal maps of the interval with negative Schwarzian derivative and no inflection points, see also [Mi]. J. C. Yoccoz, [Y], proved the non-existence of wandering intervals for C^∞ homeomorphisms of the circle having only non-flat critical points. He combines techniques of Denjoy away from the critical points with some analytical estimates near the critical points which are related to the Schwarzian derivative. In [MS1] the same result was proven for smooth unimodal maps (not necessarily having negative Schwarzian derivative) with a non-flat critical point and also for maps satisfying the so-called Misiurewicz condition. In [MS1] the main tool is the control of the distortion of the cross-ratio under iterates. This control implies that under some disjointness assumptions the diffeomorphic inverse branches of iterates of a smooth map behaves very much like univalent holomorphic maps. This similarity is clear from the minimum principle and the Koebe distortion principle for real maps, see § 2 and § 4 of this paper. In 1988, Blokh and Lyubich proved in [L], [BL] the non-existence of wandering intervals for smooth maps whose only critical points are turning points. They introduced some very nice and powerful new topological tools generalizing those of [Gu1] and used the analytical tools developed in [MS1]. Our proof of the first part of the Main Theorem combines the analytical tools developed in [MS1] and [MMMS] (which allow for inflection points) with an extension of the topological ingredients of [L] and [BL]. In fact many of the ideas of § 6–§ 9 in this paper are simplifications and modifications of results contained in [L] and [BL]. We should note however that for the proof in [L] and [BL] it is necessary that the maps are C^2 and that our proof of part (1) of the Main Theorem also works for piecewise linear maps (see § 7). To be more specific our proof applies to all diffeomorphisms of the circle for which Denjoy's results hold. So Theorem A is a natural extension of Denjoy's original ideas to maps with critical points.

The first contribution to an analogue of part (2) of the Main Theorem for rational maps on the Riemann sphere is due to Julia. He proved in [J] (see also [Fa]) that if the orbit of a periodic domain contains an attracting point then it must also contain a critical point. Hence the number of orbits of these periodic domains must be bounded by the number of critical points. In [Si], Singer introduced for the first time the Schwarzian derivative in one dimensional dynamics and proved the same result of Julia for maps with negative Schwarzian derivative. Mañé, using estimates related to Denjoy and Schwartz proved that (2) holds for maps of the circle without critical points. Instead of (2) we will prove a still stronger result. We show that for each smooth map satisfying the hypothesis of the Main Theorem there exists $\varrho>0$ such that if p is a periodic points of sufficiently high period n then $|Df^n(p)|\geq 1+\varrho$. This last estimate is new even for maps from the family $x \to ax(1-x)$.

Let us finish this introduction by stating a corollary of the Main Theorem.

COROLLARY. (For the proof see [Me].) *In the space* $\mathcal{U}^r$ *of* C^r *unimodal maps of the interval* $[-1, 1]$ *endowed with the* C^r *topology,* $r\geq 3$ *the set of structurally stable maps is open and dense.*

The above corollary is the analogue of Mañé–Sad–Sullivan Theorem on the density of structural stable rational maps on the Riemann sphere. Note however that it does not follow that structurally stable maps satisfy the Axiom A condition.

COROLLARY. *Let $f: [-1, 1] \to [-1, 1]$ be a C^2 unimodal map which is infinitely renormalisable and whose turning point is non-flat. Then f has an attracting Cantor set and its basin contains intervals.*

Proof. By Theorem B the turning point of f cannot be approximated by periodic attractors. Therefore after renormalizing a finite number of times, the new renormalized map has no periodic attractors. So for this new map all points, except those which are eventually mapped into periodic orbits, tend to the closure of the forward orbit of the turning point. Q.E.D.

Many of D. Sullivan's ideas were an important source of inspiration for our work. We are grateful to him for inviting the last two authors to C.U.N.Y. and his interest in this work. Also we would like to thank the mathematics department of the University of Warwick at which this paper was finished.

§ 1. Definitions and statement of the theorems

Let N be a smooth compact 1-dimensional manifold (i.e., a finite union of closed intervals or circles). We say that $p \in N$ is a *periodic point* of f of period n if $f^n(p)=p$ and $f^i(p) \neq p$ for $0<i<n$. $J \subset N$ is a *periodic interval* of period n if $f^n(J) \subset J$ and $f^i(J) \cap J = \emptyset$ for $0<i<n$. An interval $J \subset N$ is a *wandering interval* for f if (i) $f^i(J) \cap f^j(J) = \emptyset$ for all $0 \leq i < j$ and (ii) the ω-limit of J, $\omega(J) = \{x; \exists n_i \to \infty$ and $y \in J$ such that $f^{n_i}(y) \to x\}$, is not a periodic orbit. (In fact, if it is a periodic orbit then it necessarily has to be a (possibly one-sided) attracting periodic orbit.) If f is not a homeomorphism and has a finite number of turning points then a component of J of the Fatou set of f is either eventually periodic (i.e., eventually mapped into a periodic component of $F(f)$) or a wandering interval. If U is a periodic component of period $n>0$ of the Fatou set of f (i.e. $f^n(U) \subset U$), then either the orbit of U contains a critical point of f or $f^n|U$ is monotone and f^n has a non-repelling fixed point in the closure of U. Therefore our Main Theorem follows from the following two theorems.

THEOREM A. *Let $f: N \to N$ be a C^∞ map whose critical points are non-flat. Then f has no wandering interval.*

THEOREM B. *Let $f: N \to N$ be a C^∞ map whose critical points are non-flat. Then* *there exists n_0 and $\varrho>0$ such that for every periodic points of f of period $n\geq n_0$ one has*

$$|Df^n(p)|\geq 1+\varrho.$$

The assumptions in these theorems are satisfied for analytic maps $f\colon N\to N$. In fact Theorem B implies that analytic maps can have at most a finite number of non-repelling periodic orbits (unless f or f^2 is the identity map).

The non-flatness conditions in Theorems A and B cannot be dropped. Indeed if one considers C^∞ maps with a flat critical point then Theorems A and B are false. Such maps may have wandering intervals and an infinite number of attractors. The first example of such a map was given by Hall, [Ha], but see also [SI], [Me]. Moreover, Theorem A and B are actually proved under much weaker smoothness assumptions. Theorem A is for example also proved for general continuous piecewise linear maps (in fact it is proved under the same smoothness conditions as the result of [D] for circle diffeomorphism).

More precisely we will prove

THEOREM A′. *Let $f\colon N\to N$ be in the clase $\mathscr{A}$ (defined presently). Then f has no wandering interval.*

Here $\mathscr{A}$ is the class of absolutely continuous maps $f\colon N\to N$ such that the following two conditions are satisfied:

(A1) There exists a finite set K_f such that for each neighbourhood of U of the set K_f, the map $N\setminus U\ni x\mapsto \log|Df(x)|$ (which exists almost everywhere because f is absolutely continuous) extends to a map of bounded variation on $N\setminus U$;

(A2) For each $x_0\in K_f$ there exists $\alpha\geq 1$, a neighbourhood $U(x_0)$ of x_0 and a homeomorphism $\phi\colon U(x_0)\to(-1,1)$ such that $\phi(x_0)=0$, ϕ, ϕ^{-1} are absolutely continuous, $(-1,1)\ni x\mapsto \log D\phi(x)$ (which exists almost everywhere) can be extended to a map of bounded variation, and such that

$$f(x)=\pm|\phi(x)|^\alpha+f(x_0),\quad \forall x\in U(x_0).$$

This number α is called the *order* of $x_0\in K_f$.

Notice that condition (A2) gives a non-flatness condition at the set of critical points. (If x_0 is an inflection point we can also allow that $f(x)=\pm|\phi(x)|^{\alpha_1}+f(x_0)$ on one side of x_0 and $f(x)=\pm|\phi(x)|^{\alpha_2}+f(x_0)$ on the other side, where $\alpha_1,\alpha_2\geq 1$ need not be the same. On the other hand, if x_0 is a turning point then we need the order or flatness to be the same on both sides; indeed, otherwise the involution τ from Section 4 below would not be Lipschitz.) This class $\mathcal{A}$ contains the class of C^∞ maps with non-flat critical points (and therefore Theorem A′ implies Theorem A) and also for example the class of continuous piecewise linear maps. For these piecewise linear maps even the non-existence of wandering intervals whose orbits stays away from turning points was previously unknown. The classical proof based on Denjoy–Schwartz requires $\log|Df|$ to be Lipschitz, see [Sc], and also [CE], [Str2]. Our proof relies on a disjointness result for backward iterates of certain intervals, see Theorem 6.4. In this sense our proof is closer to the original ideas of Denjoy than the later developments of Schwartz.

Instead of Theorem B we will prove the following stronger result:

THEOREM B′. *Let $f: N \to N$ be in the clase $\mathcal{B}$ (defined presently). Then there exists n_0 and $\varrho > 0$ such that for every periodic point of f of period $n \geq n_0$ one has*

$$|Df^n(p)| \geq 1 + \varrho.$$

The class $\mathcal{B}$ is somewhat smaller than the class $\mathcal{A}$. Indeed, $f \in \mathcal{B}$ if and only if:

(B1) f is C^2;

(B2) Let K_f be the set of critical points of f. Then for every $x_0 \in K_f$, there exist $\alpha > 1$ and a C^2 coordinate system $\phi: U(x_0) \to (-1, 1)$ on a neighbourhood $U(x_0)$ of x_0 such that $\phi(x_0) = 0$ and

$$f(x) = \pm|\phi(x)|^\alpha + f(x_0), \quad \forall x \in U(x_0).$$

Clearly $\mathcal{B}$ includes the class of C^∞ maps with non-flat critical points (and therefore Theorem B′ implies Theorem B). Moreover, it is not difficult to prove that condition (B2) is satisfied if for each critical point c there exists $k \geq 2$ such that f is C^{k+1} at c and $Df^k(c) \neq 0$. It is not clear whether we can weaken the condition that f is C^2 in the proof of Theorem B′.

The numbers n_0 and ϱ from Theorem B obviously depend in an essential way on the maps f. Indeed, there exists a sequence of analytic maps f_n converging to an analytic map f such that f_n has an attracting periodic orbit of period $>n$. (Take for example f to be a quadratic map with an eventually periodic critical point.) Nevertheless, our proof of Theorem B gives more uniform estimates. More precisely, if $\mathcal{K}$ is a compact family of maps in $\mathcal{B}$ then there exists $\varrho > 0$ and $n_0 \in \mathbf{N}$ such that if $f \in \mathcal{K}$ and p is a periodic point of period $n \geq n_0$ of f then one of the following possibilities hold.

(i) p is an attracting periodic point whose immediate basin of attraction contains a critical point;

(ii) p is in the boundary of the immediate basin of a periodic attractor which attracts a critical point;

(iii) $|Df^n(p)| \geqslant 1+\varrho$.

In particular the number of periodic orbits of maps in $\mathcal{K}$ of type (i) and (ii) is bounded by the number of critical points.

Notation. We will use the following notation. J_n will denote $f^n(J)$. If I, J are intervals in the same component of N, let $[I, J]$ be the (smallest) convex hull of I and J. (Even if this component of N is equal to S^1 it will always be clear which interval we mean. If I and J are in different components then we define $[I, J]=N$.) $(I, J]$ denotes the set $[I, J]\setminus I$. Similarly define $[I, J)$ and (I, J). The Lebesgue measure of a measurable set $I \subset N$ is denoted by $|I|$.

2. Analytical estimates on diffeomorphic branches of f^n when $f \in \mathcal{B}$: the Koebe and the minimum principle

In Denjoy's theory for circle diffeomorphisms the main technical tool is the control of the distortion of iterates of the map restricted to some interval under some disjointness assumptions on the iterates of this interval. Here the distortion of a differentiable map f on an interval T is defined as the maximal ratio of the absolute values of the derivative in two different points. This number measures the non-linearity of the map. Another way to present the same concept is to consider pairs of intervals $L, R \subset T$, intersecting at a common boundary point, and the distortion of the ratio $D(L, R)=|L|/|R|$ by the map f, i.e., the number $D(f, L, R)=D(f(L), f(R))/D(L, R)$. It is easy to see that the distortion of a differentiable map f in the interval N is bounded if and only if there is an upperbound for $D(f, L, R)$ for any pair of intervals $L, R \subset T$.

If a map f has critical points we cannot hope to get a bound for its non-linearity. Hence, instead of the distortion of a pair of consecutive intervals, or three consecutive points, we have to analyze the distortion of a more complicate configuration called the cross-ratio of four points.

2.1. *Definition.* Let $J \subset T$ be open and bounded intervals in N such that $T \setminus J$ consists of intervals L and R. Define the cross ratio of these intervals as

$$D(T, J)=\frac{|J|\,|T|}{|L|\,|R|},$$

(where $|I|$ denotes the length of an interval I). If $g: T \rightarrow N$ is continuous and monotone define

$$B(g,T,J)=\frac{D(g(T),g(J))}{D(T,J)}.$$

We notice that if $f^n|T$ is monotone and continuous then

$$B(f^n,T,J)=\prod_{i=0}^{n-1}B(f,f^i(T),f^i(J)).$$

Of course this cross-ratio is related to the hyperbolic metric. Indeed, let T be an open and bounded interval on N. For $x,y\in T$ let

$$\varrho_T(x,y)=\frac{1}{2}\operatorname{Log}\frac{|L\cup J|\,|J\cup R|}{|L|\,|R|}=\frac{1}{2}\operatorname{Log}(1+D(T,J))$$

where J is the interval bounded by the points x,y. Then ϱ_T is a metric in T and the group of isometries of this metric is exactly the group $\mathcal{M}_T$ of all Möbius transformations that map T onto T. Furthermore, the group $\mathcal{M}_T$ acts transitively on T, namely, given $x,y\in T$, there exists an isometry $\phi\in\mathcal{M}_t$ such that $\phi(x)=y$. If we take $T=(-1,1)$ and $x,y\in T$ then $\varrho_T(x,y)$ is exactly the hyperbolic distance between the two points $(x,0)$, $(y,0)$ in the unit disc. Therefore we shall call ϱ_T the *hyperbolic metric* of the interval T.

As is well known, holomorphic maps of the unit disc contract the hyperbolic metric. Moreover holomorphic maps of the disc have many very powerful properties: for example there is a universal bound for the non-linearity of these maps on a smaller disc (this follows from the Koebe Lemma). We shall show that diffeomorphisms of the interval T which contract the hyperbolic metric ϱ_T satisfy similar properties.

One way to check that a map contracts the hyperbolic metric on an interval is through the Schwarzian derivative.

2.2. *Definition.* Let $g\colon T\to N$ be a C^3-map on the interval $T\subset N$. Then

$$Sg(x)=\frac{g'''(x)}{g'(x)}-\frac{3}{2}\left(\frac{g''(x)}{g'(x)}\right)^2$$

is called the Schwarzian derivative of g.

2.3. PROPOSITION. *Let $g\colon T\to N$ be a C^3 diffeomorphism on the open interval $T\subset N$ such that for every $x\in T$, $Sg(x)<0$. Let* $\mathrm{cl}(I)\subset\mathrm{int}(T)$. *Then*

$$B(g,T,I)>1.$$

Furthermore if $Sg(x)\leq 0$, $\forall x\in T$, then $B(g,T,I)\geq 1$.

Proof. The proof is well known and can be found in for example [MS].

In particular a diffeomorphism $g: T \to T'$ having negative Schwarzian derivative expands the hyperbolic metric: $\varrho_{T'}(g(x), g(y)) > \varrho_T(x, y)$ for all $x, y \in T$ and $x \neq y$. Therefore the inverse of g contracts these metrics.

So our aim is to show that inverses of diffeomorphisms $g: T \to T'$ with negative Schwarzian derivative behave in many ways as holomorphic maps on a disc in the complex plane.

Unfortunately the condition that a map has negative Schwarzian derivative is not very natural: it is not preserved under coordinate changes, has no dynamic interpretation and it excludes a large class of maps.

Therefore we will not only consider maps with negative Schwarzian derivative but maps with some smoothness properties. By studying the distortion of the cross-ratio under iteration, we will first show in this section that the iterates of a smooth map restricted to an interval where it is a diffeomorphism, does not contract the metric too much provided some disjointness assumption on the iterates of the interval under consideration. Then we will show that in this case one obtains analogues of the Koebe and Maximum Principle for conformal maps (the analogy is with respect to the inverse of the maps). These results will hold for maps $f \in \mathcal{B}$. In § 4 the case is considered that f^n is a homeomorphism which may have critical points of inflection type and when $f \in \mathcal{A}$.

In the following theorem we give a lower bound for $B(f^n, T, J)$ if $f \in \mathcal{B}$, see also in [MS1] or in [Str3], see also [Str2] and [MS2] (where also a different cross-ratio is considered).

2.4. THEOREM. *Let $f \in \mathcal{B}$. Then there exists a bounded continuous function $\sigma: [0, \infty) \to \mathbf{R}_+$ such that $\sigma(t) \to 0$ as $t \to 0$ with the following property. If T is an interval such that $f^m|T$ is diffeomorphism then for any interval $J \subset T$ with* $\mathrm{cl}(J) \subset \mathrm{int}(T)$:

$$B(f^m, T, J) \geq \exp\{-\sigma(\tau) \cdot \sum_{i=0}^{m-1} |f^i(T)|\}. \tag{2.1}$$

Here $\tau = \max_{i=0,\dots,m-1} |f^i(T)|$.

Proof. Since $B(f^n, T, J) = \Pi_{i=0}^{n-1} B(f, f^i(T), f^i(J))$, it suffices to show that there exists a constant $C_0 \in (0, \infty)$ and an increasing continuous function $\sigma: [0, \infty) \to [0, C_0)$ with $\lim_{t \to 0} \sigma(t) = 0$ such that

$$B(f, T, J) \geq \exp\{-\sigma(|T|) \cdot |T|\}$$

for all intervals $J \subset T \subset N$ such that $Df(x) \neq 0$ for all $x \in T$. It is enough to prove that

$$B(f,T,J)-1 \geqslant -\sigma(|T|)\cdot|T| \tag{2.2}$$

for some (other) function σ as above. Let $V \subset \mathrm{cl}(V) \subset U$ be a neighbourhood of K_f such that each component U' of U contains a unique critical point and such that f is of the form $f(x) = \pm|\phi(x)|^\alpha + f(x_0)$, for all $x \in U'$ where $\alpha \geqslant 1$ and $\phi: U' \to (-1,1)$ is a diffeomorphism.

Case 1. $T \subset N \setminus V$. In this case write $T=[a,d]$ and $M=[b,c]$. Then

$$B(f,T,J)-1 = \frac{\dfrac{f(c)-f(b)}{c-b}\cdot\dfrac{f(d)-f(a)}{d-a} - \dfrac{f(b)-f(a)}{b-a}\cdot\dfrac{f(d)-f(c)}{d-c}}{\dfrac{f(b)-f(a)}{b-a}\cdot\dfrac{f(d)-f(c)}{d-c}}$$

$$\geqslant \frac{-1}{K^2}\cdot\left|\frac{f(c)-f(b)}{c-b}\cdot\frac{f(d)-f(a)}{d-a} - \frac{f(b)-f(a)}{b-a}\cdot\frac{f(d)-f(c)}{d-c}\right|,$$

where $K = \inf\{|Df(x)|; x \in N \setminus V\}$. Writing

$$f(a+x) = f(a) + \mu(a,x)\cdot x$$

one gets

$$\frac{f(d)-f(a)}{d-a} = \mu(a,d-a),$$

$$\frac{f(b)-f(a)}{b-a} = \mu(a,b-a).$$

Also

$$\begin{aligned} f(c)-f(b) &= \mu(a,c-a)\cdot(c-a) - \mu(a,b-a)\cdot(-c+b-a+c) \\ &= [\mu(a,c-a)-\mu(a,b-a)](c-a) + \mu(a,b-a)\cdot(c-b) \end{aligned}$$

and therefore

$$\frac{f(c)-f(b)}{c-b} = \mu(a,b-a) + \frac{\mu(a,c-a)-\mu(a,b-a)}{c-b}\cdot(c-a)$$

and similarly

$$\frac{f(d)-f(c)}{d-c} = \mu(a,d-a) + \frac{\mu(a,d-a)-\mu(a,c-a)}{d-c}\cdot(c-a).$$

Hence

$$B(f,T,J)-1 \geqslant \frac{-1}{K^2}\cdot \Bigg|\mu(a,d-a)\left(\mu(a,b-a)+\frac{\mu(a,c-a)-\mu(a,b-a)}{c-b}\cdot(c-a)\right)$$
$$-\mu(a,b-a)\cdot\left(\mu(a,d-a)+\frac{\mu(a,d-a)-\mu(a,c-a)}{d-c}\cdot(c-a)\right)\Bigg|$$
$$=\frac{-1}{K^2}\cdot|c-a|\cdot\Bigg|\mu(a,d-a)\cdot\frac{\mu(a,c-a)-\mu(a,b-a)}{c-b}$$
$$-\mu(a,b-a)\cdot\frac{\mu(a,d-a)-\mu(a,c-a)}{d-c}\Bigg|.$$

Hence

$$\frac{B(f,T,J)-1}{|T|}\geqslant\frac{-1}{K^2}\cdot\Bigg|\mu(a,d-a)\cdot\frac{\mu(a,c-a)-\mu(a,b-a)}{c-b}$$
$$-\mu(a,b-a)\cdot\frac{\mu(a,d-a)-\mu(a,c-a)}{d-c}\Bigg|.$$

Since f is C^2

$$\sigma(t)=\sup_{|T|\leqslant t}\left|\mu(a,d-a)\cdot\frac{\mu(a,c-a)-\mu(a,b-a)}{c-b}-\mu(a,b-a)\cdot\frac{\mu(a,d-a)-\mu(a,c-a)}{d-c}\right|$$

is a bounded increasing function with $\sigma(t)\to 0$ as $t\to 0$. From this the result follows.

Case 2. $T\subset U$. Since $\alpha\geqslant 1$, the map $\phi_\alpha(x)=x^\alpha$ has Schwarzian derivative $\leqslant 0$. Then Proposition 2.3 implies $B(\phi_\alpha,\phi(T),\phi(U))\geqslant 1$. Therefore

$$B(f,T,U)=B(\phi_\alpha,\phi(T),\phi(U))\cdot B(\phi,T,U)\geqslant B(\phi,T,U).$$

Since ϕ is C^2 and $\inf\{|D\phi(x)|;x\in T\}$ is bounded from below, $B(\phi,T,U)\geqslant 1-\sigma(|T|)\cdot|T|$ follows as in Case 1. Combining this gives (2.2).

Case 3. If T contains a component of $U\setminus V$ then $(B(f,T,M)-1)/|T|$ is bounded from below since $|T|$ is bounded from below. Combining Cases 1, 2 and 3 this finishes the proof of Theorem 2.4. Q.E.D.

Remark. We should emphasise that it is really essential in this theorem that f is C^2. It is *not* sufficient that f' is Lipschitz. This is in contrast with the usual bounded non-linearity results in the theory of A. Denjoy and of A. Schwartz. If in addition f'' is

Lipschitz then it is proved in [MS2] that there exists $C_0>0$ such that

$$B(f^m,T,J)\geqslant\exp\left\{-C_0\cdot\sum_{i=0}^{m-1}|f^i(T)|^2\right\}\geqslant\exp\left\{-C_0\cdot\tau\cdot\sum_{i=0}^{m-1}|f^i(T)|\right\}.$$

The previous theorem shows how to find lower bounds for $B(f^m,T,J)$. From the next results the usefulness of these lower bounds will become clear. The first of these are similar to well known properties of (quasi)-conformal mappings.

2.5. "Minimum Principle." *Let $T\subset N$ and $g\colon T\to g(T)\subset N$ be a C^1 diffeomorphism with $T=[a,b]\subset N$. Let $x\in(a,b)$. If for any $J^*\subset T^*\subset T$.*

$$B(g,T^*,J^*)\geqslant C>0$$

then

$$|Dg(x)|\geqslant C^3\min\{|Dg(a)|,|Dg(b)|\}. \tag{2.3}$$

Proof. The proof of this principle is given already in [MS1]. In order to be self-contained let us give it here too. Let us consider the following two operators:

$$B_0(g,T^*)=\frac{|g(T^*)|^2}{|T^*|^2}\frac{1}{|Dg(a^*)|\,|Dg(b^*)|}$$

$$B_1(g,T,x)=\frac{|Dg(x)|\dfrac{|g(T)|}{|T|}}{\dfrac{|g(L)|}{|L|}\dfrac{|g(R)|}{|R|}},$$

where $T^*=[a^*,b^*]\subset T$ and L and R are the connected components of $T-\{x\}$. Observe that

$$B_0(g,T^*)=\lim_{J\to T^*}B(g,T^*,J),\quad B_1(g,T,x)=\lim_{J\to x}B(g,T,J).$$

Hence $B_0(g,L),B_0(g,R),B_1(g,T,x)\geqslant C>0$. Since

$$B_0(g,L),B_0(g,R)\geqslant C$$

we have:

$$\left(\frac{|g(L)|}{|L|}\right)^2\geqslant C|Dg(a)|\,|Dg(x)|,\quad\left(\frac{|g(R)|}{|R|}\right)^2\geqslant C|Dg(x)|\,|Dg(b)|.$$

Since $B_1(g, T, x) \geq C > 0$ we have

$$|Dg(x)| \frac{|g(T)|}{|T|} \geq C \frac{|g(L)|}{|L|} \frac{|g(R)|}{|R|}.$$

Since $g|T$ is a diffeomorphism,

$$\min\left\{\frac{|g(L)|}{|L|}, \frac{|g(R)|}{|R|}\right\} \leq \frac{|g(T)|}{|T|} \leq \max\left\{\frac{|g(L)|}{|L|}, \frac{|g(R)|}{|R|}\right\}.$$

Then

$$|Dg(x)|^2 \geq C^2 \left(\frac{|g(L)|/|L| \cdot |g(R)|/|R|}{|g(T)|/|T|}\right)^2 \geq C^2 \min\left\{\left(\frac{|g(L)|}{|L|}\right)^2, \left(\frac{|g(R)|}{|R|}\right)^2\right\}$$
$$\geq C^3 \min\{|Dg(a)|\,|Dg(x)|, |Dg(b)|\,|Dg(x)|\}.$$

Hence $|Dg(x)| \geq C^3 \min\{|Dg(a)|\,|Dg(b)|\}$. Q.E.D.

For maps with negative Schwarzian derivative one version of the next principle was first used and proved in [Str1] and later rediscovered by S. Johnson and J. Guckenheimer, see [Gu2]. See also [Str2] and [Str3].

In order to state this principle it is convenient to introduce the following terminology. Let $U \subset V$ be two intervals. We say that V is a *δ-scaled neighbourhood of U* if each component of $V \setminus U$ has length $\delta|U|$. Similarly we define when V *contains a δ-scaled neighbourhood of U.*

2.6. "Koebe Principle." *For each $C, \tau > 0$ there exists $K(C, \tau) < \infty$ with the following property. Let $g: T \to g(T) \subset N$ be a C^1 diffeomorphism where T is a subinterval of N. Assume that for any intervals J^* and T^* with $J^* \subset T^* \subset T$ one has*

$$B(g, T^*, J^*) \geq C > 0.$$

Let $M \subset T$ and assume that $g(T)$ contains a τ-scaled neighbourhood of $g(M)$ then

$$\frac{1}{K(C, \tau)} \leq \frac{|g'(x)|}{|g'(y)|} \leq K(C, \tau), \quad \forall x, y \in M. \tag{2.4}$$

Proof. See [Str3] for a more general statement of this principle. In order to be complete we will include a proof of (2.4) here. By rescaling we may assume that $M = g(M) = [0, 1]$ and that g is increasing. Let $a, b \in T$ be such that $a < 0 < 1 < b$ and that

$g(a)=-\tau$ and $g(b)=1+\tau, L=[a,0]$, $M=[0,1]$ and $R=[1,b]$. As in the proof of (2.3) one has

$$|Dg(0)|\,|Dg(1)| \leq \frac{1}{C}\left(\frac{|g(M)|}{|M|}\right)^2. \tag{2.5}$$

Similarly

$$|Dg(0)| \geq C \cdot \frac{|g(L)|/|L| \cdot |g(M)|/|M|}{|g(L\cup M)|/|L\cup M|}.$$

Using $|g(M)|=|M|=1$ and

$$\frac{|g(L)|}{|L|}\frac{|L\cup M|}{|g(L\cup M)|} = \frac{\tau}{|L|}\frac{|L|+|M|}{1+\tau} \geq \frac{\tau}{1+\tau},$$

this gives

$$|Dg(0)| \geq \frac{C\tau}{1+\tau}. \tag{2.6}$$

Similarly

$$|Dg(1)| \geq \frac{C\tau}{1+\tau}. \tag{2.7}$$

Combining (2.5)–(2.7) gives that there exists $K'<\infty$ such that

$$\frac{1}{K'} \leq |Dg(0)|,\ |Dg(1)| \leq K'. \tag{2.8}$$

Using the Minimum Principle one obtains that for each $x \in [0,1]$,

$$|Dg(x)| \geq \frac{C^3}{K'}. \tag{2.9}$$

Let $U=[0,x]$ and $V=[x,1]$. Since g is a diffeomorphism one has either $|g(U)|/|U| \leq |g(M)|/|M|=1$ or $|g(V)|/|V| \leq |g(M)|/|M|=1$. If the former holds then

$$\frac{[|g(U)|/|U|]^2}{|Dg(0)|\,|Dg(x)|} \geq C$$

gives

$$|Dg(x)|\,|Dg(0)| \leq \frac{1}{C}\cdot 1.$$

Using (2.6) this gives

$$|Dg(x)| \leq \frac{1}{C}\frac{1+\tau}{C\tau}.$$

From this and (2.9) it follows that there exists $K''<\infty$ such that

$$\frac{1}{K''}\leqslant Dg(x)|\leqslant K''. \tag{2.10}$$

Therefore

$$\frac{1}{(K'')^2}\leqslant\frac{|Dg(x)|}{|Dg(y)|}\leqslant(K'')^2,\quad \forall x,y\in M.$$

Q.E.D.

The next two results will play an important role in proving that the periodic points of high period of a map in $\mathcal{B}$ are repelling and are concerned with the situation that $B(g,T,M)-1$ is positive and bounded from below. The next result states that whenever an interval is mapped monotonically over itself with expansion of the cross-ratios then the map is really 'bending' and therefore at some point expanding.

2.7. "EXPANSION PRINCIPLE." *Let T be an interval in N and $g\colon T\to g(T)\subset N$ a C^1 diffeomorphism. For every $\varepsilon>0$ and $\delta>0$ there exists $\varrho>0$ such that the following holds. Let T,M be intervals such that both components of $T\setminus M$ have at least length $\delta\cdot|T|$. If $B(g,T,M)\geqslant 1+\varepsilon$ and $g(T)\supset T$ then there exists $\theta\in T$ with $Dg(\theta)\geqslant 1+\varrho$.*

Proof. Let $\xi>0$ be so small that $(1+\varepsilon)(1-\xi)^2\geqslant 1+\frac{1}{2}\varepsilon$.

Case 1. First suppose that $|g(L)|/|L|\geqslant 1-\xi$ and $|g(R)|/|R|\geqslant 1-\xi$. Then, using $B(g,T,M)\geqslant 1+\varepsilon$, we get

$$\frac{|g(T)|}{|T|}\cdot\frac{|g(M)|}{|M|}\geqslant(1+\varepsilon)(1-\xi)^2\geqslant 1+\frac{1}{2}\varepsilon.$$

So at least one of the terms $|g(T)|/|T|, |g(M)|/|M|$ is greater or equal than $\sqrt{1+\varepsilon/2}$. Using the mean value theorem we obtain in either case $\theta\in T$ so that

$$|Dg(\theta)|\geqslant\sqrt{1+\varepsilon/2}. \tag{2.11}$$

Case 2. Suppose that $|g(L)|/|L|<1-\xi$. The case that $|g(R)|/|R|<1-\xi$ is proved similarly. Then, using $|L|,|R|\geqslant\delta|T|$, we get

$$\begin{aligned}\frac{|g(M\cup R)|}{|M\cup R|}&=\frac{|g(T)|-|g(L)|}{|M|+|R|}\geqslant\frac{|T|-|g(L)|}{|M|+|R|}\\&>\frac{|T|-(1-\xi)|L|}{|M|+|R|}\geqslant 1+\delta\xi.\end{aligned}$$

The mean value theorem then gives $\theta \in M \cup R$ such that

$$|Dg(\theta)| \geq 1+\delta\varepsilon. \tag{2.12}$$

Together, Case 1 and Case 2 prove the result. Q.E.D.

We will need to use the previous lemma in the case where g is an iterate f^n of f. Hence we will need that $B(f^n, T, J)$ is strictly bigger than one. In order to get such a lower bound for $B(f^n, T, J)$ we will use the following result which is based on the non-flatness of the critical point. (This lemma does not hold for piecewise linear maps.)

2.8. LEMMA. *Let $g \in \mathcal{B}$ and c a critical point of g. Then for every $\tau>0$ there exists a neighbourhood U of c and a number $\xi>0$ such that if T is an interval in $U \setminus \{c\}$ with $|T| \geq \tau |[c, T)|$ and M the middle third interval in T then*

$$B(g, T, M) \geq 1+\xi. \tag{2.13}$$

Proof. We may assume that $N=[-1,1]$. Let $\hat{B}(g, T)=B(g, T, M)$ where M is the middle third interval in T. Let $b \neq c$, k the order of the critical point and define

$$\phi_b\colon [c, b] \to [0, 1] \quad \text{by} \quad \phi_b(x) = \frac{|x-c|}{|b-c|},$$

$$\psi_b\colon [g(c), g(b)] \to [0, 1] \quad \text{by} \quad \psi_b(y) = \frac{|y-g(c)|}{|g(b)-g(c)|},$$

$$g_b\colon [0, 1] \to [0, 1] \quad \text{by} \quad g_b(x) = \psi_b \circ g \circ \phi_b^{-1}(x)$$

and $g_c(x)=x^k$. Since ϕ_b and ψ_b are affine transformations and the crossratio are invariant under affine transformations, one has for any interval $T=[a, b] \subset [c, b]$

$$\hat{B}(g, T) = \hat{B}\left(g_b, \left[\frac{|a-c|}{|b-c|}, 1\right]\right). \tag{2.14}$$

Let $\Phi\colon [c, b] \times [0, 1/(1+\tau)] \to \mathbf{R}$ be defined by

$$\Phi(x, y) = \hat{B}(g_x, [y, 1]).$$

Since $g(x)=\pm|\phi(x)|^k+g(x_0)$ where ϕ is a C^2 diffeomorphism and $k>1$ one has that g_b depends continuously on b and g_b tends in the C^0 topology to g_c as $b \to c$ where g_c is the function $g_c(x)=x^k$. Therefore Φ is continuous.

Notice that since $k \geq 2$, the map g_c has negative Schwarzian derivative and therefore $B(g_c, V, U)>1$ for any intervals $U\subset V\subset[0,1]$ with $\mathrm{cl}(U)\subset\mathrm{int}(V)$. Therefore the map $[0,1/(1+\tau)]\ni y\to\hat{B}(g_c,[y,1])-1$ has a positive minimum, say $2\xi>0$. So $\Phi(0,y)\geq 1+2\xi$ for all $y\in[0,1/(1+\tau)]$. Let y_0 be such that

$$\Phi(x,y)\geq 1+\xi,\quad \forall(x,y)\in[c,y_0]\times\left[0,\frac{1}{1+\tau}\right]. \tag{2.15}$$

Now let U be a neighbourhood of c with diameter $\leq y_0$. If $I=(a,b)$ is an interval in $U\setminus\{c\}$ and $|I|/|[c,I)|\geq\tau$, then $|b-a|/|c-a|\geq\tau$. Therefore $|a-c|/|b-c|\in[0,1/(1+\tau)]$ and from (2.15),

$$\hat{B}(g,I)=\hat{B}\left(g_b,\left[\frac{|a-c|}{|b-c|},1\right]\right)=\Phi\left(b,\frac{|a-c|}{|b-c|}\right)\geq 1+\xi.$$

This finishes the proof of the lemma. Q.E.D.

3. Analytical estimates on monotone branches of f^n when $f\in\mathcal{A}$: the Macroscopic Koebe Principle

In this section we will analyze the distortion of the cross-ratio for iterates of a map $f\in\mathcal{A}$. We will face two new difficulties. First we have less differentiability as in the previous section and, therefore, we will have to rely on more disjointness assumptions in order to get good bounds for the distortion of the cross ratio of iterates. The second type of problems is that we will need later some estimates on the cross-ratios when the iterates of the interval may contain critical points of inflection type. So we will prove here a version of the Koebe Principle for branches of f^n which are only monotone, and not necessarily diffeomorphic.

The basic strategy is the same as before: we need to estimate the contraction of the cross ratio under high iterates of f on an interval T. Because in this section $f\in\mathcal{A}$, we only have that $\log Df$ has bounded variation. Therefore we need to split up the iterates of T into collections of disjoint intervals. So we start by discussing some disjointness properties of families of intervals.

3.1. *Definition.* The *intersection multiplicity* of a finite collection of intervals in N is the maximal number of intervals in this collection whose interior has a non-empty intersection.

3.2. Proposition. *Let $\mathcal{W}$ be a finite collection of intervals in N with intersection multiplicity at most p then there exists a partition of the collection*

$$\mathcal{W} = A_1 \cup A_2 \cup \ldots \cup A_{2p},$$

such that A_k consists of mutually disjoint intervals for $k=1,2,\ldots,2p$.

Proof. Clearly we may assume that N is connected. So we only need to consider the cases that N is equal to an interval or a circle.

Proof if $N=[-1,1]$. Of course we may assume that all the intervals in $\mathcal{W}$ are open. We claim that if $N=[-1,1]$ then there are p classes

$$A_1, A_2, \ldots, A_p,$$

which form the desired partition of $\mathcal{W}$. Indeed let $\mathcal{I}$ be some collection of intervals in $[-1,1]$. For $I \in \mathcal{I}$ let next$\{I, \mathcal{I}\}$ be some interval $I' \in \mathcal{I}$ such that (i) I' does not intersect I and is to the right of I and (ii) there is no interval $J \in \mathcal{I}$ satisfying (i) which is closer to I (if there is no interval to the right of I take next$\{I, \mathcal{I}\}$ to be the empty set). Similarly $\mathcal{R}(\mathcal{I})$ is some interval in $\mathcal{I}$ such that there is no interval in $\mathcal{I}$ which has points to the right of $\mathcal{R}(\mathcal{I})$. For $k=1,2,\ldots,p$ we define inductively A_k as follows. Let $A_0=\emptyset$ and suppose that we have defined by induction A_{k-1}. If $k \leq p$, and $\mathcal{B}_k = \mathcal{W} \setminus \bigcup_{i=1,\ldots,k-1} A_i = \emptyset$ then let $A_k=\emptyset$. If $\mathcal{B}_k$ is non-empty then take an interval $I_k \in \mathcal{B}_k$ so that there is no interval in $\mathcal{B}_k$ containing points to the left of I_k. Then define

$$A_{k,1} = \{I_k\},$$

$$A_{k,n+1} = A_{k,n} \cup \text{next}\{\mathcal{R}(A_{k,n}), (\mathcal{W} \setminus (A_0 \cup \ldots \cup A_{k-1} \cup A_{k,n}))\},$$

and

$$A_k = \bigcup_{n \geq 1} A_{k,n}.$$

Then A_k is a collection of intervals with disjoint interiors and the collections $A_1,\ldots,A_p$ are mutually disjoint. Now we will show that $\mathcal{W}=A_1 \cup \ldots \cup A_p$. Suppose this is not the case and there exists an interval I in the collection $\mathcal{W} \setminus (A_1 \cup \ldots \cup A_p)$. Then

$$I \in \mathcal{W} \setminus (A_0 \cup \ldots \cup A_{k-1} \cup A_k)$$

for each $k=1,\ldots,p$ and since $I \notin A_k$ it follows from the inductive definition above that there exist for $i=1,2,\ldots,p$, $T_i \in A_i$ such that the left boundary point of T_i is to the left of (or equal to) the left boundary point x of I. But then since all these intervals are open, points just to the right of this boundary point x are contained in I as well as in T_k,

$k=1,2,\ldots,p$. This contradicts the assumption that the intersection multiplicity of $\mathscr{W}$ is at most p.

Proof if $N=S^1$. If $N=S^1$ then choose some $x\in S^1$ and let $I_1,\ldots,I_r$ be intervals in $\mathscr{W}$ which contain x. Then $r\leqslant p$. Since $S^1\setminus\{x\}\simeq(-1,1)\subset[-1,1]$ it follows from the previous case that $\mathscr{W}\setminus\{I_1,\ldots,I_r\}$ can be disjointly decomposed into collections $A_1,\ldots,A_p$ of disjoint intervals. Then $\mathscr{W}=A_1\cup\ldots\cup A_p\cup\bigcup_{k=1,\ldots,r}\{I_k\}$ has the desired properties. Q.E.D.

The main result of this section is a version of the Koebe Principle for maps f^n restricted to an interval T such that $f^n|T$ is monotone (but may have critical points of inflection type) and such that the collection $T,f(T),\ldots,f^{n-1}(T)$ has low intersection multiplicity.

3.3. Macroscopic Koebe Principle. *Let $f\in\mathscr{A}$. Then there exists a strictly positive function B_0: $(0,1)\to(0,1)$ such that for any pair of intervals $M\subset T$, any $n\geqslant 0$ and any $0<\varepsilon<1$, satisfying the following conditions:*

(a) *$f^n(T)$ contains an ε-scaled neighbourhood of $f^n(M)$;*

(b) *the intersection multiplicity of $\{T,f(T),\ldots,f^{n-1}(T)\}$ is at most* 17;

(c) *$f^i(M)\cap K_f=\varnothing$, $i\in\{0,1,\ldots,n-1\}$*

one has:

T is a $B_0(\varepsilon)$-scaled neighbourhood of M.

For the proof of this theorem we will need a number of lemmas. Let $f\in\mathscr{A}$. Let $\mathscr{V}\subset\mathrm{cl}(\mathscr{V})\subset\mathrm{int}(\mathscr{U})$ be interval neighbourhoods of K_f such that the number of components of $\mathscr{V}$ and $\mathscr{U}$ are both the same as $\#K_f$.

3.4. Lemma. *Let f be a map in $\mathscr{A}$ and $\Xi=\{T_1,T_2,\ldots,T_n\}$ a collection of intervals in N and $M_i\subset T_i$. Assume that the intersection multiplicity of Ξ is at most* 17 *and that none of the intervals T_i contains a component of $\mathscr{U}\setminus\mathscr{V}$ or contains points of K_f. Then there exists $V<\infty$ such that*

$$\sum_{i=1}^{n}\log B(f_i,T_i,M_i)\geqslant -V. \tag{3.1}$$

Proof. Since $f\in\mathscr{A}$, the restriction of $\log|Df|$ to $N\setminus\mathscr{V}$ exists almost everywhere and $\log|Df|$ has bounded variation on this set. Let

$$V_f=\mathrm{Var}(\log[|Df|(N\setminus\mathscr{V})])<\infty.$$

For each $x_i \in K_f$ let U_i be the component of $\mathcal{U}$ which contains x_i. Since $f \in \mathcal{A}$, $f(x) = \phi_{\alpha_i} \circ \phi_i(x) + f(x_i)$, where $\phi_{\alpha_i}(x) = \pm |x|^{\alpha_i}$ and $\alpha_i \geq 1$. Here $\phi_i \colon U_i \to (-1, 1)$ is a homeomorphism and $\log |D\phi_i|$ exists almost everywhere and has bounded variation. Hence $V' = \Sigma_i \operatorname{Var}(\log |D\phi_i|)$ is finite. Let $V = 34(V_f + V')$.

Let $I_1 = \{i; T_i \cap \mathcal{V} = \emptyset\}$ and $I_2 = \{i; T_i \subset \mathcal{U}\}$. Since the intervals T_i never contain a component of $\mathcal{U} \setminus \mathcal{V}$, $I_1 \cup I_2 = \{1, 2, \ldots, n\}$.

First assume that $i \in I_1$. Let L_i and R_i be the components of $T_i \setminus M_i$. For $u_i, v_i \in T_i$ let (u_i, v_i) be the open interval connecting u_i and v_i. Because Df exists almost everywhere there exist $m_i \in M_i$, $l_i \in L_i$, $r_i \in R_i$ and $\tau_i \in T_i$ such that Df exists in these points and such that $|f(M_i)| / |M_i| \geq |Df(m_i)|$, $|f(T_i)| / |T_i| \geq |Df(\tau_i)|$, $|f(L_i)| / |L_i| \leq |Df(l_i)|$ and $|f(R_i)| / |R_i| \leq |Df(r_i)|$. Using this in the definition of $B(f, T_i, M_i)$ we find that there exist $l_i \in L_i$, $r_i \in R_i$, $m_i \in M_i$ and $\tau_i \in T_i$ such that

$$\log B(f, T_i, M_i) \geq \log\left(\frac{|Df(m_i)|\,|Df(\tau_i)|}{|Df(l_i)|\,|Df(r_i)|}\right), \tag{3.2}$$

and

$$m_i \in (l_i, r_i). \tag{3.3}$$

From (3.2) and the choice of the points l_i, m_i, r_i, τ_i one has

$$\log B(f, T_i, M_i) \geq -\{|\log |Df(m_i)| - \log |Df(l_i)|| + |\log |Df(\tau_i)| - \log |Df(r_i)||\} \tag{3.4}$$

and also

$$\log B(f, T_i, M_i) \geq -\{|\log |Df(m_i)| - \log |Df(r_i)|| + |\log |Df(\tau_i)| - \log |Df(l_i)||\}. \tag{3.5}$$

Rename the points l_i, m_i, r_i, τ_i in increasing order a_i, b_i, c_i, d_i. From (3.3) one gets that either $(l_i, m_i) \cap (\tau_i, r_i) = \emptyset$ or $(\tau_i, l_i) \cap (m_i, r_i) = \emptyset$, and so we can use either (3.4) or (3.5) and get

$$\begin{aligned} \log B(f, T_i, M_i) &\geq -\{|\log |Df(b_i)| - \log |Df(a_i)||\} + |\log |Df(d_i)| - \log |Df(c_i)||\} \\ &\geq -\operatorname{Var}(\log |Df|T_i|) \end{aligned} \tag{3.6}$$

and therefore

$$\log B(f, T_i, M_i) \geq -\operatorname{Var}(\log |Df|T_i|). \tag{3.7}$$

Now consider $i \in I_2$. Then T_i is contained in some component U_i of $\mathcal{U}$ (and does not intersect K_f) and so f has the form $f(x)=\phi_{\alpha_i}\circ\phi_i$. Hence

$$B(f, T_i, M_i) = B(\phi_{\alpha_i}, T_i', M_i')\times B(\phi_i, T_i, M_i).$$

Here $T_i'=\phi_i(T_i)$ and $M_i'=\phi_i(M_i)$. Since the Schwarzian derivative of ϕ_{α_i} is less or equal to 0 (because $\alpha_i\geq 1$) one gets $B(\phi_{\alpha_i}, T_i', M_i')\geq 1$. Hence, as above,

$$\begin{aligned}\log B(f, T_i, M_i) &\geq 0+\log B(\phi_i, T_i, M_i)\\ &\geq -\mathrm{Var}(\log|D\phi_i|T_i|).\end{aligned} \tag{3.8}$$

Since the intersection multiplicity of Ξ is at most 17, using Proposition 3.2, one can write $\Xi=A_1\cup A_2\ldots\cup A_{34}$ where A_j consists of a collection of mutually disjoint intervals. Hence from (3.7) and (3.8) one gets

$$\sum_{i=1}^{n}\log B(f, T_i, M_i)\geq -34\cdot(V_f+V'). \tag{3.9}$$

The lemma follows. Q.E.D.

3.5. Lemma. *Let $f\in\mathcal{A}$. Then there exists $A_0>0$ such that if $I\subset W$ are intervals in N such that*

(a) *$W\setminus I$ consists of one component H;*

(b) $|H|\leq|I|$;

(c) $I\cap K_f=\varnothing$,

then

$$\frac{|f(I)|}{|I|}\geq A_0\cdot\frac{|f(H)|}{|H|}. \tag{3.10}$$

Proof. The proof of this lemma is elementary and can be found in [MMMS]. Q.E.D.

3.6. Lemma. *Let $f\in\mathcal{A}$ and I, T be intervals with $\mathrm{cl}(I)\subset\mathrm{int}(T)$ and L and R be the components of $T\setminus I$. Let $\gamma\in(0,1)$. If*

$$\frac{|f(I)|}{|f(R)|}\geq\gamma\cdot\frac{|I|}{|R|}\quad\textit{and}\quad\frac{|f(I)|}{|f(L)|}\geq\gamma\cdot\frac{|I|}{|L|},$$

then

$$B(f, T, I)\geq\gamma^2.$$

Proof. Let

$$d=\left(\frac{|I|\,|I|}{|R|\,|L|}\right)\Big/\left(\frac{|I|}{|R|}+\frac{|I|}{|L|}\right).$$

Then

$$B(f,T,I)=\frac{\dfrac{|f(I)|}{|f(R)|}+\dfrac{|f(I)|}{|f(R)|}\dfrac{|f(I)|}{|f(L)|}+\dfrac{|f(I)|}{|f(L)|}}{\dfrac{|I|}{|R|}+\dfrac{|I|}{|R|}\dfrac{|I|}{|L|}+\dfrac{|I|}{|L|}}\geqslant\frac{\gamma+\gamma^2\cdot d}{1+d}=\gamma\cdot\frac{1+\gamma\cdot d}{1+d}\geqslant\gamma^2.$$

The last inequality follows since $\gamma\in(0,1)$ and therefore $(1+\gamma x)/(1+x)$ is always greater than γ for $x>0$. The result follows. Q.E.D.

3.7. Lemma. *Let $f\in\mathcal{A}$. Then there exists $A_1>0$ such that if I,T are intervals with* $\mathrm{cl}(I)\subset\mathrm{int}(T)$ *and L and R the components of $T\setminus I$ such that*

(a) $|L|\leqslant|I|$ *or* $|R|\leqslant|I|$;

(b) $I\cap K_f=\varnothing$;

then

$$B(f,T,I)\geqslant A_1.$$

Proof. Let A_0 be the number from Lemma 3.5. We may assume that $A_0\in(0,1)$. We will prove the lemma for $A_1=\frac{1}{3}\cdot(A_0)^2$. By possibly renaming L and R, we may consider the case that $|R|\leqslant|I|$. Then from Lemma 3.5 we get

$$\frac{|f(I)|\,/\,|I|}{|f(R)|\,/\,|R|}\geqslant A_0, \tag{3.11}$$

and hence

$$B(f,T,I)\geqslant\frac{|f(T)|}{|f(L)|}\cdot\frac{|L|}{|T|}\cdot A_0\geqslant A_0\cdot\frac{|L|}{|T|}.$$

If $|L|\geqslant|I|$ then it follows from this and $|R|\leqslant|I|$ that

$$B(f,T,I)\geqslant A_0\cdot\frac{|L|}{|L|+|I|+|R|}\geqslant A_0\cdot\frac{|L|}{|L|+|L|+|L|}=A_0\cdot\frac{1}{3}\geqslant A_1$$

and the lemma is proved. So assume that $|L|\leqslant|I|$. Then applying Lemma 3.5 again we get

(3.12) $$\frac{|f(I)|/|I|}{|f(L)|/|L|} \geq A_0,$$

and it follows from (3.11), (3.12) and Lemma 3.6 that $B(f,T,I) \geq A_0 \cdot A_0 > A_1$. Q.E.D.

Proof of Theorem 3.3. Let $\varepsilon \in (0,1)$. Let $m \leq n$ be the smallest number such that $|f^m(L)| \geq \varepsilon |f^m(M)|$ and $|f^m(R)| \geq \varepsilon |f^m(M)|$. Let A_1 be the number from Lemma 3.7 and assume $A_1 < 1$. Let $V > 0$ be the number from Lemma 3.4. Let

$$B_1 = [A_1]^{51\#K_f+1} \cdot e^{-V}.$$

We claim that

$$B(f^m, T, M) \geq B_1.$$

Indeed let $t(1) < t(2) < \ldots < t(s) < m$ be the integers $t < m$ such that $f^t(T)$ contains a component of $\mathcal{U} \setminus \mathcal{V}$ or such that $f^t(T) \cap K_f \neq \varnothing$. Since the intersection multiplicity of $\{T, f(T), \ldots, f^{n-1}(T)\}$ is at most 17 one gets $s < 51\#K_f$. From the choice of m either $|f^{t(i)}(L)| \leq \varepsilon |f^{t(i)}(M)| \leq |f^{t(i)}(M)|$ or $|f^{t(i)}(R)| \leq \varepsilon |f^{t(i)}(M)| \leq |f^{t(i)}(M)|$. So from Lemma 3.7

$$B(f, f^{t(i)}(T), f^{t(i)}(M)) \geq A_1.$$

From Lemma 3.4

$$\sum_{j \leq m-1, j \notin \{t(1), \ldots, t(s)\}} \log B(f, f^j(T), f^j(M)) \geq -V.$$

Hence the claim follows. From the claim and the definition of the operator B the theorem easily follows. Q.E.D.

4. Some simplifications and the induction assumption

The proof of Theorem A will go by induction on the number of turning points of f. So let $\mathcal{A}^d$ be the collection of all endomorphisms of N in $\mathcal{A}$ with $f(\partial N) \subset \partial N$ and with precisely d turning points. So we will prove inductively that

(Ind$_d$) maps in $\bigcup_{i=0}^{d} \mathcal{A}^i$ have no wandering intervals.

Because the proof of Theorem A goes by induction on the number of turning points we have to consider a more general situation: the manifold N is not necessarily connected (but does consist of a finite number of components). This does not give, however, a more general result. In fact if a map $f: N \to N$ has a wandering interval then one of the connected components N' of N is periodic of period s and the map $f^s: N' \to N'$ has a wandering interval. Therefore, if we prove the theorem for connected manifolds we get also the same theorem for non-connected manifolds. However, if f has d turning points, the map f^s may have more than d turning points. This is the main reason to start with a disconnected manifold. So if Ind_{d-1} holds then we may use the following fact: *if there exists a finite disjoint union of intervals which is invariant by f and contains a wandering interval then the union of these intervals contains at least d turning points.*

By extending f to a slightly bigger interval we may assume that

$$f(\partial N) \subset \partial N.$$

It suffices to prove the theorem for maps f such that none of the turning points is contained in (the closure) of a wandering interval. Indeed, otherwise f also has a wandering interval W containing a turning point c in its interior. Now modify the map f in a small neighbourhood $V \subset W$ of c to a map $g \in \mathscr{A}$ such that $f=g$ on $N-V$ for which $g|V$ has a unique turning point in c and $g(c)$ is not contained in a wandering interval. Then $f(W)$ is still a wandering interval for g (the forward iterates of $f(W)$ under f and g are the same because these iterates never enter W) and $g(c)$ is not contained in a wandering interval of g. Repeating this procedure for every wandering interval which contains a turning point in its closure we get a map g which still has wandering intervals but such that none of its turning points is contained in the closure of a wandering interval. So it suffices to prove the theorem for g.

Furthermore, for each turning point c of f there exists a neighbourhood S_c of c and a continuous involution $\tau: S_c \to S_c$ (which is not the identity) such that $f(\tau(x))=f(x)$ for all $x \in S_c$. From the non-flatness conditions for maps in $\mathscr{A}$ the map τ is Lipschitz.

5. The pullback of space: the Koebe/Contraction Principle

In this section we will start the proof of the non-existence of wandering intervals for maps in $\mathscr{A}$. This will be proved by contradiction. More precisely, suppose that J is a wandering interval and that J is not contained in a larger wandering interval. The strategy of the proof is to show that there exists necessarily an interval I which strictly contains J such that $\inf_{n\geq 0}|f^n(I)|=0$. Using the next principle, which also can be found in [L], it follows that I is a wandering interval (since $f^n(J)$ does not converge to a periodic orbit the same holds for $f^n(I)$), contradicting the maximality of J.

5.1. CONTRACTION PRINCIPLE. *Suppose I is an interval such that*

$$\inf_{n\geq 0} |f^n(I)| = 0$$

then I is a wandering interval or there exists a periodic orbit $\mathcal{O}$ such that $f^k(I)\to\mathcal{O}$ as $k\to\infty$. In particular, if I contains a wandering interval J then it is also a wandering interval.

Proof. Let $\mathcal{I}=\bigcup_{n\geq 0} f^n(\text{int}(I))$. Then $\mathcal{I}$ is forward invariant.

First suppose that there exists a component U of $\mathcal{I}$ and $n>0$ such that $f^n(U)\cap U\neq\emptyset$. Since $\mathcal{I}$ is forward invariant this implies $f^n(U)\subset U$. There are three cases.

(1) U is an interval which contains a fixed point p of f^n: $U\to U$ in its interior. In this case some iterate of I contains this fixed point of f^n in its closure and since $\inf_{k\geq 0}|f^k(I)|=0$ this fixed point of f^n must attract I, $f^k(I)\to O(p)$ as $k\to\infty$. So we are finished in this case.

(2) U is an interval and there exists no fixed point as in (1). Then $\text{cl}(U)$ contains in its boundary an attracting fixed point p of $f^n\colon \text{cl}(U)\to\text{cl}(U)$. If $f^n(U)\neq U$ then every point in $\text{cl}(U)$ is asymptotic to $O(p)$, $\omega(I)=O(p)$. If $f^n(U)=U$ then the boundary point $\{q\}=\partial U\setminus\{p\}$ is a repelling periodic point and $\inf_{k\geq 0}|f^k(I)|=0$ implies that no iterate of I contains q in its closure. Since every point in $\text{int}(U)$ is asymptotic to $O(p)$ this implies that $f^k(I)\to O(p)$. Again the result follows.

(3) U is a circle. Then there exists a finite collection $n_1<\ldots<n_r$ of positive integers such that $\bigcup_{i=1}^r f^{n_i}(\text{int}(I))$ covers S^1. But since $\inf_{n\geq 0}|f^n(I)|=0$ this implies that there exists an integer $n>n_r$ such that $f^n(I)$ is strictly contained in $f^{n_i}(\text{int}(I))$ for some $i=1,\ldots,r$. But then f^{n-n_i} has an attracting fixed point in $f^{n_i}(\text{int}(I))$ which attracts I. So again the result follows.

Now assume that for every component U of $\mathcal{I}$ one has $f^n(U)\cap U=\emptyset$ for all $n\geq 1$. Since $\mathcal{I}$ is forward invariant and this holds for each component, this implies that $f^n(U)\cap f^m(U)=\emptyset$ for all $n>m\geq 0$. It follows that U and therefore I is a wandering interval (or asymptotic to a periodic orbit). Q.E.D.

5.2. *Definition*. The *pullback* of $P_n\supset J_n$ is the sequence of intervals

$$\{P_i | i=0,1,\ldots,n\}$$

where P_{i-1} is the maximal interval containing J_{i-1} such that

$$f(P_{i-1})\subset P_i$$

for each $i=1,\dots,n$. The integers i for which P_i contains a turning point in its closure are called the *cutting times*. It will turn out to be useful to call n also a cutting time. This pullback is *monotone* if $f^n|P_0$ is monotone and $f^n(P_0)=P$. Furthermore it is said to be *diffeomorphic* if $f^n|P_0$ is a diffeomorphism and $f^n(P_0)=P$ and *unimodal* if P_i contains at most one turning point for each i.

All the pullbacks we will consider are unimodal because of the following lemma:

5.3. LEMMA. *There exists $\eta>0$ such that if $P_n\supset J_n$ and $|P_n|\leq\eta$ then the pullback of P_n is unimodal.*

Proof. Follows from the Contraction Principle and since by assumption no wandering interval contains two turning points. Q.E.D.

5.4. CONTRACTION/KOEBE PRINCIPLE. *For each $\varepsilon>0$ and each $p\in\mathbf{N}$ there exists $N_0(\varepsilon,p)$ with the following properties. Let $P_n\supset J_n$ contain an ε-scaled neighbourhood of J_n. If the pullback of P_n has intersection multiplicity $\leq p$ then $n\leq N_0(\varepsilon,p)$.*

Proof. Let $\varrho=\min(\varepsilon,1/2)$ and $\hat{P}_n\subset P_n$ be a ϱ-scaled neighbourhood of J_n. Let $m(0)<m(1)<\dots<m(l)=n$ be the cutting times of the pullback of P_n. Since the intersection multiplicity of the pullback is $\leq p$, $l\leq p\cdot d$. Let $\hat{P}_i^{\pm}$ be the components of $\hat{P}_i\setminus J_i$. For $t=0,\dots,l-1$, let $P^+_{m(t)}$ be the component which contains the turning point. Now the map

$$f^{n-m(l-1)+1}:\hat{P}_{m(l-1)+1}\to\hat{P}_n$$

is monotone. Since $\hat{P}_n$ is a ϱ-scaled neighbourhood of J_n it follows from the Macroscopic Koebe Principle that both components of $\hat{P}_{m(l-1)+1}\setminus J_{m(l-1)+1}$ have length at least $B_0(\varrho)$. From this and the non-flatness of the turning points we get that there exists a universal constant $C\in(0,1)$ such that

$$|\hat{P}^-_{m(l-1)}|\geq C\cdot B_0(\varrho)\cdot|J_{m(l-1)}|.$$

Because $\hat{P}_{m(l-1)}$ is symmetric around a turning point, we also have that $|\hat{P}^+_{m(l-1)}|\geq|J_{m(l-1)}|$. Therefore

$$|\hat{P}^{\pm}_{m(l-1)}|\geq C\cdot B_0(\varrho)\cdot|J_{m(l-1)}|.$$

Repeating this $l\leq p\cdot d$ times we get, letting $g(x)=C\cdot B_0(x)$,

$$(*)\qquad |\hat{P}_0^{\pm}|\geq g^l(\varrho)\cdot|J|.$$

Hence there exists an interval P which does not depend on n and which *strictly* contains the wandering interval J such that $|f^n(P)| \leq (1+2\varrho)|J_n| \leq 2|J_n|$ for each n as above. Now if there exists no upperbound $N_0(\varepsilon, p)$ for n then we would get $|f^{n_i}(P)| \leq 2|J_{n_i}| \to 0$ for some sequence $n_i \to \infty$ and since P contains J this would imply from the Contraction Principle that P is also a wandering interval. But this contradicts the maximality of J. Q.E.D.

6. Disjointness of orbits of intervals

In this section we will give some upperbounds on the intersection multiplicity of orbits of intervals. These bounds are needed in order to apply the Macroscopic Koebe Principle. Let $J_n = f^n(J)$. We will define a natural neighbourhood T_n of J_n such that its monotone pullback has good disjointness properties. (For those familiar with the circle homeomorphisms these neighbourhoods will coincide with the neighbourhood $[f^{q_{k-2}}(J), f^{q_{k-2}+(a_k-1)q_{k-1}}(J)]$ of $f^{q_k}(J)$ when $n = q_k$ and with

$$[f^{q_{k-2}+(i-1)q_{k-1}}(J), f^{q_{k-2}+(i+1)q_{k-1}}(J)]$$

of $f^{q_{k-2}+iq_{k-1}}(J)$ when $n = q_{k-2} + iq_{k-1}$ and $1 \leq i \leq a_k - 1$.)

6.1. *Definition.* We say that J_{n_1} and J_{n_2} have the *same orientation* if the forward iterate of f which sends one of these intervals in the other is orientation preserving. If $n \in \mathbb{N}$, we say that J_k is a *predecessor* of J_n if $0 \leq k < n$, if J_k and J_n have the same orientation and if $J_s \subset (J_k, J_n)$ and $0 \leq s < n$ implies that J_s and J_n have different orientations. If J_n has a predecessor to its left (right) then we denote the corresponding iterate by $L(n)$ (respectively $R(n)$). J_n has a *successor* J_{n+a} if

(1) J_{n-a} is predecessor of J_n (with $0 < a \leq n$);

(2) $f^a | [J_{n-a}, J_{n+a}]$ is monotone, orientation preserving, and its image contains no predecessor of J_n (if $L(n)$ and $R(n)$ both exist and if for example $n-a = L(n)$ then this implies that $f^t[J_{n-a}, J_{n+a}] \subset [J_n, J_{R(n)})$;

(3) if $J_k \subset (J_n, J_{n+a})$ and $k = 0, 1, \ldots, n+a-1$ then the intervals J_k and J_{n+a} have different orientations.

Next we define the *natural neighbourhood* of J_n to be the biggest open interval containing J_n which contains no neighbourhood of a predecessor or successor.

Remark. Of course J_n can have at most one predecessor on each side and it has a predecessor to, say, its right if there exists an interval J_s with $s < n$ to the right of J_n with the *same orientation* as J_n. Moreover, as we will see in the lemma below an interval J_n

has at most one successor; so denote this successor by $s(n)$. Therefore, if J_n has two predecessors and no successor then the natural neighbourhood T_n of J_n is equal to $T_n=[J_{L(n)}, J_{R(n)}]$ and if it has a successor then it is equal to

$$T_n=[J_{L(n)}, J_{s(n)}] \quad \text{or} \quad T_n=[J_{s(n)}, J_{R(n)}].$$

6.2. LEMMA. *For every $n\in\mathbf{N}$, J_n can have at most one successor.*

Proof. Suppose that J_n had two successors. Then it also has two predecessors, $J_{L(n)}$ and $J_{R(n)}$. By definition $f^{n-L(n)}\colon [J_{L(n)}, J_n]\to[J_n, J_{R(n)}]$ and $f^{n-R(n)}\colon [J_n, J_{R(n)}]\to[J_{L(n)}, J_n]$ are both monotone. Hence J is attracted to a periodic point, a contradiction. Q.E.D.

6.3. LEMMA. *Assume that the interval J_n has two predecessors $J_{L(n)}, J_{R(n)}$, and a successor $J_{s(n)}$. If this successor is to the right of J_n then the predecessors of $J_{s(n)}$ are J_n and $J_{R(n)}$ and if $J_{s(n)}$ has a successor then this successor is between $J_{s(n)}$ and $J_{R(n)}$.*

Proof. The left predecessor of $s(n)$ is n by the definition of $s(n)$. The right predecessor of $s(n)$ is certainly defined because $J_{R(n)}$ and $J_{s(n)}$ have the same orientation and $R(n)<n<s(n)$. Let us show that $R(n)$ is this predecessor. If this was not the case then there exists $k<s(n)$ so that $J_k\subset(J_{s(n)}, J_{R(n)})$ and so that J_k and $J_{s(n)}$ have the same orientation. Because of the definition of $R(n)$ this implies certainly that $k>n$. Because $k<s(n)$ this implies that $0\leq k-n<s(n)-n=n-L(n)$ and $L(n)+k-n<n$. From the first of these two inequalities and the definition of $s(n)$ it follows that f^{k-n} is monotone and orientation preserving on $H=[J_{L(n)}, J_n]$. In particular $J_{L(n)+k-n}$ has the same orientation as J_n. From this, the definition of $L(n)$ and $R(n)$ and the second of these inequalities it follows $J_{L(n)+k-n}$ cannot be between $J_{L(n)}$ and $J_{R(n)}$. It follows that

$$f^{k-n}(H)\supset[J_{L(n)}, J_k]\supset[J_n, J_{s(n)}]=f^{n-L(n)}(H).$$

In particular, $f^{(n-L(n))-(k-n)}$ maps $f^{k-n}(H)$ monotonically into itself, and hence J would be attracted to a periodic attractor, a contradiction.

Let us finally show that $s(n)$ cannot have a successor to its left. Indeed if it did, then by definition $f^{s(n)-R(n)}$ would map $[J_{s(n)}, J_{R(n)}]$ monotonically into $[J_n, J_{s(n)}]$. From the definition of $s(n)$, $f^{s(n)-n}$ maps $[J_n, J_{s(n)}]$ monotonically into $[J_{s(n)}, J_{R(n)}]$. Combining this gives that J is attracted to a periodic attractor. With this contradiction the proof of this lemma is completed. Q.E.D.

Remark. The previous lemma implies that if J_n has a successor $J_{s(n)}$ and $J_{s(n)}$ also has a successor $J_{s(s(n))}$ then $s(n)-n=a=s(s(n))-s(n)$ and $J_{s(n)}$ is between J_n and $J_{s(s(n))}$. So continuing this there exists a maximal integer k such that $J_{s^{i+1}(n)}$ is a successor of $J_{s^i(n)}$ for $i=0,1,\ldots,k-1$. In this case the intervals

$$J_{s(n)},\ J_{s(s(n))},\ \ldots,\ J_{s^k(n)}(n)$$

lie ordered and $f^a|[J_n, J_{s^k(n)}]$ is monotone. So f^a acts as a translation on these intervals.

6.4. THEOREM. *Let $n\in\mathbb{N}$ and assume that J_n has two predecessors $J_{R(n)}$ and $J_{L(n)}$. Let M_n be an open interval contained either in $[J_n, J_{R(n)}]$ or in $[J_{L(n)}, J_n]$. Assume that $\{M_{t_0}, M_{t_0+1},\ldots, M_n\}$ is a monotone pullback of M_n. If the intersection multiplicity of this piece is at least $2p$ and $p\geq 2$ then there exists $t\in\{t_0,\ldots,n\}$ such that*

(1) $J_{s(t)}, J_{s^2(t)},\ldots, J_{s^{2p-2}(t)}$ are defined;

(2) $n=s^p(t)$ and $J_{s^j(t)}$ is contained in M_n for $j=p,\ldots,2p-2$.

COROLLARY. *If the pullback of an interval $T\supset J_n$ with T contained in the natural neighbourhood T_n is monotone then the intersection multiplicity of this pullback is at most* 11. *Similarly, if $T\supset J_n$ and $s^k(n)$ does not exist then the monotone pullback of T has at most intersection multiplicity $2k+4$.*

Proof of Corollary. Consider the pullback of $T\cap[J_n, J_{R(n)}]$ and $T\cap[J_{L(n)}, J_n]$ separately. If the intersection multiplicity of T is at least 12 then the pullback of either $T\cap[J_n, J_{R(n)}]$ or $T\cap[J_{L(n)}, J_n]$ has intersection multiplicity ≥ 6. So take $p=3$ and the previous theorem implies that $p+1\leq 2p-2$ and $J_{s(n)}$ is contained in $T\cap[J_n, J_{R(n)}]$, which is impossible since $T\subset T_n$. The second statement follows also immediately. (Note that $2k+4\geq 6$ for $k\geq 1$.) Q.E.D.

Proof of Theorem 6.4. In order to be definite assume that $M_n\subset[J_n, J_{R(n)}]$. By assumption there exists a point y which is contained in $2p$ of the intervals $M_{t_0},\ldots, M_n$. Of course this implies that for at least p of these intervals M_i the corresponding intervals J_i all lie on the same side of y. So let $N\geq p$ be the maximal number of distinct integers $t\leq i(1),\ldots,i(N)\leq n$ such that each of the intervals $M_{i(1)},\ldots, M_{i(N)}$ contains this point y, $J_{i(j)}$ all lie on one side of y and that these are labeled so that

$$[J_{i(1)}, y]\supset[J_{i(2)}, y]\supset\ldots\supset[J_{i(N)}, y].$$

Since M_i has a common endpoint with J_i this implies that the intervals $J_{i(1)},\ldots, J_{i(N)}$ all have the same orientation.

Claim 1. $i(1)<i(2)<\ldots<i(N)$ and we may assume that $i(N)=n$. Furthermore $J_{i(1)}$ cannot be contained in $f^t(M_{i(1)})$ for $t=1,\ldots,i(N)-i(1)$ if f^t is orientation preserving on

$M_{i(1)}$. In particular if we take $a=i(N)-i(N-1)$ then f^a maps $[J_{i(1)}, J_{i(N-1)}]$ (which is contained in $M_{i(1)}$) monotonically into $(J_{i(1)}, J_{i(N)}]$.

Proof of Claim 1. For $j \in \{2,3,\dots,N\}$ the interval $J_{i(j)}$ is contained in $[J_{i(j-1)}, y] \subset M_{i(j-1)}$ and $J_{i(j)}$ and $J_{i(j-1)}$ have the same orientation. Since $f^{n-i(j-1)}: M_{i(j-1)} \to M_n \subset [J_n, J_{R(n)}]$ is monotone this implies that $J_{i(j)+n-i(j-1)}$ has the same orientation as J_n and is between J_n and $J_{R(n)}$. Since $J_{R(n)}$ is a predecessor of J_n this implies that $i(j)+n-i(j-1)>n$ and therefore $i(j)>i(j-1)$. Let us show that we may assume that $i(N)=n$. Indeed from what we have shown $f^{n-i(N)}$ is monotone on $M_{i(j)}$ for $j=1,\dots,N$. In particular letting $i'(j)=i(j)+(n-i(N))$, and taking the images of these intervals under this map we get that $M_{i'(1)},\dots,M_{i'(N)}$ all contain one point y', $J_{i'(j)}$ all lie on one side of y' and that these are labeled so that $[J_{i'(1)}, y] \supset [J_{i'(2)}, y] \supset \dots \supset [J_{i'(N)}, y]$. Since $i'(N)=n$ we may as well assume that $i(N)=n$. The first statement of the claim follows and it follows that f^a is monotone and orientation preserving on $M_{i(1)} \supset [J_{i(1)}, J_{i(N-1)}]$. Now $J_{i(1)}$ cannot be contained in $f^t(M_{i(1)})=M_{i(1)+t}$ for $t=1,\dots,i(N)-i(1)$ because otherwise $J_{i(N)-t}$ would be contained in $M_{i(N)}=M_n$ and would have the same orientation as J_n. But this is impossible because $M_n \subset [J_n, J_{R(n)}]$ and $J_{R(n)}$ is a predecessor of J_n. Q.E.D.

Claim 2. If $k<n$, J_k and J_n have the same orientation and $J_k \subset [J_{i(1)}, J_n]$ then $k \in \{i(1), i(2), \dots, i(N)\}$. Furthermore if we let $a=i(N)-i(N-1)$ then $i(j+1)-i(j)=a$ for $j=1,2,\dots,N-1$.

Proof. Let $j<n$ be maximal such that $J_k \subset [J_{i(j)}, y] \subset M_{i(j)}$. Therefore $f^{n-i(j)}$ maps J_k into M_n. Because J_k, $J_{i(j)}$ and J_n all have the same orientation, it follows that $J_{k+n-i(j)}$ also has the same orientation. This implies that $k \geq i(j)$. Suppose $k>i(j)$. If $M_k \not\supset J_n$ then $M_k \subset M_{i(j)}$ and $f^{k-i(j)}$ maps $M_{i(j)}$ monotonically into $M_k \subset M_{i(j)}$. This implies that J is attracted by a periodic orbit, contradiction. Hence, $M_k \not\supset J_n$ and by the maximality of N, $k \in \{i(0), \dots, i(N)\}$. This proves the first statement of the claim. From Claim 1, f^a maps $[J_{i(1)}, J_{i(N-1)}]$ monotonically and orientation preserving into $(J_{i(1)}, J_{i(N)}]$. It follows from this and the first part of this claim that $i(j)+a=i(j+1)$ for $j=1,\dots,N-2$. Thus Claim 2 is proved.

Define $i(N+j)=n+j \cdot a$ for $j=1,\dots,N$. As we have shown in the previous claim this formula holds for j negative. So we get

$$i(N+j)=n+j \cdot a \quad \text{for} \quad j=-N+1, -N+2, \dots, N-1, N.$$

Now consider the interval

$$H=[J_{i(1)}, J_{i(N)}].$$

The map $f^{n-i(1)}=f^{(N-1)a}$ maps $H\subset M_{i(1)}$ monotonically and orientation preserving into $M_n\subset[J_n, J_{R(n)}]$. In particular f^a maps $[J_{i(1)}, J_{i(2N-2)}]$ monotonically and orientation onto $[J_{i(2)}, J_{i(2N-1)}]$. Therefore $J_{i(j)}$ lies between $J_{i(j-1)}$ and $J_{i(j+1)}$ for $|j|\leqslant N-2$. Furthermore if $f^t(H)\cap[J_{i(1)}, J_{R(n)}]\neq\emptyset$ for some $t=1,\dots,(N-1)a$ then $f^t|H$ is either orientation reversing or t is a multiple of a. Indeed, assume that $f^t|H$ is orientation preserving and that this intersection is non-empty. By Claim 1, $f^t(H)\subset f^t(M_{i(1)})$ does not contain $J_{i(1)}$. So $f^t(J_{i(1)})$ must be contained in $[J_{i(1)}, J_{R(n)}]$. But $f^t(J_{i(1)})\subset f^t(H)$ cannot be contained in $[J_n, J_{R(n)}]$ since $J_{R(n)}$ is a predecessor of J_n. So $f^t(J_{i(1)})\subset[J_{i(1)}, J_n]$ and it follows from the previous claim that t must be a multiple of a. Furthermore consider J_k with $k<n$ and with the same orientation as J_n. This interval cannot be contained in $[J_{i(1)}, J_n]$ (see Claim 2) and neither in $[J_n, J_{R(n)}]$ from the definition of $J_{R(n)}$. Combining all this shows that the only intervals J_k with $k\leqslant i(2N-1)$ inside $[J_{i(1)}, J_{R(n)}]$ with the same orientation as J_n are intervals of the form $k=i(1)+j\cdot a$. It follows that $J_{i(j+1)}$ is a successor of $J_{i(j)}$ for $j=1,\dots,2N-2$. Q.E.D.

7. Wandering intervals accumulate on turning points

In this section we are going to prove that the ω-limit set of a wandering interval contains at least one turning point and thus prove that Ind_0 holds. From this is follows in particular that maps without turning points, e.g. circle homeomorphisms in $\mathscr{A}$, cannot have wandering intervals. So the proof in this section includes the classical proof of Denjoy.

Suppose we have a map $f\colon N\to N$ which has a wandering interval J which stays away from the turning points of f. But then we can modify the map f near these turning points without affecting the orbit of J. So change f so that each maximal interval on which f is monotone is mapped by f onto a component of N. Once we have done this we may assume that every pullback is monotone.

7.1. Proposition. *There exists n_0 such that if J_n, $n\geqslant n_0$, has two predecessors $J_{L(n)}$ and $J_{R(n)}$ and $|J_{R(n)}|>|J_n|$ and $|J_{L(n)}|>|J_n|$ then J_n has a successor and $|J_{s(n)}|<|J_n|$.*

Remark. Suppose for example that $J_{s(n)}$ is to the right of J_n. Then Lemma 6.3 implies that the interval $J_{s(n)}$ has two predecessors, namely J_n and $J_{R(n)}$ and that it has no left successor. From the conclusion of this proposition it follows that the assumptions of this proposition are again satisfied for $J_{n'}$ where $n'=s(n)$. So one can apply the proposition infinitely often!

Proof. Let $n_0=N_0(1;11)$ be as in the Koebe/Contraction Principle and let $n\geq n_0$. Suppose $s(n)$ is not defined and let $T_n=[J_{L(n)}, J_{R(n)}]$ be the natural neighbourhood of J_n. By the corollary to Theorem 6.4 the monotone pullback of T_n has intersection multiplicity bounded by 11. Because the intervals $|J_{L(n)}|, J_{R(n)}|\geq|J_n|$ this contradicts the Koebe/Contraction Principle. Hence $s(n)$ is defined. From Lemma 6.3 $L(s(n))=n$ and $R(s(n))=R(n)$. Furthermore $s(n)$ has no left successor. Now the natural neighbourhood of J_n is $T_n=[J_{L(n)}, J_{s(n)}]$ and again by the corollary to Theorem 6.4 the monotone pullback of T_n has intersection multiplicity bounded by 11. Since $|J_{L(n)}|>|J_n|$ it follows from the Koebe/Contraction Principle that $|J_{s(n)}|<|(J_n, J_{s(n)}]|<|J_n|$. Q.E.D.

7.2. Theorem. *If $f\in\mathcal{A}$ and f has a wandering interval the ω-limit set of J contains a turning point.*

Proof. Let us first show that $\omega(J)$ cannot be finite. Indeed otherwise $\omega(J)$ contains a periodic point p of, say, period k and there exists a neighbourhood U of p such that $U\cap\omega(J)=\{p\}$. Furthermore there exist a neighbourhood $V\subset U$ of p such that $f^k(V)\subset U$ and an integer n' for which $J_{n'}\subset V$ and such that $J_n\subset V$ whenever $n\geq n'$ and $J_n\cap U\neq\emptyset$. Since $f^k(V)\subset U$ this implies by induction that $J_n\subset V\cup f(V)\cup\ldots\cup f^{k-1}(V)$ for all $n\geq n'$ and therefore that $\omega(J)=O(p)$. Hence $\omega(J)$ is attracted to a periodic orbit, a contradiction.

Therefore, and since all intervals J_i are disjoint, there exists arbitrarily large integers $l, r<n$ such that J_l, J_r, J_n are in he same component of N, have the same orientation, $J_n\subset(J_l, J_r)$, such that

$$|J_n|\leq\min(|J_l|, |J_r|)$$

and, for $i=0,1,\ldots,n-1$,

$$J_i\cap(J_l, J_r) \text{ implies that } J_i \text{ has a different orientation.}$$

Assume that l, r, n are bigger than the number n_0 from above. It follows that J_l and J_r are the predecessors of J_n. So we can apply Proposition 7.1 and hence J_n has infinitely many successors $J_{s^k(n)}, k=1,2,\ldots$. From the description in Lemma 6.3, all these successors are contained in $[J_l, J_r]$, they either all lie to the right of the previous one or all to the left. Moreover $s^k(n)-s^{k-1}(n)$ is independent of k. It follows that as k tends to infinity these intervals $J_{s^k(n)}$ converge to a fixed point of f^a where $a=s(n)-n$. Hence this fixed point is an attracting fixed point with J in its basin, a contradiction. Q.E.D.

Now we know that iterates of wandering intervals tend to some turning point. In order to analyze the metric properties of iterates of a wandering interval we will pay special attention to the moments where the iterates get closest to some turning point. This is formalised in the following definition.

7.3. *Definition.* If c is a turning point in $\omega(J)$, $J_n \subset S_c$ and $m \geq n$, then we say that J_n is the *m-closest approach* to c if $(J_n, \tau(J_n)) \cap (\cup_{i \leq m} J_i) = \emptyset$. Similarly J_n is the *closest approach* to c if it is the n-closest approach to c. Now fix a turning point c in $\omega(J)$ and let $N(c)$ be the collection of integers $i \in \mathbf{N}$ with

$$J_i \subset S_c \quad \text{and} \quad [J_i, \tau(J_i)] \cap (\bigcup_{0 \leq j < i} J_j) = \emptyset.$$

From now on let

$$N(c) = \{n(1), n(2), \dots\}$$

where $n(1) < n(2) < \dots$ We call $J_{n(1)}, J_{n(2)}, \dots$ the *sequence of closest approach to c.*

7.4. Lemma. *If $J_{n(k)}$ has a successor $J_{s(n(k))}$ then $s(n(k)) = n(k+1)$ and $J_{n(k+1)}$ is between $J_{n(k)}$ and c. Furthermore, if there exists an integer j such that $s(j)$ and $s^2(j))$ are both defined and such that $j < n(k) < s(j)$ then*

$$n(k+1) = s(n(k)).$$

Proof. In order to be definite assume that $J_{n(k)}$ is to the left of c. Since $J_{n(k)}$ is a closest interval to c, any predecessor of $J_{n(k)}$ to its right must also be to the right of c. Therefore $c \in [J_{n(k)}, J_{R(n(k))}]$ and there can be no successor of $J_{n(k)}$ to its left. Hence if $J_{n(k)}$ has a successor then it must be to its right and because $f^{s(n(k))-n(k)}|[J_{L(n)}, J_{s(n)}]$ is monotone it even must be between $J_{n(k)}$ and c. So if $s(n(k)) \neq n(k+1)$ then $n(k+1) < s(n(k))$. Consider $H = [J_{L(n(k))}, J_{n(k)}]$ and let $a = n(k) - L(n(k)) = s(n(k)) - n(k)$. Then f^{2a} is monotone on H and $a > n(k+1) - n(k)$. Since $L(n(k)) + n(k+1) - n(k) < n(k)$ the interval $J_{L(n(k))+n(k+1)-n(k)} \subset f^{n(k+1)-n(k)}(H)$ is not contained in $[J_{n(k)}, \tau(J_{n(k)})]$ whereas by assumption $J_{n(k+1)} \subset f^{n(k+1)-n(k)}(H)$ is contained in this interval $[J_{n(k)}, \tau(J_{n(k)})]$. It follows that either $J_{n(k)}$ or $\tau(J_{n(k)})$ is contained in $f^{n(k+1)-n(k)}(H)$. Hence for $t = a - (n(k+1) - n(k))$ one has $0 < t < a$ and

$$(*) \qquad f^t(J_{n(k)}) \subset f^a(H) = [J_{n(k)}, J_{s(n(k))}] \subset [J_{n(k)}, c].$$

Since $J_{s(n(k))}$ is the successor of $J_{n(k)}$ this implies that f^t is orientation reversing on H. So $J_{L(n(k))+t}$ is to the right of $J_{n(k)}$. But since $L(n(k)) + t < n(k)$ this interval cannot be contained in $[J_{n(k)}, \tau(J_{n(k)})]$. Therefore, and because of $(*)$, $f^t(H)$ contains c; this contradicts the monotonicity of $f^a|H$. This proves the first statement of this lemma.

Let us now prove the second statement. According to the first part it is enough to show that $s(n(k))$ is defined. Now let $a=s(j)-j$ then $L(j)=j-a$, $s(j)=j+a$ and $s(s(j))=j+2a$. Since f^a is monotone on $[L_{j-a}, L_{j+2a}]$ and $n(k)-j<s(j)-j=a$ it follows that $J_{n(k)}$ is contained in $T=[J_{n(k)-a}, J_{n(k)+a}]$ and that f^a is monotone on T. Furthermore there is no predecessor of $J_{n(k)}$ in $f^a(T)$ because otherwise there would be a predecessor of J_{j+a} in $f^a([L_j, L_{j+2a}])$, contradicting that $s^2(j)$ exists. So property (1) and (2) of the definition of successor of $J_{n(k)}$ hold. Finally there is also no interval J_t in $[J_{n(k)}, J_{n(k)+a}]$ with $t<n(k)+a$ and with the same orientation as $J_{n(k)}$ because $J_{t+(s(j)-n(k))}$ would have the same orientation as $J_{s(j)}$ and be contained in $[J_{s(j)}, J_{s(j)+a}]$, contradicting the definition of $s^2(j)$. Q.E.D.

8. Topological properties of unimodal pullback's

From now on we will assume

$$(\mathrm{Ind}_{d-1}) \qquad \text{maps in } \bigcup_{i=0}^{d} \mathscr{A}^i \text{ have no wandering intervals}$$

and try to show that this implies Ind_d. Throughout this section we will consider properties of pullbacks of two intervals. The first of these intervals is $Q_{n(k)} \supset J_{n(k)}$: this is the interval in $M \setminus \{c\}$ such that

$$f(Q_{n(k)}) = [f(J_{n(k-1)}), f(J_{n(k+1)})].$$

Similarly let

$$\hat{Q}_{n(k)} = Q_{n(k)} \cup [J_{n(k+1)}, \tau(J_{n(k+1)})].$$

In this section we will describe the structure of the unimodal pullback $P_0, \ldots, P_{n(k)}$ of intervals $P_{n(k)}$ which are contained in $Q_{n(k)}$ or in $\hat{Q}_{n(k)}$.

In the next result it is shown that the intervals from the pullback of $\hat{Q}_{n(k)}$ meet the turning points in a periodic way.

8.1. STRUCTURE THEOREM. *Let $P_{n(k)} \supset J_{n(k)}$ be an interval which is contained in $\hat{Q}_{n(k)}$. Let $m(0)<m(1)\ldots<m(l)=n(k)$ be the cutting times of its unimodal pullback $P_0, \ldots, P_{n(k)}$ and let c_j denote the turning point in $P_{m(j)}$. Then we have the following properties.*

(1) *If $i \in \{0, \ldots, l-d\}$ then $J_{m(i)}$ is a $m(i+d)-1$ closest approach to c_i;*

(2) *If $i \in \{l-d+1, \ldots, l\}$ then $J_{m(i)}$ is a $n(k)$ closest approach to c_i;*

(3) $c_i = c_{i+d}$ *for* $i=0,\dots,l-d$ *and* $\{c_{l-d+1},\dots,c_l\}$ *are distinct;*

(4) *If* $J_{m(i)+j} \subset (J_{m(i)}, \tau(J_{m(i)}))$, $i\in\{0,\dots,l-1\}$ *and* $m(i)<m(i)+j\le n(k)$ *then*

$$P_{m(i)+j}\subset [J_{m(i)},\tau(J_{m(i)})];$$

(5) $P_{m(i+d)}\subset[J_{m(i)},\tau(J_{m(i)})]\subset P_{m(i)}$ *for* $i=0,1,\dots,l-d$ *and therefore* $f^{m(i+d)-m(i)}$ *maps* $P_{m(i)}$ *into itself.*

Proof of (4). If property (4) does not hold then the closure of $J_{m(i)}$ is contained in the interior of $P_{m(i)+j}$. So the closure of $J_{m(i)+n(k)-(m(i)+j)}$ is contained in the interior of $P_{n(k)}$. From the definition of $P_{n(k)}$ this implies $m(i)+n(k)-m(i)-j\ge n(k)$ and therefore $j\le 0$, a contradiction.

Proof of (1) *and* (2). Let us just prove (1). Statement (2) is proved in exactly the same way. Suppose by contradiction that there exists $l\in\{0,\dots,m(i+d)-1\}$ such that $J_l\subset(J_{m(i)},\tau(J_{m(i)}))$. Then $J_{l+n(k)-m(i)}=f^{n(k)-m(i)}(J_l)\subset f^{n(k)-m(i)}(P_{m(i)})\subset P_{n(k)}$ and $l\neq m(i)$. Hence $l>m(i)$. From statement (4) we know that $P_l\subset P_{m(i)}$. So $f^{l-m(i)}$ maps $P_{m(i)}$ into $P_l\subset P_{m(i)}$. Because $l<m(i+d)$ the map $f\colon \bigcup_{t=0}^{l-m(i)-1} f^t(P_{m(i)})\to\bigcup_{t=0}^{l-m(i)-1} f^t(P_{m(i)})$ has at most $d-1$ turning points. Since $J_{m(i)}$ is a wandering interval of this map, we get a contradiction with the induction hypothesis.

Proof of (3). Suppose there are $i-d<j\le i\le s$ with $c_i=c_j$. From (1) we get that $J_{m(i)}$ and $J_{m(j)}$ are both $m(j+d)$ closest approaches to $c_i=c_j$. Hence because $m(i), m(j)< m(j+d)$ this implies $i=j$. Since f has precisely d turning points one gets that $c_{l-d+1},\dots,c_l$ are distinct and that $c_i=c_{i+d}$ for $i\in\{0,\dots,l-d\}$.

Proof of (5). The proof of statement (5) follows immediately from the other statements. Q.E.D.

The following theorem shows that we can even take monotone pullbacks of intervals which contain topologically rather large sets. A unimodal version of this theorem was already used by J. Guckenheimer for his proof of the non-existence of wandering intervals for unimodal maps with negative Schwarzian derivative.

8.2. Monotone Extension Theorem. *Let* $P_{n(k)}=\hat{Q}_{n(k)}$ *and* $\{P_0,P_1,\dots,P_{n(k)}\}$ *be its unimodal pullback. Let* $m(0)<m(1)<\dots<m(l)=n(k)$ *be the cutting times and* c_i *the turning point in* $P_{m(i)}$. *Let* $H_{m(i)}\supset J$ *be the maximal interval such that* $f^{m(i)}$ *is monotone on* $H_{m(i)}$. *Let* $R_{m(i)}$ *be the component of* $P_{m(i)}\setminus J_{m(i)}$ *which contains* c_i *and* $L_{m(i)}$ *the other component. If the number* l *of cutting times of the pullback is at lest* $d+1$ *then*

$$f^{m(i)}(H_{m(i)}) \supset [L_{m(i)}, c_i],$$

$$f^{m(i-1)}(H_{m(i)}) \supset [L_{m(i-1)}, c_{i-1})$$

and

$$f^{m(i)-m(i-1)}(L_{m(i-1)}) \subset L_{m(i)}$$

for $i=0,1,\ldots,l-d$.

Proof. Let us first show that

$$(*) \qquad f^{m(i+1)-m(i)}(P_{m(i)}) \supset [J_{m(i+1)}, c_{i+1}]$$

for $i=0,1,2,\ldots,l-d$. Suppose by contradiction that there exists $i \in \{0,1,\ldots,l-d\}$ with $P_{m(i+1)} \supset f^{m(i+1)-m(i)}(P_{m(i)}) \not\ni c_{i+1}$. By statement (5) of the previous theorem $f^{m(i+d)-m(i)}$ maps $P_{m(i)}$ into itself. Now $f^{m(i+1)-m(i)}(P_{m(i)}) \not\ni c_{i+1}$ implies that

$$T = \bigcup_{t=0}^{m(i+d)-m(i)-1} f^t(P_{m(i)})$$

does not contain c_{i+1}. Hence f maps T into itself and has at most $d-1$ turning points. Since $J \subset P_{m(i)}$ it follows from the induction hypothesis that J is not a wandering interval, a contradiction. This proves (*). Furthermore

$$(**) \qquad f^{m(i+1)-m(i)}: L_{m(i)} \to L_{m(i+1)} \text{ is monotone and onto}$$

for $i=0,1,2,\ldots,l-d$ because otherwise there exists such an integer i with

$$f^{m(i+1)-m(i)}(R_{m(i)}) = L_{m(i+1)}$$

and then as before

$$T = \bigcup_{t=0}^{m(i+d)-m(i)-1} f^t(R_{m(i)} \cup J_{m(i)})$$

contains at most $d-1$ turning points and f maps this interval into itself. Since J is contained in T this contradicts the induction hypothesis. It follows from (*) and (**) that $f^{m(i+1)-m(i)}$ maps $[L_{m(i)}, c_i]$ monotonically over $[L_{m(i+1)}, c_{i+1}]$. The theorem clearly follows. Q.E.D.

Next we give two results about the disjointness of unimodal pullbacks of intervals in $Q_{n(k)}$ and in $\hat{Q}_{n(k)}$. The first result deals with the unimodal pullback of $Q_{n(k)}$.

8.3. THEOREM. *Let $m(0)<m(1)<\ldots<m(l)=n(k)$ be the cutting times of the unimodal pullback of $Q_{n(k)}$. Then:*

(1) $l\leq d-1$.

(2) *For every $0\leq j\leq l$ the second successor $s^2(m(j))$ is not defined.*

(3) *The intersection multiplicity of the unimodal pullback of $Q_{n(k)}$ is universally bounded (in fact by $12d$).*

Proof. Suppose by contradiction that $l\geq d$. By statement (5) of Theorem 8.1 it follows that $Q_{m(l)}=Q_{m(k)}$ is contained in $Q_{m(l-d)}$. Hence f maps

$$\bigcup_{t=0}^{m(l)-m(l-d)-1} f^t(Q_{m(l-d)})$$

into itself and since $Q_{n(k)}$ contains no turning point this map has at most $d-1$ turning points. This contradicts Ind_{d-1}. So let us prove statement 2 by assuming by contradiction that there exists $j\in\{0,1,\ldots,l\}$ for which $s(m(j))$ and $s(s(m(j)))$ are defined. Because $m(j)$ is $n(k)$-closest we get $s(m(j))>n(k)$. Hence from Lemma 7.4 we get that $n(k+1)=s(n(k))$ and $s(n(k))-n(k)=s(j)-j$. Because the closure of $J_{s(m(j))}$ is contained in $Q_{m(j)}$ we get that the closure of $J_{n(k+1)}=f^{n(k)-m(j)}(J_{s(m(j))})$ is contained in $f^{n(k)-m(j)}(Q_{m(j)})\subset Q_{n(k)}$ which contradicts the definition of $Q_{n(k)}$. So let us prove statement (3). From statement (2) and the corollary of Theorem 6.4 it follows that the intersection multiplicity of $\{P_{m(j)},\ldots,P_{m(j+1)}\}$ for $j=-1,0,1,\ldots,l-1$ (where we let $m(-1)=0$) is bounded by 11. Since $l\leq d-1$ the theorem follows. Q.E.D.

8.4. THEOREM. *Assume $n(k)>n(k-1)+(n(k-1)-n(k-2))$. Let $m(0)<m(1)<\ldots<m(l)=n(k)$ be the cutting times of the unimodal pullback of $\hat{Q}_{n(k)}$ and $l\geq 2d$. Then $s(n(k-1))=s(m(l-d))$ and $s^l(m(j))$ are both not defined if $j\in\{l-ld,\ldots,l-ld+d-1\}$ and $l\geq 2$. Furthermore, if $j\in\{l-ld,\ldots,l-ld+d-1\}$, the intersection multiplicity of*

$$\{\hat{Q}_{m(j)},\ldots,\hat{Q}_{m(j+1)}\}$$

is bounded by $4+2l$. Similarly the intersection multiplicity of $\{\hat{Q}_0,\hat{Q}_1,\ldots,\hat{Q}_{m(0)}\}$ is bounded by $5+2[l/d]$. (Here $[l/d]$ is the smallest integer smaller than l/d.)

Proof. If $s(n(k-1))$ were defined then $J_{n(k-2)}\subset[J_{L(n(k-1))},J_{n(k-1)})$ and $n(k)=s(n(k-1))$. Therefore, for $a=n(k)-n(k-1)=s(n(k-1))-n(k-1)$,

$$J_{n(k-2)+a}\subset[J_{n(k-1)},J_{n(k)}).$$

But since $n(k-2)+a<n(k-1)+a=n(k)$ this implies that $n(k-2)+(n(k)-n(k-1))=n(k-2)+a$ must be equal to $n(k-1)$ and this contradicts the assumption of the theorem.

Let us show that if $l\geqslant 2$ and $j\in\{l-ld,\dots,l-ld+d-1\}$ then $s^l(m(j))$ is not defined. So suppose by contradiction $s^l(m(j))$ exists. We claim that then $s^{i+1}(m(j))>n(k-1)>s^i(m(j))$ for some $i=0,1,\dots,l-1$. Indeed, since $m(0)<n(k-1)$ we may otherwise assume that $s^{l-1}(m(j))<n(k-1)$. But then $J_{m(j+(l-1)d)}$ is the $(l-1)$th successor of $J_{m(j)}$ and, again by Lemma 7.4, the successor of $J_{m(j+(l-1)d)}$ must be between $J_{m(j+(l-1)d)}$ and c_j. By statement (2) of Theorem 8.1 this implies that $s^l(m(j))>n(k-1)$. This proves the claim. Hence Lemma 7.4 and $s^{i+1}(m(j))>n(k-1)>s^i(m(j))$ imply that $s(n(k-1))$ exists and so we get a contradiction.

The disjointness statements immediately follow from the Corollary of Theorem 6.4. Q.E.D.

9. The non-existence of wandering intervals

In Section 7 we have proved Ind_0 and so Theorem A follows from

$$\text{Ind}_{d-1} \Rightarrow \text{Ind}_d.$$

So let us assume that Ind_{d-1} holds and that there exists a map $f\in\mathscr{A}_d$ which has a wandering interval J. Asume that J is maximal in the sense that J is not contained in any strictly larger wandering intervals. From the Contraction Principle this implies that

$$\frac{|H_n|}{|J|}\to 1 \quad \text{if} \quad n\to\infty$$

where H_n is the maximal interval containing J on which f^n is monotone.

From Section 7 we know that J accumulates at a turning point, say c. Consider the sequence of closest approach to c, $\{J_{n(k)}\}_{k\geqslant 0}$.

9.1. THEOREM. *There exists k_0 such that for all $k\geqslant k_0$*

$$n(k)-n(k-1)\leqslant n(k-1)-n(k-2).$$

COROLLARY. *J is not a wandering interval.*

Proof of Corollary. Since $n(k)-n(k-1)\leqslant n(k-1)-n(k-2)$ it follows that $n(k)-n(k-1)$ is eventually equal to some integer a for all k sufficiently large. In particular since $J_{n(k-1)}$ and $J_{n(k)}$ tends to c it follows that c is an attractive fixed point of f^a which attracts J. Hence J is not a wandering interval. Q.E.D.

Proof of Theorem 9.1. Since the intervals $J_{n(k)}$ are disjoint there exist arbitrarily large k such that $|J_{n(k-1)}|>|J_{n(k)}|$. Let $Q_{n(k)}$ be as before. If $|J_{n(k)}|<|J_{n(k+1)}|$ then $|J_{n(k\pm 1)}|>|J_{n(k)}|$. Because f is non-flat in the critical point c this implies that $Q_{n(k)}$ is a 1/2-scaled neighbourhood of $J_{n(k)}$ (if k is large). By Theorem 8.3 the intersection multiplicity of the unimodal pullback of $Q_{n(k)}$ is at most $12d$ and therefore the Koebe/Contraction Principle gives a contradiction when $n(k)$ is large. So we have shown that $k\mapsto|J_{n(k)}|$ is monotone decreasing for k large.

Let us now show that $n(k)\leq n(k-1)+(n(k-1)-n(k-2))$ for k large. So assume by contradiction $n(k)>n(k-1)+(n(k-1)-n(k-2))$. Consider the unimodal pullback of $\hat{Q}_{n(k)}$. If $l\leq 2d$ then from Theorem 8.4 the intersection multiplicity of $\{\hat{Q}_0, \ldots, \hat{Q}_{n(k)}\}$ is uniformly bounded. Because $\hat{Q}_{n(k)}$ contains a 1/2-scaled neighbourhood of $J_{n(k)}$ this gives a contradiction with the Koebe/Contraction Principle.

This implies that for k large the number of cutting times l is at least $2d+1$. Now let L_i and R_i be as in Theorem 8.2. From Theorem 8.2 and for $j=1,2,\ldots l-d$, there exists an interval $H\supset J$ which is mapped by $f^{m(j)}$ monotonically onto $[L_{m(j)}, c_{j-d}]$. We claim that $f^{m(j)}(H)$ does not contain $\tau(J_{m(j)})$. Indeed, since Theorem 8.2 gives

$$f^{m(j)}(H)\supset[L_{m(j)}, c_{m(j)}]$$

one would otherwise have

$$f^{m(j)}(H)\supset[L_{m(j)}, \tau(J_{m(j)})].$$

If we take $j\in\{l-2d, \ldots, l-d\}$ then $s^2(m(j))$ does not exist by Theorem 8.4, and since $f^{m(j)}$ is monotone on H we can apply the Corollary of Theorem 6.4 and the intersection multiplicity of $H, f(H), \ldots, f^{m(j)}(H)$ is at most 8. Furthermore, from the definition $\hat{Q}_{n(k)}$, $[L_{m(j)}, c_{j-d}]$ contains $[J_{m(j-d)}, c_j]$ or it contains $[\tau(J_{m(j-d)}), c_j]$. Since the length of the closest approach intervals decreases one has $|J_{m(j-d)}|>|J_{m(j)}|$ and therefore $f^{m(j)}(H)$ contains a 1/2-scaled neighbourhood of $f^{m(j)}(J)$. Therefore we get a contradiction with the Koebe/Contraction Principle and this proves the claim.

Now let $F_{m(j)}=[J_{m(j)}, \tau(J_{m(j)})]$. By definition $f^{m(j)}|H$ is monotone and by Theorem 8.2 $f^{m(j-1)}(H)\supset[J_{m(j-1)}, c_{j-1})$ and $f^{m(j)-m(j-1)}$ maps $[J_{m(j-1)}, c_{j-1})$ monotonically over $[J_{m(j)}, c_j]$. By the previous claim, the image of $[J_{m(j-1)}, c_{j-1})$ under this map is contained in $F_{m(j)}$ and therefore we get

$$f^{m(j)-m(j-1)}(F_{m(j-1)})\subset F_{m(j)}$$

for all $j \in \{l-2d, \ldots, l-d\}$. In particular

$$f^{n(k-1)-n(k-2)}(F_{n(k-2)}) \subset F_{n(k-1)}.$$

Therefore

$$J_{n(k-1)+(n(k-1)-n(k-2))} \subset [J_{n(k-1)}, \tau(J_{n(k-1)})].$$

Since $n(k-1)+(n(k-1)-n(k-2))>n(k-1)$ and $J_{n(k)}$ is a closest approach this gives $n(k-1)+(n(k-1)-n(k-2)) \geq n(k)$, a contradiction. Q.E.D.

10. The proof of Theorem B: finiteness of attractors

In this section we will prove Theorem B. If $f: N \to N$ is a diffeomorphism then the period of periodic orbits of f is bounded. So Theorem B holds trivially. So from now on assume that f is not a diffeomorphism and that $f \in \mathcal{B}$. Of course some points in $\mathcal{K}_f$ may be attracted by periodic orbits so let $\hat{n}$ be an upper bound for the period of this attracting orbits.

In this section we have to show that one has some expansion near periodic orbits. For this it will be convenient to consider the orientation preserving period of a periodic orbit. More precisely let $\mathcal{O}$ be a periodic orbit of period $k>\max(\hat{n}, 300)$. Then $p \in \mathcal{O}$ implies $Df^k(p) \neq 0$. Let $n=2k$ if $Df^k(p)<0$ and $n=k$ otherwise.

The main idea of the proof of Theorem B is to choose $p \in \mathcal{O}$ and get on both sides of p points θ^1 and θ^2, very close to p, with $Df^n(\theta^i) \geq 1+2\varrho$. Using the Minimum Principle we will get $Df^n(p) \geq 1+\varrho$ for large n and we are done.

For $p \in \mathcal{O}$ define T_p to be the maximal open interval such that both components of $T_p \setminus \{p\}$ contain at most one point of $\mathcal{O}$. (So the closure of T_p contains at most 5 points of $\mathcal{O}$.) The interval $T_q, q \in \mathcal{O}$, is a *direct neighbours* of T_p if $T_q \cap T_p = \emptyset$ and $\mathrm{cl}(T_p)$ and $\mathrm{cl}(T_q)$ have one point in common.

10.1. LEMMA. *There exists a number $\tau>0$ such that for each periodic orbit $\mathcal{O}$ of period $\geq \max(\hat{n}, 300)$ there exists $p \in \mathcal{O}$ such that:*

(1) *T_p has direct neighbours on both sides;*

(2) *$|T_{q_1}| \geq 2\tau |T_p|$ and $|T_{q_2}| \geq 2\tau |T_p|$ where T_{q_1} and T_{q_2} are the two direct neighbours of T_p.*

Proof. Let $s \in \mathcal{O}$ be such that (i) $\#(\mathrm{cl}(T_s) \cap \mathcal{O})=5$ and (ii) $|T_s| \leq |T_q|$ for all q with $\#(\mathrm{cl}(T_q) \cap \mathcal{O})=5$. If T_s has neighbours on both sides then we take $p=s$ and we are done.

Otherwise $N=[-1,1]$, and then let T^l and T^r be the smallest open intervals containing points of respectively $\{-1\}$ and $\{1\}$ such that $\#(\mathrm{cl}(T^l)\cap\mathcal{O})=\#(\mathrm{cl}(T^r)\cap\mathcal{O})=5$. Because T_s has no two neighbours $T_s\subset T^l$ or $T_s\subset T^r$. Since the interval $\mathrm{cl}(T_s)$ contains five points of $\mathcal{O}$, and n is at most twice the period of p, there are at most $5\times2\times(5+5)=100$ integers t, $0\leq t\leq n$ such that f^t maps a point in $\mathrm{cl}(T_s)\cap\mathcal{O}$ into a point of $\mathrm{cl}(T^r)\cup\mathrm{cl}(T^l)\cap\mathcal{O}$. So there exists a $0\leq t\leq 101$ such that $f^t(\mathrm{cl}(T_s)\cap\mathcal{O})$ is between T^l and T^r. Let p be the middle point of $f^t(\mathrm{cl}(T_s)\cap\mathcal{O})$. Let $S=\max\{1,\sup_{x\in N}|Df(x)|\}$. Since $T_p\subset f^t(T_s)$, we get

$$|T_p|\leq|f^t(T_s)|\leq\left(\sup_{x\in N}|Df(x)|\right)^t|T_s|\leq S^t|T_s|.$$

As $t\leq 101$ and $|T_s|\leq|T_{q_i}|$, $i=1,2$ the lemma is proved. Q.E.D.

Let $p\in\mathcal{O}$ be the point from Lemma 10.1 and as before let n be the period or twice the period of p. Let U_n be the interval around T_p such that

(10.1) $$U_n \text{ is a } 2\tau\text{-scaled neighbourhood of } T_p.$$

So $U_n\subset T_{q_1}\cup T_p\cup T_{q_2}$. Let J be the maximal interval around p satisfying:

(10.2) $$f^n(J)\subset U_n$$

and

(10.3) $$f^n|J \text{ is an orientation preserving diffeomorphism.}$$

Let J^l and J^r be the components of $J\setminus\{p\}$ and let U_n^i be the component of $U_n\setminus\{p\}$ which contains J^i for $i\in\{l,r\}$. Let $\hat{n}$ be as in the beginning of this section.

10.2. LEMMA. *If $n>\hat{n}$ we have*

(10.4) $$J\subset T_p,$$

(10.5) $$f^n(J^i)\supset J^i,\quad i=l,r.$$

Proof. We claim that $J\cap P$ consists of at most 2 points. Indeed, take I to be the maximal interval containing p such that $\partial I\subset P$ and such that $f^n|I$ is a diffeomorphism. Since f is not a diffeomorphism, I is an interval (i.e. not equal to S^1). Then $f^k|I$ is a diffeomorphism for all $k\geq0$ and since I is maximal, $f^i(I)\cap I\neq\emptyset$ implies that $f^i(I)=I$. In particular the boundary points of I (which are in P) cannot be mapped into $\mathrm{int}(I)$ and so I contains at most two points of P. (This argument also shows that if I contains two points of P then f^n interchanges these two points.) This proves statement (10.4). If

(10.5) does not hold then $f^n(J^i)\subset J^i$ and a critical point of f is attracted by a periodic orbit of period n. This implies $n\leq\hat{n}$. Q.E.D.

The main step in the proof of Theorem B is the proof of the following:

10.3. Proposition. *There exist a number $\varrho>0$ and an integer n_0 such that if the number n corresponding to p is greater than n_0 then there exist $\theta^i\in J^i$, $i=l, r$, such that*

$$Df^n(\theta^i)\geq 1+2\varrho.$$

Before proving Proposition 10.3 we will make a few remarks. Let τ be the number from Lemma 10.1. If there exists $\theta^i\in J^i$ with $Df^n(\theta^i)\geq 1+\tau$ then we are done with J^i. So from now on we may assume that

$$0<Df^n(x)<1+\tau, \quad \forall x\in J^i. \tag{10.6}$$

By the previous lemma, for $n>\hat{n}$ and $i=l, r$, $J^i\subset f^n(J^i)\subset U^i$. Then $f^n(J^i)=U^i$ is impossible because otherwise

$$\frac{|f^n(J^i)|}{|J^i|}\geq\frac{2\tau|T_p|+|J^i|}{|J^i|}\geq(1+2\tau),$$

a contradiction with (10.6). Hence in this case

$$f^n(J^i)\subsetneq U^i.$$

Let $\{U_0, \ldots, U_n\}$ be the diffeomorphic pullback of $U_n\supset f^n(J)$: U_i is the maximal interval containing $f^i(J)$ which is mapped by f diffeomorphically into U_{i+1}. Since $f^n|J$ is a diffeomorphism this is well defined and $U_k\supset f^k(J)$. Furthermore from the maximality of J one has $U_0=J$.

10.4. Lemma. (i) *There is a universal upperbound for the intersection multiplicity of $\{U_0, U_1, \ldots, U_n\}$ (in fact it is at most 74);*

(ii) *For every $\varepsilon>0$ there exists n_0 such that if $n>n_0$ then $|U_k|\leq\varepsilon$ for all $k=0, 1, \ldots, n$.*

Proof. Let $U_{k(1)}\cap\ldots\cap U_{k(r)}\ni x$ with $k(1)<\ldots<k(r)\leq n$. Because $f^{n-k}(U_k)\subset\text{int}[T_{q_1}, T_{q_2}]$ and $f(U_k)\subset U_{k+1}$ we get that

$$f^{n-k(r)}(x)\in U_{k(1)+n-k(r)}\cap\ldots\cap U_n.$$

Therefore $\#\{k; U_k \cap \mathrm{int}[T_{q_1}, T_{q_2}] \neq \emptyset\} \geq r$. Hence statement (i) follows from

$$\#\{k | U_k \cap \mathrm{int}[T_{q_1}, T_{q_2}] \neq \emptyset\} \leq 74. \tag{10.7}$$

Let us prove (10.7). Notice that $\mathrm{int}[T_{q_1}, T_{q_2}]$ contains 11 points of $\mathcal{O}$. So there are at most 22 integers $0 \leq k < n$ such that $f^k(p) \in \mathrm{int}[T_{q_1}, T_{q_2}]$. Hence there are at most 22 integers $0 \leq k < n$ such that $U_k \subset \mathrm{int}[T_{q_1}, T_{q_2}]$. Now let a and b be the boundary points of $[T_{q_1}, T_{q_2}]$, so $a, b \in \mathcal{O}$. If U_k is not contained in $\mathrm{int}[T_{q_1}, T_{q_2}]$ but has a non-empty intersection with this set then $f^{n-k}(a)$ or $f^{n-k}(b) \in \mathrm{int}[T_{q_1}, T_{q_2}]$. Because $\mathrm{cl}([T_{q_1}, T_{q_2}])$ contains only 13 points of $\mathcal{O}$, there exist at most $2\times 2\times 13 = 52$ integers $0 \leq k < n$ with this property. This implies inequality (10.7) and finishes the proof of (i).

Statement (ii) follows from the Contraction Principle and the fact that U_k contains at most 5 points of $\mathcal{O}$. Q.E.D.

Proof of Proposition 10.3. Let $C \in (0, 1)$ be so that if $T \supset I$ are intervals, $f|T$ is an diffeomorphism and $f(T)$ is an ε-scaled neighbourhood of $f(I)$ then T is a $C\varepsilon$-scaled neighbourhood of I. Since f is non-flat at its critical points such a constant exists.

Let $m(1) < m(2) < \ldots < m(l)$ be the 'cutting' times of the pullback, i.e., the integers for which U_j contains a turning point in its closure. Let L_j and R_j be the components of $U_j \setminus f^j(J^i)$ and $J^i_j = f^j(J^i)$ and let $R_{m(k)}$ be the component which contains c_k in its boundary. Now U_n contains a τ scaled neighbourhood of $f^n(J^i)$. Since $f^{n-m(l)+1}\colon U_{m(l)+1} \to U_n$ is a diffeomorphism it follows from the Macroscopic Koebe Principle that there exists a positive function B_0 (which only depends on f and the intersection multiplicity 74 from Lemma 10.6) such that $U_{m(l)+1}$ contains a $B_0(\tau)$-scaled neighbourhood of $J^i_{m(l)+1}$. Now let $g(x) = C \cdot B_0(x)$. If $U_{m(l)}$ contains a $C \cdot B_0(\tau) = g(\tau)$-scaled neighbourhood of $J^i_{m(l)}$ we repeat this procedure and we get from the Macroscopic Koebe principle again that $U_{m(l-1)+1}$ contains a $B_0(g(\tau))$-scaled neighbourhood of $J^i_{m(l-1)+1}$. If $U_{m(l-1)}$ contains a $g^2(\tau) = C \cdot B_0(g(\tau))$-scaled neighbourhood of $J^i_{m(l-1)}$ then we repeat this procedure again. Since $U = J$ this procedure must stop however. Say it stops at $m(r)$ where $r \leq l$ and then (by the definition of C above)

$$|R_{m(r)}| < g^r(\tau) |J^i_{m(r)}|,$$

where, since intersection multiplicity is at most 74, one has $r \leq 74 \cdot \#K_f$.

Now let M' be the middle third interval of $J^i_{m(r)}$ and $M \subset J^i$ be such that $f^{m(r)}(M) = M'$. From the Macroscopic Koebe principle J^i is δ-scaled neighbourhood of M. From Lemma 10.4, $|U_k|$ is small if n is large. So we can apply Lemma 2.8 and get some universal constant $\xi>0$ such that

$$B(f,f^{m(r)}(J^i),M')\geq 1+\xi.$$

Let λ be such that $(1-\lambda)^2(1+\xi)\geq 1+\frac{1}{2}\xi$. From the disjointness property of the orbit of J^i we get

$$B(f^{m(r)},J^i,M)\geq 1-\lambda,$$

$$B(f^{n-m(r)-1},f^{m(r)+1}(J^i),f(M'))\geq 1-\lambda,$$

for n large enough. Hence

$$B(f^n,J^i,M)\geq 1+\frac{1}{2}\xi,$$

for n large enough. Because of the First Expansion Principle, see Theorem 1.3, it suffices to show that the length of both components of $J^i\setminus M$ is at least $\delta\cdot|J^i|$. But since $f^{n-m(r)-1}$ has bounded distortion on $f(J^i_{m(r)})$, since f is non-flat at critical points and since the components of $J^i_{m(r)}\setminus M'$ have the same length as M', there exists a universal constant β such that the length of both components of $f^n(J^i\setminus M)=f^{n-m(r)}(J^i_{m(r)}\setminus M')$ is at least $\beta\cdot|f^n(J^i)|$. However, by Lemma 10.2, $f^n(J^i)\supset J^i$ and by assumption $|Df^n|\leq 1+\tau$ on J^i. It follows that both components of $J^i\setminus M$ have at least size $(\beta/(1+\tau))\cdot|J^i|$. Q.E.D.

Proof of Theorem B. Let $\mathcal{O}$ and as before let n bet he period or twice the period of $\mathcal{O}$. Assume that $n>n_0$ where n_0 is as in Proposition 10.3. So there exist two points θ^1,θ^2 such that $p\in T=[\theta^1,\theta^2]$,

$$Df^n(\theta^i)\geq 1+2\varrho,\quad i=l,r.$$

For n large, $|f^i(T)|$ is small for all $i\in\{0,1,\dots,n\}$ and the orbits has intersection multiplicity ≤ 74. Therefore we get

$$[B(f^n,T^*,J^*)]^3\geq\frac{1+\varrho}{1+2\varrho}$$

for all intervals $J^*\subset T^*\subset T$, provided n is large. Now we apply the Minimum Principle. Then

$$|Df^n(x)|\geq[\inf B(f^n,T^*,J^*)]^3(1+2\varrho)\geq 1+\varrho$$

for all $x\in T$. So $Df^n(p)\geq 1+\varrho$. This proves Theorem B. Q.E.D.

References

[BL] Blokh, A. M. & Lyubich, M. Ju., Non-existence of wandering intervals and structure of topological attractors of one dimensional dynamical systems 2. The smooth case. *Ergodic Theory Dynamical Systems*, 9 (1989), 751–758.

[CE] Collet, P. & Eckmann, J., *Iterated Maps of the Interval as Dynamical Systems*. Birkhäuser, 1980.

[D] Denjoy, A., Sur les courbes definie par les equations differentielles a la surface du tore. *J. Math. Pures Appl. 9,* 11 (1932), 333–375.

[Fa] Fatou, P., Sur les equations fonctionnelles. *Bull. Soc. Math. France,* 47 (1919), 161–271; 48 (1920), 33–94; 48 (1920), 208–314.

[Gu1] Guckenheimer, J., Sensitive dependence on initial conditions for one dimensional maps. *Comm. Math. Phys.*, 70 (1979), 133–160.

[Gu2] — Limit sets of S-unimodal maps with zero entropy. *Comm. Math. Phys.*, 110 (1987), 655–659.

[Ha] Hall, C. R., A C^∞ Denjoy counterexample. *Ergodic Theory Dynamical Systems*, 1 (1981), 261–272.

[J] Julia, G., Memoire sur l'iteration des fonctions rationnelles. *J. de Math.*, 8 (1918), 47–245.

[L] Lyubich, M. Ju., Non-existence of wandering intervals and structure of topological attractors of one dimensional dynamical systems 1. The case of negative Schwarzian derivative. *Ergodic Theory Dynamical Systems*, 9 (1989), 737–750.

[Ma] Mañé, R., Hyperbolicity, sinks and measure in one dimensional dynamics. *Comm. Math. Phys.*, 100 (1985), 495–524; Erratum. *Comm. Math. Phys.*, 112 (1987), 721–724.

[Me] de Melo, W., A finiteness problem for one dimensional maps. *Proc. Amer. Math. Soc.*, 101 (1987), 721–727.

[Mi] Misiurewicz, M., Absolutely continuous measure for certain maps of an interval. *Publ. Math. I.H.E.S.*, 53 (1981), 17–51.

[MS1] de Melo, W. & van Strien, S. J., A structure theorem in one-dimensional dynamics. *Ann. of Math.*, 129 (1989), 519–546.

[MS2] — Schwarzian derivative and beyond. *Bull. Amer. Math. Soc.*, 18 (1988), 159–162.

[MMMS] Martens, M., de Melo, W., Mendes, P. & van Strien, S., Cherry flows on the torus: towards a classification. *Ergodic Theory Dynamical Systems*, 10 (1990), 531–554.

[MT] Milnor, J. & Thurston, W., On iterated maps of the interval: I, II. Preprint Princeton, 1977.

[Si] Singer, D., Stable orbits and bifurcations of maps of the interval. *SIAM J. Appl. Math.*, 35 (1978), 260–267.

[Sc] Schwartz, A., A generalization of a Poincaré–Bendixon theorem to closed two dimensional manifolds. *Amer. J. Math.*, 85 (1963), 453–458.

[SI] Sharkovskii, A. N. & Ivanov, A. F., C^∞-mappings of an interval with attracting cycles of arbitrary large periods. *Ukrain. Mat. Zh.*, 4 (1983), 537–539.

[Str1] van Strien, S., On the bifurcations creating horseshoes, in *Dynamical Systems and Turbulence,* Warwick, 1980. Lecture Notes in Mathematics, 898 (1981), 316–351. Springer-Verlag, Berlin.

[Str2] — Smooth dynamics on the interval, in *New Direction in Chaos*. Cambridge University Press, Cambridge, 1987, pp. 57–119.

[Str3] — Hyperbolicity and invariant measures for general C^2 interval maps satisfying the Misiurewicz condition. *Comm. Math. Phys.*, 128 (1990), 437–496.

[Su] Sullivan, D., Quasiconformal homeomorphisms and dynamics I: a solution of

Fatou–Julia problem on wandering domains. *Ann. of Math.*, 122 (1985), 401–418; II: Structural stability implies hyperbolicity for Kleinian groups. *Acta Math.*, 155 (1985), 243–260.

[Y] Yoccoz, J. C., Il n'y a pas de contre-examples de Denjoy analytique. *C. R. Acad. Sci. Paris Sér. I Math.*, 298 (1984), 141–144.

Received January 19, 1989
Received in revised form January 7, 1991

The Cost of Computing Integers

W. de Melo and B.F. Svaiter

Proc. Amer. Math. Soc. **124**(5), 1377–1378 (1996)

ABSTRACT. We analyse the growth rate of a number theoretic function related to the operational complexity of integers

The purpose of this note is to answer a question raised by Smale on the cost of computing integers using arithmetic operations. More precisely, let $\tau\colon \mathbf{N} \to \mathbf{N}$ be the function that associates to each number n the minimum number of arithmetic operations (addition, subtraction and multiplication) one needs to obtain n starting from 1 and 2. Although 2 is obtainable from 1 in one operation, we have included it as a "starting number" (like 1) to simplify our formulas and induction.

Definition. An *allowable list* of length k is a list of k integers $n_1, n_2, \dots, n_k$ such that for each $l \leq k$, there exist integers $-1 \leq i, j < l$ such that $n_l = op(n_i, n_j)$, where op is either addition, subtraction or multiplication and $n_{-1} = 1, n_0 = 2$.

It follows that $\tau(n) \leq k$ if and only if there exists an allowable list of length k, $\{n_1, \dots, n_k\}$ with $n_k = n$. Also, $\tau(n) = k$ if $\tau(n) \leq k$ but $\tau(n)$ is not less than or equal to $k - 1$.

Proposition 1. (a) $\log\log(n) \leq \tau(n) \leq 2\log(n)$, *where* log *is the logarithm in base* 2.

(b) $\tau(2^{2^k}) = k = \log(\log(2^{2^k}))$.

Proof. Suppose that $\tau(n) = k$. Then there exists an allowable list $\{n_1, \dots, n_k\}$ with $n_k = n$. Let us consider the allowable list $\{m_1, \dots, m_k\}$, where $m_l = m_{l-1} \times m_{l-1}$. By induction we have that $n_l \leq m_l$ for every $l \leq k$ because $m_i \leq m_j$ for $i \leq j$. Therefore, $n \leq m_k = 2^{2^k}$. Thus, $\log(\log(n)) \leq k = \tau(n)$. This proves (b) and the first inequality in (a). To prove the second inequality we consider the binary expansion $n = 2^{k_1} + 2^{k_2} + \cdots + 2^{k_l}$, with $0 \leq k_1 < \cdots < k_l$. The following is an allowable sequence:

$$\{2^2, 2^3, \dots, 2^{k_l}, 2^{k_l} + 2^{k_{l-1}}, \dots, 2^{k_l} + \cdots + 2^{k_1} = n\}.$$

Hence, $\tau(n) \leq k_l + l \leq 2\log(n)$.

Remark. $\tau(2^n) \leq 2\log\log(2^n)$. In fact, if $n = 2^{k_1} + \cdots + 2^{k_l}$, then

$$\{2, 2^2, 2^{2^2}, \dots, 2^{2^{k_l}}, 2^{2^{k_l}} \times 2^{2^{k_{l-1}}}, \dots, 2^{2^{k_l} + \cdots + 2^{k_1}} = n\}$$

is an allowable list and, therefore, $\tau(n) \leq k_l + l \leq 2\log\log(2^n)$.

Received by the editors May 31, 1994 and, in revised form, October 24, 1994.
1991 *Mathematics Subject Classification.* Primary 11N56, 11A25, 11Y16.

Originally published in *Proc. Amer. Math. Soc.* **124**(5), 1377–1378 (1996).

Lemma 1. *Let $B(k) = \{n \in \mathbf{N}; \tau(n) \leq k\}$. Then the cardinality $\#B(k) \leq 3^k \times ((k+1)!)^2$.*

Proof. Let us consider the space $\mathcal{S}_k = \{s = (s_1, \ldots, s_k)\}$, where each $s_l = (op_l, i_l, j_l)$ and $op_l \in \{+, \times -\}$, i_l, j_l are integers smaller than l. To each point $s \in \mathcal{S}_k$ we can associate an allowable sequence $n_1, \ldots, n_k$ by taking $n_l = op_l(n_{i_l}, n_{j_l})$, starting with $n_{-1} = 1$ and $n_0 = 2$. In particular we have a mapping $\phi: \mathcal{S}_k \to B(k)$ which associates to s the integer n_k constructed above. Since ϕ is onto, it follows that the cardinality of $B(k)$ is at most equal to the cardinality of $\mathcal{S}_k$ which is equal to $3^k \times ((k+1)!)^2$.

Definition. A property P holds for almost all integers if the number of integers smaller than n that do not satisfy P is $n \times o(n)$.

Theorem. *If $\epsilon > 0$, then almost all integers n satisfy the property:*

$$\tau(n) \geq \frac{\log(n)}{(\log\log(n))^{1+\epsilon}}.$$

Proof. Suppose, by contradiction, that this is not true. Let

$$\psi(n) = \frac{\log(n)}{(\log\log(n))^{1+\epsilon}}.$$

Then, there exists $0 < \rho < 1$ such that, for infinitely many values of m, the cardinality of the set

$$C_m = \{n \leq m; \tau(n) \leq \psi(n)\}$$

is bigger than $\rho \times m$. If $\psi(m) \leq k < \psi(m) + 1$, then $C_m \subset B_k$. Therefore, by the lemma, $\rho \times m \leq 3^k((k+1)!)^2$ for infinitely many values of m. Thus,

$$\rho \times m \leq 3^{\psi(m)+1}(\psi(m) + 2)^{2(\psi(m)+2)}$$

which is a contradiction because a straightforward calculation shows that the above inequality cannot hold for m big enough.

The above theorem answers negatively Smale's first question: does there exist a polynomial p such that $\tau(n) \leq p(\log\log(n))$?

Smales's question 2. Is $\tau(k!) \leq p(\log k)$ for some universal polynomial p?

Smale's question 3. Does there exist a polynomial p such that for each k there exists an m satisfying $\tau(m \times k!) \leq p(\log k)$? In [SS], Shub and Smale proved that a negative answer to this question implies that one cannot find an algorithm having polynomial cost to decide whether a family of polynomials have a common zero, and, by the results of [BSS], this implies that $N \neq NP$ over the complex numbers.

References

[SS] M. Shub and S. Smale, *On the Intractability of Hilbert's Nullestellensatz and an algebraic version of "$NP \neq P$?"*, preprint.

[BSS] L. Blum, M. Shub, and S. Smale, *On a theory of computation and complexity over the real numbers: NP-completeness, recursive functions and universal machines*, Bull. Amer. Math. Soc. **21** (1989), 1–46. MR **90a:**68022

Instituto de Matematica Pura e Aplicada, Estrada Dona Castorina 110, Jardim Botanico, Rio de Janeiro, Brazil

E-mail address: demelo@impa.br

E-mail address: benar@impa.br

Rigidity of Critical Circle Mappings I

Edson de Faria and Welington de Melo

J. Eur. Math. Soc. **1**, 339–392 (1999)

Abstract. We prove that two C^3 critical circle maps with the same rotation number in a special set $\mathbb{A}$ are $C^{1+\alpha}$ conjugate for some $\alpha > 0$ provided their successive renormalizations converge together at an exponential rate in the C^0 sense. The set $\mathbb{A}$ has full Lebesgue measure and contains all rotation numbers of bounded type. By contrast, we also give examples of C^∞ critical circle maps with the same rotation number that are not $C^{1+\beta}$ conjugate for any $\beta > 0$. The class of rotation numbers for which such examples exist contains Diophantine numbers.

1. Introduction

The purpose of this paper is to study certain rigidity questions concerning critical circle mappings. This study is continued in [5].

In the qualitative theory of smooth dynamical systems, the notions of *rigidity* and *flexibility* play an important role. The smooth systems are usually classified according to the equivalence relation given by topological conjugacies: two smooth maps f and g are topologically equivalent if there exists a homeomorphism h of the ambient space such that $h \circ f = g \circ h$. Such a homeomorphism maps orbits of f onto orbits of g. One can also consider a stronger equivalence relation given by smooth conjugacies. This leads to a quantitative or geometric classification of smooth dynamical systems, since a smooth conjugacy, being essentially affine at small scales, preserves the small-scale geometric properties of the dynamics. Hence each topological equivalence class is "foliated" by the smooth conjugacy classes and the quotient space is the *moduli* or *deformation space* of the dynamics. The moduli space describes the *flexibility* of the dynam-

E. de Faria: Instituto de Matemática e Estatística, Universidade de São Paulo, Rua do Matão 1010, Butantã, CEP05508-900 São Paulo SP, Brasil, e-mail: edson@ime.usp.br

W. de Melo: Instituto de Matemática Pura e Aplicada, Estrada Dona Castorina 110, Jardim Botânico, CEP22460 Rio de Janeiro RJ, Brasil, e-mail: demelo@impa.br

This work has been partially supported by the Pronex Project on Dynamical Systems, by FAPESP Grant 95/3187-4 and by CNPq Grant 30.1244/86-3.

Mathematics Subject Classification (1991): 58F03, 58F23

Originally published in *J. Eur. Math. Soc.* **1**, 339–392 (1999).

ics. When this space reduces to a single point, we are in the presence of *rigidity*.

In general, since eigenvalues at the periodic points are smooth conjugacy invariants, we can hope to find rigidity only in the absence of periodic points. From this viewpoint, the simplest case to consider is that of circle diffeomorphisms. If f is a circle diffeomorphism without periodic points then f is combinatorially equivalent to a rigid rotation $R_\rho : x \mapsto x + \rho \pmod 1$, in the sense that for each N, the first N elements of an orbit of f are ordered in the circle in the same way as the first N elements of an orbit of R_ρ. From Denjoy's theorem it follows that if f is C^2 (or at least C^1 and its derivative has bounded variation) then f is topologically conjugate to R_ρ. By a fundamental result of Herman [11], improved by Yoccoz [20], if the rotation number ρ satisfies a Diophantine condition such as

$$\left|\rho - \frac{p}{q}\right| \geq \frac{C}{q^{2+\beta}} ,$$

for all rationals p/q, with $C > 0$ and $0 < \beta < 1$, and if f is C^r, $r \geq 3$, then the conjugacy is C^1 (it is in fact $C^{r-1-\beta-\varepsilon}$ for every $\varepsilon > 0$). On the other hand, Arnold proved that some such condition on the rotation number is essential: there exist real analytic circle diffeomorphisms with irrational rotation number such that the conjugacy with a rigid rotation is not even absolutely continuous with respect to Lebesgue measure.

Maps with periodic points cannot be rigid, but we can analyze the rigidity of some relevant invariant set, such as an attractor of the map. This is the situation studied by Sullivan and McMullen in the context of unimodal maps of the interval. They considered the so-called infinitely renormalizable maps of bounded combinatorial type. For such maps, almost all orbits are asymptotic to a Cantor set which is the closure of the critical orbit. They proved that if two such maps are smooth enough and have the same combinatorics then there exists a $C^{1+\alpha}$ diffeomorphism of the real line that conjugates the restriction of the maps to the corresponding Cantor attractors. The tools they developed have been of fundamental importance for the proofs of our results.

Perhaps the most famous rigidity result in Geometry is the celebrated *Mostow rigidity theorem*. A special case of this theorem states that two compact hyperbolic manifolds of dimension at least 3 which have the same homotopy type are in fact isometric. Here a hyperbolic manifold is the quotient space $\mathbb{H}^n/\Gamma$ of the hyperbolic space $\mathbb{H}^n$ by a discrete group Γ of isometries. The hypothesis of the theorem implies the existence of a quasiconformal homeomorphism of the sphere at infinity that conjugates the actions of the two groups there. Such *a-priori* step may be regarded as a *pre-rigidity* result. The rigidity is then obtained by proving that this qc-homeomorphism is in fact conformal, *i.e.* a Moebius transformation.

The situation for critical circle mappings fits perfectly into this framework. A critical circle mapping is a homeomorphism $f : S^1 \to S^1$ that is of class C^r, $r \geq 3$, and has a unique critical point c around which, in some C^r coordinate system, f has the form $x \mapsto x^p$, where $p \geq 3$ is an odd integer called the *power law* of f. Yoccoz proved in [21] that a critical circle mapping without periodic points is topologically conjugate to an irrational rotation. Later, in an unpublished work, he proved that the conjugacy between two critical circle mappings with the same rotation number is in fact quasisymmetric, *i.e.* there exists a constant $K \geq 1$ such that, for all pairs of adjacent intervals I_1, I_2 of equal length $|I_1| = |I_2|$, we have

$$\frac{1}{K} \leq \frac{|h(I_1)|}{|h(I_2)|} \leq K .$$

This is in contrast with the diffeomorphism case where, without restriction on the rotation number, the conjugacy may fail to be quasisymmetric (see[16], p. 75). Yoccoz's result, whose proof we present in §4 and Appendix B, is the exact analogue of the pre-rigidity step in the proof of Mostow's theorem.

Rigidity Conjecture. *If f, g are C^3 critical circle mappings with the same irrational rotation number of bounded type and the same power-law at the critical point, then there exists a $C^{1+\alpha}$ conjugacy h between f and g for some universal $\alpha > 0$.*

So far we have succeeded in proving this conjecture only when the maps are real-analytic. Our proof involves real techniques developed in this paper, and deformation of complex structures, developed in [5].

1.1. Summary of results

We now present a quick summary of our results. As already mentioned, we prove two main new theorems concerning critical circle homeomorphisms.

The first theorem brings forth the connection between renormalization and rigidity in the context of circle maps. It also indicates that the above Rigidity Conjecture might be true for a much larger class of rotation numbers than bounded type. The proof is given in §4.4.

First Main Theorem. *There exists a set $\mathbb{A}$ of rotation numbers, having full Lebesgue measure and containing all numbers of bounded type, for which the following holds. Let f and g be topologically conjugate C^3 critical* *circle maps, and let h be the conjugacy between f and g that maps the critical point of f to the critical point of g. If their common rotation number belongs to* $\mathbb{A}$, *and if their renormalizations converge together exponentially fast in the* C^0*-topology, then h is* $C^{1+\alpha}$ *for some* $\alpha > 0$.

The second theorem shows that we cannot expect the above Rigidity Conjecture to be true without restriction on the rotation number. The proof occupies §5 in its entirety.

Second Main Theorem. *There exists an uncountable set* $\mathbb{B}$ *of rotation numbers such that for any* $\rho \in \mathbb{B}$ *there exist* C^∞ *critical circle maps f and g with rotation number* ρ *with the property that the conjugacy between f and g sending the critical point of f to the critical point of g is not* $C^{1+\beta}$ *for any* $\beta > 0$.

The set $\mathbb{B}$ is very small: its Hausdorff dimension is not greater than $1/2$. But it does contain Diophantine numbers, in somewhat remarkable contrast with the case of circle diffeomorphisms. The saddle-node surgery procedure we develop here is quite general, and can be used to produce similar counterexamples to the rigidity of infinitely renormalizable unimodal maps with special unbounded combinatorics.

All estimates performed in this paper rely heavily on the *real a-priori bounds* of M. Herman [12] and G. Świątek [18]. These bounds are revisited in §3. Several technical consequences of the real bounds needed in this paper, such as the C^{r-1} boundedness of the renormalizations of a C^r critical circle map, are gathered in Appendix A.

2. Preliminaries

We have three goals in this section. First, we present some of the basic notations commonly used when studying circle maps. Second, we present the notions of commuting pair and renormalization in the context of circle maps, and discuss their relationship. Third, we state the distortion tools that are necessary for proving the real bounds in §3.

2.1. Critical circle mappings

Following the tradition in this subject, we identify the unit circle S^1 with the one-dimensional torus $\mathbb{R}/\mathbb{Z}$. The obvious advantage of such identification is that it allows us to use additive notation when dealing with circle mappings.

We briefly recall some standard facts concerning circle mappings. Given a homeomorphism $f : S^1 \to S^1$, we denote its rotation number by $\rho(f)$. It can be expressed as a continued fraction

$$\rho(f) = [a_0, a_1, \dots, a_n, \dots] = \cfrac{1}{a_0 + \cfrac{1}{a_1 + \cfrac{1}{\cfrac{\cdots}{a_n + \cfrac{1}{\cdots}}}}},$$

which can be finite or infinite, depending on whether $\rho(f)$ is rational or irrational, respectively. The positive integers a_n are the *partial quotients* of $\rho(f)$. They give rise to a sequence of *return times* for f, recursively defined by $q_0 = 1$, $q_1 = a_0$ and $q_{n+1} = a_n q_n + q_{n-1}$ for all $n \geq 1$ (for which a_n exists – an assumption that will be implicit henceforth). Given $x \in S^1$ and $n \geq 1$, we denote by $J_n(x)$ the closed interval containing x whose endpoints are $f^{q_n}(x)$ and $f^{q_{n-1}}(x)$. We also let $I_{n-1}(x) \subseteq J_n(x)$ be the closed interval whose endpoints are x and $f^{q_{n-1}}(x)$. Observe that $J_n(x) = I_n(x) \cup I_{n-1}(x)$ for all $n \geq 1$.

From the dynamics standpoint, we are not interested in *all* circle homeomorphisms, but only in those that possess a unique critical point in S^1, being local diffeomorphisms everywhere else. These are the so-called *critical circle maps*. More precisely, let $f : S^1 \to S^1$ be a C^r homeomorphism, for some $r \geq 1$. We say that f is a critical circle map if there exists $c \in S^1$ (the critical point) such that $f'(c) = 0$ and $f'(x) \neq 0$ for all $x \neq c$. Moreover, we require f to have a *power-law* at c. This means that in a suitable C^r coordinate system around the critical point, our f is represented by a map of the form $x \mapsto x|x|^{p-1} + a$, for some real number $p > 1$ called the *power-law exponent* of f. The class of all C^r critical circle maps will be denoted by $\mathrm{Crit}^r(S^1)$.

Since the critical point c of a critical circle map is a distinguished point on the circle, we will write I_n and J_n throughout, instead of $I_n(c)$ and $J_n(c)$, respectively.

2.2. *Commuting pairs*

We will study the successive renormalizations of a critical circle map f. Here, as in many other settings in dynamics, the word renormalization is taken to mean a (suitably normalized) Poincaré first return map of f to some neighborhood of its critical point. Abstracting the essential features of such first return maps yields the notion of *commuting pair*, due to O. Lanford [13] and D. Rand [17]. We formulate this notion as follows.

Definition. A C^r *commuting pair* consists of two mappings $f_- : [\lambda, 0] \to \mathbb{R}$, where $\lambda < 0$, and $f_+ : [0, 1] \to \mathbb{R}$, satisfying the following conditions.

[P_1] Both f_- and f_+ are C^r orientation-preserving homeomorphisms onto their images.
[P_2] We have $f_-(0) = 1$, $f_+(0) = \lambda$ and $0 < f_-(\lambda) = f_+(1) < 1$.
[P_3] We have $Df_-(x) > 0$ for all $\lambda \leq x < 0$, and $Df_+(x) > 0$ for all $0 < x \leq 1$.
[P_4] For each $1 \leq k \leq r$, we have $D^k(f_+ \circ f_-)(0) = D^k(f_- \circ f_+)(0)$.

A *critical* commuting pair is a commuting pair such that $Df_-(0) = 0 = Df_+(0)$.

Although it is more customary to use the symbols ξ and η instead of f_- and f_+, respectively, the present notation will be more convenient for our purposes in this paper. It can be proved that, in the presence of the other conditions, P_4 is equivalent to the following.

[P'_4] There exist open intervals $\Delta_- \supseteq [\lambda, 0]$ and $\Delta_+ \supseteq [0, 1]$, and C^r homeomorphic extensions $F_- : \Delta_- \to \mathbb{R}$ and $F_+ : \Delta_+ \to \mathbb{R}$ of f_- and f_+ respectively, satisfying $F_+ \circ F_-(x) = F_- \circ F_+(x)$ for all $x \in \Delta_- \cap \Delta_+$ such that $F_{\pm}(x) \in \Delta_{\mp}$ (the set of such x is an open interval around 0).

This justifies the name *commuting pair*. The class of all C^r critical commuting pairs will be denoted by $\mathbf{P}^r$. We shall henceforth identify a commuting pair (f_-, f_+) with a single map $f : [\lambda, 1] \to [\lambda, 1]$, called the *shadow* of the commuting pair, defined as follows,

$$f(x) = \begin{cases} f_-(x), & \text{when } \lambda \leq x \leq 0 \\ f_+(x), & \text{when } 0 < x \leq 1. \end{cases}$$

To each commuting pair f we associate an element $a \in \mathbb{N} \cup \{\infty\}$ called the *height* of f, in the following way. If there exists $n \geq 1$ such that $f^{n+1}(1) < 0 \leq f^n(1)$, then we set $a = n$; otherwise we set $a = \infty$. It is clear that f has infinite height if and only if there exists $0 < \bar{x} < 1$ such that $f(\bar{x}) = \bar{x}$.

2.3. Renormalizing a commuting pair

Every commuting pair f with finite height a such that $f^a(1) > 0$ can be *renormalized*, in the following sense. Let $\Lambda : \mathbb{R} \to \mathbb{R}$ be the linear map

$x \mapsto \lambda x$, let $\lambda' = f^a(1)/\lambda < 0$, and let $\mathcal{R}f : [\lambda', 1] \to [\lambda', 1]$ be the map defined by

$$\mathcal{R}f(x) = \begin{cases} \Lambda^{-1} \circ f \circ \Lambda(x), & \text{when } \lambda' \le x \le 0 \\ \Lambda^{-1} \circ f^{a+1} \circ \Lambda(x), & \text{when } 0 < x \le 1. \end{cases}$$

This map is (the shadow of) a commuting pair $(\mathcal{R}f_-, \mathcal{R}f_+)$, called the *first renormalization* of f. Equivalently,

$$\begin{cases} \mathcal{R}f_-(x) = \Lambda^{-1} \circ f_+ \circ \Lambda(x), & \text{for all } \lambda' \le x \le 0 \\ \mathcal{R}f_+(x) = \Lambda^{-1} \circ f_+^a \circ f_- \circ \Lambda(x), & \text{for all } 0 \le x \le 1. \end{cases}$$

The class of all C^r critical commuting pairs which are renormalizable in this sense will be denoted $\mathbf{P}^r_1$. In this way, we have a well-defined map $\mathcal{R} : \mathbf{P}^r_1 \to \mathbf{P}^r$, the so-called *renormalization operator*. More generally, for all $n \ge 1$ we write $\mathbf{P}^r_n = \mathcal{R}^{-n}(\mathbf{P}^r)$ for the set of all C^r critical commuting pairs which can be renormalized n times. We have $\mathbf{P}^r_{n+1} \subseteq \mathbf{P}^r_n$ for all n. We are especially interested in the set of all *infinitely renormalizable* critical commuting pairs, namely

$$\mathbf{P}^r_\infty = \bigcap_{n \ge 1} \mathbf{P}^r_n .$$

Given $f \in \mathbf{P}^r$, let $a_0 = a$ be its height, and for each $n \ge 1$ such that $f \in \mathbf{P}^r_n$, let a_n be the height of $\mathcal{R}^n(f)$. This can be a finite or infinite sequence; in any case, using the convention $1/\infty = 0$, we define the *rotation number* of f to be

$$\rho(f) = [a_0, a_1, \ldots, a_n, \ldots] = \cfrac{1}{a_0 + \cfrac{1}{a_1 + \cfrac{1}{\cfrac{\cdots}{a_n + \cfrac{1}{\cdots}}}}}.$$

In particular, $\rho(\mathcal{R}f) = [a_1, a_2, \ldots]$, that is, the renormalization operator acts as the Gaussian shift on continued fractions.

2.4. Renormalizing a critical circle map

Let f be a critical circle map with critical point c, and for each $k \ge 0$ such that $f^{q_k}(c) \ne c$, let $A_k : \mathbb{R} \to \mathbb{R}/\mathbb{Z}$ be the affine covering map such that $A_k([0, 1]) = I_k$, with $A_k(0) = c$ and $A_k(1) = f^{q_k}(c)$. For each $n \ge 1$ such that $f^{q_k}(c) \neq c$ for all $0 \leq k \leq n$, consider the Poincaré first return map $f_n : I_n \cup I_{n-1} \to I_n \cup I_{n-1}$, namely

$$f_n(x) = \begin{cases} f^q(x) & \text{when } x \in I_n \\ f^Q(x) & \text{when } x \in I_{n-1}, \end{cases}$$

where $q = q_{n-1}$ and $Q = q_n$. Define λ_n to be the largest negative number such that $A_{n-1}(\lambda_n) = f^Q(c)$ (one sees in fact that $\lambda_n = -|I_n|/|I_{n-1}|$). Then $A_{n-1}([\lambda_n, 0]) = I_n$ and $A_{n-1}([0, 1]) = I_{n-1}$, and we can consider the map $\boldsymbol{f}_n : [\lambda_n, 1] \to [\lambda_n, 1]$ given by $\boldsymbol{f}_n = A_{n-1}^{-1} \circ f_n \circ A_{n-1}$. Here, it is implicit that A_{n-1}^{-1} is the inverse branch that maps $I_n \cup I_{n-1}$ onto $[\lambda_n, 1]$. This defines (the shadow of) a C^r critical commuting pair called *n-th renormalization of f*. It is well-defined provided the rotation number of f has a continued-fraction development of length at least $n + 1$ (in particular it is well-defined for all n when the rotation number of f is irrational). It is easy to see in this case that $\boldsymbol{f}_{k+1} = \mathcal{R}\boldsymbol{f}_k$ for all $1 \leq k \leq n-1$. Moreover, if $a_0, a_1, \ldots$ are the partial quotients of the rotation number of f then, from the recurrence relations satisfied by the sequence of return times q_k, we see at once that the height of $\boldsymbol{f}_k$ is equal to a_k, and that $\rho(\boldsymbol{f}_k) = [a_k, a_{k+1}, \ldots]$.

Remark. Note that the largest interval containing I_n on which $f^{q_{n-1}}$ is a diffeomorphism is $[\alpha_n, c]$ where $f^{a_{n-1}q_{n-1}}(\alpha_n) = f^{q_n}(c)$, that is, $\alpha_n = f^{q_{n-2}}(c)$. Similarly, the largest interval containing I_{n-1} on which f^{q_n} is a diffeomorphism is $[c, \beta_n]$ where $f^{q_n}(\beta_n) = f^{q_{n-1}}(c)$.

2.5. The C^k metrics

The following is only one of several equivalent ways of defining a C^k distance between commuting pairs. We normalize our commuting pairs to be defined on $[0, 1]$, using for each $\boldsymbol{f}$ a fractional linear transformation that maps $\lambda, 0, 1$ respectively to $0, \frac{1}{2}, 1$, and then use the C^k norm of the difference of the normalized pairs. The C^k distance between $\boldsymbol{f}$ and $\boldsymbol{g}$ defined in this fashion is denoted by $d_k(\boldsymbol{f}, \boldsymbol{g})$.

Let us be a bit more precise. If a function $\varphi : [0, 1] \to \mathbb{R}$ has a jump discontinuity at $x = 1/2$ but is elsewhere k times continuously differentiable, let $\|\varphi\|_k = \max\{\|\varphi^-\|_k, \|\varphi^+\|_k\}$ where φ^- is the restriction of φ to $[0, \frac{1}{2}]$ and φ^+ is the restriction of φ to $[\frac{1}{2}, 1]$. Given two elements $\boldsymbol{f} : [\lambda, 1] \to \mathbb{R}$ and $\boldsymbol{g} : [\mu, 1] \to \mathbb{R}$ of $\mathbf{P}^r$, and given $0 \leq k \leq r$, let A_λ be the fractional linear transformation which maps $0, \lambda, 1$ to $0, \frac{1}{2}, 1$, respectively, and let A_μ be similarly defined. Then write

$$d_k(\boldsymbol{f}, \boldsymbol{g}) = \max\left\{|\lambda - \mu|, \|A_\lambda \boldsymbol{f} A_\lambda^{-1} - A_\mu \boldsymbol{g} A_\mu^{-1}\|_k\right\} .$$

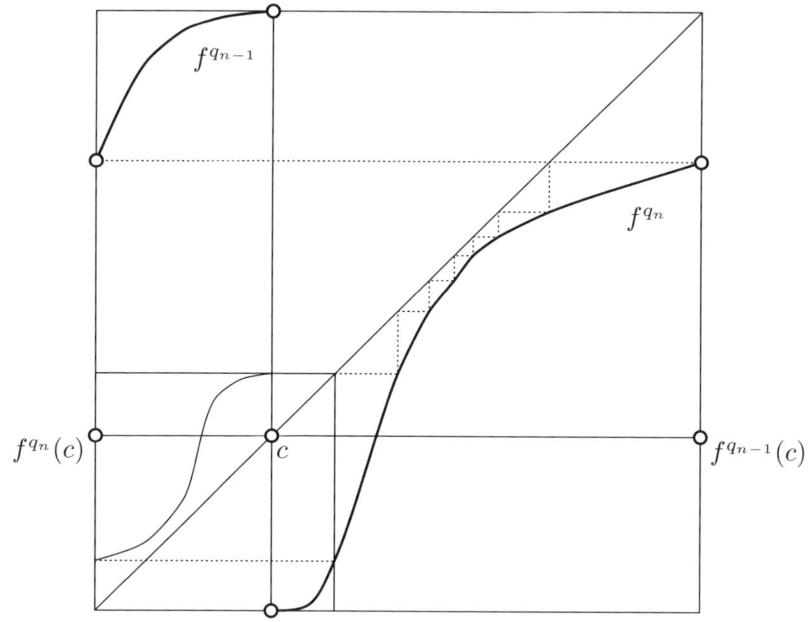

Fig. 1. Two consecutive renormalizations of f

Lemma 2.1. *For each $0 \leq k \leq r$, d_k is a metric.*

Proof. The only thing not entirely obvious is that $d_k(\boldsymbol{f}, \boldsymbol{g}) = 0$ implies $\boldsymbol{f} = \boldsymbol{g}$. But if $d_k(\boldsymbol{f}, \boldsymbol{g}) = 0$ then on one hand $\lambda = \mu$, so that $A_\lambda = A_\mu$, and on the other hand $A_\lambda \boldsymbol{f} A_\lambda^{-1} - A_\mu \boldsymbol{g} A_\mu^{-1} = 0$, so that $\boldsymbol{f} = A_\lambda^{-1} A_\mu \boldsymbol{g} A_\mu^{-1} A_\lambda = \boldsymbol{g}$. □

Proposition 2.2. *Let $\boldsymbol{f} : [\lambda, 1] \to [\lambda, 1]$ and $\boldsymbol{g} : [\mu, 1] \to [\mu, 1]$ be elements of $\mathbf{P}^r_\infty$, and suppose there exists a C^r diffeomorphism $h : [\lambda, 1] \to [\mu, 1]$ such that $h \circ \boldsymbol{f} = \boldsymbol{g} \circ h$. Then for all $k \leq r - 1$ the distances $d_k\left(\mathcal{R}^n(\boldsymbol{f}), \mathcal{R}^n(\boldsymbol{g})\right)$ converge to 0 at an exponential rate.*

Proof. Let $\boldsymbol{f}_n = \mathcal{R}^n \boldsymbol{f}$ and $\boldsymbol{g}_n = \mathcal{R}^n \boldsymbol{g}$. Then $\boldsymbol{f}_n = h_n^{-1} \circ \boldsymbol{g}_n \circ h_n$, where h_n is obtained from h by restriction and affine rescaling. We will see below (after we prove the real bounds for critical circle maps, *cf.* Theorem 3.1) that $\{h_n\}$ converges exponentially in the C^r sense to the space of *affine* maps. Therefore, we have that $d_{r-1}(\boldsymbol{f}_n, \boldsymbol{g}_n) \to 0$ exponentially fast. □

2.6. Distortion tools

In §3 we will need some distortion tools to get real bounds for critical circle maps. The most basic is the notion of *cross-ratio distortion*. Given intervals

$M \subseteq T$ on the line or circle, their cross-ratio is defined as

$$D(M,T) \;=\; \frac{|M|\,|T|}{|L|\,|R|}\,,$$

where L and R are the left and right components of $T \setminus M$. The cross-ratio distortion of a map f (whose domain contains T) on the pair of intervals (M,T) is

$$B(f;\,M,T) \;=\; \frac{D(f(M),\,f(T))}{D(M,T)}\,.$$

Cross-ratios are always increased by a map with negative Schwarzian derivative. More precisely, if f is C^3 and $Sf<0$ then $B(f;\,M,T)>1$.

Lemma 2.3. *(Cross-ratio distortion principle)*
Given a map f as above, $m \geq 1$ and intervals $M \subseteq T$ such that $f^m|T$ is a diffeomorphism onto its image, we have

$$B(f^m;\,M,T) \;\geq\; \exp\Big\{-\sigma\sum_{j=0}^{m-1}|f^j(T)|\Big\}\,,$$

where $\sigma>0$ depends on f and $\max_{0\leq j\leq m-1}|f^j(T)|$.

For a proof of (a much more general version of) this principle, see [16], p. 287. This fact will be used in combination with the following classical distortion principle. For intervals $M \subseteq T$ as above we define the *space* of M inside T to be the smallest of the ratios $|L|/|M|$ and $|R|/|M|$.

Lemma 2.4. *(Koebe distortion principle)*
Given $\ell, \tau>0$ and a map f as above, there exists $K=K(\ell,\tau,f)>1$ of the form

$$K \;=\; \left(1+\frac{1}{\tau}\right)^2 \exp C\ell\,,$$

where C is a constant depending only on f, with the following property. If T is an interval such that $f^m|T$ is a diffeomorphism and if $\sum_{j=0}^{m-1}|f^j(T)| \leq \ell$, then for each interval $M \subseteq T$ for which the space of $f^m(M)$ inside $f^m(T)$ is at least τ and for all $x,y \in M$ we have

$$\frac{1}{K} \;\leq\; \left|\frac{Df^m(x)}{Df^m(y)}\right| \;\leq\; K\,.$$

Once again, see [16], p. 295, for a proof. Used in combination with Lemma 2.3, the Koebe distortion principle allows one to propagate space around under fairly general circumstances.

3. The real a-priori bounds

In this section we establish real *a-priori* bounds for critical circle maps, obtaining as a corollary the fact that their renormalizations are pre-compact in the C^1 topology. The results are well-known, and the reader will not fail to notice the overlap with some of the material in [19] and [10].

Let $f: S^1 \to S^1$ be a critical circle homeomorphism with critical point c. The iterates of c are denoted by $c_i = f^i(c)$. Let I_n be the interval with endpoints c and c_{q_n} that contains $c_{q_{n+2}}$, as defined in Sect. 2. For simplicity, we write $I_n^j = f^j(I_n)$ for all j and n. The most basic combinatorial fact to be remembered here is that the collection of intervals

$$\mathcal{P}_n = \left\{I_{n-1}, I_{n-1}^1, \ldots, I_{n-1}^{q_n-1}\right\} \cup \left\{I_n, I_n^1, \ldots, I_n^{q_{n-1}-1}\right\}$$

constitutes a partition of S^1 modulo endpoints, called the *dynamical partition of level n* associated to f. In order to get an actual partition we exclude from each interval in $\mathcal{P}_n$ its right endpoint, say, according to the standard choice of orientation of S^1. Let $\mathcal{P}_n(x)$ denote the atom of the partition $\mathcal{P}_n$ that contains x (in particular, $\mathcal{P}_n(c)$ is either I_n or I_{n-1} according to the parity of n).

Theorem 3.1. *(Real Bounds) Let $f \in Crit^r(S^1)$ be a map with irrational rotation number. There exist constants $C_0 > 1$ and $0 < \mu_0 < \mu_1 < 1$ depending only on f such that*

(a) If I and J are any two adjacent atoms of $\mathcal{P}_n$, then $C_0^{-1}|J| < |I| < C_0|J|$;
(b) For every $x \in S^1$, we have $|\mathcal{P}_n(x)| < \mu_1|\mathcal{P}_{n-1}(x)|$;
(c) If the rotation number of f is of bounded type then $|\mathcal{P}_n(x)| > \mu_0^n/C_0$;
(d) If the rotation number of f is of bounded type then $|\mathcal{P}_n(x)| > |\mathcal{P}_{n-1}(x)|/C_0$;
(e) If $0 < i \leq j \leq q_n$ then the distortion of the restriction of f^{j-i} to $I_{n-1}^i = f^i(I_{n-1})$ is bounded by C_0.

In particular, the critical commuting pairs $\mathcal{R}^n f$ form a bounded sequence in the C^1 topology.

Later in this section we will see that the bounds in this theorem are eventually universal.

3.1. Bounding space

In what follows, two positive numbers a and b are said to be *comparable modulo f*, or simply *comparable*, if there exists a constant $C > 1$, depending only on our map f, such that $C^{-1}b \leq a \leq Cb$. This relation is denoted by $a \asymp b$. It is also convenient to write $a \preccurlyeq b$ to indicate that $a \leq Cb$. Comparability modulo f is reflexive and symmetric, but not transitive since the constants multiply. Hence, if $b_1 \asymp b_2 \asymp \cdots \asymp b_N$, we can only say that $b_1 \asymp b_N$ if N is bounded (by a constant depending only on f).

Lemma 3.2. *For each $n \geq 0$ there exist $z_1, z_2, z_3, z_4, z_5 \in S^1$ with $z_{j+1} = f^{q_n}(z_j)$ such that $|z_1 - z_2| \asymp |z_2 - z_3| \asymp |z_3 - z_4| \asymp |z_4 - z_5|$.*

Proof. Let $z \in S^1$ be a point such that $|f^{q_n}(z) - z| \leq |f^{q_n}(x) - x|$ for all $x \in S^1$. From Koebe's principle applied successively to f^{-q_n}, f^{-2q_n} and f^{-3q_n}, we have

$$|z-f^{q_n}(z)| \succcurlyeq |f^{-q_n}(z)-z| \succcurlyeq |f^{-2q_n}(z)-f^{-q_n}(z)| \succcurlyeq |f^{-3q_n}(z)-f^{-2q_n}(z)|.$$

Moreover, by our choice of z we have $|z - f^{q_n}(z)| \preccurlyeq |f^{-3q_n}(z) - f^{-2q_n}(z)|$. Therefore we can take $z_5 = f^{q_n}(z)$, $z_4 = z, \ldots, z_1 = f^{-3q_n}(z)$ as the desired five points. □

Lemma 3.3. *Let $z_1, z_2, \ldots, z_5$ and $w_0, w_1, \ldots, w_5$ be points on the circle such that $z_{j+1} = f^{q_n}(z_j)$ and $w_{j+1} = f^{q_n}(w_j)$, and such that w_1 lies on the interval of endpoints z_1 and z_2 in the partition of S^1 determined by the z_i's. If $|z_1 - z_2| \asymp |z_2 - z_3| \asymp |z_3 - z_4| \asymp |z_4 - z_5|$, then*

$$|w_0 - w_1| \succcurlyeq |w_1 - w_2| \preccurlyeq |w_2 - w_3| \,. \tag{1}$$

Proof. Let $\ell = \min |z_j - z_{j+1}|$. If there is a j with $1 \leq j \leq 3$ such that $|w_j - w_{j+1}| \leq \ell/2$, then we must have $|w_{j-1} - w_j| \geq \ell/2$ and $|w_{j+1} - w_{j+2}| \geq \ell/2$ also. But then $[w_j, w_{j+1}]$ has space on both sides inside $[w_{j-1}, w_{j+2}]$. Applying $f^{-(j-1)q_n}$ to these points and using the Koebe principle, we get (1). If on the other hand there is no j with that property, then $|w_1 - w_2| \asymp |w_2 - w_3| \asymp |w_3 - w_4|$. Again, applying f^{-q_n} and using Koebe we get (1). □

Lemma 3.4. *For all $n \geq 0$ and all $x \in S^1$, we have $|f^{q_n}(x) - x| \asymp |x - f^{-q_n}(x)|$.*

Proof. To show that $|x - f^{-q_n}(x)| \geq C^{-1}|f^{q_n}(x) - x|$, let $i \leq q_n$ be such that $f^i(x) \in [z_1, z_2]$, where $z_1, z_2, \ldots$ are the points given by Lemma 3.2. Then let $w_0 = f^{i-q_n}(x)$, $w_1 = f^i(x)$, etc. We know from Lemma 3.3 that $|w_0 - w_1| \succcurlyeq |w_1 - w_2| \preccurlyeq |w_2 - w_3|$. Applying f^{-i} to these points and using the Koebe distortion principle, we find a definite *space* around $[x, f^{q_n}(x)]$ inside $[f^{-q_n}(x), f^{2q_n}(x)]$. Therefore $|x - f^{-q_n}(x)| \succcurlyeq |f^{q_n}(x) - x|$. The proof of the opposite inequality is similar. □

We arrive at the following fundamental fact first proved by Świątek [18] and Herman [12].

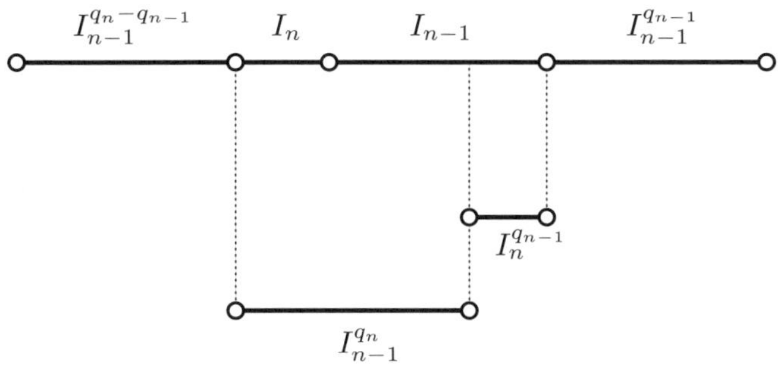

Fig. 2. These six intervals are pairwise comparable

Lemma 3.5. *Any two adjacent intervals in the dynamical partition of level n of f are comparable.*

Proof. First we prove that all intervals in Fig. 2 are pairwise comparable, through the following steps.

(a) From Lemma 3.4, we know that $|I_{n-1}| \asymp |I_{n-1}^{q_{n-1}}|$ and $|I_{n-1}^{q_n-q_{n-1}}| \asymp |I_{n-1}^{q_n}|$.

(b) Since the dynamical symmetric of I_n, namely the interval $I_n^{-q_n}$, is contained in I_{n-1}, we also have $|I_n| \preccurlyeq |I_{n-1}|$.

(c) Since the dynamical symmetric of I_{n-1}, namely $I_{n-1}^{-q_{n-1}}$ is contained in $I_n \cup I_{n-1}^{q_n-q_{n-1}}$, we have $|I_{n-1}| \preccurlyeq |I_{n-1}^{q_n-q_{n-1}}|$. Moreover, since $I_{n-1}^{q_n} \subseteq I_n \cup I_{n-1}$, items ($a$) and ($b$) yield $|I_{n-1}^{q_n-q_{n-1}}| \preccurlyeq |I_{n-1}|$. Therefore $|I_{n-1}| \asymp |I_{n-1}^{q_n-q_{n-1}}|$.

(d) Next, we claim that $|I_n| \asymp |I_n^{q_{n-1}}|$. To see why, consider the diffeomorphism

$$f^{q_n-q_{n-1}} : I_{n-1} \cup I_{n-1}^{q_{n-1}} \to I_{n-1}^{q_n-q_{n-1}} \cup I_{n-1}^{q_n} .$$

By the cross-ratio inequality (Lemma 2.3) applied to $M = I_n^{q_{n-1}}$ and $T = I_{n-1} \cup I_{n-1}^{q_{n-1}}$, we have $|I_n^{q_{n-1}}| \preccurlyeq |I_n^{q_n}| \asymp |I_n|$. Conversely, considering the diffeomorphism

$$f^{q_{n-1}} : I_{n-1}^{q_n-q_{n-1}} \cup I_n \to I_{n-1}^{q_n} \cup I_n^{q_{n-1}}$$

and applying the cross-ratio inequality to $M = I_n^{q_n}$ and $T = I_{n-1}^{q_n-q_{n-1}} \cup I_n$, we get

$$|I_n| \asymp |I_n^{q_n}| \preccurlyeq |I_{n-1}^{q_n+q_{n-1}}| \asymp |I_n^{q_{n-1}}| .$$

This proves our claim.

(e) Finally, we claim that $|I_{n-1}| \preccurlyeq |I_n|$, thereby reversing the inequality in (b). It is here that we use the critical point in a crucial way. Let $\theta_n = |I_n|/|I_{n-1}|$; we already know that $\theta_n \preccurlyeq 1$. Look at the intervals I_{n-1}^1, I_n^1 and $I_{n-1}^{q_n-q_{n-1}+1}$, all near the critical value of f. By an argument similar to the one in (c), we have $|I_{n-1}^1| \asymp |I_{n-1}^{q_n-q_{n-1}+1}|$. Moreover, $|I_n^1| \asymp \theta_n^p |I_{n-1}^1|$, where $p > 1$ is the power-law of f at the critical point. Hence these three intervals have a cross-ratio comparable to θ_n^p. On the other hand the map $f^{q_{n-1}-1}$ carries them diffeomorphically onto $I_{n-1}^{q_{n-1}}$, $I_n^{q_{n-1}}$ and $I_{n-1}^{q_n}$, respectively, whose cross-ratio is comparable to $|I_n^{q_{n-1}}|/|I_{n-1}^{q_n}|$, which in turn is comparable to θ_n. Applying the Koebe distortion principle, we see that $\theta_n^p \succcurlyeq \theta_n$, and so $\theta_n \succcurlyeq 1$ as claimed.

This proves that all six intervals in Fig. 2 are comparable. To derive the remaining comparability relations, propagate this information using Koebe's distortion principle. □

Proof of Theorem 3.1. Part (a) is Lemma 3.5 above. The remaining statements are straightforward consequences of (a). □

3.2. Beau property of renormalization

The bounds obtained in the proof of Theorem 3.1 depended on f, more precisely on the space that each atom of $\mathcal{P}_n$ enjoys relative to its two neighbors in $\mathcal{P}_n$. We now concentrate in proving that such bounds eventually become universal. It suffices to prove that the space in question is eventually universal. Bounds of this type are called *beau* by Sullivan.

Lemma 3.6. *There exists $n_0 = n_0(f)$ such that for all $n \geq n_0$ the first return map $f_n : J_n \to J_n$ satisfies $Sf_n(x) < 0$ for all $x \in J_n$.*

Proof. This is proved in Theorem A.4 of Appendix A. □

Lemma 3.7. *Given $\varepsilon > 0$, there exists $n_1 = n_1(f, \varepsilon) > n_0(f)$ such that the following holds for all $n \geq n_1$. Let $\Delta \in \mathcal{P}_n$, let $k \geq 1$ be an integer such that $f^j(\Delta)$ is contained in an element of $\mathcal{P}_n$ for all $1 \leq j \leq k$, and let Δ^* be the union of Δ with its left and right neighbors in $\mathcal{P}_n$. Then we have $f^k|\Delta^* = \phi_1 \circ \phi_2 \circ \phi_3$ where ϕ_1 and ϕ_3 are diffeomorphisms with distortion bounded by $1 + \varepsilon$ and ϕ_2 is either the identity or a map with negative Schwarzian derivative. In particular, if ε is small enough and if $I_{n-1} \neq \Delta \neq I_n$, then the distortion of $f^k|\Delta$ is bounded from below by one-half.*

Proof. Let $n_1 > n_0$ be such that $\mu_0^{n_1-n_0} \ll \varepsilon$, where μ_0 is the constant of Theorem 3.1. For $n \geq n_1$, Δ and k as in the statement, let $J \in \mathcal{P}_{n_0}$ be such that $\Delta \subseteq J$, let J^* be the union of J with its two neighbors in $\mathcal{P}_{n_0}$, and note that the space of Δ^* inside J^* is bounded from below by $C|J^*|/|\Delta^*|$, for some constant $C > 0$. Let $m \geq 0$ be the smallest integer such that

$f^m(J) \subseteq J_{n_0}$. Then for all $j \leq m$ the map $f^j|J^*$ is a diffeomorphism onto its image and, by Theorem 3.1 (b) and the Koebe distortion principle, its distortion on Δ^* is bounded by

$$\left(1 + C\frac{|\Delta^*|}{|J^*|}\right)^2 \exp\left\{C\frac{|\Delta^*|}{|J^*|}\right\} \leq \exp\{C\mu_0^{n_1-n_0}\} \leq 1+\varepsilon\,.$$

Now, there are two possibilities. The first is that $m \geq k$; in this case we can take $\phi_1 = f^k|\Delta^*$ and $\phi_2 = \phi_3 =$ identity map. The second is that $m < k$. In this case we consider the first return map $f_{n_0} : J_{n_0} \to J_{n_0}$ and let $\ell \geq 0$ be the *largest* such that

$$f^k = f^{k_1} \circ f_{n_0}^{\ell} \circ f^{k_3}\,,$$

where $k_1 \geq 0$ and $k_3 \geq 0$. We then take $\phi_1 = f^{k_1-1}$, $\phi_2 = f \circ f_{n_0}^{\ell}$ and $\phi_3 = f^{k_3}|\Delta^*$ (if $k_1 = 0$ we take instead $\phi_2 = f_{n_0}^{\ell}$ and $\phi_1 =$ identity). By Lemma 3.6, $S\phi_2 < 0$, and by the above remarks the distortions of both ϕ_1 and ϕ_3 are bounded by $1+\varepsilon$ in the appropriate domains. □

Proposition 3.8. *All bounds in Theorem 3.1 are beau. In other words, there exist universal constants $K_0 > 0$ and $0 < \lambda_0 < \lambda_1 < 1$ and some $\overline{n} = \overline{n}(f) > 0$ such that for all $n \geq \overline{n}$ the constants C_0, μ_0 and μ_1 in Theorem 3.1 can be replaced by K_0, λ_0 and λ_1, respectively.*

Proof. This is straightforward from Lemma 3.7. □

Remark. From now on, whenever we say that a constant "depends only on the real bounds", we mean that the said constant is a universal function of constants K_0, λ_0 and λ_1 given by this proposition.

4. How smooth is the conjugacy?

Now we turn to the first main result in this paper. The theorem states that if the successive renormalizations of two critical circle maps with the same rotation number ρ converge together at an exponential rate, then such maps are $C^{1+\alpha}$ conjugate for some $\alpha > 0$, provided ρ belongs to a certain full-measure set $\mathbb{A}$ of rotation numbers (defined in §4.4). First, in order to get bounds that do not depend on the maximum of the partial quotients of the rotation number, we need to perform some "saddle-node" estimates and constructions.

4.1. Saddle-node geometry

Let a be a positive integer and $\Delta_1, \Delta_2, \ldots, \Delta_{a+1}$ be consecutive intervals on the line or circle. By an *almost parabolic map* of length a and fundamental domains Δ_j, $1 \leq j \leq a$, we mean a negative-Schwarzian diffeomorphism

$$f : \Delta_1 \cup \Delta_2 \cup \cdots \cup \Delta_a \to \Delta_2 \cup \Delta_3 \cup \cdots \cup \Delta_{a+1}$$

such that $f(\Delta_j) = \Delta_{j+1}$.

The basic geometric estimate on almost parabolic maps is due to J.-C. Yoccoz.

Yoccoz's Lemma. *Let $f : I \to J$ be an almost parabolic map of length a and fundamental domains Δ_j, $1 \leq j \leq a$. If $|\Delta_1| \geq \sigma|I|$ and $|\Delta_a| \geq \sigma|I|$, then*

$$\frac{1}{C_\sigma}\frac{|I|}{\min\{j, a-j\}^2} \leq |\Delta_j| \leq C_\sigma\frac{|I|}{\min\{j, a-j\}^2},$$

where $C_\sigma > 1$ depends only on σ.

For a proof, see Appendix B. We will use Yoccoz's estimates to compare two almost parabolic maps.

Proposition 4.1. *Let f and g be two almost parabolic maps with the same length a defined on the same interval. Then for all $x \in \Delta_1(f) \cap \Delta_1(g)$ and all $0 \leq k \leq a/2$ we have*

$$\left|f^k(x) - g^k(x)\right| \leq C\,\|f-g\|_0 k^3 . \tag{2}$$

Proof. First note, using the mean-value theorem, that

$$\begin{aligned}\left|f^k(x) - g^k(x)\right| &= \left|\sum_{j=0}^{k-1}\left(f^{k-j-1}(f(g^j(x))) - f^{k-j-1}(g^{j+1}(x))\right)\right| \\ &\leq \sum_{j=0}^{k-1}\left|Df^{k-j-1}(\xi_j)\right|\,\left|f(g^j(x)) - g(g^j(x))\right| ,\end{aligned}$$

where ξ_j lies between $f(g^j(x))$ and $g^{j+1}(x)$. Hence we have

$$\left|f^k(x) - g^k(x)\right| \leq \|f-g\|_0 \sum_{j=0}^{k-1}\left|Df^{k-j-1}(\xi_j)\right| . \tag{3}$$

Let us estimate each summand in the right-hand side of (3). Let $m = m(j)$ be such that $\xi_j \in \Delta_{j+m}(f)$, and assume also that $j + m \leq a/2$. This last condition is always satisfied if the central fundamental domain of g lies to the left of the central fundamental domain of f (if this is not the case, then reverse the roles of f and g in (3) and throughout). Using Yoccoz's Lemma, we see that

$$\left|Df^{k-j-1}(\xi_j)\right| \asymp \frac{(j+m)^2}{(a-k-m+1)^2} \leq \left(\frac{j+m}{j+1}\right)^2 . \tag{4}$$

Hence, it suffices to estimate m as a function of j. For this purpose, let $n = n(j)$ be such that $g^{j+1}(x) \in [f^{j+n-1}(x), f^{j+n}(x)]$. We claim that $m \leq n+1$. There are two possibilities. The first is that $f(g^j(x)) \geq g^{j+1}(x)$: in this case we see easily that

$$\xi_j \in \left[g^{j+1}(x), f(g^j(x))\right] \subseteq \left[f^{j+n-1}(x), f^{j+n+1}(x)\right]$$

and so $m \leq n+1$. The second is that $f(g^j(x)) < g^{j+1}(x)$. In this case we have $\xi_j < g^{j+1}(x) < f^{j+n}(x) \in \Delta_{j+n+1}(f)$, so once again $m \leq n+1$. This proves our claim.

So now we must bound n as a function of j. Again, there are two cases to consider.

(*a*) We have $[g^{j+1}(x), g^{j+2}(x)] \subseteq [f^{j+n-1}(x), f^{j+n}(x)]$. In this case (Fig. 3*a*) Yoccoz's Lemma gives us

$$\frac{1}{j^2} \leq \frac{C}{(j+n)^2} ,$$

which implies $n \leq Cj$.

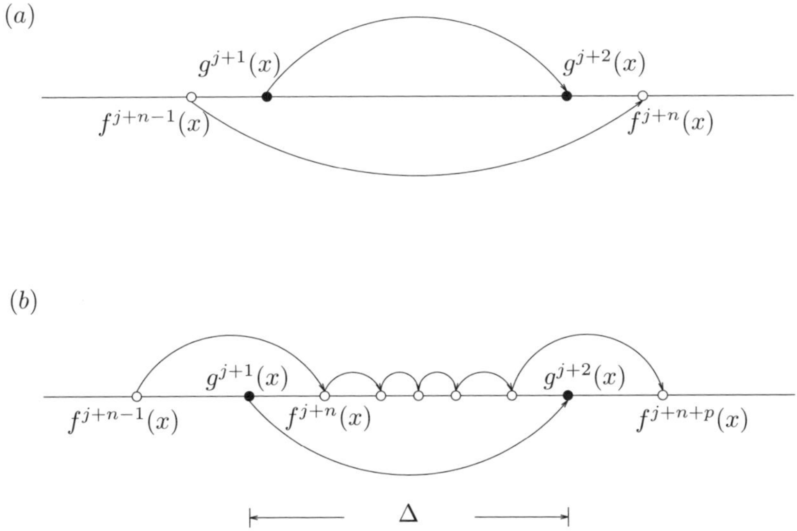

Fig. 3. Bounding n as a function of j

(*b*) We have $g^{j+2}(x) > f^{j+n}(x)$. In this case (Fig. 3*b*), $f^{j+n}(x)$ is the first point in the f-orbit of x that lands inside the interval $\Delta = [g^{j+1}(x), g^{j+2}(x)]$. Let p be such that $f^{j+n+i}(x) \in \Delta$ for $i = 0, 1, \ldots, p-1$ but $f^{j+n+p}(x) \notin \Delta$. Then we have $\Delta \subseteq [f^{j+n-1}(x), f^{j+n+p}(x)]$, and this time Yoccoz's Lemma gives us

$$\frac{1}{j^2} \leq C\left(\frac{1}{(j+n)^2} + \frac{1}{(j+n+1)^2} + \cdots + \frac{1}{(j+n+p)^2}\right) \leq \frac{C}{j+n} .$$

Therefore $n \leq Cj^2$ in this case.

In either case we see that $m \leq Cj^2$. Carrying this information back to (4), we deduce that

$$\left|Df^{k-j-1}(\xi_j)\right| \leq C j^2 . \tag{5}$$

Substituting (5) into (3), we arrive at (2), and the proof is complete. □

4.2. *A criterion for smoothness*

One key ingredient in the proof of our First Main Theorem is a slight extension of a result originally due to Carleson [2], namely Proposition 4.3 below. To formulate it, we need an auxiliary definition.

Definition. A *fine grid* is a sequence $\{\mathcal{Q}_n\}_{n\geq 0}$ of finite partitions of S^1 which satisfies

(*a*) Each $\mathcal{Q}_{n+1}$ is a strict refinement of $\mathcal{Q}_n$;
(*b*) There exists $a > 0$ such that each $I \in \mathcal{Q}_n$ is the disjoint union of at most a atoms of $\mathcal{Q}_{n+1}$;
(*c*) There exists $c > 0$ such that $c^{-1}|I| \leq |J| \leq c|I|$ for each pair of adjacent atoms $I, J \in \mathcal{Q}_n$.

For example, the dynamical partitions $\{\mathcal{P}_n\}$ of a critical circle map with rotation number of bounded type always form a fine grid, by Theorem 3.1. We note the following easy lemma concerning a fine grid $\{\mathcal{Q}_n\}$.

Lemma 4.2. *If $I \in \mathcal{Q}_n$, $J \in \mathcal{Q}_{n+1}$ and $J \subseteq I$, then $(1 + c^{-1})|J| \leq |I| \leq ac^a|J|$. In particular, there exist $C_0 > 1$ and $0 < \lambda_0 < \lambda_1 < 1$ such that $C_0^{-1}\lambda_0^n \leq |I| \leq C_0\lambda_1^n$, for all $I \in \mathcal{Q}_n$.* □

The constants $a, c, C_0, \lambda_0, \lambda_1$, are the *fine constants* of $\{\mathcal{Q}_n\}$.

Proposition 4.3. *Let $h : S^1 \to S^1$ be a homeomorphism and let $\{\mathcal{Q}_n\}_{n\geq 0}$ be a fine grid.*

(a) *If there exists* $C > 0$ *such that*

$$\left|\frac{|I|}{|J|} - \frac{|h(I)|}{|h(J)|}\right| \leq C ,$$

for each pair of adjacent atoms $I, J \in \mathcal{Q}_n$*, for all* $n \geq 0$*, then* h *is quasisymmetric.*

(b) *If there exist constants* $C > 0$ *and* $0 < \lambda < 1$ *such that*

$$\left|\frac{|I|}{|J|} - \frac{|h(I)|}{|h(J)|}\right| \leq C\lambda^n , \tag{6}$$

for each pair of adjacent atoms $I, J \in \mathcal{Q}_n$*, for all* $n \geq 0$*, then* h *is a* $C^{1+\alpha}$*-diffeomorphism for some* $\alpha > 0$.

The proof of Proposition 4.3 will depend on the following fact from elementary real analysis. If ϕ is a real-valued function in an interval or oriented arc on the circle, let $D^+\phi(x) = \lim_{t\downarrow 0}(\phi(x+t) - \phi(t))/t$ be the *right derivative* of ϕ at x, if the limit exists.

Lemma 4.4. *Let* $\phi_n : [0,1] \to \mathbb{R}$ *be a sequence of continuous, right differentiable mappings such that* $D^+\phi_n$ *converges uniformly to an* α*-Hölder continuous function* $\varphi : [0,1] \to \mathbb{R}$*, and such that each* $D^+\phi_n$ *is Riemann-integrable. If* ϕ_n *converges uniformly to* ϕ*, then* ϕ *is* $C^{1+\alpha}$ *and* $D\phi = \varphi$. □

Proof of Proposition 4.3. We will prove (b) only, the proof of (a) being somewhat easier. Let ϕ_n be the piecewise affine C^0-approximations to h determined by the vertices of $\mathcal{Q}_n$. Then ϕ_n is differentiable on the right, and $D^+\phi_n$ is a step function. First we show that $\{D^+\phi_n\}_{n\geq 0}$ is a uniform Cauchy sequence, and then that the limit is Hölder continuous. Take an atom I of $\mathcal{Q}_n$, and consider the decomposition

$$I = J_1 \cup J_2 \cup \cdots \cup J_p ,$$

with $J_k \in \mathcal{Q}_{n+1}$ consecutive and pairwise disjoint and $p \leq a$. Then $D^+\phi_n$ is constant on I and $D^+\phi_{n+1}$ is constant on each J_k, say

$$\begin{cases} D^+\phi_n(t) = \sigma = \dfrac{|\phi_n(I)|}{|I|} & (t \in I) \\ D^+\phi_{n+1}(t) = \sigma_k = \dfrac{|\phi_{n+1}(J_k)|}{|J_k|} & (t \in J_k) \end{cases} .$$

Thus, we have

$$\sigma|I| = \sum_{k=1}^{p} \sigma_k |J_k| ,$$

and in particular $\sigma' = \min \sigma_k \le \sigma \le \max \sigma_k = \sigma''$. Also, $\sigma'/\sigma'' \le \sigma/\sigma_k \le \sigma''/\sigma'$ for all k. Since by assumption $|1 - (\sigma_{k+1}/\sigma_k)| \le C\lambda^{n+1}$, an easy telescoping trick gives us

$$\frac{\sigma''}{\sigma'} \le (1 + C\lambda^{n+1})^a \le 1 + C\lambda^{n+1} .$$

A similar lower bound holds true for σ'/σ''. Therefore we have

$$1 - C\lambda^n \le \frac{\sigma}{\sigma_k} \le 1 + C\lambda^n , \tag{7}$$

for all $k = 1, 2, \ldots p$. This shows that the sequence $\left\{D^+\phi_n\right\}_{n\ge 0}$ is uniformly bounded, and moreover that for all $m \ge n \ge 0$ and all $t \in S^1$, we have

$$\left|D^+\phi_m(t) - D^+\phi_n(t)\right| \le C\sum_{j=n}^{m-1} \lambda^j < \frac{C}{1-\lambda}\lambda^n . \tag{8}$$

Hence $\left\{D^+\phi_n\right\}_{n\ge 0}$ is a uniform Cauchy sequence as claimed. Let $\varphi = \lim D^+\phi_n$, and let $\alpha > 0$ be such that $\lambda_0^\alpha = \lambda$. We prove φ is α-Hölder as follows. It suffices to consider points $x, y \in S^1$ whose distance is smaller than $\inf_{I\in\mathcal{Q}_0} |I|$. Take the smallest n such that x and y belong to distinct elements of $\mathcal{Q}_n$. Then either $n = 0$ or x and y lie in a common element of $\mathcal{Q}_{n-1}$. Either way we have by (7)

$$\left|D^+\phi_n(x) - D^+\phi_n(y)\right| \le C\lambda^n . \tag{9}$$

Combining (8) and (9), we deduce that

$$\begin{aligned}
|\varphi(x) - \varphi(y)| &\le \left|\varphi(x) - D^+\phi_n(x)\right| + \left|D^+\phi_n(x) - D^+\phi_n(y)\right| \\
&\quad + \left|D^+\phi_n(y) - \varphi(y)\right| \\
&\le \frac{C}{1-\lambda}\lambda^n + C\lambda^n + \frac{C}{1-\lambda}\lambda^n \le C\lambda_0^{n\alpha} \\
&\le C|x-y|^\alpha ,
\end{aligned}$$

and so φ is α-Hölder as claimed. □

Remark. In the language of conditional expectations, the sequence $\{D^+\phi_n\}_{n\ge 0}$ satisfies $\mathbb{E}\left(D^+\phi_n \mid \mathcal{B}_n\right) = D^+\phi_{n+1}$, where $\mathcal{B}_n$ is the σ-algebra generated by $\mathcal{Q}_n$, and therefore constitutes a *martingale*. Thus, the existence of a pointwise *a.e.* limit φ, merely as an integrable function, is a special case of J. Doob's *martingale convergence theorem*, see [1], p. 490.

4.3. A suitable fine grid

The dynamical partitions $\mathcal{P}_n$ of a critical circle map f do *not* determine a fine grid, unless the rotation number of f is of bounded type. We will however use these dynamical partitions to build a fine grid $\{\mathcal{Q}_n\}$ for our map f. The construction requires some preliminary definitions.

An element $I \in \mathcal{P}_n$ is a *saddle-node* atom if it is the disjoint union of some number $a \geq 1000$ of atoms of $\mathcal{P}_{n+1}$.

Given two atoms $\mathcal{P}_{n+1} \ni J \subseteq I \in \mathcal{P}_n$, the *order* of J inside I is one plus the smallest number of atoms of $\mathcal{P}_{n+1}$ on the right and left components of $I \setminus J$.

Note that inside a saddle-node atom $I \in \mathcal{P}_n$ there are exactly two atoms of $\mathcal{P}_{n+1}$ of order k for each $k \leq a/2$. Let $N \geq 0$ be largest with the property that $2^{N+1} < a/2$. For each $0 \leq i \leq N$, we define M_i, the *i-th central interval* of I, to be the convex-hull $[J, J^*] \subseteq I$ of the union of both atoms J, J^* of order 2^i. Note that these central intervals are nested (see Fig. 4). The left and right components of $M_i \setminus M_{i+1}$, respectively L_i and R_i, are the *lateral intervals* of I. The central interval M_N is also called the *final interval* of I. The lateral intervals together with the final interval form a special partition of I, the *balanced* partition of I.

Remark. It follows from Yoccoz's lemma that $|L_i| \asymp |M_{i+1}| \asymp |R_i|$ for all i.

Now we define an auxiliary partition $\widetilde{\mathcal{P}}_n$, for each $n \geq 1$. The atoms of $\widetilde{\mathcal{P}}_n$ are all atoms of $\mathcal{P}_n$ which are not saddle-node, together with the atoms

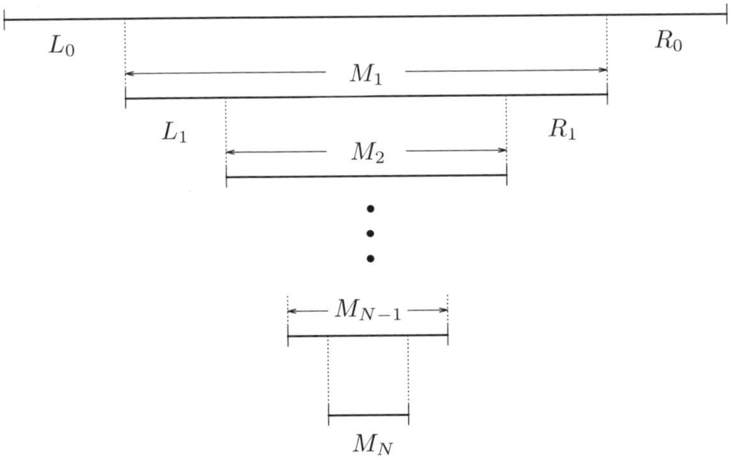

Fig. 4. Central and lateral intervals of a saddle-node atom

of the balanced partitions of all saddle-node atoms of $\mathcal{P}_n$. The partition $\mathcal{Q}_n$ that we want is constructed from $\widetilde{\mathcal{P}}_n$ and $\mathcal{P}_n$ as follows.

Proposition 4.5. *There exists a fine grid* $\{\mathcal{Q}_n\}$ *in* S^1 *with the following properties.*

(a) *Every atom of* $\mathcal{Q}_n$ *is the union of at most 3 atoms of* $\mathcal{Q}_{n+1}$.
(b) *Every atom* Δ *of* $\mathcal{Q}_n$ *is a union of atoms of* $\mathcal{P}_m$ *for some* $m \leq n$, *and there are four possibilities:*
  (b₁) Δ *is a single atom of* $\mathcal{P}_m$*;*
  (b₂) Δ *is a central interval of* $\widetilde{\mathcal{P}}_m$*;*
  (b₃) Δ *is the union of at least two atoms of* $\mathcal{P}_{m+1}$ *contained in a single atom of* $\widetilde{\mathcal{P}}_m$.
  (b₄) Δ *is a union of intervals which are simultaneously atoms of* $\mathcal{P}_m$ *and* $\widetilde{\mathcal{P}}_m$.

Proof. The proof is by induction on n. The first partition $\mathcal{Q}_1$ consists of all atoms of $\mathcal{P}_1$ which are not saddle-node atoms together with the intervals L_0, M_1 and R_0 of each saddle-node interval $I \in \mathcal{P}_1$ ($I = L_0 \cup M_1 \cup R_0$). It is clear that each atom of $\mathcal{Q}_1$ falls within one of the categories (b_1)-(b_4) above.

Assuming $\mathcal{Q}_n$ defined, define $\mathcal{Q}_{n+1}$ as follows. Take an atom $I \in \mathcal{Q}_n$ and consider the four cases below.

(1) If I is a single atom of $\mathcal{P}_m$ then one of two things can happen:
   (i) I is a saddle-node atom: In this case write $I = L_0 \cup M_1 \cup R_0$ as above and take L_0, R_0 and M_1 as atoms of $\mathcal{Q}_{n+1}$. Note that the lateral intervals L_0 and R_0 are atoms of type (b_1), while the central interval M_1 is of type (b_2).
   (ii) I is not a saddle-node atom: In this case write $I = L \cup M \cup R$ where L and R are the atoms of $\mathcal{P}_{m+1}$ adjacent to the endpoints of I and M is the union of the other atoms of $\mathcal{P}_{m+1}$ inside I. Add these three intervals to $\mathcal{Q}_{n+1}$, noting that L and R are of type (b_1), while M is of type (b_4).
(2) If I is a central interval of $\widetilde{\mathcal{P}}_m$ which is not the final interval, consider the next central interval of $\widetilde{\mathcal{P}}_m$ inside I, say M, and the two corresponding lateral intervals L and R such that $I = L \cup M \cup R$, and declare L, R and M members of $\mathcal{Q}_{n+1}$. Note that L and R are of type (b_3), while M is of type (b_2).
(3) If I is a union of $p \geq 2$ consecutive atoms $\Delta_1, \ldots, \Delta_p$ of $\widetilde{\mathcal{P}}_{m+1}$ inside a single atom of $\mathcal{P}_m$, divide it up into three approximately equal parts. More precisely, write $p = 3q + r$ and, when $r = 0$ or 1, consider

$I = L \cup M \cup R$ where

$$L = \bigcup_{j=1}^{q} \Delta_j \,, \quad M = \bigcup_{j=q+1}^{p-q} \Delta_j \,, \quad R = \bigcup_{j=p-q+1}^{p} \Delta_j \,.$$

When $r = 2$, consider $I = L \cup M \cup R$ where

$$L = \bigcup_{j=1}^{q+1} \Delta_j \,, \quad M = \bigcup_{j=q+2}^{p-q-1} \Delta_j \,, \quad R = \bigcup_{j=p-q}^{p} \Delta_j \,.$$

Note that M is empty when $p = 2$. In any case, we obtain two or three new atoms of $\mathcal{Q}_{n+1}$ which are either single atoms of $\mathcal{P}_{m+1}$, and therefore of type (b_1), or once again intervals of type (b_3).

(4) If I is a union of intervals which are simultaneously atoms of $\mathcal{P}_m$ and $\widetilde{\mathcal{P}}_m$, divide it up exactly as in (3), obtaining either two or three new atoms of $\mathcal{Q}_{n+1}$ which are either single atoms of $\mathcal{P}_m$, and therefore of type (b_1), or once again intervals of type (b_4).

This completes the induction. That $\{\mathcal{Q}_n\}_{n \geq 0}$ constitutes a fine grid follows easily from the real bounds and the remark preceding this proposition. □

An immediate consequence of the mere existence of such a fine grid is the fact that any two critical circle maps with the same rotation number are quasisymmetrically conjugate.

Corollary 4.6. *Let f and g be critical circle maps with the same irrational rotation number, and let h be the conjugacy between f and g that maps the critical point of f to the critical point of g. Then h is quasisymmetric.*

Proof. Apply Proposition 4.3 (a) to the fine grid constructed above. □

4.4. Proof of the First Main Theorem

We will proceed according to the following strategy. Given two critical circle maps f and g with the same rotation number, consider the special partitions $\mathcal{Q}_n = \mathcal{Q}_n(f)$ and $\widetilde{\mathcal{Q}}_n = \mathcal{Q}_n(g)$ given by Proposition 4.5. The conjugacy h between f and g is an isomorphism between the corresponding fine grids. We want to show that the coherence property (6) holds for h and $\{\mathcal{Q}_n\}$. The stated $C^{1+\alpha}$ smoothness of h will then follow from Proposition 4.3.

To achieve our goal, we need to impose certain conditions on the common rotation number of f and g. Let us consider the set $\mathbb{A} \subseteq [0, 1]$ of rotation numbers whose partial quotients (a_n) satisfy

$$\limsup_{n \to \infty} \frac{1}{n} \sum_{j=1}^{n} \log a_j \;<\; \infty \,, \tag{10}$$

as well as

$$\lim_{n\to\infty} \frac{1}{n} \log a_n = 0 , \tag{11}$$

and which also satisfy

$$\frac{1}{n} \sum_{j=k+1}^{k+n} \log a_j \leq \omega\left(\frac{n}{k}\right) , \tag{12}$$

for all $0 < n \leq k$, where $\omega(t)$ is a positive function (that depends on the rotation number) defined for $t > 0$ such that $t\omega(t) \to 0$ as $t \to 0$. This is the class of rotation numbers for which we shall prove our theorem. Note that all numbers of bounded type satisfy (10), (11) and (12). The number whose partial quotients are given by $a_n = k$ if $n = 2^k$ with $k \geq 1$ and $a_n = 1$ otherwise is an explicit element of $\mathbb{A}$ that is not of bounded type. This number satisfies (12) with $\omega(t) = 1/\sqrt{t}$. It is a well-known consequence of the fact that the Gauss map preserves a measure absolutely continuous with respect to Lebesgue measure (the Gauss measure) and Birkhoff's ergodic theorem that both (10) and (11) hold Lebesgue almost everywhere (see [3], page 175). Condition (12) also holds for Lebesgue almost all numbers if we take $\omega(t) = C(1 - \log t)$, where $C > 0$ depends on the number (this last fact was communicated to us by C.G. Moreira – for a proof, see Appendix C). Therefore our set $\mathbb{A}$ has full Lebesgue measure in [0, 1].

In what follows we use the notation $x_n = x_n(f) = f^{q_n}(c)$. We also assume that the critical point c is the same for both maps. If the renormalizations $\boldsymbol{f}_n$ and $\boldsymbol{g}_n$ converge together exponentially fast then $|x_n(f) - c|/|x_n(g) - c|$ converges to a limit exponentially fast also. More precisely, we have the following lemma.

Lemma 4.7. *If* $\|\boldsymbol{f}_n - \boldsymbol{g}_n\|_0 \leq C\mu^k$ *for some* $0 < \mu < 1$ *and all* $n \geq 0$, *then the ratio* $|x_n(f) - c|/|x_n(g) - c|$ *converges to a limit exponentially fast. Moreover, for all* $m, k \geq 1$ *we have*

$$\left| \frac{|I_m(f)|}{|I_k(f)|} - \frac{|I_m(g)|}{|I_k(g)|} \right| \leq C\mu^{\min\{m,k\}} \frac{|I_m(f)|}{|I_k(f)|} . \tag{13}$$

Proof. The hypothesis tells us that

$$\left| \frac{|I_{n+1}(f)|}{|I_n(f)|} - \frac{|I_{n+1}(g)|}{|I_n(g)|} \right| \leq C_1\mu^n ,$$

for all $n \geq 1$. Writing $\alpha_n = |x_n(f) - c|/|x_n(g) - c| = |I_n(f)|/|I_n(g)|$, and taking into account that $C_2^{-1}|I_n(g)| \leq |I_{n+1}(g)| \leq C_2|I_n(g)|$ by the real bounds (for some $C_2 > 1$), we see that the above inequality is equivalent to

$$\left| \frac{\alpha_{n+1}}{\alpha_n} - 1 \right| \leq C_3\mu^n .$$

This is the same as $\alpha_{n+1} = (1+\varepsilon_n)\alpha_n$ where $|\varepsilon_n| \le C_3\mu^n$. Therefore $\alpha_n = \alpha_1 \prod_{j=1}^{n-1}(1+\varepsilon_j)$, and this shows that $\lim \alpha_n$ exists. Finally, note that if $m > k \ge 1$ then

$$\left|\frac{\alpha_m}{\alpha_k} - 1\right| \le \left|\prod_{j=k}^{m-1}(1+\varepsilon_j) - 1\right|$$

$$\le C_4 \sum_{j=k}^{m-1} \varepsilon_j < C_5\mu^k ,$$

and similarly for $|1 - \alpha_k/\alpha_m|$, and these facts clearly imply (13). □

Remark. Having established this lemma, we may assume, after conjugating one of the maps (say g) by a suitable smooth diffeomorphism, that the limit of the ratios $|I_n(f)|/|I_n(g)|$ is in fact equal to one. This will be our standing hypothesis from now on (used at the end of the proof of Lemma 4.10 below).

Definition. Let $f_m : J_m(f) \to J_m(f)$ be the m-th first return map of f and let $k \neq 0$ be an integer such that $|k| \le \lceil a_m/2 \rceil$ (where $\lceil x \rceil$ denotes the smallest integer $\ge x$). The *restricted domain* of f_m^k, denoted $D_{m,k}$, is defined as follows.

$$D_{m,k} = \begin{cases} I_m \cup [f_m^{\lceil \frac{a_m}{2} \rceil - k}(x_{m-1}), x_{m-1}], & \text{when } k > 0 \\ [f_m(x_{m+1}), f_m^{\lceil \frac{a_m}{2} \rceil - k}(x_{m+1})], & \text{when } k \le -1 . \end{cases}$$

In less precise terms, the restricted domain $D_{m,k}$ is the the set of points in J_m which can be iterated k times by f_m without ever going *across* the central fundamental domain of f_m in $J_m(f) \setminus J_{m+1}(f)$.

Lemma 4.8. *For all $x \in D_{m,k}$ we have $|Df_m^k(x)| \le K$, where $K \ge 1$ depends only on the real bounds.*

Proof. Use Theorem 3.1 and Yoccoz's Lemma. □

Lemma 4.9. *Let v be a vertex of $\mathcal{P}_{k+p}(f)$ such that $v \in J_k(f)$. Then there exist $k \le m \le k+p$ and $1 \le N \le p$ such that v can be represented in the form*

$$v = \phi_1 \circ \phi_2 \circ \cdots \circ \phi_N(x_m) ,$$

where $\phi_j = f_{m_j}^{k_j}$ for some $k \le m_j \le k+p$ and $|k_j| \le \lceil a_{m_j}/2 \rceil$, and where the point $\phi_{j+1} \circ \cdots \circ \phi_N(x_m)$ belongs to the restricted domain of ϕ_j for each j.

Proof. For simplicity of notation, we write $J_i = J_i(f)$ in this proof. Let $k \le m_1 \le k+p$ be largest with the property that $v \in J_{m_1} \setminus J_{m_1+1}$, and let $0 < i \le a_{m_1}$ be such that $f^i_{m_1}(v) \in J_{m_1+1}$. If $i \le \lceil a_{m_1}/2 \rceil$ then let $k_1 = -i$; otherwise let $k_1 = a_{m_1} - i$. We get $\phi_1 = f^{k_1}_{m_1}$ and a new vertex $v_1 = f^{-k_1}_{m_1}(v) \in J_{m_1+1}$. If $v_1 \in J_{k+p}$ then $v_1 = f_{k+p}(x_{k+p-1})$ necessarily, and we can stop. On the other hand, if $v_1 \notin J_{k+p}$, then once again there exists m_2 in the range $m_1 < m_2 < k+p$ such that $v_1 \in J_{m_2} \setminus J_{m_2+1}$, and we can proceed inductively. At the end of this process we get sequences $m_1 < m_2 < \cdots < m_N \le k+p$ (so $N \le p$) and $v_1, v_2, \ldots, v_N$ with $v_j \in J_{m_j} \setminus J_{m_j+1}$, and for each j an integer k_j with $|k_j| \le \lceil a_{m_j}/2 \rceil$ such that $v_{j+1} = f^{-k_j}_{m_j}(v_j)$. The last vertex v_N is necessarily x_m for some $m \le k+p$. Hence it suffices to take $\phi_j = f^{k_j}_{m_j}$ to get the desired representation. □

From now on, we assume that the corresponding sucessive renormalizations of f and g approach each other exponentially, in other words $\|\boldsymbol{f}_n - \boldsymbol{g}_n\|_0 \le C\mu^n$ for some $0 < \mu < 1$ and all $n \ge 0$, just as stated in the hypothesis of Lemma 4.7.

Lemma 4.10. *There exists a constant $0 < \mu_* < 1$ for which the following holds. Let $v \in J_k(f)$ be a vertex of $\mathcal{P}_{k+p}(f)$ and let $w = h(v) \in J_k(g)$ be the corresponding vertex of $\mathcal{P}_{k+p}(g)$. If $\rho(f)$ satisfies condition (11), then we have*

$$|v - w| \ \le \ C|J_k(f)|K^p\mu_*^k \,, \tag{14}$$

where $K \ge 1$ is the constant of Lemma 4.8.

Proof. By Lemma 4.9 above, there exist points $x_m = x_m(f)$, $y_m = x_m(g)$ and a number $N \le p$ such that

$$|v - w| \ = \ |\phi_1 \circ \phi_2 \circ \cdots \circ \phi_N(x_m) - \psi_1 \circ \psi_2 \circ \cdots \circ \psi_N(y_m)| \ ,$$

where $\phi_j = f^{k_j}_{m_j}$ and $\psi_j = g^{k_j}_{m_j}$, with $k \le m_j \le k+p$ and $|k_j| \le \lceil a_{m_j}/2 \rceil$. For each $i \ge 1$, let $\Lambda_{i,f}$ be the affine map $x \mapsto c + |I_i(f)|x$, and define $\Lambda_{i,g}$ in the same way. For each $i \ge k$, let $A_{i,f} = \Lambda^{-1}_{k,f} \circ \Lambda_{i,f}$ and $A_{i,g} = \Lambda^{-1}_{k,g} \circ \Lambda_{i,g}$. In order to estimate $|v - w|$, we shall estimate $|v^* - w^*|$, where $v^* = \Lambda^{-1}_{k,f}(v)$ and $w^* = \Lambda^{-1}_{k,g}(w)$. To do this, for each $i \ge k$ consider the map $f_i^* : \Lambda^{-1}_{k,f}(J_i(f)) \to \Lambda^{-1}_{k,f}(J_i(f))$ given by

$$f_i^* \ = \ \Lambda^{-1}_{k,f} \circ f_i \circ \Lambda_{k,f} \ = \ A_{i,f} \circ \boldsymbol{f}_i \circ A^{-1}_{i,f} \,,$$

and let g_i^* be similarly defined.

First we claim that for all $x \in \Lambda^{-1}_{k,f}(J_i(f)) \cap \Lambda^{-1}_{k,g}(J_i(g))$ we have

$$\left| f_i^*(x) - g_i^*(x) \right| \ \le \ C_1\mu^k \frac{|I_i(f)|}{|I_k(f)|} \,. \tag{15}$$

To see why, note that by inequality (13) of Lemma 4.7 we have, for all z in the domain of both renormalizations $\boldsymbol{f}_i$ and $\boldsymbol{g}_i$,

$$\begin{aligned} \left|A_{i,f}(z) - A_{i,g}(z)\right| &= \left|\frac{|I_i(f)|}{|I_k(f)|} - \frac{|I_i(g)|}{|I_k(g)|}\right| |z| \\ &\leq C_2\mu^k \frac{|I_i(f)|}{|I_k(f)|} . \end{aligned}$$

Similarly, for all $x \in \Lambda_{k,f}^{-1}(J_i(f)) \cap \Lambda_{k,g}^{-1}(J_i(g))$ we have, again by (13),

$$\begin{aligned} \left|A_{i,f}^{-1}(x) - A_{i,g}^{-1}(x)\right| &= \left|\frac{|I_k(f)|}{|I_i(f)|} - \frac{|I_k(g)|}{|I_i(g)|}\right| |x| \\ &\leq C_3\mu^k \frac{|I_k(f)|}{|I_i(f)|} |x| \leq C_4\mu^k . \end{aligned}$$

Here we have used that $|x| \leq |J_i(f)|/|I_k(f)| \leq C_5|I_i(f)|/|I_k(f)|$ (recall from the real bounds that $|J_i(f)| \asymp |I_i(f)|$). Also, by hypothesis we have $\|\boldsymbol{f}_i - \boldsymbol{g}_i\|_0 \leq C_6\mu^k$. Combining these three estimates with a standard telescoping trick, we get (15), and the claim is proved.

Now let $\phi_j^* = A_{m_j,f} \circ \phi_j \circ A_{m_j,f}^{-1}$ and $\psi_j^* = A_{m_j,g} \circ \psi_j \circ A_{m_j,g}^{-1}$. Applying (15) with $i = m_j$ and using Proposition 4.1, we have

$$\left|\phi_j^*(x) - \psi_j^*(x)\right| \leq C_7 \, |k_j|^3 \mu^k \frac{|I_{m_j}(f)|}{|I_k(f)|} . \tag{16}$$

By the real bounds, there exists $0 < \lambda_1 < 1$ such that $|I_{m_j}(f)|/|I_k(f)| \leq C_8\lambda_1^{m_j-k}$. Taking $\lambda = \max\{\mu, \lambda_1\}$, we deduce from (16) that

$$\left|\phi_j^*(x) - \psi_j^*(x)\right| \leq C_9 \, a_{m_j}^3 \lambda^{m_j} . \tag{17}$$

We can at last start our estimate of $|v^* - w^*|$. First, note that $x_m = \Lambda_{m,f}(1)$ and $y_m = \Lambda_{m,g}(1)$. Writing $x_m^* = \Lambda_{k,f}^{-1}(x_m)$ and $y_m^* = \Lambda_{k,g}^{-1}(y_m)$, we see after a simple computation that $|x_m^* - y_m^*| \leq C_{10}\lambda^m$. Combining this fact with (17) and using Lemma 4.8, we get

$$\begin{aligned} \left|\phi_N^*(x_m^*) - \psi_N^*(y_m^*)\right| &\leq \left|\phi_N^*(x_m^*) - \psi_N^*(x_m^*)\right| + \left|\psi_N^*(x_m^*) - \psi_N^*(y_m^*)\right| \\ &\leq C_9 a_{m_N}^3 \lambda^{m_N} + C_{10}K\lambda^m . \end{aligned}$$

From this, and since

$$\begin{aligned} &\left|\phi_{N-1}^*(\phi_N^*(x_m^*)) - \psi_{N-1}^*(\psi_N^*(y_m^*))\right| \leq \\ &\left|\phi_{N-1}^*(\phi_N^*(x_m)) - \psi_{N-1}^*(\phi_N^*(x_m^*))\right| + \left|\psi_{N-1}^*(\phi_N^*(x_m^*)) - \psi_{N-1}^*(\psi_N^*(y_m^*))\right| , \end{aligned}$$

we deduce that

$$\left|\phi_{N-1}^*(\phi_N^*(x_m^*)) - \psi_{N-1}^*(\psi_N^*(y_m^*))\right| \le C_9\left(a_{m_{N-1}}^3 \lambda^{m_{N-1}} + K a_{m_N}^3 \lambda^{m_N}\right) + C_{10}K^2\lambda^m .$$

Proceeding inductively in this fashion, we get in the end

$$|v^* - w^*| \le C_9 \sum_{j=1}^{N} K^{j-1} a_{m_j}^3 \lambda^{m_j} + C_{10}K^N\lambda^m .$$

Using that $N \le p$ and taking $C_{11} = \max\{C_9, C_{10}\}$, we arrive at

$$|v^* - w^*| \le C_{11}K^p \left(\lambda^m + \sum_{j=1}^{N} a_{m_j}^3 \lambda^{m_j}\right) . \tag{18}$$

We have of course $\lambda^m \le \lambda^k$. Moreover, since $k \le m_j < m_{j+1}$ for all j, we have

$$\sum_{j=1}^{N} a_{m_j}^3 \lambda^{m_j} < \sum_{n=k}^{\infty} a_n^3 \lambda^n .$$

But since (a_n) satisfies condition (11), we know that $\lim (a_n^3)^{1/n} = 1$. In particular, if $\varepsilon > 0$ is such that $(1+\varepsilon)\sqrt{\lambda} = 1$, there exists $C_{12} = C_{12}(\varepsilon) > 0$ such that $a_n^3 < C_{12}(1+\varepsilon)^n$ for all n. Therefore

$$\sum_{n=k}^{\infty} a_n^3 \lambda^n < C_{12} \sum_{n=k}^{\infty} (\sqrt{\lambda})^n = \frac{C_{12}}{1-\sqrt{\lambda}} (\sqrt{\lambda})^k .$$

Taking this back to (18) yields $|v^* - w^*| \le C_{13}K^p(\sqrt{\lambda})^k$. Therefore, noting that under the assumption given in the remark after Lemma 4.7 we have

$$|v - w| = |\Lambda_{k,f}(v^*) - \Lambda_{k,g}(w^*)| \le |I_k(f)| \left(|v^* - w^*| + C_{14}\mu^k\right) ,$$

and taking $\mu_* = \sqrt{\lambda}$, we get (14) as desired. □

Lemma 4.11. *There exists a constant $M > 0$ depending only on the real bounds such that if $\Delta^* \in \mathcal{P}_k(f)$ and $\Delta \in \mathcal{P}_{k+p}(f)$ is contained in Δ^*, then*

$$|\Delta| \ge \frac{M^p}{\left(a_{k+1}a_{k+2}\cdots a_{k+p}\right)^2} |\Delta^*| .$$

Proof. This again follows from Yoccoz's Lemma and a simple inductive argument. □

Definition. The *level* of an atom $\Delta \in \mathcal{Q}_n(f)$, denoted $\ell(\Delta)$, is the largest $m \leq n$ such that Δ is contained in an atom of $\mathcal{P}_m(f)$.

Lemma 4.12. *If $\mathcal{Q}_n(f)$ contains an atom of level m, then*

$$n \leq c_0 \sum_{j=1}^{m} \log(1 + a_j) \tag{19}$$

for some absolute constant $c_0 > 0$. In particular, if the partial quotients of $\rho(f)$ satisfy (10), then $m \geq c_1 n$ for some constant $0 < c_1 < 1$ that depends only on $\rho(f)$.

Proof. Let $\Delta \in \mathcal{Q}_n(f)$ be an atom of level m. Let $\Delta_1 \supseteq \Delta_2 \supseteq \cdots \supseteq \Delta_n = \Delta$ be such that $\Delta_k \in \mathcal{Q}_k(f)$, and note that $1 = \ell(\Delta_1) \leq \ell(\Delta_2) \leq \cdots \leq \ell(\Delta_n) = m$. Given $1 \leq l \leq m$, let i and s (maximal) be such that

$$\ell(\Delta_{i+1}) = \ell(\Delta_{i+2}) = \cdots = \ell(\Delta_{i+s}) = l\,.$$

Then there exists $I \in \mathcal{P}_l(f)$ such that each Δ_j with $i + 1 \leq j \leq i + s$ is a union of atoms of $\mathcal{P}_{l+1}(f)$ inside I. From the very construction of the partitions $\mathcal{Q}_j(f)$ (Proposition 4.5), we see that the number of atoms of $\mathcal{P}_{l+1}(f)$ inside Δ_j is at least *twice* the number of such atoms inside Δ_{j+1}, for each $i + 1 \leq j \leq i + s - 1$. Moreover, Δ_{i+s} contains at least two such atoms (otherwise its level would be $l + 1$). Since the total number of atoms of $\mathcal{P}_{l+1}(f)$ that lie inside I is at most $1 + a_l$, it follows that $2^s \leq 1 + a_l$, whence $s \leq \log_2(1 + a_l)$. This proves (19) with $c_0 = 1/\log 2$.

Now, if $\rho(f)$ satisfies (10), then there exists $B > 0$ depending on $\rho(f)$ such that $\sum_{j=1}^{m} \log a_j \leq Bm$. Therefore

$$n \leq c_0 \sum_{j=1}^{m} \log(1 + a_j) \leq c_0\,(B + \log 2)\,m\,,$$

which proves the last assertion, with $c_1 = c_0^{-1}(B + \log 2)^{-1}$. □

Lemma 4.13. *If $\rho(f)$ satisfies (11) and (12) then there exists $0 < \beta < 1$ with the following property. If L and R are adjacent atoms of $\mathcal{Q}_n(f)$ and we have $\ell(L) \geq m$ and $\ell(R) \geq m$, then*

$$\left| \frac{|L|}{|R|} - \frac{|h(L)|}{|h(R)|} \right| \leq C\beta^m\,. \tag{20}$$

Proof. Write $m = k + p$ with $p = \lceil \sigma k \rceil$ where $\sigma > 0$ is a small constant (its size will be determined in the course of the argument). We may assume that $L \cup R$ is contained in a single atom Δ of $\mathcal{P}_k(f)$. There are two cases to consider.

(a) If $L \cup R \subseteq J_k(f)$, then the required coherence estimate (20) follows from Lemma 4.10 and Lemma 4.11. To see this, let $v_1, v_2, v_3 \in \mathcal{P}_{k+p}(f)$ be the endpoints of L and R, v_2 being their common endpoint. Let w_1, w_2, w_3 be the corresponding endpoints of $h(L)$ and $h(R)$. Then by Lemma 4.10 we have $|v_i - w_i| \le C_0 |J_k(f)| \theta^k$, where $\theta = K^\sigma \mu_* < 1$ if σ is small enough. On the other hand, condition (12) tells us that

$$a_{k+1} a_{k+2} \cdots a_{k+p} \le \exp\{p\omega(p/k)\} \le \exp\{p\omega(\sigma)\} .$$

Combining this fact with Lemma 4.11, we get

$$|v_1 - v_2| \ge \frac{M^p}{\left(a_{k+1} a_{k+2} \cdots a_{k+p}\right)^2} |J_k(f)| \ge \frac{M^p}{e^{2p\omega(\sigma)}} |J_k(f)| .$$

The same lower bound holds for $|v_2 - v_3|$. From these facts, we deduce after some simple computations that

$$\begin{aligned} \left| \frac{|L|}{|R|} - \frac{|h(L)|}{|h(R)|} \right| &= \left| \frac{|v_1 - v_2|}{|v_2 - v_3|} - \frac{|w_1 - w_2|}{|w_2 - w_3|} \right| \\ &\le C_1 \frac{\theta^k e^{2p\omega(\sigma)}}{M^p} \le C_2 \left(\frac{\theta e^{2\sigma\omega(\sigma)}}{M^\sigma} \right)^k \le C_3 \beta_1^m , \end{aligned}$$

where $\beta_1 = \left(\theta e^{2\sigma\omega(\sigma)} / M^\sigma\right)^{1/(1+\sigma)}$. Since $\theta < 1$ and $\sigma\omega(\sigma) \to 0$ as $\sigma \to 0$, we see that $\beta_1 < 1$ if σ is small enough.

(b) If $L \cup R$ is not contained in $J_k(f)$, there exists $j < q_{k+1}$ such that f^j is a diffeomorphism on an interval containing Δ and its two neighbors in $\mathcal{P}_k(f)$ and such that $f^j(\Delta) \subseteq J_k(f)$. By the Koebe principle and the real bounds, the distortion of f^j on $L \cup R$ is bounded by $\exp(C_4 \mu_0^p)$ (where $0 < \mu_0 < 1$ is the beau constant of Theorem 3.1). Therefore we have

$$\left| \frac{|L|}{|R|} - \frac{|f^j(L)|}{|f^j(R)|} \right| \le C_5 \mu_0^p \le C_6 \mu_1^m , \tag{21}$$

where $\mu_1 = \mu_0^{\sigma/(1+\sigma)}$. Working similarly with $h(L), h(R) \in \mathcal{Q}_n(g)$, we get also

$$\left| \frac{|h(L)|}{|h(R)|} - \frac{|g^j(h(L))|}{|g^j(h(R))|} \right| \le C_7 \mu_1^m . \tag{22}$$

Putting (21) and (22) together and using (a) we get inequality (20) with the constant $\beta = \max\{\mu_1, \beta_1\}$.

Hence in both cases (20) is established, and we are done. □

The proof of our First Main Theorem is now almost complete. If L and R are adjacent atoms of $\mathcal{Q}_n(f)$ as above, then combining Lemma 4.12 with Lemma 4.13 we deduce that the coherence condition (6) is satisfied with $\lambda = \beta^{c_1}$. Therefore by Proposition 4.3 the conjugacy h is indeed $C^{1+\alpha}$ for some $\alpha > 0$.

5. Counterexamples to $C^{1+\alpha}$ rigidity

Our purpose now is to construct C^∞ counterexamples to the conjectured $C^{1+\alpha}$ rigidity of critical circle maps. We will consider critical circle maps whose rotation number $\rho(f) = [a_0, a_1, \dots, a_n, \dots]$ satisfies

$$\begin{cases} \limsup \dfrac{1}{n} \log a_n = \infty \\ a_n \geq 2 \qquad\qquad\qquad \text{for all } n. \end{cases} \tag{23}$$

The class of all rotation numbers satisfying (23) will be denoted by $\mathbb{B}$. It can be shown that the Hausdorff dimension of $\mathbb{B}$ is less than or equal to $1/2$, see [7]. On the other hand, $\mathbb{B}$ contains Diophantine numbers: for example, the number ρ whose partial quotients are $a_n = 2^{2^n}$ is Diophantine and satisfies (23).

Theorem 5.1. *For every $\rho \in \mathbb{B}$ there exist C^∞ critical circle maps f, g with $\rho(f) = \rho(g) = \rho$ such that f and g are not $C^{1+\beta}$ conjugate for any $\beta > 0$.*

The proof will make use of a C^∞ surgery procedure that we explain below. These counterexamples have one additional feature: their successive renormalizations do converge together at an exponential rate. This will be clear from the construction.

5.1. Saddle-node surgery

Given f as above and a fixed $n \geq 1$, let $J_n = J_n(f) = [f^{q_n}(c), f^{q_{n-1}}(c)] \subseteq S^1$ be the n-th renormalization interval of f. When a_n is very large, the first return map $f_n : J_n \to J_n$ is an almost parabolic map of length a_n.

Let $\Delta_1^{(n)}$ be the fundamental domain of this almost parabolic map which is adjacent to $x_{n-1} = f^{q_{n-1}}(c)$, and let $\Delta_j^{(n)} = f_n^{j-1}(\Delta_1^{(n)})$, $j \leq a_n$. Let $z_n \in \Delta_1^{(n)}$ be the point such that $f_n^{a_n}(z_n) = x_{n+2} = f^{q_{n+2}}(c)$, that is, $z_n = f^{q_{n+2}-a_n q_n}(c)$. Note that since $a_n \geq 2$, x_{n+2} is not an endpoint of $f_n^{a_n}(\Delta_1^{(n)})$, and so by the real bounds it splits $f_n^{a_n}(\Delta_1^{(n)})$ into two intervals of comparable lengths. Hence the same holds for z_n. Namely, z_n splits $\Delta_1^{(n)}$ into two intervals L_n, R_n with $|L_n| \asymp |R_n|$. In particular we have $\tau|\Delta_1^{(n)}| \leq |L_n| \leq (1-\tau)|\Delta_1^{(n)}|$ (and similarly for R_n) for some constant τ depending on the real bounds. We use this fact in the proof of Proposition 5.2 below.

Consider now another critical circle map $\tilde{f}$ with the same rotation number as f, the interval $\tilde{J}_n = J_n(\tilde{f})$, the first return map $\tilde{f}_n : \tilde{J}_n \to \tilde{J}_n$, the point $\tilde{z}_n = \tilde{f}^{q_{n+2}-a_n q_n}(\tilde{c})$ and the corresponding intervals $\tilde{L}_n$, $\tilde{R}_n$. Also, let $N_n = \lceil a_n/2 \rceil$.

Definition. The number

$$\left| \frac{|f_n^{N_n-1}(L_n)|}{|f_n^{N_n-1}(R_n)|} - \frac{|\tilde{f}_n^{N_n-1}(\tilde{L}_n)|}{|\tilde{f}_n^{N_n-1}(\tilde{R}_n)|} \right|$$

is called the n-th order discrepancy between f and $\tilde{f}$.

Proposition 5.2. *Given a C^∞ critical circle map f with $\rho(f) \in \mathbb{B}$, consider a function $\sigma(n) \to \infty$ such that*

$$\limsup \frac{1}{n\sigma(n)} \log a_n = \infty .$$

Then for all $n \geq 1$, there exists a critical circle map $\tilde{f} = F(n; f)$ with the same rotation number and critical point as f and having the following properties.

(a) We have $\tilde{f}^j(c) = f^j(c)$ for $0 \leq j \leq q_{n+1}$; in particular, $J_n(\tilde{f}) = J_n = J_n(f)$.
(b) We have $\tilde{f} = \Phi \circ f$, where Φ is a C^∞ diffeo such that

$$\left\| \Phi^{\pm 1} - id_{S^1} \right\|_{C^k} \leq B_k |J_n|^{\sigma(n)-k+1}$$

for all k, where $B_k > 0$ is constant depending only on k.
(c) The n-th order discrepancy between f and $\tilde{f}$ is $\geq C|J_n|^{2\sigma(n)}$.
(d) We have $J_{n+1}(\tilde{f}) = J_{n+1}(f)$ and $\tilde{f}_{n+1} = f_{n+1}$; in particular, m-th order discrepancy between f and $\tilde{f}$ is equal to zero for all $m > n$.

Proof. We modify f inside $f^{-1}(\Delta_1^{(n)})$ using a C^∞ bump function so as to move z_n by a distance $\geq C|\Delta_1^{(n)}|^{1+\sigma(n)}$ inside $\Delta_1^{(n)}$. This we do as follows.

Let $\varphi : [0, 1] \to [0, 1]$ be a C^∞ perturbation of the identity such that $|\varphi(x) - x| \geq |\Delta_1^{(n)}|^{\sigma(n)}$ for all $\tau \leq x \leq 1 - \tau$ (and τ as above), and such that $|D^k\varphi(x)| \leq B_k|\Delta_1^{(n)}|^{\sigma(n)}$ for all $0 \leq x \leq 1$ and all $k \geq 2$. Define $\phi_n : \Delta_1^{(n)} \to \Delta_1^{(n)}$ by $\phi_n = A_n \circ \varphi \circ A_n^{-1}$ where A_n is the affine orientation-preserving map that carries $[0, 1]$ onto $\Delta_1^{(n)}$. Note that $|\phi_n(z_n) - z_n| \geq |\Delta_1^{(n)}|^{1+\sigma(n)}$. Moreover, since $D^k\phi_n = |\Delta_1^{(n)}|^{1-k} D^k\varphi$, we have

$$\left\| \phi_n^{\pm 1} - \mathrm{id}_{\Delta_1^{(n)}} \right\|_{C^k} \leq B_k \left| \Delta_1^{(n)} \right|^{\sigma(n)-k+1}$$

for all k. Define $\psi_n : \Delta_{a_n}^{(n)} \to \Delta_{a_n}^{(n)}$ as the conjugate of ϕ_n^{-1} by the diffeo $f_n^{a_n-1} : \Delta_1^{(n)} \to \Delta_{a_n}^{(n)}$, namely

$$\psi_n = f_n^{a_n-1} \circ \phi_n^{-1} \circ (f_n^{a_n-1})^{-1} . \tag{24}$$

Using the C^m Approximation Lemma (see Appendix A), we see from (24) that

$$\left\| \psi_n^{\pm 1} - \mathrm{id}_{\Delta_{a_n}^{(n)}} \right\|_{C^{k-1}} \leq C \left\| \phi_n^{\pm 1} - \mathrm{id}_{\Delta_1^{(n)}} \right\|_{C^k} \leq B_k |\Delta_1^{(n)}|^{\sigma(n)-k+1} .$$

Define $\Phi : S^1 \to S^1$ to be equal to ϕ_n on $\Delta_1^{(n)}$, to ψ_n on $\Delta_{a_n}^{(n)}$ and to the identity everywhere else. The critical circle map we look for is $\tilde{f} = \Phi \circ f$. Note that $\|\Phi^{\pm 1} - \mathrm{id}_{S^1}\|_{C^k} \leq B_k |\Delta_1^{(n)}|^{\sigma(n)-k+1}$ for all k; since $|\Delta_1^{(n)}| \asymp |J_n|$ by the real bounds, this proves (b). It is also clear from the construction that property (a) holds too. It follows in particular that the first $n+1$ partial quotients of the rotation number of $\tilde{f}$ agree with those of f. More remarkable is that, because what ϕ_n does is undone by ψ_n, we have

$$\begin{cases} \tilde{f}^{q_n} | I_{n+1} = f^{q_n} | I_{n+1} \\ \tilde{f}^{q_{n+1}} | I_n = f^{q_{n+1}} | I_n \end{cases} .$$

In other terms, $\tilde{f}_n = f_n$, the n-th renormalizations agree. Therefore all subsequent renormalizations agree as well. This shows that $\rho(\tilde{f}) = \rho(f)$ and also proves (d).

It remains to prove (c), so we estimate the n-th order discrepancy between f and $\tilde{f}$ from below. Since $|z_n - \tilde{z}_n| \geq |\Delta_1^{(n)}|^{1+\sigma(n)}$, a simple calculation yields

$$\left| \frac{|L_n|}{|R_n|} - \frac{|\tilde{L}_n|}{|\tilde{R}_n|} \right| \geq C |\Delta_1^{(n)}|^{\sigma(n)} \geq C |J_n|^{2\sigma(n)} , \tag{25}$$

provided n is sufficiently large. Since, by the real bounds, the map $f_n^{N-1} : \Delta_1^{(n)} \to \Delta_{N_n}^{(n)}$ has bounded distortion, and since $\tilde{f}_n = f_n$, inequality (25) gives us

$$\left| \frac{|f_n^{N_n-1}(L_n)|}{|f_n^{N_n-1}(R_n)|} - \frac{|\tilde{f}_n^{N_n-1}(\tilde{L}_n)|}{|\tilde{f}_n^{N_n-1}(\tilde{R}_n)|} \right| \geq C |J_n|^{2\sigma(n)} ,$$

and this proves (c). □

5.2. *The counterexamples*

We now iterate the procedure given by Proposition 5.2 to prove our Second Main Theorem (that is, Theorem 5.1). We start with a C^∞ map f with $\rho(f) \in \mathbb{B}$ as before and select $n_1 < n_2 < \cdots$ such that

$$\lim_{i \to \infty} \frac{1}{n_i \sigma(n_i)} \log a_{n_i} = \infty , \tag{26}$$

where $\sigma(n)$ is as in Proposition 5.2. Now we generate a sequence g_0, $g_1, \dots, g_i, \dots$ recursively, starting with $g_0 = f$, and taking, for all $i \geq 0$, $g_{i+1} = F(n_{i+1}, g_i)$, where $F(\cdot, \cdot)$ is as given in Proposition 5.2. Each g_i is a C^∞ critical circle map with $\rho(g_i) = \rho(f)$, and $g_{i+1} = \Phi_{i+1} \circ g_i$, where Φ_{i+1} is a C^∞ diffeo with

$$\left\| \Phi_{i+1}^{\pm 1} - \mathrm{id}_{S^1} \right\|_{C^k} \leq B_k \theta^{n_i(\sigma(n_i)-k+1)} , \tag{27}$$

for all k, where $0 < \theta < 1$ is a constant depending only on the real bounds. From (27) it follows that $\Phi = \lim \Phi_i \circ \cdots \circ \Phi_1$ exists as a C^∞ diffeo, and therefore so does $g = \lim g_i = \Phi \circ f$ as a critical circle map.

Using properties (*c*) and (*d*) of Proposition 5.2 for each g_i, we deduce that the n_i-th order discrepancy between f and g satisfies

$$\left| \frac{|f_{n_i}^{N_i - 1}(L_{n_i})|}{|f_{n_i}^{N_i - 1}(R_{n_i})|} - \frac{|g_{n_i}^{N_i - 1}(\tilde{L}_{n_i})|}{|g_{n_i}^{N_i - 1}(\tilde{R}_{n_i})|} \right| \geq C |J_{n_i}|^{2\sigma(n_i)} , \tag{28}$$

where $N_i = \lceil a_{n_i}/2 \rceil$, etc.

Now, let $h : S^1 \to S^1$ be the conjugacy between f and g mapping the critical point c to itself. Suppose h were $C^{1+\beta}$ for some $\beta > 0$. Then the left-hand side of (28) would be $\leq C|f_{n_i}^{N_i - 1}(\Delta_1^{(n_i)})|^\beta$, where $\Delta_1^{(n_i)} = L_{n_i} \cup R_{n_i}$. But by Yoccoz's Lemma, we have

$$\left| f_{n_i}^{N_i - 1}(\Delta_1^{(n_i)}) \right| \asymp \frac{1}{N_i^2} |J_{n_i}| \asymp \frac{1}{a_{n_i}^2} |J_{n_i}| . \tag{29}$$

Combining the above with (28) and (29), we would get the inequality

$$a_{n_i}^{2\beta} |J_{n_i}|^{2\sigma(n_i) - \beta} \leq C .$$

But by the real bounds $|J_n| \geq C\mu^n$ for all n, where $0 < \mu < 1$. Therefore, taking logarithms, we would have

$$\limsup \frac{\log a_{n_i}}{n_i \sigma(n_i)} \leq \frac{1}{\beta} \log \frac{1}{\mu} , \tag{30}$$

but this clearly contradicts (26). □

Remark. A closer look at the construction performed above, especially at expressions (27) and (30), reveals that if

$$\limsup \frac{1}{n} \log a_n > \frac{k}{\beta_0} \log \frac{1}{\mu}$$

then one can construct a pair of C^k critical circle maps (whose renormalizations converge exponentially fast) that are not $C^{1+\beta}$ conjugate for any $\beta \geq \beta_0$.

Appendix A. Compactness of renormalizations

The real a-priori bounds proved in the Sect. 3 have produced a very important corollary, namely, that the renormalizations of an arbitrary C^3 critical circle map are uniformly bounded in the C^1 topology. In this appendix we will use further a-priori estimates, this time involving the Schwarzian derivative, to prove that such renormalizations are uniformly bounded in the C^{r-1} topology when the critical circle map is C^r. Some technical tools are necessary.

A.1. The C^m-Approximation Lemma

In what follows, $m \geq 1$ will be a fixed integer and $I, J \subseteq \mathbb{R}$ fixed closed intervals. We denote by $C^m(I)$ the Banach space of C^m-mappings $f : I \to \mathbb{R}$ with the norm $\|f\|_m = \max\{\|D^i f\|_0 : 0 \leq i \leq m\}$, where $\|\phi\|_0 = \sup_{x\in I} |\phi(x)|$. Sometimes, when we need to emphasize the domain of f, we write $\|f\|_{I,m}$ instead of $\|f\|_m$. We consider also the closed, convex subset $C^m(I, J) \subseteq C^m(I)$ consisting of those f's such that $f(I) \subseteq J$.

Recall Leibnitz's formula for the k-th derivative of a product of two functions,

$$D^k(uv) = \sum_{j=0}^{k} \binom{k}{j} D^j u \, D^{k-j} v ,$$

from which it is clear that

$$\|uv\|_m \leq 2^m \|u\|_m \|v\|_m \tag{31}$$

whenever $u, v \in C^m(I)$. Something similar holds for the composition of two C^m mappings. Namely, we have Faa-di-Bruno's formula (cf. [11], p. 42)

$$D^k(f \circ g) = \sum_{j=1}^{k} B_{j,k}(D^1 g, D^2 g, \ldots , D^j g) \, D^{k-j+1} f \circ g ,$$

where each $B_{j,k}$ is a homogeneous polynomial of degree $k - j + 1$ on j variables whose coefficients are non-negative numbers depending only on k and j. It readily follows from this formula that if $\psi \in C^m(I, J)$ and $\phi \in C^m(J)$ then

$$\|\phi \circ \psi\|_m \leq A(m)\|\phi\|_m \sum_{k=1}^{m} \|\psi\|_m^k , \tag{32}$$

where $A(m) = \max_{1\leq k\leq m} \max_{1\leq j\leq k} B_{j,k}(1, 1, \ldots , 1)$.

Another well-known fact we will need below is the following (*cf.* [6], Th. 3.1). Suppose $m > 1$ and consider the composition operator

$(f, g) \mapsto f \circ g$ as a map $\Theta : C^m(J) \times C^{m-1}(I, J) \to C^{m-1}(I)$. Then Θ is C^1 and its Fréchet derivative is given by

$$D\Theta(f, g)\,(u, v) \;=\; u \circ g + v\, Df \circ g\,. \tag{33}$$

Note that $C^m(J) \times C^{m-1}(I, J) \subseteq C^m(J) \times C^{m-1}(I)$; we consider this last product endowed with the norm

$$\big|(f, g)\big|_{I,J,m} = \max\{\|f\|_{J,m}, \|g\|_{I,m-1}\}\,.$$

Lemma A.1. *For each $M > 0$, there exists $c(M) > 0$ such that, if $f_1, g_1 \in C^m(J)$ and $f_2, g_2 \in C^{m-1}(I, J)$ and if $|(f_1, f_2)|_{I,J,m} < M$ and $|(g_1, g_2)|_{I,J,m} < M$, then*

$$\|f_1 \circ f_2 - g_1 \circ g_2\|_{m-1} \;\le\; c(M)\big|(f_1 - g_1, f_2 - g_2)\big|_{I,J,m}\,.$$

Proof. By the mean value theorem,

$$\|f_1 \circ f_2 - g_1 \circ g_2\|_{m-1} \;\le\; \sup_{(\phi,\psi)} \|D\Theta(\phi, \psi)\|\; |(f_1 - g_1, f_2 - g_2)|_{I,J,m}\,,$$

where the supremum is taken over all (ϕ, ψ) in the line segment joining (f_1, f_2) to (g_1, g_2) inside $C^m(J) \times C^{m-1}(I, J)$, and where

$$\|D\Theta(\phi, \psi)\| \;=\; \sup\big\{\|D\Theta(\phi, \psi)(u, v)\|_{m-1} : |(u, v)|_{I,J,m} \le 1\big\}$$

is the operator-norm of $D\Theta(\phi, \psi)$. Using (33), and then (31) and (32), we have

$$\begin{aligned}\|D\Theta(\phi, \psi)(u, v)\|_{m-1} \;&\le\; \|u \circ \psi\|_{m-1} + \|v\, D\phi \circ \psi\|_{m-1}\\ &\le \big(\|u\|_{m-1} + 2^{m-1}\|v\|_{m-1}\|D\phi\|_{m-1}\big)\; A(m-1) \sum_{k=1}^{m-1} \|\psi\|_{m-1}^k\,.\end{aligned}$$

From this, and taking into account that $\|u\|_{m-1} \le \|u\|_m \le |(u, v)|_{I,J,m}$ as well as $\|v\|_{m-1} \le |(u, v)|_{I,J,m}$, we deduce that

$$\|D\Theta(\phi, \psi)\| \;\le\; A(m-1)\; \big(1 + 2^{m-1}\|D\phi\|_{m-1}\big) \sum_{k=1}^{m-1} \|\psi\|_{m-1}^k\,.$$

Finally, since $\|D\phi\|_{m-1} \le \|\phi\|_m$ and $|(\phi, \psi)|_{I,J,m} < M$, we get

$$\sup_{(\phi,\psi)} \|D\Theta(\phi, \psi)\| \;\le\; A(m-1)\; \big(1 + 2^{m-1}M\big) \sum_{k=1}^{m-1} M^k \;=\; c(M)\,. \qquad \square$$

Let us denote by $\mathbb{B}^m(I; M)$ the ball of radius M centered at the origin in $C^m(I)$.

Lemma A.2. *(The C^m-Approximation Lemma)*
For each $M > 0$, there exist constants $\varepsilon_M > 0$ and $C_M > 0$ such that the following holds for all $\varepsilon \le \varepsilon_M$. Let $\Delta_1, \Delta_2, \dots, \Delta_{n+1}$ be closed intervals on the line or on the circle, and for each $1 \le i \le n$ let $f_i, g_i \in C^m(\Delta_i, \Delta_{i+1})$ be such that

(*a*) *For all $1 \le j \le k \le n$, we have $f_k \circ f_{k-1} \circ \cdots \circ f_j \in \mathbb{B}^m(\Delta_j; M)$;*
(*b*) *We have $\sum_{i=1}^n \|f_i - g_i\|_m < \varepsilon$.*

Then for all $k \le n$ we have $g_k \circ g_{k-1} \circ \cdots \circ g_1 \in \mathbb{B}^{m-1}(\Delta_1; 2M)$, and moreover

$$\|f_k \circ f_{k-1} \circ \cdots \circ f_1 - g_k \circ g_{k-1} \circ \cdots \circ g_1\|_{m-1} \le C_M \sum_{j=1}^{k} \|f_j - g_j\|_m \,.$$

Proof. In the notation of Lemma A.1, let us write

$$C_M = \max\{1,\, c(2M),\, c(2M)c(3M)\}$$

and $\varepsilon_M = M/C_M$. We proceed by induction on k. When $k = 1$, we have $\|f_1 - g_1\|_m \le \varepsilon$ and there is nothing to prove. Suppose the assertion is valid for all $j < k$, and write (omitting the composition symbols)

$$\|f_k f_{k-1} \cdots f_1 - g_k g_{k-1} \cdots g_1\|_{m-1} \le \sum_{j=1}^{k} \|f_k \cdots f_{j+1} g_j g_{j-1} \cdots g_1 - f_k \cdots f_{j+1} f_j g_{j-1} \cdots g_1\|_{m-1} \,. \quad (34)$$

Since $|(f_j, g_{j-1} \circ \cdots \circ g_1)|_{\Delta_1, \Delta_j, m} < 2M$ and also $|(g_j, g_{j-1} \circ \cdots \circ g_1)|_{\Delta_1, \Delta_j, m} < 2M$, it follows from Lemma A.1 that

$$\|f_j g_{j-1} \cdots g_1 - g_j g_{j-1} \cdots g_1\|_{m-1} \le c(2M) \|f_j - g_j\|_m \,,$$

for $j = 1, \dots, k$. In particular, by the induction hypothesis, we have for all $1 \le j \le k-1$

$$\|f_j g_{j-1} \cdots g_1\|_{m-1} \le \|g_j g_{j-1} \cdots g_1\|_{m-1} + \varepsilon_M c(2M) < 3M \,.$$

Taking this back to (34) and applying Lemma A.1 again, we get

$$\begin{aligned} &\|f_k f_{k-1} \cdots f_1 - g_k g_{k-1} \cdots g_1\|_{m-1} \\ &\qquad \le c(2M)\|f_k - g_k\|_m + c(2M)c(3M) \sum_{j=1}^{k-1} \|f_j - g_j\|_m \\ &\qquad \le C_M \sum_{j=1}^{k} \|f_j - g_j\|_m \,, \end{aligned}$$

and this shows also that $\|g_k g_{k-1} \cdots g_1\|_{m-1} \le M + \varepsilon_M C_M < 2M$, thereby completing the induction. □

A.2. Koebe principle revisited

We present a generalization of the classical Koebe non-linearity principle. This principle states that if a C^3 diffeomorphism has non-negative Schwarzian derivative on an open interval, then its non-linearity on any smaller closed subinterval with *space* on both sides is bounded. The generalized version below seems to be new. We denote by $S\phi$ the Schwarzian derivative of ϕ.

Lemma A.3. *Given positive constants B and τ, there exists $K_{\tau,B} > 0$ such that the following holds. If ϕ is a C^3-diffeomorphism of an interval $I \supseteq [-\tau, 1+\tau]$ into the reals and if $S\phi(t) \geq -B$ for all $t \in I$, then for all $t \in [0,1]$ we have*

$$\left|\frac{\phi''(t)}{\phi'(t)}\right| \leq K_{\tau,B} \ .$$

Proof. Writing $y = \phi''/\phi'$, so that $S\phi = y' - \frac{1}{2}y^2$, we have the differential inequality

$$y' \geq \frac{1}{2}y^2 - B \ . \tag{35}$$

Let $0 \leq t_0 \leq 1$ be a point where $|y(t)|$ attains its maximum in $[0,1]$ and suppose $y_0 = y(t_0)$ is such that $|y_0| > \sqrt{2B} = \beta$. If $z(t)$ is the solution of the differential equation corresponding to (35) with initial condition $z(t_0) = y_0$, then by a well-known comparison theorem we must have $y(t) \geq z(t)$ for all $t \geq t_0$ and $y(t) \leq z(t)$ for all $t \leq t_0$. Now, if $y_0 > \beta$ then integration of the ODE leads to

$$z(t) = \beta\frac{(y_0+\beta) + (y_0-\beta)e^{\beta(t-t_0)}}{(y_0+\beta) - (y_0-\beta)e^{\beta(t-t_0)}} \ .$$

Since this solution explodes at time

$$t_1 = t_0 + \frac{1}{\beta}\log\left(\frac{y_0+\beta}{y_0-\beta}\right) ,$$

so does $y(t)$. Hence $t_1 \notin I$, i.e. $t_1 - t_0 > \tau$, which gives us

$$\frac{\phi''(t_0)}{\phi'(t_0)} = y_0 < \beta\,\frac{e^{\beta\tau}+1}{e^{\beta\tau}-1} \ .$$

If instead $y_0 < -\beta$, then we get

$$z(t) = \beta\frac{(\beta+y_0) - (\beta-y_0)e^{\beta(t-t_0)}}{(\beta+y_0) + (\beta-y_0)e^{\beta(t-t_0)}} ,$$

and arguing as before for $t \leq t_0$ gives us

$$\frac{\phi''(t_0)}{\phi'(t_0)} = y_0 > -\beta\,\frac{e^{\beta\tau}+1}{e^{\beta\tau}-1} \ .$$

Therefore the lemma is proved if we take

$$K_{\tau,B} = \beta \frac{e^{\beta\tau}+1}{e^{\beta\tau}-1} . \qquad \square$$

Remark. As $B \to 0$, $K_{\tau,B} \to 2/\tau$ and we recover the classical Koebe principle.

A.3. Bounding the C^2 norms

As before, let $f \in \mathrm{Crit}^r(S^1)$, $r \geq 3$, be a critical circle map with critical point c of power-law $p > 1$. Conjugating f by a suitable C^r-diffeomorphism, we may assume that there exists a neighborhood $\mathcal{U} \subseteq \mathbb{R}/\mathbb{Z}$ of c such that

$$f(x) \;=\; (x-c)|x-c|^{p-1} + a$$

for all $x \in \mathcal{U}$, where a is a constant. This will be our standing hypothesis on f, and we will sometimes say that f is a *canonical* circle map. Note in this case that for all $x \in \mathcal{U} \setminus \{c\}$, the Schwarzian derivative of f equals

$$Sf(x) \;=\; -\frac{p^2-1}{2(x-c)^2} . \tag{36}$$

We are interested in the maps $f^{q_{n-1}-1}\colon I_n^1 \to I_n^{q_{n-1}}$ and $f^{q_n-1}\colon I_{n-1}^1 \to I_{n-1}^{q_n}$, for a fixed $n \geq 1$. They extend as diffeomorphisms to maximal open intervals $J_{n,1}^- \supseteq I_n^1$ and $J_{n,1}^+ \supseteq I_{n-1}^1$ respectively. When linearly rescaled to unit size, these diffeomorphisms are called the *coefficients* of the n-th renormalization of f.

Let us be more precise. Consider the n-th renormalization of f, namely the commuting pair $f_n : [\lambda_n, 1] \to \mathbb{R}$ defined in Sect. 3. We write $J_{n,i}^- = f^{i-1}(J_{n,1}^-)$ for each $1 \leq i \leq q = q_{n-1}$ and $J_{n,j}^+ = f^{j-1}(J_{n,1}^+)$ for each $1 \leq j \leq Q = q_n$. We also write $J_{n,0}^- = f^{-1}(J_{n,1}^-)$ and $J_{n,0}^+ = f^{-1}(J_{n,1}^+)$. For each $0 \leq j \leq Q$, let $\Lambda_j\colon \mathbb{R} \to \mathbb{R}/\mathbb{Z}$ be the affine (orientation-preserving) covering map such that $\Lambda_j([0,1]) = I_{n-1}^j$. Let Δ_n^- be the component of $\Lambda_1^{-1}(J_{n,1}^-)$ that contains the interval $[\lambda_n, 0]$, and let Δ_n^+ be the component of $\Lambda_1^{-1}(J_{n,1}^+)$ that contains the interval $[0,1]$. Then define

$$\begin{cases} \mathcal{F}_n^- = \Lambda_0^{-1} \circ f^{q-1} \circ \Lambda_1 : \Delta_n^- \to \mathbb{R} \\ \mathcal{F}_n^+ = \Lambda_0^{-1} \circ f^{Q-1} \circ \Lambda_1 : \Delta_n^+ \to \mathbb{R} \end{cases} .$$

These are the n-th renormalization coefficients of f. Consider also the so-called *folding factors* of f_n, namely the maps

$$\begin{cases} \varphi_n^- = \Lambda_1^{-1} \circ f \circ \Lambda_0 : \Lambda_0^{-1}(J_{n,0}^-) \to \mathbb{R} \\ \varphi_n^+ = \Lambda_1^{-1} \circ f \circ \Lambda_0 : \Lambda_0^{-1}(J_{n,0}^+) \to \mathbb{R} \end{cases} .$$

Each of these maps is a homeomorphism with a unique critical point at zero. One verifies at once that the maps $\boldsymbol{F}_n^- = \mathcal{F}_n^- \circ \varphi_n^-$ and $\boldsymbol{F}_n^+ = \mathcal{F}_n^+ \circ \varphi_n^+$ are C^r extensions of $\boldsymbol{f}_n^-$ and $\boldsymbol{f}_n^+$, respectively.

It will be useful to express the coefficients $\mathcal{F}_n^\pm$ as long compositions of rescaled diffeomorphisms in the following way. We will give the explicit decomposition for $\mathcal{F}_n^+$. A similar decomposition can be worked out for $\mathcal{F}_n^-$. Let us denote by $\Delta_{n,j}^+$ the component of $\Lambda_j^{-1}(J_{n,j}^+)$ containing the unit interval. Note in particular that $\Delta_n^+ = \Delta_{n,1}^+$. For each j in the range $0 \leq j \leq Q-1$, let

$$f_j = \Lambda_{j+1}^{-1} \circ f \circ \Lambda_j : \Delta_{n,j}^+ \to \Delta_{n,j+1}^+ .$$

We call such maps the *elementary factors* of $\mathcal{F}_n^+$. Each f_j is a C^r diffeomorphism such that $f_j([0,1]) = [0,1]$ (see Fig. 5). We have of course $\varphi_n^+ = f_0$, but more importantly

$$\mathcal{F}_n^+ = \left(\Lambda_0^{-1} \circ \Lambda_Q\right) \circ \left(f_{Q-1} \circ \cdots \circ f_j \circ \cdots f_1\right) . \tag{37}$$

We note also that for all $t \in \Delta_{n,j}^+$

$$Sf_j(t) = Sf\left(\Lambda_j(t)\right)\left[D\Lambda_j(t)\right]^2 = Sf\left(\Lambda_j(t)\right)|I_{n-1}^j|^2 , \tag{38}$$

by the chain rule for the Schwarzian derivative.

Notation. Given $J = [a,b] \subseteq \mathbb{R}$ and $\tau > 0$, we denote by J^τ the interval $[c,d] \supseteq J$ such that $(a-c)/(b-a) = (d-b)/(b-a) = \tau$. Note that J has *space* equal to τ inside J^τ.

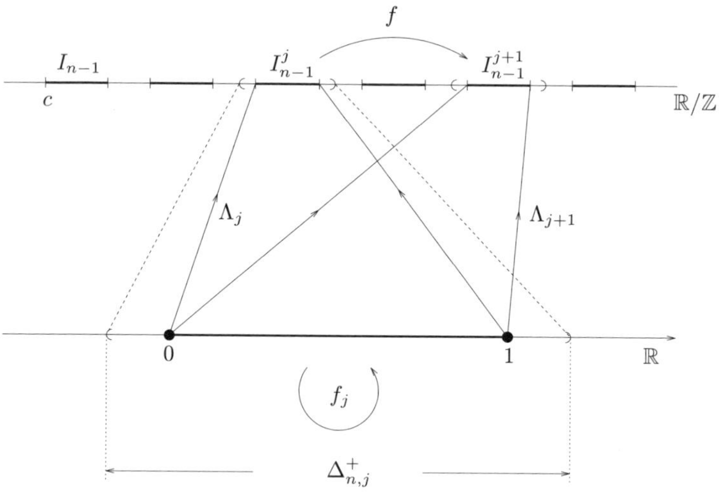

Fig. 5. The elementary factors of $\mathcal{F}_n^+$

Theorem A.4. *(The C^2 bounds)*
Let $f \in Crit^3(S^1)$ be a critical circle map with arbitrary irrational rotation number, let $\boldsymbol{f}_n : [\lambda_n, 1] \to \mathbb{R}$ be the n-th renormalization of f, and let $\mathcal{F}_n^{\pm} : \Delta_n^{\pm} \to \mathbb{R}$ be the coefficients of $\boldsymbol{f}_n$. Also, let $f_j : \Delta_{n,j}^+ \to \Delta_{n,j+1}^+$ be the elementary factors of $\mathcal{F}_n^+$. There exist positive constants B and τ depending only on the real bounds for f such that the following statements hold for all $n \geq 1$.

(a) *We have $\Delta_n^- \supseteq [\lambda_n, 0]^\tau$ and $\Delta_n^+ \supseteq [0, 1]^\tau = [-\tau, 1+\tau]$.*
(b) *For all $0 \leq j \leq Q$, we have $\Delta_{n,j}^+ \supseteq [0, 1]^\tau$.*
(c) *We have $|S\mathcal{F}_n^-(t)| \leq B$ for all $t \in \Delta_n^-$ and $|S\mathcal{F}_n^+(t)| \leq B$ for all $t \in \Delta_n^+$.*
(d) *More generally, for all $1 \leq j < k \leq Q$, we have $|S(f_k \circ \cdots \circ f_j)(t)| \leq B$ for all $t \in \Delta_{n,j}^+$.*
(e) *The C^2 norms of the restrictions $\mathcal{F}_n^- | [\lambda_n, 0]^{\tau/2}$ and $\mathcal{F}_n^+ | [0, 1]^{\tau/2}$ are bounded by B.*
(f) *More generally, for all $1 \leq j < k \leq Q$, the C^2 norm of the restriction of $f_k \circ \cdots \circ f_j$ to the interval $f_{j-1} \circ \cdots \circ f_1([0, 1]^{\tau/2})$ is bounded by B.*
(g) *The C^2 norms of $\boldsymbol{f}_n^-$ and $\boldsymbol{f}_n^+$ are bounded by B.*

Moreover, if n is sufficiently large then both coefficients have negative Schwarzian derivatives at all points of their respective domains.

The proof will use the following lemma concerning the dynamical partitions $\mathcal{P}_n$. Let us denote by $d(c, I)$ the distance between an interval $I \subseteq S^1$ and the critical point c. For each $n \geq 1$, let

$$S_n = \sum_{I \in \mathcal{P}_n \setminus \{I_{n-1}, I_n\}} \left(\frac{|I|}{d(c, I)} \right)^2 . \tag{39}$$

Lemma A.5. *The sequence S_n is bounded (by a constant depending only on f).*

Proof. Recall that $\mathcal{P}_{n+2}$ is a strict refinement of $\mathcal{P}_n$. From the real bounds, we know that there exists a constant $0 < \lambda < 1$ depending only on f such that, if I is in $\mathcal{P}_n$ and $J \subseteq I$ is in $\mathcal{P}_{n+2}$, then $|J| \leq \lambda |I|$. Hence

$$\sum_{I \supseteq J \in \mathcal{P}_{n+2}} |J|^2 \leq \left(\max_{I \supseteq J \in \mathcal{P}_{n+2}} |J| \right) |I| \leq \lambda |I|^2 .$$

Since we also have $d(c, J) \geq d(c, I)$ whenever $J \subseteq I$, it follows that

$$S_{n+2} \leq \lambda S_n + \sum_{\mathcal{P}_{n+2} \ni J \subseteq I_{n-1} \setminus I_{n+1}} \left(\frac{|J|}{d(c, J)} \right)^2 + \sum_{\mathcal{P}_{n+2} \ni J \subseteq I_n \setminus I_{n+2}} \left(\frac{|J|}{d(c, J)} \right)^2$$

$$\leq \lambda S_n + \lambda \left(\frac{|I_{n-1}|}{|I_{n+1}|} \right)^2 + \lambda \left(\frac{|I_n|}{|I_{n+2}|} \right)^2 .$$

From this and the facts that $|I_{n-1}| \asymp |I_{n+1}|$ and $|I_n| \asymp |I_{n+2}|$, we get $S_{n+2} \leq \lambda S_n + \mu$, where μ is a constant depending only on f. But then, by induction,

$$S_{2n} \leq \lambda^{n-1} S_2 + \frac{\mu}{1-\lambda}, \quad S_{2n+1} \leq \lambda^n S_1 + \frac{\mu}{1-\lambda},$$

and therefore S_n is bounded as claimed. □

Proof of Theorem A.4. It is enough to prove this theorem under the assumption that f is canonical. The existence of $\tau > 0$ such that (a) and (b) hold is a consequence of the real bounds. Hence we proceed to prove (c) for $\mathcal{F}_n^+$, the proof for $\mathcal{F}_n^-$ being completely similar. Making τ smaller if necessary and using the classical Koebe non-linearity principle, we can assume that there exists $C > 0$ depending only on the real bounds for f such that

$$|D(f_j \cdots f_1)(t)| \leq C, \tag{40}$$

for all $t \in [-\tau, 1+\tau]$ and all $j = 1, \ldots, Q-1$.

Let $\mathcal{V} \subseteq S^1$ be an open set whose closure does not contain c and such that $\mathcal{U} \cup \mathcal{V} = S^1$. Also, let $M = \sup_{x \in \mathcal{V}} |Sf(x)|$. We assume that n is so large that the largest interval in $\mathcal{P}_n$ has length smaller than the Lebesgue number of the covering $\{\mathcal{U}, \mathcal{V}\}$. Together with (37) and (38), iterated use of the chain rule for the Schwarzian yields

$$\begin{aligned} S\mathcal{F}_n^+(t) &= S(f_{Q-1} \cdots f_j \cdots f_1)(t) \\ &= \sum_{j=1}^{Q-1} Sf_j(f_{j-1} \cdots f_1(t)) \left[D(f_{j-1} \cdots f_1)(t)\right]^2 \\ &= \sum_{j=1}^{Q-1} Sf(\Lambda_j f_{j-1} \cdots f_1(t)) \left|I_{n-1}^j\right|^2 \left[D(f_{j-1} \cdots f_1)(t)\right]^2 . \end{aligned}$$

We split this last sum into $\Sigma_1(t) + \Sigma_2(t)$, where $\Sigma_1(t)$ is the sum over all j's such that $I_{n-1}^j \subseteq \mathcal{U}$ and $\Sigma_2(t)$ is the sum over the remaining terms (i.e. those with $I_{n-1}^j \subseteq \mathcal{V}$). Then we have on one hand

$$|\Sigma_2(t)| \leq C^2 M \sum_{I_{n-1}^j \subseteq \mathcal{V}} \left|I_{n-1}^j\right|^2 \leq C^2 M \max_{1 \leq j \leq Q-1} \left|I_{n-1}^j\right| . \tag{41}$$

On the other hand, since $d(c, J_{n,j}^+) \asymp d(c, I_{n-1}^j)$ for all j, we have by (36)

$$|\Sigma_1(t)| \leq C^2 \sum_{I_{n-1}^j \subseteq \mathcal{U}} \frac{\left|I_{n-1}^j\right|^2}{\left[d(c, J_{n,j}^+)\right]^2} \leq C' S_n , \tag{42}$$

where C' is another constant depending only on f and S_n is given by (39). From (41) and (42) it follows that $|S\mathcal{F}_n^+(t)|$ is uniformly bounded, and this proves (c). Moreover, since by (41) $\Sigma_2(t)$ goes to zero with n while $\Sigma_1(t)$ is always negative and bounded away from zero, we deduce that $S\mathcal{F}_n^+(t) < 0$ for all n sufficiently large. The proof of (d) is entirely analogous.

To prove (e), let B_0 be the upper-bound that we have just obtained for $|S\mathcal{F}_n^+|$. Applying Lemma A.3 to $\mathcal{F}_n^+$, we get for all $t \in [0,1]^{\tau/2}$

$$\left|\frac{D^2\mathcal{F}_n^+(t)}{D\mathcal{F}_n^+(t)}\right| \leq K_{\tau_0,B_0},$$

where $\tau_0 = \tau/2(1+\tau)$ is the space of $[0,1]^{\tau/2}$ inside $[0,1]^{\tau}$. Therefore $\|D^2\mathcal{F}_n^+\|_0 \leq K_{\tau_0,B_0}\|D\mathcal{F}_n^+\|_0 \leq CK_{\tau_0,B_0}$, by (40) above. This shows that the C^2 norm of $\mathcal{F}_n^+$ is bounded as claimed. A similar argument proves (f). Finally, (g) follows from (e) and the fact that the folding factors $\varphi_n^{\pm}$ are linear blow-ups of a fixed power-law map. The theorem is therefore proved if we take B to be the largest of all the upper-bounds obtained in the argument.□

Remark. We can go a bit further in (e), (f) and (g) and bound also the C^3 norms. For this purpose, it suffices to note for instance that

$$D^3\mathcal{F}_n^+(t) = D\mathcal{F}_n^+(t)\left(S\mathcal{F}_n^+(t) + \frac{3}{2}\left[\frac{D^2\mathcal{F}_n(t)}{D\mathcal{F}_n(t)}\right]^2\right),$$

and then use (c) and (e). However, this argument does not generalize to get bounds for higher derivatives. Our bootstrap argument in the next section will follow a different route, based on the C^m Approximation Lemma.

A.4. Bounding the C^{r-1} norms

We will show that the sequence of renormalizations of a C^r critical circle map is bounded in the C^{r-1} sense. The limits fall into (a compact subset of) a special family of analytic critical commuting pairs known as the Epstein class. Moreover, we will prove that such limits are attained at an exponential rate in the C^{r-1} topology. The rate of convergence turns out to depend only on the rotation number of the given critical circle map.

An *Epstein map* is a homeomorphism $\varphi : I \to J$ between closed intervals on the real line such that φ^{-1} is the restriction of an analytic univalent map $\Phi : \mathbb{C}(J') \to \mathbb{C}(I')$, where $I' \supseteq I$ and $J' \supseteq J$ are open intervals. Here we use the notation $\mathbb{C}(\Delta) = (\mathbb{C} \setminus \mathbb{R}) \cup \Delta$. For example, every fractional linear transformation in $PSL_2(\mathbb{R})$ is an Epstein map when restricted to an interval on the line which does not contain any of its poles. Further examples include polynomial or rational diffeomorphisms with real coefficients.

Definition. A commuting pair $\boldsymbol{f}$ is said to be an *Epstein* commuting pair if $\boldsymbol{f}^+ = \varphi^+ \circ Q$ and $\boldsymbol{f}^- = \varphi^- \circ Q$, where φ^+, φ^- are Epstein maps and Q is the power-law map $x \mapsto x^p$ (for some $p > 1$).

Theorem A.6. *Let $r \geq 3$ and let f be a C^r critical circle map with arbitrary irrational rotation number. Then the sequence of renormalizations $\{\mathcal{R}^n(f)\}$ is bounded in the C^{r-1} metric and converges C^{r-1} exponentially fast to the Epstein class.*

The idea behind the proof of Theorem A.6 is quite simple. In the long composition defining the n-th renormalization of a critical circle map, we replace the factors away from the critical point by suitable fractional linear approximations, which are all Epstein maps. The factors which are close to the critical point are already Epstein because the map is assumed to be a power-law there. Therefore the entire new composition is an Epstein map. The Moebius approximations have to be carefully chosen, however, so that the total error involved, estimated with the help of the C^m Approximation Lemma, be exponentially small in n (the step of renormalization). We now present the technical result which is needed.

Lemma A.7. *Given $r \geq 3$ and an orientation preserving C^r-diffeomorphism $\phi : I \to \mathbb{R}$ of a closed interval I onto its image, there exist constants $\ell_\phi > 0$ and $K_\phi > 0$ with the following property. For each closed interval $\Delta \subseteq I$ of length $|\Delta| \leq \ell_\phi$, there exists a fractional linear transformation $T_\Delta \in PSL_2(\mathbb{R})$ with $T_\Delta(\Delta) = \phi(\Delta)$ such that,*

(a) $\sup_{x\in\Delta} \left|D^k\phi(x) - D^kT_\Delta(x)\right| \leq K_\phi|\Delta|^{3-k}$ *for* $k = 0, 1, 2$.
(b) $\sup_{x\in\Delta} \left|D^kT_\Delta(x)\right| \leq K_\phi$ *for all* $1 \leq k \leq r$.

Proof. Let ℓ_ϕ be the constant

$$\ell_\phi = \min\left\{1, \inf_{x\in I}\left|\frac{\phi'(x)}{\phi''(x)}\right|\right\} .$$

Take any closed interval $\Delta \subseteq I$ with $|\Delta| \leq \ell_\phi$, and let x_0 be the left endpoint of Δ. Let T be the unique fractional linear transformation with the same 2-jet as ϕ at x_0. Thus, if $T(x) = (a(x - x_0) + b)/(c(x - x_0) + d)$ with $ad - bc = 1$, then the coefficients are uniquely determined by the conditions

$$\begin{aligned} T(x_0) &= \frac{b}{d} = \phi(x_0) , \\ T'(x_0) &= \frac{1}{d^2} = \phi'(x_0) , \\ T''(x_0) &= \frac{-2c}{d^3} = \phi''(x_0) . \end{aligned} \tag{43}$$

Moreover, for all $k \geq 1$,

$$D^k T(x) = \frac{(-1)^{k+1} k! c^{k-1}}{[c(x-x_0)+d]^{k+1}} . \tag{44}$$

Since $|x - x_0| \leq |\Delta| \leq \ell_\phi \leq |\phi'(x_0)|/|\phi''(x_0)| = |d|/2|c|$ for all $x \in \Delta$, we have

$$\frac{1}{2}|d| \leq |c(x-x_0)+d| \leq \frac{3}{2}|d| \tag{45}$$

for each such x. Combining (44) with the lower bound in (45), we get

$$\left|D^k T(x)\right| \leq \frac{2^{k+1} k! |c|^{k-1}}{|d|^{k+1}} = \frac{4k! |\phi''(x_0)|^{k-1}}{|\phi'(x_0)|^{k-2}} ,$$

for all $x \in \Delta$ and all $k \geq 1$, and consequently

$$\sup_{x\in\Delta} \left|D^k T(x)\right| \leq C_0 = \max_{1\leq k\leq r} \sup_{x\in I} \left\{ \frac{4k! |\phi''(x)|^{k-1}}{|\phi'(x)|^{k-2}} \right\} , \tag{46}$$

when $1 \leq k \leq r$. In particular, from

$$D^2\phi(x) - D^2 T(x) = \int_{x_0}^{x} D^3\phi(t)\, dt - \int_{x_0}^{x} D^3 T(t)\, dt ,$$

we deduce that

$$\begin{aligned}\left|D^2\phi(x) - D^2 T(x)\right| &\leq \|D^3\phi\|_0 \,|x-x_0| + \frac{24|\phi''(x_0)|^2}{|\phi'(x_0)|} |x-x_0| \\ &\leq \left(\|D^3\phi\|_0 + C_0\right) |\Delta| ,\end{aligned}$$

for all $x \in \Delta$. Integrating this inequality twice, using (43), we get

$$\sup_{x\in\Delta} \left|D^k\phi(x) - D^k T(x)\right| \leq C_1 |\Delta|^{3-k} , \tag{47}$$

for $k = 0, 1, 2$, where $C_1 = C_0 + \|D^3\phi\|_0$.

Looking at (46) and (47), we see that T is almost what we want, but not quite because in general it does not map Δ onto $\phi(\Delta)$. To correct this flaw, we replace T by $T_\Delta = A \circ T$, where A is the unique affine, orientation-preserving map that carries $T(\Delta)$ onto $\phi(\Delta)$. We have

$$A(t) - t = \left[\frac{|\phi(\Delta)|}{|T(\Delta)|} - 1\right] (t - T(x_0)) , \tag{48}$$

for all $t \in T(\Delta)$, because $\phi(x_0) = T(x_0)$. Let $\mu = |\phi(\Delta)|/|T(\Delta)|$. Since by (47) we have $||\phi(\Delta)| - |T(\Delta)|| \leq 2C_1|\Delta|^3$, and since by the upper-bound in (45) we have

$$\frac{|T(\Delta)|}{|\Delta|} \geq \inf_{x\in\Delta} \frac{1}{[c(x-x_0)+d]^2} \geq \frac{4}{9d^2} = \frac{4}{9} |\phi'(x_0)| ,$$

it follows that

$$|\mu - 1| \leq \frac{9C_1}{2|\phi'(x_0)|} |\Delta|^2 .$$

Thus we see that, for all $t \in T(I)$,

$$|A'(t) - 1| = |\mu - 1| \leq \frac{9C_1}{2\inf_{x\in I}|\phi'(x)|} |\Delta|^2 = C_2|\Delta|^2 ,$$

On the other hand, since $|T(\Delta)| \leq \|D\phi\|_0|\Delta| + 2C_1|\Delta|^3$, and since $|\Delta| \leq 1$, it follows from (48) that

$$|A(t) - t| \leq C_2 \left(\|D\phi\|_0 + 2C_1\right) |\Delta|^3 = C_3|\Delta|^3 .$$

Therefore

$$|\phi(x) - T_\Delta(x)| \leq |\phi(x) - T(x)| + |T(x) - A(T(x))| \leq (C_1 + C_3)\,|\Delta|^3 , \quad (49)$$

and moreover, using the fact that $D^kT_\Delta(x) = \mu D^kT(x)$ for all k,

$$\begin{aligned} |D^k\phi(x) - D^kT_\Delta(x)| &\leq |D^k\phi(x) - D^kT(x)| + |\mu - 1|\,|D^kT(x)| \\ &\leq C_1|\Delta|^{3-k} + C_0C_2|\Delta|^2 \\ &\leq (C_1 + C_0C_2)\,|\Delta|^{3-k} , \end{aligned} \quad (50)$$

for all $x \in \Delta$ and $k = 1, 2$. Finally, for all $k \geq 1$ we have

$$|D^kT_\Delta(x)| \leq \left(1 + C_2|\Delta|^2\right) |D^kT(x)| \leq (1 + C_2)\,C_0 . \quad (51)$$

Part (a) now follows from (49) and (50), while part (b) follows from (51), provided we take $K_\phi = \max\{C_1 + C_3, C_1 + C_0C_2, (1 + C_2)C_0\}$. □

Proof of Theorem A.6. We now expand the outline given above and present a complete proof of Theorem A.6. In the proof, we will denote by $C_0, C_1, \ldots$ positive constants depending only on the real bounds for f. As before, we may assume from the start that f is *canonical*, and accordingly we consider the covering $\{\mathcal{U}, \mathcal{V}\}$ of S^1 defined in the proof of Theorem A.4. Since the folding factors of $\boldsymbol{f}_n$ are power-law maps, and therefore already Epstein, it suffices to prove that the *coefficients* of $\boldsymbol{f}_n$ can be approximated by Epstein maps, up to an error exponentially small in n in the C^{r-1} topology. We will do this for $\mathcal{F}_n^+$, the proof for $\mathcal{F}_n^-$ being the same.

As in the previous section, let $f_j : \Delta_{n,j}^+ \to \Delta_{n,j+1}^+$, $1 \leq j \leq Q - 1$, be the elementary factors of $\mathcal{F}_n^+$. For each $1 \leq j \leq Q$ we define

$$\Delta_j = f_{j-1} \circ \cdots \circ f_2 \circ f_1 \left([0, 1]^{\tau/4}\right) \subseteq \Delta_{n,j}^+ ,$$

where τ is the constant of Theorem A.4. Note that $f_j(\Delta_j) = \Delta_{j+1}$. Let $\Delta_j' = \Lambda_j(\Delta_j)$, and observe also that $I_{n-1}^j \subseteq \Delta_j' \subseteq J_{n,j}^+$.

We introduce individual Epstein approximations g_j to each f_j. There are two cases to consider. It may happen that $\Delta'_j \subseteq \mathcal{U}$, in which case we simply take $g_j = f_j$. Otherwise, we have $\Delta'_j \subseteq \mathcal{V}$. In this case, we let $T_j : \Delta'_j \to \Delta'_{j+1}$ be the Moebius approximation to $f|\Delta'_j$ that we get applying Lemma A.7 to the restriction of f to $\mathcal{V}$, and then take $g_j = \Lambda_{j+1}^{-1} \circ T_j \circ \Lambda_j$. Note that $g_j(\Delta_j) = \Delta_{j+1}$.

Claim 1. We have $\|f_j - g_j\|_r \leq C_0 |I_{n-1}^j|^2$ for all j.

This is obvious when $I_{n-1}^j \subseteq \mathcal{U}$. When $I_{n-1}^j \subseteq \mathcal{V}$, we have $|I_{n-1}^j| \asymp |I_{n-1}^{j+1}|$, because the derivative of f on $\mathcal{V}$ is bounded away from zero, and we also have $|\Delta'_j| \asymp |I_{n-1}^j|$. Moreover, for all $1 \leq s \leq r$ and all $x \in \Delta_j$,

$$D^s f_j(x) - D^s g_j(x) = \frac{\left|I_{n-1}^j\right|^s}{\left|I_{n-1}^{j+1}\right|} \left(D^s f(\Lambda_j(x)) - D^s T_j(\Lambda_j(x))\right) .$$

Therefore the claim follows from Lemma A.7 (treat the cases $s = 1, 2$ separately).

Now, recall from Theorem A.4 that for all $1 \leq j < k \leq Q - 1$ we have

$$\|f_k \circ \cdots \circ f_j\|_2 \leq B .$$

Claim 2. If n is sufficiently large then for all $1 \leq j < k \leq Q - 1$ we have

$$\|f_k \circ \cdots \circ f_j - g_k \circ \cdots \circ g_j\|_1 \leq C_1 \max_{0 \leq i \leq Q} \left|I_{n-1}^i\right| . \tag{52}$$

Take n_0 so large that $C_0 \max |I_{n_0-1}^j| < \varepsilon_B$, where ε_B is the constant given by Lemma A.2 when we take $M = B$. Then from Claim 1 and (52), the hypotheses of Lemma A.2 are satisfied, and we get for all $n \geq n_0$

$$\begin{aligned}\|f_k \circ \cdots \circ f_j - g_k \circ \cdots \circ g_j\|_1 &\leq C_B \sum_{i=j}^{k} \|f_i - g_i\|_2 \leq C_0 C_B \sum_{i=j}^{k} \left|I_{n-1}^i\right|^2 \\ &\leq C_0 C_B \max_{0 \leq i \leq Q} \left|I_{n-1}^i\right| ,\end{aligned}$$

where C_B is the constant of Lemma A.2 for $M = B$. This proves the claim.

In order to bootstrap these C^1 estimates up to C^{r-1} estimates, we apply the C^m Approximation Lemma once more, this time reversing the roles of f_j and g_j, and with $m = r$. Thus, we need to verify the hypotheses of that lemma in this new situation.

Claim 3. For all $1 \leq j < k \leq Q - 1$, we have $\|g_k \circ \cdots \circ g_j\|_r \leq C_2$.

For brevity, write $G_{jk} = g_k \circ \cdots \circ g_j$. Then G_{jk}^{-1} is univalent on $\mathbb{C}(\Delta_{jk})$, where Δ_{jk} is an interval containing $G_{jk}(\Delta_j)$ with definite space on both sides, by our choice of τ. Using Koebe's one-quarter theorem, it is not difficult to see that the domain $\Omega_{jk} = G_{jk}^{-1}(\mathbb{C}(\Delta_{jk}))$ contains a rectangle $W_j = \Delta_j^\alpha \times [-\beta, \beta]$, and that $d(\partial W_j, \partial\Omega_{jk}) \geq \gamma$ where α, β and γ are positive constants depending only on τ and the real bounds for f. Hence, from the complex Koebe's distortion theorem, we get

$$\left|\frac{G'_{jk}(z)}{G'_{jk}(w)}\right| \leq \exp\left\{\frac{4}{\gamma}\operatorname{diam}(W_j)\right\} \leq C_3 ,$$

for all $z, w \in W_j$. This together with the mean-value theorem gives us $|G'_{jk}(z)| \leq C_4$, and therefore also $|G_{jk}(z)| \leq C_5$, for all $x \in W_j$. Now we use Cauchy's integral formula to bound all higher derivatives of G_{jk}. We have for all $x \in \Delta_j$ and all $s \geq 1$

$$|D^s G_{jk}(x)| = \frac{s!}{2\pi}\left|\int_{\partial W_j}\frac{G_{jk}(z)}{(z-x)^{s+1}}\,dz\right| \leq \frac{C_5 s!}{\pi}(\beta+(1+2\alpha)|\Delta_j|)\,\delta_j^{-s-1} ,$$

where $\delta_j = \inf_{x\in\Delta_j} d(x, \partial W_j) = \min\{\alpha|\Delta_j|, \beta\} \geq \delta = \min\{a, b\}$. Therefore $|D^s G_{jk}(x)| \leq C_6 s!\delta^{-s-1}$. This shows that $\|G_{jk}\|_r$ is bounded as claimed.

From Claims 1 and 3, the hypotheses of Lemma A.2 are therefore satisfied, and we have

$$\begin{aligned}\|f_k \circ \cdots \circ f_j - g_k \circ \cdots \circ g_j\|_{r-1} &\leq C_{C_2}\sum_{i=j}^{k}\|f_i - g_i\|_r \\ &\leq C_0 C_{C_2}\sum_{i=j}^{k}\left|I_{n-1}^i\right|^2 \\ &\leq C_0 C_{C_2}\max_{0\leq i\leq Q}\left|I_{n-1}^i\right| ,\end{aligned}$$

this time for all n large enough so that $C_0 \max|I_{n-1}^j| < \varepsilon_{C_2}$, where C_{C_2} and ε_{C_2} are the constants of Lemma A.2 for $M = C_2$. Since $\max|I_{n-1}^j|$ decreases exponentially with n, we are done. □

Appendix B. Proof of Yoccoz's Lemma

The main geometric idea behind the proof of Yoccoz's Lemma is to use the negative Schwarzian property of f to *squeeze* the graph of f between the graphs of two Moebius transformations. The required estimate for f will then follow from the corresponding estimate for Moebius transformations, which we now state and prove.

Consider the fractional linear transformation $T(x) = x/(1+x)$, and given $\varepsilon > 0$, let $T_\varepsilon(x) = T(x) - \varepsilon$. We are interested in certain quantitative aspects of the orbit $x_n = T_\varepsilon^n(x_0)$ for $x_0 = 1$. Observe that this sequence is strictly decreasing.

Lemma B.1. *Let $N > 0$ be such that $x_{N+1} \leq 0 < x_N$. Then we have $N \asymp 1/\sqrt{\varepsilon}$ and moreover $x_n - x_{n+1} \asymp 1/n^2$ for $n = 0, 1, \ldots, N$.*

Proof. Writing $\delta_n = T^n(x_0) - T_\varepsilon^n(x_0)$, we have

$$\delta_n = \varepsilon + \frac{\delta_{n-1}}{\left(1+\frac{1}{n}\right)\left(1+\frac{1}{n}-\delta_{n-1}\right)} \tag{53}$$

for all $n = 1, 2, \ldots, N+1$. We claim that

$$\frac{n\varepsilon}{6} \leq \delta_n \leq n\varepsilon\,. \tag{54}$$

The last inequality is clear. To prove the first, we note from (53) that

$$\delta_n \geq \varepsilon + \left(\frac{n}{n+1}\right)^2 \delta_{n-1}\,.$$

By induction, this gives us

$$\delta_n \geq \frac{\varepsilon}{(n+1)^2}\left(1^2+2^2+\cdots+n^2\right) = \frac{\varepsilon}{(n+1)^2}\,\frac{n(n+1)(2n+1)}{6} \geq \frac{n\varepsilon}{6}\,,$$

which proves the claim. Now, from the fact that $x_{N+1} \leq 0 < x_N$ we have the inequalities

$$\delta_N < \frac{1}{N+1}\,, \quad \delta_{N+1} \geq \frac{1}{N+2}\,.$$

Then, using (54), we get

$$\frac{1}{(N+1)(N+2)} \leq \varepsilon < \frac{6}{N(N+1)}\,, \tag{55}$$

which proves the first assertion.

Next, note that since $[x_{N+1}, x_N] \subseteq \left[T_\varepsilon(0), T_\varepsilon^{-1}(0)\right] = [-\varepsilon, \varepsilon/(1-\varepsilon)]$, we have

$$\varepsilon < x_N - x_{N+1} < 3\varepsilon \tag{56}$$

Hence, by (55), we get $x_N - x_{N+1} \asymp 1/N^2$ and the second assertion is proved when $n = N$. To prove it in general using this information, observe that

$$x_n - x_{n+1} = \frac{x_{n-1}-x_n}{(1+x_{n-1})(1+x_n)} = \frac{x_{n-1}-x_n}{\left(1+\frac{1}{n}-\delta_{n-1}\right)\left(1+\frac{1}{n+1}-\delta_n\right)}$$

implies

$$x_n - x_{n+1} \geq \frac{n}{n+2}(x_{n-1} - x_n)\,.$$

By induction, this gives on one hand

$$x_n - x_{n+1} \geq \frac{2}{(n+1)(n+2)}(x_0 - x_1) \geq \frac{1}{(n+1)(n+2)}\,,$$

and on the other hand, using (55) and (56),

$$x_n - x_{n+1} \leq (x_N - x_{N+1}) \prod_{j=1}^{N-n} \left(\frac{n+j+2}{n+j}\right) < \frac{54}{(n+1)(n+2)}\,.$$

This proves the second assertion in all cases. □

Now recall that $f : \Delta_1 \cup \Delta_2 \cup \cdots \cup \Delta_a \to \mathbb{R}$ satisfies $f(\Delta_j) = \Delta_{j+1}$ for all j. Without loss of generality, we can assume that $f(x) < x$ for all x. Thus, if we call x_0 the right endpoint of Δ_1 and write $x_j = f^j(x_0)$, we have $\Delta_j = [x_j, x_{j-1}]$ for all j. Since our map f is a negative-Schwarzian diffeomorphism, there exists a unique z in the domain of f such that $\varepsilon = |f(z) - z| \leq |f(x) - x|$ for all x. Since the statement we want to prove is invariant under affine changes of coordinates, we may assume also that $z = 0$ and $x_0 = 1$. In this setting, we want to prove that $|\Delta_j| \asymp 1/j^2$ for all j such that $\Delta_j \subseteq [0, 1]$. Note that $f'(0) = 1$.

Next, let A be the Moebius transformation on the line such that $A(x_0) = f(x_0)$ and $A(0) = f(0)$ and $A'(0) = f'(0) = 1$. This determines A uniquely, and in fact

$$A(x) = \frac{x}{1 + \lambda x} - \varepsilon\,,$$

for some $\lambda > 0$. Since $Sf < 0$, we see that $A(x) \leq f(x)$ for all $x \in [0, 1]$.

Likewise, let B be the Moebius transformation such that $B(x_a) = f(x_a)$, $B(0) = f(0)$ and $B'(0) = f'(0) = 1$. This determines B uniquely, and in fact

$$B(x) = \frac{x}{1 + \mu x} - \varepsilon\,,$$

for some $\mu > 0$. This time, since $x_a < 0$ and $Sf < 0$, we have $f(x) \leq B(x)$ for all $x \in [0, 1]$. In particular, $\lambda > \mu$. It is easy to see that $\lambda/\mu \leq c_\sigma$, where c_σ depends only on the constant σ in the statement.

Lemma B.2. *Let $x \in [0, 1]$ and $k > 0$ be such that $A(x) < B^k(x)$. Then $k \leq 1 + \lambda/\mu$.*

Proof. By induction we have

$$B^k(x) \leq \frac{x}{1+(k-1)\mu x} - \varepsilon\,.$$

Therefore $A(x) < B^k(x)$ implies $(k-1)\mu x < \lambda x$. □

Now, let us write $\alpha_n = A^n(x_0)$ and $\beta_n = B^n(x_0)$. By Lemma B.2, the number of β_j's inside each interval of the form $[\alpha_{n+1}, \alpha_n]$ is bounded independently of n. Moreover, since $\alpha_n < x_n < \beta_n$ for all n, the number of x_j's inside each $[\alpha_{n+1}, \alpha_n]$ is also bounded independently of n. To prove that $|\Delta_j| \asymp 1/j^2$, we proceed as follows. Let $\ell > 0$ be such that $\beta_{\ell+1} \leq x_j \leq \beta_\ell \leq x_{j-1}$. Then Lemma B.2 says that $\ell \leq Cj$, and we have also

$$|\beta_{\ell+1} - \beta_\ell| < |B(x_{j-1}) - x_{j-1}| < |x_j - x_{j-1}|\,.$$

Since by Lemma B.1 we have

$$|\beta_{\ell+1} - \beta_\ell| \asymp \frac{1}{\ell^2} \geq \frac{1}{Cj^2}\,,$$

it follows that $|\Delta_j| = |x_j - x_{j-1}| \geq 1/Cj^2$.

To prove an inequality in the opposite direction, let m be the largest integer such that $\alpha_m > x_{j-1}$. Then, again by Lemma B.2, we have $j \leq Cm$. Since $A(x) < f(x) < x$ for all x, we also have $\Delta_j \subseteq [\alpha_{m+2}, \alpha_m]$. Using Lemma B.1 once more, we deduce that

$$|\Delta_j| \leq \frac{C}{m^2} \leq \frac{C}{j^2}\,.$$

This completes the proof of Yoccoz's Lemma. □

Appendix C. A full-measure condition on rotation numbers

We present here our account of C.G. Moreira's probabilistic argument showing that condition (12) holds for a set of full-measure in [0, 1]. His original probability estimates were done for Lebesgue measure in [0, 1]. We prefer to use instead a probability measure which is invariant under the Gauss map, namely the *Gauss measure*. These two measures are mutually absolutely continuous, the density of the latter with respect to the former being $dx/(1+x)\log 2$. With respect to the Gauss probability measure (the only one we will use from now on), the partial quotients $a_n(\theta)$ of $\theta \in [0, 1]$ are identically distributed random variables. We warn the reader that these are *not* independent random variables; they are only *asymptotically* independent, a fact that is perhaps best expressed by saying that the Gauss map is *mixing*.

Now, the probability that the n-th partial quotient $a_n(\theta)$ be equal to a given integer $m \geq 1$ is

$$p_m = \mathbb{P}[a_n(\theta) = m] = \frac{1}{\log 2} \log\left(1 + \frac{1}{m(m+2)}\right) < \frac{2}{m^2} . \quad (57)$$

From this, we see that the probability that $a_n(\theta)$ be at least m is smaller than $4/m$. These facts form the basis for our estimates, and yield the following lemma.

Lemma C.1. *There exists a full-measure set $E \subseteq [0, 1]$ such that for each $\theta \in E$ we have $a_n(\theta) < n^2$ for all sufficiently large n.*

Proof. Let $E = \liminf([0, 1] \setminus A_n)$, where A_n is the set of all $\theta \in [0, 1]$ such that $a_n(\theta) \geq n^2$. Then $\mathbb{P}(A_n) < 4/n^2$, and since the series $\sum 4/n^2$ converges, we deduce by the Borel-Cantelli lemma that $\mathbb{P}(E) = 1$. □

We shall prove the following result, which establishes condition (12) for almost all numbers $\theta \in [0, 1]$ with $\omega(t) = C_\theta(1 - \log t)$, where $C_\theta > 0$.

Proposition C.2. *Almost all numbers $\theta \in E$ satisfy, for all $1 \leq n \leq k$,*

$$\frac{1}{n} \sum_{j=k+1}^{k+n} \log a_j(\theta) \leq C_\theta \left(1 + \log \frac{k}{n}\right) , \quad (58)$$

for some constant $C_\theta > 0$.

Proof. In attempting to prove the inequality (58) for a given $\theta \in E$, we may assume that k is so large that $a_j(\theta) < j^2$ for all $j \geq k$. The remaining cases, corresponding to the remaining finitely many pairs (n, k), are taken care of by a suitable choice of the constant C_θ.

Given (n, k), there are two possibilities to consider. The first possibility is that $n^2 < k$. In this case we simply observe that

$$\frac{1}{n} \sum_{j=k+1}^{k+n} \log a_j(\theta) \leq 2 \log(k+n) < 5 \log \frac{k}{n} ,$$

for all sufficiently large k.

The second possibility is that $n \leq k \leq n^2$. Here, we shall prove that *with probability one* the left-hand side of (58) is bounded by 10. For this purpose, let us consider the following pathologies.

(a) For a given $m \geq 1$, there are more than $2np_m$ partial quotients $a_{k+i}(\theta)$ with $1 \leq i \leq n$ such that $a_{k+i}(\theta) = m$ (where p_m is as defined in (57)).

By an elementary combinatorial argument, we see that this occurs with probability at most

$$\sum_{j=\lceil 2np_m\rceil}^{n} \binom{n}{j} p_m^j (1-p_m)^{n-j} < \left(\frac{e}{4}\right)^{np_m} . \qquad (59)$$

The probability that this happens for *some* m in the range $1 \le m \le n^{1/3}$ is therefore smaller than

$$n^{1/3} \times \left(\frac{e}{4}\right)^{2n^{1/3}} < \frac{1}{n^4}$$

if n is sufficiently large.

(b) There are more than $n^{2/3}$ partial quotients $a_{k+i}(\theta)$ with $1 \le i \le n$ such that $a_{k+i}(\theta) > n^{1/3}$. By a similar reasoning to the one used in (a), we see that this occurs with probability smaller than

$$\left(\frac{e}{4}\right)^{n^{1/3}} < \frac{1}{n^4}$$

if n is sufficiently large.

Therefore, fixing n sufficiently large, the probability that there exists k in the range $n \le k \le n^2$ such that one of the above pathologies occurs for (n, k) is certainly less than $n^2 \times (2/n^4) = 2/n^2$. Since the series $\sum 2/n^2$ converges, again by Borel-Cantelli we deduce that with probability one there are no pathologies for (n, k) if k (and hence n) is sufficiently large.

Now, if there are no pathologies for (n, k), and noting that for $1 \le i \le n$ we have

$$a_{k+i}(\theta) < (k+i)^2 \le (n^2+n)^2 \le 4n^4$$

if k is sufficiently large, we deduce that

$$\frac{1}{n}\sum_{j=k+1}^{k+n} \log a_j(\theta) \le \frac{1}{n}\sum_{m=1}^{\lfloor n^{1/3}\rfloor} (2np_m)\log m + \frac{n^{2/3}}{n}\log(4n^4)$$

$$< \sum_{m=1}^{\lfloor n^{1/3}\rfloor} \frac{4\log m}{m^2} + \frac{1}{n^{1/3}}(2\log 2 + 4\log n) ,$$

which is less than 10 if n is sufficiently large. This completes the proof. □

Acknowledgements. We are grateful to Dennis Sullivan for suggesting many interesting ideas, and to Curt McMullen for explaining to us various aspects of his beautiful work on renormalization and rigidity. We want to thank the referee for very keen remarks leading to a considerable improvement of our final results. We also want to thank Jean-Christophe Yoccoz for providing us with a sketch of the proof of his unpublished theorem on quasisymmetric conjugacies between critical circle maps. We are indebted also to Marco Martens,

Jacek Graczyk, Oscar Lanford, Rafael de la Llave, Michael Lyubich, Henri Epstein and C.G. Moreira for several useful discussions on these and related matters. This work has been financially sponsored by PRONEX, CNPq and FAPESP. We also would like to thank IHES, MSRI, ETH-Zürich, IMPA and SUNY at Stony Brook for their warm hospitality and generous support.

References

1. P. Billingsley: Probability and measure. New York: John Wiley & Sons 1986
2. L. Carleson: On mappings conformal at the boundary. J. Anal. Math. **19**, 1–13 (1967)
3. I. Cornfeld, S. Fomin, Ya. Sinai: Ergodic Theory. New York: Springer 1982
4. E. de Faria: Asymptotic rigidity of scaling ratios for critical circle mappings. Ergod. Th. & Dynam. Sys. **19**(4), 995–1035 (1999)
5. E. de Faria, W. de Melo: Rigidity of critical circle mappings II. To appear in J. Amer. Math. Soc., IMS Stony Brook preprint 97/17 (1997)
6. J. Franks: Manifolds of C^r mappings and applications to differentiable dynamics. Stud. in Anal. Adv. Math. Suppl. Stud. **4**, 271–290 (1979)
7. I. J. Good: The fractional dimension theory of continued fractions. Proc. Cambridge Phil. Soc. **37**, 199–228 (1941)
8. J. Graczyk: Harmonic scalings for smooth families of diffeomorphisms of the circle. Nonlinearity **4**, 935–954 (1990)
9. J. Graczyk: Hölder classes for circle maps. Proc. Amer. Math. Soc. (to appear)
10. J. Graczyk, G. Świątek: Critical circle maps near bifurcation. Commun. Math. Phys. **176**, 227–260 (1996)
11. M. Herman: Sur la conjugaison differentiable des difféomorphismes du cercle a des rotations. Publ. Math. IHES **49**, 5–234 (1979)
12. M. Herman: Conjugaison quasi-simétrique des homéomorphismes du cercle a des rotations. 1988 (manuscript)
13. O. Lanford: Renormalization group methods for critical circle mappings with general rotation number. VIIIth International Congress on Mathematical Physics, pp.532–536 (Marseille, 1986). Singapore: World Sci. Publishing 1987
14. M. Lyubich: Combinatorics, geometry and attractors of quasi-quadratic mappings. Ann. of Math. **140**, 347–404 (1994)
15. C. McMullen: Renormalization and 3-manifolds which fiber over the circle. Annals of Math Studies **142**. Princeton: Princeton University Press 1996
16. W. de Melo, S. van Strien: One dimensional dynamics. Berlin, New York: Springer 1993
17. D. Rand: Global phase-space universality, smooth conjugacies and renormalization: the $C^{1+\alpha}$ case. Nonlinearity **1**, 181–202 (1988)
18. G. Świątek: Rational rotation numbers for maps of the circle. Commun. Math. Phys. **119**, 109–128 (1988)
19. G. Świątek: One-dimensional maps and Poincaré metric. Nonlinearity **4**, 81–108 (1992)
20. J.C. Yoccoz: Conjugaison differentiable des diffeomorphismes du cercle dont le nombre de rotation verifie une condition Diophantienne. Ann. Sci. de l'Ec. Norm. Sup. **17**, 333–361 (1984)
21. J.C. Yoccoz: Il n'y a pas de contre-example de Denjoy analytique. C. R. Acad. Sci. Paris **298**, 141–144 (1984)

Rigidity of Critical Circle Mappings II

Edson de Faria and Welington de Melo

J. Amer. Math. Soc. **13**(2), 343–370 (1999)

1. Introduction

A celebrated rigidity theorem of M. Herman [5] states that any two smooth diffeomorphisms of the circle with the same Diophantine rotation number are smoothly conjugate. Herman used purely real-variable techniques to prove this theorem. In recent years, other fundamental rigidity results have been established in the context of one-dimensional dynamics with the help of complex-analytic techniques. Thus, in the context of circle diffeomorphisms, J.-C. Yoccoz [16] used conformal surgery to describe the optimal class of rotation numbers for which Herman's theorem holds in the real-analytic category. In the context of interval maps, D. Sullivan [13] and C. McMullen [10] used a combination of real-variable techniques with deep results from Teichmüller theory and hyperbolic geometry to establish the rigidity of the Cantor attractor of a unimodal map with bounded combinatorics. The conjugacy between two such maps with the same combinatorics is shown to be $C^{1+\alpha}$ for some $\alpha > 0$ depending only on the combinatorics. In sharp contrast with the case of circle diffeomorphisms, in this case one does not expect the conjugacy to be much better than C^1, even when the unimodal maps are analytic.

In this paper we prove the following strong rigidity theorem for real-analytic critical circle maps.

Theorem 1.1. *Any two real-analytic critical circle maps with the same irrational rotation number ρ of bounded type are $C^{1+\alpha}$ conjugate for some $0 < \alpha < 1$ depending only on the least upper-bound of the coefficients of the continued fraction expansion of ρ.*

The proof of this theorem uses both real and complex-analytic tools. We use the *real a-priori bounds* of Świątek and Herman (see [14], [6] and [3]) in the form stated in §2. The complex-analytic part we develop here combines some of the powerful new ideas on conformal rigidity and renormalization recently developed by McMullen in [9], [10], [11] with the basic theory of holomorphic commuting pairs introduced in [2] and a generalization of the Lyubich-Yampolsky approach to the complex bounds given in [8], [15].

The general outline of the paper is as follows. First we prove that after a *finite* number of renormalizations, every real-analytic critical circle map with arbitrary irrational rotation number gives rise to a *holomorphic commuting pair* with good

Received by the editors November 9, 1998 and, in revised form, September 20, 1999.
2000 *Mathematics Subject Classification.* Primary 37F25; Secondary 37E10, 30D05, 37F40.
Key words and phrases. Holomorphic pairs, complex bounds, uniform twist, rigidity.
This work has been partially supported by the Pronex Project on Dynamical Systems, by FAPESP Grant 95/3187-4 and by CNPq Grant 30.1244/86-3.

Originally published in *J. Amer. Math. Soc.* **13**(2), 343–370 (1999).

geometric bounds. Next, we prove that such a holomorphic pair (with good geometric control) possesses McMullen's *uniform twisting* property [10]. We also prove that its critical point is δ-*deep* for some universal $\delta > 0$. The general machinery developed by McMullen in [9], [10] permits us to deduce from these facts that the conjugacy between any two real-analytic critical circle maps with rotation number of type bounded by N is $C^{1+\varepsilon(N)}$-conformal at the critical point. This implies that the successive renormalizations of both circle maps around their critical points are converging together *exponentially fast* in the C^0 sense (even $C^{1+\beta}$ for some $\beta > 0$), with rate of convergence given by a universal function of ε. By a general proposition proved in [3], this implies at last that the conjugacy between both maps is $C^{1+\alpha(N)}$ for some $\alpha(N) > 0$. Although in our proof the Hölder exponent depends on N, we believe that this should not be the case.

It is to be noted that in [3] we constructed C^∞ counterexamples (even with Diophantine rotation numbers) to this form of $C^{1+\alpha}$ rigidity. One may ask whether such counterexamples exist in the analytic category. We do not know the answer to this question. Nevertheless, we conjecture that in the general case the conjugacy is always $C^{1+\alpha}$ at the critical point, for some universal α independent of the rotation number. A possible approach to this conjecture is to use M. Lyubich's recent work on quadratic-like maps and universality of unimodal maps, but it is still unclear at this time how to adapt his methods to holomorphic pairs and critical circle maps.

2. Preliminaries

In this section, we introduce some basic concepts and notations. We denote by $|I|$ the length of an interval I on the line or the circle. We use $\mathrm{dist}(\cdot,\cdot)$ for Euclidean distance, and $\mathrm{diam}(\cdot)$ for Euclidean diameter. All bounds achieved in this paper will depend on certain *a-priori* constants (which can ultimately be traced back to the *real a-priori bounds* given by Theorem 2.1 below). In this context, two positive quantities a and b are said to be *comparable* if there exists a constant $C > 1$ depending only on the a-priori bounds such that $C^{-1}b \leq a \leq Cb$. We write $a \asymp b$ to denote that a is comparable to b.

Critical circle maps. We identify the unit circle S^1 with the one-dimensional torus $\mathbb{R}/\mathbb{Z}$. A homeomorphism $f : S^1 \to S^1$ which is at least C^1 and satisfies $f'(c) = 0$ and $f'(x) \neq 0$ for all $x \neq c$ is called a *critical circle map*. The point c is the *critical point* of f. In this paper f will be real-analytic, and the critical point will be cubic: this means that near c we have $f = \phi \circ Q \circ \psi$, where ϕ, ψ are real analytic diffeomorphisms and Q is the map $x \mapsto x^3$.

We are interested in the geometry of orbits of f only when there are no periodic points. In this case the rotation number of f is irrational and can be represented as an infinite continued fraction

$$\rho(f) = [a_0, a_1, \dots, a_n, \dots] = \cfrac{1}{a_0 + \cfrac{1}{a_1 + \cfrac{1}{\cfrac{\cdots}{a_n + \cfrac{1}{\cdots}}}}}.$$

When the partial quotients a_n are bounded, we say that $\rho(f)$ is a number of *bounded type*. We also refer to $\sup a_n$ as the *combinatorial type* of $\rho(f)$.

The denominators of the convergents of $\rho(f)$, defined recursively by $q_0 = 1$, $q_1 = a_0$ and $q_{n+1} = a_n q_n + q_{n-1}$ for all $n \geq 1$, are the *closest return times* of the orbit of any point to itself. We denote by Δ_n the closed interval containing c whose endpoints are $f^{q_n}(c)$ and $f^{q_{n+1}}(c)$. We also let $I_n \subseteq \Delta_n$ be the closed interval whose endpoints are c and $f^{q_n}(c)$. Observe that $\Delta_n = I_n \cup I_{n+1}$. The most important fact to remember when studying the geometry of a circle map is that for each n the collection of intervals

$$\mathcal{P}_n = \left\{I_n, f(I_n), \ldots, f^{q_{n+1}-1}(I_n)\right\} \cup \left\{I_{n+1}, f(I_{n+1}), \ldots, f^{q_n-1}(I_{n+1})\right\}$$

constitutes a partition of the circle (modulo endpoints), called *dynamical partition of level n* of the map f. Note that, for all n, $\mathcal{P}_{n+1}$ is a refinement of $\mathcal{P}_n$.

Of course, these definitions make sense for an arbitrary homeomorphism of the circle. For a rigid rotation, we have $|I_n| = a_{n+1}|I_{n+1}| + |I_{n+2}|$. Therefore, if a_{n+1} is very large, then I_n is much longer than I_{n+1}. It is a remarkable fact, first proved by Świątek and Herman, that this never happens for a critical circle map! The dynamical partitions $\mathcal{P}_n$ have *bounded geometry*, in the sense that adjacent atoms have comparable lengths.

Theorem 2.1 (The real bounds). *Let $f : S^1 \to S^1$ be a critical circle map with arbitrary irrational rotation number. There exists $n_0 = n_0(f)$ such that for all $n \geq n_0$ and every pair I, J of adjacent atoms of $\mathcal{P}_n$ we have $K^{-1}|J| \leq |I| \leq K|J|$, where $K > 1$ is a universal constant.*

These *real a-priori bounds* are essential in all the estimates that we perform in this paper. For proofs, see [4] and §3 of [3]. An important consequence of this theorem is the fact that every critical circle map with rotation number of bounded type is conjugate to the rotation with the same rotation number by a *quasisymmetric* homeomorphism. Another consequence is the following lemma.

Lemma 2.2. *Let f be a critical circle map with arbitrary rotation number, let $n \geq 1$, and let $J_{-i} = f^{q_{n+1}-i}(I_n)$ for $0 \leq i < q_{n+1}$. Given $m < n$, let $i_1 < i_2 < \cdots < i_\ell$ be the moments in the backward orbit $\{J_{-i}\}$ of I_n before the first return to I_{m+1} such that $J_{-i_k} \subseteq I_m$. Then $\ell = a_{m+1}$, and we have*

$$J_{-i_k} \subseteq f^{q_m+(a_{m+1}-k+1)q_{m+1}}(I_{m+1}) \ .$$

Moreover, given an integer $M \geq 1$, there exists $C_M > 1$ (depending only on M) such that for all sufficiently large n we have $C_M^{-1}|I_n| \leq |J_{-i_k}| \leq C_M|I_n|$, provided $1 \leq k \leq M$ or $a_{m+1} - M + 1 \leq k \leq a_{m+1}$.

Proof. The largest $j < q_{n+1}$ such that $f^j(I_n) \subseteq I_{m+1}$ is easily computed as $j = q_{n+1} - q_{m+2}$. Since $q_{m+2} = q_m + a_{m+1}q_{m+1}$, there are exactly a_{m+1} subsequent moments $j < i < q_{n+1}$ such that $f^i(I_n) \subseteq I_m$. The rest follows from Theorem 2.1 and the Koebe distortion principle (see [12]). □

Commuting pairs and renormalization. Let f be a critical circle map as before, and let $n \geq 1$. The first return map $f_n : \Delta_n \to \Delta_n$ to $\Delta_n = I_n \cup I_{n+1}$, called the *$n$-th renormalization of f without rescaling*, is determined by a pair of maps, namely $f^{q_n} : I_{n+1} \to \Delta_n$ and $f^{q_{n+1}} : I_n \to \Delta_n$. This pair of maps is what we call a *critical commuting pair*. Note that f_n has a well-defined rotation number (identifying the endpoints of Δ_n, we recover a circle map). Each f_{n+1} is by definition the *renormalization without rescaling* of f_n. These objects can be normalized to unit size to yield the *n-th renormalization of* f, denoted $\mathcal{R}^n(f)$, in the following way. Consider the exponential covering map $\exp : \mathbb{R} \to S^1$ given by $\exp(x) = e^{2\pi i x}$, and let T denote the translation $x \mapsto x+1$. We assume that the critical point of f is $c = 1 \in S^1$. There is a unique lift $\hat{f} : \mathbb{R} \to \mathbb{R}$ of f to the real line through the exponential map such that $0 \leq \hat{f}(0) < 1$; henceforth referred to as the *standard lift* of f. Let $\widetilde{I}_n \subset \mathbb{R}$ be the interval whose endpoints are 0 and $\lambda_n = T^{-p_n} \circ \hat{f}^{q_n}(0)$, where p_n is the numerator of the n-th convergent of $\rho(f)$. Note that $\exp(\widetilde{I}_n) = I_n$. Write $\widetilde{\Delta}_n = \widetilde{I}_n \cup \widetilde{I}_{n+1}$. Let $\xi_n = T^{-p_n} \circ \hat{f}^{q_n} : \widetilde{I}_{n+1} \to \widetilde{\Delta}_n$ and $\eta_n = T^{-p_{n+1}} \circ \hat{f}^{q_{n+1}} : \widetilde{I}_n \to \widetilde{\Delta}_n$. This pair of maps projects down to the pair $(f^{q_n}, f^{q_{n+1}})$ on the circle. Finally, let $\hat{\xi}_n = \Lambda_n^{-1} \circ \xi_n \circ \Lambda_n$ and $\hat{\eta}_n = \Lambda_n^{-1} \circ \eta_n \circ \Lambda_n$, where Λ_n is the linear map $x \mapsto \lambda_n x$. The resulting critical commuting pair $(\hat{\xi}_n, \hat{\eta}_n)$ is by definition the (normalized) n-th renormalization $\mathcal{R}^n(f)$. For more on commuting pairs, including the abstract definition, see [3].

Holomorphic pairs. The concept of *holomorphic commuting pair* was introduced in [2], and will play a crucial role in this paper. We recall the definition and some of the relevant facts about these objects; henceforth called simply *holomorphic pairs.* Assume we are given a configuration of four simply-connected domains $\mathcal{O}_\xi, \mathcal{O}_\eta, \mathcal{O}_\nu, \mathcal{V}$ in the complex plane, called a *bowtie*, such that

(a) each $\mathcal{O}_\gamma$ is a Jordan domain whose closure is contained in $\mathcal{V}$;
(b) we have $\overline{\mathcal{O}}_\xi \cap \overline{\mathcal{O}}_\eta = \{0\} \subseteq \mathcal{O}_\nu$;
(c) the sets $\mathcal{O}_\xi \setminus \mathcal{O}_\nu$, $\mathcal{O}_\eta \setminus \mathcal{O}_\nu$, $\mathcal{O}_\nu \setminus \mathcal{O}_\xi$ and $\mathcal{O}_\nu \setminus \mathcal{O}_\eta$ are non-empty and connected.

Definition. A holomorphic pair Γ with domain $\mathcal{U} = \mathcal{O}_\xi \cup \mathcal{O}_\eta \cup \mathcal{O}_\nu$ and co-domain $\mathcal{V}$ is the dynamical system generated by three holomorphic maps $\xi : \mathcal{O}_\xi \to \mathbb{C}$, $\eta : \mathcal{O}_\eta \to \mathbb{C}$ and $\nu : \mathcal{O}_\nu \to \mathbb{C}$ satisfying the following conditions (see Figure 1).

(H_1) Both ξ and η are univalent onto $\mathcal{V} \cap \mathbb{C}(\xi(J_\xi))$ and $\mathcal{V} \cap \mathbb{C}(\eta(J_\eta))$ respectively, where $J_\xi = \mathcal{O}_\xi \cap \mathbb{R}$ and $J_\eta = \mathcal{O}_\eta \cap \mathbb{R}$. (Notation: $\mathbb{C}(I) = (\mathbb{C} \setminus \mathbb{R}) \cup I$.)
(H_2) The map ν is a 3-fold branched cover onto $\mathcal{V} \cap \mathbb{C}(\nu(J_\nu))$, where $J_\nu = \mathcal{O}_\nu \cap \mathbb{R}$, with a unique critical point at 0.
(H_3) We have $\mathcal{O}_\xi \ni \eta(0) < 0 < \xi(0) \in \mathcal{O}_\eta$, and the restrictions $\xi|[\eta(0), 0]$ and $\eta|[0, \xi(0)]$ constitute a critical commuting pair.
(H_4) Both ξ and η extend holomorphically to a neighborhood of zero, and we have $\xi \circ \eta(z) = \eta \circ \xi(z) = \nu(z)$ for all z in that neighborhood.
(H_5) There exists an integer $m \geq 1$, called the *height* of Γ, such that $\xi^m(a) = \eta(0)$, where a is the left endpoint of J_ξ; moreover, $\eta(b) = \xi(0)$, where b is the right endpoint of J_η.

The interval $J = [a, b]$ is called the *long dynamical interval* of Γ, whereas $\Delta = [\eta(0), \xi(0)]$ is the *short dynamical interval* of Γ. They are both forward invariant under the dynamics. The *rotation number* of Γ is by definition the rotation number of the critical commuting pair of Γ (condition H_3).

Examples of holomorphic pairs with arbitrary rotation number and arbitrary height were carefully constructed in §4 of [2], with the help of the family of entire maps given by

$$z \mapsto z + \theta - \frac{1}{2\pi} \sin 2\pi z$$

where θ is real, the so-called *Arnold family.* Many interesting properties of holomorphic pairs can be proved with these examples at hand.

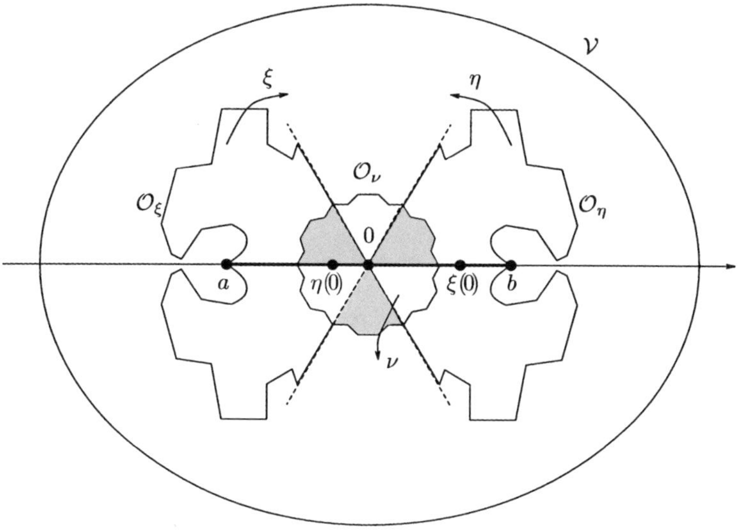

FIGURE 1.

The *shadow* of a holomorphic pair is the map $F : \mathcal{O}_\xi \cup \mathcal{O}_\eta \cup \mathcal{O}_\nu \to \mathcal{V}$ given by

$$F(z) = \begin{cases} \xi(z), & \text{when } z \in \mathcal{O}_\xi, \\ \eta(z), & \text{when } z \in \mathcal{O}_\eta, \\ \nu(z), & \text{when } z \in \mathcal{O}_\nu \setminus (\mathcal{O}_\xi \cup \mathcal{O}_\eta). \end{cases}$$

The shadow captures the essential dynamical features of a holomorphic pair. It is a simple fact (see [2], Prop. 2.4) that every Γ-orbit is an F-orbit and conversely.

The *limit set* of a holomorphic pair Γ, denoted $\mathcal{K}_\Gamma$, is the closure of the set of points in $\mathcal{U}$ which never escape from $\mathcal{U}$ under iteration by F. The annulus $\mathcal{V} \setminus \mathcal{U}$ is a fundamental domain for the dynamics off the limit set, and for this reason it will be called the *fundamental annulus* of Γ.

Holomorphic pairs can be renormalized. In other words, the first renormalization of the critical commuting pair of Γ extends in a natural way to a holomorphic pair $\mathcal{R}(\Gamma)$ with the same co-domain $\mathcal{V}$. For the careful construction of $\mathcal{R}(\Gamma)$, see Prop. 2.3 in [2]. There is also a pull-back theorem for holomorphic pairs.

Definition. We say that Γ has *geometric boundaries* if $\partial\mathcal{U}$ and $\partial\mathcal{V}$ are quasicircles.

Theorem 2.3. *Let Γ and Γ' be holomorphic pairs with geometric boundaries, and let $h : J \to J'$ be a quasisymmetric conjugacy between $F|J$ and $F'|J'$. Then there exists a quasiconformal conjugacy $H : \mathcal{V} \to \mathcal{V}'$ between Γ and Γ' which is an extension of h.*

For a proof of this theorem, see Th. 3.1 in [2].

Conformal distortion. We will need some classical facts about conformal maps; see [1] and [9]. Consider a univalent map $\phi : \Omega \to \mathbb{C}$ on a domain $\Omega \subseteq \mathbb{C}$. The *distortion* of ϕ on a compact set $E \subseteq \Omega$ is the largest ratio $|\phi'(z)|/|\phi'(w)|$ with $z, w \in E$. When E is convex (a disk in most cases), the distortion on E is bounded by $\exp N_\phi(E)$, where $N_\phi(E) = \sup_{z\in E} |\phi''(z)|/|\phi'(z)|$ is the *total non-linearity* of ϕ on E. The *Koebe distortion theorem* states that $N_\phi(E) \leq 2/\operatorname{dist}(E, \partial\Omega)$. *Koebe's one-quarter theorem* states that if $D \subseteq \Omega$ is a disk centered at z, then $\phi(D)$ contains a disk centered at $\phi(z)$ of radius $\frac{1}{4}\operatorname{diam}(D)|\phi'(z)|$.

Almost parabolic maps. In order to adapt Yampolsky's argument for the complex bounds in §3, we will need to know some general facts about complex analytic maps that are very close to maps with a parabolic fixed-point. Given $J \subseteq \mathbb{R}$ and $0 < \theta < \pi$, we denote by $\mathbb{P}_\theta(J) \subseteq \mathbb{C}$ the set of all z in the complex plane that *view* J under an angle $\geq \theta$. This set is the *Poincaré neighborhood* of J in $\mathbb{C}(J)$ with angle θ.

Definition. Let $J \subseteq \mathbb{R}$ be an interval, and let $\theta > 0$. A holomorphic univalent map $\phi : \mathbb{P}_\theta(J) \to \mathbb{C}$ is called *almost parabolic* if the following conditions are satisfied.

(a) ϕ is symmetric about the real axis, that is, $\phi(\overline{z}) = \overline{\phi(z)}$ for all $z \in \mathbb{P}_\theta(J)$.
(b) $\phi|J$ is monotone without fixed points.
(c) ϕ has positive Schwarzian derivative on J.
(d) $J \cap \phi(J)$ is non-empty.

If Δ_ϕ is the interval $J \setminus \phi(J)$, the largest $a = a(\phi) > 0$ such that $\phi^{a-1}(\Delta_\phi) \subseteq J$ is called the *length* of ϕ. The number $\theta = \theta(\phi)$ is the *angle* of ϕ.

The most important example of an almost parabolic map is the inverse branch of a high first return f^{q_n} of a critical circle map f (or one of its renormalizations), in the situation where a_n is large, that is, the rotation number is "almost rational".

Given $0 < \sigma < 1$, we denote by $\mathcal{F}_\sigma$ the family of all almost parabolic maps ϕ such that $|\Delta_\phi| \geq \sigma|J|$ and $|\phi^{a-1}(\Delta_\phi)| \geq \sigma|J|$, and also normalized so that

$$[0,1] = \Delta_\phi \cup \phi(\Delta_\phi) \cup \cdots \cup \phi^{a-1}(\Delta_\phi) .$$

Now we have the following fundamental lemma due to Yoccoz (see [3] for a proof).

Yoccoz's Lemma. *There exists $C_\sigma > 1$ such that for each $\phi \in \mathcal{F}_\sigma$ and for each $0 \leq j \leq a-1$ we have*

$$\frac{1}{C_\sigma m(j)^2} \leq |\phi^j(\Delta_\phi)| \leq \frac{C_\sigma}{m(j)^2} ,$$

where $m(j) = \min\{j+1, a-j\}$.

Every member of $\mathcal{F}_\sigma$ whose length is sufficiently large has two fixed points, symmetric about the real axis. These fixed points are necessarily attracting, due to the positive Schwarzian property of the maps in $\mathcal{F}_\sigma$. More precisely, we have the following fact. Let us denote by $\mathbb{H}$ the upper half-plane.

Lemma 2.4. *Given $0 < \sigma < 1$, there exist $C > 1$, $a_0 > 0$ and $\theta_0 > 0$ such that, if $\phi \in \mathcal{F}_\sigma$ has length $a = a(\phi) > a_0$ and angle $\theta(\phi) < \theta_0$, then there exist two attracting fixed points $z_+ \in \mathbb{H} \cap \operatorname{Dom}(\phi)$ and $z_- = \overline{z}_+$ with*

$$\frac{1}{Ca} \leq \operatorname{Im} z_+ \leq \frac{C}{a} .$$

Moreover, if $|z - z_+| \leq C/a$, then $|z - \phi(z)| \leq C/a^2$.

Proof. The lemma follows from Yoccoz's Lemma, the saddle-node bifurcation and a normality argument. □

The family $\mathcal{F}_\sigma$ is normal in the sense of Montel, and every limit map *not in* $\mathcal{F}_\sigma$ is a map with a parabolic (indifferent) fixed point on the real axis, whose multiplier is necessarily equal to one.

Lemma 2.5. *Given a compact set $W \subset \mathbb{H}$ and an open set $D \supset [0,1]$ in the plane, there exist $N_* > 0$, $\theta_* > 0$ and $a_* > 0$ with the following property. For each $\phi \in \mathcal{F}_\sigma$ such that $a(\phi) \geq a_*$ and $\theta(\phi) < \theta_*$, the domain of ϕ contains W, and for each $z \in W$ there exists $n < N_*$ such that $\phi^n(z) \in D$.*

Proof. If the statement is false, we find sequences $N_k \to \infty$, $a_k \to \infty$ and $\theta_k \to 0$, maps $\phi_k \in \mathcal{F}_\sigma$ with $\theta(\phi_k) = \theta_k$ and $a(\phi_k) = a_k$ (whose domains contain W), and points $z_k \in W$ such that $\phi_k^n(z_k)$, whenever defined, does not belong to D for all $n \leq N_k$. Since $\mathcal{F}_\sigma$ is normal and W is compact, we may assume that the sequence ϕ_k converges uniformly on compact subsets of $\mathbb{H}$ to a map $\phi : \mathbb{H} \to \mathbb{H}$ and that $z_k \to z \in W$. Applying Lemma 2.4 to each ϕ_k, we deduce that ϕ has a fixed-point $x_0 \in [0,1]$. By the Denjoy-Wolff theorem, $\phi^n(z) \to x_0$ as $n \to \infty$. Hence there exists N such that $\phi^N(z) \in D$. But then $\phi_k^N(z_k) \in D$ also, for all sufficiently large k, a contradiction. □

3. Complex bounds for real-analytic circle maps

We will extend Yampolsky's proof of complex bounds for circle maps in the Epstein class [15] to the more general case of *real-analytic* circle maps. More precisely, we will prove the following result.

Theorem 3.1 (The complex bounds). *Let $f : S^1 \to S^1$ be a real-analytic critical circle map with arbitrary irrational rotation number. Then there exists $n_0 = n_0(f)$ such that for all $n \geq n_0$ the n-th renormalization of f extends to a holomorphic pair with geometric boundaries (see definition in §2) whose fundamental annulus has conformal modulus bounded from below by a universal constant.*

The proof of Theorem 3.1 will be given at the end of this section. The main step in the proof is to show that for all sufficiently large n the appropriate inverse branch of $f^{q_{n+1}}$ maps a sufficiently large disk around the n-th renormalization domain $I_n \cup I_{n+1}$ well within itself.

We consider the unit circle $S^1 = \mathbb{R}/\mathbb{Z}$ embedded in the infinite cylinder $\mathbb{C}/\mathbb{Z}$, and in all the geometric considerations that follow we use on $\mathbb{C}/\mathbb{Z}$ the conformal metric induced from the standard Euclidean metric $|dz|$ on $\mathbb{C}$ via the exponential map $\exp : \mathbb{C} \to \mathbb{C}/\mathbb{Z}$ given by $\exp(z) = e^{2\pi i z}$. Note that $\operatorname{Im} z$ is well defined for every $z \in \mathbb{C}/\mathbb{Z}$ (as the imaginary part of any one of its pre-images under the exponential).

In what follows, we will fix $f : S^1 \to S^1$ as in the statement of Theorem 3.1. We will assume wherever necessary that f is normalized so that its critical point is $c = 1 \in S^1$. Since f is real-analytic, it extends to a holomorphic map $f : A_R \to \mathbb{C}/\mathbb{Z}$, where A_R is the annulus $\{z \in \mathbb{C}/\mathbb{Z} : |\operatorname{Im} z| < R\}$. Making R smaller if necessary, we may assume that f has no critical points outside S^1. Using Koebe's distortion theorem, it is easy to see that there exists $R_0 > 0$ such that, if $z \in S^1$ and $f(z)$ is at a distance $> R_0$ from the critical value of f, then the inverse branch f^{-1} which maps $f(z)$ back to z is well defined and univalent on the disk $D(f(z), R_0)$.

On an intuitive level, the key to the proof of Theorem 3.1 is to show that for all sufficiently large n the n-th renormalization of f satisfies an inequality of the form $|\mathcal{R}^n(f)(z)| \geq C|z|^3$ on a neighborhood of the origin, where C is a universal constant. Thus, the relevant inverse branches of $\mathcal{R}^n(f)$ behave as cube roots, mapping a large disk about the origin well within itself, giving rise to a holomorphic pair. The proof depends on the following crucial proposition, which is the analogue of Lemma 3.1 in [15]. Following [15], for each $m \geq 1$ we define $D_m \subseteq \mathbb{C}/\mathbb{Z}$ to be the disk of diameter the interval $[f^{q_{m+1}}(c), f^{q_m - q_{m+1}}(c)] \subseteq S^1$ containing the critical point c. Note that diam (D_m) is comparable with $|I_m|$: this follows from the real a-priori bounds (Theorem 2.1).

Proposition 3.2. *There exist universal constants B_1 and B_2 and for each $N \geq 1$ there exists $n(N)$ such that for all $n \geq n(N)$ the inverse branch $f^{-q_{n+1}+1}$ taking $f^{q_{n+1}}(I_n)$ back to $f(I_n)$ is well defined and univalent over $\Omega_{n,N} = (D_{n-N} \setminus S^1) \cup f^{q_{n+1}}(I_n)$, and for all $z \in \Omega_{n,N}$ we have*

$$\frac{\operatorname{dist}\left(f^{-q_{n+1}+1}(z), f(I_n)\right)}{|f(I_n)|} \leq B_1 \left(\frac{\operatorname{dist}(z, I_n)}{|I_n|}\right) + B_2 \,. \tag{1}$$

Yampolsky's proof of this fact in [15] is greatly facilitated by his assumption that the circle map has the Epstein property (namely, its lift to $\mathbb{C}$ has univalent inverse branches globally defined in the upper half-plane). Thus, the Poincaré neighborhoods $\mathbb{P}_\theta(J)$ are automatically invariant in the sense that $f^{-1}(\mathbb{P}_\theta(J)) \subseteq \mathbb{P}_\theta(f^{-1}J)$. In our setting this is no longer true, but as we look at smaller and smaller scales we see that it is asymptotically true, as the following lemma shows.

Lemma 3.3. *For every small $a > 0$, there exists $\theta(a) > 0$ satisfying $\theta(a) \to 0$ and $a/\theta(a) \to 0$ as $a \to 0$, such that the following holds. Let $F : \mathbb{D} \to \mathbb{C}$ be univalent and symmetric about the real axis, and assume $F(0) = 0$, $F(a) = a$. Then for all $\theta \geq \theta(a)$ we have $F(\mathbb{P}_\theta([0, a])) \subseteq \mathbb{P}_{(1-a^{1+\delta})\theta}([0, a])$, where $0 < \delta < 1$ is an absolute constant.*

Proof. There exists a Moebius transformation G such that $G(0) = 0$, $G(a) = a$ and $|D^j F(x) - D^j G(x)| \leq C_0 a^{3-j}$ for $j = 0, 1, 2$ and all $x \in [0, a]$, where C_0 is an absolute constant (consider the Moebius transformation with the same 2-jet as F at zero, post-composed with a linear map to meet the normalization condition $G(a) = a$). This G has no pole in a disk $D_0 \subseteq \mathbb{D}$ of definite radius around zero. Let $\varphi(z) = F(z) - G(z)$. Then $\varphi(z) = b_1 z + b_2 z^2 + \cdots$ for all $z \in D_0$, where $|b_1| \leq C_0 a^2$ and $|b_2| \leq C_0 a$. Take a small number $\varepsilon > 0$ and consider the disk $D_1 = D(0, a^{1-\varepsilon}) \subseteq D_0$. In this disk we have the estimate

$$|\varphi(z)| \leq |b_1||z| + |b_2||z|^2 + C_1|z|^3 \leq C_2 a^{2-2\varepsilon}|z| \,.$$

At the same time, using the fact that $F'(\zeta) = 1$ for some $\zeta \in [0, a]$ and the Koebe distortion lemma, we see that $|F(z)| \geq C_3|z|$ and $|G(z)| \geq C_3|z|$ for all $z \in D_1$. Therefore, for every such z the triangle with vertices at 0, $F(z)$ and $G(z)$ has an angle at zero $\leq C_4 a^{2-2\varepsilon}$; similarly for the angle at a in the triangle with vertices a, $F(z)$ and $G(z)$. Now suppose $z \in \mathbb{P}_\theta([0, a])$, for $\theta \geq a^\varepsilon$. Since G preserves this neighborhood, $G(z)$ forms an angle $\geq \theta$ with $(-\infty, 0]$, and the same holds for the angle $G(z)$ forms with $[a, +\infty)$. It follows that $F(z) \in \mathbb{P}_{\theta'}([0, a])$, where $\theta' = \theta - C_4 a^{2-2\varepsilon} \geq \theta(1 - a^{1+\delta})$ for some $0 < \delta < 1$ depending only on ε. This proves the lemma with $\theta(a) = a^\varepsilon$. □

The main consequence of this lemma is the following.

Lemma 3.4. *For each $n \geq 1$ there exist $K_n \geq 1$ and $\theta_n > 0$, with $K_n \to 1$ and $\theta_n \to 0$ as $n \to \infty$, such that for all $\theta \geq \theta_n$ and all $1 \leq j \leq q_{n+1}$ the inverse branch f^{-j+1} mapping $f^j(I_n)$ back to $f(I_n)$ is well defined over $\mathbb{P}_\theta(f^j(I_n))$ and maps this neighborhood univalently into $\mathbb{P}_{\theta/K_n}(f(I_n))$.*

Proof. Let $d_n = \max_{1\leq j\leq q_{n+1}} |f^j(I_n)|$; from the real bounds, these numbers go to zero exponentially with n. Take $\delta > 0$ as in Lemma 3.3, and let K_n be given by

$$K_n^{-1} = \prod_{j=1}^{q_{n+1}} \left(1 - |f^j(I_n)|^{1+\delta}\right) .$$

Then define $\theta_n = K_n\theta(d_n)$, where $\theta(\cdot)$ is the function in Lemma 3.3. Note that

$$\log K_n \leq C \sum_{j=1}^{q_{n+1}} |f^j(I_n)|^{1+\delta} \leq C d_n^\delta .$$

Therefore $K_n \to 1$ and $\theta_n \to 0$ as required. Also, $d_n/\theta_n \to 0$.

Now fix j as in the statement and suppose $\theta \geq \theta_n$. Define inductively $\vartheta_0 = \theta$ and $\vartheta_{i+1} = (1-|f^{j-i}(I_n)|^{1+\delta})\vartheta_i$ for $i = 0, 1, \ldots, j-2$, and note that $\vartheta_{j-1} \geq \theta/K_n$. Moreover,

$$\text{diam}\left(\mathbb{P}_{\vartheta_i}\left(f^{j-i}(I_n)\right)\right) = \frac{|f^{j-i}(I_n)|}{\sin\vartheta_i} \leq \frac{C'd_n}{\theta(d_n)} \ll R_0 .$$

Therefore f^{-1} is well defined and univalent over $\mathbb{P}_{\vartheta_i}(f^{j-i}(I_n))$, and by Lemma 2.4 we have the inclusion $f^{-1}(\mathbb{P}_{\vartheta_i}(f^{j-i}(I_n))) \subseteq \mathbb{P}_{\vartheta_{i+1}}(f^{j-i-1}(I_n))$. This completes the proof. □

Remark. The same result holds true if we replace I_n by any interval $J \supseteq I_n$ such that the map $f^{q_{n+1}-1} : f(J) \to f^{q_{n+1}}(J)$ is a diffeomorphism.

We will need four lemmas concerning the sequence $\{D_m\}$ introduced earlier. The first is an easy consequence of Lemma 3.4 and the above remark.

Lemma 3.5. *There exists $m_0 \geq 1$ such that for all $m \geq m_0$ the inverse branch f^{-q_m+1} taking $f^{q_m}(I_m)$ back to $f(I_m)$ is well defined and univalent in D_m, and*

$$\frac{\text{diam}(f^{-q_m+1}(D_m))}{|I_m|} \leq C\frac{\text{diam}(f(D_m))}{|f(I_m)|} .$$

The second is the analogue of Lemma 4.1 in [15].

Lemma 3.6. *There exist $\varepsilon_1 > 0$ and $m_1 \geq m_0$ such that for all $m \geq m_1$ and each $w \in f^{-q_{m+1}}(D_m) \setminus D_m$ we have* (a) $\text{dist}\,(w, I_m) \leq C|I_m|$ *and* (b) *for each $x \in I_m$, $\varepsilon_1 < |\arg(w - x)| < \pi - \varepsilon_1$.*

Proof. The same proof given in [15] applies here. Invariance of Poincaré neighborhoods is replaced by *quasi-invariance*, using Lemma 3.4. □

The third is the analogue of Lemma 4.4 in [15]. It provides us with the tools we need for the inductive step in the proof of Proposition 3.2 (see Figure 2).

Lemma 3.7. *There exist $\varepsilon_2 > 0$ and $m_2 \geq m_0$ such that the following holds for all $m \geq m_2$. Let $\zeta \in D_m \setminus D_{m+1}$ be a point not on the circle, and let $\zeta' = f^{-q_m}(\zeta)$ and $\zeta'' = f^{-q_{m+2}}(\zeta')$. Then either $\zeta'' \in D_{m+1}$, or else* $\text{dist}\,(\zeta'', I_{m+1}) \leq C|I_m|$ *and $\varepsilon_2 < \arg(\zeta'' - x) < \pi - \varepsilon_2$ for all $x \in I_m \cup I_{m+1}$.*

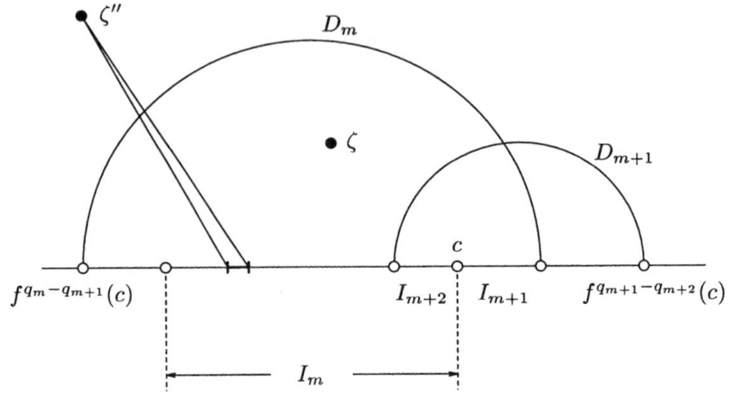

FIGURE 2.

Proof. Once again, the proof of Lemma 4.4 in [15] can be repeated here, *mutatis mutandis.* □

Notation. Given a point $\zeta \in \mathbb{C}$ and an interval $J = (a, b) \subseteq \mathbb{R}$, we denote by angle$(\zeta, J)$ the smallest of the angles $\pi - \arg(\zeta - a)$ and $\arg(\zeta - b)$.

The fourth is a consequence of Lemma 2.5.

Lemma 3.8. *There exist universal constants $N_* > 0$ and $a_* > 0$ and some $m_3 > 0$ with the following property. For all $m \geq m_3$ such that $a_{m+1} > a_*$ and each $w \in V_m = f^{-q_{m+1}}(D_m) \setminus D_m$, there exists $1 < i < N_*$ such that the iterate $(f^{-q_{m+1}})^i(w)$ is well defined and belongs to D_m.*

Proof. Lift $f^{-q_{m+1}}$ to the real line and normalize it so that $I_m \setminus I_{m+2}$ becomes the interval $[0, 1]$ to get an almost parabolic map ϕ_m. Note that $\phi_m \in \mathcal{F}_\sigma$ for some σ depending only on the real bounds. Let W_m be the image of V_m under such normalization. It is an easy matter to check that $f^{-q_{m+1}}(D_m) \cap S^1 \subseteq D_m \cap S^1$, so that V_m does not intersect S^1, and that $W_m^+ = W_m \cap \mathbb{H}$ is compactly contained in $\mathbb{H}$. Therefore, by Lemma 3.6, there exists a fixed compact set $W \subset \mathbb{H}$ such that $W_m^+ \subseteq W$ for all sufficiently large m. Similarly, the normalized copies of D_m contain a fixed open set $D \supseteq [0, 1]$ for all sufficiently large m. Hence we can take N_* and a_* as given by Lemma 2.5. □

Proof of Proposition 3.2. We will start with a point z in the disk D_{n-N}. For the argument to work, n will have to be sufficiently large. We start taking $n > N + \max\{m_1, m_2, m_3\}$, where m_1, m_2 and m_3 are given respectively by Lemma 3.6, Lemma 3.7 and Lemma 3.8.

In keeping with Yampolsky's notation in [15], let us denote by J_{-i} the interval $f^{q_{n+1}-i}(I_n)$. Also, given z, let $z_{-i} = f^{-i}(z)$ be the corresponding pre-images of z.

The proof runs by finite induction in the range $n - N \leq m \leq n$. Let m be the largest with the property that $z \in D_m$, and keep in mind that dist$(z, I_n) \asymp |I_m|$. Consider those moments $i_1 < i_2 < \cdots < i_\ell$ in the backward orbit $\{J_{-i}\}$ before the first return to I_{m+1} such that $J_{-i_k} \subseteq I_m$. Then, there are two possibilities.

The first possibility is that $z_{-i_\ell} \notin D_m$. In this case there exists a smallest $k \leq \ell$ such that $z_{-i_s} \notin D_m$ for $s = k, k+1, \ldots, \ell$. We claim that $|J_{-i_k}| \asymp |I_n|$. This is clear from the real bounds if $\ell = a_{m+1} \le a_*$, where a_* is given by Lemma 3.8. If on the other hand $\ell > a_*$, then again by Lemma 3.8 we must have $\ell - k < N_*$, and the claim follows from Lemma 2.2. Therefore, by Lemma 3.6,

$$\frac{\operatorname{dist}(z_{-i_k}, J_{-i_k})}{|J_{-i_k}|} \le C\frac{|I_m|}{|J_{-i_k}|} \le C'\frac{\operatorname{dist}(z, I_n)}{|I_n|}. \tag{2}$$

Moreover, angle$\,(z_{-i_k}, J_{-i_k}) \ge \varepsilon_1$, so there exists $\theta = \theta(\varepsilon_1, N)$ such that $z_{-i_k} \in \mathbb{P}_\theta(J_{-i_k})$. Now, if n is sufficiently large, $\theta_n < \theta$ and we can use Lemma 3.4 to get that $z_{-q_{n+1}+1} \in \mathbb{P}_{\theta/K_n}(f(I_n))$. This gives us

$$\frac{\operatorname{dist}(z_{-q_{n+1}+1}, f(I_n))}{|f(I_n)|} \le C''K_n\frac{\operatorname{dist}(z_{-i_k}, J_{-i_k})}{|J_{-i_k}|}, \tag{3}$$

and this together with (2) yields the proposition in this case.

The second possibility is that $\zeta = z_{-i_\ell} \in D_m$, and we can assume that $\zeta \notin D_{m+1}$ (otherwise the induction step is complete). In this case, consider $\zeta' = f^{-q_m}(\zeta)$ and $\zeta'' = f^{-q_{m+2}}(\zeta')$ and the corresponding interval $J'' = f^{-q_m-q_{m+2}}(J_{-i_\ell})$, and apply Lemma 3.7. Then either $\zeta'' \in D_{m+1}$, in which case the induction step is complete, or else dist$\,(\zeta'', I_{m+1}) \le C|I_m|$ and angle$\,(\zeta'', J'') \ge \varepsilon_1$, in which case we can apply the same argument leading to (2) and (3).

If the backward orbit survives all the steps of the induction, this means that in the end $z_{-q_{n+1}+q_{n-1}} \in D_{n-1}$. By Lemma 3.5, the image of D_{n-1} under $f^{-q_{n-1}+1}$ has diameter comparable to $|f(I_n)|$, so the first member of (3) is simply bounded by an absolute constant. So in any case we have (1). □

Proof of Theorem 3.1. First we remark that, since f is a cubic critical circle map, there exists a neighborhood Ω of the critical point of f such that the restriction $f : \Omega \to f(\Omega)$ is of the form $f = \psi \circ Q$ where ψ is a univalent map with universally bounded distortion and Q is the map $z \mapsto z^3$.

Let B_1, B_2 be the constants of Proposition 3.2 and let us fix a large integer N. How large N must be will be determined in the course of the argument. By Proposition 3.2, if $n \ge n(N)$, then inequality (1) holds true for all $z \in \Omega_{n,N}$. Making n larger still if necessary, we have $f^{-q_{n+1}+1}(D_{n-N}) \subseteq f(\Omega)$. By the above remark, the branch of f^{-1} mapping $f(I_n)$ back to I_n is the composition of a cube root with a univalent map with bounded distortion. Using this fact and inequality (1), we have

$$\frac{\operatorname{diam}(f^{-q_{n+1}}(D_{n-N}))}{|I_n|} \le C\sqrt[3]{B_1\frac{\operatorname{diam}(D_{n-N})}{|I_n|} + B_2}, \tag{4}$$

for some universal constant $C > 0$.

Now, as we know from the real bounds (Theorem 2.1), there exist universal constants $K_2 > K_1 > 1$ such that, for all sufficiently large n,

$$K_1^N \le \frac{\operatorname{diam}(D_{n-N})}{|I_n|} \le K_2^N. \tag{5}$$

Taking $N \ge 1$ to be the smallest integer greater than $3\log(2C\sqrt[3]{B_1 + B_2})/2\log K_1$, we can check from (4) and the first inequality in (5) that

$$\frac{\operatorname{diam}(f^{-q_{n+1}}(D_{n-N}))}{|I_n|} < \frac{1}{2}\frac{\operatorname{diam}(D_{n-N})}{|I_n|}.$$

Note that N depends only on B_1, B_2, K_1, C, and is therefore universal. From these facts, we see at once that for all sufficiently large n the topological disk $f^{-q_{n+1}}(D_{n-N})$ is compactly contained in D_{n-N}, and moreover

$$\text{mod}\,(D_{n-N} \setminus f^{-q_{n+1}}(D_{n-N})) \geq \mu\ , \tag{6}$$

where $\mu > 0$ is a universal constant. With these basic geometric bounds at hand, we can easily construct the holomorphic pair to which $\mathcal{R}^n(f)$ extends, in the following way. For more details on this construction, see §4 of [2].

First, let $\hat{f}$ be the standard lift of f to the real line. For each n, let $\lambda_n = T^{-p_n} \circ \hat{f}^{q_n}(0)$ and denote by Λ_n the linear map $x \mapsto \lambda_n x$. Take the topological disks $\widetilde{\mathcal{V}}_n = D_{n-N}$ and $\widetilde{\mathcal{O}}_{\eta_n} = f^{-q_{n+1}}(D_{n-N})$ in the cylinder and consider their lifted and normalized copies in $\mathbb{C}$, namely $\mathcal{V}_n = \Lambda_n^{-1}(\exp^{-1}(\widetilde{\mathcal{V}}_n))$ and $\mathcal{O}_{\eta_n} = \Lambda_n^{-1}(\exp^{-1}(\widetilde{\mathcal{O}}_{\eta_n}))$ (here, $\exp^{-1}$ denotes the inverse branch of the exponential that maps the critical point $c = 1 \in S^1$ of f to the origin). Then consider the map

$$\eta_n = \Lambda_n^{-1} \circ T^{-p_{n+1}} \circ \hat{f}^{q_{n+1}} \circ \Lambda_n : \mathcal{O}_{\eta_n} \to \mathbb{C}\ .$$

The geometric estimates proved above show that $\mathcal{O}_{\eta_n}$ is compactly contained in $\mathcal{V}_n$, while $\eta_n(\mathcal{O}_{\eta_n}) \subseteq \mathcal{V}_n$ holds by construction. We define the domains $\mathcal{O}_{\xi_n}, \mathcal{O}_{\nu_n}$ and the maps $\xi_n : \mathcal{O}_{\xi_n} \to \mathbb{C}$, $\nu_n : \mathcal{O}_{\nu_n} \to \mathbb{C}$ in a similar way. We obtain in this fashion a holomorphic pair Γ_n whose underlying critical commuting pair is precisely the n-th renormalization $\mathcal{R}^n(f)$. In addition, (6) shows that the conformal modulus of Γ_n is bounded from below by μ. Finally, it is straightforward to check that all the above topological disks have piecewise analytic boundaries, consisting of finitely many analytic arcs meeting at definite angles, so Γ_n has geometric boundaries. This completes the proof of the theorem. □

Remark. Note from the second inequality in (5) that $\text{diam}(\mathcal{V}_n) \leq K_2^N$ for all sufficiently large n. In other words, the size of the co-domain of Γ_n is universally bounded from above. This will be very useful later (*cf.* condition G_2 in §4 below).

4. Geometric control of holomorphic pairs

In this section, we show that if a holomorphic pair satisfies certain geometric constraints, then one can bound from below how much its dynamics expands the hyperbolic metric in the complement of the real axis. From the renormalization viewpoint, these constraints are quite natural, being a consequence of the complex bounds.

We start with the definition of geometric control. For brevity, let us say that a finite set in the plane is K-bounded ($K > 1$), if any two non-zero distances d_1 and d_2 between points in the set satisfy $d_1 \leq K d_2$. We denote by Q the map $z \mapsto z^3$ in the complex plane.

Definition. A holomorphic pair Γ is controlled by $K > 1$, or simply K-controlled, if the following geometric conditions are satisfied.

(G_1) The set $\{0, \xi(0), \eta(0), \nu(0), a, b\}$ is K-bounded.
(G_2) We have $\text{diam}(\mathcal{V}) \leq K|J|$.
(G_3) On the real axis, $|\gamma'(x)| \leq K$ for $\gamma = \xi, \eta, \nu$.
(G_4) ξ has holomorphic extensions $\xi_a : D(a, K^{-1}|J|) \to \mathbb{C}$ and $\xi_0 : D(0, K^{-1}|J|) \to \mathbb{C}$ such that $\xi_a = \psi_a \circ Q \circ \phi_a$ and $\xi_0 = \psi_\xi \circ Q$ respectively, where ψ_a, ϕ_a and ψ_ξ are univalent maps with distortion bounded by K, and $\phi_a(a) = 0$.

(G_5) η has holomorphic extensions $\eta_b : D(b, K^{-1}|J|) \to \mathbb{C}$ and $\eta_0 : D(0, K^{-1}|J|) \to \mathbb{C}$ such that $\eta_b = \psi_b \circ Q \circ \phi_b$ and $\eta_0 = \psi_\eta \circ Q$ respectively, where ψ_b, ϕ_b and ψ_η are univalent maps with distortion bounded by K, and $\phi_b(b) = 0$.
(G_6) The open sets $\mathcal{O}_\xi \cap D(a, K^{-1}|J|)$ and $\mathcal{O}_\eta \cap D(b, K^{-1}|J|)$ are connected.
(G_7) We have $D(0, K^{-1}|J|) \subseteq \mathcal{O}_\nu$.
(G_8) We have mod $(\mathcal{V} \setminus \mathcal{U}) \geq K^{-1}$.

Remark. Condition G_6 guarantees that the shadow of a K-controlled holomorphic pair extends to a single-valued map

$$F_* : \mathrm{Dom}(\Gamma) \cup D(a, K^{-1}|J|) \cup D(b, K^{-1}|J|) \to \mathbb{C} .$$

We call this map the *extended shadow* of our holomorphic pair.

Now a strengthened version of the complex bounds can be stated as follows.

Theorem 4.1. *There exists a universal constant $K_0 > 1$ such that the following holds. If Γ is a holomorphic pair with arbitrary rotation number, there exists $n_0 = n_0(\Gamma)$ such that for all $n \geq n_0$ the holomorphic pair $\mathcal{R}^n\Gamma$ restricts to a holomorphic pair Γ_n controlled by K_0.*

Proof. This is straightforward from the real bounds given by Theorem 2.1 and the complex bounds given by Theorem 3.1. For condition G_2, use the remark following the proof of Theorem 3.1. For conditions G_4 and G_5, use also Koebe distortion. □

Let us now formulate the tools from function theory that we need to prove expansion of the hyperbolic metric by a controlled holomorphic pair. Given a hyperbolic Riemann surface X, we write ρ_X for the hyperbolic density of X, d_X for the hyperbolic metric of X, and $\mathrm{diam}_X(E)$ for the diameter of $E \subseteq X$ in that metric. If $f : \mathcal{O} \subseteq X \to Y$ is a holomorphic map between two hyperbolic Riemann surfaces ($\mathcal{O}$ an open subset of X), we denote by

$$\|f'(z)\|_{X,Y} = \frac{\rho_Y(f(z))}{\rho_X(z)} \, |f'(z)|$$

the derivative of f at $z \in X$ measured with respect to the hyperbolic metrics in X and Y, and simplify the notation a bit to $\|f'(z)\|_Y$ when $X = Y$.

The key to expansion is the following lemma, stated as Prop. 4.9 in [10].

Lemma 4.2. *There exists a universal positive function $C(s)$ decreasing to zero as s decreases to zero such that the following holds. If $f : X \subset Y$ is the strict inclusion of a hyperbolic Riemann surface into another, then we have the inequality $\|f'(z)\|_{X,Y} \leq C(d_Y(z, Y \setminus f(X))) < 1$.*

The second tool we need, to be used in combination with the above, is the following.

Lemma 4.3. *Let Ω be a doubly-connected region in the plane, let E be the bounded component of $\mathbb{C} \setminus \Omega$, and let $V = E \cup \Omega$. Then we have*

$$\mathrm{diam}_V(E) \leq \frac{C}{\mathrm{mod}\ (\Omega)} , \tag{7}$$

for some constant $C > 0$, as well as

$$\mathrm{mod}\ (\Omega) \geq \frac{4}{\pi} \left[\frac{\delta}{\mathrm{diam}(\Omega)} \right]^2 , \tag{8}$$

where δ is the Euclidean distance between the boundary components of Ω.

Proof. For (7), see [9], Th. 2.4, where in fact a more precise estimate is given . For (8), use the definition of mod (Ω) as the extremal length of the family $\mathcal{F}$ of curves in Ω joining the two components of $\partial\Omega$ to get

$$\text{mod }(\Omega) \geq \frac{1}{A(\Omega)} \inf_{\gamma\in\mathcal{F}} L^2(\gamma) ,$$

where $L(\gamma)$ is the Euclidean length of γ and $A(\Omega)$ is the Euclidean area of Ω. Since $A(\Omega) \leq \pi(\operatorname{diam}\Omega)^2/4$, whereas $L(\gamma) \geq \delta$ for all $\gamma \in \mathcal{F}$, this proves (8). □

While inequality (8) gives a lower bound for the modulus of an annulus in terms of the minimum separation between boundary components, an upper bound is provided by Teichmüller's inequality

$$\text{mod }(\Omega) \leq \Phi\left(\frac{\delta}{\operatorname{diam}\Omega}\right) , \tag{9}$$

where $\Phi(x)$ is a certain universal function, which is continuous and monotone increasing, and which tends to zero as $x \to 0$; *cf.* [1].

Later in this section it will also be very important to be able to control the variation of holomorphic distortion of an analytic map between hyperbolic Riemann surfaces. When the time comes, we will resort to the following consequence of Koebe's distortion theorem, as stated by McMullen in [9], Cor. 2.27.

Lemma 4.4. *Let $f : X \to Y$ be an analytic map between hyperbolic Riemann surfaces whose derivative vanishes nowhere. Then for all $x_1, x_2 \in X$ we have*

$$\|f'(x_1)\|_{X,Y}^{\alpha} \leq \|f'(x_2)\|_{X,Y} \leq \|f'(x_1)\|_{X,Y}^{1/\alpha} ,$$

where $\alpha = \exp\{C_0 d_X(x_1, x_2)\}$ and $C_0 > 0$ is a universal constant.

Besides these very general tools, we will need the following specific consequence of geometric control. Given $E \subseteq \mathbb{C}$ and $\varepsilon > 0$, we write $V(E, \varepsilon) = \{z : d(z, E) < \varepsilon \operatorname{diam}(E)\}$.

Lemma 4.5. *Given $K > 1$, there exist $\varepsilon_0 > 0$ and $M > 1$ such that the following holds for all $0 < \varepsilon \leq \varepsilon_0$. If Γ is a K-controlled holomorphic pair and F_* is its extended shadow, then*

(a) *$V(J, \varepsilon) \subseteq \operatorname{Dom}(F_*)$;*
(b) *for all $z \in V(J, \varepsilon)$ we have $|F_*'(z)| \leq M$;*
(c) *there exists $0 < \tau = \tau(\varepsilon, K) < \varepsilon$ such that $F_*(V(J, \tau)) \subseteq V(F_*(J), \varepsilon)$;*
(d) *$F_*(V(J, \varepsilon)) \subseteq \operatorname{Dom}(\Gamma)$.*

Proof. Everything boils down to controlling the map F_* on a neighborhood of J. Let $\Delta = F_*(J) = [\xi(a), \eta(b)] = [\xi(a), \xi(0)]$. To prove (a), we note that

$$\text{mod }(\mathcal{V} \setminus \Delta) > \text{mod }(\mathcal{V} \setminus \operatorname{Dom}(\Gamma)) \geq K^{-1} ,$$

by condition G_8. Hence, there exists $\varepsilon_1 > 0$ depending only on K such that $V(\Delta, \varepsilon_1) \subseteq \mathcal{V}$, by Teichmüller's inequality (9). Using condition G_4 and making ε_1 smaller if necessary, we find $0 < \varepsilon_2 < K^{-1}$ and $0 < \varepsilon_3 < \varepsilon_1$, both depending only on K, such that

$$\begin{cases} D(\xi(a), \varepsilon_3|\Delta|) \subseteq \xi_a(D(a, \varepsilon_2|J|)) \subseteq V(\Delta, \varepsilon_1) , \\ \\ D(\xi(0), \varepsilon_3|\Delta|) \subseteq \xi_0(D(0, \varepsilon_2|J|)) \subseteq V(\Delta, \varepsilon_1) . \end{cases} \tag{10}$$

Now, consider $a < x < 0$. If $|\xi(x) - \xi(a)| < \varepsilon_3|\Delta|$, then by (10) we have $x \in D(a, \varepsilon_2|J|)$. Similarly, if $|\xi(x) - \xi(0)| < \varepsilon_3|\Delta|$, then $x \in D(0, \varepsilon_2|J|)$. If neither of these happen, then

$$\xi(a) + \varepsilon_3|\Delta| < \xi(x) < \xi(0) - \varepsilon_3|\Delta| .$$

In this case the disk $D_x = D(\xi(x), \varepsilon_3|\Delta|) \subseteq V(\Delta, \varepsilon_1)$ is contained in the image of ξ. Since $|\xi'(x)| \leq K$ by condition G_3, and since $|\Delta| \geq K^{-1}|J|$ by condition G_1, we deduce from Koebe's one-quarter theorem that

$$\xi^{-1}(D_x) \supseteq D\left(x, \frac{\varepsilon_3|J|}{4K^2}\right) .$$

Putting these facts together, and denoting the extended ξ by ξ_*, we get

$$\xi_*^{-1}(V(\Delta, \varepsilon_1)) \supseteq V([a, 0], \varepsilon') , \tag{11}$$

where $\varepsilon' = \min\{\varepsilon_2, \varepsilon_3/4K^2\}$. Proceeding similarly with η replacing ξ and condition G_5 replacing condition G_4, and noting that $\Delta \supseteq [\eta(0), \eta(b)]$, we obtain $\varepsilon'' > 0$ depending only on K such that

$$\eta_*^{-1}(V(\Delta, \varepsilon_1)) \supseteq V([0, b], \varepsilon'') , \tag{12}$$

where η_* is the extension of η. Taking $\varepsilon_0 = \min\{|a|\varepsilon', |b|\varepsilon''\}/|J|$, we deduce from (11) and (12) that $V(J, \varepsilon_0) \subseteq \mathrm{Dom}(F_*)$, and this proves (a).

The proof of the remaining assertions is more of the same, so we omit the details. Using condition G_3 and Koebe distortion, we see that the maps defining F_* have distortion on $V(J, \varepsilon_0/2)$ bounded only in terms of K. Therefore (b) holds for some constant M depending only on K if we replace ε_0 by $\varepsilon_0/2$. Making this new ε_0 still smaller if necessary, we can assume also that $V(\Delta, \varepsilon_0) \subseteq \mathrm{Dom}(\Gamma)$. Then, (c) follows for some τ as stated if we use (b) and apply the mean value theorem to F_* in $V(J, \tau)$. Finally, replacing ε_0 by $\tau(\varepsilon_0, K)$, we see that (d) follows from (c). □

Next, let Γ be a holomorphic pair and let $Y = \mathcal{V} \setminus \mathbb{R}$. How do we compare the hyperbolic metric of Y with the hyperbolic metric of the upper half-plane $\mathbb{H}$? The answer is given by the following lemma.

Lemma 4.6. *Let $\rho_Y(z)$ and $\rho(z) = 1/|\operatorname{Im} z|$ be the hyperbolic densities of Y and $\mathbb{C} \setminus \mathbb{R}$, respectively. Then for all $z \in \mathrm{Dom}(\Gamma)$ we have $\rho(z) < \rho_Y(z) \leq c_M \rho(z)$, where $c_M > 1$ is a constant depending only on $M = \mathrm{mod}\ (\mathcal{V} \setminus \mathrm{Dom}(\Gamma))$.*

Proof. By monotonicity of hyperbolic densities, $\rho_Y > \rho$. Let δ be the Euclidean distance between $\partial\mathcal{V}$ and $\partial\,\mathrm{Dom}(\Gamma)$. By Teichmüller's inequality (9), we have

$$\delta \geq \Phi^{-1}(M)\ \mathrm{diam}\,\mathcal{V} \geq \Phi^{-1}(M)\,|\operatorname{Im} z| .$$

Hence the largest disk D contained in Y and centered at z has radius

$$R \geq \min\{\delta, |\operatorname{Im} z|\} \geq 2c_M^{-1}|\operatorname{Im} z| ,$$

where $c_M^{-1} = \frac{1}{2}\min\{1, \Phi^{-1}(M)\}$. Therefore, again by monotonicity of hyperbolic densities, we get

$$\rho_Y(z) \leq \rho_D(z) = \frac{2}{R} \leq c_M \frac{1}{|\operatorname{Im} z|} = c_M\,\rho(z) ,$$

and this completes the proof. □

We are ready for the first expansion result.

Proposition 4.7. *Let Γ be a K-controlled holomorphic pair with co-domain $\mathcal{V}$, and let $Y = \mathcal{V} \setminus \mathbb{R}$. Then for each $\varepsilon > 0$ we have the following.*

(a) *There exists $\varrho = \varrho(\varepsilon, K) > 0$ satisfying $\|F'(z)\|_Y \geq 1 + \varrho$ for all z in the domain of Γ such that $|\operatorname{Im} z| \geq \varepsilon|J|$.*

(b) *There exists $\lambda = \lambda(\varepsilon, K) > 1$ such that $\|F'(z)\|_Y \geq \lambda$ for all $z \in \operatorname{Dom}(\Gamma) \cap V(J,\varepsilon)$ such that $F(z) \notin V(J,\varepsilon)$.*

Moreover, if Γ is the renormalization of a holomorphic pair Γ^* with co-domain $\mathcal{V}^*$ and $Y^* = \mathcal{V}^* \setminus \mathbb{R}$, then the same statements are true with $\|\cdot\|_{Y^*}$ replacing $\|\cdot\|_Y$ throughout.

Proof. It suffices to consider only $\varepsilon \leq \varepsilon_0$, where ε_0 is given by Lemma 4.5. By condition G_8 and (3), there exists $\varepsilon_1 = \varepsilon_1(K) > 0$ such that the Euclidean distance between $\partial \operatorname{Dom}(\Gamma)$ and $\partial\mathcal{V}$ is at least $\varepsilon_1|J|$.

Let E be the set of points in $\operatorname{Dom}(\Gamma)^+ \subseteq \mathcal{V}^+$ such that $\operatorname{Im} z \geq \varepsilon|J|$, and let δ be the Euclidean distance between the boundary components of the annulus $\mathcal{V}^+ \setminus E$. Then we have

$$\delta \geq |J| \min\{\varepsilon_1, \varepsilon\} \ . \tag{13}$$

Moreover, by condition G_2,

$$\operatorname{diam}(\mathcal{V}^+ \setminus E) \leq \operatorname{diam}(\mathcal{V}) \leq K|J| \ . \tag{14}$$

Combining (13) and (14) with inequalities (7) and (8) of Lemma 4.3, we get an upper bound for the diameter of E in the hyperbolic metric of $\mathcal{V}^+$, namely

$$\operatorname{diam}_{\mathcal{V}^+}(E) \leq \frac{C_0 \pi K^2}{4\left[\min\{\varepsilon_1, \varepsilon\}\right]^2} = s \ . \tag{15}$$

Now let $z \in E$ and suppose that $F(z) = \xi(z)$, *i.e.* that $z \in \mathcal{O}_\xi^+$ (other cases being treated in the same way). If w is any point in the non-empty set $\partial\mathcal{O}_\xi^+ \cap \partial E$, we have

$$d_{\mathcal{V}^+}(z, \mathcal{V}^+ \setminus \mathcal{O}_\xi^+) \leq d_{\mathcal{V}^+}(z, w) \leq \operatorname{diam}_{\mathcal{V}^+}(E) \ . \tag{16}$$

Viewed as a (univalent) map from $\mathcal{V}^+$ into itself, ξ^{-1} is the composition of the hyperbolic isometry $\xi^{-1} : \mathcal{V}^+ \to \mathcal{O}_\xi^+$ with the hyperbolic contraction given by the inclusion $\mathcal{O}_\xi^+ \subseteq \mathcal{V}^+$. Therefore, by Lemma 4.2 and (16),

$$\|(\xi^{-1})'(\xi(z))\|_Y \leq C\left(\operatorname{diam}_{\mathcal{V}^+}(E)\right) < 1 \ ,$$

which put together with (15) gives us

$$\|F'(z)\|_Y = \|\xi'(z)\|_Y \geq \frac{1}{C(s)} > 1 \ .$$

This proves part (a) with $\varrho(\varepsilon, K) = C(s)^{-1} - 1 > 0$.

To prove part (b) we note, using parts (c) and (d) of Lemma 4.5, that if $z \in \operatorname{Dom}(\Gamma) \cap V(J,\varepsilon)$ is such that $F(z) \notin V(J,\varepsilon)$, then $z \notin V(J, \tau(\varepsilon, K))$. In particular, $\operatorname{Im} z \geq \tau(\varepsilon, K)|J|$, and therefore by (a) we have

$$\|F'(z)\|_Y \geq 1 + \varrho(\tau(\varepsilon, K), K) = \lambda(\varepsilon, K) \ .$$

This completes the proof of both assertions for $\|\cdot\|_Y$. The proof for $\|\cdot\|_{Y^*}$ is similar. $\square$

We now turn to our second expansion result.

Proposition 4.8. *Let* Γ *be a* K*-controlled holomorphic pair with arbitrary rotation number. Then for each* $\mu > 1$ *there exists* $n_* = n_*(\mu, \Gamma) > 0$ *with the following property. If* z *is a point in the domain of* Γ *such that* $F^j(z) \in \mathrm{Dom}(\Gamma)$ *for all* $0 \le j < n$ *but* $F^n(z) \notin \mathrm{Dom}(\Gamma)$ *for some* $n > n_*$*, then* $\|(F^n)'(z)\|_Y \ge \mu$.

Proof. Let $\Delta = [\xi(0), \eta(0)]$ be the short dynamical interval of Γ, and let $f : \Delta \to \Delta$ be the associated circle map. Also, let K_0 and $n_0 = n_0(\Gamma)$ be given by Theorem 4.1. Then for all $n \ge n_0$ the n-th renormalization of Γ is a holomorphic pair with the same co-domain $\mathcal{V}$ that restricts to a K_0-controlled holomorphic pair Γ_n. We denote by $\mathcal{U}_n = \mathrm{Dom}(\Gamma_n)$ the domain and $\mathcal{V}_n$ the co-domain of Γ_n, by F_n its shadow, by J_n its long dynamical interval and by I_n its short dynamical interval. We also consider the protection bands $\mathcal{W}_n = V(J_n, \varepsilon_0)$, where ε_0 is the constant in Lemma 4.5. Recall that $F_n(\mathcal{W}_n) \subseteq \mathrm{Dom}(\Gamma_n)$.

Since $|J_n| \to 0$ as $n \to \infty$, and since $\mathrm{diam}\,\mathcal{V}_n \le K_0|J_n|$ by condition G_2, there exists a sequence $n_0 < n_1 < n_2 < \cdots < n_j < \cdots$ along which domains and co-domains are nested in the sense that

$$\mathcal{V}_{n_{j+1}} \subseteq \mathcal{U}_{n_j} \cap \mathcal{W}_{n_j} \tag{17}$$

holds true for all $j \ge 1$. Now choose an integer k such that $\lambda^{k-1} > \mu$, where $\lambda = \lambda(\varepsilon_0, K_0)$ is given by Proposition 4.7 (b). □

Claim. If $z \in \mathcal{U}_{n_k}$ and $n > 0$ are such that $F^j(z) \in \mathrm{Dom}(\Gamma)$ for $j = 0, 1, \ldots, n-1$ but $F^n(z) \notin \mathrm{Dom}(\Gamma)$, then $\|(F^n)'(z)\|_Y > \mu$.

To prove this claim, we proceed as follows. Since $z \in \mathcal{U}_{n_k}$ and the F-orbit of z eventually exits the domain of Γ, there exists $s_k \ge 1$ such that $F^i_{n_k}(z) \in \mathcal{U}_{n_k}$ for all $0 \le i < s_k$ but $z_{k-1} = F^{s_k}_{n_k}(z) \notin \mathcal{U}_{n_k}$. By the nesting condition (17), we must have $z_{k-1} \in \mathcal{U}_{n_{k-1}} \cap \mathcal{W}_{n_{k-1}}$. Proceeding inductively in this fashion until the F-orbit of z exits $\mathcal{U}_{n_1}$, we obtain positive integers $s_k, s_{k-1}, \ldots, s_1$ and points $z_k = z, z_{k-1}, \ldots, z_0$ satisfying

$$z_{j-1} = F^{s_j}_{n_j}(z_j) \in \mathcal{U}_{n_{j-1}} \cap \mathcal{W}_{n_{j-1}} ,$$

and such that $F^i_{n_j}(z_j) \in \mathcal{U}_{n_j}$ for all $0 \le i < s_j$ while $z_{j-1} \notin \mathcal{U}_{n_j}$, for all $1 \le j \le k$. Note that since the F-orbit of z exits each $\mathcal{U}_{n_j}$, it exits each protection band $\mathcal{W}_{n_j}$ *a fortiori*. Therefore, by Proposition 4.7 (b), we have $\|(F^{s_j}_{n_j})'(z_j)\|_Y \ge \lambda$ for $j = 1, 2, \ldots, k-1$. When the F-orbit of z reaches z_0, there are still, say, r iterates to go before it finally exits the domain of Γ, and we see that

$$F^n(z) = F^r \circ F^{s_1}_{n_1} \circ F^{s_2}_{n_2} \circ \cdots \circ F^{s_k}_{n_k}(z) .$$

Therefore by the chain rule we have

$$\|(F^n)'(z)\|_Y = \|(F^r)'(z_0)\|_Y \prod_{j=1}^{k} \|(F^{s_j}_{n_j})'(z_j)\|_Y > \lambda^{k-1} > \mu ,$$

which proves the claim.

The next step (*cf.* Figure 3) is to cover J with pre-images of $\mathcal{U}_{n_k}$ under F. More precisely, since $\mathcal{U}_{n_k} \supseteq I_{n_k}$ and

$$\Delta = \bigcup_{j=0}^{q_{n_k}} f^{-j}(I_{n_k}) ,$$

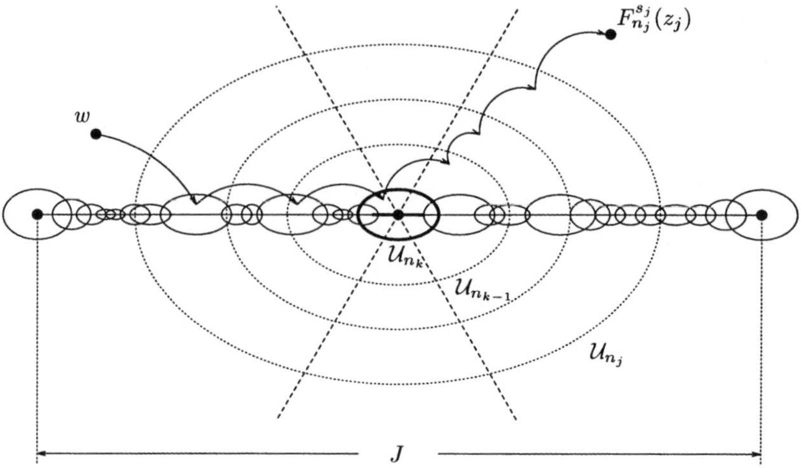

FIGURE 3.

it follows that Δ is contained in the open set

$$\mathcal{O} = \bigcup_{j=0}^{q_{n_k}} F^{-j}(\mathcal{U}_{n_k}) \ .$$

Here we have used the fact that f is simply the restriction of F to the short dynamical interval of Γ. Now recall that $J = \Delta \cup \eta^{-1}(\Delta) \cup \xi^{-1}(\Delta) \cup \cdots \cup \xi^{-m}(\Delta)$, where m is the *height* of Γ. Hence J is contained in the open set

$$\mathcal{O}' \ = \ \bigcup_{j=0}^{m} F_*^{-j}(\mathcal{O}) \ ,$$

where F_* is the extended shadow of Γ (we use F_* instead of F so we can cover the endpoints of J also). This means that for some $\varepsilon > 0$ small enough $V(J,\varepsilon) \subseteq \mathcal{O}'$. All points in the domain of Γ that are not in $V(J,\varepsilon)$ lie at a definite positive distance from the real axis, in other words,

$$\varepsilon' = \inf \left\{ |\operatorname{Im} z| : z \in \operatorname{Dom}(\Gamma) \setminus V(J,\varepsilon) \right\} \ > \ 0 \ .$$

Hence, let n_* be such that $(1+\varrho)^{n_*} > \mu$, where $\varrho = \varrho(\varepsilon', K)$ is given by Proposition 4.7 (a). If $w \in \operatorname{Dom}(\Gamma)$ and $n > n_*$ are such that $F^j(w) \in \operatorname{Dom}(\Gamma)$ for all $0 \leq j < n$ but $F^n(w) \notin \operatorname{Dom}(\Gamma)$, then *either* there exists j such that $F^j(w) \in V(J,\varepsilon)$, in which case there exists $i \geq j$ such that $z = F^i(w) \in \mathcal{U}_{n_k}$ and therefore by the Claim above

$$\|(F^n)'(w)\|_Y \geq \|(F^{n-i})'(z)\|_Y \geq \mu \ ,$$

or $F^j(w) \notin V(J,\varepsilon)$ for all $j < n$, in which case $|\operatorname{Im} F^j(w)| \geq \varepsilon'|J|$, whence by Proposition 4.7 (a) and the chain rule we get $\|(F^n)'(w)\|_Y \geq (1+\varrho)^n > \mu$ also. This completes the proof.

In order to prove our next result about holomorphic pairs, we shall need the following property of maps in the Arnold family.

Lemma 4.9. *Let $f : \mathbb{C}^* \to \mathbb{C}^*$ be a map in the Arnold family. If the rotation number of $f|\partial\mathbb{D}$ is irrational, then the Julia set of f is $\mathbb{C}^*$.*

Proof. A well-known generalization of Sullivan's no-wandering domains theorem states that a holomorphic self-map of $\mathbb{C}^*$ with finitely many singular values has no wandering domains (see [7] for a proof). The lemma follows from this and the fact that f has no non-repelling cycles or Herman rings, since its unique critical point has an orbit dense in the unit circle. □

Definition. A *generalized* holomorphic pair is a conformal copy of a holomorphic pair, in other words, the conjugate of a holomorphic pair by a univalent map. The image of a bowtie under a univalent map is called a *generalized* bowtie.

Proposition 4.10. *Let Γ be a holomorphic pair with arbitrary rotation number, and let $\mathcal{K}_\Gamma$ be its limit set. Then*

$$\mathcal{K}_\Gamma = \overline{\bigcup_{n\geq 0} F^{-n}(J)} \, .$$

Proof. Since the statement we want to prove is purely topological, we can assume, using the pull-back theorem for holomorphic pairs (Theorem 2.3) and Theorem 4.1, that Γ satisfies all the hypotheses of Proposition 4.8. For the same reasons, we also know that Γ is conjugate to a deep renormalization of a suitable element f of the Arnold family. More precisely, there exist domains $\mathcal{V}^0, \mathcal{O}^0_\xi, \mathcal{O}^0_\eta, \mathcal{O}^0_\nu \subseteq \mathbb{C}^*$, symmetric about $\partial\mathbb{D}$ with respect to inversion, forming a (generalized) bowtie with center at $1 \in \partial\mathbb{D}$, and a positive integer s such that the maps

$$\xi_0 = f^{q_s}|\mathcal{O}^0_\xi, \ \eta_0 = f^{q_{s+1}}|\mathcal{O}^0_\eta, \ \nu_0 = f^{q_s+q_{s+1}}|\mathcal{O}^0_\nu$$

determine a generalized holomorphic pair Γ^0, and moreover there exists a quasiconformal homeomorphism $H : \mathcal{V} \to \mathcal{V}^0$ conjugating Γ to Γ^0. Let F_0 be the shadow of Γ^0, and let $J_0 = H(J) \in \partial\mathbb{D}$ be the long dynamical interval of Γ^0. The distinctive feature of Γ^0 is the fact that, for each $w \in \mathrm{Dom}(\Gamma^0)$ and each $k > 0$ for which $F_0^k(w)$ is defined, there exist non-negative integers m, n with $k \leq m + n$ such that

$$F_0^k(w) \;=\; f^{mq+nQ}(w) \, , \tag{18}$$

where $q = q_s$ and $Q = q_{s+1}$. □

Claim. The set $\Lambda = \mathcal{K}_\Gamma \setminus \bigcup_{n\geq 0} F^{-n}(J)$ has empty interior.

To prove this, we argue by contradiction. Suppose $D \subseteq \Lambda$ is a non-empty open disk, and let $D_0 = H(D)$. Since points in D can be iterated by F forever, $F_0^k(w)$ exists for all $k \geq 0$, for each $w \in D_0$, and so by (18) there exist infinitely many $n > 0$ such that $f^n(w) \in \mathrm{Dom}(\Gamma_0)$. From Lemma 4.9, we know that the Julia set of f is $\mathbb{C}^*$, so by Montel's theorem there exist $n_0 > 0$ and $w \in D_0$ such that $f^{n_0}(w) \in \partial\mathbb{D}$. But then $f^n(w) \in \partial\mathbb{D}$ for all $n \geq n_0$, and so there exists $k > 0$ such that $F_0^k(w) \in \partial\mathbb{D} \cap \mathrm{Dom}(\Gamma_0) = J_0$. We have thus found a point $z = H^{-1}(w)$ in D which belongs to $F^{-k}(J)$, a contradiction that proves the claim.

Now, let x be an arbitrary point of Λ. We want to show that for every $\delta > 0$ there exists a point $p \in D(x, \delta)$ such that $F^m(p) \in J$ for some $m > 0$. Since $F^n(x)$ exists for all $n \geq 0$ and lies in the *interior* of $\mathrm{Dom}(\Gamma)$, it follows from the claim above that for all $\delta > 0$ and all sufficiently large k there exists $y \in D(x, \delta)$ such that $F^j(y) \in \mathrm{Dom}(\Gamma)$ for $0 \leq j < k$ but $F^k(y) \notin \mathrm{Dom}(\Gamma)$. By an easy connectedness argument we can assume also that $F^k(y) \in \partial\,\mathrm{Dom}(\Gamma)$. We call such y a *(k,δ)-escaping point for x.*

Next, observe that $C_1 = \sup\{d_Y(z, F^{-1}(J)\cap Y) : z \in \partial\,\mathrm{Dom}(\Gamma)\} < \infty$. Given $\delta > 0$, let $\mu > 1$ be such that

$$\mu > \left(\frac{2C_1}{\delta}\right)^{e^{C_0C_1}}, \tag{19}$$

where C_0 is the universal constant of Lemma 4.4. Let y be an $(n,\delta/2)$-escaping point for x with $n > n_*$, where n_* is given by Proposition 4.8, and let $w = F^n(y) \in \partial\,\mathrm{Dom}(\Gamma)$. Take any point $w^* \in F^{-1}(J)$ such that $d_Y(w,w^*) \le C_1$. Denote by F^{-n} the inverse branch of F^n that maps w back to y, and let $p = F^{-n}(w^*)$.

To estimate the Euclidean distance between x and p, we first estimate the hyperbolic distance in Y between y and p. For this purpose, let σ be the hyperbolic geodesic in Y joining w to w^*, and let ρ_Y be the hyperbolic density of Y. The distance $d_Y(y,p)$ is certainly not greater than the hyperbolic length $L_Y(F^{-n}\sigma)$ of the arc $F^{-n}\sigma$ joining y to p. On the other hand, by Proposition 4.8 we have $\|(F^{-n})'(w)\|_Y \le \mu^{-1}$, and applying Lemma 4.4 we get

$$\|(F^{-n})'(z)\|_Y \le \frac{1}{\mu^{1/\alpha}}, \tag{20}$$

for all $z \in \sigma$, where $\alpha = \exp\{C_0 d_Y(w,z)\} \le e^{C_0C_1}$. Hence, using (19) and (20), we see that

$$\begin{aligned} L_Y(F^{-n}\sigma) &= \int_\sigma \|(F^{-n})'(z)\|_Y\, \rho_Y(z)|dz| \\ &\le \frac{1}{\mu^{1/\alpha}} \int_\sigma \rho_Y(z)|dz| = \frac{1}{\mu^{1/\alpha}}\, d_Y(w,w^*) \\ &\le C_1\mu^{-e^{-C_0C_1}} < \frac{\delta}{2}\,. \end{aligned}$$

This gives us $|y-p| \le d_Y(y,p) < \delta/2$, and therefore $|x-p| < \delta$ as desired.

The main consequence of Proposition 4.10 is one of the key points we shall use in §6 to prove that the limit set of a holomorphic pair is uniformly twisting. Let z be a point in the limit set $\mathcal{K}_\Gamma$ of a holomorphic pair Γ. For each tangent vector v at z, we write $\ell_Y(v)$ for the length of v measured in the hyperbolic metric of Y, and put $\ell_Y(v) = \infty$ if z happens to lie in the real axis. For each $k \ge 0$, we let $v_k = DF^k(z)\,v$.

Theorem 4.11. *Let Γ be a holomorphic pair with arbitrary rotation number, and let z be a point in the limit set $\mathcal{K}_\Gamma$. Then, for each tangent vector v at z, we have*

$$\ell_Y(v_0) \le \ell_Y(v_1) \le \cdots \le \ell_Y(v_k) \;\to\; \infty$$

as $k \to \infty$.

Proof. The sequence is clearly non-decreasing, so we concentrate on proving that it diverges to infinity. Assume $z \in \mathcal{K}_\Gamma \cap Y$, otherwise there is nothing to prove, and fix $k > 0$. Let $\mathcal{O}_k \subseteq \mathrm{Dom}(\Gamma)$ be the connected component of $F^{-k}(Y)$ that contains z. We know that F^k maps $\mathcal{O}_k$ univalently onto one of the two connected components of Y. For definiteness, let us assume that $\mathcal{O}_k \subseteq \mathcal{V}^+$ and that $F^k(\mathcal{O}_k) = \mathcal{V}^+$. Thus, viewed as a map from $\mathcal{V}^+$ into itself, F^{-k} is the composition of an *isometry* between $\mathcal{V}^+$ and $\mathcal{O}_k$, with their respective hyperbolic metrics, and a *contraction* given by the inclusion $\mathcal{O}_k \subseteq \mathcal{V}^+$. In particular,

$$\|DF^{-k}(F^k(z))\|_Y = \frac{\rho_Y(z)}{\rho_{\mathcal{O}_k}(z)} . \tag{21}$$

Now, if D_k is the largest disk centered at z and contained in $\mathcal{O}_k$, then

$$\rho_{\mathcal{O}_k}(z) \asymp \rho_{D_k}(z) = \frac{2}{R_k} ,$$

where R_k is the radius of D_k. We also saw in Lemma 4.6 that $\rho_Y(z) \asymp 1/|\operatorname{Im} z|$. Combining these facts with (21) and the chain rule, we get

$$\ell_Y(v_k) = \|(F^k)'(z)\|_Y \, \ell_Y(v) \asymp \frac{2|\operatorname{Im} z|}{R_k} \ell_Y(v) .$$

Since by Proposition 4.10 we have $R_k \to 0$ as $k \to \infty$, we are done. □

5. The critical point is deep

Now we use the expansion results established in the previous section to prove that the critical point of a holomorphic pair Γ is a δ-deep point of the limit set $\mathcal{K}_\Gamma$ of Γ, for some $\delta > 0$.

Definition. Let $\Lambda \subseteq \mathbb{C}$ be compact. We say that a point $p \in \Lambda$ is a *δ-deep point* of Λ if for every $r > 0$ the largest disk contained in $D(p,r)$ which does not intersect Λ has radius $\leq r^{1+\delta}$.

In particular, if p is δ-deep for Λ, then linear blow-ups of Λ around p fill-out the plane at an exponential rate in the Hausdorff metric.

Theorem 5.1. *Let Γ be a holomorphic pair with arbitrary rotation number and limit set $\mathcal{K}_\Gamma$. Then there exists $\delta > 0$ such that the critical point of Γ is a δ-deep point of $\mathcal{K}_\Gamma$.*

Proof. In the course of the proof, we write C_0 for the constant of Lemma 4.4, and $C_1, C_2, \ldots$ for constants that depend only on the complex bounds. Let Γ_n be the n-th renormalization of Γ. By Theorem 4.1 we may assume that Γ_n is K_0-controlled for all $n \geq n_0$. As in the proof of Proposition 4.8, we choose $n_0 < n_1 < \cdots n_i < \cdots$ so that $\mathcal{V}_{n_{i+1}} \subseteq \mathcal{U}_{n_i} \cap \mathcal{W}_{n_i}$, where the $\mathcal{W}_n$ are the protection bands defined in that proposition. Since $\operatorname{diam}(\mathcal{U}_n) \asymp \operatorname{diam}(\mathcal{V}_n) \asymp |J_n|$ and $|J_n| \to 0$ exponentially as $n \to \infty$, we can choose the n_i so that $n_{i+1} - n_i$ is bounded (by a constant depending only on the real bounds). This entails $n_i \leq C_1 i$ for all i.

We want to show that for every z in the domain of Γ such that $z \notin \mathcal{K}_\Gamma$ we have $\operatorname{dist}(z, \mathcal{K}_\Gamma) \leq C_2 |z|^{1+\delta}$ for some $\delta > 0$. Let k be the largest integer such that $z \in \mathcal{U}_{n_k}$. Then $\operatorname{diam}(\mathcal{U}_{n_k}) \asymp |z|$, and by the real bounds

$$n_k \asymp \log\left(\frac{1}{|J_{n_k}|}\right) \asymp \log\left(\frac{1}{|z|}\right) .$$

Since $n_k \leq C_1 k$, we get $k \geq C_3 \log(1/|z|)$. Now, since $z \notin \mathcal{K}_\Gamma$, its forward orbit eventually leaves $\mathcal{U}$, and so it leaves each $\mathcal{U}_{n_i}$, and each $\mathcal{W}_{n_i}$ as well. Let n be such that $F^n(z) \in \mathcal{U}_{n_0} \cap \mathcal{W}_{n_0}$ but $F^{n+1}(z) \notin \mathcal{W}_{n_0}$. Then, by the same argument used in the proof of the Claim in Proposition 4.8, we have

$$\|DF^n(z)\|_Y \;\geq\; \lambda^k \;\geq\; \left(\frac{1}{|z|}\right)^{C_3 \log \lambda} \;=\; \frac{1}{|z|^\beta} \,.$$

Moreover, by Lemma 4.5 (c), the point $z' = F^n(z)$ satisfies $\operatorname{Im} z' \geq \tau(\varepsilon_0, K_0)|J_{n_0}|$, and so we can find a point $w' \in \mathcal{K}_\Gamma \cap \partial \mathcal{W}_{n_0}$ such that $d_Y(z', w') \leq C_4$. Let γ' be the geodesic arc in Y joining z' to w' (so that $L_Y(\gamma') \leq C_4$). Then the arc $\gamma = F^{-n}(\gamma')$ joins z to $w = F^{-n}(w') \in \mathcal{K}_\Gamma$. Using Lemma 4.4, we see that $\|DF^{-n}(\zeta)\|_Y \leq \|DF^{-n}(z)\|_Y^\alpha$ for all $\zeta \in \gamma'$, where $\alpha = e^{C_0 C_4}$. Therefore $L_Y(\gamma) \leq C_5|z|^\delta$, where $\delta = \alpha\beta$. Since the hyperbolic density of Y at points in $\mathcal{U}$ is comparable to the hyperbolic density of the upper half-plane $\mathbb{H}$, it follows that $L_{\mathbb{H}}(\gamma) \leq C_6|z|^\delta$. Thus, w belongs to the hyperbolic disk $\Omega \subseteq \mathbb{H}$ of center z and radius $C_6|z|^\delta$. Since the point of $\partial\Omega$ farthest from z in the Euclidean metric lies vertically above z, let us say $z + iL$, we get

$$\log\left(1 + \frac{L}{\operatorname{Im} z}\right) \;\leq\; C_6|z|^\delta \,.$$

But since $|z|$ is bounded, the left-hand side of this inequality is $\geq C_7 L/\operatorname{Im} z$, and we finally have

$$|w - z| \;\leq\; L \;\leq\; \frac{C_6}{C_7}|z|^\delta(\operatorname{Im} z) \;\leq\; C_8|z|^{1+\delta} \,.$$

We deduce that the largest Euclidean disk centered at $z \notin \mathcal{K}_\Gamma$ which does not meet $\mathcal{K}_\Gamma$ has radius $\leq C_8|z|^{1+\delta}$, and this proves the theorem. □

6. Small limit sets everywhere

In this section we use Theorem 4.11 to prove an important property of holomorphic pairs that will imply (§7) that the limit set is *uniformly twisting* in the sense of McMullen [10]. The precise statement is given in Theorem 6.3 below. We will need the following property of critical circle maps.

Lemma 6.1. *Let $f : S^1 \to S^1$ be a critical circle map for which the real a-priori bounds hold true. For each $n \geq 1$, let Δ_n be the interval of endpoints $f^{q_n}(c)$ and $f^{q_{n-1}}(c)$ containing the critical point c. Then $\Delta_n, f^{-1}(\Delta_n), \ldots, f^{-q_n+1}(\Delta_n)$ are pairwise disjoint and each $f^{-j}(\Delta_n)$ has definite and bounded space on both sides inside the largest interval containing it that does not meet any of the others.*

Proof. The endpoints of $f^{-j}(\Delta_n)$ are $f^{q_n - j}(c)$ and $f^{q_{n+1}-j}(c)$, and both belong to the dynamical partition $\mathcal{P}_n$ of level n of f. □

We will also need an easy property of controlled holomorphic pairs.

Lemma 6.2. *Let Γ be a controlled holomorphic pair. Then there exist a disk $D' \subseteq \mathcal{O}_\nu \setminus \mathbb{R}$ and a disk $D'' \subseteq \mathcal{U}$ centered at the origin, with* $\operatorname{diam}(D') \asymp |J| \asymp \operatorname{diam}(D'')$ *and* $\operatorname{dist}(D', \mathbb{R}) \asymp |J|$, *such that ν is univalent in D' and $\nu(D') \supseteq D''$.*

Proof. Control implies the existence of a disk $D \subseteq \mathcal{U}$ of definite radius around 0 and at a definite distance from $\nu(0)$. In any one of the two components of $\nu^{-1}(D)$ disjoint from $\mathbb{R}$ we find, by Koebe distortion, a disk D' of definite radius around the corresponding pre-image of 0. To obtain D'' inside $\nu(D')$, apply Koebe again! □

Definition. We say that a holomorphic pair Γ is *super-controlled* if Γ is K-controlled and there exist $R = R(K) > 0$, domains $\widetilde{\mathcal{O}}_\xi, \widetilde{\mathcal{O}}_\eta, \widetilde{\mathcal{O}}_\nu$, and complex-analytic extensions $\widetilde{\xi} : \widetilde{\mathcal{O}}_\xi \to \mathbb{C}$, $\widetilde{\eta} : \widetilde{\mathcal{O}}_\eta \to \mathbb{C}$, $\widetilde{\nu} : \widetilde{\mathcal{O}}_\nu \to \mathbb{C}$ of ξ, η, ν respectively, such that for $\gamma = \xi, \eta, \nu$ and for all $z \in \partial\mathcal{O}_\gamma \cap \mathcal{K}_\Gamma$ we have $D(z, R|J|) \subseteq \widetilde{\mathcal{O}}_\gamma$ and the restriction of $\widetilde{\gamma}$ to $D(z, R|J|)$ has distortion bounded by K.

To go from control to super-control, one renormalization suffices. In other words, if Γ is K-controlled, then its first renormalization $\mathcal{R}(\Gamma)$ is a super-controlled holomorphic pair.

Theorem 6.3. *Let Γ be a super-controlled holomorphic pair with rotation number of bounded type, with limit set $\mathcal{K}_\Gamma$ and shadow F. Then for each $z_0 \in \mathcal{K}_\Gamma$ and each $r > 0$ there exists a pointed domain (U, y) with $|z_0 - y| \asymp r$ and* $\mathrm{diam}(U) \asymp r$, *and there exists $k > 0$ such that F^k maps (U, y) onto a pointed domain $(V, 0)$ univalently with bounded distortion. In particular, U contains a conformal copy of some renormalization of Γ whose limit set has size commensurable with r.*

Proof. The proof is divided into several steps.

Step I. First we show that the statement is true when $z_0 \in [\xi(0), \eta(0)]$. Let $f = F|[\xi(0), \eta(0)]$ and for each $n \geq 1$ let Δ_n be as in Lemma 6.1. Each Δ_n is the short dynamical interval of the n-th renormalization Γ_n of Γ. Let $\mathcal{V}_n \supseteq \Delta_n$ be the co-domain of Γ_n; by the complex bounds, we can assume that $\mathrm{diam}(\mathcal{V}_n) \asymp |\Delta_n|$. For each $0 < i \leq q_n - 1$, let $V_{n,i} = f^{-i}(\mathcal{V}_n)$; note that by Koebe's distortion theorem we have $\mathrm{diam}(V_{n,i}) \asymp |f^{-i}(\Delta_n)|$. Now, choose n so that the interval of $\mathcal{P}_n$ which contains z_0 has length $\asymp r$. Then choose k so that $V_{n,k}$ is closest to z_0. This value of k and the pointed domain $(U, y) = (V_{n,k}, f^{-k}(0))$ will do.

Step II. Next, the statement is true for all $z_0 \in (a, b)$, the long dynamical interval of Γ. Here, let p be the *height* of Γ and consider the domains $V_{n,i}$ together with $\xi^{-1}(V_{n,i}), \xi^{-2}(V_{n,i}), \ldots, \xi^{-p}(V_{n,i})$ and $\eta^{-1}(V_{n,i})$. Then apply the same argument as in Step I.

Step III. Now we prove the more difficult case when z_0 is not on the real axis. Here is where we use Theorem 4.11. This case breaks down into further sub-cases. We start with a vector v_0 at z_0 and Euclidean norm $|v_0| \asymp r$, and we iterate: $z_k = F^k(z)$, $v_k = DF^k(z)v$. Note that $|F'(z_k)| = |v_{k+1}|/|v_k|$ for all k. We also know from Theorem 4.11 that $\ell_Y(v_k) \to \infty$ monotonically as $k \to \infty$. There are two possibilities.

Step IIIa. There exists k such that $\ell_Y(v_k) \ll 1$ but $\ell_Y(v_{k+1}) \gg 1$. More precisely, $\ell_Y(v_k) < \varepsilon$ while $\ell_Y(v_{k+1}) > 1/\varepsilon$, for some constant ε that will be determined in the course of the argument. To be definite, assume z_k lies in the domain of ξ, so $z_{k+1} = \xi(z_k)$. Since at all points z of the domain of Γ the hyperbolic density of Y is comparable to $1/\operatorname{Im} z$, we have

$$\frac{|v_k|}{\operatorname{Im} z_k} \asymp \ell_Y(v_k) < \varepsilon .$$

Hence $\operatorname{Im} z_k \geq C|v_k|/\varepsilon \gg |v_k|$. This means that ξ is univalent in a disk $D(z_k, R)$ of radius $R \asymp \operatorname{Im} z_k$ (provided it is well defined there – it may happen that z_k is too close to the boundary of the domain of ξ, in which case we must remember that ξ extends holomorphically to a definite neighborhood of such boundary because the holomorphic pair Γ is super-controlled). By Koebe's one-quarter theorem,

$\xi(D(z_k, R))$ contains the disk $D(z_{k+1}, R')$, where $R' = R|v_{k+1}|/4|v_k| \gg |v_{k+1}|$. But $\ell_Y(v_{k+1}) \asymp |v_{k+1}|/\operatorname{Im} z_{k+1}$ is large ($> 1/\varepsilon$), so $|v_{k+1}| \gg \operatorname{Im} z_{k+1}$. Let $\zeta = \operatorname{Re} z_{k+1}$ be the point on the real axis closest to z_{k+1}. By Step II, there exists a pointed domain (U'', y''), which is mapped univalently by some iterate of F to a pointed domain $(V, 0)$ around the critical point, such that $\operatorname{diam}(U'') \asymp |v_{k+1}|$ and $|\zeta - y''| = O(|v_{k+1}|)$, so $|z_{k+1} - y''| = O(|v_{k+1}|)$ also. Now, if ε is chosen small enough, we have $U'' \subseteq D(z_{k+1}, R'/2)$. Take $U' = \xi^{-1}(U'') \subseteq D(z_k, R)$ and $y' - \xi^{-1}(y'') \in U'$. Again by Koebe's distortion theorem we have $\operatorname{diam}(U') \asymp |v_k|$ and $|z_k - y'| = O(|v_k|)$. Since $D(z_k, R) \subseteq Y$, the inverse branch F^{-k} mapping z_k back to z_0 is well defined and univalent in $D(z_k, R)$. Thus, take $U = F^{-k}(U')$ and $y = F^{-k}(y')$. Once again by Koebe's distortion theorem, $\operatorname{diam}(U) \asymp |v_0| \asymp r$, and $|y - z_0| = O(|v_0|) = O(|r|)$, so we are done in this case.

Step IIIb. There exists k such that $\ell_Y(v_k) \asymp 1$. As in McMullen's original argument, this case is the most delicate. Here, we have $|v_k| \asymp \operatorname{Im} z_k$. As before, let $\zeta = \operatorname{Re} z_k$ be the point on the real axis closest to z_k. Using Step II, we find some $V_{n,i}$ and $y_{n,i} \in V_{n,i}$ such that $\operatorname{diam}(V_{n,i}) \asymp \operatorname{dist}(\zeta, y_{n,i}) \asymp \operatorname{Im} z_k$ (note that it is here that we use the bounded type assumption) and such that F^i maps $(V_{n,i}, y_{n,i})$ univalently onto $(\mathcal{V}_n, 0)$. Applying Lemma 6.2 to the controlled holomorphic pair Γ_n, we get a disk D'. Let $U' = F^{-i}(D') \subseteq V_{n,i}$. Using Koebe's distortion theorem, we see that $\operatorname{diam}(U') \asymp \operatorname{dist}(U', \mathbb{R}) \asymp \operatorname{diam}(V_{n,i})$. It follows at once that if γ is the geodesic arc in Y joining z_k to U', then $\operatorname{diam}(\gamma \cup U') \asymp \operatorname{Im} z_k$. Hence we can find $r_k \asymp \operatorname{Im} z_k$ and points $\zeta_1 = z_k, \zeta_2, \ldots, \zeta_s \in \gamma \cup U'$ such that

$$\gamma \cup U' \subseteq \Omega = \bigcup_{j=1}^{s} D(\zeta_j, r_k/2) ,$$

and such that $D(\zeta_j, r_k) \subseteq \mathcal{U} \setminus \mathbb{R}$ for all j. But now the inverse branch $\Phi = F^{-k}$ mapping z_k back to z_0 is well defined over the union of the disks $D(\zeta_j, r_k)$. Applying Koebe's distortion to each of these disks we see that for all $z \in \Omega$ we have $|\Phi'(z)| \asymp |\Phi'(z_k)| \asymp |v_0|/|v_k|$. But then we see that $U = \Phi(U')$ satisfies $\operatorname{diam}(U) \asymp |v_0| = r$. □

7. Proof of Theorem 1.1

The proof of Theorem 1.1 will combine the results of the previous sections with McMullen's powerful dynamic inflexibility theorem for holomorphic dynamical systems. We recall very briefly the relevant definitions and then the statement of McMullen's theorem.

Let us denote by $\mathcal{V}(\widehat{\mathbb{C}} \times \widehat{\mathbb{C}})$ the set of all analytic hypersurfaces of $\widehat{\mathbb{C}} \times \widehat{\mathbb{C}}$. For us, a holomorphic dynamical system is a subset $\mathcal{F} \subseteq \mathcal{V}(\widehat{\mathbb{C}} \times \widehat{\mathbb{C}})$. The elements of $\mathcal{F}$ are its holomorphic relations. A topology in $\mathcal{V}(\widehat{\mathbb{C}} \times \widehat{\mathbb{C}})$ can be defined as follows. If $F \subseteq \widehat{\mathbb{C}} \times \widehat{\mathbb{C}}$ is a hypersurface, its boundary $\partial F = \overline{F} \setminus F$ is closed in $\widehat{\mathbb{C}} \times \widehat{\mathbb{C}}$. Hence, given $F \in \mathcal{V}(\widehat{\mathbb{C}} \times \widehat{\mathbb{C}})$ and a sequence $F_i \in \mathcal{V}(\widehat{\mathbb{C}} \times \widehat{\mathbb{C}})$, declare $F_i \to F$ if

(a) $\partial F_i \to \partial F$ in the Hausdorff metric on closed subsets of $\widehat{\mathbb{C}} \times \widehat{\mathbb{C}}$;

(b) for each open set $U \subseteq \widehat{\mathbb{C}} \times \widehat{\mathbb{C}}$ there exist $f, f_i : U \to \mathbb{C}$ (for all sufficiently large i) such that $U \cap F = f^{-1}(0)$, $U \cap F_i = f_i^{-1}(0)$, each f, f_i vanishes to order one on F, F_i respectively, and the sequence f_i converges uniformly to f on compact subsets of U.

Then declare a set to be closed in $\mathcal{V}(\widehat{\mathbb{C}} \times \widehat{\mathbb{C}})$ if it contains the limits of all its convergent sequences. As McMullen shows in Ch. 9 of [10], the space $\mathcal{V}(\widehat{\mathbb{C}} \times \widehat{\mathbb{C}})$ with this topology is separable and metrizable.

The *geometric topology* on the space of all closed holomorphic dynamical systems is by definition the Hausdorff topology on the space of closed subsets of $\mathcal{V}(\widehat{\mathbb{C}} \times \widehat{\mathbb{C}})$. The geometric topology is typically non-Hausdorff (thus non-metrizable), but nevertheless it is sequentially compact.

We may now formulate the following notions, after McMullen [10].

Saturation. Given a holomorphic dynamical system $\mathcal{F}$, we define its saturation $\mathcal{F}^{\mathrm{sat}}$ to be the closure in $\mathcal{V}(\widehat{\mathbb{C}} \times \widehat{\mathbb{C}})$ of the set whose elements are the intersections $F \cap U$, where $F \in \mathcal{F}$ and $U \subseteq \widehat{\mathbb{C}} \times \widehat{\mathbb{C}}$ is open.

Non-linearity. A holomorphic dynamical system $\mathcal{F} \subseteq \mathcal{V}(\widehat{\mathbb{C}} \times \widehat{\mathbb{C}})$ is said to be non-linear if it does not leave invariant a parabolic line field in $\widehat{\mathbb{C}}$.

Twisting. A (closed) holomorphic dynamical system $\mathcal{F} \subseteq \mathcal{V}(\widehat{\mathbb{C}} \times \widehat{\mathbb{C}})$ is said to be twisting if every holomorphic dynamical system quasiconformally conjugate to $\mathcal{F}$ is non-linear.

Uniform twisting. A family $\{\mathcal{F}_\alpha\}$ of holomorphic dynamical systems is said to be uniformly twisting if every geometric limit of the family of saturations $\{\mathcal{F}_\alpha^{\mathrm{sat}}\}$ is a twisting dynamical system.

The family $(\mathcal{F}, \Lambda)$. Given $\mathcal{F} \subseteq \mathcal{V}(\widehat{\mathbb{C}} \times \widehat{\mathbb{C}})$ and a compact set Λ in the Riemann sphere, we define a family $(\mathcal{F}, \Lambda)$ of holomorphic dynamical systems in the following way. For each baseframe ω in the convex-hull $\mathrm{ch}(\Lambda)$ of Λ in hyperbolic 3-space, let T_ω be the fractional linear transformation that sends ω onto the standard baseframe ω_0 at $(0,1) \in \mathbb{C} \times \mathbb{R}_+ \equiv \mathbb{H}^3$. Define $(\mathcal{F}, \omega)$ to be the dynamical system $T_\omega^*(\mathcal{F})$, the pull-back of $\mathcal{F}$ by T_ω. Then let $(\mathcal{F}, \Lambda)$ be the family of all $(\mathcal{F}, \omega)$ as ω ranges through the baseframes in $\mathrm{ch}(\Lambda)$.

$C^{1+\alpha}$-conformality. A map $\phi : \widehat{\mathbb{C}} \to \widehat{\mathbb{C}}$ is said to be *$C^{1+\alpha}$-conformal* at $p \in \widehat{\mathbb{C}}$ (for some $\alpha > 0$) if the complex derivative $\phi'(p)$ exists and we have

$$\phi(z) = \phi(p) + \phi'(p)(z - p) + O(|z - p|^{1+\alpha})$$

for all z near p.

With these notions at hand, we are now in a position to state McMullen's fundamental dynamic inflexibility theorem. For a proof, see [10], page 166.

Theorem 7.1. *Let $\mathcal{F} \subseteq \mathcal{V}(\widehat{\mathbb{C}} \times \widehat{\mathbb{C}})$ be a holomorphic dynamical system, and let $\Lambda \subseteq \widehat{\mathbb{C}}$ be a compact set. If $(\mathcal{F}, \Lambda)$ is uniformly twisting and $\phi : \widehat{\mathbb{C}} \to \widehat{\mathbb{C}}$ is a K-quasiconformal conjugacy between $\mathcal{F}$ and another holomorphic dynamical system $\mathcal{F}'$, then for each δ-deep point $p \in \Lambda$ the map ϕ is $C^{1+\alpha}$-conformal at p, for some $\alpha > 0$.*

Remark. The number α depends only on the positive numbers K, δ and on a very specific measure of the non-linearity of $\mathcal{F}$ defined by McMullen in Ch. 9 of [10].

To be able to apply McMullen's theorem, we need to show that our case falls within his framework. The definition of holomorphic dynamical system adopted by McMullen is sufficiently broad to include our case. Indeed, we can define the *full dynamics* of a holomorphic pair Γ as follows.

Intuitively, the full dynamics of a holomorphic pair Γ is the class of all holomorphic maps whose graphs are contained in the holomorphic relations of Γ. These holomorphic relations are defined using the maps $\xi : \mathcal{O}_\xi \to \mathbb{C}$, $\eta : \mathcal{O}_\eta \to \mathbb{C}$, $\nu : \mathcal{O}_\nu \to \mathbb{C}$ in their open domains. Let Ω be the set of all *finite* words in the alphabet $\{\xi, \eta, \nu\}$. Given $w \in \Omega$, let $\mathrm{Dom}(w)$ be the set of all $z \in \mathrm{Dom}(\Gamma)$ such that $w(z)$ exists. Then each $\mathrm{Dom}(w) \subseteq \mathbb{C}$ is an open set. If $\mathrm{Dom}(w)$ is non-empty, we say that w is admissible for Γ, and the set of all such words is denoted by Ω_Γ. Now, for each pair $(w_1, w_2) \in \Omega_\Gamma \times \Omega_\Gamma$, let

$$G(w_1, w_2) \;=\; \big\{(z_1, z_2) \in \mathrm{Dom}(w_1) \times \mathrm{Dom}(w_2) : w_1(z_1) = w_2(z_2)\big\} \;\subseteq \widehat{\mathbb{C}} \times \widehat{\mathbb{C}}\ .$$

Each $G(w_1, w_2)$ is an analytic hypersurface in $\widehat{\mathbb{C}} \times \widehat{\mathbb{C}}$. Now consider the set $\mathcal{F}(\Gamma)$ of all holomorphic maps $g : U \to \mathbb{C}$ for which there exists $(w_1, w_2) \in \Omega_\Gamma \times \Omega_\Gamma$ such that $\mathrm{gr}(g) = \{(z, g(z)) : z \in U\} \subseteq G(w_1, w_2)$. This set is the *full dynamics* of Γ. Due to the presence of the critical point, $\mathcal{F}(\Gamma)$ is certainly non-linear.

Lemma 7.2. *Let Γ be a super-controlled holomorphic pair with rotation number of bounded type, and let $\mathcal{K}_\Gamma$ be its limit set. Then $(\mathcal{F}(\Gamma), \mathcal{K}_\Gamma)$ is uniformly twisting.*

Proof. Using Theorem 6.3, we prove that $\mathcal{F}(\Gamma)$ is uniformly twisting by the same argument as in the proof of Theorem 9.18 in [10].

Let ω_n be a sequence of baseframes in the convex-hull $\mathrm{ch}(\mathcal{K}_\Gamma) \subseteq \mathbb{H}^3$, each ω_n being based at a point $(z_n, t_n) \in \mathbb{H}^3$. Then the disk in the plane with center z_n and radius t_n intersects $\mathcal{K}_\Gamma$ – otherwise the hemisphere in $\mathbb{H}^3$ (hyperbolic half-plane) having the boundary of this disk as equator would separate (z_n, t_n) from $\mathcal{K}_\Gamma$, so that (z_n, t_n) would not belong to $\mathrm{ch}(\mathcal{K}_\Gamma)$. By Theorem 6.3, there exists a holomorphic pair Γ_n in $\mathcal{F}(\Gamma)$, whose fundamental annulus has conformal modulus uniformly bounded from below, such that $d(z_n, \mathcal{K}_{\Gamma_n}) = O(t_n)$ and $\mathrm{diam}(\mathcal{K}_{\Gamma_n}) \asymp t_n$. Conjugating Γ_n by an affine Moebius transformation that takes ω_n onto a baseframe ω_n' based at $(0, 1) \in \mathbb{H}^3$, we obtain a new holomorphic pair Γ_n' with $d(0, \mathcal{K}_{\Gamma_n'}) = O(1)$ and $\mathrm{diam}(\mathcal{K}_{\Gamma_n'}) \asymp 1$. Since each ω_n' differs from the standard baseframe at $(0, 1)$ by an elliptic rotation about $(0, 1)$ in $\mathbb{H}^3$, we deduce that Γ_n' belongs to $(\mathcal{F}(\Gamma), \omega_n)$ up to conjugation by such an elliptic rotation. It follows that any geometric limit of the form $\mathcal{G} = \lim(\mathcal{F}(\Gamma)^{\mathrm{sat}}, \omega_n)$ contains the graph of a limiting holomorphic pair (possibly conjugated by a rotation of the Riemann sphere), and so each such $\mathcal{G}$ is twisting. Therefore $(\mathcal{F}(\Gamma), \mathcal{K}_\Gamma)$ is uniformly twisting as asserted. □

Theorem 7.3. *Let Γ be a holomorphic pair with rotation number of bounded type, and let $H : \Gamma \simeq \Gamma'$ be a quasiconformal conjugacy between Γ and another holomorphic pair Γ'. Then H is $C^{1+\varepsilon}$-conformal at 0, where $\varepsilon > 0$ is universal.*

Proof. By Lemma 7.2, $(\mathcal{F}(\Gamma), \mathcal{K}_\Gamma)$ is uniformly twisting. Since by Theorem 5.1 the critical point of Γ is δ-deep for some $\delta > 0$, the desired result follows at once from McMullen's inflexibility Theorem 7.1 applied to $\phi = H$ and $p = 0$. □

Corollary 7.4. *Let f and g be real-analytic critical circle maps with the same rotation number of bounded type, and let $h : f \simeq g$ be the conjugacy that maps* *the critical point of f to the critical point of g. Then h has a quasiconformal extension which is $C^{1+\varepsilon}$-conformal at the critical point. In particular, the successive renormalizations of f and g converge together exponentially fast in the C^0 sense.*

Proof. We know from Theorem 2.1 that h is quasisymmetric. By Theorem 3.1, if n is sufficiently large, then the n-th renormalizations $\mathcal{R}^n f$ and $\mathcal{R}^n g$ extend to holomorphic pairs $\Gamma_n(f)$ and $\Gamma_n(g)$, respectively. By the pull-back theorem for holomorphic pairs (Theorem 2.3), the restriction of h to the long dynamical interval of $\Gamma_n(f)$ extends to a quasiconformal conjugacy between $\Gamma_n(f)$ and $\Gamma_n(g)$. But then Theorem 7.3 tells us that H is $C^{1+\varepsilon}$-conformal at the critical point. The fact that $\mathcal{R}^n f$ and $\mathcal{R}^n g$ converge together at an exponential rate in the C^0 topology follows easily from this and Theorem 2.1. □

The proof of Theorem 1.1 now follows from this corollary and the fact, proved in [3], that exponential convergence of renormalizations in the C^0-topology implies $C^{1+\alpha}$ conjugacy for some $0 < \alpha < 1$ depending on the rate of convergence. See also Th. 9.4 in [12], where a similar result is proved in the context of unimodal maps.

Acknowledgments

We wish to thank D. Sullivan for sharing with us some of his deep insights on renormalization. We are grateful to C. McMullen for explaining to us various aspects of his beautiful results on renormalization and rigidity.

References

[1] L. Ahlfors, *Conformal invariants*, McGraw-Hill, 1973. MR **50:**10211

[2] E. de Faria, *Asymptotic rigidity of scaling ratios for critical circle mappings*, Ergod. Th. & Dynam. Sys. **19** (1999), 995–1035.

[3] E. de Faria and W. de Melo, *Rigidity of critical circle mappings I*, J. Eur. Math. Soc. **1**.

[4] J. Graczyk and G. Świątek, *Critical circle maps near bifurcation*, Commun. Math. Phys. **176** (1996), 227–260. MR **96j:**58053

[5] M. Herman, *Sur la conjugaison differentiable des difféomorphismes du cercle a des rotations*, Publ. Math. IHES **49** (1979), 5–234. MR **81h:**58039

[6] ———, *Conjugaison quasi-simétrique des homéomorphismes du cercle a des rotations*, Manuscript, 1988.

[7] L. Keen, *Dynamics of holomorphic self-maps of* $\mathbb{C}^*$, Holomorphic Functions and Moduli I, Eds. D. Drasin *et al.*, Springer-Verlag, New York, 1988. MR **90e:**58075

[8] M. Lyubich and M. Yampolsky, *Dynamics of quadratic polynomials: complex bounds for real maps*, Ann. Inst. Fourier (Grenoble) **47** (1997), 1219–1255. MR **98m:**58113

[9] C. McMullen, *Complex dynamics and renormalization*, Annals of Math. Studies **135** (1994). MR **96b:**58097

[10] ———, *Renormalization and 3-manifolds which fiber over the circle*, Annals of Math. Studies **142** (1996). MR **97f:**57022

[11] ———, *Self-similarity of Siegel disks and Hausdorff dimension of Julia sets*, Acta Math. **180** (1998), 247–292. MR **99f:**58172

[12] W. de Melo and S. van Strien, *One-dimensional Dynamics*, Springer-Verlag, Berlin and New York, 1993. MR **95a:**58035

[13] D. Sullivan, *Bounds, quadratic differentials and renormalization conjectures*, Mathematics into the Twenty-First Century, Amer. Math. Soc. Centennial Publication, vol.2, Amer. Math. Soc., Providence, RI, 1991. MR **93k:**58194

[14] G. Świątek, *Rational rotation numbers for maps of the circle*, Commun. Math. Phys. **119** (1988), 109–128. MR **90h:**58077

[15] M. Yampolsky, *Complex bounds for critical circle maps*, Ergod. Th. & Dynam. Sys. **19** (1999), 227–257. CMP 99:09

[16] J.-C. Yoccoz, *Conjugaison analytique des difféomorphismes du cercle*, Manuscript, 1989.

Instituto de Matemática e Estatística, Universidade de São Paulo, Rua do Matão 1010, CEP05508-900 São Paulo SP - Brasil
E-mail address: `edson@ime.usp.br`

Instituto de Matemática Pura e Aplicada, Estrada Dona Castorina 110, Jardim Botânico, CEP22460-320 Rio de Janeiro RJ - Brasil
E-mail address: `demelo@impa.br`

Regular or Stochastic Dynamics in Real Analytic Families of Unimodal Maps

Artur Avila, Mikhail Lyubich and Welington de Melo

Invent. math. **154**, 451–550 (2003)

To Jacob Palis on his 60th birthday

Abstract. In this paper we prove that in any non-trivial real analytic family of quasiquadratic maps, almost any map is either regular (i.e., it has an attracting cycle) or stochastic (i.e., it has an absolutely continuous invariant measure). To this end we show that the space of analytic maps is foliated by codimension-one analytic submanifolds, "hybrid classes". This allows us to transfer the regular or stochastic property of the quadratic family to any non-trivial real analytic family.

Contents

Mathematics Subject Classification (2000): 37E05, 37F45, 30D05

Originally published in *Invent. math.* **154**, 451–550 (2003).

1. Introduction

In this paper we will consider the dynamics of unimodal maps of an interval I, i.e., smooth endomorphisms of I with a unique critical point that will be assumed to be quadratic. The simplest and most famous example of this kind is given by the real quadratic family:

$$Q_\lambda : [0, 1] \to [0, 1], \quad Q_\lambda(x) = \lambda x(1 - x),$$

where λ is a real parameter between 1 and 4. An 1976 article by R. May [May] had a big impact on the scientific community by demonstrating that this simple mathematical model exhibits a very interesting and complex dynamical behavior.

The interest in this special family grew further when Milnor and Thurston [MT] showed the *qualitative universality* of this family: any unimodal map has essentially the same dynamics as some quadratic map Q_λ. This statement becomes particularly complete if we restrict ourselves to maps with negative Schwarzian derivative: any such map is *topologically conjugate* to some quadratic map, as was shown by Guckenheimer and Misiurewicz in late 1970's. This suggests that a typical one parameter family $\{f_t\}$ of unimodal maps should have a similar qualitative dynamical evolution as the quadratic family. Discovery of Feigenbaum, Coullet and Tresser (made approximately at the same time) of *quantitative universality* of the quadratic family raised its significance even further.

In this paper we will describe a picture of an appropriate space of real analytic unimodal maps, which will give a justification for the special role of the quadratic family. We will use it to transfer some important dynamical properties from the quadratic family to any non-trivial real analytic family of *quasiquadratic maps* (a class which includes maps with negative Schwarzian derivative, see §2.8 for the precise definition).

A unimodal map is called *regular* if its critical point belongs to the basin of a hyperbolic periodic attractor and all its periodic orbits are hyperbolic. It is called *stochastic* if it has an invariant measure absolutely continuous with respect to the Lebesgue measure. Given a smooth one-parameter family $\{f_t\}$ of unimodal maps, we refer to a parameter t as regular or stochastic if the corresponding map f_t is such. The set of regular parameter values is always open. The set of stochastic parameter values has positive Lebesgue measure for an open set of families containing the quadratic family [J,BC]. In fact, in the case of the quadratic family $\{Q_\lambda\}$ much more is known:

- The set of regular parameter values λ is open and dense in the quadratic family [L4,GS2].
- The set of stochastic parameter values λ has full Lebesgue measure in the complement of the regular parameters [L7].

The former result was extended by Kozlovski [K1] to any non-trivial real analytic family $\{f_t\}$ of real analytic unimodal maps: For an open and dense set of parameter values t in such a family, the map f_t has a finite number of periodic attractors whose basin has full Lebesgue measure. One of our main theorems, Theorem B, extends the latter result to all non-trivial real analytic families $\{f_t\}$ of unimodal maps with negative Schwarzian derivative: *In such a family the map f_t is either regular or stochastic for a set of parameter values t of full Lebesgue measure.* Note that this result fits nicely to the general program of studying attractors in finite parameter families of dynamical systems (in all dimensions) formulated by Palis [Pa].

Let us now describe the picture which allows us to transfer the results from the quadratic family to other families of quasiquadratic maps. We consider an appropriate Banach space $\mathcal{B}$ of real analytic quasiquadratic maps of an interval and describe its partition into topological classes (i.e., the classes of topologically conjugate maps). One of our main results, Theorem A, states that each topological class is either an open set (in the regular case) or a codimension-one Banach submanifold. Different codimension-one classes fit together nicely giving a lamination structure in a neighborhood of any map that does not have a parabolic periodic point. This lamination is transversally quasisymmetric.

Any non-trivial real analytic one parameter family $\{f_t\} \subset \mathcal{B}$ is transverse to this lamination except possibly on a closed countable set of parameter values. Moreover, the quadratic family $\{Q_\lambda\}$ is a global transversal to this lamination. Thus, the bifurcation locus of the quadratic family (i.e., the complement of the set of regular parameters) has a *universal quasisymmetric structure*: outside a closed countable set of parameters, the bifurcation locus in any non-trivial real analytic family $\{f_t\} \subset \mathcal{B}$ is locally quasisymmetrically equivalent to that in the quadratic family. However, since quasisymmetric maps are not necessarily absolutely continuous, measure-theoretical applications of this result involve some extra work.

If we had the lamination structure in a neighborhood of parabolic maps as well, we would have a stronger result: the set of tangencies would be discrete. This would also imply the following conjecture. There is a C^∞ open and dense set of analytic families $\{f_t\} \subset \mathcal{B}$ which are transverse to all topological classes and satisfy the following property: for any nearby family $\{g_t\}$ there is a quasisymmetric homeomorphism ϕ between the parameter intervals such that f_t is quasisymmetrically conjugate to $g_{\phi(t)}$.

Problem 1.1. Is it true that codimension-one topological classes form a lamination in the space of quasiquadratic maps (near parabolic maps as well)?[1]

Though this paper is concerned with *real* unimodal maps, it is mostly based on the *complex* methods. The complex tools which are particularly important for us are the theory of holomorphic motions and the Pullback Argument, especially, its *infinitesimal* version introduced in this paper. It allows us to carry out an infinitesimal analysis of topological classes of unimodal maps which yields the above lamination structure.

[1] One can show that this fails without the quasiquadratic assumption.

After describing the dynamical and analytic background (§2), we state the main results of the paper (§3.1) and then give an outline of the main ideas of the proofs (§3.2). We encourage the reader to read this part first before plunging into the ocean of technical details. The structure of the rest of the paper containing the proofs of the main results should be clear from the Table of Contents.

Remark 1.1. We would like to draw the reader's attention to §A.6 of Appendix A which implies a short proof of Yoccoz's Rigidity Theorem.

Further development: The main results of this paper are still valid without the negative Schwarzian derivative assumption, see [Av].

The regular or stochastic dichotomy in families of unimodal maps has been recently refined, giving a better statistical description of the dynamics of typical stochastic parameters: they satisfy the "Collet-Eckmann condition", among other nice properties. Those results, first obtained in the context of the quadratic family ([AM1]), were generalized in [AM2], [Av] to non-trivial analytic families using results and methods of this paper, and then to generic smooth families.

The lamination constructed in this paper fails to be absolutely continuous in a rather dramatic way, see [AM3].

Acknowledgements. We are grateful to A. Douady and M. Yampolsky for helpful discussions and suggestions, and to the referee for many detailed comments. We are also thankful to SUNY at Stony Brook, IMPA, the Clay Institute, and the University of Toronto for their kind hospitality. This work has been partially supported by the PRONEX Project on Dynamical Systems, FAPERJ Grant E-26/151.462/99, CNPq Grant 460110/00-4, and the NSF grant DMS-9803242.

2. Preliminaries

2.1. General terminology and notations. As usual, $\mathbb{N} = \{1, 2, \dots\}$ stands for the set of natural numbers; $\mathbb{R}$ stands for the real line; $\mathbb{C}$ stands for the complex plane, and $\overline{\mathbb{C}} = \mathbb{C} \cup \{\infty\}$ stands for the Riemann sphere.

Let $\mathbb{D}_r(x) = \{z \in \mathbb{C} : |z - x| < r\}$, $\mathbb{D}_r = \mathbb{D}_r(0)$, and let $\mathbb{D} = \mathbb{D}_1$.

$\overline{X}$ or cl X denotes the closure of a set X; int X denotes its interior.

$U \Subset V$ means that U is *compactly contained in* V, i.e., $\overline{U}$ is a compact set contained in V.

For an open set $U \subset \mathbb{C}$ and a point $z \in U$, let $U(z)$ stand for the connected component of U containing z.

For two sets X and Y in $\mathbb{C}$, let

$$\operatorname{dist}(X, Y) = \inf_{x \in X,\, y \in Y} |x - y|.$$

If S is a hyperbolic Riemann surface, we consider the hyperbolic metric dist_S in it and for $X, Y \subset S$ we define

$$\operatorname{dist}_S(X, Y) = \inf_{x,y} \operatorname{dist}_S(x, y)$$

and

$$\overline{\operatorname{dist}_S}(X, Y) = \max\{\sup_{x \in X} \operatorname{dist}_S(x, Y), \sup_{y \in Y} \operatorname{dist}_S(y, X)\},$$

which is the standard Hausdorff distance.

We will reserve notation I for the interval $[-1, 1]$. For $a > 0$ let

$$\Omega_a = \{z \in \mathbb{C} : \operatorname{dist}(z, I) < a\}.$$

A *topological disk* is a simply connected domain in $\mathbb{C}$; a *Jordan disk* is a topological disk bounded by a Jordan curve;

a *topological annulus* is a doubly connected domain in $\mathbb{C}$.

The Lebesgue measure of a set $X \subset \mathbb{R}$ will be denoted by $|X|$; notation $\operatorname{meas}(X)$ will be reserved for the planar Lebesgue measure of a set $X \subset \mathbb{C}$.

A set $X \subset \mathbb{C}$ is called $\mathbb{R}$-*symmetric* if it is invariant under the conjugacy $z \mapsto \bar{z}$. A function, or vector field, or differential defined on an $\mathbb{R}$-symmetric set will be called $\mathbb{R}$-symmetric if it commutes with the conjugacy. A set X is called 0-*symmetric* if it is invariant under the 0-symmetry $z \mapsto -z$.

For a bounded function, or a vector field, or a differential, $\|\cdot\|_\infty$ will denote its sup-norm.

Given a bounded open set $V \subset \mathbb{C}$, let $\mathcal{B}_V$ be the Banach space of holomorphic functions $f : V \to \mathbb{C}$ which are continuous up to the boundary endowed with the sup-norm.

The tangent space to a manifold M at a point x is denoted by $T_x M$.

Given a map $f : X \to X$ on some metric space X, f^n will denote its *iterates*, $n = 0, 1, 2, \ldots$.

For $x \in X$, $\operatorname{orb}(x) \equiv \operatorname{orb}_f(x) = \{f^n x\}_{n=0}^\infty$ will denote the *forward orbit* or *trajectory* of x.

We will also use this notation for partially defined maps so that $\operatorname{orb}_f(x)$ consists of those points $f^n(x)$ which are well defined.

$\omega(x) \equiv \omega_f(x)$ is the *limit set* of $\operatorname{orb}(x)$:

$$\omega(x) = \bigcap_{n=0}^{\infty} \overline{\operatorname{orb}(f^n(x))}.$$

A point x is called *recurrent* if $x \in \omega(x)$.

A point x is called *fixed* if $f(x) = x$; it is called *periodic* if $f^n(x) = x$ for some $n \in \mathbb{N}$; the smallest $n \in \mathbb{N}$ with this property is called the *period* of x; The orbit of a periodic point is also called a *cycle*.

id stands for the identity map, $\operatorname{id}(x) \equiv x$.

Given a set $Y \subset X$, the *first return map* to Y is defined as follows: For $y \in Y$, let $F(y) = f^{l(y)}(y)$, where $l = l(y) \in \mathbb{N}$ is the first moment when $f^l(y) \in Y$. Such a moment may or may not exist, so that the first return map is only partially defined on Y. Somewhat abusing notations we will still write $F : Y \to Y$.

The *first landing map* $L : X \to Y$ is defined as follows: For $x \in X$, let $L(x) = f^{l(x)}(x)$, where $l = l(x) \in \mathbb{N} \cup \{0\}$ is the first moment such that $f^l(x) \in Y$. (Note that $L|Y = \text{id}$.) Again, this map is only partially defined on X.

Let $X \subset X'$ and $Y \subset Y'$. Two maps $f : X \to X'$ and $g : Y \to Y'$ are called *topologically conjugate* or *topologically equivalent* if there is a homeomorphism $h : X' \to Y'$ such that $h(X) = Y$ and

$$h(f(z)) = g(h(z)), \quad z \in X. \tag{2.1}$$

Classes of topologically conjugate maps are called *topological classes.*

We will also say that h is *equivariant* (with respect to the actions of f and g). This terminology will also be used in the case when h is only partially defined. Then it means that (2.1) is satisfied whenever it makes sense.

Given a diffeomorphism $\phi : J \to J'$ between two real intervals, its *distortion* or *non-linearity* is defined as

$$\sup_{x,y \in J} \log \frac{|D\phi(x)|}{|D\phi(y)|}.$$

Its Schwarzian derivative is given by the formula:

$$S\phi = \frac{D^3\phi}{D\phi} - \frac{3}{2}\left(\frac{D^2\phi}{D\phi}\right)^2.$$

The condition of negative Schwarzian derivative plays an important role in one-dimensional dynamics. This condition is preserved under composition.

Let $U \subset \mathbb{C}$ be a bounded open set. We say that a holomorphic function $f\colon U \to \mathbb{C}$ belongs to class $A^1(U)$ if f and its derivative f' admit a continuous extension to the closure $\overline{U}$. We will use the same notations f and f' for the extensions. We supply $A^1(U)$ with the seminorm

$$\|f\|_1 = \max_{z \in \overline{U}} |f'(z)|. \tag{2.2}$$

If $f \in A^1(U)$, $f|\overline{U}$ is a homeomorphism onto its image and f' does not vanish on $\overline{U}$, we say that $f|\overline{U}$ is a diffeomorphism (onto the image).

Remark 2.1. Notice that if U is a bounded *connected* open set then $\|\cdot\|_1$ is a Banach *norm* in the subspace $\Lambda_z \subset A^1(U)$ of functions vanishing at a given point $z \in \overline{U}$.

2.2. Hyperbolic metric. A domain $D \subset \mathbb{C}$ is called *hyperbolic* if its universal covering is conformally equivalent to the unit disc. This happens if and only if $\mathbb{C} \setminus D$ consists of at least two points. Hyperbolic domains possess the *hyperbolic (or Poincaré) metric* ρ_D of constant negative curvature. This metric is obtained by pushing down the Poincaré metric $d\rho_{\mathbb{D}} = |dz|/(1 - |z|^2)|$ from the unit disc $\mathbb{D}$.

In the case of a simply connected hyperbolic domain D ("conformal disk"), $d\rho_D = p_D(z)|dz|$ is the pull-back of the $\rho_{\mathbb{D}}$ by the Riemann mapping $D \to \mathbb{D}$. In this case the density $p_D(z)$ is comparable with $\text{dist}(z, \partial D)^{-1}$:

$$\frac{1}{4}\,\text{dist}(z, \partial D)^{-1} \le p_D(z) \le \text{dist}(z, \partial D)^{-1}.$$

The main virtue of the hyperbolic metric is that it is contracted under holomorphic maps (Schwarz Lemma) and hence is conformally invariant.

Consider an open interval $J \subset \mathbb{R}$. A special role in the complex dynamics of real maps belongs to the slit plane $\mathbb{C}_J = \mathbb{C} \setminus (\mathbb{R} \setminus J)$ endowed with the hyperbolic metric. By symmetry, J is a hyperbolic geodesic in $\mathbb{C}_J$. Let us consider hyperbolic r-neighborhoods consisting of points $z \in \mathbb{C}_J$ whose hyperbolic distance to J is at most r. It is easy to see that such a neighborhood is the union of two symmetric disk sectors based on J. For $\phi \in (0, \pi)$, let $D_\phi(J)$ denote such a neighborhood whose boundary curve meets J at angle ϕ. (One can easily work out the explicit relation between r and ϕ.) Note that for $\phi = \pi/2$ we obtain the round disk with diameter J.

2.3. Quasiconformal maps and Beltrami differentials. We assume that the reader is familiar with the basic theory of quasiconformal maps (see, e.g., [A,LV]). The goal of this section is to fix terminology and notations, and to state a few basic facts particularly important for this paper.

A homeomorphism $h : U \to V$ between two open sets $U, V \subset \mathbb{C}$, is called a *quasiconformal map*, or briefly a *qc* map, if it has locally integrable distributional derivatives ∂h, $\bar{\partial} h$, and $|\bar{\partial} h/\partial h| \le k < 1$ almost everywhere. As this local definition is conformally invariant, one can define qc homeomorphisms between Riemann surfaces.

One can associate to a qc map an analytic object called the *Beltrami differential* of h,

$$\mu = \frac{\bar{\partial} h}{\partial h}\frac{d\bar{z}}{dz},$$

with $\|\mu\|_\infty < 1$. (We will identify the Beltrami differential of a map $h : \mathbb{C} \to \mathbb{C}$ with the function $\bar{\partial} h/\partial h$.) The corresponding geometric object is a measurable family of infinitesimal ellipses (defined up to dilatation), pull-backs by Dh of the field of infinitesimal circles. The eccentricities of these ellipses are ruled by $|\mu|$, and are uniformly bounded almost everywhere, while the orientation of the ellipses is ruled by the arg μ. The *dilatation*

$$\mathrm{Dil}(h) = \frac{1 + \|\mu\|_\infty}{1 - \|\mu\|_\infty}$$

of h is the essential supremum of the eccentricities of these ellipses. A qc map h is called K-qc if $\mathrm{Dil}(h) \le K$.

Weyl's Lemma. *A 1-qc map is holomorphic.*

One of the most remarkable facts of analysis is that *any Beltrami differential with* $\|\mu\|_\infty < 1$ *(or rather a measurable field of ellipses with essentially bounded eccentricities) is locally generated by a qc map*, unique up to post-composition with a conformal map. Thus, such a Beltrami differential on a Riemann surface S induces a conformal structure quasiconformally equivalent to the original structure of S. Together with the Riemann Mapping Theorem this leads to the following result:

Measurable Riemann Mapping Theorem. *Let μ be a Beltrami differential on $\overline{\mathbb{C}}$ with $\|\mu\|_\infty < 1$. Then there is a qc map $h : \overline{\mathbb{C}} \to \overline{\mathbb{C}}$ which solves the Beltrami equation: $\bar\partial h / \partial h = \mu$. This solution is unique if it is normalized to fix three points in $\overline{\mathbb{C}}$. The normalized solution h_μ depends holomorphically*[2] *on μ.*

A map $f : U \to V$ between domains in $\overline{\mathbb{C}}$ is called *quasiregular* if it is a composition of a holomorphic map and a qc homeomorphism. Beltrami differentials can be naturally pulled back by quasiregular maps $\mu \mapsto f^*\mu$. A Beltrami differential (defined on an open set containing $U \cup V$) is called *f-invariant* if $f^*\mu = \mu$ a.e. in U.

Assume that a quasiregular map $f : U \to V$ admits an invariant Beltrami differential μ defined on $\overline{\mathbb{C}}$. Let us solve the Beltrami equation

$$\frac{\bar\partial h_\lambda}{\partial h_\lambda} = \lambda\mu, \quad |\lambda| < a \equiv \frac{1}{\|\mu\|_\infty},$$

by means of qc maps $h_\lambda : \mathbb{C} \to \mathbb{C}$ fixing two given points. Then the maps $f_\lambda = h_\lambda \circ f \circ h_\lambda^{-1}$ preserve the standard conformal structure and hence are holomorphic (by Weyl's Lemma). The dependence of f_λ on λ is also holomorphic (that is, the map $(\lambda, z) \mapsto f_\lambda(z)$, $|\lambda| < a$, $z \in h_\lambda(U)$ is holomorphic). This family of maps is called the *Beltrami disk* through f in the direction of μ. If we restrict λ to the real interval $(-a, a)$, we obtain the Beltrami *path* through f in the direction of μ.

One more fundamental property of qc maps exploited in this paper is compactness:

[2] The space of Beltrami differentials is the unit ball in the complex Banach space $L^\infty(\overline{\mathbb{C}})$, and thus it is endowed with the natural complex structure. Holomorphic dependence of h_μ on μ is understood in the pointwise sense: for any $z \in \overline{\mathbb{C}}$, $\mu \mapsto h_\mu(z)$ is holomorphic.

First Compactness Lemma. *The space of K-qc maps $h : \mathbb{C} \to \mathbb{C}$ fixing two points is compact in the uniform topology on the Riemann sphere.*

A useful consequence is the following:

Lemma 2.1. *Let $S \subset \mathbb{C}$ be a hyperbolic domain, and let $H : S \to S$ be a K-qc map homotopic to the identity rel the boundary. Then* $\operatorname{dist}_S(x, H(x)) \le C(K)$, *$x \in S$, where $C(K) \to 0$ as $K \to 1$.*

Proof. If $S = \mathbb{D}$, then the situation can be normalized so that $x = 0$, and the statement follows from the First Compactness Lemma. In general, cover S by the unit disk and lift H to a qc homeomorphism of $\mathbb{D}$ homotopic to id rel boundary (see Theorem 2.2 of [EM]). □

Let $L : \mathbb{R}_+ \to \mathbb{R}_+$. A map $h : X \to Y$ between two metric spaces is called a *L-quasi-isometry* if for any $\varepsilon > 0$,

$$\operatorname{dist}(h(x), h(y)) \le \max\{L(\varepsilon)\operatorname{dist}(x, y), \varepsilon\}, \quad x, y \in X.$$

Quasiconformal maps are quasi-isometries with respect to the hyperbolic metric:

Lemma 2.2. *For every $K \ge 1$ there exists $L_K : \mathbb{R}^+ \to \mathbb{R}^+$ such that if $h : S \to \tilde{S}$ is a K-qc map between two hyperbolic Riemann surfaces then h is a L_K quasi-isometry in the hyperbolic metric. Furthermore, for every $\varepsilon > 0$, $\lim_{K\to 1} L_K(\varepsilon) = 1$.*

Proof. Lifting H to the universal covering, we reduce the situation to the case when $S = \mathbb{D}$. Again, the conclusion follows from compactness argument. □

Lemma 2.3. *Given $M > m > 0$, there is a constant δ with the following property. Let $S, \tilde{S} \subset \overline{\mathbb{C}}$ be hyperbolic Riemann surfaces and $h_1, h_2 : S \to \tilde{S}$ be $(1+\delta)$-qc maps homotopic rel boundary. Let X and Y be subsets of S. If* $\operatorname{dist}_S(X, Y) > M$ *then* $\operatorname{dist}_{\tilde{S}}(h_1(X), h_2(Y)) > m$.

Proof. Let $L = L_{1+\delta}(m)$ be the quasi-isometric constant for h_2^{-1} (see Lemma 2.2). Let $H = h_2^{-1} \circ h_1$. Assume there is a point $x \in X$ such that $\operatorname{dist}_{\tilde{S}}(h_1(x), h_2(Y)) \le m$. Applying h_2^{-1} we conclude that $\operatorname{dist}_S(H(x), Y) \le \max\{Lm, m\} < (M+m)/2$, provided δ is sufficiently small. Hence $\operatorname{dist}_S(H(x), x) > (M-m)/2$, which for sufficiently small δ contradicts Lemma 2.1. □

A qc map $h : \mathbb{C} \to \mathbb{C}$ will be called *normalized* if it fixes points -2 and 2 (this gives a little bit of space for the dynamical interval $[-1, 1]$). We will use the notation μ_h for the Beltrami differential of the qc map h.

A homeomorphism $h : X \to h(X) \subset \mathbb{C}$ *of a closed set* $X \subset \mathbb{C}$ will be called quasiconformal if it admits a qc extension to $\mathbb{C}$.

If X is a measurable set and $h : X \to \overline{\mathbb{C}}$ is a homeomorphism which admits qc extensions h_1 and h_2 to neighborhoods of X, it is easy to see that $\mu_{h_1}|X = \mu_{h_2}|X$ up to sets of zero Lebesgue measure. This allows us to define the dilatation of h, $\mathrm{Dil}(h)$ as the essential supremum over X of the eccentricities of the corresponding ellipses field (or 0 if meas $X = 0$).

A *quasidisk* (or *quasiarc*) is the image of $\mathbb{D}$ (or $[-1, 1]$) by a qc map of $\mathbb{C}$.

2.4. Quasisymmetric maps. A homeomorphism $f : \mathbb{R} \to \mathbb{R}$ is called *quasisymmetric* (briefly: "qs") if there exists a constant $\kappa \geq 1$ such that for any $h > 0$ and any $x \in \mathbb{R}$ we have

$$\frac{1}{\kappa} \leq \frac{f(x+h) - f(x)}{f(x) - f(x-h)} \leq \kappa.$$

The *dilatation* $\mathrm{Dil}(f)$ of a qs map is defined as the smallest such κ. A map f is called κ-qs if $\mathrm{Dil}(f) \leq \kappa$.

Qs maps are important because of their relation to qc maps:

Theorem 2.4 (Ahlfors-Beurling). *Let $F : \mathbb{C} \to \mathbb{C}$ be a K-qc homeomorphism preserving the real line. Then the restriction $f|\mathbb{R}$ is $\kappa(K)$-qs. Vice versa, any κ-qs homeomorphism $f : \mathbb{R} \to \mathbb{R}$ admits a $K(\kappa)$-qc extension to the complex plane.*

Quasisymmetric maps fixing 0 and 1 are Hölder continuous with an absolute constant and an exponent depending only on the dilatation. Since the dilatation is invariant under compositions with affine maps, we obtain:

Lemma 2.5. *Let $f : \mathbb{R} \to \mathbb{R}$ be a κ-qs homeomorphism. Then there exist constants $C > 0$ and $\delta > 0$ such that for any two nested intervals $J \subset T \subset \mathbb{R}$ we have:*

$$\frac{|f(J)|}{|f(T)|} \leq C \left(\frac{|J|}{|T|} \right)^{\delta}.$$

A map f *on a set* $X \subset \mathbb{R}$ is called quasisymmetric if f extends to a qs map of $\mathbb{R}$.

2.5. Holomorphic motions and codimension-one laminations. Given a domain $\mathcal{V}$ in a complex Banach space E with a base point $*$ and a set $X_* \subset \mathbb{C}$, a *holomorphic motion* of X_* over $\mathcal{V}$ is a family of injections $h_\lambda : X_* \to \mathbb{C}$, $\lambda \in \mathcal{V}$, such that $h_* = \mathrm{id}$ and $h_\lambda(z)$ is holomorphic in λ for any $z \in X_*$. Let $X_\lambda = h_\lambda X_*$.

We will summarize fundamental properties of holomorphic motions which are usually referred to as the λ-*lemma*. It consists of two parts: extension of the motion and transversal quasiconformality, which will be stated separately. The first extension result was obtained in [L1], [MSS], and states that any holomorphic motion of a subset of $\overline{\mathbb{C}}$ extends to a holomorphic motion of the closure. A much more involved result was obtained by [BR], [ST]. Let $B_r \subset E$ stand for the Banach ball of radius r centered at $*$.

Extension Lemma ([BR], [ST]). *A holomorphic motion $h_\lambda : X_* \to X_\lambda$ of a set $X_* \subset \mathbb{C}$ over a Banach ball B_r admits an extension to a holomorphic motion $H_\lambda : \mathbb{C} \to \mathbb{C}$ of the whole complex plane over the ball $B_{r/3}$.*

Remark 2.2. Assume that a complex Banach space E is supplied with an anti-linear isometric involution conj : $E \to E$. Let

$$E^{\mathbb{R}} = \{\lambda \in E \mid \operatorname{conj}(\lambda) = \lambda\}$$

and $* \in E^{\mathbb{R}}$. Let us say that a holomorphic motion of an $\mathbb{R}$-symmetric set $X \subset \mathbb{C}$ over B_r is $\mathbb{R}$-symmetric if $\overline{h_{\operatorname{conj}\lambda}(\bar{z})} = h_\lambda(z)$. Then the above Extension Lemma actually provides an $\mathbb{R}$-symmetric extension of this motion over $B_{r/3}$. This follows from the fact that the extension constructed by [BR] is canonical.

In what follows, this remark will be applied to certain spaces of holomorphic functions on some $\mathbb{R}$-symmetric domains, and $(\operatorname{conj} f)(z) = \overline{f(\bar{z})}$.

Remark 2.3. If a holomorphic motion $h_\lambda : X_* \to X_\lambda$ is defined over a *simply connected hyperbolic domain $D \subset \mathbb{C}$*, then Slodkowski's Theorem [Sl] gives the existence of an extension to $H_\lambda : \mathbb{C} \to \mathbb{C}$ defined over the full parameter space D.

Quasiconformality Lemma [MSS], [BR]. *Let $h_\lambda : U_* \to U_\lambda$ be a holomorphic motion of a domain $U_* \subset \mathbb{C}$ over a hyperbolic domain $D \subset \mathbb{C}$. Then the maps h_λ are $K(r)$-qc, where r is the hyperbolic distance between $*$ and λ in D. Moreover, $K(r) = 1 + O(r)$ as $r \to 0$.*

Remark 2.4. Combining the Extension Lemma and the Quasiconformality Lemma, we obtain the following statement: If $h_\lambda : X_* \to X_\lambda$ is a holomorphic motion of a subset $X_* \subset \mathbb{C}$ over a hyperbolic domain D then h_λ admits a $K(r)$-qc extension to the whole $\mathbb{C}$, where r is the hyperbolic distance between $*$ and λ.

Indeed, it is enough to work in the universal cover (notice that we do not require the extensions to fit together in a holomorphic motion). If $r \le 1/4$, a direct application of the Extension Lemma and the Quasiconformality Lemma will do. In the general case, we can cover a geodesic path linking $*$ to λ by at most $4r + 1$ balls of hyperbolic diameter bounded by $1/4$, so we just have to apply the previous case at most $4r + 1$ times. See also [BR], Theorem 1, which provides optimal bounds for the dilatation of the extensions.

A holomorphic motion $h_\lambda : X_* \to X_\lambda$ over $\mathcal{V}$ can be viewed as a complex codimension-one lamination on $\mathcal{V} \times \mathbb{C}$, whose leaves are graphs of the functions $\lambda \mapsto h_\lambda(z)$, $z \in \mathcal{V}$. More generally, a codimension-one *holomorphic lamination* $\mathcal{L}$ on a complex Banach manifold $\mathcal{M}$ is a family of disjoint codimension-one Banach submanifolds of $\mathcal{M}$, called the *leaves* of the lamination that locally looks like a holomorphic motion:

- For any point $p \in \mathcal{M}$, there exists a holomorphic local chart $\Phi : \mathcal{W} \to \mathcal{V} \oplus \mathbb{C}$ (where $\mathcal{W}$ is an open neighborhood of p in $\mathcal{M}$ and $\mathcal{V}$ is an open

set in some complex Banach space) such that for any leaf L and any connected component L_0 of $L \cap \mathcal{W}$, the image $\Phi(L_0)$ is a graph of a holomorphic function $\mathcal{V} \to \mathbb{C}$.
The neighborhood $\mathcal{W}$ in the above definition is called a *flow box*, and the connected components L_0 are called *local leaves* in this flow box.

Note that by the Extension Lemma, any holomorphic lamination extends locally, near any point $p \in \mathcal{M}$, to a lamination whose leaves fill out a full Banach neighborhood of p.

A one-dimensional holomorphic submanifold of $\mathcal{M}$ which only has transversal intersections with the leaves of $\mathcal{L}$ is called a *transversal* to $\mathcal{L}$. Given two transversals X and Y within one flow box $\mathcal{W}$, we have a partially defined local holonomy map $H : X \to Y$, $H(p) = q$ iff p and q belong to the same local leaf in $\mathcal{W}$. This definition can be extended to local transversals X and Y through two points p and q on the same leaf connected by some path γ. To this end, cover the path with finitely many flow boxes $\mathcal{W}_i$ and and define the local holonomy $H : X \to Y$ as the composition of local holonomies within the $\mathcal{W}_i$.

A map $H : X \to Y$ is called *locally qc* at $p \in X$ if it admits a qc extension to some neighborhood of p. We say that a lamination $\mathcal{L}$ is *transversally quasiconformal* if the holonomy between any two transversals is locally qc. The λ-lemma implies (see e.g., [L6, Appendix 2]):

Corollary 2.6 (Transverse qc structure). *Any codimension-one holomorphic lamination is transversally quasiconformal.*

A holomorphic motion $h_\lambda : \mathbb{C} \to \mathbb{C}$ will be called *normalized* if it fixes points 2 and -2.

The Measurable Riemann mapping Theorem can be interpreted in terms of holomorphic motions in the following way. Let μ_λ, $\lambda \in \mathbb{D}$, be a holomorphic family of Beltrami differentials on $\mathbb{C}$ such that $\|\mu_\lambda\|_\infty < 1$ for all $\lambda \in \mathbb{D}$, and $\mu_0 = 0$. Then there exists a unique normalized holomorphic motion h_λ of $\mathbb{C}$ based at 0 such that $\mu_{h_\lambda} = \mu_\lambda$. The converse is also true:

Theorem 2.7 (see [BR], Theorem 2). *Let h_λ be a holomorphic motion of an open set of $\mathbb{C}$. Then μ_{h_λ} is a holomorphic family of Beltrami differentials.*

Holomorphic motions also enjoy the compactness properties of qc maps:

Lemma 2.8. *Let $X \subset \overline{\mathbb{C}}$ be a set containing 3 distinct points $\{a, b, c\}$ and let $\mathcal{V}$ be an open subset of a separable Banach space. Consider a holomorphic motion of X over $\mathcal{V}$ as a map from $\mathcal{V}$ to the space of continuous maps from X to $\overline{\mathbb{C}}$ endowed with the uniform metric. The space of all holomorphic motions $h_\lambda : X \to \overline{\mathbb{C}}$, $\lambda \in \mathcal{V}$, fixing $\{a, b, c\}$ is compact in the uniform topology over compacts of $\mathcal{V}$.*

For a proof, see [D1].

2.6. Infinitesimal deformations

2.6.1. Quasiconformal vector fields. A continuous vector field $v \equiv v(z)/dz$ on an open set $U \subset \overline{\mathbb{C}}$ is called *K-quasiconformal* (or briefly, K-qc) if it has locally integrable distributional partial derivatives ∂v and $\overline{\partial} v$, and $\|\overline{\partial} v\|_\infty \leq K$. A vector field is *quasiconformal* if it is K-qc for some K.

The theory of qc vector fields has many parallels with the theory of qc maps. Given $\mu \in L^\infty(\mathbb{C})$ one obtains a qc vector field α with $\overline{\partial}\alpha = \mu$. (Global) solutions to this problem are obtained from local ones, and those can be explicitly given (see [AB]) using the Cauchy transform

$$-\frac{1}{\pi}\int \frac{\mu(\zeta)}{z-\zeta} d\zeta \wedge d\overline{\zeta}.$$

The Cauchy transform also implies that local solutions have modulus of continuity $\phi(x) = -x\ln(x)$ (see [Mc2], Theorem A.10).

Two qc vector fields α and $\tilde{\alpha}$ such that $\overline{\partial}\alpha = \overline{\partial}\tilde{\alpha}$ differ by a conformal vector field (this is another instance of Weyl's Lemma). A conformal vector field on $\overline{\mathbb{C}}$ which vanishes at three given points vanishes on the whole sphere.

Second Compactness Lemma. *The space of K-qc vector fields of the Riemann sphere $\overline{\mathbb{C}}$ vanishing at three given points is compact in the topology of uniform convergence on $\overline{\mathbb{C}}$.*

(See Corollary A.11, p. 199 of [Mc2] for the proof.)

Corollary 2.9. *For any $L > 0$, there exists a $C > 0$ such that if α is a L-qc vector field on $\overline{\mathbb{C}}$ that vanishes at ∞ and on the boundary of some interval $T \subset \mathbb{R}$, then $|\alpha(z)| < C|T|$, for all $z \in T$.*

Proof. Let $A : [0, 1] \to T$ be an affine transformation and let $\beta = A^*\alpha$. Then the vector field β is L-qc and vanishes at 0, 1 and ∞. By the Second Compactness Lemma, β is bounded by some universal C. Hence α is bounded by $C|T|$. □

A qc vector field will be called *normalized* if it vanishes on $\{-2, 2, \infty\}$.

A continuous vector field v *on a closed set* $X \subset \mathbb{C}$ is called *quasiconformal* if it extends to a qc vector field on $\mathbb{C}$. If a vector field v on a closed set X admits a normalized qc extension to $\overline{\mathbb{C}}$ (this is always the case when X does not intersect $\{-2, 2, \infty\}$) then we let

$$\|v\|_{\mathrm{qc}} = \inf \|\overline{\partial}\beta\|_\infty, \tag{2.3}$$

where β runs over all normalized qc extensions of v.

Notice that if α_1 and α_2 are qc vector fields that coincide in some measurable set X then $\overline{\partial}\alpha_1 = \overline{\partial}\alpha_2$ on X up to sets of zero Lebesgue measure. If we define α as the restriction of those vector fields to X, the object $\overline{\partial}\alpha \in L^\infty(X)$ is well defined.

Quasiconformal vector fields are the infinitesimal counterparts of qc maps. More precisely, they are tangent at identity to holomorphic motions.

Lemma 2.10. *Let $h_\lambda : X \to \mathbb{C}$, $\lambda \in \mathbb{D}$, be a holomorphic motion with base point* 0*. Then*

$$\alpha \equiv \frac{d}{d\lambda} h_\lambda \Big|_{\lambda=0}$$

is a qc vector field on X. Moreover, if X is an open set,

$$\bar{\partial}\alpha = \frac{d}{d\lambda} \mu_{h_\lambda} \Big|_{\lambda=0}. \tag{2.4}$$

Proof. Consider an extension of h_λ, which we still denote h_λ. By Theorem 2.7, μ_{h_λ} depends holomorphically on λ, and (2.4) follows from the proof of Lemma 19 of [AB]. □

2.6.2. Equivariant vector fields. Let $f : \Omega \to \mathbb{C}$ be a holomorphic map and let v be a holomorphic vector field on Ω. A vector field α is called *equivariant* on some set $X \subset \Omega$ (with respect to the pair (f, v)) if for any $z \in X$,

$$v(z) = \alpha(f(z)) - f'(z)\alpha(z). \tag{2.5}$$

Note that this equation can also be written in the form

$$f^*\alpha - \alpha = \frac{v}{f'}. \tag{2.6}$$

This equation tells us that α is an "infinitesimal conjugacy" between f and its "infinitesimal deformation" v. It is obtained by linearizing the following commutative diagram:

$$\begin{array}{ccc} \Omega & \xrightarrow[\mathrm{id}+\varepsilon\alpha]{} & \Omega_\varepsilon \\ {\scriptstyle f}\downarrow & & \downarrow{\scriptstyle f+\varepsilon v} \\ \mathbb{C} & \xrightarrow[\mathrm{id}+\varepsilon\alpha]{} & \mathbb{C} \end{array} \tag{2.7}$$

Let $X \subset \Omega$ and let α be a vector field on $Y \equiv f(X)$. A vector field β is called the *lift* of α to X by (f, v) if $v = \alpha \circ f - f'\beta$. This equation is obtained by linearization of the following commutative diagram:

$$\begin{array}{ccc} X & \xrightarrow[\mathrm{id}+\varepsilon\beta]{} & X_\varepsilon \\ {\scriptstyle f}\downarrow & & \downarrow{\scriptstyle f+\varepsilon v} \\ \mathbb{C} & \xrightarrow[\mathrm{id}+\varepsilon\alpha]{} & \mathbb{C} \end{array}$$

Note that if 0 is a critical point of f, then a "liftable" vector field α must necessarily satisfy condition $v(0) = \alpha(f(0))$ (this condition is also sufficient if 0 is a simple critical point, see Lemma 6.4).

Obviously, a vector field is equivariant if and only if it is equal to its lift.

Assume now that the set X is open and the vector field α is quasiconformal. Since $\beta = f^*\alpha - v/f'$ where v/f' is holomorphic,

$$\bar\partial\beta = \bar\partial(f^*(\alpha|Y)) = f^*\bar\partial(\alpha|Y),$$

where the former pullback acts on vector fields while the latter one acts on Beltrami differentials. Hence

$$\|\bar\partial\beta\|_\infty = \|f^*\bar\partial(\alpha|Y)\|_\infty = \|\bar\partial(\alpha|Y)\|_\infty,$$

i.e., *lifts preserve the qc norm of vector fields.*

2.6.3. Variational formula. Let us consider now the iteration operator $S_n : f \mapsto f^n$ acting between some spaces of (real or complex) analytic functions. Linearizing the expression $(f+\varepsilon v)^n$, we obtain (by induction) the following formula for the differential of S_n:

(2.8)

$$v^n \equiv DS_n(f)v = Df^{n-1} \circ f \sum_{k=0}^{n-1} \frac{v \circ f^k}{Df^k \circ f} = Df^n \sum_{k=0}^{n-1} (f^k)^* \left(\frac{v}{f'}\right).$$

(Though f is implicit in the notation v^n, it should not lead to a confusion.)

Note that if f_t is a one-parameter family of analytic maps such that

$$\left.\frac{d}{dt} f_t \right|_{t=0} = v$$

then

$$\left.\frac{d}{dt} f_t^n \right|_{t=0} = v^n. \tag{2.9}$$

Applying to (2.6) the iterates of f^* and summing up, we see that if α is equivariant with respect to (f, v) on $\cup_{k=0}^{n-1} f^k(X)$, then it is equivariant with respect to (f^n, v^n) on X:

$$(f^n)^*\alpha - \alpha = \sum_{k=0}^{n-1} (f^k)^*(f^*\alpha - \alpha) = \sum_{k=0}^{n-1} (f^k)^* \left(\frac{v}{f'}\right) = \frac{v^n}{Df^n}. \tag{2.10}$$

(Another way of seeing it is to linearize the "iteration" of the diagram (2.7).)

The variational formula makes it clear that if α is a bounded vector field equivariant on $\mathrm{orb}_f(x)$ and $Df^k(x) \to \infty$ then (using (2.8) and (2.10)) we have

$$\alpha(x) = -\sum_{j=0}^{\infty} \frac{v \circ f^j}{Df^{j+1}}.$$

Note finally that if β is obtained from α by n consecutive lifts by (f, v), then β is the lift of α by (f^n, v^n).

2.7. Markov maps and expanding Cantor sets.

Definition 2.1. *Consider two open sets $U \Subset \tilde{U}$ and a smooth map $f : \tilde{U} \to \mathbb{C}$. The map $f : U \to \mathbb{C}$ is called Markov if:*

(i) U is the union of finitely many Jordan disks U_i with piecewise smooth boundary and disjoint closures;
(ii) the restrictions $f|\overline{U_i}$ are diffeomorphism onto the image;
(iii) for any i and k, the curves $f(\partial U_k)$ and $f(\partial U_i)$ are either disjoint or coincide;
(iv) for any i, j, the image $f(U_i)$ is either disjoint from U_j or contains U_j (Markov property).

A Markov map will be called *strictly* Markov if the Markov property (iv) is strengthened to the strict Markov property: $fU_i \Supset U_j$.

The "Julia set" of a Markov map is defined as

$$K(f) = \{z : f^n z \in U, \ n = 0, 1, 2, \dots\}.$$

The map f and its Julia set are called *expanding* if $K(f)$ is compact and there exist constants $C > 0$ and $\rho > 1$ such that

$$|Df^n(z)v| \geq C\rho^n|v|, \quad z \in K(f), \ v \in T_z U, \ n = 0, 1, 2, \dots.$$

We will be mostly concerned with holomorphic (on $\tilde{U}$) Markov maps, though we will sometimes encounter a more general situation. It follows from the Schwarz Lemma that the Julia set of a holomorphic strictly Markov map is an expanding Cantor set.

Recall that $\mathcal{B}_V$ stands for the Banach space of holomorphic functions $f : V \to \mathbb{C}$ which are continuous up to the boundary.

Proposition 2.11. *Let $f : U \to \mathbb{C}$ be a holomorphic strictly Markov map as above. Let V be an open set such that $U \Subset V \Subset \tilde{U}$. Then there exists a holomorphic motion $h_g : \mathbb{C} \to \mathbb{C}$ over some neighborhood $\mathcal{V} \subset \mathcal{B}_V$ of f conjugating $f : U \to \mathbb{C}$ to $g : U_g \to \mathbb{C}$, where $U_g = h_g(U)$.*

Proof. We will give a proof of this well-known statement which illustrates a simplest version of the so called "pull-back argument". By property (iii) of Definition 2.1, the image $\Gamma = f(\partial U)$ is a 1-cycle, i.e., a finite union of disjoint Jordan curves. Let $U^n = f^{-n}U$. Since the restrictions $f|U_i$ are univalent, each U^n is a finite union of Jordan disks with disjoint closures. Moreover,

$$U \equiv U^0 \Supset U^1 \Supset U^2 \Supset \dots.$$

Let $\gamma = \partial U$.

Let $\gamma_g = g^{-1}\Gamma$. If g is sufficiently close to f in $\mathcal{B}_V$, then γ_g is a 1-cycle moving holomorphically with g, where the motion $h_g^0 : \gamma \to \gamma_g$ is defined by the requirement: $g(h_g^0(z)) = f(z)$, $z \in \gamma$. Moreover, γ_g bounds some domain U_g and the map $g : U_g \to \mathbb{C}$ is a strictly Markov map.

Extend h_g^0 to Γ as the identity. By the λ-lemma, this motion further extends to a holomorphic motion of the whole plane over some neighborhood $\mathcal{V} \subset \mathcal{B}_V$. Let us keep the same notation h_g^0 for the extension. By definition, h_g^0 conjugates $f : \gamma \to \Gamma$ to $g : \gamma_g \to \Gamma$.

Let $U_g^n = g^{-n}U$.

Since the restrictions $f|U_i$ are univalent, we can lift h_g^0 to U as follows:

$$g \circ h_g^1 \big| U = h_g^0 \circ f \big| U.$$

Since h_g^0 is equivariant on γ, it matches with h_g^1 on γ. Thus, we can let $h_g^1 = h_g^0$ in $\mathbb{C} \setminus U$. We obtain a holomorphic motion of the whole plane which conjugates $f : U \setminus U^1 \to \mathbb{C}$ to $g : U_g \setminus U_g^1 \to \mathbb{C}$.

Similarly, pulling this motion back to U^2, we obtain a motion h_g^2 conjugating $f : U \setminus U^2 \to \mathbb{C}$ to $g : U_g \setminus U_g^2 \to \mathbb{C}$, etc. Since the motions h_g^l, $l \geq n$, coincide with h_g^n on $\mathbb{C} \setminus U^n$, at the end we obtain a motion $h_g : \mathbb{C} \setminus K(f) \to \mathbb{C} \setminus K(g)$ conjugating $f : U \setminus K(f) \to \mathbb{C}$ to $g : U \setminus K(g) \to \mathbb{C}$. By the λ-lemma, this motion admits an extension through the Julia sets. Since the Julia sets are nowhere dense, this extension conjugates $f : U \to \mathbb{C}$ to $g : U_g \to \mathbb{C}$. □

Remark 2.5. The same result is valid for (non-strictly) Markov maps for which one can construct a holomorphic motion h_g^0 of $\gamma \cup \Gamma$, provided its Julia set has empty interior. The rest of the argument carries to this more general situation without changes.

Given a smooth map $f : U \to \mathbb{C}$ and a Riemannian metric ν on $\mathbb{C}$, we say that ν is *expanded* by f on an invariant set $Q \subset U$ if there is a constant $\lambda > 1$ such that $\|Df(x)v\|_\nu \geq \lambda \|v\|_\nu$ for all $x \in Q$ and $v \in T_x\mathbb{C}$. The following useful fact is well-known:

Lemma 2.12. *Assume that a smooth map $f : U \to \mathbb{C}$ is expanding on a compact invariant set $Q \subset U$. Then there exists a Riemannian metric ν which is expanded by f on Q. If f is holomorphic then the metric ν can be selected to be conformal.*

2.8. Unimodal maps. We refer to the book of de Melo & van Strien [MS] for the general background in one-dimensional dynamics.

A smooth map $f : I \to I$ of the interval $I = [-1, 1]$ is called *unimodal* if it has a single critical point and this point is an extremum. We always assume that the critical point is located at the origin. Let $\mathcal{U}^3$ be the space of C^3 unimodal maps $f : I \to I$ with quadratic critical point, which are even (symmetric), that is $f(x) = f(-x)$, $Df(0) = 0$ and $D^2(f(0)) \neq 0$. We normalize the maps so that -1 is a fixed point and $f(1) = -1$. We endow $\mathcal{U}^3$ with the C^3 topology. If $Df(-1) < 1$, either the dynamics is trivial (-1 is the global attractor) or the map has a proper unimodal restriction. For this reason we will assume further that $Df(-1) \geq 1$.

The symmetry assumption above was introduced to simplify the proofs, but is not essential. In fact the general case reduces to the symmetric case, as we show in Appendix C.

We let $c_n = f^n(0)$, $n = 0, 1, 2, \dots$.

Basic examples of unimodal maps are given by quadratic maps

$$q_\tau : I \to I, \quad q_\tau(x) = \tau - 1 - \tau x^2, \tag{2.11}$$

where $\tau \in [1/2, 2]$ is a real parameter.

A map $f \in \mathcal{U}^3$ is *quasiquadratic* if any nearby map $g \in \mathcal{U}^3$ is topologically conjugate to some quadratic map. We denote by $\mathcal{U} \subset \mathcal{U}^3$ the space of quasiquadratic maps. By the theory of Milnor-Thurston [MT] and Guckenheimer [G], a map $f \in \mathcal{U}^3$ with negative Schwarzian derivative is quasiquadratic, so the quadratic family is contained in $\mathcal{U}$.

Let us consider a periodic point q of some period n and its cycle $\overline{q} = \{f^k(q)\}_{k=0}^{n-1}$. Let $\lambda = (Df^n)(q)$ be its *multiplier*. The point q and its cycle $\overline{q}$ are called *attracting, parabolic*, or *repelling* depending on whether $|\lambda| < 1$, $|\lambda| = 1$, or $|\lambda| > 1$. A periodic point is called *superattracting* if $\lambda = 0$, which means that the cycle of this point contains 0.

The *basin of attraction* $D(\overline{q})$ of an attracting cycle $\overline{q}$ is defined as $\{x \in I : f^n(x) \to \overline{q}\}$. The basin of attraction $D(\overline{q})$ of a parabolic cycle $\overline{q}$ is defined similarly, except that the orbits landing at the cycle itself are not considered to be in the basin (this makes the basin open).

A quasiquadratic map f is called *hyperbolic* or *regular* if it has an attracting cycle $\overline{q}$. In this case the orbit of the critical point 0 converges to $\overline{q}$, hence a quasiquadratic map can have at most one attracting cycle [Fa,Si]. Moreover, if it has one then almost all orbits converge to this cycle (this follows from a result by Guckenheimer and Mañé, see [M]).

A quasiquadratic map f is called *parabolic* if it has a parabolic cycle $\overline{q}$. Similarly to the hyperbolic case, in the parabolic case the critical point 0 belongs to the basin $D(\overline{q})$ and this basin has full Lebesgue measure in I. Thus, a quasiquadratic map can have at most one parabolic cycle (and the map cannot be simultaneously hyperbolic and parabolic).

Remark 2.6. Let us denote by $D^0(\overline{q})$ the union of connected components of $D(\overline{q})$ whose closure intersects $\overline{q}$ (the *immediate basin* of $\overline{q}$).

There is a simple criteria to decide if a unimodal map $f \in \mathcal{U}^3$ is topologically equivalent to a quadratic map: any non-repelling periodic orbit must contain the critical point in its immediate basin. This is a consequence of Milnor-Thurston Theory and non-existence of wandering intervals for maps in $\mathcal{U}^3$ (see [MS], Theorem 6.2, p. 156 and Theorem 6.4, p. 162).

To prove that a map is quasiquadratic, we need a robust version of the above criteria given in the following:

Lemma 2.13. *If $f \in \mathcal{U}^3$ is topologically conjugate to a quadratic map and has no parabolic periodic orbit then f is quasiquadratic.*

Proof. We only have to show that the criteria of Remark 2.6 is valid for all g in a neighborhood of f.

By the result of Guckenheimer and Mañé (see also Proposition 2.15 below), if J is a neighborhood of the critical point then $f|I \setminus J$ is expanding. This property is persistent, so we conclude that for any J there is a neighborhood $\mathcal{V} \subset \mathcal{U}^3$ of f such that for $g \in \mathcal{V}$ all periodic orbits which do not intersect J are repelling. This implies the result if f has an attracting periodic orbit, so let us assume that all periodic orbits are repelling. In this case, the (automatically repelling) periodic cycles of f accumulate on 0.[3]

By Theorem A of [K2], there exists a symmetric interval J_f containing the critical point such that the first return map to $f(J_f)$ has negative Schwarzian derivative. Shrinking the interval J_f we may suppose that $f(J_f) = [p_f, f(0)]$ where p_f is a repelling periodic point and $\overline{p}_f \cap f(J_f) = \{p_f\}$ (this condition implies the nice condition, to be defined in the next section).

The interval J_f has a continuation J_g for $g \in \mathcal{U}^3$ near f such that $g(\partial J_g) = p_g$ is the continuation of p_f, and by the proof in [K2], it is clear that there exists a neighborhood $\mathcal{V} \subset \mathcal{U}^3$ of f such that if $g \in \mathcal{V}$ then the first return map to $g(J_g)$ still has negative Schwarzian derivative. By the argument of Singer [Si], it follows that any non-repelling periodic orbit for g which intersects J_g must contain the critical point in its immediate basin. Shrinking $\mathcal{V}$ we may assume that there exists an interval J which is contained in $\bigcap_{g \in \mathcal{V}} J_g$. Shrinking $\mathcal{V}$ again, we conclude (by the previous argument) that all periodic orbits contained in $I \setminus J$ are repelling, and the result follows. □

In what follows, all unimodal maps under consideration will be assumed to be quasiquadratic.

If the fixed point -1 is repelling, then f has a unique fixed point in the interior of I. This point will always be denoted by α. If the α-point is repelling, then it is orientation reversing, that is, $f'(\alpha) < 0$.

A unimodal map is called *preperiodic* if the orbit of the the critical point 0 lands at a repelling cycle. Simplest examples of preperiodic maps are provided by *Ulam-Neumann* (or *Chebyshev*) maps which are defined by the condition $f(0) = 1$ (so that the second iterate of 0 lands at the fixed point -1, which is automatically repelling in this case). For instance, the quadratic map q_2 is Chebyshev.

A unimodal map f is *renormalizable* if there exists an interval J containing the critical point and an integer $n \geq 2$ such that $f^n(J) \subset J$ and the intervals $J, f(J), \ldots, f^{n-1}(J)$ have pairwise disjoint interior. The smallest such n is called the *renormalization period*.

Let n be the renormalization period and $J \ni 0$ be the maximal periodic interval of period n as above. This interval is bounded by a periodic point p

[3] To see this, first notice that by the No Wandering Intervals Theorem, the preimages of the critical point are dense on I. By the Intermediate Value Theorem, if $n \geq 1$ and $f^n(x) = 0$ then there exists $p \in [-x, x]$ with $f^n(p) = p$.

(of period n if $n \geq 3$ and period 1 if $n = 2$) and the symmetric point. The restriction $f^n|J$ is called the *pre-renormalization* of f.

Let $A : I \to J$ be the affine scaling mapping -1 to p. Then the map

$$R(f) \equiv A^{-1} \circ f^n \circ A : I \to I$$

is called the *renormalization* of f.

Remark 2.7. The renormalization of a quadratic map has negative Schwarzian derivative and hence is quasiquadratic. Since the renormalization operator is an open continuous map in the topology of $\mathcal{U}^3$ and respects topological equivalence, it acts on the space of quasiquadratic maps.

If the renormalization $R(f)$ is in turn renormalizable, then the map f is *twice renormalizable*. In this way we can define *n times renormalizable* maps for any $n = 1, 2, \ldots,$ including $n = \infty$.

A hyperbolic or parabolic map with an attracting/parabolic cycle of period $n > 1$ is renormalizable but at most finitely many times (the last renormalization of this map has an attracting/parabolic fixed point). Preperiodic maps are at most finitely renormalizable. In fact, infinitely renormalizable maps have a recurrent critical point.

A unimodal map is called *Yoccoz* if it is not infinitely renormalizable and has all periodic orbits repelling. A Yoccoz map with a non-recurrent critical point is called *Misiurewicz*.

2.9. Spaces of unimodal maps. Let $\mathcal{UN}$ stand for the set of Ulam-Neumann maps $f \in \mathcal{U}^3$. It is a domain in an affine hyperplane of the Banach space C^3.

Unimodal maps $f \in \mathcal{U}^3$ with negative Schwarzian derivative form an open subset of $\mathcal{U}$ containing the quadratic family $\{q_\tau : \tau \in [1/2, 2]\}$. We will be interested in the intersection of $\mathcal{U}$ with some Banach spaces of real analytic unimodal maps.

Let $a > 0$, and let $\mathcal{E}_a \subset \mathcal{B}_{\Omega_a}$ be the complex Banach space of holomorphic maps $v : \Omega_a \to \mathbb{C}$ continuous up to the boundary which are 0-symmetric (that is, $v(z) = v(-z)$) and such that $v(-1) = v(1) = 0$, endowed with the sup-norm $\|v\|_a = \|v\|_\infty$. It contains the real Banach space $\mathcal{E}_a^{\mathbb{R}}$ of "real maps" v, i.e, holomorphic maps symmetric with respect to the real line: $v(\bar{z}) = \overline{v(z)}$.

The complex affine subspace $q_2 + \mathcal{E}_a$ will be denoted as $\mathcal{A}_a$.

If $f \in \mathcal{A}_a$, we denote the postcritical set $\overline{\text{orb}(f(0))}$ by O_f.

Let $\mathcal{U}_a = \mathcal{U} \cap \mathcal{A}_a$. Note that $\mathcal{U}_a$ is the union of an open set in the affine subspace $\mathcal{A}_a^{\mathbb{R}} = q_2 + \mathcal{E}_a^{\mathbb{R}}$ and a codimension-one space of Ulam-Neumann maps.

2.10. Real puzzle. Let us start with some combinatorial preparation, which will be used throughout this paper. *We assume that both fixed points of f, -1 and α, are repelling.*

A symmetric interval J containing 0 is called *nice* in the sense of Martens if the orbits of its boundary points do not intersect int J.

The *real Yoccoz puzzle* $\mathcal{P}^{\mathbb{R}}$ for a quasiquadratic map f is a collection of closed intervals P_i^n, $n \in \mathbb{N} \cup \{0\}$, called *real Yoccoz puzzle pieces* such that $P_0^0 = [-\alpha, \alpha]$ and the P_i^n, $n > 0$, are the components of $f^{-n} P_0^0$. Intervals of the Yoccoz puzzle containing the critical point are called *critical* and are labeled as P_0^n. Any critical Yoccoz puzzle piece P_0^n is nice. Moreover,

- any non-critical Yoccoz puzzle piece P_i^n is diffeomorphically mapped onto some other puzzle piece $P_{k(i)}^{n-1}$;
- any critical Yoccoz puzzle piece P_0^n, $n > 0$, is folded into the Yoccoz puzzle piece P_1^{n-1} containing the critical value c_1 in such a way that $f(\partial P_0^n) \subset \partial P_1^{n-1}$.

Take now a critical Yoccoz puzzle piece $J_0 \in \mathcal{P}^{\mathbb{R}}$ and consider the first landing map L to it. The domain of this map consists of a family $\mathcal{J}$ of disjoint Yoccoz puzzle pieces $J_i \in \mathcal{P}^{\mathbb{R}}$, $i \in \mathbb{N}$, satisfying the following properties (see [Ma]): Any J_i, $i > 0$, is diffeomorphically mapped by f onto some other interval $J_{k(i)} \in \mathcal{J}$, and there exists $n_i \in \mathbb{N}$ such that the branch $L|J_i = f^{n_i}|J_i$ diffeomorphically maps J_i onto J_0.

More generally, a similar description for the landing map applies to any nice interval J_0. In this case, we will loosely say that the collection $\{J_i\}$ (of pairwise disjoint intervals) is the real puzzle associated to J_0, and we will call the J_i real puzzle pieces.

The following statement explains the role of nice intervals.

Theorem 2.14 **([Ma]).** *Let f be a quasiquadratic map. Let us consider two symmetric intervals $J_0 \subset T_0$ such that the orbit of $f(\partial J_0)$ does not return to* int T_0 *(in particular, J_0 is nice). Let $f^{n_i} : J_i \to J_0$ be a branch of the landing map L to J_0. Then f^{n_i} diffeomorphically maps some interval $T_i \supset J_i$ onto T_0. Moreover, the distortion of $L|J_i$ is $O(|J_0|/|T_0|)$.*

The following is a consequence of the result by Guckenheimer and Mañé.

Proposition 2.15 (see [MS], Corollary 1, p. 248)**.** *Let $f : I \to I$ be a quasiquadratic map with all periodic orbits repelling and let $L : \cup J_i \to J_0$ be the first landing map to a nice interval J_0. Then the complement $Q = I \setminus \cup$* int *J_i is an expanding set.*

Remark 2.8. The above expanding set is usually a Cantor set (for instance, if the boundary points of J_0 are not periodic). If ∂J_0 is contained in the preorbit of a periodic point q, then this Cantor set admits a real Markov partition $\{M_j^{\mathbb{R}}\}$ constructed with the help of finitely many preimages of q (see the proof of Lemma 7.10 for an explicit construction of a real Markov partition). If f is analytic, this partition can be further refined and thickened to become a (complex) strictly Markov partition $\{M_j\}$.

For a given unimodal map f, let us construct the *principal nest*

(2.12) $$[-\alpha, \alpha] = T^0 \supset T^1 \supset T^2 \supset \dots$$

of real puzzle pieces. Consider the first return map $g_1 : \cup T_i^1 \to T^0$ to the interval T^0. It is defined on the union of disjoint closed intervals T_i^1 contained in T^0. Assuming the critical point 0 returns to T^0, one of these intervals contains the critical point. Call it $T^1 \equiv T_0^1$. Otherwise we stop and the principal nest consists only of T_0.

We define inductively the first return maps

(2.13) $$g_n : \cup T_i^n \to T^{n-1}$$

to T^{n-1}, and we let $T^n \equiv T_0^n$ be the critical interval of the domain of g_n. If there is n such that the critical point never returns to T^n, the principal nest consists of the intervals T^0, ..., T^n. In this case f is necessarily non-renormalizable and has a non-recurrent critical point. Otherwise the principal nest is an infinite sequence of intervals. If f is non-renormalizable, the intervals T^n shrink to zero. Otherwise they shrink to the domain of the pre-renormalization of f (see [L3]).

The *scaling factors* of f are defined as follows: $\lambda_n = |T^n|/|T^{n-1}|$.

We will now reformulate Theorem 2.14 in a way convenient for further references.

Theorem 2.16. *Let f be a non-renormalizable quasiquadratic map with recurrent critical point such that $f(0) \neq 0$. Let $J_0 = T^{n+1}$, $n \in \mathbb{N}$, and let $L : \cup J_i \to J_0$ be the corresponding first landing map. Let $L|J_i = f^{n_i}$ and $\tilde{J}_i$ be the monotonicity interval of f^{n_i} containing J_i. Then $f^{n_i}(\tilde{J}_i) \supset T^n$ and hence the distortion of $L|J_i$ is $O(\lambda_{n+1})$.*

It is important to distinguish two combinatorial possibilities for the returns of the critical point: central and non-central returns. The return to level $n-1$ (and the level $n-1$ itself) is called *central* if $g_n(0) \in T^n$. Let $\{n_k - 1\}$ be the sequence of non-central levels in the principal nest. (Under the assumption that the principal nest consists of infinitely many intervals, the sequence $\{n_k - 1\}$ is infinite if and only if f is non-renormalizable.)

The following result will provide us with a big space around certain intervals of the principal nest.

Theorem 2.17 ([L3]). *Let f be a non-renormalizable quasiquadratic map with non-trivially recurrent critical point (that is, $f(0) \neq 0$). Then there exist constants $C > 0$ and $\rho \in (0, 1)$ such that*

$$\lambda_{n_k+1} \leq C\rho^k.$$

Combining the last two theorems, we see that the branches of the first landing map $L : \cup J_i \to J_0$ become almost linear if J_0 is selected sufficiently deep in the principal nest.

Remark 2.9. Theorems 2.14 and 2.17 were proven in the quoted papers for quasiquadratic maps with negative Schwarzian derivative. The general case was reduced to this one by Kozlovski [K2], who has proven that for sufficiently big n, the first return map to T^{n-1} has negative Schwarzian derivative, and this is enough to obtain good properties for g_n.

Lemma 2.18. *Let us use the notations of Theorem 2.16. There is a function $C(\lambda) \to \infty$ as $\lambda \to 0$ with the following property. For any n, there exist topological disks $U_i \supset J_i$ such that $\operatorname{mod}(U_0 \setminus J_0) = C(\lambda_{n+1})$ and the first landing map $L|_{J_i}$ extends to a univalent map from U_i onto U_0.*

Proof. Fix some $\phi \in (0, \pi)$ and let U_0 be the hyperbolic neighborhood $D_\phi(T^n)$ of T^n (see §2.2). □

2.11. Stochastic maps. A unimodal map $f : I \to I$ is called *stochastic* if it has an invariant measure μ which is absolutely continuous with respect to the Lebesgue measure on I (such a measure will be abbreviated as a.c.i.m.).

Existence of an a.c.i.m. is related to the rate of expansion along the orbit of the critical value c_1. It was shown by Collet & Eckmann [CE] (with a complement by Nowicki [N]) that the map is stochastic if the expansion rate is exponential:

$$Df^n(c_1) \geq C\lambda^n, \quad C > 0,\ \lambda > 1.$$

This criterion was improved by Nowicki & van Strien [NS] who replaced the exponential rate with the summability condition:

$$\sum |Df^n(c_1)|^{-1/2} < \infty. \tag{2.14}$$

Since the strongest contraction occurs near the critical point 0, one should expect that the rate of *expansion* along the critical orbit is related to the rate of *recurrence* of the critical orbit. Here is an efficient criterion of this kind:

Theorem 2.19 (Martens & Nowicki [MN]). *Let f be a non-renormalizable quasiquadratic map and let λ_n be its scaling factors. If*

$$\sum \sqrt{\lambda_n} < \infty$$

then f is stochastic (in fact, f satisfies (2.14)).

This result together with Theorem 2.17 implies the following combinatorial criterion:

Theorem 2.20. *Let f be a non-renormalizable quasiquadratic map. If all but finitely many levels in its principal nest are non-central then f is stochastic.*

2.12. Quadratic-like maps. A holomorphic map $f : U \to U'$ is called *quadratic-like* if it is a double branched covering between topological disks U, U' such that $U \Subset U'$. It has a single critical point which is assumed to be located at the origin 0, unless otherwise is stated. We will also make the following technical assumptions:

- U is symmetric with respect to the origin and f is even, i.e., $f(-z) = f(z)$.
- The domains U and U' are bounded by piecewise smooth curves.

Quadratic-like maps are considered up to affine conjugacy. We will say that a quadratic-like map f is *normalized* at 0 if it has the following expansion at 0:

$$f(z) = c + z^2 + O(z^3). \tag{2.15}$$

Note that this normalization is not the same as the normalization we use for unimodal maps.

A quadratic-like map is called *real* if the domains U and U' are $\mathbb{R}$-symmetric and f preserves the real line: $f(U \cap \mathbb{R}) \subset \mathbb{R}$.

The *filled Julia set* of a quadratic-like map is defined as the set of non-escaping point:

$$K(f) = \{z : f^n z \in U,\ n = 0, 1 \dots\}.$$

Its boundary is called the *Julia set*, $J(f) = \partial K(f)$. The sets $K(f)$ and $J(f)$ are connected if and only if the critical point itself is non-escaping: $0 \in K(f)$. Otherwise these sets are Cantor.

If f is a real quadratic-like map with connected Julia set, then $K(f) \cap \mathbb{R}$ is an interval $[-\beta, \beta]$, where β is a fixed point of f. Since f is considered up to affine conjugacy, we can normalize it so that $\beta = -1$.

The *fundamental annulus* of a quadratic-like map $f : U \to U'$ is the annulus between the domain and the range of f, $A = U' \setminus \overline{U}$.

Two quadratic-like maps f and g are called *topologically equivalent* if they are topologically conjugate in some neighborhoods of their Julia sets. They are called *hybrid equivalent* if they are conjugate by a qc map h with $\bar\partial h = 0$ a.e. on $K(f)$. Note that in the hyperbolic case (when f has an attracting cycle), the hybrid class of f consists of topologically equivalent quadratic-like maps with the same multiplier of the attracting cycle (Douady & Hubbard).

By the Douady-Hubbard Straightening Theorem [DH1], every hybrid class with connected Julia set intersects the quadratic family

$$\{P_c : z \mapsto z^2 + c\}_{c\in\mathbb{C}}$$

at a single point c of the Mandelbrot set. (Recall that the Mandelbrot set is defined as the set of parameter values $c \in \mathbb{C}$ for which the Julia set $J(P_c)$ is connected.) It follows that given a real quadratic-like map f without parabolic points, the restriction $f|I$ is quasiquadratic (this is still true when parabolic points are allowed, but requires an extra argument).

Theorem 2.21 ([L4,GS2]). *Let us consider two real non-hyperbolic quadratic-like maps f and g with connected Julia set. If f and g are topologically equivalent then they are hybrid equivalent. Thus, there exists a unique quadratic polynomial q_τ, $\tau \in [1/2, 2]$, in the topological class of f.*

Let us now consider the projectivized *real* tangent bundle $\mathcal{L} \to \mathbb{C}$ over the plane $\mathbb{C}$. An *invariant line field* on the Julia set $J(f)$ is a measurable section $X \to \mathcal{L}$ invariant under the action of f, where $X \subset J(f)$ is a measurable invariant set of positive (plane) Lebesgue measure. In other words, it is a measurable function $\theta : X \to \mathbb{R} \bmod \pi\mathbb{Z}$ such that for a.e. $z \in X$,

$$\theta(fz) = \theta(z) + \arg Df(z) \bmod \pi\mathbb{Z}.$$

Associated to an invariant line field on the Julia set, there is family of f-invariant Beltrami differentials

$$\mu_\lambda(z) = \lambda e^{2i\theta(z)}, \quad |\lambda| < 1,$$

on X (extended by 0 to the whole complex plane). Hence any invariant line field generates a Beltrami disk $f_\lambda = h_\lambda \circ f \circ h_\lambda^{-1}$, $|\lambda| < 1$, of the map f, where h_λ is the solution of the Beltrami equation $\bar\partial h_\lambda = \mu_\lambda \partial h_\lambda$. This deformation is non-trivial on the Julia set. Thus, Theorem 2.21 yields the following result due to Yoccoz (in the finitely renormalizable case, see [H]) and McMullen [Mc1]:

Theorem 2.22. *A real quadratic-like map with connected Julia set does not have invariant line fields on the Julia set.*

2.13. Real hybrid classes. Theorem 2.21 can be carried to the class of real analytic maps:

Theorem 2.23 (see §B.3). *If two non-hyperbolic real analytic unimodal maps*
$f, g \in \mathcal{U}_a$ are topologically conjugate then they are quasisymmetrically conjugate.

(For a a stronger statement in the Yoccoz case, see Theorem B.1.)

The above discussion motivates the following definition. Two quasi-quadratic maps f and g of class $\mathcal{U}_a$ are called *(real) hybrid equivalent* if they are topologically equivalent and (in the hyperbolic case) their attracting cycles have the same multiplier. (Because of Theorem 2.21, two quasiquadratic maps which have quadratic-like extensions belong to the same real hybrid class if and only if they belong to the same hybrid class as quadratic-like maps.)

We denote by $\mathcal{H}_f^{\mathbb{R}} \equiv \mathcal{H}_{f,a}^{\mathbb{R}} \subset \mathcal{U}_a$ the real hybrid class of f (for simplicity we often omit a in the notation). By Theorem 2.21, each hybrid class intersects the quadratic family $\{q_\tau\}_{\tau\in[1/2,2]}$ at a single point. Hence we can consider the *straightening* map $\chi : \mathcal{U} \to [1/2, 2]$, which associates to a quasiquadratic map $f \in \mathcal{U}$ the hybrid equivalent quadratic polynomial.

Remark 2.10. By the Milnor-Thurston theory, two quasiquadratic maps belong to the same hybrid class if and only if they have the same *kneading sequence* and the multipliers of any non-repelling cycles are the same for both maps. This is a consequence of non-existence of wandering intervals for quadratic maps, and can be easily obtained by the argument of [MS], Corollary of Theorem 6.2, p. 157.

This has the following nice consequence for the action of renormalization in the space of quasiquadratic maps. Let f and g be renormalizable of the same type. It is clear that they have the same kneading sequence if and only if their renormalizations do, so f and g belong to the same hybrid class if and only if their renormalizations do. So each renormalization operator (corresponding to some renormalization combinatorics) acts on the hybrid classes injectively.

2.14. A priori bounds. A relation between general real analytic and quadratic-like maps is provided by the renormalization: an appropriate renormalization of a real analytic map is quadratic-like. This statement is usually encoded as *a priori* bounds:

Theorem 2.24 ([LS1,LY]). *Let f be an infinitely renormalizable real analytic map of class $\mathcal{U}_a$. Then some renormalization $R^n f$ admits a quadratic-like extension to the complex plane.*

Note that this property is robust: if it holds for some map $f_0 \in \mathcal{U}_a$ then it also holds, with the same n, for nearby maps $f \in \mathcal{U}_a$.

2.15. Parameter geometry of the quadratic family. The real quadratic family $\{q_\tau\}$ can be partitioned according to the combinatorics of the first return maps [L5]. According to this construction, the set $\mathcal{N}$ of non-renormalizable quadratic maps with both fixed points repelling[4] is covered with countable unions $\mathcal{D}^n$ of intervals Δ_i^n each of which is endowed with a family of first return maps $g_{\tau,l} : \cup T^l_{\tau,j} \to T^{l-1}_\tau$ of a certain level $l = l(n, i)$. Each interval Δ_i^n contains a subinterval Π_i^n corresponding to the central return of the critical point: $g_{\tau,l}(0) \in T^l_{\tau,0}$.

Theorem 2.25 ([L5]). *There exist constants $C > 0$ and $\rho \in (0, 1)$ such that*

$$\frac{|\Pi_i^n|}{|\Delta_i^n|} \le C\rho^n. \tag{2.16}$$

Thus, the probability of the central return on level n is exponentially small in n. By the Borel-Cantelli Lemma, the probability of infinitely many central returns is equal to 0. By Theorem 2.20, almost all quadratic maps q_τ with $\tau \in \mathcal{N}$ are stochastic.

[4] $\mathcal{N}$ will also denote the corresponding set of parameter values $\tau \in [1/2, 2]$, and the same convention applies to other sets which appear below: $\mathcal{F}$, $\mathcal{I}$, etc.

By means of the renormalization, this result extends to quadratic maps which are not infinitely renormalizable:

Theorem 2.26 ([L5]). *Almost all Yoccoz quadratic maps q_τ, $\tau \in [1/2, 2]$, are stochastic maps whose last renormalization satisfies the Martens-Nowicki criterion for existence of a.c.i.m.*

Recently, Avila and Moreira (see [AM1]) have strengthened the result above by showing that almost every Yoccoz map satisfies the Collet-Eckmann condition.

2.16. Regular or Stochastic Theorem. Let $\mathcal{I}$ stand for the set of infinitely renormalizable quadratic maps q_τ, $\tau \in [1/2, 2]$.

We say that a set $X \subset \mathbb{R}$ has *definite gaps everywhere* if there exists a $C > 0$ such that for any $x \in X$ and any $\varepsilon > 0$ there exists an interval $J \subset (x - \varepsilon, x + \varepsilon) \setminus X$ such that

$$C^{-1} \operatorname{dist}(x, J) \leq |J| \leq C \operatorname{dist}(x, J).$$

By the Lebesgue Density Theorem, such a set has zero measure. But unlike the measure zero property, the property to have definite gaps everywhere is preserved by quasisymmetric maps.

Theorem 2.27 ([L7], §4.1). *The set $\mathcal{I}$ has definite gaps everywhere and hence has zero Lebesgue measure in the parameter interval* $[1/2, 2]$.

Putting together the last two theorems, we obtain:

Theorem 2.28. *Almost any real quadratic map q_τ is either regular or stochastic.*

2.17. Invariant line fields and equivariance. In this work, the non-existence of invariant line fields on the set of non-escaping points of a holomorphic dynamical system will be a recurrent hypothesis. It will allow us to estimate the global dilatation of equivariant qc maps or vector fields in terms of their "external dilatation". The following two Lemmas will clarify this relation.

Lemma 2.29. *Let $f : U \to \mathbb{C}$, $\tilde{f} : \tilde{U} \to \mathbb{C}$ be two non-constant holomorphic maps such that there exists a qc map $h : \mathbb{C} \to \mathbb{C}$, $h(U) = \tilde{U}$, equivariant on U. Let $K(f)$ be the set of non-escaping points of f. Then*

$$\operatorname{Dil}(h) = \max\{\operatorname{Dil}(h|\mathbb{C} \setminus U), \operatorname{Dil}(h|K(f))\}.$$

Furthermore, if $\operatorname{Dil}(h|K(f)) > 1$ *then f has an invariant line field on $K(f)$.*

Proof. Equivariance implies that $f^*(\mu_h) = \mu_h$, so

$$\operatorname{Dil}(h|\mathbb{C} \setminus K(f)) = \max \operatorname{Dil}(h|f^{-n}(\mathbb{C} \setminus U)) = \operatorname{Dil}(h|\mathbb{C} \setminus U).$$

If $\operatorname{Dil}(h|K(f)) > 1$, μ_h gives an invariant line field on $K(f)$. □

The same argument gives the following infinitesimal version:

Lemma 2.30. *Let $f : U \to \mathbb{C}$ be a non-constant holomorphic map and let v be a holomorphic vector field on U such that there exists a qc vector field α on the plane equivariant on U. Let $K(f)$ be the set of non-escaping points. Then $\|\bar{\partial}\alpha\|_\infty = \max\{\|\bar{\partial}\alpha|\mathbb{C} \setminus U\|_\infty, \|\bar{\partial}\alpha|K(f)\|_\infty\}$. Furthermore, if $\|\bar{\partial}\alpha|K(f)\|_\infty > 0$ then f has an invariant line field on $K(f)$.*

2.18. Some generalities about Banach spaces. Let E denote a complex Banach space.

We say that a set $\mathcal{K} \subset E \setminus \{0\}$ is a *cone* if $v \in \mathcal{K}$ implies $\lambda v \in \mathcal{K}$ for all $\lambda \in \mathbb{C} \setminus \{0\}$. We say that a codimension-one subspace F is *transverse* to $\mathcal{K}$ if $F \cap \mathcal{K} = \emptyset$.

Lemma 2.31. *Let F be a codimension-one subspace transverse to an open cone $\mathcal{K}$ and $v \in \mathcal{K}$. There exists $C > 0$ such that if $w \in E \setminus \mathcal{K}$ and $w = w_1 + \lambda v$ with $w_1 \in F$ then $|\lambda| \leq C\|w_1\|_E$.*

For a sequence of subspaces $F_n \subset E$, let

$$\limsup F_n = \mathrm{Lin}\{v \in E |\, \liminf \, \mathrm{dist}_E(v, F_n) = 0\}.$$

Lemma 2.32. *If the F_n are codimension-one subspaces, then* $\limsup F_n$ *is either E or a codimension-one subspace.*

Proof. Let G be any subspace of E of dimension 2. Since each F_n has codimension one, $F_n \cap G$ contains a unitary vector v_n. So there exists a subsequence converging to some unitary vector $v \in \limsup F_n$. Since $\limsup F_n$ is a subspace which intersects any 2-dimensional subspace of E, we conclude that $\limsup F_n$ is either E or a codimension-one subspace. □

3. Results and methods

3.1. Statement of the results. We are now ready to formulate the main results of this paper.

Fix some $a > 0$. Note that the affine space $\mathcal{A}_a$ has a natural involution around its real subspace $\mathcal{A}_a^{\mathbb{R}}$. A subset in $\mathcal{A}_a$ is called $\mathbb{R}$-*symmetric* if it is invariant under this involution.

Theorem A. *Every real hybrid class $\mathcal{H}_f^{\mathbb{R}}$, $f \in \mathcal{U}_a$, is an embedded codimension-one real analytic Banach submanifold of $\mathcal{U}_a$. Furthermore, the hybrid classes laminate a neighborhood of any non-parabolic map $f \in \mathcal{U}_a$. More precisely, any non-parabolic map $f \in \mathcal{U}_a$ has an $\mathbb{R}$-symmetric neighborhood $\mathcal{V}$ in the complex affine space $\mathcal{A}_a$ endowed with a codimension-one $\mathbb{R}$-symmetric holomorphic lamination such that for any $g \in \mathcal{V} \cap \mathcal{U}_a^{\mathbb{R}}$, the intersection of the leaf through g with $\mathcal{A}_a^{\mathbb{R}}$ coincides with $\mathcal{H}_g^{\mathbb{R}} \cap \mathcal{V}$.*

A real analytic one-parameter family $\{f_\lambda\}_{\lambda\in\Lambda}$ of quasiquadratic maps is called *non-trivial* if it is not contained in a single real hybrid class.

Theorem B. *Let* $\{f_\lambda\}_{\lambda\in\Lambda} \subset \mathcal{U}_a$ *be a non-trivial one-parameter real analytic family of quasiquadratic maps. Then for Lebesgue almost all parameter values* $\lambda \in \Lambda$*, the map* f_λ *is either regular or stochastic.*

Theorem C. *Under the circumstances of the previous theorem, there is an open and dense set* $\Lambda_0 \subset \Lambda$ *of parameter values with countable complement such that the straightening map* $\chi(f_t)$ *is quasisymmetric on any compact interval contained in* Λ_0*.*

Remark 3.1. The map conjugating f_t to its straightening does not depend continuously on t (see [NPT], Theorem 3.2, p. 15, or [DH1], Proposition 15, p. 315).

3.2. Ideas of the proofs.

3.2.1. Quadratic-like maps. The entry point for this paper is the theory of quadratic-like maps $g : U \to U'$ described in §2.12. Briefly, the picture is as follows. The space $\mathcal{Q}$ of quadratic-like maps is endowed with a natural complex analytic structure based on families of Banach spaces. The connectedness locus $\mathcal{C}$ of this space is laminated by the hybrid classes $\mathcal{H}_f$ each of which is a codimension-one complex analytic submanifold in $\mathcal{Q}$. By the λ-lemma, this lamination is transversally quasiconformal (but it is not transversally smooth!). The quadratic family $\{P_c : z \mapsto z^2 + c\}_{c\in\mathbb{C}}$ is a global transversal to this lamination.

The tangent ("horizontal") space to the hybrid class $\mathcal{H}_f$ consists of holomorphic vector fields v on U, which admits representation (2.5), $v = \alpha \circ f - f'\alpha$, where α is a qc vector field on U whose $\bar\partial$-derivative vanishes a.e. on the filled Julia set $K(f)$. This equation tells us that v is an "infinitesimal hybrid deformation" of f and α is the corresponding "infinitesimal conjugacy".

A transverse ("vertical") direction to the hybrid class $\mathcal{H}_f$ can be selected as a holomorphic vector field $V^{\mathrm{tr}} \equiv V_f^{\mathrm{tr}}$ which can be represented as (2.5) in $U \setminus K(f)$, where $\alpha(z)/dz$ is a *holomorphic* vector field on $\overline{\mathbb{C}} \setminus K(f)$ vanishing at ∞. This explicit description of a vertical direction exploits essentially the "external structure" of quadratic-like maps (existence of the fundamental annulus $U' \setminus U$). Lack of this structure for general real analytic maps is the source of major difficulties addressed in this paper.

3.2.2. Horizontal space. Theorem 2.23 motivates the following definition. Assume that $f \in \mathcal{U}_a$ is not hyperbolic, and let v be a holomorphic vector field in the neighborhood Ω_a. We say that it is *horizontal* if there is a qc vector field $\alpha(z)$ on $\mathbb{C}$ satisfying (2.5) on orb(0). (This definition is designed in such a way that it will still make sense for complex perturbations of f.)

Clearly, the horizontal vector fields form a linear space, which will be denoted T_f. Moreover, it is easy to see (by the Implicit Function Theorem) that for simple combinatorics, like parabolic or preperiodic, the hybrid class $\mathcal{H}_f^{\mathbb{R}}$ is a real analytic codimension-one submanifold in $\mathcal{U}_a$, whose tangent space coincides with T_f. For general combinatorics, this is still true but requires much finer analysis; it will be outlined below.

3.2.3. Infinitely renormalizable case. As usual in one-dimensional dynamics, the analysis depends on the combinatorial properties of the maps under consideration. It turns out that for our problem, the infinitely renormalizable case is easier to handle. The reason is that by means of renormalization, it can be reduced to the quadratic-like case.

Take some infinitely renormalizable map $f \in \mathcal{U}_a$. By *a priori* bounds (Theorem 2.24) some renormalization $R^n f$ is a quadratic-like map, and the same is true for all (complex) $\tilde{f} \in \mathcal{A}_a$ near f. Moreover, all maps $R^n \tilde{f}$ belong to some Banach ball $\mathcal{V}$ of quadratic-like maps.

The classical Runge Theorem implies that the differential $DR^n(f)$ has a dense image in $\mathcal{V}$. By the Implicit Function Theorem, the pullback of the hybrid lamination near $g = R^n f$ in $\mathcal{V}$ is a holomorphic lamination near f in $\mathcal{A}_a$. It is easy to see that the real slices of the leaves of this lamination are local real hybrid classes.

3.2.4. Puzzle maps. To handle the non-renormalizable case (the finitely renormalizable case is completely analogous), we need to consider a special class of piecewise holomorphic maps. A *puzzle map* is defined on a disjoint union $\cup_{i\geq 0} U_i$ of topological disks U_i called "puzzle pieces" such that the "critical" piece U_0 contains 0 and is symmetric with respect to 0, while any noncritical puzzle piece U_i is univalently mapped by f onto some other puzzle piece $U_{j(i)}$ (see Fig. 1). Moreover, we require that for any i, either $f(U_0)$ contains U_i or $\overline{f(U_0)}$ is disjoint from $\overline{U_i}$ (and there are also some other technical requirements). Let U_1 denote the puzzle piece containing the critical value $c_1 = f(0)$.

The *filled Julia set* $K(f)$ of the puzzle map is defined as the set of all non-escaping (from $\cup U_i$) points. It is not necessarily compact.

Let us fix from now on a non-renormalizable unimodal map $f \in \mathcal{U}_a$ with repelling fixed points and recurrent critical point. Any such map can be restricted to a puzzle map in the following way. Select a nice critical interval $J_0 \equiv T^n$ in the principal nest (see §2.10), and let U_0 be an appropriate hyperbolic neighborhood of J_0 in the slit complex plane $\mathbb{C} \setminus (\mathbb{R} \setminus J_0)$. Consider the first landing map $L : \cup_{j\geq 0} J_j \to J_0$ to J_0, where we denote by J_1 the component containing the critical value c_1 (notice that $f^{-1}(J_1) = T^{n+1}$). Then $U_i \supset J_i$ are defined as the preimages of U_0 under L. The last requirement on the puzzle map can be ensured by selecting the interval $J_0 = T^n$ sufficiently deep in the principal nest to make the corresponding scaling factor λ_n so small (by Theorem 2.17) that $\overline{f^{-1}(U_1)} \subset U_0$ (see Lemma 5.5).

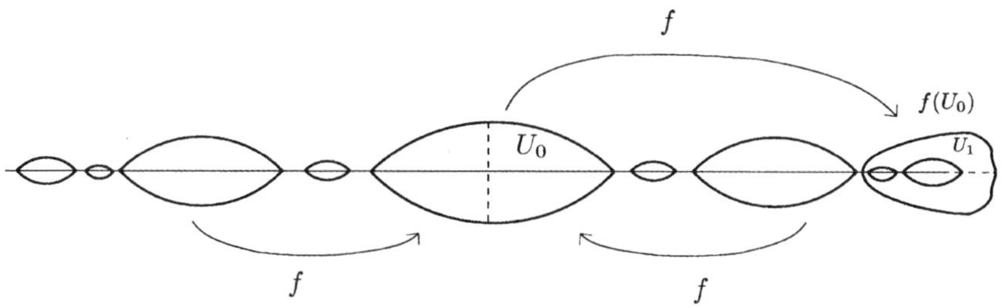

Fig. 1. Puzzle map

This puzzle structure of unimodal maps $f \in \mathcal{U}_a$ will serve as a substitute for the external structure of quadratic-like maps. A crucial property of the puzzle structure (see §5.2) is that it is *persistent* under perturbations of the map, even *complex*, and moreover *moves holomorphically* under the perturbation.

3.2.5. Key estimate. By definition, a horizontal vector field v for a non-hyperbolic puzzle map is a holomorphic vector field on $\cup U_i$ satisfying (2.5) for some qc vector field α on orb(0). In the hyperbolic case, we also require that the multiplier of the corresponding periodic attractor does not change "infinitesimally" along the direction v (that is, v satisfies (5.1)) .

The key estimate says that, for any map $g \in \mathcal{V}$ in some neighborhood $\mathcal{V} \ni f$ which either does not have invariant line fields on $K(g)$ or is hyperbolic, the dependence of α on v is *uniformly bounded*:

$$\|\alpha\|_{\text{qc}} \leq L\|v\|_a, \quad v \in T_g. \tag{3.1}$$

(The norms $\|\cdot\|_{\text{qc}}$ and $\|\cdot\|_a$ are defined in §2.6.1 and 2.9.) Recall that the no-invariant line fields assumption is satisfied for all real maps which are not hyperbolic or parabolic (see Theorem 2.22 and Lemma A.24). It is also satisfied for complex preperiodic puzzle maps for simpler reasons (see Appendix B).

The proof goes as follows. Select an $\varepsilon > 0$ so that the puzzle structure is persistent in the 2ε-neighborhood $\mathcal{V}$ of f, and let g belong to the ε-neighborhood of f. Consider a holomorphic curve $g_\lambda = g + \lambda v \in \mathcal{V}$ tangent to v, $|\lambda| < \varepsilon/\|v\|_a$. Let $h_\lambda : (\mathbb{C}, \cup U_i) \to (\mathbb{C}, \cup U_i^\lambda)$ be a holomorphic motion of the puzzle structure, and let

$$\alpha_0 = \left.\frac{dh_\lambda}{d\lambda}\right|_{\lambda=0}.$$

Since h_λ is equivariant on the boundary of the puzzle, α_0 is equivariant with respect to (g, v) on ∂U. Moreover,

$$\bar\partial\alpha_0 = \left.\frac{d\mu_\lambda}{d\lambda}\right|_{\lambda=0},$$

where μ_λ is the Beltrami differential of h_λ. Applying the Schwarz Lemma to the map $\lambda \mapsto \mu_\lambda$, we obtain: $\|\bar{\partial}\alpha_0\|_\infty \le \varepsilon^{-1}\|v\|_a$.

We exploit now a new tool, the "Infinitesimal Pullback Argument" (§6.2), to construct a qc vector field β equivariant with respect to (g, v) on U, coinciding with α on orb(0), and coinciding with α_0 on $\mathbb{C}\setminus U$. Moreover, if g has no invariant line fields on the filled Julia set, then $\|\bar{\partial}\beta\|_\infty \le \|\bar{\partial}\alpha_0\|_\infty$, and the key estimate follows with $L = \varepsilon^{-1}$.

3.2.6. Transverse direction. We are now approaching the most delicate issue, construction of a "transverse vector field", i.e., a non-horizontal vector field belonging to $T_f(\mathcal{U}_a)$. The idea is first to construct a *smooth* transverse vector field v and then to approximate it with a holomorphic one using Mergelyan's Theorem. In fact, v is going to be holomorphic on the critical value puzzle piece U_1 and vanishing on all other puzzle pieces U_i, $i \neq 1$. Moreover, v can be selected in such a way that it has a definite hyperbolic length (with respect to the hyperbolic metric of U_1) at any given point $q \in J_1 = U_1 \cap \mathbb{R}$, no matter how close this point to is ∂J_1 (it is essential here that v is required to be only C^1). It turns out that if the nice interval J_0 was originally selected sufficiently deep, then the vector field v corresponding to $q = c_1$ cannot be horizontal. Assuming otherwise, we use the infinitesimal pullback argument to construct a qc vector field α equivariant on orb(0) and vanishing on the boundary of the real puzzle. Let us then consider the first landing point p of the orb(q) in the domain $W = f^{-1}U_1 \Subset U_0$ (notice that by construction, α must vanish in $\partial W \cap \mathbb{R}$). Equation (2.5) allows us to bound from below the hyperbolic length of $\alpha(p)$ in U_0 via the hyperbolic length of v in U_1. So, the former length is also definite. Moreover, using the ideas of the proof of the Key estimate we are able to bound from above the qc norm of α uniformly.

But the hyperbolic diameter of W in U_0 is very small, provided the nice interval J_0 was selected deep enough (note that $W \cap \mathbb{R}$ is the puzzle piece of the principal nest following J_0). Hence the length of $\alpha(p)$ is very big compared with the diam W. But this contradicts Corollary 2.9, since α vanishes at the boundary points of $W \cap \mathbb{R}$ and has uniformly bounded qc norm.

Next we approximate v in the union of two appropriate sectors by polynomial vector fields v_n vanishing to the first order at the boundary point of J_1 where the sectors touch. We claim that the v_n are eventually not horizontal (and hence they represent transverse directions at f in $\mathcal{U}_a$). Assuming contrary, we prove the key estimate for the corresponding vector fields α_n. It follows the same lines as described above, though technically more involved. The key estimate allows us to pass to a limit in equation (2.5) and to conclude that v is horizontal – contradiction.

Remark 3.2. In this construction we control uniformly the $\|\cdot\|_1$ norm of v. The scaling invariance of the $\|\cdot\|_1$ norm is essential for this argument, as it is for the C^1 Closing Lemma of Pugh [Pu]. In fact, the construction of the transverse direction can be seen as the infinitesimal counterpart to the C^1 Connecting Lemma of Hayashi [Ha] (we want to close the critical orbit at the infinitesimal level). The key estimate, based heavily on complex analysis, is what allows us to promote a C^1 perturbation to an infinitesimal holomorphic perturbation.

3.2.7. Transverse cone field. Thus, the tangent space at f can be decomposed in the direct sum of the horizontal and transverse spaces, $T_f\mathcal{A}_a = T_f \oplus V_f^{\mathrm{tr}}$, where $V_f^{\mathrm{tr}} = \mathrm{Lin}\{V^{\mathrm{tr}}\}$ and V^{tr} is the transverse vector field constructed above. Since $\mathcal{A}_a$ is an affine space, we can use this decomposition as a "coordinate system" in it. Let $\theta \in (0, \pi/2)$. For a complex map $g \in \mathcal{A}_a$ near f, let us consider a tangent cone in $T_g\mathcal{A}_a$,

$$\mathcal{K}_g^\theta = \{v \in T_g\mathcal{A}_a : \|v^h\| < \tan\theta\,\|v^{\mathrm{tr}}\|\},$$

where v^h and v^{tr} are the projections of v to T_f and V_f^{tr} respectively.

If the angle θ is sufficiently small and g (possibly complex) is sufficiently close to f and does not have invariant line fields on $K(g)$, or is hyperbolic, then

$$\mathcal{K}_g^\theta \cap T_g = \emptyset. \tag{3.2}$$

Otherwise there would exist a sequence of maps $g_n \to f$, either without invariant line fields on $K(g_n)$ or hyperbolic, and a sequence of unitary horizontal vector fields $v_n \in T_{g_n}\mathcal{A}_a$ converging to V^{tr}. Let α_n be a a qc vector field equivariant with respect to (g_n, v_n) on $\mathrm{orb}_{g_n}(0)$. By (3.1) and the Second Compactness Lemma, the sequence $\{\alpha_n\}$ admits a subsequence converging to a qc vector field α equivariant on $\mathrm{orb}_f(0)$ with respect to (f, V^{tr}). It follows that the vector field V^{tr} is horizontal – contradiction.

3.2.8. Local laminations. Let $\mathcal{V} \subset \mathcal{A}_a$ be a neighborhood of f where (3.2) holds. We can select this neighborhood as a product $\mathcal{V}^h \times \Sigma^{\mathrm{tr}}$ where $\mathcal{V}^h \subset T_f$ and $\Sigma^{\mathrm{tr}} \subset V_f^{\mathrm{tr}}$ is a transverse segment. Exploiting a macroscopic version of the Key estimate, we show that for Σ^{tr} small enough, each hybrid class may intersect Σ^{tr} only at a single point (the estimate we use essentially implies that if two nearby maps g_1 and g_2 are hybrid conjugate then $g_2 - g_1$ is almost tangent to the hybrid class of g_1, so g_1 and g_2 cannot be both in the transverse segment Σ^{tr}). This implies that preperiodic and hyperbolic maps are dense in Σ^{tr} (thus, at this stage we obtain a new proof of Kozlovski's Theorem [K1]).

If $g \in \mathcal{V}$ is preperiodic or hyperbolic then the hybrid class $\mathcal{H}_g$ is a complex analytic submanifold in $\mathcal{A}_a$ whose tangent space coincides with the horizontal space T_g. By the cone transversality (3.2), this submanifold has a bounded slope in the coordinate system $T_f \oplus V_f^{\mathrm{tr}}$. This implies (together with the existence of uniform qc conjugacies for $\mathcal{H}_g \cap \mathcal{V}$, obtained using the Macroscopic Pullback Argument) that it is a graph with a bounded slope over the whole neighborhood $\mathcal{V}^h$. For these simple combinatorics (preperiodic or hyperbolic), it is easy to see that the local hybrid classes $\mathcal{H}_g \cap \mathcal{V}$ of different maps cannot intersect, so we have indeed constructed the lamination through a dense subset of Σ^{tr}. An application of the Extension Lemma promotes it to a lamination through the whole transversal Σ^{tr}.

Since we already dealt with infinitely renormalizable maps, the remaining case of the construction of local laminations is for hyperbolic maps, which is much easier. If f is hyperbolic, there is a neighborhood $\mathcal{V}^{\mathbb{R}} \subset \mathcal{U}_a$ of f consisting of hyperbolic maps. The analytic map that associates to each $g \in \mathcal{V}^{\mathbb{R}}$ the multiplier of its attracting periodic orbit is a submersion. This implies that the real hybrid classes in $\mathcal{V}^{\mathbb{R}}$ form a transversally analytic foliation.

3.2.9. Connectivity of the hybrid classes. In the Appendix B, we prove that any hybrid class $\mathcal{H}^{\mathbb{R}}_{f,a}$ of a Yoccoz map f is connected. To this end we complement the puzzle structure of f with a Markov structure in such a way that the combinatorics of this pattern depends on the combinatorics of f only, while the geometry of the pattern is definite. In other words, these patterns are qc equivalent for any two maps f and $\tilde{f} \in \mathcal{H}^{\mathbb{R}}_f$. By the pullback argument, f and $\tilde{f}$ are qc equivalent in a complex neighborhood of the interval I.

In particular, we can select $\tilde{f}$ as the quadratic polynomial q_τ in the hybrid class of f. Let h be an $\mathbb{R}$-symmetric qc map conjugating f to $\tilde{f}$ near I, and let μ be its Beltrami differential. Let $\mu_t = t\mu$, $0 \leq t \leq 1$, and let $h_t : \mathbb{C} \to \mathbb{C}$ be the solution of the corresponding Beltrami equation fixing 0 and 1. Then the Beltrami path $f_t = h_t \circ f \circ h_t^{-1}$ is a real analytic family of unimodal maps connecting f to $\tilde{f}$ in $\mathcal{H}^{\mathbb{R}}_{f,a'}$ for some a'.

Starting with this path f_t in $\mathcal{H}^{\mathbb{R}}_{f,a'}$ we build a two parameter family $f_{t_1,t_2} = f_{t_1} + t_2 v_{t_1}$ in $\mathcal{U}_{a'}$ using a transverse direction v_t along the path f_t. This family can be approximated by a two parameter family in $\mathcal{U}_a$, passing through f and $\tilde{f}$. By the Implicit Function Theorem, the intersection of this family with $\mathcal{H}^{\mathbb{R}}_{f,a}$ yields a path in $\mathcal{H}^{\mathbb{R}}_{f,a}$ linking f and $\tilde{f}$.

3.2.10. Regular or Stochastic Theorem. We will outline now a proof of Theorem B. Take some map $f \in \mathcal{U}_a$ which is not parabolic, and consider a real analytic family $f_t \in \mathcal{U}_a$, $f_0 = f$, transverse to the hybrid class $\mathcal{H}^{\mathbb{R}}_f$. We wish to prove that almost all maps f_t near f are either regular or stochastic. Of course, we can assume that f is not hyperbolic. Moreover, since maps with non-recurrent critical point have absolutely continuous invariant measures (Misiurewicz [Mi]), it is enough to consider the recurrent case.

Assume first that f is at most finitely renormalizable. Since the real hybrid class $\mathcal{H}^{\mathbb{R}}_f$ is connected, there is a local holonomy from the family $\{f_t\}$ to the quadratic family $\{q_\tau\}$. By Corollary 2.6, this holonomy is quasisymmetric.

Quasisymmetric maps are not in general absolutely continuous. However, by Lemma 2.5, they respect the property of exponential decay of the parapuzzle geometry (Theorem 2.25). This property implies that almost any map f_t near f, which is not hyperbolic and is at most finitely renormalizable, satisfies the Martens-Nowicki criterion for existence of a.c.i.m. Hence almost any map f_t near f which is at most finitely renormalizable is either regular or stochastic.

If f is infinitely renormalizable then some renormalization $\{g_t = R^n f_t\}$ is a quadratic-like family near $g = R^n f$ transverse to the real hybrid class $\mathcal{H}_g^{\mathbb{R}}$ (since the image of $DR^n(f)$ is dense, R^n is transversally non-singular). Since the hybrid classes form a holomorphic lamination with connected leaves in the space of quadratic-like germs, the holonomy from $\{g_t\}$ to the quadratic family $\{q_\tau\}$ is quasisymmetric. By Theorem 2.27 the set $\mathcal{I}$ of infinitely renormalizable quadratic maps has definite gaps everywhere. This property carries to the family $\{g_t\}$ by means of the quasisymmetric holonomy. Since R^n analytically maps $\{f_t\}$ to $\{g_t\}$, the same is true for the family $\{f_t\}$. By the Lebesgue Density Theorem, the set of infinitely renormalizable maps f_t near f has zero Lebesgue measure.

Consider now an arbitrary non-trivial real analytic family $\{f_t\} \subset \mathcal{U}_a$. We show in §9.1 that this family contains at most countably many tangencies with the real hybrid classes and at most countably many parabolic points. At all other points it is transverse to the hybrid classes. It follows that almost all parameter values in this family are either regular or stochastic. The set of transverse points is an open set with countable complement, so this argument also proves Theorem C.

4. Infinitely renormalizable case

4.1. Hybrid lamination in the space of quadratic-like maps

4.1.1. Banach balls of quadratic-like maps. Let U be a 0-symmetric domain, and let $\mathcal{B}_U^0$ be the space of normalized at 0 even holomorphic functions $f \in \mathcal{B}_U$. This is an affine subspace of $\mathcal{B}_U$. The ball or radius ε centered at $f \in \mathcal{B}_U^0$ is denoted as $\mathcal{B}_U^0(f, \varepsilon)$.

Each class of affinely conjugate quadratic-like maps contains a unique representative normalized at 0. Let $\mathcal{Q}$ stand for the space of normalized at 0 quadratic-like maps. The *connectedness locus* $\mathcal{C}$ is the set of quadratic-like maps $f \in \mathcal{Q}$ with connected Julia set. The *hybrid class* $\mathcal{H}_f \subset \mathcal{Q}$ is the set of quadratic-like maps $g \in \mathcal{Q}$ which are hybrid equivalent to f on some neighborhoods of the filled Julia sets.

Take some quadratic-like map $f : U \to U'$ in $\mathcal{Q}$. Consider a 0-symmetric Jordan domain $V \Subset U$ with a piecewise smooth boundary such that $V \supset K(f)$ and $fV \Supset V$. Then any normalized at 0 even holomorphic function $g \in \mathcal{B}_U^0$ which is sufficiently close to f restricts to a quadratic-like map $g : V \to V'$. Thus, we have an embedding $j_V : \mathcal{B}_U^0(f, \varepsilon) \to \mathcal{Q}$. Its image $\mathcal{Q}_V(f, \varepsilon)$ endowed with the topology induced from $\mathcal{B}_U$ will be called a *Banach ball* (centered at f of radius ε) of quadratic-like maps.

Given a set $X \subset \mathcal{Q}$, the intersection $X \cap \mathcal{Q}_V(f, \varepsilon)$ will be called a *Banach slice* of X (by the ball $\mathcal{Q}_V(f, \varepsilon)$).

4.1.2. Hybrid lamination. Below we will summarize the results of [L6] (especially §4) about the hybrid lamination in the space of quadratic-like maps.

Theorem 4.1. *Let $f : U \to U'$ be a normalized at 0 quadratic-like map with connected Julia set. Then there exists a Banach ball $\mathcal{V} = \mathcal{Q}_V(f, \varepsilon)$ such that for any $g \in \mathcal{V} \cap \mathcal{C}$, the slice of the hybrid class $\mathcal{H}_g$ by the ball $\mathcal{V}$ is a complex codimension-one analytic submanifold in $\mathcal{V}$. These submanifolds form a holomorphic lamination in $\mathcal{V}$.*

In the setting of the above theorem, let us consider a map $\tilde{f} \in \mathcal{V}$ which is hybrid equivalent to f and in the same connected component of $\mathcal{V} \cap \mathcal{H}_f$. Take two transversals $\mathcal{T} \ni f$ and $\tilde{\mathcal{T}} \ni \tilde{f}$ to $\mathcal{H}_f \cap \mathcal{V}$. Then we have a well defined local holonomy $h : \mathcal{T} \cap \mathcal{C} \to \tilde{\mathcal{T}} \cap \mathcal{C}$ along the leaves of the lamination. Moreover, by the λ-lemma, h admits a qc extension to a neighborhood of f in $\mathcal{T}$. (We will formulate this property briefly by saying that "the holonomy is locally quasiconformal" or that the "the lamination is transversally quasiconformal"). In fact, these holonomies can be extended to the whole hybrid class of f:

Proposition 4.2. *Let $f \in \mathcal{C}$ be a normalized quadratic-like map with connected Julia set. If $\tilde{f} \in \mathcal{H}_f$ is hybrid equivalent to f then it can be joined to f with a path $\{f_t\} \subset \mathcal{H}_f$ covered with finitely many Banach balls from Theorem 4.1. Thus, there is a well-defined local qc holonomy between any two transversals $\mathcal{T} \ni f$ and $\tilde{\mathcal{T}} \ni \tilde{f}$ to $\mathcal{H}_f$.*

The role of the quadratic family is partly explained by the following statement:

Proposition 4.3. *The quadratic family $\{q_\tau\}$ is transverse to the hybrid lamination (in any Banach ball from Theorem 4.1).*

The tangent space T_f to the above Banach slice of $\mathcal{H}_f$ consists of vector fields $v \in \mathcal{B}_U$ that admit representation (2.5), $v = \alpha \circ f - f'\alpha$, where α is a qc vector field on U whose $\bar{\partial}$-derivative vanishes a.e. on the filled Julia set $K(f)$. Such vector fields are called *horizontal.*

A transverse, *vertical* direction to the hybrid class $\mathcal{H}_f$ can be selected as a holomorphic vector field V_f^{tr} which can be represented as (2.5) in $U \setminus K(f)$, where $\alpha(z)/dz$ is a *holomorphic* vector field on $\overline{\mathbb{C}} \setminus K(f)$ vanishing at ∞. The latter condition means that near ∞, $\alpha(z) = az + h(z)$, where h is a bounded holomorphic function. If f is $\mathbb{R}$-symmetric, the transverse direction can also be chosen $\mathbb{R}$-symmetric.

4.2. From real analytic to quadratic-like. Let us consider an infinitely renormalizable quasi-quadratic map $f \in \mathcal{U}_a$. By the *a priori bounds*, there is a quadratic-like renormalization of f. This means that there exist $\mathbb{R}$-symmetric Jordan domains $U \equiv U_0, \dots, U_{n-1}, U_p \equiv U'$ such that:

- $U_i \Subset \Omega_a$, $i = 0, 1, \dots, n-1$.
- $U_0 \ni 0$ is 0-symmetric and $f : U_0 \to U_1$ is a branched double covering;
- the maps $f : U_i \to U_{i+1}$ are conformal isomorphisms, $i = 1, \dots, n-1$;
- $U' \Supset U$.

We call the map

$$P(f) \equiv f^n : U \to U'$$

a *quadratic-like pre-renormalization* of f. If $g \in \mathcal{U}_a$ is sufficiently close to f, then the restriction $g^n | U$ gives a quadratic-like pre-renormalization of g. Normalizing these maps at the origin, we obtain a renormalization $R : \mathcal{V} \to \mathcal{Q}_W^{\mathbb{R}}$, where $\mathcal{V} \subset \mathcal{U}_a$ is some neighborhood of f and $W \ni 0$ is an $\mathbb{R}$-symmetric and 0-symmetric Jordan disk obtained from U by little shrinking and rescaling.

To carry out the infinitesimal analysis of the renormalization operator, we will make use of the variational formula (2.8).

Lemma 4.4. *The renormalization operator $R : \mathcal{V} \to \mathcal{Q}_W^{\mathbb{R}}$ is real analytic.*

Proof. The pre-renormalization of f analytically depends on f, since it is a restricted iterate of f (with the differential explicitly given by (2.8)). The normalization of a function f is the rescaling by factor $f''(0)$ analytically depending on f as well. □

We want to prove that the derivative of the renormalization operator is transversally non-singular, that is, the image of $DR(f)$ contains transverse vectors to $\mathcal{H}_{R(f)}^{\mathbb{R}}$. Let us deal first with an easier case.

Lemma 4.5. *Assume $\overline{U}_i$ are disjoint, $0 \le i \le n-1$. Then the image of the infinitesimal renormalization $DR(f) : T_f \mathcal{A}_a^{\mathbb{R}} \to T_{R(f)} \mathcal{Q}_W^{\mathbb{R}}$ is dense.*

Proof. The operator R is a composition of the pre-renormalization operator P from $\mathcal{A}_a^{\mathbb{R}}$ to $\mathcal{B}_U^{\mathbb{R}}$ and a rescaling operator from $\mathcal{B}_U^{\mathbb{R}}$ to $\mathcal{Q}_W^{\mathbb{R}}$. It is easy to see that the derivative of the rescaling operator has dense image, so we just have to show that $DP(f) : T_f \mathcal{A}_a^{\mathbb{R}} \to T_{R(f)} \mathcal{B}_U^{\mathbb{R}}$ has dense image. Since $P(f)$ is the restriction of the iterate f^n, the differential $DP(f)v$ is given by the variational formula (2.8).

Take an even holomorphic vector field $w \in \mathcal{B}_U^{\mathbb{R}}$. Let us define a holomorphic vector field $\tilde{v}$ on $\cup U_i$ such that $DP(f)\tilde{v} = w$. First let $\tilde{v} = 0$ on $\cup_{i=1}^{n-1} U_i$. Using (2.8) with w in the left-hand side, extend this vector field to U:

$$\tilde{v}(z) = \frac{w(z)}{Df^{n-1}(f(z))} \tag{4.1}$$

(notice that in (2.8) the only non-vanishing term corresponds to $k = 0$).

Then $\tilde{v}$ is an even vector field on U, and the pre-renormalization of $\tilde{v}$ is equal to w. By the Mergelyan Polynomial Approximation Theorem (see [R], Theorem 20.5, p. 423), $\tilde{v}$ can be approximated by a polynomial vector field v with which is even and $\mathbb{R}$-symmetric. Then the pre-renormalization of v approximates w. Rescaling the domain U, we obtain the assertion. □

The above situation occurs whenever the last renormalization is not period doubling. In the doubling case, the intersections between the U_i are unavoidable, since the little Julia sets (the forward images of $J(f^{2n}|U)$, where $2n$ is the period of renormalization) intersect.

In this case, we can still select the U_i with the following properties:

- If the closure of U_i intersects the closure of U_j and $i < j$ then $j = i + n$ and $U_i \cup U_j$ is simply connected;
- $\overline{U}_0 \cap \overline{U}_n$ is contained in a small neighborhood V of a repelling periodic point q of period n;
- V is a topological disk such that $f^n|V$ is a diffeomorphism and $\overline{V} \subset f^n(V)$; hence $f^n|V$ is linearizable.

Lemma 4.6. *Under the circumstances just described, $DR(f) : T_f\mathcal{A}_a^{\mathbb{R}} \to T_{R(f)}\mathcal{Q}_W^{\mathbb{R}}$ has dense image.*

We will make use of the following lemma.

Lemma 4.7. *Let $|\lambda| > 1$, $L : s \mapsto \lambda s$, and let S be a bounded connected open set such that $0 \in S \subset L(S)$. Consider two holomorphic functions $a, b \in \mathcal{B}_S$ such that $b(0) = \lambda$. Then there exists a holomorphic function $u \in \mathcal{B}_S$ such that*

$$a(s) = u(s) + b(s)u(\lambda^{-1}s), \quad s \in S. \tag{4.2}$$

Proof. Let us first consider the case where $S = \mathbb{D}_\varepsilon$ and $\|b\|_{\mathcal{B}_S} < |\lambda|^2$. Let $a(s) = \sum_{k=0}^{\infty} a_k s^k$, $b(s) = \sum_{k=0}^{\infty} b_k s^k$. Then (4.2) has a formal solution $u(z) = u_k z^k$ whose coefficients are recursively found from the equations:

$$(\lambda + \lambda^k)u_k = a_k - \sum_{j=0}^{k-1} b_{k-j}u_j.$$

To prove convergence of the formal solution, let us consider the linear jet of u, $\hat{u}(s) = u_0 + u_1 s$, and let $\hat{a}(s) = \hat{u}(s) + b(s)\hat{u}(\lambda^{-1}s)$. A simple calculation gives $a(s) - \hat{a}(s) = O(|s^2|)$.

Hence it is enough to solve (4.2) in the space of functions u with vanishing linear jet (assuming that a is also in the same space). In terms of $w(s) = u(s)/s^2$ and $c(s) = a(s)/s^2$, (4.2) assumes the form

$$c(s) = w(s) + \lambda^{-2}b(s)w(\lambda^{-1}s) \equiv (\mathrm{id} + T)w(s),$$

where $T : \mathcal{B}_S \to \mathcal{B}_S$ is a contracting linear operator (since $\|T\| = \lambda^{-2}\|b\| < 1$). Hence $\mathrm{id} + T$ is invertible, and the conclusion follows.

In the general case, we first notice that for ε sufficiently small we still have $a, b \in \mathcal{B}_{\mathbb{D}_\varepsilon}$ and $\|b\|_{\mathcal{B}_{\mathbb{D}_\varepsilon}} < |\lambda|^2$. By the previous consideration, we obtain $u \in \mathcal{B}_{\mathbb{D}_\varepsilon}$ satisfying (4.2) for $s \in \mathbb{D}_\varepsilon$. Then we find a $k \geq 0$ such that $L^{-k}(S) \subset \mathbb{D}_\varepsilon$ and use the functional equation k times to extend u to S. □

Proof of Lemma 4.6. As before, we just have to show that the image of $DP(f)$ is dense.

Arguing as in Lemma 4.5, let $w \in \mathcal{B}_U^{\mathbb{R}}$ be an arbitrary even polynomial vector field (note that even polynomial vector fields are dense in $\mathcal{B}_U^{\mathbb{R}}$). We claim that there exists a holomorphic vector field v_{lin} in $f^n(V)$ such that

(4.3)
$$w(z) = Df^{n-1}(f^{n+1}(z))\, v_{\mathrm{lin}}(f^n(z)) + Df^{2n-1}(f(z))\, v_{\mathrm{lin}}(z), \quad z \in V.$$

(Note that the right-hand side of this equation is formally equal to $DP(f)v_{\mathrm{lin}}$ assuming that v_{lin} is extended outside $f^n(V)$ by zero.) To solve (4.3), let us linearize f^n in V. Let $\lambda = Df^n(q)$, $L(s) = \lambda s$, S be a neighborhood of 0 such that $L(S) \supset S$, and let $\phi : (L(S), 0) \to (f^n(V), q)$ be a conformal map satisfying $\phi \circ L = f^n \circ \phi$ for $s \in S$.

In this linear coordinate s the above equation can be written in the form (4.2), with $a = w \circ \phi$, $b = Df^n \circ f^n \circ \phi$ (so that $b(0) = \lambda$), and

$$u = (v_{\mathrm{lin}} \circ \phi \circ L) \cdot (Df^{n-1} \circ f \circ \phi \circ L).$$

Using Lemma 4.7, we obtain a solution $u \in \mathcal{B}_S$ of (4.2). Passing back to V we obtain a solution v_{lin} of (4.3).

Let v_n be an even polynomial vector field which is close to v_{lin} on $f^n(V)$, and let v_0 be an even polynomial vector field close to

$$\frac{w(z) - Df^{n-1}(f^{n+1}(z))\, v_n(f^n(z))}{Df^{2n-1}(f(z))}$$

in a neighborhood of $\overline{U_0 \cup V}$. Notice that v_0 is close to v_n on V.

Let us now construct a vector field interpolating between v_0 and v_n. Since we want even vector fields, we first symmetrize the domains. Let $-U_n$ and $-V$ be the central reflections of respectively U_n and V about the origin, and let

$$U_n^{\mathrm{sym}} = U \cup (-U), \qquad V^{\mathrm{sym}} = V \cup (-V)$$

(see Fig. 2). Let $K \subset V^{\mathrm{sym}}$ be a compact 0-symmetric neighborhood of $\overline{U}_0 \cap \overline{U_n^{\mathrm{sym}}}$.

Using a partition of unity, construct an even smooth (C^∞) vector field v_{sm} on $\mathbb{C}$ with the following properties: $v_{\mathrm{sm}} = v_0$ on $\overline{U}_0 \setminus K$, $v_{\mathrm{sm}} = v_n$ on $U_n^{\mathrm{sym}} \setminus K$, and v_{sm} is C^1 close to v_n on K (this is possible because v_0 is C^1 close to v_n on $K \subset V$).

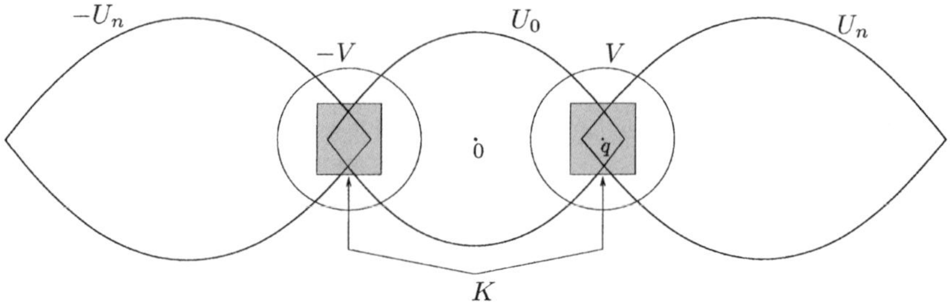

Fig. 2.

Consider now a normalized qc vector field α such that $\bar{\partial}\alpha | U_0 \cup U_n^{\text{sym}} = \bar{\partial} v_{\text{sm}}$ and $\bar{\partial}\alpha | \mathbb{C} \setminus (U_0 \cup U_n^{\text{sym}}) = 0$ (see §2.6.1). Notice that $\|\bar{\partial}\alpha\|_\infty$ is close to 0 and so by the Second Compactness Lemma, α is close to 0. Then $v_{\text{sm}} - \alpha \in \mathcal{B}^0_{U_0 \cup U_n^{\text{sym}}}$ interpolates between v_0 and v_n.

By the Mergelyan Theorem, there is an even polynomial vector field v which is close to v_0 on U_0, to v_n on U_n and to 0 on the U_i, $i \neq 0, n$. Then $DP(f)v$ is close to w in $\mathcal{B}_U$. □

Lemmas 4.5 and 4.6 imply:

Lemma 4.8. *DR is transversally non-singular.*

4.3. Local laminations near infinitely renormalizable maps.

Theorem 4.9. *Let $f \in \mathcal{U}_a$ be an infinitely renormalizable map. Then some neighborhood $\mathcal{V} \subset \mathcal{U}_a$ of f is foliated by the real hybrid classes, which are codimension-one real analytic submanifolds in $\mathcal{V}$.*

Proof. Consider the above renormalization operator $R : \mathcal{V} \to \mathcal{Q}^{\mathbb{R}}_W$. Let $g = R(f)$. By Theorem 4.1, $\mathcal{Q}^{\mathbb{R}}_W = T_g \oplus V^{\text{tr}}_g$, where T_g is the tangent space to the real hybrid class at g and V^{tr}_g is the transverse vertical line. Moreover, g has a neighborhood $\mathcal{W} \subset \mathcal{Q}^{\mathbb{R}}_W$ foliated by the real hybrid classes of quadratic-like maps, which are graphs of real analytic functions $T_g \to V^{\text{tr}}_g$.

If the neighborhood $\mathcal{W}$ is sufficiently small, then the hybrid class $\mathcal{H}^{\mathbb{R}}_g \cap \mathcal{W}$ is the zero-set of some real analytic submersion $\phi : \mathcal{W} \to \mathbb{R}$. By Lemma 4.8, there exists a vector $v \in \mathcal{A}^{\mathbb{R}}_a$ such that $w \equiv DR(f)v \notin T_g$. Hence the composition $\phi \circ R : \mathcal{V} \to \mathbb{R}$ is a submersion at f. By the Implicit Function Theorem, its zero-set $X_f = R^{-1}\mathcal{H}_g$ is a codimension-one submanifold near f.

Since R is real analytic and the horizontal space T_G varies continuously for $G \in \mathcal{W}$, $DR(F)v \notin T_{R(F)}$ for all $F \in \mathcal{V}$ sufficiently close to f. By shrinking $\mathcal{V}$, we can assume that this is valid for all $F \in \mathcal{V}$. Applying the Implicit Function Theorem as above we see that the preimages $X_F = R^{-1}\mathcal{H}_{R(F)}$, $F \in \mathcal{V}$, are real analytic codimension-one submanifolds in $\mathcal{V}$.

Moreover, these submanifolds are closed in $\mathcal{V}$ and are transverse to v at any point $F \in \mathcal{V}$. Hence they form a real analytic lamination of $\mathcal{V}$.

Observe finally that the leaves X_F of this lamination coincide with the real hybrid classes. Indeed, if two non-hyperbolic maps F and $\tilde{F}$ in $\mathcal{V}$ are topologically equivalent on the real line then so are their renormalizations $R(F)$ and $R(\tilde{F})$. By Theorem 2.21, these two maps are hybrid equivalent. If F and $\tilde{F}$ are hyperbolic with the same multiplier then so are their renormalizations.

Vice versa, if $R(F)$ and $R(\tilde{F})$ are topologically equivalent on the real line then the maps F and $\tilde{F}$ have the same kneading sequence. Hence, by Remark 2.10, if $R(F)$ and $R(\tilde{F})$ are hybrid equivalent, then so do F and $\tilde{F}$. □

Remark 4.1. Since the hybrid class of an infinitely renormalizable map f has just been shown to be codimension-one submanifolds, the transverse non-singularity of the renormalization operator at f (and the fact that hybrid classes are preserved by the renormalization) implies that the image by DR of any vector transverse to $\mathcal{H}_f^{\mathbb{R}}$ is transverse to $\mathcal{H}_{R(f)}^{\mathbb{R}}$.

4.4. Regular or stochastic property near infinitely renormalizable parameter values.

Theorem 4.10. *Let $\{f_t\} \subset \mathcal{U}_a$ be a one-parameter real analytic family of quasi-quadratic maps such that $f \equiv f_{t_0}$ is infinitely renormalizable. If $\{f_t\}$ is transverse to the real hybrid class $\mathcal{H}_f^{\mathbb{R}}$ then for almost all t near t_0, the map f_t is either regular or stochastic.*

Proof. Let us consider the above renormalization operator $R : \mathcal{V} \to \mathcal{Q}_W^{\mathbb{R}}$. By Remark 4.1, $DR(f)v \notin T_{R(f)}$ if v is not tangent to $\mathcal{H}_f^{\mathbb{R}}$. Since R is analytic, $\{g_t = R(f_t)\}$ is a real analytic family of quadratic-like maps transverse to the hybrid class $\mathcal{H}_g^{\mathbb{R}} \cap \mathcal{Q}_W^{\mathbb{R}}$, where $g = g_0$.

By Theorem 4.1, the foliation by the real hybrid classes in $\mathcal{Q}^{\mathbb{R}}$ is transversally quasisymmetric with connected leaves, so the straightening $g_t \mapsto q_{\chi(t)}$ is quasisymmetric near 0. By Theorem 2.28 and the argument of §3.2.10 (detailed in §9.4), this implies the Regular or Stochastic property for the family $\{g_t\}$ near 0.

Since the map g_t is regular or stochastic if and only if the corresponding map f_t is, the conclusion follows. □

4.5. Straightening near infinitely renormalizable parameters.

Theorem 4.11. *Let $\{f_t\}$ be a one-parameter real analytic family of unimodal maps such that $f = f_{t_0}$ is infinitely renormalizable. If $\{f_t\}$ is transverse to $\mathcal{H}_f^{\mathbb{R}}$ at f then the straightening χ of this family is quasisymmetric near f.*

Proof. Consider the renormalization operator $R : \mathcal{V} \to \mathcal{Q}$ in a neighborhood $\mathcal{V}$ of f described in §4.2. Since R is transversally non-singular, it diffeomorphically maps the family $\{f_t\}$ near f onto its image $\{Rf_t\}$.

Let $q = \chi(f)$ be the straightening of f. By Proposition 4.3, the quadratic family q_τ is transverse to the hybrid lamination. For the same reason as above, the renormalization R diffeomorphically maps the quadratic family near q onto its image $\{Rq_\tau\} \subset \mathcal{Q}$.

By Proposition 4.2, there is a well-defined quasisymmetric holonomy h from the family $\{Rf_t\}$ to the family $\{Rq_\tau\}$ in the space $\mathcal{Q}$ of quadratic-like maps. Since $\chi = R^{-1} \circ h \circ R$ (the renormalization operator acts on hybrid classes, see Remark 2.10), we conclude that χ is quasisymmetric as well. □

We will consider next the case of at most finitely renormalizable maps.

5. Tangent space and puzzle maps

5.1. Tangent space. To prove that a hybrid class is a Banach submanifold we have to find a candidate for its tangent space. We will now associate to each map f a complex vector space whose intersection with the real slice will be the tangent space to the hybrid class whenever the map is real and unimodal.

As in the unimodal case, a map $g \in \mathcal{A}_a$ is called preperiodic if there exists k such that c_k is a repelling periodic orbit. It is called hyperbolic if $\omega(0)$ is an attracting cycle.

Remark 5.1. For the class of maps that we are considering, there could be (complex) attracting cycles that do not attract the critical point. However, whenever we refer to an attracting cycle, we mean the one which does attract the critical point.

The following three propositions will motivate the definition of the tangent space to a hybrid class.

Proposition 5.1. *Let $f \in \mathcal{A}_a$ and $v \in \mathcal{B}_a$. Let $\{f_t\}$ be a smooth curve in $\mathcal{A}_a$ such that $f_0 = f$ and*

$$\left.\frac{d}{dt} f_t\right|_{t=0} = v.$$

Then, for each n, the curve $c_n(t) = f_t^n(0) \in \mathbb{C}$ is smooth in a neighborhood of $t = 0$. If there is an equivariant vector field α on $\operatorname{orb}_f(0)$ *then*

$$\alpha(c_n) = \left.\frac{d}{dt} c_n(t)\right|_{t=0} = v^n(0).$$

Proof. The second equation follows from (2.9). Since α is also equivariant by (f^n, v^n) (see §2.6.2), we have:

$$\alpha(c_n) = \alpha(c_n) - Df^n(0)\alpha(0) = v^n(0).$$

□

Remark 5.2. If the orbit of zero is infinite, there is always a (perhaps not even continuous) vector field α on $\mathrm{orb}_f(0)$ satisfying (2.5). If the critical orbit of f is periodic or preperiodic, the set T_f of vector fields v for which there exists an α satisfying (2.5) is a codimension-one subspace. Moreover, α on the orbit of the critical value is uniquely determined by v on the orbit of the critical point.

Recall that O_f stands for the postcritical set $\overline{\mathrm{orb}(c_1)}$.

Definition 5.1. *Let us consider a map $f \in \mathcal{U}_a$. We denote by $\mathcal{G}_f$ the set of qc vector fields on the closed set O_f endowed with the norm $\|\cdot\|_{\mathrm{qc}}$.*

Proposition 5.2. *Let f_λ be a holomorphic family in $\mathcal{A}_a$ through $f \equiv f_0$, with non escaping critical orbit. Assume there is a holomorphic motion h_λ satisfying the equation $h_\lambda \circ f \circ h_\lambda^{-1} = f_\lambda$ on the critical orbit. Let*

$$v = \frac{d}{d\lambda} f_\lambda \Big|_{\lambda=0}$$

and

$$\alpha = \frac{d}{d\lambda} h_\lambda \Big|_{\lambda=0}.$$

Then α is a qc vector field that satisfies equation (2.5) on the critical orbit.

Proof. Differentiating the equation

$$h_\lambda \circ f(c_n) = f_\lambda \circ h_\lambda(c_n)$$

at $\lambda = 0$ we conclude that

$$\alpha(c_{n+1}) = f'(c_n)\alpha(c_n) + v(c_n)$$

for every $n \geq 0$, which is clearly equivalent to equation (2.5). Quasiconformality of α follows from Lemma 2.10. □

If $f \in \mathcal{A}_a$ is a preperiodic map we denote by $\mathcal{C}_f \subset \mathcal{A}_a$ the connected component of f in the set of preperiodic maps with the same relation on the critical orbit. If f is hyperbolic we denote by $\mathcal{C}_f$ the connected component of f in the set of hyperbolic maps with the same multiplier of the attractor as f.

Definition 5.2. *Let f be either a non-hyperbolic map in $\mathcal{U}_a$ or a complex map in $\mathcal{A}_a$ with preperiodic or periodic critical point. We denote by T_f the set of vectors $v \in T\mathcal{A}_a$ such that there exists a qc vector field α equivariant on the critical orbit.*

By Remark 5.2, for each $v \in T_f$, there exists a unique qc vector field α on O_f which extends to a qc vector field equivariant on the critical orbit. It is clear that the correspondence $L_f : T_f \to \mathcal{G}_f$ that to each v associates such an α is a linear map.

Remark 5.3. We will later see that L_f is a *continuous* linear map. This is part of the content of the Key estimate (Lemma 6.12) and is one of the main results of our analysis.

Proposition 5.3. *Let $f \in \mathcal{A}_a$ be a preperiodic map. Then $\mathcal{C}_f$ is a complex codimension-one submanifold in $\mathcal{A}_a$ whose tangent space at f is T_f.*

Proof. If f is a preperiodic then there exists minimal $p, q > 0$ such that $c_p = c_{p+q}$. By the Implicit Function Theorem, the solutions of the equation $c_p = c_{p+q}$ form a codimension-one submanifold whose tangent space at f is T_f. □

With this motivation we define T_f for a hyperbolic but not superattracting map $f \in \mathcal{A}_a$ as the tangent space to $\mathcal{C}_f$, which is a codimension-one complex submanifold by the Implicit Function Theorem. The following simple proposition shows that in this case, T_f cannot be described by Definition 5.2.

Proposition 5.4. *If $f \in \mathcal{A}_a$ is hyperbolic but not superattracting and $v \in \mathcal{B}_a$, then there is a qc vector field on $\mathbb{C}$ which satisfies the equation (2.5) on the critical orbit.*

Proof. Since f is hyperbolic but not superattracting, for λ small enough f is conjugate to $f_\lambda \equiv f + \lambda v$ in a neighborhood V of the attractor. Since there are only a finite number of n such that $c_n \notin V$ we conclude that f is conjugate to f_λ on the critical orbit. We can now consider a holomorphic motion h_λ such that $h_\lambda(f^n(0)) = f_\lambda^n(0)$. The conclusion follows from Proposition 5.2. □

If f is hyperbolic but not superattracting and $v \in T_f$ we define $L_f(v) = \alpha \in \mathcal{G}_f$ satisfying (2.5) which exists by the above proposition (and is unique by Remark 5.2).

It is easy to see that if f is a hyperbolic map with a periodic attractor p of period n then vector fields v tangent to $\mathcal{C}_f$ at f satisfy the following formula,

$$Dv^n(p)(Df^n(p) - 1) = v^n(p)D^2 f^n(p), \tag{5.1}$$

since this formula implies that the multiplier of the continuation of p along $f + tv$ (given infinitesimally by $p + tv^n(p)(1 - Df^n(p))^{-1}$) does not change infinitesimally.

5.2. Puzzle maps. Let $D \subset \mathbb{C}$ be a topological disk. We say that D has L-bounded shape around $x \in D$ if there exists $r > 0$ such that

$$\mathbb{D}_r(x) \subset D \subset \mathbb{D}_{Lr}(x).$$

Let $U \subset \mathbb{C}$ be an open set and let $X \subset U$ be measurable. We say that X is *thin* in U if there exist $L, \varepsilon > 0$ with the following property: any $x \in X$ has a neighborhood $D \subset U$ with L-bounded shape around x, such that $\operatorname{mod}(U \setminus D) > \varepsilon$ and $\operatorname{meas}(D \setminus X)/\operatorname{meas} D > \varepsilon$. Notice that this notion is qc invariant: if X is thin in U and $f : U \to \mathbb{C}$ is a qc map then $f(X)$ is thin in $f(U)$. It is also invariant with respect to the lifting by branched coverings: if $f : V \to U$ is a holomorphic finite branched covering and X is thin in U then $f^{-1}(X)$ is thin in V.

Definition 5.3. *A holomorphic map $f\colon U \to \mathbb{C}$ of class $A^1(U)$ is called a puzzle map if:*

- *U is a countable union of quasidisks U_i, $i \geq 0$, called puzzle pieces, with pairwise disjoint closures, and $U_0 \ni 0$;*
- *For $i > 0$, f is a diffeomorphism of the closure of U_i onto the closure of some U_j;*
- *There exists a sequence $n_i \geq 0$ such that $f^{n_i}|U_i$ is a diffeomorphism onto U_0;*
- *0 is a critical point of f and f' does not vanish on ∂U_0;*
- *$f|U_0$ is a double covering onto the image;*
- *For any i, either $\overline{U_i}$ is contained in $f(U_0)$ or it does not intersect $\overline{f(U_0)}$.*

And furthermore the collection U_i satisfy the following geometric conditions:

(1) $\inf_{U_i \subset f(U_0)} \operatorname{mod}(f(U_0) \setminus U_i) > 0$;
(2) $\overline{\cup U_i} \cap f(U_0)$ *is thin in* $f(U_0)$;
(3) $\lim_{i\to\infty} \operatorname{diam}(U_i) = 0$.

We will use notation $U_1 \equiv U(c_1)$ for the puzzle piece containing the critical value c_1 (whenever $c_1 \in U$). We will also use the notation $\mathcal{P} = \mathcal{P}_f$ for the collection of puzzle pieces $\{U_i\}$.

Definition 5.4. *The filled Julia set $K(f)$ of a puzzle map $f\colon U \to \mathbb{C}$ is the set of points which do not escape U.*

We say that a puzzle map $f : U \to \mathbb{C}$ is hyperbolic if there is an attracting periodic cycle $\overline{p}$ contained in U. For hyperbolic puzzle maps, the immediate basin of attraction of $\overline{p}$ is compactly contained in U and contains the critical orbit.

Definition 5.5. *Let $f \in \mathcal{U}_a$. A geometric puzzle for f is a countable collection of topological disks U_i (puzzle pieces) with piecewise smooth boundary such that:*

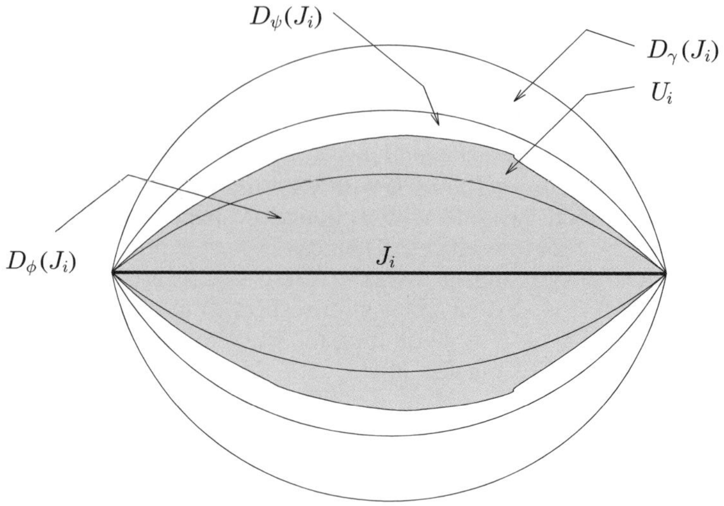

Fig. 3. Geometric domains

- *f restricted to the union of the puzzle pieces is a puzzle map;*
- *There is a real puzzle P such that the real slices $J_i = U_i \cap \mathbb{R}$ are puzzle pieces of P;*
- *The U_i are $\mathbb{R}$-symmetric;*
- *The closure of the union of the J_i contains a neighborhood of the interval $[-1, c_1]$ in $[-1, 1]$.*

And furthermore, there exist $0 < \phi < \psi < \gamma < \pi/2$ and $k > 0$ such that

(1) $D_\phi(J_i) \subset U_i \subset D_\psi(J_i)$;
(2) For each i, either the closure of $D_\gamma(J_i)$ is contained in $f(U_0)$ or it does not intersect the closure of $f(U_0)$;
(3) If U_i is contained in $f(U_0)$ then $\mathrm{mod}(f(U_0) \setminus U_i) > k$;
(4) There exists $\tilde{U}_0 \supset U_0$ with $\mathrm{mod}(\tilde{U}_0 \setminus U_0) > k$, such that if f^{n_i} maps U_i diffeomorphically onto U_0 then there is an $\tilde{U}_i \supset U_i$ which is mapped diffeomorphically onto $\tilde{U}_0$. Furthermore $\tilde{U}_i \subset f(U_0)$ or $\tilde{U}_i \cap f(U_0) = \emptyset$.

We will call (ϕ, ψ, γ, k) the geometric parameters of the puzzle.

Remark 5.4. The geometric conditions 1–4 of Definition 5.5 of a geometric puzzle map automatically imply the geometric conditions 1–3 of Definition 5.3 of a puzzle map.

If U is a geometric puzzle for f, then $f : U \to \mathbb{C}$ is naturally called a geometric puzzle map.

Definition 5.6. *We say that the puzzle $\mathcal{P}_f$ persists in a neighborhood $\mathcal{V}$ of f in $\mathcal{A}_a$ if there exists a normalized holomorphic motion $H_g : \mathbb{C} \to \mathbb{C}$, $g \in \mathcal{V}$,*

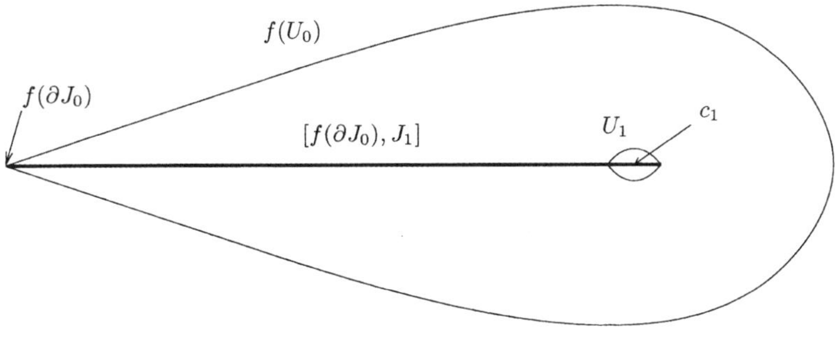

Fig. 4. Domain $f(U_0)$

such that $g(H_g(z)) = H_g(f(z)), \forall z \in \partial U^f$. *We denote* $H_g(U^f) = U^g$. *Notice that* $g|U^g$ *is automatically a puzzle map.*

Lemma 5.5. *Let* $0 < \phi < \psi < \gamma < \pi/2$ *and let* $k > 0$ *be arbitrarily big. If* $f \in \mathcal{U}_a$ *is a Yoccoz map, then there exists a puzzle for* f *with geometric parameters* (ϕ, ψ, γ, k).

Proof. Let $g = f^n\colon J \to J$ be the last pre-renormalization of f and let p be the fixed point of g in the interior of J. Let T^i be the principal nest for g.

Assume first that the critical point is recurrent. By Theorem 2.17, $|T^{i-1}|/|T^i|$ is unbounded. Let i_0 be such that $|T^{i_0-1}|/|T^{i_0}|$ is big. Letting $J_0 = T^{i_0}$, consider the associated real puzzle $\{J_i\}$. Let $\tilde{J}_0 = T^{i_0-1}$. In what follows, we discard the puzzle pieces to the right of the puzzle piece J_1. We still have that $\overline{\cup J_i}$ is a neighborhood of $[-1, c_1]$ in $[-1, 1]$ (since the set of points that never enter J_0 is a Cantor set, see Remark 2.8).

By Theorem 2.16, there are intervals $\tilde{J}_i \supset J_i$ such that f^{n_i} is a diffeomorphism from $\tilde{J}_i$ to $\tilde{J}_0$ and $\tilde{J}_i \cap J_0 = \emptyset$ for $i \neq 0$.

Let $\theta = (\phi + \psi)/2$ and let $U_0 = D_\theta(J_0)$. By Lemma 2.18, there exist topological disks $U_i \supset J_i$ such that for each i, f^{n_i} maps U_i onto U_0 and extends to a univalent map onto some $\tilde{U}_0$ with big $\mathrm{mod}(\tilde{U}_0 \setminus U_0)$. By the Koebe Distortion Theorem, each U_i satisfies $D_\phi(J_i) \subset U_i \subset D_\psi(J_i)$. In particular the domains U_i are pairwise disjoint.

Let us show that if U_i intersects $V \equiv f(U_0)$, then U_i is well inside V (see Fig. 4). Note first that in this case, $J_i \subset V \cap \mathbb{R}$. Otherwise there would be an interval J_k such that $f(J_k) = J_i$, $J_k \cap J_0 = \emptyset$, but $U_k \cap U_0 \neq \emptyset$ contradicting disjointness of the U_j. It follows that J_i is contained in the convex hull of $f(\partial J_0)$ and J_1 (since we assume there are no J_i to the right of J_1).

Notice also that there exists a constant ρ that only depends on θ and the initial bounds on f such that V contains $D_\rho(V \cap \mathbb{R})$. It follows that there is a κ (depending only on the same data) such that for any $x \in [f(\partial J_0), c_1]$, $\mathrm{dist}(x, f(\partial J_0)) < \kappa \,\mathrm{dist}(x, \partial V)$.

Furthermore, the interior of $\tilde{J}_i$ does not intersect $f(\partial J_0)$ since the orbit of $f(\partial J_0)$ never returns to the interior of $\tilde{J}_0$. Since $\operatorname{dist}(J_i, \partial \tilde{J}_i)/|J_i|$ is big when $|\tilde{J}_0|/|J_0|$ is big, we conclude that $|J_i|$ is much smaller than $\operatorname{dist}(J_i, f(\partial J_0))$. It follows that $|J_i|$ is much smaller than $\operatorname{dist}(J_i, \partial V)$. Thus, U_i is well inside V as was asserted. All other properties of a geometric puzzle are easily supplied.

Suppose now that the critical point is non-recurrent. Let T^i be the last interval of the principal nest, so that the iterates of 0 never return to the interior of T^i. It follows that the set of points in T^i which do not return to the interior of T^i accumulate on 0. Let us consider a small nice interval J_0 whose endpoints do not return to T^i and the associated real puzzle $\{J_i\}$.

Let $\tilde{J}_0 = T^i$. By Theorem 2.14, there are intervals $\tilde{J}_i \supset J_i$ such that f^{n_i} is a diffeomorphism from $\tilde{J}_i$ onto $\tilde{J}_0$. Notice that c_1 is accumulated from the right by intervals J_i, so we can discard all J_i outside a small neighborhood of $[-1, c_1]$ and still get a neighborhood of $[-1, c_1]$ in $[-1, 1]$. We can now argue exactly as before. □

Remark 5.5. Notice that the above construction can be adapted to construct a (non-geometric) puzzle for f with any central puzzle piece U_0 with a reasonable shape. More precisely, given $0 < \phi < \psi < \pi$, if $|T_n|/|T_{n-1}|$ is small enough, then any U_0 trapped in between $D_\phi(T_n)$ and $D_\psi(T_n)$ (i.e., $D_\phi(T_n) \subset U_0 \subset D_\psi(T_n)$) generates a puzzle $\{U_j\}$ whose real trace $\{J_j\}$ form a real puzzle for f.

Lemma 5.6. *Let $0 < \phi < \psi < \gamma < \pi/2$ and let $k > 0$ be arbitrarily big. If $f \in \mathcal{U}_a$ is a Yoccoz map, then there exists a puzzle for f with geometric parameters (ϕ, ψ, γ, k) which persists on a neighborhood $\mathcal{V} \subset \mathcal{A}_a$ of f.*

Proof. We will show that the puzzle given by Lemma 5.5 is a persistent puzzle. We will keep the notation of that lemma. We show how to define a holomorphic motion H_g in a neighborhood of f.

We first observe that the Cantor set Q of points which never enter J_0 is contained in a persistent Markov family $\{M_j\}$, that is, f restricted to $\cup M_j$ is strictly Markov (see Remark 2.8). So by Proposition 2.11 there is a holomorphic motion $h_g : \mathbb{C} \to \mathbb{C}$ over a neighborhood of f such that $h_g \circ f = g \circ h_g$ on $\cup M_j$.

Let us show that there is a neighborhood $\mathcal{V}$ of f such that for any i there is a holomorphic motion h^i_g of $Q \cup U_i$ such that $h^i_g|Q = h_g$ and $h_g \circ f^{n_i+1} = g^{n_i+1} \circ h^i_g$ in U_i. It is clear that for any fixed i we can get a neighborhood $\mathcal{V}_i$ where such a holomorphic motion is defined.

To deal with all i at the same time we notice that all but finitely many U_i are compactly contained in $\cup M_j$ since the U_i accumulate on Q and Q is compactly contained in $\cup M_j$. Let $\mathcal{I}$ be the set of U_i which are not compactly contained in $\cup M_j$. Let $\mathcal{J}$ be the set of U_i which are compactly contained in $\cup M_j$ but $f(U_i) \in \mathcal{I}$ (in particular $\mathcal{J}$ is also finite). Shrinking the neighborhood $\mathcal{V}$ if needed, we may suppose that for $g \in \mathcal{V}$, h_g^i is defined for $U_i \in \mathcal{I} \cup \mathcal{J}$, and $h_g^i(U_i) \subset \cup h_g(M_j)$ for $U_i \in \mathcal{J}$.

Now, if $U_i \notin \mathcal{I} \cup \mathcal{J}$, there is a unique k such that $f^k(U_i) = U_j$ belongs to $\mathcal{J}$. We then define $h_g^i | U_i$ so that $g^k \circ h_g^i = h_g^j \circ f^k$ in U_i. These are the desired holomorphic motions h_g^i.

We extend each of the holomorphic motions h_g^i to normalized holomorphic motions of $\mathbb{C}$.

The property that $\text{orb}(f(\partial J_0))$ never enters int $\tilde{J}_0$ implies that $\tilde{J}_j \cap J_i = \emptyset$, for all $i \neq j$ with $n_i \leq n_j$ (compare the proof of the previous Lemma). Hence $\text{dist}(J_j, \partial J_i)$ is much bigger than $|J_j|$. By the condition 1 in the definition of geometric parameters, we have for all k, $U_k \subset D_{\pi/2}(J_k)$, hence

$$\inf_{i \neq j} \text{dist}_S(U_i, U_j) > 0, \tag{5.2}$$

where $S = \mathbb{C} \setminus Q$ (use the Schwarz Lemma and $\partial J_j \subset Q$). Moreover, by condition 4 in the definition of geometric parameters, for all k there exists $\tilde{U}_i$ with $\tilde{U}_i \cap \partial V = \emptyset$ and $\text{mod}(\tilde{U}_i \setminus U_i) > k$, so by similar considerations,

$$\inf_i \text{dist}_S(U_i, \partial V) > 0, \quad \text{where } V \equiv f(U_0). \tag{5.3}$$

By the Quasiconformality Lemma (see §2.5), $\mathcal{V}$ can be shrunk so that the dilatation of the motions h_g and h_g^i will be close to 1. By Lemma 2.3,

$$\begin{aligned} &\inf_{i \neq j} \text{dist}_{h_g(S)} \left(h_g^i(U_i), h_g^j(U_j)\right) > 0, \\ &\inf_i \text{dist}_{h_g(S)} \left(h_g^i(U_i), h_g(\partial V)\right) > 0. \end{aligned} \tag{5.4}$$

So we can define a holomorphic motion H_g which agrees with h_g on $Q \cup \partial V$ and with h_g^i on U_i. This concludes the proof of the lemma. □

Remark 5.6. The above proof also shows that the holomorphic motion H_g, $g \in \mathcal{V}$ associated to the persistent puzzle $\mathcal{P}_f$ may be taken $\mathbb{R}$-symmetric. More precisely, we can choose $\mathcal{V}$ to be a small ball around f and construct H_g with the property that $H_g(\overline{z}) = \overline{H_{\text{conj}(g)}(z)}$, where $\text{conj}(g)(\overline{z}) = \overline{g(z)}$. To see this, notice that all the constructive (dynamical) steps in the above proof respect the $\mathbb{R}$-symmetry. If we also choose all neighborhoods of f to be small balls centered at f, then the successive applications of the Extension Lemma in the above argument yield $\mathbb{R}$-symmetric holomorphic motions by Remark 2.2.

From now on, when considering a holomorphic motion associated to a persistent geometric puzzle we will always assume it is $\mathbb{R}$-symmetric. In particular, if $g \in \mathcal{U}_a \cap \mathcal{V}$ then H_g is an $\mathbb{R}$-symmetric qc map.

Remark 5.7. The above construction is refined in Lemma 7.9. For a related but different construction see also Lemma 12.5 of [LS2].

6. Pullback arguments and the key estimate

6.1. Macroscopic pullback argument. Let $f : U^f \to \mathbb{C}$ and $\tilde{f} : U^{\tilde{f}} \to \mathbb{C}$ be two puzzle maps, and let h be a homeomorphism of $\mathbb{C}$ such that $h(U^f) = U^{\tilde{f}}$, equivariant on ∂U^f. If $h(f(0)) = \tilde{f}(0)$, then there is a unique homeomorphism h_1 coinciding with h on $\mathbb{C}\setminus U^f$ and such that $h \circ f = \tilde{f} \circ h_1$ on U^f. It is called the *lift* of h.

Definition 6.1. *We say that a homeomorphism $h : \mathbb{C} \to \mathbb{C}$ is a combinatorial equivalence between f and $\tilde{f}$ if it is equivariant on ∂U^f and the lift h_1 of h is homotopic to h rel $\partial U^f \cup \operatorname{orb}_f(0)$.*

Lemma 6.1. *Let h be a qc combinatorial equivalence between f and $\tilde{f}$. Then the lift h_1 of h is a qc combinatorial equivalence between f and $\tilde{f}$ and* $\operatorname{Dil}(h_1) \le \operatorname{Dil}(h)$.

Proof. Let H^t be a homotopy between h and h_1 rel $\partial U^f \cup \operatorname{orb}_f(0)$, and let H_1^t be the lift of H^t. Then H_1^t is a homotopy rel $\partial U^f \cup \operatorname{orb}_f(0)$ between h_1 and its lift h_2.

Define a sequence of homeomorphisms ψ_k as follows

$$\psi_k = \begin{cases} h_1 & \text{on } \cup_{j=0}^k U_j^f \\ h & \text{otherwise.} \end{cases}$$

We notice that ψ_k is quasiconformal on $\mathbb{C} \setminus (\cup_{j=0}^k \partial U_j \cup \{0\})$, since it coincides with h on the complement of $\cup_{j=0}^k U_j$ and is the conformal lift of h on $\cup_{j=0}^k U_j \setminus \{0\}$. Since quasiarcs and points are qc removable, ψ_k is quasiconformal on the whole complex plane. Since quasiarcs have zero Lebesgue measure, the estimate $\operatorname{Dil}(\psi_k) \le \operatorname{Dil}(h)$ follows from the fact that conformal lifts preserve the norm of Beltrami differentials.

Since $\psi_k \to h_1$ pointwise, the result follows from the First Compactness Lemma. □

The Pullback Argument in the context of quadratic-like maps was formulated by Sullivan, see [MS], Chap. 6, Sect. 4. The following result adapts it to the context of puzzle maps.

Theorem 6.2. *Let us consider two puzzle maps f and $\tilde{f}$ with all periodic orbits hyperbolic. Let h be a qc combinatorial equivalence between f and $\tilde{f}$. Then there is a qc homeomorphism $H : \mathbb{C} \to \mathbb{C}$ such that $H \circ f = \tilde{f} \circ H$ on U^f, $H = h$ on $\mathbb{C} \setminus U^f$, and* $\operatorname{Dil}(H) \le \operatorname{Dil}(h)$. *If there are no invariant line fields on $K(f)$ or if f and $\tilde{f}$ are hyperbolic maps with the same multiplier, then* $\operatorname{Dil}(H) \le \operatorname{Dil}(h|\mathbb{C} \setminus U^f)$.

Proof. Assume first that all periodic orbits are repelling. In this case $K(f)$ has empty interior by Lemma A.21.

Let $h_0 = h$ and define by induction h_{k+1} as the lift of h_k. By the previous lemma all h_k are qc maps with $\mathrm{Dil}(h_k) \leq \mathrm{Dil}(h)$. By the First Compactness Lemma, there exists an accumulation point H of h_k which is quasiconformal and whose dilatation is bounded by the dilatation of h. Since the $h_k(z)$ are eventually the same for any $z \in \mathbb{C} \setminus K(f)$, H is a conjugacy between f and $\tilde{f}$ on $U^f \setminus K(f)$. Since $K(f)$ has empty interior, H is a conjugacy between f and $\tilde{f}$ on U^f.

Moreover, $\mathrm{Dil}(H|\mathbb{C} \setminus K(f) \leq \mathrm{Dil}(h|\mathbb{C} \setminus U^f)$. Hence, if there are no invariant line fields on $K(f)$, then the dilatation of H on the entire complex plane has the same bound.

If f and $\tilde{f}$ are hyperbolic maps with the same multiplier we can use Lemma A.26 to modify h inside $K(f)$ in order to have $\overline{\partial} h|K(f) = 0$. □

Remark 6.1. We will show in the Appendix A that the assumption that a puzzle map f has only hyperbolic periodic orbits is satisfied for all complex preperiodic or hyperbolic maps (Lemma A.23), as well as for the puzzle extensions of unimodal maps without parabolic points on I (Lemma A.20).

Lemma 6.3. *There exists a constant $L > 0$ with the following property. Let $f \in \mathcal{U}_a$ be a Yoccoz map, and let $\mathcal{P}_f$ be a geometric puzzle which persists in an ε-neighborhood of f in $\mathcal{A}_a$ (which exists by Lemma 5.6). Let $\mathcal{V}$ be an $\varepsilon/2$-neighborhood of f. If $g \in \mathcal{U}_a \cap \mathcal{V}$ is preperiodic or hyperbolic and $\tilde{g}$ belongs to the same connected component of $\mathcal{C}_g \cap \mathcal{V}$, then there is a normalized L-qc homeomorphism $h : \mathbb{C} \to \mathbb{C}$ equivariant with respect to g and $\tilde{g}$ on U^g.*

Proof. Let H_g be the holomorphic motion of $\mathcal{P}_f$. By the λ-lemma, the dilatation of the motion restricted to $\mathcal{V}$ is bounded by some constant L.

Consider some preperiodic or hyperbolic $g \in \mathcal{U}_a \cap \mathcal{V}$ and a holomorphic path $g_\lambda, \lambda \in \mathbb{D}$, in $\mathcal{C}_g \cap \mathcal{V}$ connecting $g_0 = g$ with $\tilde{g}$ (such path exists since $\mathcal{C}_g$ is a codimension-one analytic submanifold). Let $H_\lambda = H_{g_\lambda} \circ H_g^{-1}$ denote the corresponding holomorphic motion of the puzzle $\mathcal{P}_{g_\lambda}$ with base point $\lambda = 0$.

Let $h_\lambda = H_\lambda$ on the complement of U^g and $h_\lambda(g^n(0)) = g_\lambda^n(0)$ whenever $g^n(0) \in U^g$.

Assume first that the critical orbit of g never escapes U^g. Then

$$g^n(0) \notin \partial U^g, \quad n = 0, 1, \ldots. \tag{6.1}$$

Let us consider the set W of points $\lambda_0 \in \mathbb{D}$ such that h_λ is injective in a neighborhood of λ_0. Then $0 \in W$. Indeed, by (6.1), any curve $\lambda \mapsto g_\lambda^n(0)$ does not collide with ∂U^{g_λ} for λ sufficiently close to 0 (depending on n). This makes the statement obvious in the preperiodic case. In the hyperbolic case, there exist $N \in \mathbb{N}$ and neighborhoods $V \ni 0$ and $\Lambda \ni 0$ such that $g_\lambda|V$

is injective and $g_\lambda(V) \Subset V$ for $\lambda \in \Lambda$, and $g_\lambda^N(0) \in V$. Then the points $g_\lambda^n(0)$, $n \geq N$, do not collide neither with ∂U^{g_λ}, nor with each other.

Let W_0 be the connected component of W containing 0. We claim that for each $\lambda_0 \in W_0$, h_{λ_0} extends to a qc combinatorial equivalence $\hat{h}_{\lambda_0}$ between g and g_{λ_0}. To see it, let us consider a simply connected domain $D \subset W_0$ containing 0 and λ_0. Restricting the motion h_λ to $\lambda \in D$ and using Slodkowski's Theorem (see Remark 2.3)[5], we obtain an extension $\hat{h}_\lambda : \mathbb{C} \to \mathbb{C}$, $\lambda \in D$. By the Quasiconformality Lemma, $\hat{h}_{\lambda_0}$ is qc. Moreover, since $\hat{h}_{\lambda_0}$ comes from a holomorphic motion, it is a combinatorial equivalence between g and g_{λ_0}[6]. This concludes the proof of the claim.

By Theorem 6.2 (together with Remark 6.1), there exists an L-qc map of $\mathbb{C}$ which conjugates g_λ to g on U^g. By the First Compactness Lemma, for any $\lambda \in \overline{W_0} \cap \mathbb{D}$, g_λ is L-qc conjugate to g on U^g.

Vice versa, if g_λ is qc conjugate to g on U^g, then there exists a neighborhood of λ where h_λ is injective, so $\lambda \in W$. Hence $\overline{W_0} \cap \mathbb{D} \subset W_0$, so that W_0 is open and closed. Thus, $W_0 = \mathbb{D}$, so that h_λ is always injective and g_λ is always L-qc conjugate to g on U^g.

Assume now that the critical orbit of g escapes U. Since g is real, there is a smallest n such that $g^n(0)$ belongs to the invariant Cantor set $Q \equiv \partial U^g \cap \mathbb{R}$ (see Proposition 2.15). The holomorphic motion H_λ restricts to the dynamical motion of Q given by Proposition 2.11. Hence $H_\lambda(g^m(0)) = g_\lambda^m(0)$, for $m \geq n$, so that the maps h_λ are equivariant on $\mathrm{orb}_g(0) \cup \partial U_g$.

By the same argument of the non-escaping case, the points $g_\lambda^m(0) = h_\lambda(g^m(0))$, $m < n$, do not collide with $H_\lambda(\mathbb{C} \setminus U^g)$, which yields the desired statement. □

Remark 6.2. In the above Lemma we constructed, for each $\tilde{g}$ in the connected component of g in $\mathcal{C}_g \cap \mathcal{V}$, a qc map $h_{\tilde{g}}$ which has the three following properties:

- $h_{\tilde{g}} | \mathbb{C} \setminus U^g = H_{\tilde{g}} \circ H_g^{-1}$ is a holomorphic motion of the puzzle with base point g;
- $h_{\tilde{g}}$ is equivariant on U_g;
- $\bar{\partial} h_{\tilde{g}} = 0$ on $K(g)$.

It is easy to see that such a qc map $h_{\tilde{g}}$ is uniquely defined by the above properties (once the motion $H_{\tilde{g}}$ is fixed). It follows that the map $\tilde{g} \mapsto h_{\tilde{g}} | \mathbb{C} \setminus \mathrm{int}\, K(g)$ depends holomorphically on $\tilde{g}$ along the connected component of g in $\mathcal{C}_g \cap \mathcal{V}$. Indeed, this is automatic in the case where $K(g)$ has empty interior. By Lemma A.22, since g is $\mathbb{R}$-symmetric, if $K(g)$ has non-empty interior then $\mathrm{int}\, K(g)$ is the basin of attraction of an attracting

[5] We can use the Extension Lemma instead to construct a piecewise holomorphic motion joining 0 with λ_0, see Remark 2.4.

[6] If we denote the lift of $\hat{h}_\lambda$ by $\hat{h}_{\lambda,1}$ and let $s : [0, 1] \to D$ be any path connecting 0 to λ_0, then a homotopy between $\hat{h}_{\lambda_0}$ and its lift can be written explicitly as $\hat{h}_{\lambda_0} \circ \hat{h}_{s(t)}^{-1} \circ \hat{h}_{s(t),1}$, $t \in [0, 1]$.

periodic cycle. The holomorphic dependence of the linearizing coordinate of an attracting periodic cycle implies that $h_{\tilde{g}}$ depends holomorphically in int $K(g)$ as well.

6.2. Infinitesimal Pullback Argument. The "Infinitesimal Pullback Argument" introduced in this section will allow us to reconstruct, by means of consecutive liftings (see §2.6.2), a qc vector field which is equivariant on some essential part of the dynamical domain into a qc vector field which is equivariant on the whole domain. One step of the pullback argument is given by the following lemma:

Lemma 6.4. *Let $\Omega \ni 0$ be a quasidisk. Consider a map $f \in A^1(\Omega)$ whose derivative does not vanish on $\overline{\Omega} \setminus \{0\}$. Assume that $f : \overline{\Omega} \to f(\overline{\Omega})$ is either a diffeomorphism or a double branched covering ramified at 0. Let $v \in \mathcal{B}_\Omega$. Let α and β be qc vector fields on $\mathbb{C}$ such that $\beta|\partial\Omega$ is the lift of α by (f, v). Moreover, if f is a double branched covering, we assume that $v(0) = \alpha(f(0))$. Then there exists a qc vector field γ such that $\gamma|\Omega$ is the lift of α by (f, v), $\gamma|\mathbb{C} \setminus \Omega = \beta$, and*

$$\|\bar{\partial}\gamma\|_\infty \leq \max\{\|\bar{\partial}\alpha\|_\infty, \|\bar{\partial}\beta\|_\infty\}. \tag{6.2}$$

Proof. Define a continuous vector field γ on $\mathbb{C} \setminus \{0\}$ by letting $\gamma = \beta$ on $\mathbb{C} \setminus \Omega$ and letting $\gamma = (\alpha \circ f - v)/f'$ on $\Omega \setminus \{0\}$.

If f is a diffeomorphism then γ clearly extends to 0. Assume f is a branched double covering. Since the modulus of continuity of qc vector fields is $\phi(x) = -x \ln(x)$ (see §2.6.1), we have for z near 0:

$$|\alpha(f(z)) - \alpha(f(0))| = O(\phi(|f(z) - f(0)|)) = O(\phi(|z|^2)).$$

Since $v(0) = \alpha(f(0))$ and f' has a simple root at 0, we have

$$\gamma(z) = \frac{v(0) - v(z)}{f'(z)} + O(\phi(|z|)),$$

where the first term is a regular holomorphic function. Hence γ admits a continuous extension to 0.

It is clear that γ is quasiconformal on $\mathbb{C} \setminus (\partial\Omega \cup \{0\})$. Since quasiarcs and isolated points are qc removable, γ is quasiconformal on the whole complex plane.

Since the lifts preserve the qc norm of vector fields, we have $\|\bar{\partial}\gamma\|_\infty = \|\bar{\partial}\alpha\|_\infty$ on Ω, while we have $\|\bar{\partial}\gamma\|_\infty = \|\bar{\partial}\beta\|_\infty$ on $\mathbb{C} \setminus \Omega$. Since quasiarcs are removable, (6.2) follows. □

Remark 6.3. This lemma extends to branched coverings of any degree n with some extra conditions on the derivatives of v (which assure that the critical point does not bifurcate infinitesimally along $f + tv$).

Theorem 6.5. *Let $f : U \to \mathbb{C}$ be a puzzle map whose critical point does not escape $\overline{U}$, and let v be a tangent vector field at f. Assume there exists a normalized qc vector field β on $\mathbb{C}$ which is equivariant on $\partial U \cup \text{orb}(0)$. Then there exists an equivariant (on U) qc vector field α with $\|\bar{\partial}\alpha\|_\infty \leq \|\bar{\partial}\beta\|_\infty$ which coincides with β on $\mathbb{C} \setminus U$. Furthermore, if there are no invariant line fields on $K(f)$, then $\|\bar{\partial}\alpha\|_\infty \leq \|\bar{\partial}\beta \mid \mathbb{C} \setminus U\|_\infty$.*

Proof. This proof is the essence of the "infinitesimal pullback argument".

Using Lemma 6.4, let us lift the vector field β to the puzzle piece U_0 by means of $f : U_0 \to \mathbb{C}$. We obtain a qc vector field γ_0 on $\mathbb{C}$ with $\|\bar{\partial}\gamma_0\|_\infty \leq \|\bar{\partial}\beta\|_\infty$, coinciding with β on $\mathbb{C} \setminus U_0$. Let $U^m = (f|U \setminus U_0)^{-m}(U_0)$. Define inductively a sequence of qc vector fields γ_m on $\mathbb{C}$ by letting γ_m on U^m be the lift of γ_{m-1} and letting $\gamma_m = \gamma_{m-1}$ on $\mathbb{C} \setminus U^m$. Since holomorphic lifts preserve the qc norm, $\|\bar{\partial}\gamma_n\|_\infty \leq \|\bar{\partial}\gamma_{n-1}\|_\infty$.

By The Second Compactness Lemma, this sequence of vector fields is precompact. Since it stabilizes pointwise, it converges to a qc vector field α_1. This vector field satisfies the following properties:

- $\alpha_1|U_0$ is the lift of β;
- $\alpha_1|U \setminus U_0$ is equivariant;
- $\alpha_1|\mathbb{C} \setminus U = \beta$;
- $\|\bar{\partial}\alpha_1\|_\infty \leq \|\bar{\partial}\beta\|_\infty$.

Replace now $\beta \equiv \alpha_0$ with α_1 and repeat the procedure. In this way we will construct a sequence of qc vector fields α_n with the following properties

(i) $\alpha_n|U_0$ is the lift of α_{n-1};
(ii) $\alpha_n|U \setminus U_0$ is equivariant;
(iii) $\alpha_n|\mathbb{C} \setminus U = \alpha_{n-1}$ (so by induction $\alpha_n|\mathbb{C} \setminus U = \beta$ for all n);
(iv) $\|\bar{\partial}\alpha_n\|_\infty \leq \|\bar{\partial}\alpha_{n-1}\|_\infty$.

Taking any Cesaro limit of this sequence (which exists by the Second Compactness Lemma),

$$\alpha = \lim \frac{1}{n_i} \sum_{k=0}^{n_i - 1} \alpha_k,$$

we obtain a qc vector field α with $\|\bar{\partial}\alpha\|_\infty \leq \|\bar{\partial}\beta\|_\infty$. Properties (i)–(iii) imply that α is equivariant everywhere on U.

Note that α on $\mathbb{C} \setminus K(f)$ is obtained by consecutive liftings of β on $\mathbb{C} \setminus U$. Hence

$$\|\bar{\partial}\alpha \mid \mathbb{C} \setminus K(f)\|_\infty \leq \|\bar{\partial}\beta \mid \mathbb{C} \setminus U\|_\infty.$$

If there are no invariant line fields on $K(f)$ then

$$\|\bar{\partial}\alpha\|_\infty = \|\bar{\partial}\alpha \mid \mathbb{C} \setminus K(f)\|_\infty,$$

and the last assertion follows (see Lemma 2.30). □

Remark 6.4. The same proof also applies to the case when the critical point escapes. More precisely, let f be a puzzle map and $\Omega \subset U$ be a union of puzzle pieces. Let $n = \min\{k \geq 0, c_k \notin \Omega\}$ (the escaping time of the critical point). Assuming the critical point escapes Ω, we have $n < \infty$. If β is a vector field equivariant on $\partial\Omega$ and on $\{c_k, 0 \leq k \leq n-1\}$, then there exists a vector field α equivariant on Ω coinciding with β outside of Ω and such that $\|\bar{\partial}\alpha\|_\infty \leq \|\bar{\partial}\beta\|_\infty$. By Lemma A.23, $K(f|\Omega)$ has zero Lebesgue measure, so $\|\bar{\partial}\alpha\|_\infty \leq \|\bar{\partial}\beta|\mathbb{C} \setminus U\|_\infty$.

6.2.1. Gluing qc vector fields. Assume now that we have two qc vector fields β_i on $\mathbb{C}$ which respect some dynamical data on sets $X_i \subset \mathbb{C}$, $i = 0, 1$. Let us say that these vector fields can be *glued* if there is a qc vector field on $\mathbb{C}$ coinciding with β_i on X_i. The case which will interest us is when $X_1 = O_f$, $X_0 = \mathbb{C} \setminus U$.

Note first that if X_0 and X_1 have disjoint closures, the vector fields β_i can be obviously glued using partition of unity. Together with Theorem 6.5 this implies:

Lemma 6.6. *Let f be a puzzle map with non-escaping critical point such that the postcritical set O_f is disjoint from ∂U, and let $v \in \mathcal{B}_U$. Assume that there are qc vector fields β_0 and β_1 on $\mathbb{C}$ such that β_1 is equivariant on* $\operatorname{orb}_f(0)$ *and β_0 is equivariant on ∂U. Then there exists an equivariant qc vector field α on U coinciding with β_0 on $\mathbb{C} \setminus U$ and coinciding with β_1 on O_f. If there are no invariant line fields on $K(f)$, then $\|\bar{\partial}\alpha\|_\infty \leq \|\bar{\partial}\beta_0 \,|\, \mathbb{C} \setminus U\|_\infty$.*

If the postcritical set intersects the boundary of the puzzle, the gluing method is much more delicate. It will be based on a pullback construction.

Recall that $U(x)$ is the connected component of U containing x.

Lemma 6.7. *Let f be a puzzle map such that $c_1 \in U$. Let $v \in \mathcal{B}_U$, and let β_0 be a qc vector field on $\mathbb{C}$ equivariant on ∂U. Then there are constants C_1 and C_2 with the following property. Consider a point $x \in U$ such that $0 \in \omega(x)$. If there exists a bounded vector field β equivariant on* $\operatorname{orb}(x)$ *then there exists a qc vector field α on $\mathbb{C}$ such that*

- $\|\bar{\partial}\alpha\|_\infty \leq C_1 + C_2\|\beta\|_\infty$;
- $\alpha(x) = \beta(x)$;
- *α coincides with β_0 outside of $U(x)$.*

Proof. Let $W = (f|U_0)^{-1}(U)$ (the domain of the first return map to U_0) and $W_0 = W(0)$ (the central domain of the first return map).

Let β_1 be a qc vector field on $\mathbb{C}$ coinciding with β_0 on $\mathbb{C} \setminus U_1$ and such that $v(0) = \beta_1(c_1)$. By Lemma 6.4 there exists a qc vector field ζ coinciding with β_0 on $\mathbb{C} \setminus U_0$ and such that ζ is the lift of β_1 on U_0. So ζ is equivariant on $\partial U \cup \partial W$.

Since $\overline{W}_0 \subset U_0$, there exist qc vector fields θ_0, θ_1 such that $\theta_0|W_0 = 1$, $\theta_0|\mathbb{C} \setminus U_0 = 0$, and $\theta_1|W_0 = 0$, $\theta_1|\mathbb{C} \setminus U_0 = \beta_0$. We let

$$C_1 = \max\{\|\bar{\partial}\beta_0\|_\infty, \|\bar{\partial}\zeta\|_\infty, \|\bar{\partial}\theta_1\|_\infty\} \quad \text{and} \quad C_2 = \|\bar{\partial}\theta_0\|_\infty.$$

Let s_i denote the sequence of landing moments of x, that is,

$$s_0 = \min\{k \geq 0 | f^k(x) \in U_0\} \quad \text{and} \quad s_{i+1} = \min\{k > s_i | f^k(x) \in U_0\}.$$

Let $m = \min\{k \geq 0 | f^{s_k}(x) \in W_0\}$, so that $f^{s_m}(x)$ is the first landing of x on W_0 ($m < \infty$ since $0 \in \omega(x)$). For each $k \leq m$ there exists a unique quasidisk V_k around x such that $f^{s_k}|\overline{V}_k$ is a diffeomorphism onto $\overline{U}_0$.

Let $\eta = \theta_1 + \beta(f^{s_m}(x))\theta_0$ (we identify freely vector fields and functions when defined on subsets of $\mathbb{C}$), so that

$$\eta|\mathbb{C} \setminus U_0 = \beta_0, \quad \eta(f^{s_m}(x)) = \beta(f^{s_m}(x)), \quad \|\overline{\partial}\eta\|_\infty \leq C_1 + C_2\|\beta\|_\infty.$$

Let γ_m be the lift to V_m of η by (f^{s_m}, v^{s_m}). For each $k < m$, let γ_k be the lift to V_k of ζ by (f^{s_k}, v^{s_k}). So we have constructed, for $k \leq m$, qc vector fields γ_k on V_k such that $\|\overline{\partial}\gamma_k\|_\infty \leq C_1 + C_2\|\beta\|_\infty$.

Notice that the equivariance of ζ on $\partial U \cup \partial W$ implies that, for $k < m$, γ_k and γ_{k+1} coincide on ∂V_{k+1}. It also follows that β_0 coincides with γ_0 on ∂V_0.

Let $\alpha = \beta_0$ outside V_0, $\alpha = \gamma_k$ on $V_k \setminus V_{k+1}$, $k < m$, and $\alpha = \gamma_m$ on V_m. Then α is a continuous vector field on $\mathbb{C}$ and since quasiarcs are removable, α is a qc vector field with $\|\overline{\partial}\alpha\|_\infty \leq C_1 + C_2\|\beta\|_\infty$. By equivariance of β on $\text{orb}(x)$, $\alpha(x) = \beta(x)$. □

Remark 6.5. The proof above uses that $0 \in \omega(x)$ only to assure that there exists an $m < \infty$ such that $f^{s_m}(x) \in W_0$. If this is not the case (but x is still non-escaping from U), then one can show that the conclusion still holds under the (weak) assumption $|Df^{s_k}(x)| \to \infty$.

Lemma 6.8. *Let f be a puzzle map with a recurrent critical point such that* orb(0) *intersects infinitely many puzzle pieces, and let $v \in \mathcal{B}_U$. Assume there are qc vector fields β_0 and β_1 on $\mathbb{C}$ such that β_1 is equivariant on* orb(0) *and β_0 equivariant on ∂U. Then there is a qc vector field α equivariant on U coinciding with β_0 on $\mathbb{C} \setminus U$ (and coinciding with β_1 on* orb(f(0))*). Furthermore, if there are no invariant line fields on $K(f)$, then $\|\overline{\partial}\alpha\|_\infty \leq \|\overline{\partial}\beta_0|\mathbb{C} \setminus U\|_\infty$.*

Proof. By the assumption, there exists a sequence $k(j) \to \infty$ and a sequence V_j of components of U such that

$$c_{k(j)} \in V_j \quad \text{but} \quad c_i \notin V_j, \; i < k(j).$$

Let $\Omega_j = U \setminus V_j$.

We first construct a uniformly bounded sequence of vector fields α_j equivariant only on Ω_j. This is done in three steps.

In the first, we modify the vector field β_0 inside V_j to obtain the correct value (given by β_1) on $c_{k(j)}$, uniform bounds come from Lemma 6.7. In other words, there exists a constant C and a sequence γ_j of vector fields such that $\|\overline{\partial}\gamma_j\|_\infty \leq C$, $\gamma_j(c_{k(j)}) = \beta_1(c_{k(j)})$, and γ_j coincides with β_0 on the complement of V_j.

In the second step, we prepare ourselves for a pullback argument by modifying γ_j inside Ω_j to obtain the correct values in the finite set $\{c_i, 0 \leq i < k(j)\}$. More precisely, we define a sequence ζ_j of qc vector fields on $\mathbb{C}$ such that ζ_j coincides with γ_j on $\mathbb{C} \setminus \Omega_j$ and ζ_j coincides with β on c_i, $0 \leq i \leq k(j)$. In this step we lose uniform estimates, which are recovered in the next one.

The third step is the pullback argument truncated to Ω_j. Since ζ_j is equivariant on $\partial\Omega_j$ and on c_i, $0 \leq i < k(j)$, we can apply the Infinitesimal Pullback Argument (or rather, its escaping version outlined in Remark 6.4) to obtain a qc vector field α_j equivariant on Ω_j and coinciding with ζ_j on $\mathbb{C} \setminus \Omega_j$ (hence $\alpha_j | \mathbb{C} \setminus \Omega_j = \gamma_j$). Moreover, $\|\bar{\partial}\alpha_j\|_\infty \leq C$.

To obtain the desired vector field equivariant on all of U we just need to take a limit α of the vector fields α_j (using the Second Compactness Lemma). If there are no invariant line fields, the estimate on α follows as before.

Notice that the equivariance of α and β_1 on the critical orbit determines uniquely their values on the orbit of the critical value, so $\alpha | \operatorname{orb}(f(0)) = \beta_1$. □

Remark 6.6. In the above proof, it was only used that the vector field β_1 is bounded on orb(0). No assumption of quasiconformality or even continuity is necessary.

Lemmas 6.6 and 6.8 immediately imply:

Theorem 6.9. *Let f be a puzzle map with a recurrent critical point, and let $v \in \mathcal{B}_U$. Assume there are qc vector fields β_0 and β_1 such that β_1 is equivariant on* orb(0) *and β_0 is equivariant on ∂U. Then there is a qc vector field α equivariant on U and coinciding with β_0 on $\mathbb{C} \setminus U$. Furthermore, if there are no invariant line fields on $K(f)$, then $\|\bar{\partial}\alpha\|_\infty \leq \|\bar{\partial}\beta_0 | \mathbb{C} \setminus U\|_\infty$.*

Let us finish this section with a discussion of infinitesimal deformations of hyperbolic puzzle maps.

Lemma 6.10. *Let f be a hyperbolic puzzle map. If v satisfies equation (5.1) then there exists a qc vector field α on $\mathbb{C}$ which is conformal and equivariant on a neighborhood of* $\overline{\operatorname{orb}(0)}$.

Proof. Let p be the attracting periodic point whose immediate basin of attraction contains 0, and let n stand for its period. Take some f^n-invariant topological disks $V \Subset V' \ni p$ containing $\overline{\operatorname{orb}_{f^n}(0)}$. It is enough to find a solution of the equation $v^n(z) = \alpha(f^n(z)) - \alpha(z) Df^n(z)$ which is conformal and equivariant in V, since such solution can be spread (by means of $n-1$ lifts) around the rest of the orbit of 0 and then extended arbitrarily to a qc vector field on $\mathbb{C}$.

Equation (5.1) tells us that v^n is an infinitesimal deformation of $f^n | V'$ preserving the multiplier of the attracting periodic point. Hence there is a holomorphic family $g_\lambda : V \to \mathbb{C}$, $g_0 = f^n$, tangent to v^n, such that each g_λ has an attracting fixed point p_λ (depending analytically on λ) with multiplier $Df^n(p)$. By the standard local theory, there is a holomorphic family of conformal maps

$$\psi_\lambda : (V, 0) \to (V_\lambda, 0)$$

conjugating f^n to g_λ. Then $\alpha = d\psi/\partial\lambda|_{\lambda=0}$ is the desired vector field. □

Remark 6.7. The above equation can also be solved directly using the Böttcher or linearizing coordinate near p. In these coordinates, it is reduced to either $u(s) = a(s^2) - 2sa(s)$ (superattracting case) or $u(s) = a(\lambda s) - \lambda a(s)$ (simply attracting case). Both equations can be easily analyzed by means of power expansions.

Lemma 6.11. *Let f be a hyperbolic puzzle map and let $v \in \mathcal{B}_U$ be a vector field satisfying equation (5.1). Let β_0 be a qc vector field on $\mathbb{C}$ equivariant on ∂U. Then there exists a vector field α on $\mathbb{C}$ equivariant on U such that $\|\bar\partial\alpha\|_\infty \leq \|\bar\partial\beta_0|\mathbb{C} \setminus U\|_\infty$. Moreover, α is holomorphic on the basin of attraction of the attracting cycle.*

Proof. Let β be the vector field given by Lemma 6.10. We can create a vector field γ by gluing β_0 on the complement of U with β on a neighborhood of orb(0). Applying the Infinitesimal Pullback Argument (Theorem 6.5), we obtain a vector field α which is equivariant on U, conformal on the basin of attraction int $K(f)$ and satisfies

$$\|\bar\partial\alpha|\mathbb{C} \setminus K(f)\|_\infty \leq \|\bar\partial\beta_0|\mathbb{C} \setminus U\|_\infty.$$

By Lemma A.21, $\partial K(f)$ has zero Lebesgue measure for a hyperbolic puzzle map, so the estimate follows. □

6.3. Key estimate. We say that a preperiodic or hyperbolic complex map g has *special combinatorics* with respect to $\mathcal{V}$, a complex neighborhood of g, if the connected component of g in $\mathcal{C}_g \cap \mathcal{V}$ contains a real map.

Recall that L_g is the linear map which associates to any tangent vector field $v \in T_g$ the unique qc vector field α on the postcritical set such that $v = \alpha \circ f - \alpha f'$ and $v(0) = \alpha(c_1)$.

Lemma 6.12 (Key estimate). *Let $f \in \mathcal{U}_a$ be a Yoccoz map. There exists a neighborhood $\mathcal{V}$ of f in $\mathcal{A}_a$ and a constant $C > 0$ such that, for any g with special combinatorics with respect to $\mathcal{V}$, the operator norm of L_g is bounded by C.*

Proof. Consider a persistent puzzle for f given by Lemma 5.6. Take an $\varepsilon > 0$ such that this puzzle persists in an ε-neighborhood of f, and let H_g be the associated holomorphic motion. Let $\mathcal{V}$ be an $\varepsilon/2$-neighborhood of f

and let $C = 2/\varepsilon$. Given $g \in \mathcal{V}$ with special combinatorics and $v \in T_g$ with $\|v\|_a = 1$, let $h_\lambda = H_{g+\lambda v} \circ H_g^{-1}$, $\lambda \in \mathbb{D}_{\varepsilon/2}$. Let

$$\beta_0 = \frac{d}{d\lambda} h_\lambda \bigg|_{\lambda=0}.$$

Notice that β_0 is equivariant on ∂U^g with respect to (g, v). By Theorem 2.7, μ_{h_λ} is a holomorphic function on $\mathbb{D}_{\varepsilon/2}$ with values in the unit ball of $L^\infty(\mathbb{C})$. By Lemma 2.10,

$$\overline{\partial}\beta_0 = \frac{d}{d\lambda} \mu_{h_\lambda} \bigg|_{\lambda=0}.$$

Since $\mu_{h_0} = 0$, we can apply the Schwarz Lemma to get $\|\overline{\partial}\beta_0\|_\infty \leq 2/\varepsilon$.

Assume g is preperiodic. By Lemma A.23, g has no invariant line fields on $K(g)$.

By Lemma 6.3, the special combinatorics assumption implies that $g|U^g$ is qc conjugate to $\tilde{g}|U^{\tilde{g}}$ for some real map $\tilde{g} \in \mathcal{V}$; in particular, the critical orbit does not escape $\overline{U^g}$. By Theorem 6.5, there exists a qc vector field α equivariant on U^g, coinciding with β_0 on $\mathbb{C} \setminus U$, and such that $\|\overline{\partial}\alpha\|_\infty \leq C$.

In the hyperbolic case we proceed as above using Lemma 6.11. □

Remark 6.8. It is possible to show that for each $f \in \mathcal{U}_a$, L_f has a bounded operator norm. Near infinitely renormalizable maps, we can reduce to a persistent quadratic-like renormalization, and our argument works unchanged. For the hyperbolic and parabolic case, one can obtain explicit estimates, using the formulas for the tangent space. However, in the parabolic case those explicit estimates are less stable than the ones in our arguments, and do not seem to allow to obtain estimates for nearby maps.

7. Transverse direction

In this section we will construct a transverse direction to the tangent space T_f for at most finitely renormalizable maps f. Later on, in §8.3, we will identify T_f with the genuine tangent space to $\mathcal{H}_f$. Let us start with simple cases.

7.1. Non-recurrent cases. In the hyperbolic case T_f was actually defined as the tangent space to the hybrid class $\mathcal{H}_f$, which is a codimension-one submanifold (see §5.1). This tangent space is explicitly given by equation (5.1), so that the transverse vector fields are those which satisfy the inequality

$$(7.1) \qquad Dv^n(p)(Df^n(p) - 1) - v^n(p)D^2 f^n(p) \neq 0,$$

where $\overline{p}$ is the attracting periodic orbit of period n.

In the parabolic case with $Df^n(p) = 1$, Definition 5.2 implies that $v^n(p) = 0$ for $v \in T_f$. Obviously this condition specifies a codimension-one subspace.

The parabolic case with $Df^n(p) = -1$ is more delicate:

Lemma 7.1. *Let $f \in \mathcal{U}_a$ be a parabolic map such that the multiplier of its parabolic orbit is -1 and let $v \in T_f$. Then v satisfies (5.1), that is*

$$2Dv^n(p) + v^n(p)D^2 f^n(p) = 0.$$

Proof. Let v be a vector field in T_f and let α be a qc vector field of $\mathbb{C}$, equivariant on O_f. Since 0 is attracted by $\overline{p}$, we may assume $\lim_{k\to\infty} c_{kn} = p$.

Equivariance allows us to write

$$v^n(c_{kn}) = \alpha(c_{(k+1)n}) - Df^n(c_{kn})\alpha(c_{kn}).$$

By continuity of α on O_f, $\alpha(c_{kn}) - \alpha(p) = o(1)$. Thus,

$$v^n(p) + Dv^n(p)(c_{kn} - p) =$$
$$\alpha(c_{(k+1)n}) + \alpha(c_{kn}) - D^2 f^n(p)\alpha(p)(c_{kn} - p) + o(1)(c_{kn} - p).$$

Taking the difference between two consecutive equations, we obtain:

$$(Dv^n(p)+D^2 f^n(p)\alpha(p))(c_{(k+1)n} - c_{kn}) = \alpha(c_{(k+2)n}) - \alpha(c_{kn}) + o(c_{kn} - p).$$

Since p is parabolic with multiplier -1, $c_{(k+1)n} - c_{kn} = -2(c_{kn} - p) + o(c_{kn} - p)$. Thus,

$$\lim_{k\to\infty} \frac{\alpha(c_{(k+2)n}) - \alpha(c_{kn})}{c_{kn} - p} = 2Dv^n(p) + 2D^2 f^n(p)\alpha(p).$$

Equivariance gives $v^n(p) = 2\alpha(p)$, so to obtain (5.1) we just have to show that

$$\lim_{k\to\infty} \frac{\alpha(c_{(2k+2)n}) - \alpha(c_{2kn})}{c_{2kn} - p} = 0.$$

To see this, notice that $c_{2kn} - p$ approaches p from one of the sides, that is, for big k, all $c_{2kn} - p$ have the same sign. Since p is parabolic with multiplier -1, $c_{2kn} - p \sim \eta n^{-1/2}$ for some $\eta \neq 0$ (see Lemma B.3), so $\sum_k (c_{2kn} - p)$ diverges. Since α is uniformly bounded on orb(0),

$$\sum_k \alpha(c_{(2k+2)n}) - \alpha(c_{2kn})$$

is bounded. Since the limit

$$\lim_{k\to\infty} \frac{\alpha(c_{(2k+2)n}) - \alpha(c_{2kn})}{c_{2kn} - p}$$

exists, it must be 0, proving the lemma. □

By Lemma 7.1, T_f is contained in a codimension-one subspace, and transverse directions may be obtained again by (7.1).

We will now construct the transverse direction in the Misiurewicz case. Let δ_x denote the Dirac measure concentrated on x. Let

$$\nu_n = \sum_{k=0}^{n} \frac{\delta_{c_k}}{(f^k)'(c_1)},$$

which are signed measures on the interval I or equivalently bounded functionals acting on $C^0(I)$. By Proposition 2.15, in the Misiurewicz case f is expanding on O_f, so that $|(f^n)'(c_1)|$ grows exponentially fast. Therefore, the ν_n converge in the weak-$*$ topology to a signed measure ν.

Lemma 7.2. *There exists a polynomial $p \in T_f \mathcal{A}_a^{\mathbb{R}}$ such that $\nu(p) \neq 0$.*

Proof. Since the critical point is non-recurrent, $\nu(J) = 1$ for a small interval J around 0 such that $c_j \notin J$ for $j \geq 1$. So the signed measure ν does not vanish and hence there exists a continuous function $\phi \in C^0(I)$ such that $\nu(\phi) \neq 0$. Since any continuous function can be approximated by a polynomial, the assertion follows. □

Lemma 7.3. *If $\nu(v) \neq 0$, then $v \notin T_f$.*

Proof. Let $v \in T_f$, and let α be a qc vector field equivariant on $\mathrm{orb}_f(0)$. Then $\alpha(c_{n+1}) = v^{n+1}(0) = (f^n)'(c_1)\nu_n(v)$, so that

$$\lim \frac{|\alpha(c_{n+1})|}{|(f^n)'(c_1)|} = |\nu(v)|.$$

Since α is a qc vector field, $\alpha(c_{n+1})$ is uniformly bounded, so $\nu(v) = 0$. □

The existence of the transverse direction follows from Lemmas 7.2 and 7.3. Once we show that T_f has codimension-one, it will follow that $T_f = \mathrm{Ker}(\nu)$.

Remark 7.1. More generally, this analysis allows to obtain transverse directions for maps f which satisfy the summability condition

$$\sum_{k=0}^{\infty} \frac{1}{|Df^k(c_1)|} < \infty.$$

7.2. Smooth transverse vector field. We will now proceed with analysis of puzzle maps satisfying certain geometric assumptions, which will allow us to handle the case of at most finitely renormalizable maps with a recurrent non-periodic critical point. In this section we will construct a smooth transverse vector field, which is holomorphic on the puzzle piece containing the critical value U_1 and vanishes on the other puzzle pieces.

Definition 7.1. *An equipped holomorphic family of puzzle maps over some neighborhood* $\mathcal{V} \ni 0$ *in a complex Banach space is a pair* (f_λ, h_λ), $\lambda \in \mathcal{V}$, *where* $f_\lambda : \cup U_i^\lambda \to \mathbb{C}$ *is a puzzle map and* $h_\lambda : \mathbb{C} \to \mathbb{C}$ *is a holomorphic motion of* $\mathbb{C}$ *that maps* U_i^0 *onto* U_i^λ *and such that*

$$h_\lambda \circ f_0(z) = f_\lambda \circ h_\lambda(z) \tag{7.2}$$

for $z \in \cup(\partial U_i^0)$.

Let $U \subset \mathbb{C}$ be an $\mathbb{R}$-symmetric open set which intersects the real line in some interval J. We denote by $\Upsilon(U)$ the Banach space of holomorphic vector fields of class $A^1(U)$ that vanish together with their derivatives at ∂J, endowed with the A^1 norm $\|\cdot\|_1$, that is $\|v\|_1 = \sup_{z\in U} |v'|$, see (2.2) ($\|\cdot\|_1$ is a norm in $\Upsilon(U)$ because of the boundary conditions). We will extend those vector fields to the whole real line as 0 outside J. (The case that will interest us most is $U = D_\gamma(J)$.)

Recall that for a puzzle map $f : \cup U_i \to \mathbb{C}$, we denote by J_i the real slices of the domains U_i.

Proposition 7.4. *Let* $f : \cup U_i \to \mathbb{C}$ *be a geometric puzzle map with parameters* (ϕ, ψ, γ, k). *Then there is an* $\varepsilon > 0$ *depending only on* ψ *and* γ, *and there exists an equipped holomorphic family of puzzle maps* (f_v, h_v) *over the Banach ball* $B_\varepsilon(\Upsilon(D_\gamma(J_1)))$ *such that:*

- h_v *is the identity on* U_j, $j \neq 1$, *on* $\partial(f(U_0))$, *and is normalized;*
- $(\mathrm{id} + v) \circ h_v$ *is the identity in* U_1*;*
- $f_v = f$ *in* U_j, $j \neq i$*;*
- $f_v = f \circ (\mathrm{id} + v)$ *in* $U_1^v = h_v(U_1)$.

Proof. Let $\varepsilon > 0$ and $v \in \Upsilon(D_\gamma(J_1))$ such that $\|v\|_1 < \varepsilon$. If $\varepsilon < 1$ then $I_v \equiv (\mathrm{id} + v)|\overline{D_\gamma(J_1)}$ is a diffeomorphism fixing ∂J_1 with $I_v'|\partial J_1 = 1$ (using the convexity of $D_\gamma(J)$). If ε is small (depending only on ψ and γ, then $I_v(D_\gamma(J_1))$ contains $D_\psi(J_1) \supset U_1$. Hence $I_v^{-1}(U_1) \subset D_\gamma(J_1)$. By definition of geometric parameters, the domain $I_v^{-1}(U_1)$ is disjoint from U_j, $j \neq i$, and from $\partial(f(U_0))$.

Thus, we can define f_v in $I_v^{-1}(U_1)$ as $f \circ I_v$. We then define h_v as the identity on U_j, $j \neq 1$, and on $\partial(f(U_0))$, and we let $h_v|U_1 = I_v^{-1}|U_1$. Since $U_1 \subset I_v(D_\gamma(J_1))$, $I_v^{-1}|U_1$ depends holomorphically on v. We obtain a holomorphic motion h_v on $U \cup \partial(f(U_0))$. By the Extension Lemma, it can be extended to a normalized holomorphic motion of the whole sphere over a Banach ball of radius $\varepsilon/3$. □

To motivate the next statement, note that the tangent vector to the curve $t \mapsto f_{tv}$ at $t = 0$ is given by the vector field $v(z) f'(z)$ on U_1 (and 0 elsewhere). This vector field is tangent to the hybrid class of f if there exists a qc vector field α on the orbit of the critical value c_1 such that $\alpha(c_1) = 0$ and

$$v(z) f'(z) = \alpha(f(z)) - \alpha(z) f'(z).$$

This equation can be written in the following concise form:

$$v = f^*\alpha - \alpha, \tag{7.3}$$

where $f^*\alpha$ is the pullback of α by f, compare equation (2.6).

If U is a hyperbolic domain and v is a tangent vector at a point $x \in U$, we denote $\|v\|^U_{\text{hyp}}$ the hyperbolic length of v.

Lemma 7.5. *Given (ψ, γ), there exists an $L > 0$ with the following property. Let f be a geometric puzzle map without invariant line fields on $K(f)$ with parameters (ϕ, ψ, γ, k). Let $v \in \Upsilon(D_\gamma(J_1))$ be a holomorphic vector field such that there is a qc vector field α on the closed set O_f with $\alpha(c_1) = 0$, satisfying (7.3).*

Then α has a normalized $L\|v\|_1$-qc extension, which satisfies (7.3) on U and vanishes on $\partial(U \cap \mathbb{R})$.

Proof. Without loss of generality we can assume that v is normalized: $\|v\|_1 = 1$. Take an $\varepsilon > 0$ as in the above proposition and consider the corresponding equipped holomorphic family of puzzle maps $(f_{\lambda v}, h_{\lambda v})$, $|\lambda| < \varepsilon$. Let

$$\alpha_0 = \frac{d}{d\lambda} h_{\lambda v}\bigg|_{\lambda=0}.$$

Since $h_{\lambda v}$ satisfies (7.2) for $z \in \partial U$, α_0 satisfies (7.3) for $z \in \partial U$.

As in Lemma 6.12 we conclude that $\|\alpha_0\| \leq L\|v\|_1$ with $L = 1/\varepsilon$.

Now we can use Theorem 6.9 to conclude that there exists a qc vector field β that coincides with α on the orbit of the critical value, coincides with α_0 on the complement of U and satisfies the equation (7.3) on U. Moreover, by Lemma A.24, there are no invariant line fields on $K(f)$, so we conclude that $\|\bar\partial\beta\|_\infty \leq L\|v\|_1$.

This vector fields vanishes on the $\partial(U \cap \mathbb{R})$, since $h_{\lambda v}|\partial(U \cap \mathbb{R}) = \text{id}$, $|\lambda| < \varepsilon$. □

Lemma 7.6. *Let f be a geometric puzzle map with geometric parameters (ϕ, ψ, γ, k) such that $c_1 \in U$. Then there exists a vector field $v \in \Upsilon(D_\gamma(J_1))$ such that $\|v\|_1 = 1$ and $\|v(c_1)\|^{U_1}_{\text{hyp}} > 1/7$.*

Proof. Let

$$w_n(z) = (1 - z^2)(1 - e^{-2n}) + \frac{2}{n}(e^{-n(1+z)} + e^{-n(1-z)} - e^{-2n} - 1).$$

Then $w_n \in \Upsilon(\mathbb{D})$. Also, $\|w_n\|_1 < 6$ and $w_n(z) \to 1 - z^2$ pointwise in $\mathbb{D}$ as $n \to \infty$. Hence the hyperbolic norm of $w_n(z)$ in $\mathbb{D}$ goes to the value $|1 - z^2|/(1 - |z^2|) \geq 1$. Take a big n and let w be the restriction of w_n to $D_\gamma(I)$ normalized so that $\|w\|_1 = 1$. Rescaling $D_\gamma(I)$ to $D_\gamma(J_1)$, we obtain a vector field v in the latter domain with desired properties, because both the norm $\|\cdot\|_1$ and the hyperbolic norm are scaling invariant. □

Lemma 7.7. *Given ϕ and k_0, there exists a $\kappa > 0$ with the following property. Let $f : U \to \mathbb{C}$ be a geometric puzzle map with parameters (ϕ, ψ, γ, k), with $k > k_0$, such that the critical value c_1 returns to U_1, and let $v \in \Upsilon(D_\gamma(J_1))$ be a vector field from Lemma 7.6. If there exists a qc vector field α on $\mathbb{C}$, satisfying (7.3) on U, then there exists $q \in f^{-1}(U_1) \cap \mathbb{R}$ such that $|\alpha(q)| > \kappa\, d(q, \partial J_0)$.*

Proof. Let $n > 1$ be minimal such that $c_n \in U_1$. Let us show that

$$\|\alpha(c_{n-1})\|^{U_0}_{\text{hyp}} \geq \|v(c_1)\|^{U_1}_{\text{hyp}}. \tag{7.4}$$

Since $n-2$ is the first landing time of c_1 in $f^{-1}(U_1)$, there exists a domain $W \subset U_1$ containing c_1, which is univalently mapped onto U_0 by f^{n-2} (see e.g., [L4, Lemma 3.5]). Moreover the orbit of this domain, $f^k(W)$, $k = 1, 2, \ldots, n-2$, does not intersect U_1.

Equation (7.3) implies

$$(f^{n-2})^*\alpha - \alpha = \sum_{k=0}^{n-3} (f^k)^* v.$$

Evaluating it at c_1 we obtain $(f^{n-2})^*\alpha(c_1) = v(c_1)$ (since v vanishes outside U_1 and $\alpha(c_1) = 0$). Since $f^{n-2} : W \to U_0$ is a hyperbolic isometry,

$$\|\alpha(c_{n-1})\|^{U_0}_{\text{hyp}} = \|v(c_1)\|^{W}_{\text{hyp}}.$$

This equation implies (7.4) by the Schwarz Lemma.

Let $q = c_{n-1}$. Then $[-q, q] \subset V \equiv f^{-1}(U_1)$. By condition 4 of the definition of geometric puzzle map, $\text{mod}(U_0 \setminus V) > k/2$, hence there exists a constant depending only on k which bounds $\text{dist}(q, \partial J_0)/\operatorname{diam} J_0$ from below. Since $\|\alpha(q)\|^{U_0}_{\text{hyp}} > 1/7$ and $U_0 \supset D_\phi(J_0)$, there is a constant κ depending on ϕ and k only such that $|\alpha(q)| > \kappa \,\text{dist}(q, \partial J_0)$. □

Lemma 7.8. *Given (ϕ, ψ, γ), there exists a $k > 0$ with the following property. Let $f : U \to \mathbb{C}$ be a geometric puzzle map with parameters (ϕ, ψ, γ, k) such that the critical value c_1 returns back to the domain U_1. Assume that f does not have invariant line fields on $K(f)$. Then there exists a vector field $v \in \Upsilon(D_\gamma(J_1))$ such that there are no qc vector fields α satisfying (7.3) on the critical orbit. Moreover, $v|J_1$ is real.*

Proof. Let L be the constant from Lemma 7.5 associated to ψ and γ. Fix some $k_0 > 0$ and let κ be the constant from Lemma 7.7 associated to ϕ and k_0. Let $T = f^{-1} J_1 \cap \mathbb{R}$ and let C be the constant from Corollary 2.9 corresponding to L and T.

If $k > k_0$ is big enough then $|T| \leq \kappa C^{-1}\, \text{dist}(T, \partial J)$. Let v be the vector field given by Lemma 7.7. Suppose by contradiction that equation (7.3) is satisfied on the critical orbit for some qc vector field. Then by Lemma 7.5, this equation is satisfied by some normalized L-qc vector

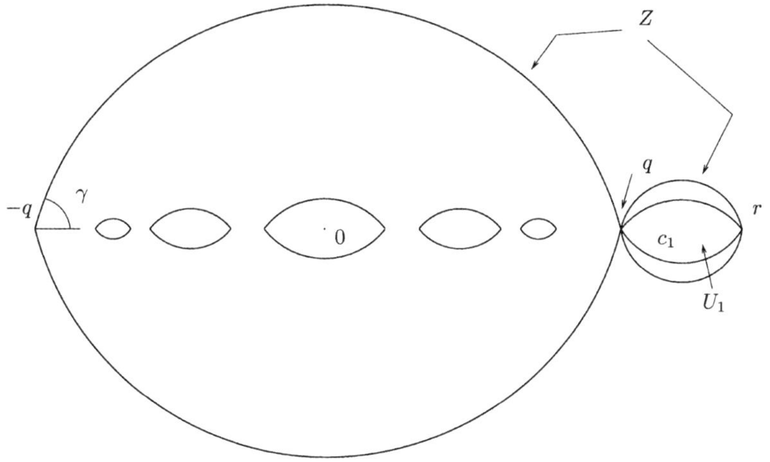

Fig. 5. Set Z and the puzzle

field α on $\cup U_j$ which vanishes at $\partial(\cup \operatorname{int} J_j)$. By (7.3), α vanishes at ∂T as well. By Corollary 2.9, $\alpha(z) \le C|T| \le \kappa \operatorname{dist}(T, \partial J_0)$ for $z \in T$. On the other hand, Lemma 7.7 yields existence of a point $q \in J$ such that $|\alpha(q)| > \kappa \operatorname{dist}(q, J_0)$. This is a contradiction. □

7.3. Puzzle motion. In this section we will show that the puzzle moves holomorphically over an appropriate neighborhood of a geometric puzzle map. First, let us introduce some notations.

Let us consider a unimodal map $f \in \mathcal{U}_a$ supplied with a geometric puzzle $\{U_i\}$ with parameters (ϕ, ψ, γ, k). Every puzzle piece U_i is univalently mapped onto U_0 by some iterate of f, $f^{n_i} : U_i \to U_0$. Recall that the (closed) real slices of the U_i are denoted by J_i and that J_1 stands for the interval containing the critical value c_1. Let $J_1 = [q, r]$, where $q < r$, and let $N = [-q, q]$. Let us consider the union of two (open) hyperbolic disks based upon the intervals N and J_1 together with their real boundary points:

$$Z \equiv Z_f = D_\gamma(N) \cup D_\gamma(J_1) \cup \{-q, q, r\}.$$

In what follows we will consider only puzzle pieces U_i which intersect Z. Notice that this family of puzzle pieces is forward invariant, except that $f(U_0)$ does not belong to it.

Let $E \equiv E_f = \operatorname{orb}(\partial J_0) \cup \{0\}$ (notice that this is a finite set). Let $\Lambda \equiv \Lambda_f$ be the space of odd vector fields $v \in A^1(Z)$ which vanish to the first order on E (where "odd" means that $v(z) = -v(-z)$ whenever both z and $-z$ belong to Z).

Notice that $\Upsilon(D_\gamma(J_1)) \subset \Lambda$ and the inclusion is an isometry.

We assume that the angle γ is so small that f is defined on some neighborhood of $\overline{Z}$. For $v \in \Lambda$, we let $f_v = f \circ (\mathrm{id} + v)$. For small enough v, this map is well defined on $\overline{Z}$, $f_v \in A^1(Z)$, and depends continuously on v.

Let $Q \equiv Q_f = \partial(\cup \operatorname{int} J_i)$ be the maximal invariant set of $f|(I \setminus \operatorname{int} J_0)$.

Lemma 7.9 (Puzzle Motion). *Let f as above be a geometric puzzle map with parameters (ϕ, ψ, γ, k) which is a restriction of a map in $\mathcal{U}_a$. Assume that $f|Q$ is hyperbolic. Then for some $\varepsilon > 0$ there is an equipped holomorphic family of puzzle maps (f_v, h_v), $v \in B_\varepsilon(\Lambda)$, such that h_v is the identity on $\partial(f(U_0)) \cup \partial Z$ and is normalized. In particular, $f_v^{n_i+1} \circ h_v = h_v \circ f^{n_i+1} = f^{n_i+1}$ on each ∂U_i.*

The proof of this lemma will be based on several lemmas. By Lemma 2.12, there exists a smooth metric $d\nu = \rho|dz|$ which is expanding on Q. By approximating ρ near Q with a piecewise constant function, we obtain an expanding piecewise Euclidean metric near Q. So, we can assume that ν is piecewise Euclidean near Q in the first place.

Lemma 7.10. *The map $f_v = f \circ (\mathrm{id} + v)$ has an invariant expanding Cantor set $Q_v \subset Z$, $Q_0 \equiv Q$, which moves holomorphically over some neighborhood $\mathcal{V} \subset \Lambda_f$ of 0.*

Proof. For simplicity, let us assume that f is non-renormalizable (the general case is dealt in the same way, but the notation is more complicated). Notice that the orientation reversing fixed point p of f belongs to $\operatorname{orb}_f(\partial J_0)$. In particular, p is also a fixed point for f_v with the same multiplier.

We will construct a holomorphically moving Markov partition $\mathcal{M}^v$ for f_v whose elements are contained in Z. To this end let us use the linearizing coordinate $L_v : (D, p) \to (\mathbb{C}, 0)$ near the orientation reversing fixed point $p_v \equiv p$ of f_v normalized by $L_v'(p_v) = 1$. The L_v-preimages of the straight rays landing at 0 will be called "rays" of f_v landing at p_v:

$$R_v(\zeta) = L_v^{-1}\{t\zeta : \ 0 \le t \le 1\}.$$

Since the multiplier of p_v is real, the second iterate of any ray "overflows" itself. Since the linearizing coordinate depends holomorphically on v, the ray $R_v(\zeta)$ with a given ζ moves holomorphically over some neighborhood $\mathcal{V} \subset \Lambda_f$. Moreover, this ray (viewed as a smooth arc in $\mathbb{C}$) smoothly depends on v.

Let x_v be a preimage of p_v under some iterate of f_v such that the whole $\operatorname{orb}_{f_v}(x_v)$ is close to Q (in particular, the smooth metric ν selected before is expanded along the orbit of x_v). If x_v is different from the special points q, $-q$ and r, then all sufficiently short rays landing at p_v can be pulled back to x_v providing us with "rays" landing at x_v.

Assume $x_v = q$. Let $f^l(q) = p$. Let us consider the union $S(q)$ of two $\mathbb{R}$-symmetric Euclidean sectors of size $2\gamma/3$ with vertex at q. Then the map $f_v^{n_1}$ is holomorphic on $S(q)$ near q and $f^l(S(q))$ contains the union $S(p)$ of two sectors centered at p of size $\gamma/2$. Hence we can take the preimages of the rays landing at p within the sectors $S(p)$ and obtain "rays" landing at q.

Similarly we construct rays landing at points $-q$ and r within appropriate sectors with vertices at those points (one sector for each of the points).

Clearly, all the rays we have constructed move holomorphically with v and smoothly depend on v. We will fix now $\theta < \gamma/2$ and only consider rays which make this angle with the real line.

Let us now defined "h-rays" for f as the rays truncated on the height $\pm h$ with respect to the smooth expanding metric ν fixed before. Here "height" denotes the vertical distance to the real line (recall that we chose ν such that in small scales near Q, ν is just a multiple of the Euclidean metric). For any point x_v, they are well-defined provided h is small enough. We define h-rays for f_v as the holomorphic motions of these rays. Since the metric ν is expanding, this family of rays satisfies the following *overflowing property*: If Γ is an h-ray landing at x_v at angle θ with the real line, then the image $f_v\Gamma$ strictly contains the h-ray landing at $f_v(x_v)$ at angle θ or $\pi - \theta$.

Given some n, we will now consider a real Markov covering $\{M_j\}$ of Q by partitioning $[-1, 1]$ by the set $Q \cap \cup_{k=0}^{n} f^{-k}(p)$ and taking only those (closed) intervals of the partition which intersect Q. For n big enough, this covering is contained in a small neighborhood of Q.

Let us now complexify this covering. The complex domains of the coverings will be $\mathbb{R}$-symmetric (smooth) hexagons H_j based upon the intervals M_j. Take some $M_j = [a, b]$. Consider four h-rays landing at points a and b at angle θ with the interval M_j. We obtain four sides of H_j. Join two endpoints of the rays lying in the upper half-plane with a horizontal interval on height h, and similarly in the lower half-plane. Since the h-rays are smoothly close to the straight rays for h sufficiently small, we obtain a smooth hexagon H_j (see Fig. 6). Since the h-rays move holomorphically with v, the hexagons also do (on the top and bottom sides of the hexagon the motion can be defined by linear interpolation). This defines hexagons H_j^v.

The overflowing property of the family of rays implies that the family of hexagons H_j^v is Markov family. By Proposition 2.11 (and the remark following it), the set of non-escaping points,

$$Q_v = \{z : \ f_v^n z \in \cup H_j^v, \ n = 0, 1, \dots\},$$

is a holomorphically moving invariant Cantor set.

Smooth dependence of the rays on v implies that all the hexagons are contained in Z provided v is sufficiently small. Hence $Q_v \subset Z$ as well. □

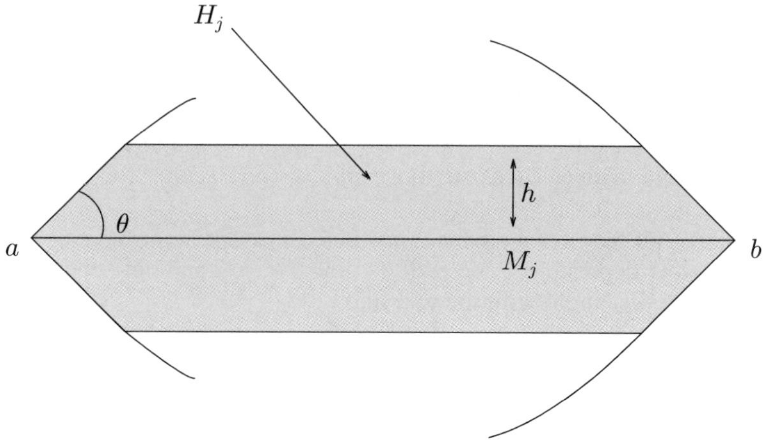

Fig. 6. A hexagonal Markov piece

Let h_v be the holomorphic motion of the Markov partition $\{H_j\}$ in the above lemma.

Lemma 7.11. *There exists a neighborhood $\mathcal{V}$ of 0 in Λ such that for each U_i there is a holomorphic motion h_v^i of $U_i \cup \partial Z \cup Q$ such that $h_v \circ f^{n_i+1} = f_v^{n_i+1} \circ h_v^i$ in U_i and $h_v^i = h_v$ in $\partial Z \cup Q$.*

Proof. Suppose first that $i = 1$. Notice that f^{n_1+1} is a ramified double covering map over a neighborhood of $f(U_0)$. Using that both f and f_v are symmetric, we conclude that for any $v \in \Lambda$ sufficiently small we can associate $w \in \Upsilon(D_\gamma(J_1))$ such that $f_v^{n_1+1} = f^{n_1+1} \circ (\mathrm{id} + w)$ and this association is continuous. We can then argue exactly like in Lemma 7.4 that for $w \in \Upsilon(D_\gamma(J_1))$ small enough, $\mathrm{id} + w$ is a diffeomorphism from $D_\gamma(J_1)$ onto a set which contains U_1, which allows us to define the motion as $(\mathrm{id} + w)^{-1}$ in U_1.

If $i \neq 1$ is fixed, it is clear that there exists a neighborhood $\mathcal{V}_i$ where such holomorphic motion is defined.

We now argue exactly as in Lemma 5.6. We notice that all but finitely many U_i are compactly contained in the domain of the persistent Markov partition $\{H_j\}$. Let $\mathcal{I}$ be the set of U_i which are not contained in any H_j. Let $\mathcal{J}$ be the set of U_i which are contained in some H_j but $f(U_i)$ is in $\mathcal{I}$. Shrinking the neighborhood if needed, we may suppose that in $\mathcal{V}$, h_v is defined, h_v^i is defined for $U_i \in \mathcal{I} \cup \mathcal{J}$ (a finite set of domains) and $h_v^i(U_i) \subset \cup h_v(H_j)$ for $U_i \in \mathcal{J}$. This allows us to construct the motion of all U_i in $\mathcal{V}$ by pulling back, as in Lemma 5.6. □

Let $S = \mathbb{C} \setminus Q$. As in Lemma 5.6 we conclude the estimates (5.2) and (5.3), which imply by Lemma 2.3 the estimates (5.4).

So we can define a holomorphic motion H_v which agrees with h_v on $Z_f \cup \partial f(U_0) \cup Q$ and with h_v^i on U_i. This conclude the proof of the Lemma 7.9.

7.4. Existence of transverse direction. Let P_f be the set of all $p \in \Lambda_f$ such that pf' is a polynomial.

Lemma 7.12. *$P_f^{\mathbb{R}}$ is dense in $\Lambda_f^{\mathbb{R}}$.*

Proof. This follows from the Mergelyan Polynomial Approximation Theorem (see [R], Theorem 20.5, p. 423). □

Lemma 7.13. *Given f as above, there exists an L with the following property. Assume $v \in \Lambda_f$ and there is a qc vector field α satisfying equation (7.3) on $\mathrm{orb}_f(0)$. Then α has a normalized $L\|v\|_1$-qc extension.*

Proof. We argue as in Lemma 6.12.

Let $\mathcal{V}$ be a neighborhood of 0 in Λ given by Lemma 7.9. Arguing as in Lemma 6.12 we conclude that $\beta = \frac{d}{dt}h_{tv}\big|_{t=0}$ is a normalized qc vector field such that $\|\beta\|_{qc} \leq L\|v\|_1$ for some constant L. Applying Theorem 6.9 and Lemma A.24 we get a normalized qc vector field α with the same estimates. □

Corollary 7.14. *The set of $v \in \Lambda_f$ satisfying equation (7.3) is closed.*

Proof. Follows from Lemma 7.13 and the Second Compactness Lemma. □

Theorem 7.15. *If $f \in \mathcal{U}_a$ is at most finitely renormalizable then there exists a vector field $v \in \mathcal{A}_a^{\mathbb{R}}$ which does not belong to T_f.*

Proof. The non-recurrent case was treated in §7.1, so let us assume that f has a recurrent critical point. By Lemma 5.6 there exists a geometric puzzle for f satisfying the assumptions of Lemma 7.8. By Lemmas 7.8, 7.12, and Corollary 7.14, there exists a real polynomial pf' such that pf' does not belong to T_f. □

Remark 7.2. Existence of the transverse direction can be proved for maps with minimal postcritical set (which includes infinitely renormalizable and some finitely renormalizable combinatorics) by different means.

This construction is based on a renormalization approach: the assumption of minimality is used to obtain a renormalization which is polynomial-like with finitely many branches. The first step is the construction of the transverse direction for the renormalization through a variation of the construction in the case of quadratic-like maps in §4.5 of [L6]. This is possible since polynomial-like maps still enjoy a tame "external structure".

The second step is to show that the derivative of the renormalization operator has dense image and is based on the construction of Lemma 4.5.

This allows to generalize this work for classes of unimodal maps without any decay of geometry and will be elaborated elsewhere (for maps such that the closure of the critical orbit is not a minimal set, there is some decay of geometry and the construction of the transverse direction we develop here can be applied).

8. Local laminations

8.1. Transverse cone field.

Lemma 8.1. *Let f be a Yoccoz map and consider two sequences f_n, g_n in $\mathcal{U}_a$ such that f_n is hybrid equivalent to g_n and* $\lim f_n = \lim g_n = f$. *Assume that*

$$\lim \frac{g_n - f_n}{\|g_n - f_n\|_a} = w.$$

Then $w \in T_f$.

Proof. Denote by $\mathcal{V}_\varepsilon$ an ε-neighborhood of f in $\mathcal{A}_a$.

Consider a geometric puzzle $\mathcal{P}_f$ for f which persists in $\mathcal{V}_\varepsilon$ (see Lemma 5.6), and let $\psi_g : \mathbb{C} \to \mathbb{C}$ be a normalized holomorphic motion of the puzzle. For $F, G \in \mathcal{V}_\varepsilon$, denote by $\mu(F, G)$ the Beltrami coefficient of $\psi_G \circ \psi_F^{-1}$. It follows from the Quasiconformality Lemma that there exists a constant C such that $\|\mu(F, G)\|_\infty \leq C\|G - F\|_a$, provided F and G belongs to $\mathcal{V}_{\varepsilon/2}$.

By Remark 5.6, $\psi_{g_n} \circ \psi_{f_n}^{-1}$ is an $\mathbb{R}$-symmetric map. By Lemma A.25, there exists a normalized qc map H_n, equivariant with respect to f_n and g_n on U^{f_n}, such that

$$\|\mu_{H_n}\|_\infty \leq \|\mu(f_n, g_n)\|_\infty \leq C\|g_n - f_n\|_a.$$

Let $\mu_{H_n} = \lambda_n \mu_n$ with $\|\mu_n\|_\infty = 1$. It follows that $f_n^*(\mu_n) = \mu_n$ on U^{f_n}. We may assume also that $\tau \equiv \lim \lambda_n / \|g_n - f_n\|_a$ exists, so that $|\tau| \leq C$.

Let $H_{n,\lambda}$ be a normalized holomorphic motion over $\mathbb{D}$ with Beltrami coefficient $\lambda\mu_n$. Let $f_{n,\lambda} : H_{n,\lambda}(U^{f_n}) \to \mathbb{C}$ be defined as

$$f_{n,\lambda} = H_{n,\lambda} \circ f_n \circ H_{n,\lambda}^{-1}.$$

Since μ_n is invariant by f_n, we conclude that $f_{n,\lambda}$ is holomorphic on $H_{n,\lambda}(U^{f_n})$. Moreover, we have $f_{n,\lambda_n}|U^{g_n} = g_n$.

By Lemma 2.8, passing to a subsequence, we may assume that the $H_{n,\lambda}$ converge to some normalized holomorphic motion H_λ uniformly on compacts of $\mathbb{D}$.

Let $f_\lambda : H_\lambda(U^f) \to \mathbb{C}$ be defined as

$$f_\lambda = H_\lambda \circ f \circ H_\lambda^{-1}.$$

It follows that for each $\lambda \in \mathbb{D}$, $f_{n,\lambda}$ converges to f_λ uniformly on compacts of $H_\lambda(U^f)$, so that f_λ is holomorphic.

For any fixed compact set $K \subset U^f$, there exists $\delta > 0$ such that $H_\lambda^{-1}(K) \subset U^f$, $\lambda \in \mathbb{D}_\delta$. It follows that $(\lambda, z) \mapsto f_{n,\lambda}(z)$ converges uniformly to $(\lambda, z) \mapsto f_\lambda(z)$ on $\mathbb{D}_{\delta/2} \times K$. By uniform convergence of derivatives of holomorphic maps, we conclude that for any $z \in U^f$,

$$\begin{aligned} \tau \left.\frac{d}{d\lambda} f_\lambda(z)\right|_{\lambda=0} &= \lim \frac{\lambda_n}{\|g_n - f_n\|_a} \lim \frac{f_{n,\lambda_n}(z) - f_{n,0}(z)}{\lambda_n} \\ &= \lim \frac{g_n - f_n}{\|g_n - f_n\|_a} \\ &= w(z). \end{aligned}$$

Since $\|w\|_a = 1$, it does not vanish identically, so that $\tau \neq 0$. Let

$$\alpha = \left.\frac{d}{d\lambda} H_\lambda\right|_{\lambda=0}.$$

By the argument of Proposition 5.2, α is a qc vector field equivariant with respect to $(f, w/\tau)$ on $\mathrm{orb}(0)$, so $w \in T_f$. □

From now on, let us fix a Yoccoz map f, and let $v \in \mathcal{A}_a^{\mathbb{R}}$ be a transverse vector field given by Lemma 7.15 such that $\|v\|_a = 1$.

Corollary 8.2. *Let $\Sigma_\varepsilon = \{f + tv | t \in (-\varepsilon, \varepsilon)\}$. Then there exists an ε such that Σ_ε intersects each hybrid class in at most one point.*

Proof. If this is not the case, there exist sequences $t_{1,n}, t_{2,n} \to 0$ such that $t_{1,n} \neq t_{2,n}$ with $f + t_{1,n}v$ hybrid conjugate to $f + t_{2,n}v$. Applying Lemma 8.1 we conclude that $v \in T_f$, which is a contradiction. □

Remark 8.1. It follows that the straightening $\chi : \Sigma_\varepsilon \to [1/2, 2]$ is a homeomorphism onto the image. Since hyperbolic maps are dense in the quadratic family, they are dense in Σ_ε as well.

Below we use the notion of special combinatorics defined in §6.3.

Lemma 8.3. *There exists a neighborhood $\mathcal{V}$ of f such that if $f_n \to f$ is a sequence of maps with special combinatorics with respect to $\mathcal{V}$, then* $\limsup T_{f_n} \subset T_f$.

Proof. Consider a geometric puzzle for f which persists in a neighborhood of f and take a neighborhood $\mathcal{V}$ of f as in Lemma 6.12. Let $f_n \to f$ be a sequence of maps with special combinatorics with respect to $\mathcal{V}$, and let $v_n \in T_{f_n}$ be a sequence of vector fields converging to some vector field v. Since $\|v_n\|_a$ is uniformly bounded, and the operator norm of L_{f_n} is uniformly bounded, there exists a qc vector field α_n, equivariant on $\mathrm{orb}_{f_n}(0)$, and such that $\|\alpha_n\|_{\mathrm{qc}} < C$. Denote by β_n some normalized extension of α_n to $\mathbb{C}$ with $\|\bar\partial \beta_n\|_\infty < C$. By the Second Compactness Lemma, we may assume that β_n converges to some vector field β. It is easy to see that β is equivariant with respect to (f, v) on $\mathrm{orb}(0)$. So $v \in T_f$, and since T_f is a vector space, $\limsup T_{f_n} \subset T_f$. □

Corollary 8.4. *There exists a neighborhood $\mathcal{V}$ of f in $\mathcal{A}_a$ and an open cone $\mathcal{K}$ such that for any $g \in \mathcal{V}$ which has special combinatorics with respect to $\mathcal{V}$, T_g is transverse to $\mathcal{K}$.*

Proof. Suppose that the statement is false. Let $\mathcal{V}_n \subset \mathcal{A}_a$ be a $1/n$-neighborhood of f and let $\mathcal{W}_n \subset T\mathcal{A}_a$ be a $1/n$ neighborhood of the transverse vector field v. Since the cone generated by $\mathcal{W}_n$ is an open cone, there exists a sequence $f_n \in \mathcal{V}_n$, with special combinatorics with respect to $\mathcal{V}_n$, such that $T_{f_n} \cap \mathcal{W}_n \neq \emptyset$. So $v \in \limsup T_{f_n}$ and thus $\limsup T_{f_n} \not\subset T_f$.

To obtain a contradiction with Lemma 8.3, we observe that, for a fixed neighborhood $\mathcal{V}$ of f, $\mathcal{V}_n \subset \mathcal{V}$ for all n big enough, so that f_n has special combinatorics also with respect to $\mathcal{V}$. □

Corollary 8.5. *T_f is a codimension-one subspace of $T\mathcal{A}_a$.*

Proof. By Remark 8.1, there exists a sequence $f_n \to f$ with special combinatorics. For each f_n, T_{f_n} is codimension-one, so by Lemma 2.32, either $\limsup T_{f_n}$ is codimension-one or is equal to $T\mathcal{A}_a$. By Lemma 8.3, $T_f \supset \limsup T_{f_n}$, so T_f must be a codimension-one subspace, since the transverse vector field v does not belong to T_f. □

8.2. Proof of Theorem A. The construction of the lamination in the infinitely renormalizable case was carried out in Theorem 4.9. The lamination near hyperbolic maps is trivial to construct and parabolic combinatorics are codimension-one submanifolds by the Implicit Function Theorem (notice that parabolic points of quasiquadratic maps must be non-degenerate). We will construct the lamination near a Yoccoz map f.

Fix a puzzle for f and let $\mathcal{V}$ and $\mathcal{K}$ be as in Lemmas 6.3, 6.12 and Corollary 8.4 and v be as in last section. Let $\Pi_1 : T\mathcal{A}_a \to T_f$, $\Pi_2 : T\mathcal{A}_a \to \mathbb{C}$ be the linear projections along T_f and v, that is, $w = \Pi_1(w) + \Pi_2(w)v$. From Lemma 2.31, there exists a constant C such that for any $w \in T\mathcal{A}_a \setminus \mathcal{K}$, $\Pi_2(w) \leq C\|\Pi_1(w)\|_a$. In particular, tangent spaces as in Corollary 8.4 are C-Lipschitz graphics from T_f to the transverse one-dimensional subspace spanned by v.

Fix ε very small and let Λ be an ε-neighborhood of 0 in T_f and let $\Sigma = \Sigma_\varepsilon$ be as in Corollary 8.2.

Let $g \in \mathcal{A}_a$. We say that $\gamma : \mathcal{W} \to \mathcal{A}_a$ is a *g-graphic* (over $\mathcal{W}$) if $\mathcal{W}$ is a neighborhood of 0 in T_f and if $\gamma(0) = g$ and $\Pi_1 \circ \gamma - \mathrm{id} = \Pi_1(g)$. We say that a g-graphic is C-Lipschitz if $\|\Pi_2 \circ D\gamma\| \leq C$. We say that a g-graphic is contained in some set X if $\gamma(\mathcal{W}) \subset X$.

We say that a set $X \subset \mathcal{A}_a$ is a *definite* g-graphic if there is a g-graphic over Λ onto X.

Lemma 8.6. *If $g \in \Sigma$ has special combinatorics and $\gamma : \mathcal{W} \to \mathcal{C}_g$, $\mathcal{W} \subset \Lambda$ is a g-graphic, then γ is C-Lipschitz.*

Proof. This is an immediate consequence of Corollary 8.4. □

Lemma 8.7. *If $g \in \Sigma$ has special combinatorics then $\mathcal{C}_g$ contains a definite g-graphic.*

Proof. Since $\mathcal{C}_g$ is a codimension-one submanifold transverse to v at g, there exists a g-graphic contained in $\mathcal{C}_g$.

Let $\mathcal{W} \subset \Lambda$ be the set of all w such that there exists a neighborhood $\mathcal{W}_w$ of $[0, w]$ in T_f and a g-graphic over $\mathcal{W}_w$ contained in $\mathcal{C}_g$.

Since $\mathcal{C}_g$ is a codimension one submanifold, it is also clear that the continuation property is valid, that is, if γ_1 is a g_1-graphic over $\mathcal{W}_1$ and γ_2 is a g_2-graphic over $\mathcal{W}_2$ which are contained in $\mathcal{C}_g$, and $\mathcal{W}_1 \cap \mathcal{W}_2$ is simply connected, then $\gamma_1|\mathcal{W}_1 \cap \mathcal{W}_2 = \gamma_2|\mathcal{W}_1 \cap \mathcal{W}_2$. It follows that there exists a g-graphic $\gamma : \mathcal{W} \to \mathcal{C}_g$. So we just have to show that $\Lambda = \mathcal{W}$.

If this is not the case, there exists $w_0 \in \Lambda \setminus \{0\}$ such that $[0, w_0) \subset \mathcal{W}$, but $w_0 \notin \mathcal{W}$. By Lemma 8.6, γ is C-Lipschitz, so that $\gamma(\mathcal{W})$ is contained in a small neighborhood of g, and $\lim_{w \to w_0} \gamma_w$ exists and will be denoted by $\gamma(w_0)$. By Lemma 6.3, for any $w \in \mathcal{W}$, $\gamma(w)$ is L-qc conjugate to g on U^g. Using the First Compactness Lemma, we conclude that $\gamma(w_0)$ is L-qc conjugate to g on U^g, so $w_0 \in \mathcal{C}_g$ and w_0 has special combinatorics with respect to $\mathcal{V}$. By Corollary 8.4, v is transverse to $T_{\gamma(w_0)}$, so there exists a $\gamma(w_0)$-graphic contained in $\mathcal{C}_g$. By the continuation property, $w_0 \in \mathcal{W}$, which is a contradiction. $\square$

Let γ_g be the definite g-graphic contained in $\mathcal{C}_g$. Let $\Delta_g = \gamma_g(\Lambda) \subset \mathcal{C}_g$.

By the continuation property, if $g_1, g_2 \in \Sigma$ have special combinatorics and are different, then $\Delta_{g_1} \cap \Delta_{g_2} = \emptyset$.

By Remark 8.1 and the Extension Lemma for holomorphic motions, there is a unique extension of the lamination Δ_g whose leaves pass through every point of Σ. Given now $g \in \Sigma$ not necessarily with special combinatorics, we let Δ_g be the leaf of the extended lamination. It is clear that each Δ_g is still a definite C-Lipschitz g-graphic.

It follows that the slices of the Δ_g by $\mathcal{U}_a$ form a lamination with codimension-one real analytic leaves. Moreover, by the First Compactness Lemma, the maps in the same leaf are L-qc conjugate, and hence the leaves coincide with the local hybrid classes. This completes the proof of Theorem A.

Remark 8.2. The present proof for the existence of the laminations given the existence of the transverse direction (see Remark 7.2) also works in the infinitely renormalizable case.

This proof also apply to a class of maps considered in [LS2], namely covering maps of the circle with a unique critical point of inflection type (for which the existence of the transverse direction is automatic). This application will be elaborated elsewhere.

8.3. Characterization of the tangent space. Let us recall that each Δ_g constructed above consists of maps $\tilde{g}$ which are L-qc conjugate to g on U^g. Moreover, the conjugacy between g and $\tilde{g}$ can be chosen to vary holomorphically inside Δ_g:

Proposition 8.8. *For each $g \in \Sigma$ there is a normalized holomorphic motion $h_{\tilde{g}}$, $\tilde{g} \in \Delta_g$ such that $h_{\tilde{g}}(U_g) = U_{\tilde{g}}$ and $h_{\tilde{g}}$ is equivariant with respect to g and $\tilde{g}$ on U_g.*

Proof. Let us fix a holomorphic motion H_g of the puzzle on $\mathcal{V}$. For g with special combinatorics, it follows from Remark 6.2 that there exists a holomorphic motion $h_{\tilde{g}}$ over Δ_g, equivariant on U_g and coinciding with $H_{\tilde{g}}$ on $\mathbb{C} \setminus U_g$.

In general, Δ_g is the limit of a sequence Δ_{g_n}, g_n with special combinatorics. By Lemma 2.8, there exists a holomorphic motion over Δ_g which is a limit of the holomorphic motions over Δ_{g_n}. By continuity, this holomorphic motion is also equivariant. □

Though the above proof only applies for our construction of the lamination in a neighborhood of a Yoccoz map, a similar statement still holds near infinitely renormalizable maps. In this case instead of using persistent puzzles, the situation can be reduced to the quadratic-like case by means of the renormalization. The hyperbolic case can be dealt in an easier way with a construction based on the persistence of the basin of attraction (in a similar argument to Proposition 5.4). Similarly, the case of parabolic maps can be dealt using the persistence of attracting petals along the submanifold where the parabolic point persists. All those cases are summarized below:

Proposition 8.9. *Let $g \in \mathcal{U}_a$. There exists a codimension-one complex submanifold $\Delta_g \subset \mathcal{A}_a$, such that $\Delta_g \cap \mathcal{U}_a \subset \mathcal{H}_g$ and a normalized holomorphic motion $h_{\tilde{g}}$, $\tilde{g} \in \Delta_g$, which is equivariant on I. Moreover, the Δ_g form a lamination near any non-parabolic map $f \in \mathcal{U}_a$.*

Recall that except in the hyperbolic but not superattracting case, the "tangent space" T_f was defined as the set of vector fields admitting a representation $v = \alpha \circ f - f'\alpha$ on $\mathrm{orb}_f(0)$ for a qc vector field α. The following proposition shows that this choice was completely justified:

Theorem 8.10. $T_f = T_f\mathcal{H}_f$.

Proof. First we notice that T_f contains the tangent space to the real hybrid class of f. Indeed, if $v \in T_f\mathcal{H}_f$ then there exists a path $f_\lambda \in \mathcal{H}_f$ through f tangent to v at f. By Proposition 8.9, f_λ can be equipped with an equivariant holomorphic motion of the interval. By Proposition 5.2, $v \in T_f$.

We obtained in §7 a transverse vector field to T_f for at most finitely renormalizable maps f. Since $T_f\mathcal{H}_f$ is codimension-one, $T_f = T_f\mathcal{H}_f$.

Let now f be infinitely renormalizable, and let R be the renormalization operator of §4, so that $R(f) : U \to U'$ is a quadratic-like map. Let v be a vector field transverse to $T_f\mathcal{H}_f$. By transverse non-singularity of R, $DR(f)v \notin T_{R(f)}\mathcal{H}_{R(f)}$ (see Remark 4.1).

By the characterization of the tangent space to the hybrid class of a quadratic-like map (see §4.1.2), there are no qc vector fields α equivariant with respect to $(R(f), DR(f)v)$ on U satisfying $\overline{\partial}\alpha | K(R(F)) = 0$.

However, this last condition is vacuous since there are no invariant line fields on $K(R(f))$ (Theorem 2.22). By the Infinitesimal Pullback Argument (in the quadratic-like setting), we see that this is equivalent to non-existence of a qc vector field α equivariant just on $\mathrm{orb}_{R(f)}(0)$. In particular, there is no qc vector field equivariant with respect to (f, v) on $\mathrm{orb}_f(0)$, so $v \notin T_f$. □

9. Regular or stochastic theorem

9.1. Tangencies between holomorphic curves and holomorphic laminations. Let us consider a codimension-one holomorphic lamination $\mathcal{F}$ in an open set $\mathcal{V}$ of a complex Banach space $\mathcal{B}$. For a point $a \in \mathcal{V}$ in the support of the lamination, denote by L_a the leaf of $\mathcal{F}$ through a and by T_a the tangent space to this leaf at a. Let $\gamma : \mathbb{D} \to \mathcal{V}$ be a holomorphic curve. We say that γ has a tangency with $\mathcal{F}$ at some parameter value λ_0 if $\gamma(\lambda_0)$ belongs to the support of the lamination and

$$\frac{d}{d\lambda}\gamma(\lambda)\bigg|_{\lambda=\lambda_0} \in T_{\gamma(\lambda_0)}.$$

The set of tangencies is clearly closed in $\mathbb{D}$.

Lemma 9.1 (A. Douady). *If the curve γ is not contained in any leaf of $\mathcal{F}$ then the set of tangencies is discrete.*

Proof. We may assume that we have flow box coordinates $\mathcal{W} \oplus \mathbb{C}$, in other words, the leaves of the lamination are graphs over $\mathcal{W}$. Let's consider a parameter λ_0 where γ has a tangency with the lamination.

By a change of coordinates we may assume $\lambda_0 = 0$ and that the leaf containing $\gamma(0)$ is the graph of the zero function from $\mathcal{W}$ to $\mathbb{C}$. If γ is not contained in a leaf, we may write for z near 0, $\gamma(z) = (\phi(z), z^n\psi(z))$ where $\psi(0) \neq 0$ and $n \geq 2$.

Let $S : \mathcal{W} \oplus \mathbb{C} \to \mathcal{W} \oplus \mathbb{C}$ be defined by $S(z, w) = (z, w^n)$ and $\tilde{\mathcal{F}}$ be the lamination in $\mathcal{W} \oplus \mathbb{C}$ whose leaves are connected components of preimages by S of leaves of $\mathcal{F}$.

It is easy to see that if $\tilde{\gamma}$ is a path in $\mathcal{W} \oplus \mathbb{C}$ and if $S \circ \tilde{\gamma}$ has a tangency at λ with $\mathcal{F}$ then either $\tilde{\gamma}$ has a tangency at λ with $\tilde{\mathcal{F}}$ or $\tilde{\gamma}(\lambda) \in \mathcal{W} \times \{0\}$.

Let then $\tilde{\gamma} = (\phi(z), z\psi(z)^{1/n})$ for z near 0. Then $\tilde{\gamma}$ is transverse to $\tilde{\mathcal{F}}$ at 0, and so there is no tangency in a neighborhood of 0. We conclude that 0 is the unique tangency of $\gamma = S \circ \tilde{\gamma}$ in a neighborhood of 0. □

9.2. Connectivity of some hybrid classes.

Theorem 9.2. *If f is a Yoccoz map, then there exists a one parameter real analytic family $\{f_t\}_{t\in[0,1]} \subset \mathcal{U}_a$ in the hybrid class of f connecting $f_0 = f$ with the quadratic map $f_1 = q_{\chi(f)}$.*

Proof. It follows from Theorem B.1 that f is qc conjugate to the quadratic polynomial $q_{\chi(f)}$ in a neighborhood of the interval I. Moreover, this conjugacy fixes 0 and 1, and commutes with the reflections with respect to 0 and $\mathbb{R}$ (we will call such maps "symmetric"). Extend it to a global symmetric qc homeomorphism $h : \mathbb{C} \to \mathbb{C}$. Let μ be its Beltrami differential. Solving the Beltrami equation with differential $\lambda\mu$, we obtain a holomorphic family

$$h_\lambda : \mathbb{C} \to \mathbb{C}, \ |\lambda| < 1/\|\mu\|_\infty,$$

of symmetric qc maps fixing 0 and 1. Let $f_\lambda : h_\lambda(\Omega_a) \to \mathbb{C}$ be defined by $f_\lambda = h_\lambda \circ f \circ h_\lambda^{-1}$. Then there is some $a' < a$ such that $\{f_t\}_{t\in[0,1]}$ is a real analytic curve in $\mathcal{U}_{a'}$ contained in the hybrid class of f.

Let us show that this curve can be approximated, in the topology of $\mathcal{U}_{a'}$ by a similar curve taking values in $\mathcal{U}_a$. To this end, let us consider a one-parameter real analytic family of vector fields $\{v_t\}_{t\in[0,1]}$ in $T\mathcal{U}_{a'}$ along the family $\{f_t\}_{t\in[0,1]}$ such that for any t, v_t is transverse to the hybrid class of f in $\mathcal{U}_{a'}$ (that is $v_t \notin T_{f_t}\mathcal{H}_{f,a'}$). Consider a 2-parameter family

$$G(t,s) = f_t + sv(t) \text{ in } \mathcal{U}_{a'},\ (t,s) \in [0,1] \times [-1,1].$$

Letting $\zeta_0 : [0,1] \to [0,1] \times [-1,1]$ be the natural inclusion $\zeta_0(t) = (t,0)$, we have $f_t = G \circ \zeta_0(t)$.

Let us now consider a real analytic family $F : [0,1] \times [-1,1] \to \mathcal{U}_a$ such that $F(0,0) = G(0,0)$ and $F(1,0) = G(1,0)$. We may also require that $\Pi_{a,a'} \circ F$ is C^1 close to G, where $\Pi_{a,a'} : \mathcal{U}_a \to \mathcal{U}_{a'}$ is the inclusion. By the Implicit Function Theorem, there exists a real analytic curve ζ : $[0,1] \to [0,1] \times [-1,1]$, C^1 close to ζ_0, such that $\Pi_{a,a'} \circ F \circ \zeta$ is contained in $\mathcal{H}_{f,a'}$. In particular, $F \circ \zeta : [0,1] \to \mathcal{U}_a$ is a real analytic path connecting f_0 to f_1 in $\mathcal{H}_{f,a}$. □

9.3. Proof of Theorem C. Let $\{f_t\}_{t\in J}$ be a real analytic family in $\mathcal{U}_a$ defined over some (open or closed) interval $J \subset \mathbb{R}$. Assume $\{f_t\}$ is non-trivial. By Theorem A each hybrid class is a real analytic codimension-one submanifold. Let $T \subset \mathbb{R}$ be the union of the set of parabolic parameters and the set of tangencies of f_t with the hybrid classes.

Lemma 9.3. *T is countable.*

Proof. We notice that the set of parabolic parameters is countable. Indeed, those are associated with countably many analytic equations of the type $f_t^k(x) = x$, $|Df_t^k(x)| = 1$, where $k > 0$ is an integer. Since $\{f_t\}$ is not contained in a leaf, each of these equations corresponds to a discrete set of parameters.

So if T is not countable there is a parameter t such that f_t is not parabolic and any neighborhood of t intersects T in infinite many points. By Theorem A, since f_t is not parabolic there exists a neighborhood of f_λ in $\mathcal{A}_a$ which is holomorphically laminated by the hybrid classes. By Lemma 9.1, the set of tangencies in a smaller neighborhood is finite. This is a contradiction. □

To conclude the proof of Theorem C, it is enough to show that each parameter in the complement of T has a neighborhood where the straightening is quasisymmetric.

Consider then a parameter $\lambda_0 \notin T$. If f_{λ_0} is hyperbolic, χ is clearly analytic in a neighborhood of λ_0. If f_{λ_0} is infinitely renormalizable, we can apply Theorem 4.11. Assume then that f_{λ_0} is a Yoccoz map. In this case, there is a path connecting it to the quadratic family and we can just use the transversality at λ_0 to conclude the quasisymmetry of χ, since codimension-one real analytic laminations are transversally quasisymmetric.

9.4. Proof of Theorem B. Let X be the complement of the set of parameters of the quadratic family which are either hyperbolic or are Yoccoz maps and have at most a finite number of central levels in the principal nest of their last renormalization. By Theorem 2.20, a map which is neither regular or stochastic is hybrid equivalent to a parameter in X. In view of Theorem C, it is enough to prove the following:

Lemma 9.4. *The image of X by any qs map has zero Lebesgue measure.*

Proof. We first decompose $X = \mathcal{I} \cup \mathcal{F} \cup \mathcal{P}$ where $\mathcal{I}$ are infinitely renormalizable parameters, $\mathcal{F}$ are Yoccoz parameters contained in X, and $\mathcal{P}$ are parabolic. Let h be a qs map.

Then $|h(\mathcal{P})| = 0$ since $\mathcal{P}$ is countable, and $|h(\mathcal{I})| = 0$ follows from Theorem 2.27, since the property of having definite gaps everywhere is preserved by quasisymmetric maps.

Parameters in $\mathcal{F}$ can be further decomposed as a countable union $\cup \mathcal{F}_j$, where parameters in each $\mathcal{F}_j$ have the same combinatorics for their smallest renormalization interval. Let us show that each $h(\mathcal{F}_j)$ has zero Lebesgue measure. For simplicity, let us consider the case $\mathcal{F}_0 \subset \mathcal{N}$ of non-renormalizable maps, the general case reduces to this one by renormalization.

According to §2.15, for each n there exists a covering of $\mathcal{N}$ by disjoint intervals $\Delta_i^n \subset [1/2, 2]$ ("real parapuzzle pieces of level n"), and each of the Δ_i^n contains a central interval Π_i^n satisfying (2.16). Furthermore, a parameter belongs to $\mathcal{F}_0$ if and only if it belongs to infinitely many Π_i^n. By Lemma 2.5, there exists constants $\tilde{C} > 0$ and $0 < \tilde{\rho} < 1$ such that

$$\frac{|h(\Pi_i^n)|}{|h(\Delta_i^n)|} \leq \tilde{C}\tilde{\rho}^n,$$

so that $\sum_n \sum_i |h(\Pi_i^n)| < \infty$. By the Borel-Cantelli Lemma, $|h(\mathcal{F}_0)| = 0$. □

Appendix A. Complex return maps

The dynamics of certain classes of complex return maps was described in the works of Branner-Hubbard [BH], Yoccoz [H], and Lyubich [L2], [L4].

The precise hypothesis on the dynamics change from work to work. In this appendix we adapt those ideas for our setting, collecting the results needed for the analysis of puzzle maps.

A.1. Definitions. Let W be a quasidisk and let $\{W_j\}$ be a family of at least 2 quasidisks inside W with pairwise disjoint closures such that $0 \in W_0$. Assume further that

$$\inf \operatorname{mod}(W \setminus \overline{W_j}) > 0, \tag{A.1}$$

that $\overline{\cup W_j}$ is thin in W (see definition in the beginning of §5.2), and $\operatorname{diam}(W_j) \to 0$.

An *R-map* is a holomorphic map $F : \cup W_j \to W$ such that for any $j \neq 0$, $F|W_j$ is a univalent map onto W, and $F|W_0$ is a double covering onto W branched at 0 ("R" stands for "Return"). We let $W^n = F^{-n}(W)$ and we define the *filled Julia set* $K(F)$ as $\cap W^n$.

The components of $F^{-n}(W)$ are called *puzzle pieces of depth n*. For $x \in F^{-n}(W)$, we let $P^n(x)$ be the puzzle piece of depth n containing x. Puzzle pieces containing 0 are called *critical*.

An R-map F is called *renormalizable* if there exists a puzzle piece $V = P^n(0)$, $n \geq 1$, and an integer $p > 0$ such that $V \subset F^p(V)$, the puzzle pieces $F^j(V)$ $1 \leq j \leq p$ are pairwise disjoint, and $F^{mp}(0) \in V$, $m > 0$. The map $R(F) = F^p|P^n(0)$ with minimal n as above will be called the *renormalization* of F. It is a quadratic-like map with connected Julia set.

Let $F : \cup W_j \to W$ and $\tilde{F} : \cup \tilde{W}_j \to W$ be two R-maps, and let h be a homeomorphism of $\mathbb{C}$ equivariant on $\cup \partial W_j$. If $h(F(0)) = \tilde{F}(0)$, then for each j there is a unique homeomorphism $\psi_j : \operatorname{cl} W_j \to \operatorname{cl} \tilde{W}_j$ coinciding with h on ∂W_j and such that $h \circ F = \tilde{F} \circ \psi_j$ on W_j. Let

$$h_1 = \begin{cases} \psi_j & \text{on } W_j \\ h & \text{on } \mathbb{C} \setminus \cup W_j. \end{cases}$$

Since $\operatorname{diam} W_j \to 0$, h_1 is a homeomorphism of $\mathbb{C}$. It is called the *lift* of h (compare §6.1).

We say that a homeomorphism $h : \mathbb{C} \to \mathbb{C}$ is a *combinatorial equivalence* between F and $\tilde{F}$ if it is equivariant on $\cup \partial W_j$ and the lift h_1 of h is homotopic to h rel $\cup \partial W_j \cup \operatorname{orb}_F(0)$.

The notions of topological/qc/hybrid equivalence between two R-maps are self-evident.

A.2. Divergence property. For $x \in K(F)$, we let

$$A^n(x) = P^n(x) \setminus \overline{W^{n+1}}.$$

We define $\operatorname{mod} A^n(x)$ as the extremal length of the family of curves in $A^n(x)$ joining $\partial P^n(x)$ to $\partial W^{n+1} \cap P^n(x)$.

Theorem A.1. *Assume that* $F : \cup_{j=1}^{m} W_j \to W$ *is a non-renormalizable R-map defined on the union of finitely many domains* W_j*. Then for any* $x \in K(F)$,

$$\sum \operatorname{mod} A^n(x) = \infty. \tag{A.2}$$

Proof. Fix $x \in K(F)$ and let $\mu_k^n = \operatorname{mod} A^n(F^k(x))$. These numbers satisfy the following rules: $\mu_k^n = \mu_{k+1}^{n-1}$ if $P^n(x)$ is not critical and $2\mu_k^n = \mu_{k+1}^{n-1}$ otherwise.

Since there are only finitely many W_j, $\overline{\cup W_j} \subset W$. This implies that μ_k^0 (which in fact does not depend on k) is positive. Since F is non-renormalizable, the "tableau" $\{P_k^n(0)\}_{n,k}$ is aperiodic in k. By the work of Branner and Hubbard, Theorem 4.3, p. 264 of [BH], these properties imply (A.2). □

A compact set $X \subset \mathbb{C}$ is called *removable* if any qc map $H : \mathbb{C} \setminus X \to \mathbb{C}$ extends to a qc homeomorphism of $\mathbb{C}$. The following remark is due to Jeremy Kahn.

Corollary A.2. *Under the assumptions of the previous theorem, the filled Julia set* $K(F)$ *is removable and hence has zero Lebesgue measure.*

Proof. Since there are only finitely many domains W_j, the filled Julia set $K(F)$ is compact. Now the result follows from the divergence property (A.2) by [SN], §1 (see also McMullen in [BH], §5.4). □

A.3. Geometry of puzzle pieces. Let us say that $x \in K(F)$ *shadows the critical orbit* if for any k and all $m \geq m(k)$, there exists $j \leq m$ such that the puzzle piece $P^{k+j}(F^{m-j}(x))$ is critical (in other words, the map $F^m : P^{k+m}(x) \to P^k(F^m(x))$ is not univalent). In particular, for $m \geq m(k)$, $P^k(F^m(x))$ intersects the critical orbit. Moreover, for all k, $\operatorname{orb}_F(x) \cap P^k(0) \neq \emptyset$.

Lemma A.3. *Let* F *be an R-map and let* $x \in K(F)$*. If*

$$\operatorname{mod}(P^n(x) \setminus \overline{P^{n+1}(x)}) \to 0$$

then x *shadows the critical orbit.*

Proof. Assume that x does not shadow the critical orbit. Then there exist k and arbitrarily big m such that the map $F^m : P^{k+m}(x) \to P^k(F^m(x))$ is univalent. Hence

$$\operatorname{mod}(P^{k+m}(x) \setminus \overline{P^{k+m+1}(x)}) = \operatorname{mod}(P^k(F^m(x)) \setminus \overline{P^{k+1}(F^m(x))}).$$

Taking k more iterates of $F^m(x)$, we conclude that

$$\operatorname{mod}(P^k(F^m(x)) \setminus \overline{P^{k+1}(F^m(x))}) \geq \frac{1}{2^k} \operatorname{mod}(W \setminus \overline{P^1(F^{m+k}(x))}).$$

By the definition of R-map, the later modulus is bounded away from 0. This is a contradiction. □

Lemma A.4. *If F is renormalizable and x shadows the critical orbit then there exists k such that $F^k(x) \in K(R(F))$.*

Proof. Let $R(F) = F^p|P^n(0)$ and let $U_i = P^n(F^i(0))$. Then $\{U_i\}_{i=0}^{p-1}$ is a disjoint cover of the orbit of 0. Let $K_j = F^j(K(R(F)))$ so that

$$K_j = \left\{x \in U_j : F^k(x) \in \cup_{i=0}^{p-1} U_i, k = 0, 1, \ldots\right\}.$$

Let $x \in K(F)$ shadow the critical orbit and let m_0 be such that

$$P^n(F^m(x)) \cap \text{orb}_F(0) \neq \emptyset \quad \text{for } m \geq m_0.$$

Hence, $P^n(F^m(x)) \subset \cup U_i$ for all $m \geq m_0$, and the conclusion follows. □

Theorem A.5. *Let F be an R-map and let $x \in K(F)$ satisfy*

$$\sum \text{mod}(P^n(x) \setminus \overline{P^{n+1}(x)}) < \infty.$$

Then F is renormalizable and $F^k(x) \in K(R(F))$ for some k.

Proof. For $x \in K(F)$, let $B_n(x) = \text{mod}(P_n(x) \setminus \overline{P^{n+1}(x)})$. Let x be a point such that $\sum B_n(x) < \infty$. By Lemma A.3, $P^k(0)$ intersects $\text{orb}_F(x)$ for every k. In particular, if m_k is minimum such that $F^{m_k}(x) \in P^{k+1}(0)$, $B_{k+m_k}(x) = B_k(0)$. Since the sequence $k + m_k$ is strictly increasing, we conclude that $\sum B_k(0) < \infty$.

Assume first that $\text{orb}_F(0)$ intersects infinitely many domains W_j. Then $F(0)$ does not shadow the critical orbit: if m is minimal with $F^m(F(0)) \in W_j$, the map $F^m : P^{m+1}(F(0)) \to P^1(F^m(F(0)))$ is a diffeomorphism. In particular, by Lemma A.3, $B_k(F(0))$ does not converge to 0, so neither does $B_k(0)$, and this contradicts $\sum B_n(0) < \infty$.

Thus, $\text{orb}_F(0)$ intersects only finitely many W_j. Let $\tilde{F}$ be the R-map obtained by restricting F to those. The sequence $B_k(0)$ does not change by taking this restriction. The Divergence Property of Theorem A.1 implies that $\tilde{F}$ is renormalizable. So F is also renormalizable, and since x shadows the critical orbit we have $F^k(x) \in K(R(F))$ for some k by Lemma A.4. □

Corollary A.6. *If F is a non-renormalizable puzzle map, for every $x \in K(F)$, $\cap P^n(x) = \{x\}$.*

Lemma A.7. *Let F be an R-map such that* $\text{int}\, K(F) \neq \emptyset$. *Then F is renormalizable and* $\text{int}\, K(F) = \cup F^{-n}(\text{int}\, K(R(F)))$. *Furthermore, $R(F)$ has a non-repelling periodic orbit.*

Proof. Let $x \in \text{int}\, K(F)$. Notice that $\partial P^n(x)$ is not contained in $K(F)$, so $\cap P^n(x)$ contains a neighborhood V of x. By Theorem A.5, F is renormalizable, and $F^k(x) \in K(R(F))$ for some k. Since $K(R(F)) = \cap P^n(0)$, we have:

$$F^k(V) \subset F^k(\cap P^n(x)) \subset K(R(F)).$$

Hence $F^k(x) \in \operatorname{int} K(R(F))$. By the classification of Fatou components, see [Mc1] p. 37, a quadratic polynomial whose filled-in Julia set has non-empty interior must have either a hyperbolic or parabolic periodic orbit or an indifferent periodic orbit associated to a Siegel disk, so in all cases it must have a non-repelling periodic orbit. The same conclusion holds for $R(F)$ via the Straightening Theorem of Douady-Hubbard (see §2.12). □

A.4. Measure of the Julia set. In this section we generalize the results of [L2] and [Sh] on the Lebesgue measure of the Julia sets of quadratic polynomials to the setting of R-maps.

Lemma A.8. *Let F be an R-map. Then almost every $x \in K(F)$ shadows the critical orbit.*

Proof. Consider a point $x \in K(F)$ which does not shadow the critical orbit. Then there exist k and a sequence $m_j \to \infty$ such that each map $F^{m_j} : P^{m_j+k}(x) \to P^k(F^{m_j}(x))$ is univalent. Applying a few more iterates of F, we will find a critical puzzle piece $P^{k-n_j}(0)$ such that the map $F^{m_j+n_j}$: $P^{m_j+k} \to P^{k-n_j}(0)$ is also univalent.

Since the property of being thin is invariant under lifts by branched coverings, the filled Julia set $K(F)$ is thin in each $P^l(0)$, $0 \leq l \leq k$. By the Koebe Distortion Theorem, $K(F)$ is also uniformly thin in all puzzle pieces $P^{m_j+k}(x)$. Since the puzzle pieces $P^{m_j+k}(x)$ shrink to x (by Lemma A.3), x is not a density point of $K(F)$. The conclusion follows from the Lebesgue Density Points Theorem. □

Corollary A.9. *If F has an escaping critical point then* meas $K(F) = 0$.

Proof. In this case there are no points which shadow the critical orbit. □

Theorem A.10. *Let F be an R-map. If* meas $K(F) > 0$*, then F is renormalizable and $K(F) \setminus \cup F^{-n} K(R(F))$ has zero Lebesgue measure.*

Proof. Assume F is non-renormalizable. Consider a point $x \in K(F)$ which shadows the critical orbit. Recall that $\operatorname{orb}_F(x)$ intersects all critical puzzle pieces.

Assume that the critical point intersect infinitely many W_j, and let $j_k \to \infty$ be a sequence such that $\operatorname{orb}_F(0) \cap W_{j_k} \neq \emptyset$. Let n_k be the first landing time of the orbit of 0 at W_{j_k}, and let m_k be the first landing time of x at $P^{n_k+1}(0)$. Notice that the maps $F^{n_k} : P^{n_k}(F(0)) \to W$ and $F^{m_k} : P^{n_k+m_k+1}(x) \to P^{n_k+1}(0)$ are univalent.

Since $\cup W_j$ is thin in W, $K(F)$ is uniformly thin in $P^{m_k}(F(0))$ by the Koebe Distortion Theorem. Pulling back by the double covering $F|P^1(0)$, we conclude that $K(F)$ is uniformly thin in $P^{m_k+1}(0)$. Pulling back again, we get that $K(F)$ is uniformly thin in $P^{n_k+m_k+1}(x)$. Since the $P^n(x)$ shrink to x (Corollary A.6), x is not a density point of $K(F)$. By the Lebesgue Density Point Theorem and Lemma A.8, $K(F)$ has zero Lebesgue measure.

Assume now that $\text{orb}_F(0)$ intersects only finitely many W_j. Let us consider the R-map $F_\#$ which is the restriction of F to the W_j visited by the critical point. It follows that $\cup F^{-n}(K(F_\#))$ contain all points $x \in K(F)$ which shadow the critical orbit. By Corollary A.2, meas $K(F_\#) = 0$, and by Lemma A.8, $K(F)$ has zero Lebesgue measure.

Thus, F is renormalizable, and the last assertion follows from Lemma A.3 and Lemma A.4. □

A.5. Periodic orbits.

Lemma A.11. *Let F be an R-map and let $\overline{p}$ be a non-repelling periodic orbit. Then for all n, $\overline{p}$ intersects $P^n(0)$.*

Proof. Let k be the period of p and assume that $\overline{p}$ does not intersect $P^n(0)$. Then $F^k|P^{n+k}(p)$ is univalent onto $P^n(p)$. Since $\overline{P^{n+k}} \subset P^n(p)$, p is repelling by the Schwarz Lemma. □

Lemma A.12. *Let F be an R-map and assume F has a non-repelling periodic orbit $\overline{p}$. Then F is renormalizable, $\overline{p}$ intersects $K(R(F))$, and $\overline{p}$ is the unique non-repelling periodic orbit of F. Furthermore, if F is $\mathbb{R}$-symmetric, $\omega(0) = \overline{p}$, so $\overline{p} \subset \mathbb{R}$.*

Proof. By the previous lemma, $\overline{p}$ intersects $\cap P^n(0)$. By Corollary A.6, if F is non-renormalizable then $\cap P^n(0) = \{0\}$, so $0 \in \text{orb}_F(p)$. Since 0 is periodic, F is renormalizable, contradiction. So F is renormalizable, and in this case $\cap P^n(0) = K(R(F))$, so $K(R(F))$ intersects $\overline{p}$. By [D2], a quadratic-like map has at most one non-repelling periodic orbit, and this yields uniqueness of $\overline{p}$.

If F is $\mathbb{R}$-symmetric, then $R(F)$ is $\mathbb{R}$-symmetric and the straightening map can also be chosen $\mathbb{R}$-symmetric. By the classical theory of Fatou and Julia, a non-repelling periodic orbit of a $\mathbb{R}$-symmetric quadratic polynomial must be contained in $\mathbb{R}$, and this yields the conclusion for F. □

Lemma A.13. *Let F be an R-map which has an attracting hyperbolic periodic orbit $\overline{p}$. Then $K(F) \setminus \text{int}\, K(F)$ has zero Lebesgue measure and $\text{int}\, K(F)$ is the basin of attraction of $\overline{p}$.*

Proof. By the previous lemma, F is renormalizable and $R(F)$ is a quadratic-like map with a unique attracting periodic orbit (the intersection of $\text{orb}_F(p)$ and $K(R(F))$). It is well known that this implies that $\text{int}\, K(R(F))$ is the basin of attraction of $\text{orb}_F(p) \cap K(R(F))$ and $\partial K(R(F))$ has zero Lebesgue measure. By Lemma A.7, it follows that $\text{int}\, K(F)$ is the basin of attraction of $\overline{p}$. By Theorem A.10, $K(F) \setminus \text{int}\, K(F)$ is equal to the union of preimages of $\partial K(R(F))$ up to some set of zero Lebesgue measure, so it has zero Lebesgue measure. □

A.6. Rigidity. The following lemma is analogous to Lemma 6.1.

Lemma A.14. *Let h be a combinatorial equivalence between two R-maps F and $\tilde{F}$. Then the lift h_1 of h is a combinatorial equivalence between F and $\tilde{F}$. If h is quasiconformal then h_1 is quasiconformal and* $\mathrm{Dil}(h_1) \le \mathrm{Dil}(h)$.

Lemma A.15. *Let $h : \mathbb{C} \to \mathbb{C}$ be a qc map which is a combinatorial equivariance between two R-maps F and $\tilde{F}$. Assume* $\mathrm{int}\, K(F) = \emptyset$. *Then there exists a qc conjugacy $H : \mathbb{C} \to \mathbb{C}$ such that $H|\mathbb{C} \setminus \cup W_j = h$ and* $\mathrm{Dil}(H) \le \mathrm{Dil}(h)$.

Proof. As in the puzzle case (Lemma 6.2) we can obtain a qc map $H : \mathbb{C} \to \mathbb{C}$ as a limit of a series of lifts of h. This qc map automatically satisfies $\mathrm{Dil}(H) \le \mathrm{Dil}(h)$ and is equivariant on $\mathbb{C} \setminus K(F)$. Since $\mathrm{int}\, K(F) = \emptyset$, the conclusion follows. □

Theorem A.16. *Let F and $\tilde{F}$ be non-renormalizable R-maps which are combinatorially equivalent and let h be a combinatorial equivalence between them. Assume that $h|\mathbb{C} \setminus \cup W_j$ extends to a qc map of $\mathbb{C}$. Then there exists a hybrid conjugacy $H : \mathbb{C} \to \mathbb{C}$ such that $H|\mathbb{C} \setminus \cup W_j = h$.*

Proof. Let us assume first that the domain of F has finitely many components W_j. We let $h_0 = h$ and define h_{i+1} be the lift of h_i. Then $h_i(z)$ is eventually constant for every $z \notin K(F)$. Let H be the limit of h_i on $W \setminus \overline{K(F)}$. Then H is a qc homeomorphism and by Corollary A.2 H extends to a qc homeomorphism of $\mathbb{C}$. Since $K(F)$ has empty interior and H is equivariant on $\mathbb{C} \setminus \overline{K(F)}$, the result follows.

Assume now that there are infinitely many W_j, but $\mathrm{orb}_F(0)$ intersects only finitely many of them (in particular this covers the case of escaping critical orbit). Consider the restrictions $F_\#$ and $\tilde{F}_\#$ to the union of those puzzle pieces. Let $\hat{h} : \mathbb{C} \to \mathbb{C}$ be a qc extension of $h|\mathbb{C} \setminus \cup W_j$. For every $j \neq 0$ let ψ_j be the lift of $\hat{h}$ to W_j. Let us define Ψ as follows

$$\Psi = \begin{cases} \psi_j & \text{on } W_j, \text{ for } W_j \cap \mathrm{orb}_F(0) = \emptyset \\ h & \text{otherwise.} \end{cases}$$

Then Ψ is a combinatorial equivalence between $F_\#$ and $\tilde{F}_\#$ which is quasiconformal on the complement of the domain of $F_\#$. By the previous case, we obtain a qc conjugacy $H_\#$ between $F_\#$ and $\tilde{F}_\#$. Moreover, $H_\# : \mathbb{C} \to \mathbb{C}$ is also a combinatorial equivalence between F and $\tilde{F}$, so by Lemma A.15, it can be turned into the desired qc conjugacy.

Now let us assume that the critical orbit intersects infinitely many puzzle pieces. Since W is a quasidisk, the qc homeomorphism $h|\mathbb{C} \setminus W$ can be extended to a qc homeomorphism $\hat{h} : \mathbb{C} \to \mathbb{C}$ taking $F(0)$ to $\tilde{F}(0)$. As above, let ψ_j be the lift of $\hat{h}$ to W_j and let

$$\Psi = \begin{cases} \psi_j & \text{on } W_j \\ h & \text{otherwise.} \end{cases}$$

For $x \in W_0$, $y \in \tilde{W}_0$, let $h_{x,y}$ be a qc map coinciding with h on $\mathbb{C} \setminus W$ and taking x to y. It is clear that we may choose $h_{x,y}$ in such way that

$$\sup_{x \in W_0, y \in \tilde{W}_0} \operatorname{Dil}(h_{x,y}) < \infty.$$

There exist $j_k, n_k \to \infty$ such that n_k is the first landing moment of the critical orbit at W_{j_k}. Let m_k be the first landing moment of $F^{n_k}(0)$ at W_0. Assume first that $m_k < \infty$. Let $V_k^j = P^j(F^{n_k}(0))$. Define

$$\psi_{j,k} : V_k^j \setminus V_k^{j+1} \to \tilde{V}_k^j \setminus \tilde{V}_k^{j+1}, \quad 0 \le j \le m_k - 1,$$

as the lift of Ψ by the pair $(F^j, \tilde{F}^j)$. Let $\phi_k : V_k^{m_k} \to \tilde{V}_k^{m_k}$ be the lift of $h_{F^{m_k}(0), \tilde{F}^{m_k}(0)}$ by the pair $(F^{m_k}, \tilde{F}^{m_k})$. Define Ψ_k as follows:

$$\Psi_k = \begin{cases} \psi_{j,k} & \text{on } V_k^j \setminus V_k^{j+1}, \quad 0 \le j \le m_k - 1 \\ \phi_k & \text{on } V_k^{m_k} \\ h & \text{on } \mathbb{C} \setminus W. \end{cases}$$

Then the Ψ_k are qc maps such that $\Psi_k(F^{n_k}(0)) = \tilde{F}^{n_k}(0)$ and $\sup \operatorname{Dil}(\Psi_k) < \infty$.

If $m_k = \infty$, define

$$\Psi_k = \begin{cases} \psi_{j,k} & \text{on } V_k^j \setminus V_k^{j+1}, \quad j \ge 0 \\ \tilde{F}^{n_k}(0) & \text{at } F^{n_k}(0) \\ h & \text{on } \mathbb{C} \setminus W. \end{cases}$$

Notice that since $\cap_j V_k^j = \{F^{n_k}(0)\}$, this map has the same properties as before.

Let now $F_k = F | \cup_{j \ne j_k} W_j$ and $\tilde{F}_k = \tilde{F} | \cup_{j \ne j_k} W_j$. Modify Ψ_k on finitely many domains W_j which contain points $f^l(0)$, $l = 0, ..., n_k - 1$, so that it becomes a qc combinatorial equivalence Φ_k between F_k and $\tilde{F}_k$. We are now in the escaping case considered above, so that we can turn Φ_k into a qc conjugacy H_k between F_k and $\tilde{F}_k$. Moreover,

$$\operatorname{Dil}(H_k) \le \operatorname{Dil}(\Phi_k | \mathbb{C} \setminus \cup W_j) \le \operatorname{Dil}(\Psi_k),$$

since $K(F_k)$ has zero Lebesgue measure. Take some limit H of the H_k. This is the desired qc conjugacy.

Since meas $K(F) = 0$ (see Theorem A.10), H is automatically a hybrid conjugacy. □

Remark A.1. Note that the above theorem leads to a simple proof of the Yoccoz Rigidity Theorem for non-renormalizable quadratic polynomials (see [H]). Indeed such a quadratic map can be renormalized in a generalized sense to an R-map (see [L4], §3). The conjugacy between the corresponding R-maps lifts to a conjugacy between the quadratic polynomials by means of the pullback argument.

The following lemma is a slight modification of well known results, see [DH1].

Lemma A.17. *Let $F : U \to U'$ and $\tilde{F} : \tilde{U} \to \tilde{U}'$ be quadratic-like maps with connected Julia set which are qc conjugate and let ψ be a qc conjugacy on a neighborhood of $K(F)$. Let $h : \mathbb{C} \to \mathbb{C}$ be a qc map equivariant on ∂U. Then there exists a qc homeomorphism $H : \mathbb{C} \to \mathbb{C}$ such that $H|\mathbb{C} \setminus U = h$ and $H|K(F) = \psi$.*

Proof. Let $\Phi : \mathbb{C}\setminus K(F) \to \mathbb{C}\setminus K(\tilde{F})$ be obtained as a limit of a series of lifts of h by the pair of unbranched double coverings $F : U \setminus K(F) \to U' \setminus K(F)$ and $\tilde{F} : \tilde{U} \setminus K(\tilde{F}) \to \tilde{U}' \setminus K(\tilde{F})$. Then Φ is quasiconformal and equivariant on $U \setminus K(F)$.

Let $V' = F^{-n}(U)$ be a small neighborhood of $K(F)$ which is contained in the domain of ψ and let $V = F^{-1}(V')$. Let $A = \overline{V}' \setminus V$. Consider a homotopy $\Phi_t : A \to \tilde{U} \setminus K(\tilde{F})$ between $\Phi|A$ and ψ which is equivariant on ∂V (the existence of such a homotopy follows from the fact that F has degree 2).

By means of successive lifts, we obtain a homotopy $\Phi_t : \overline{V}' \setminus K(F) \to \tilde{U} \setminus K(\tilde{F})$, between $\Phi|\overline{V}' \setminus K(F)$ and ψ, equivariant on $\overline{V} \setminus K(F)$.

Let us supply $\tilde{U} \setminus K(\tilde{F})$ with the hyperbolic metric. Given a point $x \in V' \setminus K(F)$, let $c(x)$ be the length of the hyperbolic geodesic joining $\Phi_0(x) = \Phi(x)$ and $\Phi_1(x) = \psi(x)$, homotopic to the path $t \to \Phi_t(x)$. By compactness, $c(x)$ is bounded on $V' \setminus V$.

Let $x \in V$ and let $F^m(x)$ be the first landing time of x on $V' \setminus V$. By the Schwarz Lemma, $c(x) \leq c(F^n(x))$, so $c(x)$ is uniformly bounded in $V' \setminus K(F)$ as well. In particular, if x is close to $K(F)$, $|\psi(x) - \Phi(x)|$ is close to 0.

Let

$$H = \begin{cases} \Phi & \text{on } \mathbb{C} \setminus K(F) \\ \psi & \text{on } K(F). \end{cases}$$

By the previous estimate, H is continuous, so by the Gluing Lemma (see [B], Lemma 2, p. 93), H is quasiconformal. It is also clearly a conjugacy. □

Lemma A.18. *Consider two renormalizable R-maps F and $\tilde{F}$. Assume that they are combinatorially equivalent and let h be a combinatorial equivalence between them such that $h|\mathbb{C} \setminus W^1$ admits a qc extension to $\mathbb{C}$. Assume $R(F)$ and $\tilde{R}(F)$ are qc (resp. hybrid) conjugate. Then there exists a qc (resp. hybrid) conjugacy H between F and $\tilde{F}$ such that $H|\mathbb{C} \setminus W^1 = h$.*

Proof. Let $R(F) = F^p|P^n(0)$. Let ψ be the $(n+p)$-fold lift of h. Then ψ is equivariant on $W^1 \setminus W^{n+p}$. Let $V_p = P^n(0)$, $V_0 = P^{n+p}(0)$. Let $\psi_p : V_p \to \tilde{V}_p$ be a qc conjugacy between $R(F)$ and $R(\tilde{F})$, homotopic to ψ rel $\mathbb{C} \setminus V_0$ (which can be obtained using Lemma A.17).

Let $V_j = P^{n+p-j}(F^j(0))$, $1 \leq j < p$. Let $\psi_j : V_j \to \tilde{V}_j$ be the lift of ψ_p by $(F^{p-j}, \tilde{F}^{p-j})$. Then ψ_0 coincides with $\psi_p|V_0$. Let

$$\Psi = \begin{cases} \psi_j & \text{on } V_j, 1 \leq j \leq p \\ \psi & \text{on } \mathbb{C} \setminus \cup V_j. \end{cases}$$

Let us show that Ψ is a combinatorial equivalence between F and $\tilde{F}$. Let $\Psi_1 : \mathbb{C} \to \mathbb{C}$ be the lift of Ψ by $(F, \tilde{F})$. Then Ψ and Ψ_1 coincide on $\mathbb{C} \setminus W^{n+p}$ by equivariance, as well as on V_j, $0 \leq j < p$ by construction. The set $W^{n+p} \setminus \cup_{j=0}^{p-1} V_j$ is a countable union of Jordan disks with shrinking diameters (by the thin condition), so Ψ is homotopic to Ψ_1 rel $\mathbb{C} \setminus (W^{n+p} \setminus \cup_{j=0}^{p-1} V_j)$. Since $\mathrm{orb}_F(0) \subset \cup_{j=0}^{p-1} V_j$, Ψ is a combinatorial equivalence between F and $\tilde{F}$.

Let ϕ be a qc map homotopic to Ψ rel $(\mathbb{C} \setminus W^1) \cup \cup_{j=1}^{p} V_j$. Then ϕ is a qc combinatorial equivalence between F and $\tilde{F}$. We can obtain a qc conjugacy H as a limit of a series of lifts as before. Notice that $H|K(R(F)) = \psi_0$, so by Theorem A.10, $\mathrm{Dil}(H|K(F)) = \mathrm{Dil}(\psi_0|K(R(F))$, and if ψ_0 is hybrid then so is H. □

A.7. Non-critical *R*-maps. Let W_j and W be as in the definition of R-maps except that $0 \notin \cup W_j$. A holomorphic map $F : \cup W_j \to W$ will be called a *non-critical R-map* if $F|W_j$ is univalent onto W for all j. The definition of puzzle pieces and filled Julia set goes as before. By a trivialization of the arguments for R-maps we obtain:

Lemma A.19. *Let F be a non-critical R-map. Then* $\mathrm{meas}\, K(F) = 0$ *and for every $x \in K(F)$, $\cap P^n(x) = \{x\}$. Given another non-critical R-map $\tilde{F}$ and a qc map $h : \mathbb{C} \setminus \cup W_j \to \mathbb{C}$ equivariant on $\cup \partial W_j$ there exists a qc conjugacy H such that $H|\mathbb{C} \setminus \cup W_j = h$.*

A.8. Application to puzzle maps. Let $f : U \to \mathbb{C}$ be a puzzle map and let $Q(f)$ be the first return map to U_0. Then its domain consist of connected components of $(f|U_0)^{-1}(U)$ and $Q(f)$ is either an R-map (if $f(0) \in U$) or a non-critical R-map (otherwise). We will now use the results on complex return maps (applied to $Q(f)$) to understand f.

It is easy to see that $K(f) = \cup (f^{n_i}|U_i)^{-1}(K(Q(f))$. As a consequence, the following three lemmas can be deduced for puzzle maps f from the corresponding statements for $Q(f)$ (respectively Lemmas A.12, A.7 and A.13).

Lemma A.20. *Let f be a puzzle map and assume it has a non-repelling periodic orbit. Then this non-repelling periodic orbit is unique and $Q(f)$ is renormalizable. If f is $\mathbb{R}$-symmetric then such an orbit is necessarily real.*

Lemma A.21. *Let f be a puzzle map. If* $\mathrm{int}\, K(f) \neq \emptyset$ *then f has a non-repelling periodic orbit.*

Lemma A.22. *Let f be a hyperbolic puzzle map. Then $K(f) \setminus \operatorname{int} K(f)$ has zero Lebesgue measure and* $\operatorname{int} K(f)$ *coincides with the basin of attraction of the attracting cycle.*

The next lemmas extend well-known properties of quadratic-like maps.

Lemma A.23. *Let f be a puzzle map whose critical point either escapes or is preperiodic but not periodic. Then* $\operatorname{meas} K(f) = 0$ *and all periodic orbits are repelling.*

Proof. Assume that $\operatorname{meas} K(f) > 0$ or that f has a non-repelling periodic orbit. By Lemmas A.12 and A.10, $Q(f)$ is renormalizable and $\operatorname{meas} K(R(Q(f))) > 0$ or $R(Q(f))$ has a non-repelling periodic orbit. Since $Q(f)$ is renormalizable, the critical point of f is non-escaping.

It is well known that if F is a quadratic-like map which has a preperiodic but not periodic critical point then $\operatorname{meas} K(F) = 0$ and all periodic orbits are repelling. In particular, $R(Q(f))$ (and hence f) cannot have a preperiodic but not periodic critical point. □

Lemma A.24. *Let f be an $\mathbb{R}$-symmetric puzzle map. Then either f has an attracting or parabolic periodic orbit or $K(f)$ has no invariant line fields.*

Proof. Since $K(f)$ is the union of preimages of $K(Q(f))$, if $K(f)$ has an invariant line field, then so does $K(Q(f))$.

In particular $Q(f)$ is renormalizable and $R(Q(f))$ is a quadratic-like map with invariant line-field. Since f is real, the result follows from Theorem 2.22. □

Lemma A.25. *Let f, g be $\mathbb{R}$-symmetric puzzle maps in the same hybrid class and h be an $\mathbb{R}$-symmetric qc homeomorphism equivariant on ∂U^f. Then there is an $\mathbb{R}$-symmetric conjugacy H between f and g on U^f coinciding with h on $\mathbb{C} \setminus U^f$ and such that $\overline{\partial} H = 0$ in $K(f)$. In particular,* $\operatorname{Dil}(H) \le \operatorname{Dil}(h)$.

Proof. Given h we first redefine it inside $\cup U_i^f$ so that it coincides with the topological conjugacy between f and $\tilde{f}$ on $\cup J_i^f$. The resulting map can still be required to be a homeomorphism (we use that h is real and $\operatorname{diam}(J_i^f) \to 0$).

We can now lift this map to U_0^f to obtain a homeomorphism Ψ. It turns out that Ψ is a combinatorial equivalence between $Q(f)$ and $Q(\tilde{f})$. Indeed, Ψ coincides with its lift Ψ_1 outside $(f|U_0)^{-1}(U) \setminus \mathbb{R}$. This set is a union of Jordan disks with diameter going to 0, so Ψ is homotopic to Ψ_1 rel its boundary.

If $Q(f)$ is renormalizable, then $R(Q(f))$ is hybrid conjugate to $R(Q(\tilde{f}))$ by Theorem 2.21. By Lemma A.18 and Theorem A.16, there exists a hybrid conjugacy between $Q(f)$ and $Q(\tilde{f})$ coinciding with h on $\partial U_0^f \setminus (f|U_0^f)^{-1}(U^f)$. Let ψ_j be the lift of this conjugacy by $(f^{n_j}|U_j^f, \tilde{f}^{n_j}|U_j^{\tilde{f}})$.

We let H be defined as

$$H = \begin{cases} \psi_j & \text{on } U_j^f \\ h & \text{on } \mathbb{C} \setminus U^f. \end{cases}$$

□

Lemma A.26. *Let $f : U \to \mathbb{C}$ and $\tilde{f} : \tilde{U} \to \mathbb{C}$ be qc conjugate hyperbolic puzzle maps with the same multiplier and let $h : \mathbb{C} \to \mathbb{C}$ be a qc conjugacy between them. Then there exists a qc conjugacy $H : \mathbb{C} \to \mathbb{C}$, such that $H|\mathbb{C} \setminus K(f) = h$ and $\overline{\partial} h|K(f) = 0$.*

Proof. In this case, $Q(f)$ is renormalizable, and h is a qc conjugacy between $Q(f)$ and $Q(\tilde{f})$. Since $R(Q(f))$ is qc conjugate to $R(Q(\tilde{f}))$ and both are hyperbolic with the same multiplier, it follows that they are hybrid equivalent. By Lemma A.18, there is a hybrid equivalence H between $Q(f)$ and $Q(\tilde{f})$ coinciding with h outside $K(Q(f))$, and this equivalence can be lifted to a hybrid equivalence between f and $\tilde{f}$ with the desired properties. □

Appendix B. Quasiconformal conjugacies for Yoccoz maps

The main aim of this appendix is to prove the following theorem:

Theorem B.1. *Let $f, \tilde{f} \in \mathcal{U}_a$ topologically conjugate Yoccoz maps. Then there is a qc map $h : \mathbb{C} \to \mathbb{C}$, symmetric with respect to the real line and* 0 *which is equivariant with respect to f and $\tilde{f}$ in a neighborhood of I.*

This is stronger than the statement that f and $\tilde{f}$ are qs conjugate on I. Difficulties arise essentially because of the lack of a nice external structure (if f and $\tilde{f}$ have quadratic-like extension, both statements are equivalent using Sullivan's pullback argument). To compensate for this, we will provide some geometrical external structure, constructed by hand. This particular construction is based on existence of small scaling factors in the principal nest, and does not immediately adapt to the case of, say, infinitely renormalizable maps of bounded type.

To simplify the exposition we will assume that the Yoccoz maps we are dealing with are in fact non-renormalizable and have a recurrent critical point. The finitely renormalizable case is analogous and the proof applies with obvious modifications to the Misiurewicz case.

B.1. Compatible external structures. We have so far concentrated mostly on the description of puzzle structures, which is suitable to analyze the set of points which eventually land in a given nice interval. We have also in Lemma 7.10 used a Markov structure to treat the points which never land on the nice interval. We will here describe how both constructions can be made compatible in the presence of good geometric parameters.

Let p be the fixed point of f. Fix a sufficiently deep stage in the principal nest in a way that $|T_n|/|T_{n-1}|$ is very small, and let $J_0 = T_n = [-q, q]$.

Let Q be the set of points in I that never land on J_0 and Q' be a truncation of Q, so that $Q' = Q \cap [-1, r]$, where r is the right endpoint of J_1. So Q' is a hyperbolic Cantor set and we may use Lemma 2.12 to obtain a real analytic conformal metric ν which is expanded by f. We assume ν to be symmetric with respect to the real line and to 0. In what follows we will only measure distances with respect to the metric ν.

Let $Q_m = Q' \cap \cup_{j=0}^{m} f^{-j}(p) \cup \{-1\}$, so Q_m is a finite forward invariant set contained in Q.

Lemma B.2. *Fix a small $\varepsilon > 0$. For all sufficiently small h, there exists a family of curves (called υ-segments) $\Gamma(x)$, $x \in Q$, with the following properties:*

(i) $\Gamma(x)$ *is an $\mathbb{R}$-symmetric continuous path;*
(ii) $\Gamma(x)\backslash x$ *is a C^∞ curve whose tangent is ε-close to the vertical direction;*
(iii) $\mathrm{dist}_\nu(\partial\Gamma(x), x) = h$;
(iv) *overflowing:* $\Gamma(f(x)) \subset f(\Gamma(x))$.

Proof. Let us consider a complex strictly Markov covering $\{V_j\}$ of Q, which is $\mathbb{R}$-symmetric and 0-symmetric. Let $V^k = \{x \mid f^j(x) \in V^0,\ 0 \le j \le k\}$, $V^\infty = \cap V^k$ (note that V^∞ contains Q but does not necessarily coincide with it).

Let us define $g = g_k : V^k \to \mathbb{C}$ which is $\mathbb{R}$-symmetric, 0-symmetric, and for each connected component V_j^k, there exists $z \in V_j^k$ such that $g(w) = f(z) + f'(z)(w - z)$, $w \in V_j^k$. It is easy to see that $g_k : V^k \to \mathbb{C}$ is a strictly Markov map provided k is big enough. The map g is a linear model for f. This lemma is obvious for g, since we can take vertical segments for υ-segments.

Let $H : \mathbb{C} \to \mathbb{C}$ be a smooth diffeomorphism, $\mathbb{R}$-symmetric, 0-symmetric, which coincides with the identity near ∂V^k and is equivariant near ∂V^{k+1}. We may select H arbitrarily close to id choosing k sufficiently big. By Proposition 2.11, we can turn H into a qc map h equivariant on V^{k+1} and coinciding with H on $\mathbb{C} \setminus V^{k+1}$.

Notice that $\Im(Df(x))/|Df(x)| = O(\Im(x))$. Since $f|V^0$ is expanding,

$$\text{(B.1)} \qquad \frac{|\Im(Dh(x))|}{|Dh(x)|} < \varepsilon/2, \quad x \in \mathbb{C} \setminus V^\infty,$$

provided k is sufficiently big. For each $x \in Q$, consider the arc $\hat{\Gamma}(x)$, the preimage by h of the vertical line through $h(x)$. Let $\Gamma(x)$ be the truncation of $\hat{\Gamma}(x)$ at height h. Properties (i) and (iii) of this family of curves are obvious. Property (ii) follows from B.1. The overflowing property (iv) follows from the corresponding property of the linear model and the expanding property of the metric ν. □

Let us now describe the central puzzle piece U_0 of the external puzzle structure. It is an $\mathbb{R}$-symmetric hexagon whose real trace is J_0. Its boundary consists of the arcs $\Gamma(-q)$, $\Gamma(q)$ (the *wall* of U_0) and four straight segments (the *roof* of U_0) constructed as follows. Let $\Gamma(-q)$ have as upper endpoint $(-x, y)$. Then one of the straight segments joins $(-x, y)$ to $(0, x + y)$ and the others are obtained by symmetries with respect to the real line and 0. Let Λ be the roof of U_0.

Since $f|Q'$ expands ν, there is a neighborhood V of q and a constant $C' > 1$ such that $f|V$ expands distances by C'. By construction, $\mathrm{dist}(f(\Lambda) \setminus f(V), [-1, r])$ is uniformly bounded from below, independent of how small h is chosen. On the other hand, $\mathrm{dist}(\Lambda, I) = h(1 + o(1))$ (since ν is conformal and continuous). So we conclude that there exists $1 < C < C'$ such that

$$\mathrm{dist}(f(\Lambda), [-1, r]) > Ch \tag{B.2}$$

for h small enough.

By Remark 5.5, U_0 generates a complex puzzle $\mathcal{P}\{U_j\}$. The boundary of the of the puzzle pieces U_j can be described in terms of walls and roofs which are taken by f^{n_j} onto the wall and the roof of U_0. By the overflowing property of Lemma B.2, the walls of U_j consist of truncated v-segments.

Let us now construct an external Markov structure. Let $\{M_j^{\mathbb{R}}\}$ be the family of those components of $[-1, r] \setminus Q_m$ which are different from the puzzle pieces J_j. We want m to be so big that the diameter of any $M_j^{\mathbb{R}}$ is smaller than $h/10$.

The complexification of each $M_j^{\mathbb{R}} = [a, b]$ is obtained by taking four arcs, $\Gamma(a)$, $\Gamma(b)$ and straight almost horizontal segments linking the upper endpoints as well as the lower endpoints. The closed cell bounded by those arcs will be denoted M_j.

Notice that if m is big enough, then $\{-q, q\} \subset Q_m$. Furthermore, whenever the boundary of some J_m^j intersects Q_m, its boundary is contained in Q_m. By the combinatorics of the construction, given $M_k^{\mathbb{R}}$ and J_j, there are two possibilities for intersection: either J_j is compactly contained in $\mathrm{int}\, M_k^{\mathbb{R}}$, or their interiors do not intersect. Similarly, given M_k and U_j, either U_j is compactly contained in $\mathrm{int}\, M_k$, or their interiors are disjoint. Furthermore, if their boundaries do intersect, the intersection is one of the walls of U_j. This follows from the property that no puzzle piece U_j has the roof crossing the boundary of some Markov piece M_k (they can only touch at an endpoint of the roof).

To see this, notice that the roof of U_j cannot cross laterals of Markov pieces by geometric considerations (using that the angle of a U-segment is nearly $\pi/4$ and a Γ-segment is nearly vertical). As a consequence, the roof of U_j cannot intersect ∂M_k if J_j is not contained in $M_k^{\mathbb{R}}$. Furthermore, the base of M_k is at least ten times smaller then its height, while the height of U_j is almost half the size of its base since it is mapped in U_0 with small

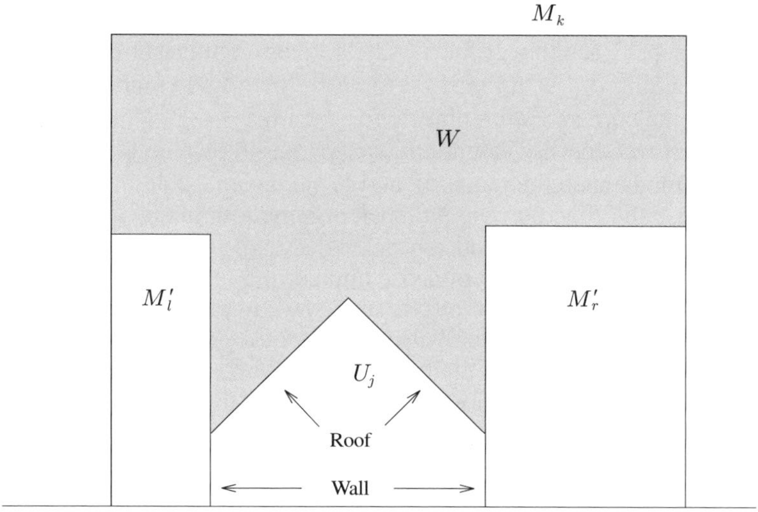

Fig. 7. Inside a Markov piece

distortion (depending on $|T_n|/|T_{n-1}|$), assuring that its roof do not intersect the top of M_k if $J_j \subset M_k$.

Observe also that all Markov pieces are in a neighborhood of $[-1, r]$ of size near $h(1 + o(1))$ (by choosing m big), so they do not intersect $f(\Lambda)$ by (B.2).

We construct the family $\{M_j'\}$ as the set of all $(f|M_k)^{-1}(M_l)$ for pieces M_k and M_l such that $M_l \cap \operatorname{int} f(M_k) \neq \emptyset$. We let $M = \cup \operatorname{int} M_j$, $M' = \cup \operatorname{int} M_j'$, and $U = \cup \operatorname{int} U_j$.

Let $W = M \setminus \overline{M' \cup U}$. This is the union of Jordan domains with piecewise smooth boundary. The maximal smooth segments contained in the boundary of those Jordan domains will be called W-segments. The boundaries of the Jordan domains can intersect, but only along a W-segment. A W-segment is either contained in ∂M, $\partial M'$ or is the roof of some U_j.

Figure 7 shows the interior of a piece M_k which contains two pieces M_l' and M_r', with a unique U_j whose interior is contained in $M_k \setminus (M_l' \cup M_r')$. The component of W inside M_k (shown shaded) is bounded by segments of ∂M_k, segments of $\partial M_l'$ and $\partial M_r'$, and U-segments of ∂U_j. The picture is skewed to show better the details, actual Markov pieces are much narrower (their bases being at least ten times smaller then the heights).

Notice in conclusion that if $x \in \overline{M} \setminus U$, then either $f^n(x) \in \overline{W}$ for some n, or $x \in Q'$. For this reason, $\overline{W}$ will work as a fundamental domain for the map $f : M' \setminus U \to M \setminus U$.

B.2. Construction of the conjugacy. Let us now consider a map $\tilde{f}$ topologically conjugate to f. Then it is possible to make the above construction simultaneously, the only care we have to take is choose n and m the same for both maps. Each object for f (say, a given segment in $\partial M \setminus \partial M'$) has a corresponding object for $\tilde{f}$, said to have "the same combinatorics". We will mark the corresponding objects for $\tilde{f}$ with a $\sim$ e.g., $\tilde{U}$, $\tilde{M}$, $\tilde{\Lambda}$.

The construction has essentially two parts: first we obtain a qc map relating the fundamental domains W and $\tilde{W}$ (and some additional structures), and then use the Macroscopic pullback argument to obtain a conjugacy on a neighborhood of the interval.

Let us first define an $\mathbb{R}$-symmetric homeomorphism h_0 on some relevant parts of the boundary of the puzzle and Markov pieces. The construction below will be $\mathbb{R}$-symmetric and also 0-symmetric "where defined" (that is, if x and $-x$ belong to the domain of h_0 then $-h_0(-x) = h_0(x)$).

Let us start the construction with the ray $\Gamma(p)$ through the fixed point p. It is well-known that conformal maps are locally qc conjugate near repelling fixed points in such a way that the conjugacy is smooth outside the fixed points themselves. In particular, f near p is qc conjugate to $\tilde{f}$ near $\tilde{p}$ so that the conjugacy is smooth in the punctured neighborhoods. Moreover, it is easy to select a conjugacy so that it maps $\Gamma(p)$ onto $\tilde{\Gamma}(\tilde{p})$. Similarly we can construct a local qc conjugacy near -1 carrying $\Gamma(-1)$ to $\Gamma(-1)$. Let h_0 stand for the restriction of the above local conjugacies to $\Gamma(p) \cup \Gamma(-1)$.

Recall that points of $Q_{m+1} \setminus \{-1\}$ are preimages of p. By the overflowing property of the family of rays $\Gamma(x)$, it is easy to extend h_0 to a homeomorphism

$$h_0 : \Delta \equiv \bigcup_{x \in Q_{m+1}} \Gamma(x) \to \bigcup_{\tilde{x} \in \tilde{Q}_{m+1}} \tilde{\Gamma}(\tilde{x})$$

which admits a qc extension to a neighborhood of Δ, is piecewise smooth on $\Delta \setminus \mathbb{R}$, and is equivariant on $\Delta \cap \overline{M}'$.

Thus, the map h_0 is defined on the laterals of M and M' (which are the arcs of Δ) and is equivariant on the laterals of M'. Extend it to the tops of M and M' so that it is equivariant on $\partial M'$. Then extend it to the roof of U_0 and to $f(\partial U_0) \setminus M$ (this set is the image of the roof of U_0 and the part of its wall that is not contained in M'), so that h_0 is now equivariant also on ∂U_0.

Notice that there are only a finite number of roofs which are not contained in M'. By a finite number of lifts, extend h_0 to those roofs. This concludes the definition of h_0. It is defined on the set Z_1 consisting of $f(\partial U_0)$, ∂M, $\partial M'$ and a finite number of roofs not contained in M'. Moreover, this map is equivariant on the set Z_2 consisting of ∂U_0, $\partial M'$ and the same union of roofs.

Notice that each embedded arc on Z_1 is a quasiarc since it consists of smooth arcs meeting at a positive angle. Hence the notion of a quasisymmetric map on Z_1 makes an obvious sense. The map h_0 being piecewise smooth is quasisymmetric on Z_1. The set $\overline{\mathbb{C}} \setminus Z_1$ is a finite union of quasidisks, so by Ahlfors-Beurling criteria (see [LV]), h_0 has a qc extension to $\mathbb{C}$, denoted by h_1, which can be required to be also $\mathbb{R}$-symmetric.

Since h_1 is equivariant on $\partial M'$, we can apply the simple version of the Macroscopic pullback argument (see Remark 2.5) to $f : M' \to M$. It provides us with a qc map (still denoted by h_1) which is equivariant on M'. In the course of this pullback procedure h_1 is not modified on Z_2. Hence it is equivariant on $\partial U \subset M' \cup Z_2$.

By Lemma A.25, there is a qc map h_2 which is equivariant on U and coincides with h_1 on $\mathbb{C} \setminus U$. Since this does not modify h_1 on $M \setminus U$, h_2 is also equivariant on $M' \setminus U$. Thus, h_2 is equivariant on $M' \cup U$. On the left, this set is bounded by the ray $\Gamma(-1)$, so it fails to be a neighborhood of the repelling fixed point -1. It is easy to modify h_2 in the complement of $M \cup U$ in order to make it equivariant on a small neighborhood V of -1 as well.

The last problem is that the set $V \cup M \cup U$ is a neighborhood of $[-1, f(0)]$ rather than of $[-1, 1]$. By lifting h_2 to the neighborhood $f^{-1}(V \cup M \cup U)$ of $[-1, 1]$, we obtain the desired map h. This concludes the proof of Theorem B.1.

B.3. Quasisymmetric conjugacies for real analytic quasiquadratic maps. We will now give a proof of Theorem 2.23. Obviously Theorem B.1 covers the case of Yoccoz maps, so we will concentrate on the remaining cases.

B.3.1. Infinitely renormalizable case. Let h be a topological conjugacy between f and $\tilde{f}$. If T is a deep enough renormalization interval (of period m), Theorem 2.24 implies that $f^m : T \to T$ and $\tilde{f}^m : h(T) \to h(T)$ extend to quadratic-like maps, and Theorem 2.21 can be applied so that $h|T$ is quasisymmetric.

Let us say that a triple (l, m, r) is κ-commensurable if

$$\kappa^{-1} \leq \frac{r - m}{m - l} \leq \kappa.$$

In this notation, h is κ-qs if and only if it takes a 1-commensurable triple to a κ-commensurable triple.

Fix a 1-commensurable triple (l, m, r), that is, m is the midpoint of the interval $J = [l, r] \subset I$. Let T' be a symmetric interval slightly smaller than T, and let $\delta = (|T| - |T'|)/2$. Let $k \geq 0$ be maximal with $f^j(J) \cap T' = \emptyset$ for $j < k$. It follows from the result of Guckenheimer [G] and Mañé [M] that $f^k|J$ and $\tilde{f}^k|h(J)$ have distortion bounded by some constant $\ln C$ independent of J. In particular $(f^k(l), f^k(m), f^k(r))$ is a C-commensurable triple. Two possibilities arise:

(1) $|f^k(J)| \geq \delta$. In this case both intervals $[f^k(l), f^k(m)]$ and $[f^k(m), f^k(r)]$ have length at least $\delta/2C$. Since h is a homeomorphism, there exists $\varepsilon > 0$ such that the image by h of any interval of length $\delta/2C$ has length at least ε. It follows that $(h(f^k(l)), h(f^k(m)), h(f^k(r)))$ is a $2\varepsilon^{-1}$-commensurable triple, and hence $(h(l), h(m), h(r))$ is a $2C\varepsilon^{-1}$-commensurable triple.

(2) $|f^k(J)| < \delta$. In this case, $f^k(J) \subset T$. Since $h|T$ is quasisymmetric, there exists C' (depending only on C, and thus independent of J) such that C-commensurable triples in T are taken by $h|T$ to C'-commensurable triples. Thus $(h(f^k(l)), h(f^k(m)), h(f^k(r)))$ is a C'-commensurable triple, and hence $(h(l), h(m), h(r))$ is CC'-commensurable.

This shows that h is κ-qs with $\kappa = \max\{2C\varepsilon^{-1}, CC'\}$.

B.3.2. Attracting case. If f and $\tilde{f}$ are in the same hyperbolic hybrid class then (using the linearizing coordinate) one obtains a topological conjugacy $h : I \to I$ which is *real analytic* on a symmetric interval T contained in the basin of attraction of the attracting periodic orbit p (here and in the parabolic case, p will denote the non-repelling periodic point which is *closest to* 0) and such that $[-p, p] \subset \operatorname{int} T$. The same argument for the infinitely renormalizable case can now be applied (where one should select a symmetric interval T' slightly smaller than T but bigger than $[-p, p]$).

B.3.3. Parabolic case. Let us first recall the standard fact concerning the asymptotics of the orbits converging to parabolic points.

Lemma B.3. *Let $g : (\mathbb{R}, 0) \to (\mathbb{R}, 0)$ be a real analytic germ near a non-degenerate*[7] *parabolic fixed point* 0. *Fix $x_0 \neq 0$ such that $x_n = g^n(x_0) \to 0$.*

(1) *If $Dg(0) = 1$ then $|x_n - x_{n+1}| \sim \eta n^{-2}$ for some constant $\eta > 0$.*
(2) *If $Dg(0) = -1$ then $|x_n - x_{n+2}| \sim \eta n^{-3/2}$ for some constant $\eta > 0$.*

If f has a parabolic point p (of period q) with multiplier -1, it is still true (as in the attracting case) that there exists a symmetric interval T with $[-p, p] \subset \operatorname{int} T$, and $f^q(T) \subset \operatorname{int} T$. This implies that we can choose a topological conjugacy which is smooth in $T \setminus \{-p, p\}$. It follows from Lemma B.3 that h is bi-Lipschitz near p. The dynamics outside T is uniformly expanding (by the result of Guckenheimer and Mañé mentioned above), so the same argument as before shows that h is quasisymmetric.

The case where f has a parabolic point p (of period q) with multiplier 1 is more delicate. For $s > q$ big, let $\hat{P}_s$ be the set of all $x \in I$ such that $f^j(x) = p$ for some $j \leq s$. Let $\{\hat{M}_j^s\}_j$ be the collection of connected components of $I \setminus \hat{P}_s$ which are not contained in the basin of attraction of p. There are exactly q elements of $\{\hat{M}_j^s\}_j$ whose closure intersects $\operatorname{orb}_f(p)$, which we will label $\hat{M}_1^s, \dots, \hat{M}_q^s$. Since $f(\hat{P}_s) \subset \hat{P}_s$, the collection $\{\hat{M}_j^s\}_j$ is a Markov partition. However, the action of f on this Markov partition is not uniformly expanding.

Let P_s be the union of $\hat{P}_s$ with the set of all $x \in I$ such that $f^k(x) \in \hat{P}_s$ for some $k \geq 1$, and $f^j(x) \in \cup_{j=1}^q \hat{M}_j^s$ for $j < k$. Notice that $P_s \setminus \hat{P}_s$ is

[7] In the case of multiplier 1 it means that $D^2 g(0) \neq 0$. In the case of multiplier -1, it means that $D^3 g^2(0) \neq 0$. Parabolic points of quasiquadratic maps are always non-degenerate.

an infinite set which accumulates on $\mathrm{orb}_f(p)$. Let $\{M_j^s\}_j$ be the collection of connected components of $\cup_j \hat{M}_j^s \setminus P_s$. Since $f(P_s) \subset P_s$, the collection $\{M_j^s\}_j$ is an "infinite Markov partition". Let us show that this infinite Markov partition is "uniformly expanding" in some sense.

Lemma B.4. *If s is sufficiently big, then for any $N > 0$, and for any $\delta > 0$ sufficiently small (depending on s and N) the following holds. Let $J \subset I$ be any interval and let $k \geq 0$ be the smallest number such that either*

(1) $f^k(J) \cap [-p, p] \neq \emptyset$, *or*
(2) $f^k(J)$ *contains N different intervals of the collection $\{M_j^s\}_j$, or*
(3) $|f^k(J)| \geq \delta$.

Then the distortion of $f^k|J$ is at most $\ln 2$.

Proof. Let $\varepsilon(s) = \sum_j |M_j^s|$. By [M], Theorem 4, the complement of the basin of attraction of p has Lebesgue measure zero. This implies that $\lim_{s\to\infty} \varepsilon(s) = 0$.

Let $\phi \equiv f^n : L \to M_j^s$ be any branch of the first return map to some M_j^s. It follows from the Markov property that ϕ is surjective, so it has a fixed point r in the closure of its domain. Notice that $\mathrm{orb}_f(p) \cap \overline{M_j^s} = \emptyset$, so r is not parabolic and by [MMS], Theorem B, $|D\phi(p)| \geq e^\rho$, where $\rho > 0$ only depends on f.

Let us choose s so big that $4C\varepsilon(s) \leq \rho$ where

$$C = \sup_{x\in I\setminus[-p,p]} \ln \frac{|D^2 f(x)|}{|Df(x)|}.$$

It is easy to see that the intervals $f^i(L)$, $0 \leq i \leq n-1$ are disjoint and contained in $\cup_j \hat{M}_j^s$. In particular, the distortion of ϕ can be bounded by

$$C \sum_{i=0}^{n-1} |f^i(L)| \leq C\varepsilon(s) \leq \frac{\rho}{4},$$

so $|D\phi| \geq e^{\rho/2}$ on its entire domain.

Let now J and k be as in the statement of the lemma. Let $k_0 \leq k$ be maximal such that for $i \leq k_0$ we have $f^i(J) \subset \cup_j \hat{M}_j^s$ (it follows that $k - k_0 \leq s$). Using that $f^{k_0-1}(J)$ is contained in the union of $N+1$ intervals $\overline{M_j^s}$, we see that

$$\sum_{i=0}^{k_0-1} |f^i(J)| \leq \frac{(N+1)}{1-e^{-\rho/2}} \sum_j \min\left\{\left|\hat{M}_j^s\right|, \delta\right\} \equiv \zeta(\delta, N, s).$$

It is clear that with N and s fixed we have $\lim_{\delta\to 0} \zeta(\delta, N, s) = 0$. Choose $\delta > 0$ so small that $C(\zeta(\delta, N, s) + \delta s) < \ln 2$. Then the distortion of $f^k|J$

is bounded by $C\sum_{i=1}^{k-1}|f^i(J)| = C(\sum_{i=0}^{k_0-1}|f^i(J)| + \sum_{i=k_0}^{k-1}|f^i(J)|) \leq C(\zeta(\delta, N, s) + \delta s) < \ln 2$ as required. □

Lemma B.5. *Let h be a topological conjugacy between f and $\tilde{f}$. For every $s > q$, there exists $N > 0$, $K > 1$, such that if (l, m, r) is a 2-commensurable triple in I and $J = [l, r]$ intersects more than N intervals of the collection $\{M_j^s\}_j$ then $(h(l), h(m), h(r))$ is a K-commensurable triple.*

Proof. It is enough to consider the case when $|J|$ is small. Since the only accumulation points of P_s are in $\mathrm{orb}_f(p)$, we may assume (by taking N big) that J is close to $\mathrm{orb}_f(p)$.

The set $P_s \cup \mathrm{orb}_f(0)$ is forward invariant and its only accumulation points are in $\mathrm{orb}_f(p)$. It follows from Lemma B.3 that $h|(P_s \cup \mathrm{orb}_f(0))$ extends to a bi-Lipschitz map $h_1 : I \to I$.

Let L, M, and R be the connected components of $I \setminus (P_s \cup \mathrm{orb}_f(0))$ containing points l, m, and r respectively (in case some of these three points lies in $P_s \cup \mathrm{orb}_f(0)$, we let the corresponding component be empty). Then $|L| + |M| + |R|$ is much smaller than $|J|$ (assuming $|J|$ is small and N is large). This implies that $h|\{l, m, r\}$ extends to a bi-Lipschitz map $I \to I$ with constants almost as good as for h_1, which implies the commensurability statement. □

Let us now fix a topological conjugacy h which is smooth in $(-p, p)$. Then it follows from Lemma B.3 that $h|[-p, p]$ is bi-Lipschitz. Let s be as in Lemma B.4, $N = N(s)$ be as in Lemma B.5, and let $\delta = \delta(s, N)$ be given by Lemma B.4. Let (l, m, r) be a 1-commensurable triple and let $J = [l, r]$. Let k be as in Lemma B.4, so that $f^k|J$ has distortion bounded by $\ln 2$ and at least one of the conditions (1), (2), or (3) hold. We may assume that the distortion of $\tilde{f}|h(J)$ is also bounded by $\ln 2$ (after some increasing of s and decreasing of δ). Let us show that in each of the three cases we can conclude that $(h(l), h(m), h(r))$ is κ-commensurable for some universal constant κ.

If (3) holds, we can argue as in the infinitely renormalizable case. If (2) holds, then we can apply Lemma B.5. If (1) holds then either $f^k(J) \subset [-p, p]$ or $\mathrm{int}\, f^k(J) \cap \{-p, p\} \neq \emptyset$. In the first case we just use that $h|[-p, p]$ is bi-Lipschitz. In the second case, $p \in \mathrm{int}\, f^{k+q}(J)$, so $f^{k+q}(J)$ intersects infinitely many intervals of the collection $\{M_j^s\}_j$, and Lemma B.5 implies the assertion.

Appendix C. Non-symmetric maps

So far we restricted our attention to a class of symmetric (i.e. even) unimodal maps. Below we will show how to extend our results to an appropriate space of asymmetric unimodal maps. The idea is to reduce one case to the other by means of a transversally non-singular projection from the space of asymmetric maps to the space of symmetric maps.

Let $\tilde{\mathcal{U}}^k$, $k \geq 3$ be the space of (not necessarily symmetric) C^k unimodal maps $f : I \to I$, that is, $f(-1) = f(1) = -1$ and f has a unique non-degenerate critical point at 0. Let $\tilde{\mathcal{U}} \subset \tilde{\mathcal{U}}^3$ be the set of quasiquadratic unimodal maps, that is, maps f which have a neighborhood $\mathcal{V}$ such that for all $g \in \mathcal{V}$, g is topologically conjugate to a quadratic map. Let $\tilde{\mathcal{E}}_a \subset \mathcal{B}_{\Omega_a}$ be the subspace of holomorphic maps v with $v(-1) = v(1) = 0$. Let $\tilde{\mathcal{A}}_a = q_2 + \tilde{\mathcal{E}}_a$ and $\tilde{\mathcal{U}}_a = \tilde{\mathcal{U}} \cap \tilde{\mathcal{A}}_a$.

Denote by c_f the critical point of f and let q_f be the quadratic unimodal map with the same critical value as f. Define $\Theta : \tilde{\mathcal{U}}^\infty \to C^\infty(I)$ which associates to each f the unique diffeomorphism of I such that $f = q_f \circ \Theta(f)$. Define the projection $\Pi : \tilde{\mathcal{U}}^\infty \to \mathcal{U}^\infty$ by $\Pi(f) = \Theta(f) \circ q_f$.

Obviously, $\Pi(f)$ is a symmetric unimodal map. It is conjugate to f, since $\Pi(f) = \Theta(f) \circ f \circ \Theta(f)^{-1}$. Moreover, any unimodal map $g \in \tilde{\mathcal{U}}^3$ near $\Pi(f)$ is conjugate to some unimodal map near f, namely, to $\Theta(f)^{-1} \circ g \circ \Theta(f) \in \tilde{\mathcal{U}}^3$. Hence $\Pi(f)$ is quasiquadratic. Clearly, if f is analytic then $\Pi(f)$ analytic as well. We conclude that Π acts from $\tilde{\mathcal{U}}$ to $\mathcal{U}$.

Moreover, if Γ is an analytic family of unimodal maps in some $\tilde{\mathcal{U}}_a$ then $\Pi \circ \Gamma$ is an analytic family of unimodal maps in some $\mathcal{U}_{a'}$. It follows that Theorems B and C can be "lifted" to the asymmetric setting.

To "lift" Theorem A we need some extra information: the derivative of Π (restricted to appropriate Banach spaces of analytic unimodal maps) has a dense image, and therefore it is transversally non-singular with respect to the hybrid lamination we constructed in the symmetric case. This is analogous to what we did in the infinitely renormalizable case in preparation for the proof of Theorem 4.9, where we used the renormalization operator instead of Π.

Lemma C.1. *Let $f \in \tilde{\mathcal{U}}_a$. Then there exist $b > 0$ and a neighborhood $\mathcal{V} \subset \tilde{\mathcal{U}}_a$ of f such that $\Pi(\mathcal{V}) \subset \mathcal{U}_b$, $\Pi : V \to \mathcal{U}_b$ is real analytic, and $D\Pi(f) : T\tilde{\mathcal{U}}_a \to T\mathcal{U}_b$ has a dense image.*

Proof. Let $a > a' > 0$ be such that $\Theta(f) \in \mathcal{B}_{\Omega_{a'}}$. Let $\mathcal{V} \subset \mathcal{U}_{a'}$ be a small neighborhood of f. If $b > 0$ is small enough, then $\Pi(\mathcal{V}) \subset \mathcal{U}_b$ and $\Pi : V \to \mathcal{U}_b$ is real analytic. Since the derivative of the inclusion from $\mathcal{U}_a$ into $\mathcal{U}_{a'}$ has a dense image, it is enough to prove that $D\Pi(f) : T\tilde{\mathcal{U}}_{a'} \to T\mathcal{U}_b$ has a dense image.

To this end, let us consider a polynomial vector field $w_0 \in T\mathcal{U}_b$, and let us try to find $v \in T\tilde{\mathcal{U}}_{a'}$ such that

$$D\Pi(f)v = w_0. \tag{C.1}$$

It is easier to compute $D\Pi(f)v$ for the case of a vector field v such that $v(c_f) = 0$. In this case the quadratic polynomial q_f does not vary infinitesimally, and we obtain:

$$w \equiv D\Theta(f)v = \frac{v}{q'_f \circ \Theta(f)}, \tag{C.2}$$
$$D\Pi(f)v = w \circ q_f.$$

Since w_0 is even, we can represent it in the form $w_0 = w \circ q_f$ where w is a polynomial vector field. Let $v = w \cdot (q'_f \circ \Theta(f))$. Then $v \in \mathcal{B}_{\Omega_{a'}}$ and $v(c_f) = 0$, and (C.1) follows from (C.2). □

Using this Lemma the argument of Theorem 4.9 can be applied to obtain Theorem A for asymmetric maps.

References

[A] Ahlfors, L.: Lectures on quasi-conformal maps. Van Nostrand Co 1966

[AB] Ahlfors, L., Bers, L.: Riemann mapping theorem for variable metrics. Ann. Math. **72**, 385–404 (1960)

[Av] Avila, A.: Bifurcations of unimodal maps: the topologic and metric picture. IMPA Thesis (2001) (http://www.math.sunysb.edu/~artur/)

[AM1] Avila, A., Moreira, C.G.: Statistical properties of unimodal maps: the quadratic family. Preprint (http://www.arXiv.org). To appear in Ann. Math.

[AM2] Avila, A., Moreira, C.G.: Statistical properties of unimodal maps: smooth families with negative Schwarzian derivative. Preprint (http://www.arXiv.org). To appear in Astérisque

[AM3] Avila, A., Moreira, C.G.: Statistical properties of unimodal maps: periodic orbits, physical measures and pathological laminations. Preprint (http://www.arXiv.org)

[BC] Benedicks, M., Carleson, L.: On iterations of $1 - ax^2$ on $(-1,1)$. Ann. Math. **122**, 1–25 (1985)

[B] Bers, L.: The moduli of Kleinian groups. Russ. Math. Surv. **29**, 86–102 (1974)

[BR] Bers, L., Royden, H.L.: Holomorphic families of injections. Acta Math. **157**, 259–286 (1986)

[BH] Branner, B., Hubbard, J.H.: The iteration of cubic polynomials. II: Patterns and parapatterns. Acta Math. **169**, 229–325 (1992)

[CE] Collet, P., Eckmann, J.-P.: Positive Liapunov exponents and absolute continuity for maps of the interval. Ergodic Theory Dyn. Syst. **3**, 13–46 (1983)

[D1] Douady, A.: Prolongement de mouvements holomorphes (d'après Slodkowski et autres). Astérisque **227**, 7–20 (1995)

[D2] Douady, A.: Systèmes dynamiques holomorphes. Bourbaki seminar, Vol. 1982/83, 39–63, Astérisque 105–106 (1983)

[DH1] Douady, A., Hubbard, J.H.: On the dynamics of polynomial-like maps. Ann. Sci. Éc. Norm. Supér., IV. Sér. **18**, 287–343 (1985)

[DH2] Douady, A., Hubbard, J.H.: Étude dynamique des polynômes complexes. Parties I et II. Publ. Math. Orsay 84–2 (1984) & 85–4 (1985)

[EM] Earle, C., McMullen, C.: Quasiconformal isotopies. In: Holomorphic Functions and Moduli I. Springer MSRI publications volume 10, 143–154 (1988)

[Fa] Fatou, P.: Sur les équations fonctionnelles. Bull. Soc. Math. Fr. **47** (1919)

[G] Guckenheimer, J.: Sensitive dependence to initial conditions for one-dimensional maps. Commun. Math. Phys. **70**, 133–160 (1979)

[GS1] Graczyk, J., Swiatek, G.: Induced expansion for quadratic polynomials. Ann. Sci. Éc. Norm. Supér., IV. Sér. **29**, 399–482 (1996)

[GS2] Graczyk, J., Swiatek, G.: Generic hyperbolicity in the logistic family. Ann. Math. **146**, 1–52 (1997)
[Ha] Hayashi, S.: Connecting invariant manifolds and the solution of the C^1 stability and Ω-stability conjectures for flows. Ann. Math. **145**, 81–137 (1997)
[H] Hubbard, J.H.: Local connectivity of Julia sets and bifurcation loci: three theorems of J.-C. Yoccoz. In: Topological Methods in Modern Mathematics, A Symposium in Honor of John Milnor's 60th Birthday, 467–511. Publish or Perish 1993
[J] Jakobson, M.: Absolutely continuous invariant measures for one-parameter families of one-dimensional maps. Commun. Math. Phys. **81**, 39–88 (1981)
[K1] Kozlovski, O.S.: Axiom A maps are dense in the space of unimodal maps in the C^k topology. Ann. Math. **157**, 1–43 (2003)
[K2] Kozlovski, O.S.: Getting rid of the negative Schwarzian derivative condition. Ann. Math. **152**, 743–762 (2000)
[LV] Lehto, O., Virtanen K.I.: Quasiconformal mappings in the plane. Springer 1973
[LS1] Levin, G., van Strien, S.: Local connectivity of Julia sets of real polynomials. Ann. Math. **147**, 471–541 (1998)
[LS2] Levin, G., van Strien, S.: Bounds for maps of an interval with one critical point of inflection type II. IHES/M/99/82. Invent. Math. **141**, 399–465 (2000)
[L1] Lyubich, M.: Some typical properties of the dynamics of rational maps. Russ. Math. Surv. **38**, 154–155 (1983)
[L2] Lyubich, M.: On the Lebesgue measure of the Julia set of a quadratic polynomial. Preprint IMS at Stony Brook, # 1991/10
[L3] Lyubich, M.: Combinatorics, geometry and attractors of quasi-quadratic maps. Ann. Math. **140**, 347–404 (1994). Note on the geometry of generalized parabolic towers. Manuscript 2000 (available at http://www.arXiv.org)
[L4] Lyubich, M.: Dynamics of quadratic polynomials, I–II. Acta Math. **178**, 185–297 (1997)
[L5] Lyubich, M.: Dynamics of quadratic polynomials, III. Parapuzzle and SBR measures. Astérisque **261**, 173–200 (2000)
[L6] Lyubich, M.: Feigenbaum-Coullet-Tresser Universality and Milnor's Hairiness Conjecture. Ann. Math. **149**, 319–420 (1999)
[L7] Lyubich, M.: Almost every real quadratic map is either regular or stochastic. Ann. Math. **156**, 1–78 (2002)
[LY] Lyubich, M., Yampolsky, M.: Dynamics of quadratic polynomials: Complex bounds for real maps. Ann. Inst. Fourier **47**, 1219–1255 (1997)
[M] Mañé, R.: Hyperbolicity, sinks and measures for one-dimensional dynamics. Commun. Math. Phys. **100**, 495–524 (1985)
[Ma] Martens, M.: Distortion results and invariant Cantor sets for unimodal maps. Ergodic Theory Dyn. Syst. **14**, 331–349 (1994)
[MMS] Martens, M., de Melo, W., van Strien, S.: Julia-Fatou-Sullivan theory for real one-dimensional dynamics. Acta Math. **168**, 273–318 (1992)
[MN] Martens, M., Nowicki, T.: Invariant measures for Lebesgue typical quadratic maps. Astérisque **261**, 239–252 (2000)
[May] May, R.M.: Simple mathematical models with very complicated dynamics. Nature **261**, 459–466 (1976)
[Mc1] McMullen, C.: Complex dynamics and renormalization. Ann. Math. Stud. **135** (1994)
[Mc2] McMullen, C.: Renormalization and 3-manifolds which fiber over the circle. Ann. Math. Stud. **142** (1996)
[Mi] Misiurewicz, M.: Absolutely continuous measures for certain maps of an interval. Publ. Math., Inst. Hautes Étud. Sci. **53**, 17–51 (1981)
[MSS] Mañé, R., Sad, P., Sullivan, D.: On the dynamics of rational maps. Ann. Sci. Éc. Norm. Supér., IV. Sér. **16**, 193–217 (1983)
[MS] de Melo W., van Strien S.: One-dimensional dynamics. Springer 1993
[MT] Milnor, J., Thurston, W.: On iterated maps of the interval. Dynamical Systems. Proc. U. Md., 1986–87, ed. J. Alexander. Lect. Notes Math. **1342**, 465–563 (1988)

[NPT] Newhouse, S., Palis, J., Takens, F.: Bifurcation and stability of families of diffeomorphisms. Publ. Math., Inst. Hautes Étud. Sci. **57**, 5–71 (1983)

[N] Nowicki, T.: A positive Liapunov exponent for the critical value of an S-unimodal mapping implies uniform hyperbolicity. Ergodic Theory Dyn. Syst. **8**, 425–435 (1988)

[NS] Nowicki, T., van Strien, S.: Invariant measures exist under a summability condition for unimodal maps. Invent. Math. **105**, 123–136 (1991)

[Pa] Palis, J.: A global view of dynamics and a conjecture of the denseness of finitude of attractors. Astérisque **261**, 335–347 (2000)

[Pu] Pugh, C.: The closing lemma. Am. J. Math. **89**, 956–1009 (1967)

[R] Rudin, W.: Real and complex analysis, second edition. McGraw-Hill 1974

[SN] Sario, L., Nakai, M.: Classification theory of Riemann surfaces. Springer 1970

[Sh] Shishikura, M.: Topological, geometric and complex analytic properties of Julia sets. In: Proceedings of the International Congress of Mathematicians (Zürich, 1994), pp. 886–895. Basel: Birkhäuser 1995

[Si] Singer, D.: Stable orbites and bifurcations of maps of the interval. SIAM J. Appl. Math. **35**, 260–267 (1978)

[ST] Sullivan, D., Thurston, W.: Extending holomorphic motions. Acta Math. **157**, 243–257 (1986)

[Sl] Slodkowski, Z.: Holomorphic motions and polynomial hulls. Proc. Am. Math. Soc. **111**, 347–355 (1991)

Global Hyperbolicity of Renormalization for C^r Unimodal Mappings

Edson de Faria*, Welington de Melo** and Alberto Pinto***

Ann. of Math. **164**, 731–824 (2006)

Abstract

In this paper we extend M. Lyubich's recent results on the global hyperbolicity of renormalization of quadratic-like germs to the space of C^r unimodal maps with quadratic critical point. We show that in this space the bounded-type limit sets of the renormalization operator have an invariant hyperbolic structure provided $r \geq 2 + \alpha$ with α close to one. As an intermediate step between Lyubich's results and ours, we prove that the renormalization operator is hyperbolic in a Banach space of real analytic maps. We construct the local stable manifolds and prove that they form a continuous lamination whose leaves are C^1 codimension one, Banach submanifolds of the ambient space, and whose holonomy is $C^{1+\beta}$ for some $\beta > 0$. We also prove that the global stable sets are C^1 immersed (codimension one) submanifolds as well, provided $r \geq 3 + \alpha$ with α close to one. As a corollary, we deduce that in generic, one-parameter families of C^r unimodal maps, the set of parameters corresponding to infinitely renormalizable maps of bounded combinatorial type is a Cantor set with Hausdorff dimension less than one.[1]

Table of Contents

*Financially supported by CNPq Grant 301970/2003-3.

**Financially supported by CNPq Grant 304912/2003-4 and Faperj Grant E-26/152.189/2002.

***Financially supported by Calouste Gulbenkian Foundation, PRODYN-ESF, POCTI and POSI by FCT and Ministério da CTES, and CMUP.

[1]There is a list of symbols used in this paper, before the references, for the convenience of the reader.

Originally published in *Ann. of Math.* **164**, 731–824 (2006).

1. Introduction

In 1978, M. Feigenbaum [10] and independently P. Coullet and C. Tresser [4] made a startling discovery concerning certain rigidity properties in one-dimensional dynamics. While analysing the transition between simple and "chaotic" dynamical behavior in "typical" one-parameter families of unimodal maps – such as the quadratic family $x \mapsto \lambda x(1-x)$ – they recorded the parameter values λ_n at which successive period-doubling bifurcations occurred in the family and found a remarkable universal scaling law, namely

$$\frac{\lambda_n - \lambda_{n-1}}{\lambda_{n+1} - \lambda_n} \to 4.669\ldots .$$

They also found universal scalings within the geometry of the post-critical set of the limiting map corresponding to the parameter $\lambda_\infty = \lim \lambda_n$ (cf. the work of E. Vul, Ya. Sinai and K. Khanin [29]). In an attempt to explain these phenomena, they introduced a certain nonlinear operator acting on the space of unimodal maps – the so-called *period doubling operator*. They conjectured that the period-doubling operator has a unique fixed point which is hyperbolic with a one-dimensional unstable direction. They also conjectured that the universal constants they found in their experiments are the eigenvalues of the derivative of the operator at the fixed point.

A few years later (1982) this conjecture was confirmed by O. Lanford [18] through a computer assisted proof. Working in a cleverly defined Banach space of real analytic maps and using rigorous numerical analysis on the computer, Lanford established at once the existence and hyperbolicity of the fixed point of the period-doubling operator. Subsequent work by M. Campanino and H. Epstein [2] (also Campanino et al. [3] and Epstein [9]) established the existence (but neither uniqueness nor hyperbolicity) of the fixed point without essential help from the computer.

It was soon realized by Lanford and others that the period-doubling operator was just a restriction of another operator acting on the space of unimodal maps – the *renormalization operator* – whose dynamical behavior is much richer. The hopes were high that the iterates of this operator would reveal the small scale geometric properties of the critical orbits of many interesting one-dimensional systems. Hence, the initial conjecture was generalized to the following.

Renormalization Conjecture. *The limit set of the renormalization operator in the space of maps of bounded combinatorial type is a hyperbolic Cantor set where the operator acts as the full shift in a finite number of symbols.*

(For a precise formulation of what is meant by bounded combinatorial type, see §2.2 below.)

In the path towards a proof of this conjecture, several new ideas were developed in the last 20 years by a number of mathematicians, especially D. Sullivan, C. McMullen and M. Lyubich. Among the deepest in Dynamical Systems, these ideas have the complex dynamics of quadratic-like maps (in the sense of Douady and Hubbard [6]) as a common thread. Sullivan proved in [28] that all limits of renormalization are quadratic-like maps with a definite modulus. Then, constructing certain Teichmüller spaces from quadratic-like maps and using a substitute of Schwarz's lemma in these spaces, Sullivan established the existence of horseshoe-like limit sets for renormalization. Later, using a different approach based on Mostow rigidity, McMullen [23] gave another proof of this result and went further by showing that the convergence (in the C^0 sense) towards the limit set is exponential.

The final breakthrough came with the work of Lyubich [20]. He endowed the space of *germs* of quadratic-like maps (modulo affine conjugacies) with a very subtle complex structure, showing that the renormalization operator is complex-analytic with respect to such a structure. In Lyubich's space, the stable sets of maps in the limit set of renormalization coincide with the very *hybrid classes* of such maps, and inherit a natural structure making them (complex codimension one) analytic submanifolds. Combining McMullen's rigidity of towers with Schwarz's lemma in Banach spaces, Lyubich proved exponential contraction along such stable leaves. To obtain expansion in the transversal directions to such leaves at points of the limit set, Lyubich argued by contradiction: if expansion fails, then one can find a map in the limit set whose orbit under renormalization is slowly shadowed by another orbit (the *small orbits theorem*, page 323 of [20]). This however contradicts another theorem of his, namely the combinatorial rigidity theorem of [21]. It follows that the limit set is indeed hyperbolic in the space of germs. Based on this result of Lyubich and using the real and complex bounds given by Sullivan, we prove in Theorem 2.4 that the attractor (for bounded combinatorics) is hyperbolic in a Banach space of real analytic maps.

In the present paper, we give the last step in the proof of the above renormalization conjecture in the (much larger) space of C^r smooth unimodal maps with r sufficiently large. The very formulation of the conjecture in this setting requires some care, because the renormalization operator is *not* differentiable in C^r. For the correct formulation, see Theorem 2.5 below. To prove the conjecture, we combine Theorem 2.4 with some nonlinear functional analysis inspired by the work of A. Davie [5]. In that work, Davie constructs the stable manifold of the fixed point of the period doubling operator in the space of $C^{2+\varepsilon}$ maps "by hand", showing it to be a C^1 codimension-one submanifold of the ambient space, even though the operator is not differentiable. To do this, he first extends the hyperbolic splitting of the derivative at the fixed point from Lanford's Banach space of real-analytic maps to the larger space of $C^{2+\varepsilon}$

maps (to which the derivative extends as a bounded linear operator). This gives him an extended codimension-one stable subspace in $C^{2+\varepsilon}$ to work with, and he views the local stable set in $C^{2+\varepsilon}$ as the graph of a function over the extended stable subspace. In attempting to prove that such function is C^1, he goes around the inherent loss of differentiability of renormalization by first noting that the local unstable manifold coming from Lanford's theorem is still there (and is still smooth in $C^{2+\varepsilon}$) and then showing that there is afterall a contraction in $C^{2+\varepsilon}$ towards that unstable manifold, whose elements are *analytic* maps. Thus, the loss of differentiability is somehow compensated by the contraction towards the unstable manifold. Davie's crucial estimates show that the renormalization operator in $C^{2+\varepsilon}$ is sufficiently well-approximated by the extension of its derivative in Lanford's space to a bounded linear operator in $C^{2+\varepsilon}$.

Our approach is based on the idea that whatever Davie can do with Lanford's Banach space relative to the fixed point, we can do with the Banach space obtained in Theorem 2.4 relative to the whole limit set. There is one fundamental difference, however. The linear and nonlinear estimates carried out by Davie rely on the special fact that the period-doubling fixed point is *concave*. This allows him to prove his main theorems in $C^{2+\varepsilon}$ for all $\varepsilon > 0$. By contrast, we cannot – and do not – rely on any such convexity assumptions. We derive our estimates (in §5 and §8) directly from the geometric properties of the postcritical set of maps in the limit set (these properties – proved in §5.2 – are a consequence of the real *a priori* bounds). As a result, our local stable manifold theorem in C^r requires $r \geq 2 + \alpha$ with α close to one.

We go beyond the conjecture in at least three respects. First, we show that the local stable manifolds form a C^0 lamination whose holonomy is $C^{1+\beta}$ for some $\beta > 0$. In particular, every smooth curve which is transversal to such a lamination intersects it at a set of constant Hausdorff dimension less than one. Second, we prove that the global stable sets are C^1 (immersed) codimension-one submanifolds in C^r provided $r \geq 3 + \alpha$ with α close to one (we globalize the local stable manifolds via the implicit function theorem, hence the further loss of one degree of differentiability). Third, we prove that in an open and dense set of C^k one-parameter families of C^r unimodal maps (for any $k \geq 2$), each family intersects the global stable lamination transversally at a Cantor set of parameters and the small-scale geometry of this intersection is the same for all nearby families. In particular, its Hausdorff dimension is strictly smaller than one.

In the path towards these results, we have made an attempt to abstract out the more general features of the renormalization operator in the form of a few properties or "axioms" – the notion of a robust operator introduced in Section 6. We prove a general local stable manifold theorem for robust operators there. It is our hope that this might be useful in other renormalization problems, for example in the case of critical circle maps (see [7] and [8]).

Acknowledgement. We wish to thank M. Lyubich and A. Avila for several useful discussions and A. Douady for his elegant proof of Lemma 9.4 (§9.2). We are greatful to the referee for his keen remarks and for pointing out several corrections. We also thank FCUP, IMPA, IME-USP, KTH, SUNY Stony Brook for their hospitality and support during the preparation of this paper.

2. Preliminaries and statements of results

In this section, we introduce the basic notions of the theory of renormalization of unimodal maps. Then we state Sullivan's theorem on the existence of topological limit sets for the renormalization operator, the exponential convergence results of McMullen, and Lyubich's theorem showing the full hyperbolicity of such limit sets in the space of germs of quadratic-like maps. Finally, we state our main results extending Lyubich's hyperbolicity theorem to the space of C^r unimodal maps with r sufficiently large.

2.1. *Quadratic unimodal maps.* We describe here two types of ambient spaces of C^r unimodal maps. These will be determined by two families of Banach spaces, denoted $\mathbb{A}^r$ and $\mathbb{B}^r$.

2.1.1. *The Banach spaces* $\mathbb{A}^r$. Let $I = [-1, 1]$ and for all $r \geq 0$ let $C^r(I)$ be the Banach space of C^r real-valued functions on I. Here r can be either a nonnegative real number, say $r = k + \alpha$ with $k \in \mathbb{N}$ and $0 \leq \alpha < 1$, in which case $C^r(I)$ is the space of C^k functions whose k^{th} derivative is α-Hölder, or else $r = k + \text{Lip}$, in which case $C^r(I)$ means the space of C^k functions whose k^{th} derivative is Lipschitz (so whenever we say that r is not an integer, we include the Lipschitz cases). Let us denote by $\mathbb{A}^r$ the space $C^r_e(I)$ consisting of all C^r functions on I which are *even* and vanish at the origin, in other words

$$\mathbb{A}^r = \{v \in C^r(I) : v \text{ is even and } v(0) = 0\} \ .$$

Then $\mathbb{A}^r$ is a closed linear subspace of $C^r(I)$ and therefore also a Banach space under the C^r norm. Now, for each $r \geq 2$, define

$$\mathbb{U}^r \subset 1 + \mathbb{A}^r \subset C^r(I)$$

to be the set of all maps $f : I \to I$ of the form $f(x) = 1 + v(x)$, where $v \in \mathbb{A}^r$ satisfies $v''(0) < 0$, which are unimodal. Then $\mathbb{U}^r$ is a Banach manifold; indeed it is an open subset of the affine space $1 + \mathbb{A}^r$. Note that for all $f \in \mathbb{U}^r$ the tangent space $T_f\mathbb{U}^r$ is naturally identified with $\mathbb{A}^r$. The elements of $\mathbb{U}^r$ are called *C^r unimodal maps with a quadratic critical point.*

2.1.2. *The Banach spaces* $\mathbb{B}^r$. We define $\mathbb{B}^r$ to be the space of functions $v : I \to \mathbb{R}$ of the form $v = \varphi \circ q$ where $q(x) = x^2$ and $\varphi \in C^r([0, 1])$ vanishes at the origin. The norm of v in this space is given by the C^r norm of φ. This makes $\mathbb{B}^r$ into a Banach space. Note that for each $s \leq r$ the inclusion map

$j : \mathbb{B}^r \to \mathbb{A}^s$ is linear and continuous (hence C^1). Now, for each $r \geq 1$, let

$$\mathbb{V}^r \subset 1 + \mathbb{B}^r$$

be the open subset of the affine space $1 + \mathbb{B}^r$ consisting of those $f = \phi \circ q$ such that $\phi([0,1]) \subseteq (-1,1]$, $\phi(0) = 1$ and $\phi'(x) < 0$ for all $0 \leq x \leq 1$. Just as before, $\mathbb{V}^r$ is a Banach manifold. Note that each $f \in \mathbb{V}^r$ is a unimodal map belonging to $\mathbb{U}^r$ when $r \geq 2$. Moreover, the inclusion of $\mathbb{V}^r$ in $\mathbb{U}^r$ is strict (for each $r \geq 2$).

2.2. *The renormalization operator.* A map $f \in \mathbb{U}^r$ is said to be *renormalizable* if there exist $p = p(f) > 1$ and $\lambda = \lambda(f) = f^p(0)$ such that $f^p|[-|\lambda|, |\lambda|]$ is unimodal and maps $[-|\lambda|, |\lambda|]$ into itself. In this case, with the smallest possible value of p, the map $Rf : [-1,1] \to [-1,1]$ given by

$$Rf(x) = \frac{1}{\lambda} f^p(\lambda x) \tag{2.2.1}$$

is called the *first renormalization* of f. We have $Rf \in \mathbb{U}^r$. The intervals $f^j([-|\lambda|, |\lambda|])$, for $0 \leq j \leq p-1$, are pairwise disjoint and their relative order inside $[-1,1]$ determines a *unimodal* permutation θ of $\{0, 1, \ldots, p-1\}$. The set of all unimodal permutations is denoted $\mathbf{P}$. The set of $f \in \mathbb{U}^r$ that are renormalizable with the same unimodal permutation $\theta \in \mathbf{P}$ is a connected subset of $\mathbb{U}^r$ denoted $\mathbb{U}^r_\theta$. Hence we have an operator

$$R : \bigcup_{\theta \in \mathbf{P}} \mathbb{U}^r_\theta \to \mathbb{U}^r\,, \tag{2.2.2}$$

the so-called *renormalization operator.*

Now let us fix a finite subset $\Theta \subseteq \mathbf{P}$. Given an infinite sequence of unimodal permutations $\theta_0, \theta_1, \ldots, \theta_n, \cdots \in \Theta$, write

$$\mathbb{U}^r_{\theta_0,\theta_1,\cdots,\theta_n,\cdots} = \mathbb{U}^r_{\theta_0} \cap R^{-1}\mathbb{U}^r_{\theta_1} \cap \cdots \cap R^{-n}\mathbb{U}^r_{\theta_n} \cap \cdots\,,$$

and define

$$\mathcal{D}^r_\Theta = \bigcup_{(\theta_0,\theta_1,\cdots,\theta_n,\cdots)\in\Theta^{\mathbb{N}}} \mathbb{U}^r_{\theta_0,\theta_1,\cdots,\theta_n,\cdots}\,.$$

The maps in $\mathcal{D}^r_\Theta$ are *infinitely renormalizable* maps with (bounded) combinatorics belonging to Θ. Note that $R(\mathcal{D}^r_\Theta) \subseteq \mathcal{D}^r_\Theta$; in fact,

$$R(\mathbb{U}^r_{\theta_0,\theta_1,\cdots,\theta_n,\cdots}) \subseteq \mathbb{U}^r_{\theta_1,\theta_2,\cdots,\theta_{n+1},\cdots}\,. \tag{2.2.3}$$

We note that if f is a renormalizable map in $\mathbb{V}^r$, then $R(f)$ belongs to $\mathbb{V}^r$ also. Hence, taking $\mathbb{V}^r_\theta = \mathbb{U}^r_\theta \cap \mathbb{V}^r$, the restriction of the renormalization operator

$$R : \bigcup_{\theta \in \mathbf{P}} \mathbb{V}^r_\theta \to \mathbb{V}^r \tag{2.2.4}$$

is well-defined.

2.3. *The limit sets of renormalization.* In [28], Sullivan established the existence of horseshoe-like invariant sets for the renormalization operator, showing that they all consist of real analytic maps of a special kind, namely, restrictions to $[-1, 1]$ of *quadratic-like maps* in the sense of Douady-Hubbard. We remind the reader that a quadratic-like map $f : V \to W$ is a holomorphic map with the property that V and W are topological disks with V compactly contained in W, and f is a proper, degree two branched covering map with a continuous extension to the boundary of V. The *conformal modulus of* f is the modulus of the annulus $W \setminus \overline{V}$.

We are interested only in quadratic-like maps that commute with complex conjugation, for which V is symmetric about the real axis. Consider the real Banach space $\mathcal{H}_0(V)$ of holomorphic functions which commute with complex conjugation and are continuous up to the boundary of V, with the C^0 norm. Let $\mathbb{A}_V \subset \mathcal{H}_0(V)$ be the closed linear subspace of functions of the form $\varphi = \phi \circ q$, where $q(z) = z^2$ and $\phi : q(V) \to \mathbb{C}$ is holomorphic with $\phi(0) = 0$. Also, let $\mathbb{U}_V$ be the set of functions of the form $f = 1 + \varphi$, where $\varphi = \phi \circ q \in \mathbb{A}_V$ and ϕ is *univalent* on some neighborhood of $[-1, 1]$ contained in V, such that the restriction of f to $[-1, 1]$ is unimodal. Then $\mathbb{U}_V$ is an open subset of the affine space $1 + \mathbb{A}_V$, which is linearly isomorphic to $\mathbb{A}_V$ via the translation by 1, and we shall regard $\mathbb{U}_V$ as an open subset of $\mathbb{A}_V$ itself via this identification. For each $a > 0$, let us denote by Ω_a the set of points in the complex plane whose distance from the interval $[-1, 1]$ is smaller than a. We may now state Sullivan's theorem as follows.

THEOREM 2.1. *Let $\Theta \subseteq \mathbf{P}$ be a nonempty finite set. Then there exist $a > 0$, a compact subset $\mathbb{K} = \mathbb{K}_\Theta \subseteq \mathbb{A}_{\Omega_a} \cap \mathcal{D}^\omega_\Theta$ and $\mu > 0$ with the following properties.*

(i) *Each $f \in \mathbb{K}$ has a quadratic-like extension with conformal modulus bounded from below by μ.*

(ii) *$R(\mathbb{K}) \subseteq \mathbb{K}$, and the restriction of R to $\mathbb{K}$ is a homeomorphism which is topologically conjugate to the two-sided shift $\sigma : \Theta^{\mathbb{Z}} \to \Theta^{\mathbb{Z}}$: in other words, there exists a homeomorphism $H : \mathbb{K} \to \Theta^{\mathbb{Z}}$ such that the diagram*

$$\begin{array}{ccc} \mathbb{K} & \xrightarrow{R} & \mathbb{K} \\ {\scriptstyle H}\big\downarrow & & \big\downarrow{\scriptstyle H} \\ \Theta^{\mathbb{Z}} & \xrightarrow[\sigma]{} & \Theta^{\mathbb{Z}} \end{array}$$

commutes.

(iii) *For all $g \in \mathcal{D}^r_\Theta \cap \mathbb{V}^r$, with $r \geq 2$, there exists $f \in \mathbb{K}$ with the property that $||R^n(g) - R^n(f)||_{C^0(I)} \to 0$ as $n \to \infty$.*

For a detailed exposition of this theorem, see Chapter VI of [26].

Later, in [23], C. McMullen established the exponential convergence of renormalization for bounded combinatorics (using rigidity of towers). His theorem forms the basis for the contracting part of Lyubich's hyperbolicity theorem in [20].

THEOREM 2.2. *If f and g are infinitely renormalizable quadratic-like maps with the same bounded combinatorial type in $\Theta \subset P$, and with conformal moduli greater than or equal to μ, then*

$$\|R^n f - R^n g\|_{C^0(I)} \leq C\lambda^n$$

for all $n \geq 0$ where $C = C(\mu, \Theta) > 0$ and $0 < \lambda = \lambda(\Theta) < 1$.

The above result was extended by Lyubich to all combinatorics. In particular it follows, in the case of bounded combinatorics, that the exponent λ and the constant C in Theorem 2 do not depend on Θ. The conclusion of the above theorem can also be improved in bounded combinatorics: for $r \geq 3$; the exponential convergence holds in the C^r topology if the maps are in $\mathbb{V}^r$ (see [24] and [25]).

In [20], Lyubich considered the space of quadratic-like germs modulo affine conjugacies in which the limit set $\mathbb{K}$ is naturally embedded. This space is a manifold modeled on a complex topological vector space (arising as a direct limit of Banach spaces of holomorphic maps). In this setting, Lyubich established in [8] the full hyperbolicity of the renormalization operator. With the help of Sullivan's real and complex bounds and Lyubich's theorem we prove the hyperbolicity of some iterate of the renormalization operator acting on a space $\mathbb{A}_{\Omega_a}$ for some $a > 0$ (see Theorem 2.4 in §2.5). Then we extend Davie's analysis for the Feigenbaum fixed point to the context of bounded combinatorics to conclude that the hyperbolic picture also holds true in the much larger space $\mathbb{U}^r$ (see Theorem 2.5 in §2.5).

2.4. *Hyperbolic basic sets.* We need to introduce the well-known concept of hyperbolic basic set for nonlinear operators acting on Banach spaces. Let us consider a Banach space $\mathcal{A}$, and an open subset $\mathcal{O} \subseteq \mathcal{A}$.

Definition 2.1. Let $T : \mathcal{O} \to \mathcal{A}$ be a smooth nonlinear operator. A *hyperbolic basic set* of T is a compact subset $\mathbb{K} \subset \mathcal{O}$ with the following properties.

(i) $\mathbb{K}$ is T-invariant and $T|\mathbb{K}$ is a topologically transitive homeomorphism whose periodic points are dense.

(ii) If $y \in \mathcal{O}$ and all T-iterates of y are defined, then $T^n(y)$ converges to $\mathbb{K}$.

(iii) There exist a continuous, DT-invariant splitting $\mathcal{A} = E^s_x \bigoplus E^u_x$, for $x \in \mathbb{K}$, and uniform constants $C > 0$ and $0 < \theta < 1$ such that

$$\|DT^n(x)\, v\| \leq C\theta^n \|v\|$$

for all $v \in E^s_x$, as well as

$$\|DT^n(x)\,v\| \geq C\theta^{-n}\|v\| \tag{2.4.1}$$

for all $v \in E^u_x$.

(iv) The dimension of E^u_x is finite and constant.

The following notions are also standard. Let $\mathcal{A}(x,\varepsilon)$ be the ball in $\mathcal{A}$ with center x and radius ε. The *local stable manifold* $W^s_\varepsilon(x)$ *of* T *at* x consists of all points $y \in \mathcal{A}(x,\varepsilon)$ such that, for all $n > 0$, we have $T^n(y) \in \mathcal{A}(T^n(x),\varepsilon)$ and

$$\|T^n(y) - T^n(x)\| \to 0 \text{ when } n \to \infty\ .$$

The *local unstable manifold* $W^u_\varepsilon(x)$ *of* T *at* x consists of all points $y \in \mathcal{A}(x,\varepsilon)$ such that, when $y_0 = y$, for all $n \geq 1$ there exists $y_n \in \mathcal{A}(T^{-n}(x),\varepsilon)$ such that $y_{n-1} = T(y_n)$ and

$$\|T^{-n}(x) - y_n\| \to 0 \text{ when } n \to \infty\ .$$

Finally the *global stable set of* T *at* x is defined as

$$W^s(x) = \{y \in \mathcal{O}\ :\ \|T^n(y) - T^n(x)\| \to 0 \text{ when } n \to \infty\}\ .$$

The question arises as to whether these sets have smooth manifold structures. We have the following general result.

THEOREM 2.3. *If* $\mathbb{K}$ *is a hyperbolic basic set of a* C^1 *operator* $T : \mathcal{O} \to \mathcal{A}$ *then*

(i) *the local stable (resp. unstable) set at* $x \in \mathbb{K}$ *is a* C^1 *Banach submanifold of* $\mathcal{A}$ *which is tangent to* E^s_x *(resp.* E^u_x*) at* x.

(ii) *If* $y \in W^s(x)$ *then*

$$\|T^n(x) - T^n(y)\| \leq C\theta^n\|x - y\|\ .$$

Moreover, $T(W^u_\varepsilon(x)) \supseteq W^u_\varepsilon(T(x))$, *the restriction of* T *to* $W^u_\varepsilon(x)$ *is one-to-one and for all* $y \in W^u_\varepsilon(x)$,

$$\left\|T^{-n}(x) - T^{-n}(y)\right\| \leq C\theta^n\|x - y\|\ .$$

(iii) *If* $y \in \mathcal{A}(x,\varepsilon)$ *is such that* $T^i(y) \in \mathcal{A}(T^i(x),\varepsilon)$ *for* $i \leq n$ *then*

$$\mathrm{dist}\,(T^n(y), W^u_\varepsilon(T^n(x))) \leq C\theta^n, \text{ as well as } \mathrm{dist}\,(y, W^s_\varepsilon(x)) \leq C\theta^n\ .$$

(iv) *The family of local stable manifolds (and also the family of local unstable manifolds) form a* C^0 *lamination: the tangent spaces to the leaves vary continuously.*

We do not prove this theorem here since we will not use it, but instead make the following comments. Using the arguments of Hirsch-Pugh in [14], we can prove that the local unstable set is a smooth manifold. The local stable set is also a smooth manifold, but a different proof is needed: one can use the ideas of Irwin in [16]. See also Theorem 2.1 on page 375 of [27]. In both cases the smoothness can be improved to C^k if the operator T is C^k.

For invertible operators the global stable set is also a smooth submanifold. In the noninvertible situation, this is not always true. However, we will prove in §9.1 that this is the case for the renormalization operator acting on $\mathbb{V}^r$, provided $r \geq 3 + \alpha$ and $\alpha > 0$ is close to one.

2.5. *Hyperbolicity of renormalization.* In the present paper we prove three main theorems. The first main theorem shows that there exists a real Banach space of analytic maps, containing the topological limit set $\mathbb{K}$ of renormalization, on which the renormalization operator R acts as a real-analytic operator and has $\mathbb{K}$ as a hyperbolic basic set. More precisely, we have the following result.

THEOREM 2.4 (Hyperbolicity in a real Banach space). *There exist $a > 0$, an open set $\mathbb{O} \subset \mathbb{A} = \mathbb{A}_{\Omega_a}$ containing $\mathbb{K} = \mathbb{K}_\Theta$ and a positive integer N with the following property. There exists a real analytic operator $T : \mathbb{O} \to \mathbb{A}$ having $\mathbb{K}$ as a hyperbolic basic set with codimension-one stable manifolds at each point, such that $T(f)|[-1,1] = R^N(f|[-1,1])$, for all $f \in \mathbb{O}$, is the Nth iterate of the renormalization operator.*

The proof of this theorem, presented in §3 (see Theorem 3.9), combines Lyubich's hyperbolicity results with Sullivan's real and complex bounds.

The second main theorem establishes the "hyperbolicity" of renormalization in $\mathbb{U}^r$. As we have mentioned before, the renormalization operator is not smooth in $\mathbb{U}^r$, so the definition of hyperbolicity of an invariant set does not even make sense. However, the hyperbolic picture holds in this situation. More precisely, we have the following theorem.

THEOREM 2.5 (Hyperbolic Picture in $\mathbb{U}^r$). *If $r \geq 2 + \alpha$, where $\alpha > 0$ is close to one, then statements* (i), (ii), (iii) *and* (iv) *of Theorem* 2.3 *hold true for the renormalization operator acting on $\mathbb{U}^r$. Furthermore,*

(i) *the local unstable manifolds are real analytic curves;*

(ii) *the local stable manifolds are of class C^1, and together they form a continuous lamination whose holonomy is $C^{1+\beta}$ for some $\beta > 0$.*

The main difficulty behind the proof of this theorem is the fact that the operator T is not Fréchet differentiable in C^r (in fact it is only continuous in a dense subset of $\mathbb{U}^r$). However, as we shall see in §8.2, it is a C^1 mapping from its domain in $\mathbb{U}^r$ into $\mathbb{U}^s$ if $s < r - 1$ (even for $s = r - 1$ if r is an integer). Hence its tangent map defines a continuous map $L\colon \mathbb{K}\times\mathbb{A}^r \to \mathbb{K}\times\mathbb{A}^s$ by $L(g, v) = (T(g), DT(g)(v)) = (T(g), L_g(v))$. The bounded linear mappings $L_g\colon \mathbb{A}^r \to \mathbb{A}^s$ extend to bounded linear operators $L_g\colon \mathbb{A}^t \to \mathbb{A}^t$ for all $0 \le t \le r$. Although L_g is not the derivative of T at g in C^r, it is nevertheless a sufficiently good linear approximation to T near g (see the properties of Definition 6.1, checked in §8).

COROLLARY 2.6 (Hyperbolic Picture in $\mathbb{V}^r$). *If $r \ge 2 + \alpha$, where $\alpha > 0$ is close to one, then statements* (i), (ii), (iii) *and* (iv) *of Theorem* 2.3 *hold true for the renormalization operator acting on $\mathbb{V}^r$. Furthermore,*

(i) *the local unstable manifolds are real analytic curves;*

(ii) *the local stable manifolds are of class C^1, and together they form a continuous lamination whose holonomy is $C^{1+\beta}$ for some $\beta > 0$.*

For the proofs of Theorem 2.5 and Corollary 2.6, see Section 8.

By an argument using the implicit function theorem and the results in [24], which *a priori* are valid only in $\mathbb{V}^r$, we shall prove in Section 9 our third main theorem, which we state as follows.

THEOREM 2.7. *If $r \ge 3+\alpha$, where $\alpha > 0$ is close to one, then the following assertions hold true for the renormalization operator acting in $\mathbb{V}^r$:*

(i) *The global stable sets are C^1 immersed submanifolds.*

(ii) *For each integer $2 \le k \le r$, there exists an open dense set of C^k one-parameter families of maps in $\mathbb{V}^r$ all of whose elements intersect the global stable lamination of (T, K_Θ) transversally.*

(iii) *In each such family, the set of parameters where the intersections occur is a Cantor set which is locally $C^{1+\beta}$ diffeomorphic to the corresponding Cantor set of the quadratic family. In particular, its Hausdorff dimension is a universal number depending only on Θ which lies strictly between zero and one if Θ has more than one element.*

It is worth emphasizing that when a generic family (in the sense of the above corollary) intersects the stable lamination at a point, then any neighborhood of this point in parameter space contains a renormalization window that is mapped under a suitable power of the renormalization operator onto a *full* transversal family.

3. Hyperbolicity in a Banach space of real analytic maps

In this section we give a proof of Theorem 2.4. Using the real and complex bounds given by Sullivan in [28], we prove in §3.1 that there is an iterate of the renormalization operator which extends as a real analytic map T to an open set $\mathbb{O}_{\Omega_a}$ of the Banach space $\mathbb{A}_{\Omega_a}$ consisting of real analytic maps whose domain is an a-neighborhood of the interval $[-1,1]$, for a suitable $a>0$. The maps $g \in \mathbb{K}$ have unique extensions belonging to $\mathbb{O}_{\Omega_a}$. In §3.2, using Lemmas 4.16 and 4.17 in Lyubich's paper [20], we show that the hybrid conjugacy classes of the maps $g \in \mathbb{K}$ form a continuous lamination of codimension one real analytic manifolds. Then in §3.4 we construct a skew-product renormalization operator that satisfies properties (W1) to (W4) on page 395 of [20] in the real-analytic case (restated in §3.5). By Theorems 8.2 and 8.8 in [20] the skew-product renormalization operator will have fiberwise stable and unstable leaves (as defined in §3.3). The local stable leaf at $g \in \mathbb{K}$ is a relatively open set of the hybrid conjugacy class of g. Then using the skew-product renormalization operator, we prove in §3.5 that $\mathbb{K}$ is a basic set for the real analytic renormalization operator $T : \mathbb{O}_{\Omega_a} \to \mathbb{A}_{\Omega_a}$.

3.1. *Real analyticity of the renormalization operator.* Using Sullivan's real and complex bounds in [28], we will show that there exists $a>0$ such that some iterate $T : \mathbb{O}_{\Omega_a} \to \mathbb{A}_{\Omega_a}$ of the renormalization operator is a (well-defined) real analytic operator with a compact derivative.

For each $f \in \mathbb{K}$, let $\mathcal{I}_f \subseteq [-1,1]$ be the postcritical set of f (the Cantor attractor of f). For each $k \geq 0$, we can write

$$R^k f(x) = \Lambda_k^{-1} \circ f^{p_k} \circ \Lambda_k(x)$$

where

$$\begin{aligned} p_k &= p(f,k) = \prod_{i=0}^{k-1} p(R^i f)\ , \\ \lambda_k &= \lambda(f,k) = \prod_{i=0}^{k-1} \lambda(R^i f)\ , \\ \Lambda_k(x) &= \Lambda(f,k)(x) = \lambda_k \cdot x\ , \end{aligned} \tag{3.1.1}$$

with $p(\cdot)$ and $\lambda(\cdot)$ as defined in §2.2. Consider the renormalization intervals $\Delta_{0,k} = \Delta_{0,k}(f) = [-|\lambda_k|, |\lambda_k|] \subset [-1,1]$, and define $\Delta_{i,k} = \Delta_{i,k}(f) = f^i(\Delta_{0,k})$ for $i = 0,1,\ldots,p_k - 1$. The collection $\mathbf{C}_k = \{\Delta_{0,k},\ldots,\Delta_{p_k-1,k}\}$ consists of pairwise disjoint intervals at level k. Moreover, $\bigcup\{\Delta : \Delta \in \mathbf{C}_{k+1}\} \subseteq \bigcup \{\Delta : \Delta \in \mathbf{C}_k\}$ for all $k \geq 0$ and we have

$$\mathcal{I}_f = \bigcap_{k=0}^{\infty} \bigcup_{i=0}^{p_k-1} \Delta_{i,k}\ .$$

Definition 3.1. The set $\mathcal{I}_f$ has *geometry bounded by* $0 < \tau < 1$ with respect to $(\mathbf{C}_k)_{k\in\mathbb{N}}$ if the following conditions are met for $k \geq 1$.

(i) If $\Delta_{j,k+1} \subset \Delta_{i,k}$ then $\tau < |\Delta_{j,k+1}| \,/\, |\Delta_{i,k}| < 1-\tau$.

(ii) If I is a *connected component* of $\Delta_{i,k} \setminus \bigcup_j \Delta_{j,k+1}$ then $\tau < |I| \,/\, |\Delta_{i,k}| < 1-\tau$.

By Sullivan's real bounds (see [28] and Section VI.2 on page 453 of [26]), there exists $\alpha > 0$, such that for every $g \in \mathbb{K}$ the set $\mathcal{I}_g$ has geometry bounded by α with respect to $(\mathbf{C}_k)_{k\in\mathbb{N}}$.

The following result is a consequence of Sullivan's complex bounds (see [28] and Section VI.5 on page 483 of [26]).

THEOREM 3.1. *There exist $\mu > 0$, $N_0 > 0$ and a neighborhood V of the dynamics with the following properties. Every $g \in \mathbb{K}$ extends to a holomorphic map $g : V \to \mathbb{C}$ and for every $N \geq N_0$ there exists a symmetric neighborhood $O_{g,N}$ of the interval $\Delta_{0,N}(g)$ such that*

(i) *the diameter of the set $g^i(O_{g,N}) \subset V$ is comparable to the length $|\Delta_{i,N}(g)|$ of the interval $\Delta_{i,N}(g)$ for every $0 \leq i \leq p = p(N,g)$;*

(ii) *the map $g^p : O_{g,N} \to g^p(O_{g,N})$ is a quadratic-like map with conformal modulus greater than $\mu > 0$.*

Applying Theorem 3.1 (ii) to $g \in \mathbb{K}$, we see that $R^N(g)$ has a quadratic-like extension to

$$U_{g,N} = \Lambda_g^{-1}(O_{g,N}) \tag{3.1.2}$$

(where $\Lambda_g = \Lambda(g,N)$) and such an extension has conformal modulus greater than $\mu > 0$.

Recall that the *filled-in Julia set $\mathcal{K}_f$ of a quadratic-like map $f : U \to U'$* is the set $\{z : f^n z \in U, n = 0, 1, \dots\}$, and its boundary is the Julia set $\mathcal{J}_f$ of f. Since all maps in $\mathbb{K}$ have conformal modulus greater than or equal to $\mu > 0$, we deduce from Proposition 4.8 on page 83 of McMullen's book [23] that there exists $b > 0$ such that for every $g \in \mathbb{K}$ we have

$$\Omega_b(\mathcal{K}_{R^N(g)}) \subset U_{g,N} \,. \tag{3.1.3}$$

Here the notation $\Omega_\varepsilon(K)$ means the set of all points whose distance from K is less than $\varepsilon/2$ times the diameter of K.

For each neighborhood U of $[-1,1]$ in $\mathbb{C}$, symmetric about the real axis, we consider the real Banach space $\mathbb{A}_U$ of holomorphic functions defined earlier. We denote by $\mathbb{A}_U(g,\delta)$ the open ball of radius δ around g. By (3.1.3), the inclusion map $i_{g,N} : \mathbb{A}_{U_{g,N}} \to \mathbb{A}_{\Omega_\alpha}$ is a well-defined compact linear operator for every $0 < \alpha < b$.

LEMMA 3.2. *Let* $\mu > 0$ *and* $N_0 > 0$ *be as in Theorem* 3.1 *and* $b > 0$ *as in* (3.1.3). *For every* $0 < \alpha < b$ *there exist* $N > N_0$ *and* $\delta_0 > 0$ *such that*

(i) *for every* $g \in \mathbb{K}$, *the operator* $T_{g,N} : \mathbb{A}_{\Omega_\alpha}(g, \delta_0) \to \mathbb{A}_{U_{g,N}}$ *is well-defined if we set*
$$T_{g,N}(f) = {\Lambda_f}^{-1} \circ f^p \circ \Lambda_f : U_{g,N} \to \mathbb{C} ,$$
where $p = p(f, N) = p(g, N)$, $\Lambda_f = \Lambda(f, N)$, *and* $T_{g,N}(f)$ *is a quadratic-like map with conformal modulus greater than* $\mu/2$;

(ii) *the operator* $T : \mathbb{O}_{\Omega_\alpha} \to \mathbb{A}_{\Omega_\alpha}$ *given by* $T = i_{g,N} \circ T_{g,N}$ *is real-analytic with a compact derivative, where*
$$\mathbb{O}_{\Omega_\alpha} = \bigcup_{g \in \mathbb{K}} \mathbb{A}_{\Omega_\alpha}(g, \delta_0) .$$

Proof. By Sullivan's real bounds, there exist $C_1 > 1$ and $0 < \nu_1 < \nu_2 < 1$ such that for all $g \in \mathbb{K}$, all $k \in \mathbb{N}$ and all $0 \le j \le p(k, g) - 1$, we have $C_1^{-1}\nu_1^k < |\Delta_{j,k}(g)| < C_1\nu_2^k$. Thus, by property (i) in Theorem 3.1, for every $\alpha > 0$ there is $N > 0$ so large that the open sets $g^j(O_{g,N})$ have diameter smaller than $\alpha/3$ for all $0 \le j \le p(N, g)$. Recall that $O_{g,N} = \Lambda_g(U_{g,N})$. By a continuity argument, there is $\delta_g > 0$ such that for every $f \in \mathbb{A}_{\Omega_\alpha}(g, \delta_g)$, the restriction $f|[-1, 1]$ is N-times renormalizable, $f^j(\Lambda_f(U_{g,N})) \subset \Omega_{\alpha/2}$ for every $0 \le j \le p = p(N, f)$, and moreover $f^p : \Lambda_f(U_{g,N}) \to f^p(\Lambda_f(U_{g,N}))$ is a quadratic-like map with conformal modulus greater than $\mu/2$. By compactness of $\mathbb{K}$ in $\mathbb{A}_{\Omega_\alpha}$, there is a finite set $\{g_i : i = 1, \ldots, l\}$ such that
$$\mathbb{K} \subset \bigcup_{i=1}^{l} \mathbb{A}_{\Omega_\alpha}(g_i, \delta_{g_i}/2) .$$
Set $\delta_0 = \min_{i=1,\ldots,l}\{\delta_{g_i}/2\}$. Then, for every $g \in \mathbb{K}$ there exists $i = i(g)$ such that $\mathbb{A}_{\Omega_\alpha}(g, \delta_0) \subset \mathbb{A}_{\Omega_\alpha}(g_i, \delta_{g_i})$. Hence $T_{g,N}(f)$ is well-defined, and it is a quadratic-like map with conformal modulus greater than $\mu/2$, for every $f \in \mathbb{A}_{\Omega_\alpha}(g, \delta_0)$ which proves (i).

Note that the real Banach space $\mathbb{A}_{\Omega_\alpha}$ is naturally embedded in the complex Banach space $\mathbb{A}_{\Omega_\alpha,\mathbb{C}}$ of maps $f : \Omega_\alpha \to \mathbb{C}$ which are holomorphic and continuous up to the boundary and that $T_{g,N}$ extends to an operator $T^{\mathbb{C}}_{g,N}$ in an open set of $\mathbb{A}_{\Omega_\alpha,\mathbb{C}}$, given by the same expression. Applying Cauchy's integral formula, we see that $T^{\mathbb{C}}_{g,N}$ is complex-analytic, and so $T_{g,N}$ is real-analytic. Since by Montel's theorem the inclusion $i_{g,N}$ is a compact linear operator, we deduce that $T : \mathbb{O}_{\Omega_\alpha} \to \mathbb{A}_{\Omega_\alpha}$ is a real-analytic operator with a compact derivative, which proves (ii). □

3.2. *Real analytic hybrid conjugacy classes.* We will introduce later (in §3.4) a skew-product renormalization operator. The fiberwise local stable manifolds of such a skew-product – which will be used to determine the stable manifolds of the real-analytic operator $T : \mathbb{O}_{\Omega_a} \to \mathbb{A}_{\Omega_a}$, for some suitable $a > 0$ –

turn out to be openly contained in the hybrid conjugacy classes of the maps in the limit set $\mathbb{K}$. Here we analyze the manifold structure of hybrid classes in more detail.

A homeomorphism $h : U \to V$, where U and V are contained in $\mathbb{C}$ or $\overline{\mathbb{C}}$, is *quasiconformal* if it has locally square integrable distributional derivatives ∂h, $\overline{\partial} h$, and there exists $\varepsilon < 1$ with the property that $|\overline{\partial} h/\partial h| \leq \varepsilon$ almost everywhere. The Beltrami differential μ_h of h is given by $\mu_h = \overline{\partial} h/\partial h$. A quasiconformal map h is K quasiconformal if $K \geq (1 + ||\mu_h||_\infty)/(1 - ||\mu_h||_\infty)$.

Two quadratic-like maps f and g are *hybrid conjugate* if there is a quasiconformal conjugacy h between f and g with the property that $\overline{\partial} h(z) = 0$ for almost every $z \in \mathcal{K}_f$. Let us denote by $\mathcal{H}(f)$ the hybrid conjugacy class of f.

By a slight abuse of notation, we will denote by $\mathbb{K} \cap \mathbb{A}_V(g, \delta)$ the set of maps $f \in \mathbb{A}_V(g, \delta)$ with the property that $f|[-1, 1]$ belongs to $\mathbb{K}$.

In the proof of the following theorem, we will need to work with the complexification of $\mathbb{A}_V$. Let $\mathbb{A}_{V,\mathbb{C}}$ be the complex Banach space of all holomorphic maps $f : V \to \mathbb{C}$ with a continuous extension to the boundary of V. Let $\mathbb{A}_{V,\mathbb{C}}(f, \delta)$ be the open ball in $\mathbb{A}_{V,\mathbb{C}}$ centered in f and with radius $\delta > 0$. Let $C : \mathbb{A}_{V,\mathbb{C}} \to \mathbb{A}_{V,\mathbb{C}}$ be the conjugation operator given by $C(f) = c \circ f \circ c$, where $c(z) = \overline{z} \in \mathbb{C}$. We note that $f \in \mathbb{A}_V$ if and only if $f \in \mathbb{A}_{V,\mathbb{C}}$ and $C(f) = f$.

THEOREM 3.3. *For every $g \in \mathbb{K}$, there exists a symmetric neighborhood $\hat{V}_g$ of the reals such that g has a quadratic-like extension to $\hat{V}_g$ (also denoted by g), $\hat{V}_g$ contains a definite neighborhood of $\mathcal{K}_g$ and for every neighborhood $V \subset \hat{V}_g$ symmetric with respect to $\mathbb{R}$ and with the property that $g|V$ is a quadratic-like map, there is $\delta_{g,V} > 0$ such that for all $f \in \mathbb{K} \cap \mathbb{A}_V(g, \delta_{g,V})$,*

$$\mathcal{H}_V(f) = \mathcal{H}(f) \cap \mathbb{A}_V(g, \delta_{g,V})$$

are codimension-one, real-analytic leaves varying continuously with f.

Proof. By Lemmas 4.16 and 4.17 on page 354 of Lyubich's paper [20], we obtain that for all $f \in \mathbb{K} \cap \mathbb{A}_{V,\mathbb{C}}(g, \delta_{g,V})$, $\mathcal{H}_{V,\mathbb{C}}(f) = \mathcal{H}(f) \cap \mathbb{A}_{V,\mathbb{C}}(g, \delta_{g,V})$ are codimension-one complex analytic leaves varying continuously with f. If $f \in \mathbb{A}_V(g, \delta_{g,V})$ then the hybrid conjugacy class of f in $\mathbb{A}_{V,\mathbb{C}}(g, \delta_{g,V})$ is invariant under the conjugation operator C. Hence, the tangent space $T_f\mathcal{H}_{V,\mathbb{C}}(f)$ at f to its hybrid conjugacy class is invariant under the conjugation operator C, and there is a one-dimensional transversal E_f to $T_f\mathcal{H}_{V,\mathbb{C}}(f)$ which is also invariant under the conjugation operator C. Locally $\mathcal{H}_{V,\mathbb{C}}(f)$ is a graph of $\mathcal{G} : Z \subset T_f\mathcal{H}_{V,\mathbb{C}}(f) \to E_f$ with the property that if $h = v + \mathcal{G}(v)$ then $C(h) = C(v) + \mathcal{G}(C(v))$. Thus, locally $\mathcal{H}_V(f)$ is also the graph of $\mathcal{G}|Z \cap \mathbb{A}_V(g, \delta_{g,V})$, and so it is a codimension-one, real-analytic leaf. Since the complex-analytic leaves $\mathcal{H}_{V,\mathbb{C}}(f)$ vary continuously with f, we deduce that the real-analytic leaves $\mathcal{H}_V(f)$ also vary continuously with f. □

3.3. *Hyperbolic skew-products.* Before going further, we pause for a moment to introduce the elementary concept of hyperbolic skew product in an abstract setting. Let $\mathbb{K}$ be a compact metric space and assume that $\mathbb{K}$ is totally disconnected. Let $\mathcal{F}$ be a finite collection of (real) Banach spaces, say $\mathcal{F} = \{\mathcal{A}_1, \mathcal{A}_2, \ldots, \mathcal{A}_N\}$, and assume we have a locally constant map $\varphi : \mathbb{K} \to \mathcal{F}$. We write $\mathcal{A}_x = \varphi(x) \in \mathcal{F}$, for all $x \in \mathbb{K}$. Let $E = \cup_{x\in\mathbb{K}}\{x\} \times \mathcal{A}_x$. We endow E with a topology as follows. If $\mathbb{K}_i = \varphi^{-1}(\mathcal{A}_i)$, then $\mathbb{K}_i$ is an open and closed set in $\mathbb{K}$, for each $i = 1, 2, \ldots, N$. Note that E is the disjoint union of $\mathbb{K}_i \times \mathcal{A}_i$, $i = 1, 2, \ldots, N$. Hence endow each factor $\mathbb{K}_i \times \mathcal{A}_i$ with the product topology and then E with the union topology. It is clear that E is metrizable also. The natural projection $E \to \mathbb{K}$ is open and continuous. We shall assume that there exists a continuous injection $\mathbb{K}_i \to \mathcal{A}_i$ for each i, and will accordingly identify each $x \in \mathbb{K}_i$ with its image in $\mathcal{A}_i$.

Now suppose $T : \mathbb{K} \to \mathbb{K}$ is a homeomorphism (in the case we are interested, T is transitive), and also that for each $x \in \mathbb{K}$ we have a real-analytic map $S_x : \mathcal{A}_x(x, \delta) \to \mathcal{A}_{T(x)}$, where $\mathcal{A}_x(x, \delta) = \{x + v \in \mathcal{A}_x : \|v\|_{\mathcal{A}_x} < \delta\}$. We define a skew-product operator $S : E(\delta) \to E$ over T, where

$$E(\delta) = \{(x, y) : x \in \mathbb{K}, y \in \mathcal{A}_x, \|y - x\|_{\mathcal{A}_x} < \delta\},$$

by $S(x, y) = (T(x), S_x(y))$.

Definition 3.2. We say that S is *fiberwise hyperbolic* if there exists a continuous splitting $\mathcal{A}_x = E^s_x \bigoplus E^u_x$ with $\dim E^u_x = 1$ which is invariant in the sense that $DS_x(E^s_x) \subseteq E^s_{T(x)}$ and $DS_x(E^u_x) \subseteq E^u_{T(x)}$, satisfying for all $v^s \in E^s_x$ and all $v^u \in E^u_x$ the inequalities

$$\left\|D\left(S_{T^{n-1}(x)} \circ \ldots S_x\right)(x)v^s\right\|_{\mathcal{A}_{T^n(x)}} \leq C\theta^n\|v^s\|_{\mathcal{A}_x}$$
$$\left\|D\left(S_{T^{n-1}(x)} \circ \ldots S_x\right)(x)v^u\right\|_{\mathcal{A}_{T^n(x)}} \geq C^{-1}\theta^{-n}\|v^u\|_{\mathcal{A}_x},$$

where $C > 1$ and $0 < \theta < 1$ are uniform constants on g.

Definition 3.3. The *fiberwise local stable manifold* $W^s_\beta(x)$ of S at x consists of all points $y \in \mathcal{A}_x(x, \beta)$ such that for all $n \geq 1$, we have $S_{T^{n-1}(x)} \circ \cdots \circ S_x(y) \in \mathcal{A}_{T^n(x)}(T^n(x), \beta)$ and

$$\left\|S_{T^{n-1}(x)} \circ \cdots \circ S_x(y) - S_{T^{n-1}(x)} \circ \cdots \circ S_x(x)\right\|_{\mathcal{A}_{T^n(x)}} \leq C\theta^n$$

where $C > 0$ and $0 < \theta < 1$ are uniform constants on $x \in \mathbb{K}$. The *fiberwise local unstable manifold* $W^u_\beta(x)$ *of* S *at* x consists of all points $y \in \mathcal{A}_x(x, \beta)$ such that when $y_0 = y$, for each $n \geq 1$ there exists $y_n \in \mathcal{A}_{T^{-n}(x)}$ such that $y_{n-1} = S_{T^{-n}(x)}(y_n)$ and $\|T^{-n}(x) - y_n\|_{\mathcal{A}_{T^{-n}(x)}} \leq C\theta^n$.

3.4. *Skew-product renormalization operator.* Our goal in this section is to build a skew-product renormalization operator that will play a central role in the proof that $\mathbb{K}$ is a basic set for $T : \mathbb{O}_{\Omega_a} \to \mathbb{A}_{\Omega_a}$, for a suitable $a > 0$. Our skew-product is constructed so as to satisfy properties (W1) to (W4) on page 395 of [20] in the real analytic case – restated in §3.5 – and therefore will have fiberwise stable and unstable manifolds, as we will explain in that section.

Using Theorem 3.1 and (3.1.3), we know that for every $0 < \alpha < b$, $\mathbb{K}$ injects continuously into $\mathbb{A}_{\Omega_\alpha}$. Hence for $f, g \in \mathbb{K}$ we define $\mathrm{dist}_{\mathbb{K}}(f,g) = \|f - g\|_{\mathbb{A}_{\Omega_\alpha}}$. We also denote by $\mathbb{K}(g,\varepsilon)$ the ball of radius ε centered at g in this metric. The metric is compatible with the natural topology of $\mathbb{K}$, independently of which α we take.

LEMMA 3.4. *The filled-in Julia set $\mathcal{K}_g$ varies continuously in the Hausdorff metric with respect to $g \in \mathbb{K}$.*

Proof. We need to show that for every $\varepsilon > 0$ there exists $\delta > 0$ such that if $\mathrm{dist}_{\mathbb{K}}(f,g) < \delta$ then (a) $\mathcal{K}_g \subset \Omega_\varepsilon(\mathcal{K}_f)$ and (b) $\mathcal{K}_f \subset \Omega_\varepsilon(\mathcal{K}_g)$. Let $U = U_{R^{-N_0}(g),N_0} \subset \mathbb{C}$ be the symmetric neighborhood of $[-1,1]$ given by Lemma 3.2. Since the operator $T_{R^{-N_0}(g),N_0}$ is continuous, every $f \in \mathbb{K}$ sufficiently close to g in $\mathbb{A}_{\Omega_\varepsilon}$ is quadratic-like on U ($f = T_{R^{-N_0}(g),N_0}(T^{-1}(f)) : U \to \mathbb{C}$) and is also close to g in $\mathbb{A}_U$.

To prove (a), cover $\mathcal{K}_g$ by finitely many disks $D(z_i(g), \varepsilon/2)$, $i = 1, 2, \ldots, m$, where each $z_i(g)$ is an expanding periodic point of g. For f sufficiently close to g, the corresponding periodic points $z_i(f) \in D(z_i(g), \varepsilon/2)$. Hence each $z \in \mathcal{K}_g$ is at distance at most ε from some $z_i(f)$, which proves (a).

To prove (b), let $n > 0$ be so large that $W = g^{-n}(U) \subset \Omega_\varepsilon(\mathcal{K}_g)$. Since f is close to g and $W \subseteq U$ is symmetric $f : W \to f(W)$ is quadratic-like also, whence $\mathcal{K}_f \subset W \subset \Omega_\varepsilon(\mathcal{K}_g)$ and so (b) is proved. □

LEMMA 3.5. *Let $g \in \mathbb{K}$ and let $V \subset \mathbb{C}$ be a symmetric neighborhood of $[-1,1]$ which is compactly contained in $\Omega_{b/2}(\mathcal{K}_g)$, where b is given by (3.1.3). Then for all $\varepsilon > 0$ sufficiently small $\mathbb{K} \cap \mathbb{A}_V(g,\varepsilon)$ is an open subset of $\mathbb{K}$.*

Proof. Take $0 < \alpha < b$ sufficiently small such that Ω_α is compactly contained in V. By Theorem 3.1 and (3.1.3), every $f \in \mathbb{K}$ is well-defined on $\Omega_b(\mathcal{K}_f)$. Since by Lemma 3.4 the map $f \mapsto \mathcal{K}_f$ is continuous in the Hausdorff metric, there exists $\varepsilon_0 > 0$ such that if $f \in \mathbb{K}$ is such that $\mathrm{dist}_{\mathbb{K}}(f,g) < \varepsilon_0$ then $\Omega_{b/2}(\mathcal{K}_g) \subset \Omega_b(\mathcal{K}_f)$. Since $\overline{V} \subset \Omega_{b/2}(\mathcal{K}_g)$, it follows that f is well-defined on V, that is, $f \in \mathbb{A}_V$. Hence there is a well-defined injection $\mathbb{K}(g,\varepsilon_0) \to \mathbb{A}_V$. Such injection is continuous. Indeed, for $f \in \mathbb{K}(g,\varepsilon_0)$, the C^0 norm of f in $\Omega_{b/2}(\mathcal{K}_g)$ is uniformly bounded, while $\|f\|_{\mathbb{A}_{\Omega_\alpha}}$ varies continuously with f. Since $\overline{\Omega_\alpha} \subset V \subset \overline{V} \subset \Omega_{b/2}(\mathcal{K}_g)$, we deduce from Hadamard's three circles theorem (see Lemma 11.5 on page 415 of [20]) that $\|f\|_{\mathbb{A}_V}$ varies continuously with f also. Therefore the map $\mathbb{K}(g,\varepsilon_0) \to \mathbb{A}_V$ is continuous as asserted. Now let $f \in \mathbb{K} \cap \mathbb{A}_V(g,\varepsilon_0)$. Since the inclusion $\mathbb{A}_V \to \mathbb{A}_{\Omega_\alpha}$ has Lipschitz constant one, we have that $f \in \mathbb{K}(g,\varepsilon_0)$. Hence, by continuity of the map $\mathbb{K}(g,\varepsilon_0) \to \mathbb{A}_V$, there exists $\varepsilon_1 > 0$ such that $\mathbb{K}(f, \varepsilon_1) \subseteq \mathbb{K} \cap \mathbb{A}_V(g, \varepsilon_0)$, which shows that this last set is open in $\mathbb{K}$. This completes the proof. □

LEMMA 3.6. *Let $b > 0$ be as defined in (3.1.3) and $\delta_0 > 0$ as in Lemma 3.2. There exist $\nu > 0$, $0 < \delta < \delta_0$, a finite set $\mathcal{V}$ of symmetric neighborhoods of $[-1, 1]$ and a locally constant map $\mathbb{K} \ni g \mapsto V_g \in \mathcal{V}$ with the following properties:*

(i) *The neighborhood V_g is compactly contained in $\Omega_{b/2}(\mathcal{K}_g)$;*

(ii) *Every $f \in \mathbb{A}_{V_g}(g, \delta)$ is a quadratic-like map with conformal modulus larger than ν;*

(iii) *If $f \in \mathbb{K} \cap \mathbb{A}_{V_g}(g, \delta)$ then $\mathcal{H}(f) \cap \mathbb{A}_{V_g}(g, \delta)$ is a codimension-one, real-analytic submanifold varying continuously with f.*

Proof. For every $g \in \mathbb{K}$, let $U_g \subset \mathbb{C}$ be a symmetric neighborood of $[-1, 1]$ where g is quadratic-like, and take $n_g > 0$ so large that $V'_g = g^{-n_g}(U_g) \subset \Omega_{b/3}(\mathcal{K}_g)$ and $V'_g \subseteq \hat{V}_g$, where $\hat{V}_g$ is as given in Theorem 3.3.

Let $\delta_g > 0$ be so small that each $f \in \mathbb{A}_{V'_g}(g, \delta_g)$ is quadratic-like in V'_g with conformal modulus greater than $\nu_g > 0$ and also so that Theorem 3.3 holds true (for V'_g and δ_g). By Lemma 3.4, making δ_g smaller if necessary, we see that $V'_g = g^{-n_g}(U_g) \subset \Omega_{b/2}(\mathcal{K}_f)$ for all $f \in \mathbb{K} \cap \mathbb{A}_{V'_g}(g, \delta_g)$.

By Lemma 3.5, each set $\mathbb{K} \cap \mathbb{A}_{V'_g}(g, \delta_g/2)$ is open in $\mathbb{K}$. Since $\mathbb{K}$ is compact, there exists a finite set $\{g_i : i = 1, \ldots, l\}$ such that

$$\mathbb{K} \subset \bigcup_{i=1}^{l} \mathbb{A}_{V'_{g_i}}(g_i, \delta_{g_i}/2) .$$

Thus we can set

$$\mathcal{V} = \left\{V'_{g_i} : i = 1, \ldots, l\right\}, \ \delta = \min_{i=1,\ldots,l} \{\delta_{g_i}/2\} \quad \text{and} \quad \nu = \min_{i=1,\ldots,l} \{\nu_{g_i}\} .$$

Therefore, since $\mathbb{K}$ is totally disconnected, there exists a locally constant map $\mathbb{K} \ni g \mapsto V_g \in \mathcal{V}$ so that properties (i), (ii) and (iii) are satisfied. □

We are now in a position to define the skew-product renormalization operator. This is accomplished in our next lemma. Let us define first its range and domain, respectively, as follows

$$E = \left\{(g, f) : g \in \mathbb{K} \text{ and } f \in \mathbb{A}_{V_g}\right\},$$
$$E(\delta) = \left\{(g, f) \in E : f \in \mathbb{A}_{V_g}(g, \delta)\right\} .$$

Let us now fix once and for all $a > 0$ so small that $\overline{\Omega_a} \subset V_g$ for every $g \in \mathbb{K}$ (this is possible because $\mathcal{V}$ in Lemma 3.6 is a finite set). The inclusion

$k_g : \mathbb{A}_{V_g} \to \mathbb{A}_{\Omega_a}$ is a well-defined compact linear operator. By (3.1.3) and Lemma 3.6 (i) we also have

$$V_{R^N(g)} \subset \Omega_{b/2}(\mathcal{K}_{R^N(g)}) \subset \Omega_b(\mathcal{K}_{R^N(g)}) \subset U_{g,N} \ .$$

Therefore the inclusion $j_{g,N} : \mathbb{A}_{U_{g,N}} \to \mathbb{A}_{V_{R^N(g)}}$ is also a well-defined compact linear operator.

LEMMA 3.7. *Let $\delta > 0$ and $V_g \in \mathcal{V}$ be as in Lemma 3.6. Let $N = N(a) > 0$, $T_{g,N}$ and $T : \mathbb{O}_{\Omega_a} \to \mathbb{A}_{\Omega_a}$ be as in Lemma 3.2.*

(i) *For every $g \in \mathbb{K}$, the operator $T_g : \mathbb{A}_{\Omega_a}(g, \delta) \to \mathbb{A}_{V_{R^N(g)}}$ given by $T_g = j_{g,N} \circ T_{g,N}$ is real analytic with a compact derivative.*

(ii) *The skew-product renormalization operator $S : E(\delta) \to E$ given by $S(g, f) = (T(g), S_g(f))$, where $S_g = T_g \circ k_g : \mathbb{A}_{V_g}(g, \delta) \to \mathbb{A}_{V_{T(g)}}$, is well-defined. Furthermore,*

$$k_{T(g)} \circ S_g = T \circ k_g \ . \tag{3.4.1}$$

Proof. The proof is similar to the proof of Lemma 3.2 (ii). □

3.5. *Hyperbolicity of the renormalization operator.* The purpose of this section is to show that $\mathbb{K}$ is a hyperbolic basic set for the operator $T : \mathbb{O}_{\Omega_a} \to \mathbb{A}_{\Omega_a}$. This will follow from the fact (Lemma 3.8 below) that the skew-product renormalization operator has fiberwise real analytic stable manifolds and fiberwise one dimensional real analytic unstable manifolds.

We start by noting that our skew-product operator satisfies the conditions W1–W4 on page 395 of Lyubich [20] in the real analytic case. Namely, we have

W1. The conformal modulus of each $g \in \mathbb{K}$ is larger than a uniform constant $\mu > 0$.

W2. There exists $\eta > 0$ such that if $\mathrm{dist}_{\mathbb{K}}(f, g) < \eta$ for some $f, g \in \mathbb{K}$, then $\mathbb{A}_{V_f} = \mathbb{A}_{V_g}$.

W3. There exists $\delta > 0$ such that $S_g(\mathbb{A}_{V_g}(g, \delta)) \subseteq \mathbb{A}_{V_{T(g)}}$.

W4. The vertical fibers $\mathcal{Z}_g$ (consisting of those normalized symmetric quadratic-like germs whose external class is the same as that of g) sit locally in $\mathbb{A}_{V_g}$ for each $g \in \mathbb{K}$.

Condition W1 is satisfied because of the complex bounds (Theorem 3.1). Condition W2 follows from Lemma 3.6. Condition W3 holds by the construction of S_g in Lemma 3.7. Condition W4 is a consequence of Lemma 3.6 (iii).

Now we have the following result.

LEMMA 3.8. *The skew-product renormalization operator* $S : E(\delta) \to E$ *defined in Lemma 3.7 is fiberwise hyperbolic. Moreover*

(i) *The local stable set* $W^s_\delta(g)$ *of* S *at* g *is a codimension-one submanifold of* $\mathbb{A}_{V_g}$ *which is relatively open in* $\mathcal{H}(g) \cap \mathbb{A}_{V_g}(g, \delta)$, *and* $W^s_\delta(g)$ *is tangent to* E^s_g *at* g.

(ii) *The local unstable set* $W^u_\delta(g) \subset \mathbb{A}_{V_g}$ *of* S *at* g *is a one-dimensional, real-analytic manifold, and* $\{g\} \times W^u_\delta(g)$ *varies continuously with* $g \in \mathbb{A}$ *in* E.

Proof. Since the operator S satisfies Lyubich's conditions W1–W4 stated above, part (i) follows from Theorem 8.2 on page 392 of [20] and Theorem 3.3, and part (ii) follows from Theorem 8.8 on page 398 of Lyubich's paper [20]. □

THEOREM 3.9. *Let* $T : \mathbb{O}_{\Omega_a} \to \mathbb{A}_{\Omega_a}$ *be the real analytic operator defined in Lemma 3.7. Then there is a continuous,* DT*-invariant splitting* $\mathbb{A}_{\Omega_a} = E^s_g \bigoplus E^u_g$, *for* $g \in \mathbb{K}$, *such that if* $v^u \in E^u_g$ *and* $v^s \in E^s_g$ *then*

$$\|DT^n(g)v^u\|_{\mathbb{A}_{\Omega_a}} \geq C^{-1}\theta^{-n}\|v^u\|_{\mathbb{A}_{\Omega_a}}, \tag{3.5.1}$$

$$\|DT^n(g)v^s\|_{\mathbb{A}_{\Omega_a}} \leq C\theta^{n}\|v^s\|_{\mathbb{A}_{\Omega_a}}, \tag{3.5.2}$$

where $C > 1$ *and* $0 < \theta < 1$ *are uniform constants on* g.

Proof. Since for every $g \in \mathbb{K}$ the map $k_g : \mathbb{A}_{V_g} \to \mathbb{A}_{\Omega_a}$ is linear and injective, it follows from Lemma 3.8 (ii) that $Z^u_g = k_g(W^u_\delta(g))$ is a real-analytic, one-dimensional manifold varying continuously with g. Let w_g be the unitary vector tangent to $W^u_\delta(g)$ at g. Then $v_g = k_g(w_g)$ is a vector tangent to Z^u_g at g and also varies continuously with g. Since k_g and $k_{T(g)}$ are linear maps we see from (3.4.1) that if λ_g is such that $DS_g(g)w_g = \lambda_g w_{T(g)}$ then $DT(g)v_g = \lambda_g v_{T(g)}$. Thus a natural candidate for E^u_g is the one-dimensional linear subspace generated by v_g. In particular, (3.5.1) is satisfied.

Let us find the natural candidate for E^s_g. We have that $DT_g(g)v_g = w_{T(g)}$ and by hypothesis $w_{T(g)}$ is transversal to the tangent space of $W^s_\delta(T(g))$. Thus, by the implicit function theorem $Z^s_g = T^{-1}_g(W^s_\delta(Tg))$ is a codimension-one manifold transversal to Z^u_g. Taking E^s_g equal to the tangent space of Z^s_g, we obtain that $E^s_g \bigoplus E^u_g = \mathbb{A}_{\Omega_a}$. By (3.4.1), we have that a neighborhood of $T(g)$ intersected with $Z^u_{T(g)}$ is contained in $T(Z^u_g)$ and a neighborhood of $T(g)$ intersected with $T(Z^s_g)$ is contained in $Z^s_{T(g)}$, which implies that the splitting $E^s_g \bigoplus E^u_g$ is invariant under DT. From assertion (i) in Lemma 3.8, we obtain that E^s_g varies continuously with g and so the splitting $E^s_g \bigoplus E^u_g$ also varies continuously with g.

Finally, let $M > 0$ be such that $\|DT_g(g)\|_{\mathbb{A}_{\Omega_a}} \leq M$ and note that $\|k_g\|_{\mathbb{A}_{\Omega_a}} \leq 1$ for all $g \in \mathbb{K}$. For all $v^s \in E^s_g$ with unit norm, let $u^s = DT_g(g)v^s \in \mathbb{A}_{V_{T(g)}}$. By

Lemma 3.8 (i) and (3.4.1) there exists $C_1 > 1$ and $0 < \theta < 1$ such that

$$\begin{aligned}\|DT^n(g)v^s\|_{\mathbb{A}_{\Omega_a}} &= \|k_{T^n(g)} \circ DS_{T^{n-1}(g)}(T^{n-1}(g)) \circ \cdots \circ DS_{T(g)}(T(g))u^s\|_{\mathbb{A}_{\Omega_a}} \\ &\leq \|DS_{T^{n-1}(g)}(T^{n-1}(g)) \circ \cdots \circ DS_{T(g)}(T(g))u^s\|_{\mathbb{A}_{V_{T^n(g)}}} \\ &\leq C_1 M\theta^{n-1} ,\end{aligned}$$

which shows that (3.5.2) is satisfied. This completes the proof. $\square$

With the above results, we have therefore established Theorem 2.4, to the effect that a suitable power of the renormalization operator is indeed hyperbolic in a suitable (real) Banach space of real analytic mappings. From now on, we shall concentrate on the problem of extending such hyperbolicity to larger ambient spaces of smooth mappings. Our journey will take us far into the wilderness of nonlinear functional analysis.

4. Extending invariant splittings

In this section we prove a certain result from functional analysis (Theorem 4.1 below) that is absolutely crucial for the stable manifold theorem that we shall prove later. This result deals with the notion of compatibility presented below and is a strong generalization of a key idea of Davie in [5]. In Section 5, we shall use the results presented here to show that the invariant splitting for the renormalization operator T in $\mathbb{A}_{\Omega_a}$ of Section 3 extends to an invariant splitting for the action of T in the larger spaces $\mathbb{A}^r$ of C^r maps.

4.1. *Compatibility.* We are interested in the answer to the following question. Given a smooth operator $T : \mathcal{O} \to \mathcal{A}$ having a hyperbolic basic set $\mathbb{K}$, and given a larger ambient space $\mathcal{B} \supseteq \mathcal{A}$ to which T extends (not necessarily smoothly), under which conditions does $\mathbb{K}$ have a hyperbolic structure in $\mathcal{B}$? To give a precise meaning to this question (and then answer it!) we introduce the following notion.

We have a natural continuous map $L : \mathbb{K} \to \mathcal{L}(\mathcal{A}, \mathcal{A})$ given by

$$\begin{aligned}\mathbb{K} \ni x \;\mapsto\; & L_x : \mathcal{A} \to \mathcal{A} \\ & L_x(v) = DT(x)\, v \,.\end{aligned}$$

We will also assume that for every $x \in \mathbb{K}$, E^u_x is a one-dimensional subspace and that we can chose a unit vector $\mathbf{u}_x \in E^u_x$ varying continuously with x so that $L_x(\mathbf{u}_x) = \delta_x \cdot \mathbf{u}_{T(x)}$ with $\delta_x > 0$. In the case of the renormalization operator there is a natural choice for the vectors $\mathbf{u}_x$: choose the unit vector pointing in the direction of increasing topological entropy.

For every $x \in \mathbb{K}$, we denote $DT^n(x) = L_{T^{n-1}(x)} \circ \cdots \circ L_x$ by $L^{(n)}_x$ and $\delta_{T^{n-1}(x)} \cdots \delta_x$ by $\delta^{(n)}_x$. By hyperbolicity of $\mathbb{K}$, there exist $C_0 > 0$ and $\lambda > 1$

such that for every $x \in \mathbb{K}$ and every $n \geq 1$ we have

$$\delta_x^{(n)} > C_0 \lambda^n \ . \tag{4.1.1}$$

We denote by $\mathcal{X}(r)$ the open ball in the Banach space $\mathcal{X}$ centered at the origin and with radius $r > 0$.

Definition 4.1. Let $\theta < \rho < \lambda$ where θ is the contraction exponent of the hyperbolic basic set $\mathbb{K}$ of the operator T and λ is as in (4.1.1). The pair $(\mathcal{B}, \mathcal{C})$ is *ρ-compatible* with $(T, \mathbb{K})$ if the following conditions are satisfied.

A1. The inclusions $\mathcal{A} \to \mathcal{B} \to \mathcal{C}$ are compact operators.

A2. There exists $M > 0$ such that each linear operator $L_x = DT(x)$ extends to a linear operator $\hat{L}_x : \mathcal{C} \to \mathcal{C}$ with

$$\left\|\hat{L}_x\right\|_{\mathcal{C}} < M,$$
$$\hat{L}_x(\mathcal{B}) \subset \mathcal{B},$$
$$\left\|\hat{L}_x(v)\right\|_{\mathcal{B}} < M \|v\|_{\mathcal{B}}.$$

A3. The map $\tilde{L} : \mathbb{K} \to \mathcal{L}(\mathcal{B}, \mathcal{C})$ given by $\tilde{L}_x = \hat{L}_x | \mathcal{B}$ is continuous.

A4. There exists $\Delta > 1$ such that $\mathcal{B}(\Delta) \cap \mathcal{A}$ is $\mathcal{C}$-dense in $\mathcal{B}(1)$.

A5. There exist $K > 1$ and a positive integer m such that

$$\left\|\hat{L}_x^{(m)}(v)\right\|_{\mathcal{B}} \leq \max\left\{\frac{\rho^m}{2(1+\Delta)}\|v\|_{\mathcal{B}}, K\|v\|_{\mathcal{C}}\right\} \ .$$

Remark 4.1. Note that neither the map $\hat{L} : \mathbb{K} \times \mathcal{C} \to \mathbb{K} \times \mathcal{C}$ given by $\hat{L}(x, v) = (T(x), \hat{L}_x(v))$ nor its restriction from $\mathbb{K} \times \mathcal{B}$ to $\mathbb{K} \times \mathcal{B}$ are necessarily continuous.

Example 4.1. As we know from Theorem 2.4, $\mathbb{K}$ is a hyperbolic basic set of the renormalization operator $T = R^N : \mathbb{O} \to \mathbb{A}$. In Section 5 (see Theorem 5.1), we will show that the pair $(\mathbb{A}^r, \mathbb{A}^0)$ is ρ-compatible for r sufficiently close to 2 and 1-compatible for $r > 2$ noninteger.

Let $\pi_x^u : \mathcal{A} \to E_x^u$ and $\pi_x^s : \mathcal{A} \to E_x^s$ be the canonical projections. We define $P_x = \pi_{T(x)}^u \circ L_x$ and $Q_x = \pi_{T(x)}^s \circ L_x$ which have the property that $L_x = P_x + Q_x$ and that $P_{T(x)} Q_x = Q_{T(x)} P_x = 0$. We also define the linear functional $\sigma_x : \mathcal{A} \to \mathbb{R}$ by $\pi_x^u(v) = \sigma_x(v) \mathbf{u}_x$, and observe that $P_x(v) = \delta_x \sigma_x(v) \mathbf{u}_{T(x)}$. We note that the map $\sigma : \mathbb{K} \to \mathcal{L}(\mathcal{A}, \mathbb{R})$ which associates to each x the linear functional σ_x is continuous.

THEOREM 4.1. *If* $(\mathcal{B},\mathcal{C})$ *is* ρ*-compatible with* $(T,\mathbb{K})$ *then each stable functional* σ_x *extends to a unique linear functional* $\hat{\sigma}_x \in \mathcal{B}^*$ *satisfying*

$$\left\|\hat{L}_x^{(n)}(v) - \delta_x^{(n)}\hat{\sigma}_x(v)\mathbf{u}_{T^n(x)}\right\|_{\mathcal{B}} \le C\hat{\theta}^n\|v\|_{\mathcal{B}} \tag{4.1.2}$$

for some $C > 0$ *and* $0 < \hat{\theta} < \rho$*. Furthermore, the map* $\hat{\sigma} : \mathbb{K} \to \mathcal{L}(\mathcal{B},\mathbb{R})$ *which associates to each* x *the linear functional* $\hat{\sigma}_x$ *is continuous.*

Proof. Let m and M be as given in Definition 4.1. Since by property **A1** the $\mathcal{C}$-closure of $\mathcal{B}(1)$ is compact, and since by property **A4** the intersection $\mathcal{A}\cap\mathcal{B}(\Delta)$ is $\mathcal{C}$-dense in $\mathcal{B}(1)$ we can find a finite set

$$\Phi \subseteq \mathcal{A}\cap\mathcal{B}(\Delta)$$

such that for each $w \in \mathcal{B}(1)$ there exists $w' \in \Phi$ such that

$$\|w - w'\|_{\mathcal{C}} < \frac{\rho^m}{4K} .$$

Now let $v \in \mathcal{B}(1)$, and let $v_0 \in \Phi$ be such that

$$\|v - v_0\|_{\mathcal{C}} < \frac{\rho^m}{2K} .$$

Since $\|v - v_0\|_{\mathcal{B}} < 1 + \Delta$, applying the inequality of property **A5** to $v - v_0$ yields

$$\begin{aligned}\left\|\hat{L}_x^{(m)}(v - v_0)\right\|_{\mathcal{B}} &\le \max\left\{\frac{\rho^m}{2(1+\Delta)}\|v - v_0\|_{\mathcal{B}}, K\|v - v_0\|_{\mathcal{C}}\right\}\\ &< \rho^m/2 .\end{aligned}$$

Therefore $\hat{L}_x^{(m)}(v) = \hat{L}_x^{(m)}(v_0) + (\rho^m/2)w_1$ for some $w_1 \in \mathcal{B}(1)$. Repeating the argument with w_1 replacing v and proceeding inductively in this fashion, we get after k steps

$$\hat{L}_x^{(km)}(v) = \sum_{j=0}^{k-1}\frac{\rho^{jm}}{2^j}L_{T^{jm}(x)}^{((k-j)m)}(v_j) + \frac{\rho^{km}}{2^k}w_k$$

for some $w_k \in \mathcal{B}(1)$ and $v_j \in \Phi$. Now recall that

$$\begin{aligned}L_{T^{jm}(x)}^{((k-j)m)}(v_j) &= P_{T^{jm}(x)}^{((k-j)m)}(v_j) + Q_{T^{jm}(x)}^{((k-j)m)}(v_j)\\ &= \delta_{T^{jm}(x)}^{((k-j)m)}\sigma_{T^{jm}(x)}(v_j)u_{T^{km}(x)} + Q_{T^{jm}(x)}^{((k-j)m)}(v_j) .\end{aligned}$$

Hence we can write

$$\begin{aligned}\hat{L}_x^{(km)}(v) = \delta_x^{(km)}&\left(\sum_{j=0}^{k-1}\frac{1}{2^j}\frac{\rho^{jm}}{\delta_x^{(jm)}}\sigma_{T^{jm}(x)}(v_j)\right)\mathbf{u}_{T^{km}(x)}\\ &+\sum_{j=0}^{k-1}\left(\frac{\rho^m}{2}\right)^j Q_{T^{jm}(x)}^{((k-j)m)}(v_j) + \frac{\rho^{km}}{2^k}w_k .\end{aligned} \tag{4.1.3}$$

The first summation in parentheses converges to a limit because, by (4.1.1), $\left|\delta_x^{(jm)}\right| \geq C_0\lambda^{jm} > C_0\rho^{jm}$ and $\{\sigma_{T^{jm}(x)}(v_j)\}$ is bounded, as the v_j run through finitely many values and $\left\|\sigma_{T^i(x)}\right\| \leq M$ for all i. We therefore define

$$\hat{\sigma}_x(v) = \lim_{k\to\infty} \sum_{j=0}^{k-1} \frac{1}{2^j} \frac{\rho^{jm}}{\delta_x^{(jm)}} \sigma_{T^{jm}(x)}(v_j) \ . \tag{4.1.4}$$

It will be clear in a moment that this extension of σ_x is independent of the choices of approximants v_j performed above, linear, continuous, and the unique extension satisfying (4.1.2). We know that

$$\left\| Q_{T^{jm}(x)}^{((k-j)m)} \right\|_{\mathcal{A}} \leq C_1 \theta^{m(k-j)}$$

for all $j < k$. Thus the second summation plus the last term in (4.1.3) add up to a vector with $\mathcal{B}$-norm bounded by

$$C_2 \left[\sum_{j=0}^{k-1} \frac{1}{2^j} \left(\frac{\theta}{\rho}\right)^{m(k-j)} + \frac{1}{2^k} \right] \rho^{km} \ .$$

This gives

$$\left\| \hat{L}_x^{(km)}(v) - \delta_x^{(km)} \hat{\sigma}_x(v) \mathbf{u}_{T^{km}(x)} \right\|_{\mathcal{B}} \leq C_3(k+1)\beta^k \rho^{km} \ , \tag{4.1.5}$$

where $\beta = \max\{1/2, \theta/\rho\} < 1$. Now choose $0 < \hat{\theta} < \rho$ so that $(k+1)\beta^k < C_4(\frac{\hat{\theta}}{\rho})^{km}$ for all k. Since by property **A2** for all $x \in \mathbb{K}$ and for all $v \in \mathcal{B}$ we have $\|\hat{L}_x(v)\|_{\mathcal{B}} < M\|v\|_{\mathcal{B}}$, writing $n = km + r$ and using the above estimates we obtain the desired inequality (4.1.2).

Let us now verify that $\hat{\sigma}_x(v)$ is the unique value satisfying (4.1.2). In particular, it does not depend on the choices of approximants v_j taken in (4.1.4). To do this we represent by $\sigma_x^*(v)$ a value satisfying (4.1.2), for instance that obtained in (4.1.4) by taking another choice of approximants. Therefore we have

$$\begin{aligned} \left\|\hat{\sigma}_x(v)\mathbf{u}_{T^{km}(x)} - \sigma_x^*(v)\mathbf{u}_{T^{km}(x)}\right\|_{\mathcal{B}} \leq & \left\| \hat{\sigma}_x(v)\mathbf{u}_{T^{km}(x)} - \left(\delta_x^{(km)}\right)^{-1} \hat{L}_x^{(km)}(v) \right\|_{\mathcal{B}} \\ & + \left\| \left(\delta_x^{(km)}\right)^{-1} \hat{L}_x^{(km)}(v) - \sigma_x^*(v)\mathbf{u}_{T^{km}(x)} \right\|_{\mathcal{B}} \\ \leq & \, 2C\hat{\theta}^{km} \left(\delta_x^{(km)}\right)^{-1} . \end{aligned}$$

Letting $k \to \infty$ in this inequality we deduce that $\hat{\sigma}_x(v) = \sigma_x^*(v)$. A similar argument shows that $\hat{\sigma}_x$ is linear. Using inequality (4.1.5) with $k = 1$, we obtain that $\|\hat{\sigma}_x\|_{\mathcal{B}}$ is bounded. Finally, the fact that $\hat{\sigma}_x$ is continuous in x can be deduced from (4.1.4) using property **A3**. □

COROLLARY 4.2. *Let $(\mathcal{B}, \mathcal{C})$ be ρ-compatible with $(T, \mathcal{A})$. Let the linear functional $\hat{\sigma}_x \in \mathcal{B}^*$ be the extension of the stable functional σ_x satisfying inequality* (4.1.2) *for all $x \in \mathbb{K}$. Then, there exists a continuous splitting $\mathcal{B} = \hat{E}_x^s \bigoplus \hat{E}_x^u$ with the following properties:*

(i) *$\hat{E}_x^u$ is the inclusion in $\mathcal{B}$ of the unstable linear space $E_x^u \subset \mathcal{A}$;*

(ii) $\hat{E}_x^s = \mathrm{Ker}(\hat{\sigma}_x)$;

(iii) *The splitting is invariant by $\hat{L}_x$;*

(iv) *There exists a constant $C > 0$ such that*

$$\left\| \hat{L}_x^m(v) \right\|_{\mathcal{B}} \geq C\lambda^m \|v\|_{\mathcal{B}}$$

for all $x \in \mathbb{K}$, for all $v \in \hat{E}_x^u$, and for all $m \in \mathbb{N}$ (where $\lambda > 1$ is the same as in (4.1.1)*);*

(v) *There exist constants $C > 0$ and $0 < \hat{\theta} < \rho < \lambda$ such that*

$$\left\| \hat{L}_x^m(v) \right\|_{\mathcal{B}} \leq C\hat{\theta}^m \|v\|_{\mathcal{B}}$$

for all $x \in \mathbb{K}$, for all $v \in \hat{E}_x^s$, and for all $m \in \mathbb{N}$. In particular, if $\rho \leq 1$ then $\hat{\theta} < 1$.

(vi) *Let $\hat{\pi}_x^s : \mathcal{B} \to \hat{E}_x^s$ and $\hat{\pi}_x^u : \mathcal{B} \to \hat{E}_x^u$ be the natural projections such that*

$$\hat{\pi}_x^s \circ \hat{\pi}_x^s = \hat{\pi}_x^s, \ \hat{\pi}_x^u \circ \hat{\pi}_x^u = \hat{\pi}_x^u \ \text{and} \ \hat{\pi}_x^s \circ \hat{\pi}_x^u = \hat{\pi}_x^u \circ \hat{\pi}_x^s = 0 \ .$$

The operators $\hat{Q}_x : \mathcal{B} \to \mathcal{B}$ and $\hat{P}_x : \mathcal{B} \to \mathcal{B}$ may be defined by $\hat{Q}_x^m = \hat{L}_x^m \circ \hat{\pi}_x^s$ and by $\hat{P}_x^m = \hat{L}_x \circ \hat{\pi}_x^u$. Then, there exists $C > 1$ such that

$$\left\| \hat{Q}_x^m \right\|_{\mathcal{B}} \leq C\hat{\theta}^m \ \text{and} \ \left\| \hat{P}^m \right\|_{\mathcal{B}} \geq C^{-1}\lambda^m$$

for all $x \in \mathbb{K}$ and for all $m \in \mathbb{N}$.

Proof. First, we observe that for all $x \in \mathbb{K}$ and for all $v \in \mathcal{B}$, we can write $v = (v - \hat{\sigma}_x(v)\mathbf{u}_x) + \hat{\sigma}_x(v)\mathbf{u}_x$, where $v - \hat{\sigma}_x(v)\mathbf{u}_x \in \mathrm{Ker}(\hat{\sigma}_x)$ and $\hat{\sigma}_x(v)\mathbf{u}_x \in \hat{E}_x^u$. Since $\hat{\sigma}_x(\mathbf{u}_x) = 1 \neq 0$, we obtain that $\mathcal{B} = \hat{E}_x^s \bigoplus \hat{E}_x^u$.

By inequality (4.1.2), there exists $C > 0$ such that

$$\left\| \hat{L}_x^{(m)}(v) \right\|_{\mathcal{B}} \leq C\hat{\theta}^m \|v\|_{\mathcal{B}} \tag{4.1.6}$$

for all $x \in \mathbb{K}$, for all $v \in \mathrm{Ker}(\hat{\sigma}_x)$, and for all $m \in \mathbb{N}$; if $v \in \mathcal{B} \setminus \mathrm{Ker}(\hat{\sigma}_x)$ then there exists $C_v > 0$ such that

$$\left\| \hat{L}_x^{(m)}(v) \right\|_{\mathcal{B}} \geq C_v\lambda^m \ .$$

Therefore, $\hat{L}_x(\mathrm{Ker}(\hat{\sigma}_x)) \subset \mathrm{Ker}(\hat{\sigma}_{T(x)})$. Since $L_x(E^u_x) = E^u_{T(x)}$ implies that $\hat{L}_x(\hat{E}^u_x) = \hat{E}^u_{T(x)}$, the splitting is invariant by $\hat{L}_x$.

By inequality (4.1.6), we obtain that property (v) and the first inequality in property (vi) are satisfied. Since $\mathbb{K}$ is compact and the map $\mathbb{K} \to \mathbb{R}^+$ which associates $\|\mathbf{u}_x\|_{\mathcal{B}}$ to each x is continuous, there is $C > 1$ such that

$$C^{-1}\|v\|_{\mathcal{A}} < \|v\|_{\mathcal{B}} < C\|v\|_{\mathcal{A}} \tag{4.1.7}$$

for all $x \in \mathbb{K}$ and for all $v \in E^u_x$. Thus, property (iv) and the second inequality in property (vi) follow from (4.1.1) and (4.1.7). □

5. Extending the invariant splitting for renormalization

Our aim in this section is to show that the invariant splitting on the limit set $\mathbb{K}$ of the operator T given by Theorem 2.4, which is an iterate of the renormalization operator, can be extended to an invariant splitting of the same operator acting in the space of C^r unimodal maps. Given the abstract results of the previous section, namely Theorem 4.1 and Corollary 4.2, all we have to do is find the appropriate compatible spaces and the corresponding compatibility constants. More precisely, we shall prove the following theorem.

THEOREM 5.1. *Let T and $\mathbb{K}$ be as above, and let λ be the expansion constant satisfying* (4.1.1).

(i) *For all $\alpha > 0$ the pair of spaces $(\mathbb{A}^{2+\alpha}, \mathbb{A}^0)$ is 1-compatible with $(T, \mathbb{K})$.*

(ii) *For all $1 < \rho < \lambda$ there exists $\alpha > 0$ sufficiently small such that $(\mathbb{A}^{2-\alpha}, \mathbb{A}^0)$ is ρ-compatible with $(T, \mathbb{K})$.*

The path towards the proof of this theorem (presented in §5.3) leads us to perform what amounts to a *spectral analysis* of the formal derivative of the renormalization operator, which in turn calls for certain estimates on the geometry of the post-critical set of each map in the limit set of renormalization. We have the following explicit formula for the derivative $L_f = DT(f)$ of T at $f \in \mathbb{K}$:

$$\begin{aligned} DT(f)v = &\frac{1}{\lambda_f}\sum_{j=0}^{p-1} Df^j(f^{p-j}(\lambda_f x))v(f^{p-j-1}(\lambda_f x)) \\ &+\frac{1}{\lambda_f}[x(Tf)'(x) - Tf(x)]\sum_{j=0}^{p-1} Df^j(f^{p-j}(0))v(f^{p-j-1}(0)) \ , \end{aligned}$$

where as before $\lambda_f = f^p(0)$ for some positive integer $p = p(f, N)$. We observe that the operator L_f extends naturally to each of the spaces $\mathbb{A}^\gamma$ for $\gamma \geq 0$.

Properties **A1**, **A2** and **A3** of Definition 4.1 are easily verified in our setting. Property **A4** follows from a general result of Hölder spaces that can be proved via smoothing operators. Hence, the heart of the matter is verifying property **A5**. This is where the geometric scaling properties of the invariant Cantor set of a map in $\mathbb{K}$ become important – see §5.2. We follow Davie's observation that $L_f^{(m)}$ is a special sort of operator – what we call an L-operator – which is amenable to analysis. The verification of the fifth property (with $(\mathcal{B}, \mathcal{C}) = (\mathbb{A}^\gamma, \mathbb{A}^0)$) – presented in §5.3 – consists in controlling the norm of a certain positive linear operator $L_{f,\gamma}^{(m)} : \mathbb{A}^0 \to \mathbb{A}^0$ associated to $L_f^{(m)}$ (see Lemma 5.3). Using the bounded distortion properties of $f \in \mathbb{K}$ and the geometry of the invariant set of f, we show that the exponential growth rate of the C^0 norm of $L_{f,\gamma}^{(m)}$ is bounded by some $\mu < \lambda$ if $\gamma = 2 - \alpha$ with $\alpha > 0$ small enough and is bounded by some $\mu < 1$ if $\gamma = 2 + \alpha$ with $\alpha > 0$.

5.1. *Hölder norms and L-operators.* First we define what we mean by an L-operator, and to each such operator L we associate another operator L_γ, acting on continuous functions. Then, we use local Hölder estimates to control the norm of compositions $L_m \circ \cdots \circ L_2 \circ L_1$ of L-operators L_i by the norm of $(L_m \circ \cdots \circ L_2 \circ L_1)_\gamma$.

Definition 5.1. An L-operator is a *bounded linear operator* $L : C^\gamma(I) \to C^\gamma(I)$ that can be written in the form

$$Lv(x) = \sum_{i=1}^{n} \phi_i(x) v(\psi_i(x)),$$

where $\phi_i \in C^{\gamma_1}(I)$ and $\psi_i \in C^{\gamma_2}(I)$ are maps such that $\psi_i(I) \subset I$ for $i = 1, \ldots, n$, and where $\gamma_1 > 0$ and $\gamma_2 \geq 1$ are such that $0 < \gamma < \gamma_1, \gamma_2$.

Example 5.1. For all $f \in \mathbb{K}$ and all $i \geq 0$, the *formal derivative* $L_f = DT(T^i(f))$ is an L-operator.

An L-operator L as above yields a positive, bounded linear operator $L_\gamma : C^0(I) \to C^0(I)$ defined by

$$L_\gamma v(x) = \sum_{i=1}^{n} |\phi_i(x)| |D\psi_i(x)|^\gamma v(\psi_i(x)) \ .$$

A straightforward computation yields the following result.

Lemma 5.2. *If* $L_1, L_2 : C^\gamma(I) \to C^\gamma(I)$ *are L-operators, then* $(L_1 \circ L_2)_\gamma = L_{1,\gamma} \circ L_{2,\gamma}$.

We remind the reader that a function $\varphi : I \to I$ is α*-Hölder continuous*, for a fixed $0 < \alpha < 1$, if there is $c > 0$ such that $|\varphi(x) - \varphi(y)| \leq c|x - y|^\alpha$ for all $x, y \in I$. Let $C^\alpha(I)$ be the Banach space of all α-Hölder continuous real functions on I, with norm

$$\|\varphi\|_\alpha = \max\left\{ \|\varphi\|_0 \,,\, \sup_{x \neq y} \frac{|\varphi(x) - \varphi(y)|}{|x-y|^\alpha} \right\} .$$

Let $C^{k+\alpha}(I)$ be the Banach space of all real functions on I for which the kth derivative is α-Hölder continuous, with norm

$$\|\varphi\|_{k+\alpha} = \max\{\|\varphi\|_0, \|D^k\varphi\|_\alpha\}.$$

LEMMA 5.3. *Let $L_i : C^\gamma(I) \to C^\gamma(I)$ be a sequence of L-operators, and assume that there exist constants $\mu > 0$ and $C > 0$ such that for all n,*

$$\|(L_n \circ \cdots \circ L_2 \circ L_1)_\gamma\|_0 \leq C\mu^n \; . \tag{5.1.1}$$

Then for all $\rho > \mu$ and all $\varepsilon > 0$ there exist $m > 0$ and $K > 0$ such that for all $v \in C^\gamma(I)$,

$$\|L_m \circ \cdots \circ L_2 \circ L_1(v)\|_\gamma \leq \max\left\{\varepsilon\rho^m\|v\|_\gamma, K\|v\|_0\right\} \; .$$

To prove the above proposition, we will use local Hölder estimates for L-operators given in our next lemma. For each $\eta > 0$ and each $\varphi \in C^\alpha(I)$, we consider an associated semi-norm

$$\|\varphi\|_{\alpha,\eta} = \sup_{0<|x-y|<\eta} \frac{|\varphi(x) - \varphi(y)|}{|x-y|^\alpha} \; .$$

The corresponding semi-norm of $\varphi \in C^{k+\alpha}(I)$ for $k > 0$ is $\|\varphi\|_{k+\alpha,\eta} = \|\varphi^{(k)}\|_{\alpha,\eta}$.

LEMMA 5.4. *Let $L : C^\gamma(I) \to C^\gamma(I)$ be an L-operator as defined above.*

(i) *For every $\varepsilon > 0$, there exists $\eta > 0$ such that*

$$\|Lv\|_{\gamma,\eta} \leq (\varepsilon + \|L_\gamma\|_{C^0(I)})\|v\|_\gamma.$$

(ii) *For every $\varepsilon > 0$ and $0 < \xi < \gamma$, there is $\eta > 0$ such that*

$$\|Lv\|_{\xi,\eta} \leq \varepsilon\|v\|_\gamma.$$

Proof. See Lemmas 1 and 2 in [5]. □

Proof of Lemma 5.3. Choosing m such that $C\mu^m < \varepsilon\rho^m/8$, we have

$$M = \|L_{m,\gamma} \circ \cdots \circ L_{1,\gamma}\|_0 < \frac{\varepsilon\rho^m}{8} \; .$$

By Lemma 5.4, given $\varepsilon' = \varepsilon\rho^m/8$, there exists $\eta > 0$ such that

$$\|L_m \circ \cdots \circ L_1(v)\|_{\gamma,\eta} \leq \left(\varepsilon' + M\right)\|v\|_\gamma \leq \frac{\varepsilon\rho^m}{4}\|v\|_\gamma \; .$$

Taking $K = 8k!\|L_m \circ \cdots \circ L_1\|_0/\eta^\gamma$, writing $\gamma = k + \alpha$, where k is an integer and $0 < \alpha < 1$, and using interpolation of norms (see Lemma 4 in [5]), we deduce that

$$\begin{aligned}\|L_m \circ \cdots \circ L_1(v)\|_\gamma &\leq 4\max\left\{\|L_m \circ \cdots \circ L_1(v)\|_{\gamma,\eta}, \frac{2k!}{\eta^\gamma}\|L_m \circ \cdots \circ L_1(v)\|_0\right\}\\ &\leq \max\{\varepsilon\rho^m\|v\|_\gamma, K\|v\|_0\}\ . \qquad \square\end{aligned}$$

5.2. *Bounded geometry.* Our aim in this section is to prove two crucial propositions concerning the geometry of the invariant Cantor set of an infinitely renormalizable map in the limit set of renormalization. They are important not only in the proof of Theorem 5.1, but also in the proof (presented in §8) that the renormalization operator is robust (in the sense of §6).

We recall our notation. For each $f \in \mathbb{K}$, let $\mathcal{I}_f \subseteq I$ be the closure of the postcritical set of f (the Cantor attractor of f). For each $k \geq 0$, we can write

$$R^k f(x) = \frac{1}{\lambda_k} \cdot f^{p_k}(\lambda_k x)$$

where $p_k = \prod_{i=0}^{k-1} p(R^i f)$ and $\lambda_k = \prod_{i=0}^{k-1} \lambda(R^i f)$. Recall that the renormalization intervals $\Delta_{0,k} = [-|\lambda_k|, |\lambda_k|] \subset [-1,1]$, and $\Delta_{i,k} = f^i(\Delta_{0,k})$ for $i = 0, 1, \ldots, p_k - 1$. The collection $\mathbf{C}_k = \{\Delta_{0,k}, \ldots, \Delta_{p_k-1,k}\}$ consists of pairwise disjoint intervals. Moreover, $\bigcup\{\Delta : \Delta \in \mathbf{C}_{k+1}\} \subseteq \bigcup\{\Delta : \Delta \in \mathbf{C}_k\}$ for all $k \geq 0$ and we have

$$\mathcal{I}_f = \bigcap_{k=0}^{\infty} \bigcup_{i=0}^{p_k-1} \Delta_{i,k}\ .$$

In our first proposition, f is a normalized, symmetric quadratic unimodal map, infinitely renormalizable, sufficiently smooth (say C^2) for Sullivan's real bounds to be true for f. But there are no restrictions on the combinatorics. We shall use the general fact, due to Guckenheimer [12], that among those renormalization intervals at the kth level the one that contains the critical point of f (namely, $\Delta_{0,k}$) is the largest (up to multiplication by a constant). This can be seen as follows. First suppose that f is also S-unimodal. If $n > 0$ is such that $f^n(x)$ belongs to the interval with endpoints $-x, x$ but $f^j(x)$ does not, for all $1 \leq j < n$, then $|Df^n(x)| > 1$ – this uses the fact that f has negative Schwarzian. From this it follows that if $J \subset [-1,1]$ is an interval that does not contain the critical point, whose iterates $f^j(J)$ are pairwise disjoint for $0 \leq j \leq n$, such that $f^n(J)$ lies in the convex-hull of J union its symmetric while the previous iterates $f^j(J)$, $1 \leq j < n$, do not, then $|f^n(J)| > |J|$. Hence, if f is renormalizable, symmetric and S-unimodal then at each renormalization level the interval that contains the critical point is the largest. If we drop the negative Schwarzian hypothesis, the same is true up to a multiplicative constant. This is because every sufficiently deep renormalization of f already has negative Schwarzian derivative.

PROPOSITION 5.5. *For each $\alpha > 0$ there exist constants C_0 and $0 < \mu < 1$ such that*

$$\sum_{i=0}^{p_k-1} \frac{|\Delta_{i,k}|^{2+\alpha}}{|\Delta_{i+1,k}|} \leq C_0 \mu^k \ . \tag{5.2.1}$$

Proof. Let $\ell(\Delta_{i,k})$ be the level of $\Delta_{i,k}$, i.e., the largest integer j such that $\Delta_{i,k} \subseteq \Delta_{0,j} \setminus \Delta_{0,j+1}$. Let $d_{i,k}$ be the distance from $\Delta_{i,k}$ to zero (the critical point). Using that $\Delta_{i,k}$ has space around itself we see that for all $i \neq 0$ and all $x \in \Delta_{i,k}$ we have $d_{i,k} \leq |x| \leq K d_{i,k}$, where $K > 1$ is a constant that depends only on the real bounds. Hence $K^{-1} \leq |x|/|y| \leq K$ whenever $x, y \in \Delta_{i,k}$. These facts are implicitly used in the estimates below.

Now, we have $|\Delta_{i,k}|/|\Delta_{i+1,k}| = 1/|f'(x_{i,k})|$ for some $x_{i,k} \in \Delta_{i,k}$. Since the critical point is quadratic, we have $|f'(x_{i,k})| \geq C_1 |x_{i,k}|$, and so

$$\frac{|\Delta_{i,k}|}{|\Delta_{i+1,k}|} \leq \frac{1}{C_1 |x_{i,k}|} \ . \tag{5.2.2}$$

Therefore, for all $0 \leq j \leq k-1$,

$$\begin{aligned}
\sum_{\ell(\Delta_{i,k})=j} \frac{|\Delta_{i,k}|^2}{|\Delta_{i+1,k}|} \ &\leq \ C_1^{-1} \sum_{\ell(\Delta_{i,k})=j} \frac{|\Delta_{i,k}|}{|x_{i,k}|} \ \leq \ C_2 \sum_{\ell(\Delta_{i,k})=j} \int_{\Delta_{i,k}} \frac{dx}{|x|} \\
&\leq \ C_3 \int_{\Delta_{0,j} \setminus \Delta_{0,j+1}} \frac{dx}{|x|} \ \leq \ C_4 \log \frac{|\Delta_{0,j}|}{|\Delta_{0,j+1}|} \ .
\end{aligned}$$

With these estimates, and using the fact proved above that $|\Delta_{0,k}| \geq C_5 |\Delta_{i,k}|$ for all $0 \leq i \leq p_k - 1$, we see that

$$\begin{aligned}
\sum_{i=0}^{p_k-1} \frac{|\Delta_{i,k}|^{2+\alpha}}{|\Delta_{i+1,k}|} \ &\leq \ C_6 \max |\Delta_{i,k}|^\alpha \left(1 + \sum_{j=0}^{k-1} \log \frac{|\Delta_{0,j}|}{|\Delta_{0,j+1}|} \right) \\
&\leq C_6 |\Delta_{0,k}|^\alpha \left(1 + \log \frac{1}{|\Delta_{0,k}|} \right) \\
&\leq K_\alpha |\Delta_{0,k}|^{\alpha/2} \ ,
\end{aligned}$$

where K_α is a positive constant depending on α. This proves (5.2.1) because $|\Delta_{0,k}|$ decays exponentially with k, with uniform rate depending only on the real bounds. □

In addition to Proposition 5.5 — valid for maps with arbitrary combinatorial type — we shall need also an estimate that seems specific for maps with *bounded* combinatorial type, namely Proposition 5.8 below; first, a couple of lemmas.

For each $f \in \mathbb{K}$, let d_f be the infimum of all positive numbers s such that

$$\sum_{i=0}^{p_k-1} |\Delta_{j,k}|^s \to 0 \ \text{ as } \ k \to \infty \ .$$

It is possible to prove, using some thermodynamic formalism, that d_f agrees with the Hausdorff dimension of $\mathcal{I}_f$, but we will not need this fact. Let $0 < D < 1$ be the supremum of d_f as f ranges through $\mathbb{K}$.

LEMMA 5.6. *For each $s > D$ there exist $C_s > 0$ and $0 < \eta_s < 1$ such that for all $f \in \mathbb{K}$,*

$$\sum_{j=0}^{p_k-1} |\Delta_{j,k}|^s < C_s \eta_s^k \ .$$

Proof. Apply bounded geometry and the compactness of $\mathbb{K}$. □

Next, let us define

$$S_{j,k}(f;s) = \sum_{\ell(\Delta_{i,k})=j} |\Delta_{i,k}|^s \ ,$$

for $j = 0, 1, \ldots, k-1$, all $k \geq 0$, and all $f \in \mathbb{K}$.

LEMMA 5.7. *For each $s > D$ and each $f \in \mathbb{K}$ we have $S_{j,k}(f;s) \leq C_s \lambda_j^s \eta_s^{k-j}$, where $C_s > 0$ and $0 < \eta_s < 1$ are the constants of Lemma 5.6, and $\lambda_j = \lambda(f,j)$ is as in (3.1.1).*

Proof. Using renormalization, we see that $S_{j,k}(f;s) = \lambda_j^s S_{0,k-j}(R^j(f);s)$. From Lemma 5.6, we know that

$$S_{0,k-j}(R^j(f);s) \leq C_s \eta_s^{k-j} \ .$$

The result follows. □

PROPOSITION 5.8. *For each $\mu > 1$ close to one, there exist $0 < \alpha < 1-D$ close to zero and $C > 0$ such that for all $f \in \mathbb{K}$,*

$$\sum_{i=0}^{p_k-1} \frac{|\Delta_{i,k}|^{2-\alpha}}{|\Delta_{i+1,k}|} \leq C\mu^k \ .$$

Proof. Using (5.2.2) and the fact that $|x_{i,k}| \geq \lambda_{j+1}$ when $\ell(\Delta_{i,k}) = j$, we have

$$\begin{aligned} \sum_{i=0}^{p_k-1} \frac{|\Delta_{i,k}|^{2-\alpha}}{|\Delta_{i+1,k}|} &= \sum_{i=0}^{p_k-1} \frac{|\Delta_{i,k}|}{|\Delta_{i+1,k}|} |\Delta_{i,k}|^{1-\alpha} \\ &\leq C \left(|\Delta_{0,k}|^{-\alpha} + \sum_{j=0}^{k-1} \lambda_j^{-1} S_{j,k}(f;1-\alpha) \right) \ . \end{aligned}$$

If $1 - \alpha > D$ then, by Lemma 5.7 with $s = 1 - \alpha$,

$$\sum_{i=0}^{p_k-1} \frac{|\Delta_{i,k}|^{2-\alpha}}{|\Delta_{i+1,k}|} \leq C C_{1-\alpha} \sum_{j=0}^{k} \lambda_j^{-\alpha} \eta_{1-\alpha}^{k-j} \leq K_\alpha \mu_\alpha^k \ ,$$

where $K_\alpha > 0$ and $\mu_\alpha = \max\{\lambda(f,1)^{-\alpha} : f \in \mathbb{K}\}$ depend on α. But if α is small enough we will have $\mu_\alpha < \mu$, and this completes the proof. □

Remark 5.1. By a continuity argument and the real bounds, we can prove that Propositions 5.5 and 5.8 remain true for maps $\tilde{f} \in \mathbb{U}^4$ sufficiently close to $\mathbb{K}$. More precisely, for each $k > 0$ there exists $\varepsilon_k > 0$ such that for all $f \in \mathbb{K}$ and all $\tilde{f} \in \mathbb{U}^4$ with $\left\|\tilde{f} - f\right\|_{C^4(I)} < \varepsilon_k$, the map $\tilde{f}$ is k-times renormalizable, and the statements of both propositions hold for $\tilde{f}$. This will be used in §8.4 only for real analytic maps in an open neighborhood of $\mathbb{K}$ in $\mathbb{A}$.

5.3. *Spectral estimates.* In this section we prove Theorem 5.1. Fixing $f \in \mathbb{K}$ and considering the Banach space $\mathcal{A}$ given by Theorem 2.4, we recall that the Fréchet derivative $L_f = DT(f) : \mathcal{A} \to \mathcal{A}$ is given by formula (5.0.1). It is clear from that formula that L_f extends to a bounded linear operator $\hat{L}_f : \mathbb{A}^0 \to \mathbb{A}^0$, and moreover $\hat{L}_f(\mathbb{A}^s) \subseteq \mathbb{A}^s$ for all $s \geq 0$ (because f is analytic).

We want to verify the compatibility properties of Definition 4.1 for the spaces $\mathcal{B} = \mathbb{A}^s$ and $\mathcal{C} = \mathbb{A}^0$ when s is close to (but different from) 2. Properties **A1** and **A2** are clearly satisfied, while **A3** follows from Lemma 8.13. Property **A4** is a consequence of the following simple fact about Hölder spaces (see [15]).

LEMMA 5.9. *There exists $\Delta > 1$ such that $\mathbb{A} \cap \mathbb{A}^s(\Delta)$ is C^0-dense in $\mathbb{A}^s(1)$.*

Proof. By Theorem A.10 on page 43 of [15], there exist a family S_t, $t > 0$, of smoothing operators preserving even functions and $C \geq 1$ such that, for all $v \in \mathbb{A}^s(1)$, we have $\|S_t v\|_{C^s} \leq C$ and $\|v - S_t v\|_{C^0} \leq C t^s$. By the Stone-Weierstrass theorem, for all small $0 < \varepsilon < C$ there is a polynomial w_t with real coefficients and vanishing at zero such that $\|w_t - S_t v\|_{C^s} < \varepsilon$. Now let $v_t(x) = \frac{1}{2}(w_t(x) + w_t(-x))$, so that $v_t \in \mathbb{A}$ and we still have $\|v_t - S_t v\|_{C^s} < \varepsilon$. Then $\|v_t\|_{C^s} < \|S_t(v)\|_{C^s} + \varepsilon < 2C$ on one hand, while $\|v_t - v\|_{C^0} < \varepsilon + Ct^s$ on the other hand. For t small enough, this gives $\|v_t - v\|_{C^0} < 2\varepsilon$ with $v_t \in \mathbb{A}^s(2C)$. □

Hence all that remains is to check that property **A5** is satisfied. By Lemma 5.3, this will be the case provided we can control the C^0 norms of $\hat{L}_{f,s}^{(m)}$. We shall prove this now, with the help of Propositions 5.5 and 5.8.

Recall that for each $m \geq 1$ the operator $\hat{L}_f^{(m)}$ is an L-operator and its associated positive, bounded linear operator $\hat{L}_{f,s}^{(m)} : \mathbb{A}^0 \to \mathbb{A}^0$ is given by

(5.3.1)
$$\hat{L}_{f,s}^{(m)}(v) = \frac{1}{\lambda_k} \sum_{j=0}^{p_k-1} |Df^j(f^{p_k-j}(\lambda_k x))||\lambda_k Df^{p_k-j-1}(\lambda_k x)|^s v(f^{p_k-j-1}(\lambda_k x)) ,$$

where $k = mN$ (recall that $T = R^N$). Now we have the following fact coming from bounded geometry:

$$|Df^j(f^{p_k-j}(\lambda_k x))| \asymp \frac{|\Delta_{0,k}|}{|\Delta_{p_k-j,k}|}\,, \tag{5.3.2}$$

for all $0 \leq j \leq p_k - 1$. Since $|Df(\lambda_k x)| \leq C\lambda_k$ for some constant $C > 0$ independent of k and uniform in $f \in \mathbb{K}$, and $|\Delta_{0,k}| = 2\lambda_k$ we have

$$|Df^{p_k-j-1}(\lambda_k x)| \leq C|\Delta_{0,k}||Df^{p_k-j-2}(f(\lambda_k x))|\,.$$

Again, by bounded geometry, for all $0 \leq j \leq p_k - 2$

$$|Df^{p_k-j-2}(f(\lambda_k x))| \asymp \frac{|\Delta_{p_k-j-1,k}|}{|\Delta_{1,k}|}\,,$$

and so

$$|Df^{p_k-j-1}(\lambda_k x)| \leq C|\Delta_{0,k}|\frac{|\Delta_{p_k-j-1,k}|}{|\Delta_{1,k}|}\,. \tag{5.3.3}$$

Using (5.3.2) and (5.3.3) in (5.3.1), we see that

$$\|\hat{L}^{(m)}_{f,s}\| \;\leq\; \frac{C}{\lambda_k}\left(\frac{\lambda_k^s|\Delta_{0,k}|}{|\Delta_{1,k}|} + \sum_{j=0}^{p_k-2}\lambda_k^{2s}\frac{|\Delta_{0,k}|}{|\Delta_{p_k-j,k}|}\frac{|\Delta_{p_k-j-1,k}|^s}{|\Delta_{1,k}|^s}\right).$$

But $|\Delta_{0,k}| = 2\lambda_k$ and since the critical point of f is quadratic, $|\Delta_{1,k}| \asymp |\Delta_{0,k}|^2 \asymp \lambda_k^2$. Therefore, we arrive at

$$\|\hat{L}^{(m)}_{f,s}\| \;\leq\; C_1\sum_{j=0}^{p_k-1}\frac{|\Delta_{p_k-j-1,k}|^s}{|\Delta_{p_k-j,k}|}\,. \tag{5.3.4}$$

The proof of part (i) of Theorem 5.1 now follows from Proposition 5.5, while the proof of part (ii) is a consequence of Proposition 5.8. This ends the proof of Theorem 5.1.

6. The local stable manifold theorem

In this section we isolate those features of the renormalization operator that are essential for the promotion of “hyperbolicity” from the Banach space $\mathbb{A}$ of Theorem 2.4 to the space $\mathbb{U}^r$. This leads us to the definition of a robust operator (see §6.1). Such definition is necessarily rather technical, since it has to account for the fact that the renormalization operator is not Fréchet differentiable in $\mathbb{U}^r$. In particular a robust operator acts simultaneously on four different Banach spaces (corresponding in the case of renormalization to the space $\mathbb{A}$ given by Theorem 2.4, $\mathbb{A}^r$, $\mathbb{A}^s$ and $\mathbb{A}^0$, where $r > 1 + s$ and s is close to 2), and satisfies several properties. The major goal of this section is to prove a local stable manifold theorem for robust operators.

6.1. *Robust operators.* Before moving on to a precise definition of a robust operator, we give the following informal description. A robust operator acts on four Banach spaces $\mathcal{A} \subset \mathcal{B} \subset \mathcal{C} \subset \mathcal{D}$. In the smaller space $\mathcal{A}$ it acts smoothly and has a hyperbolic basic set $\mathbb{K}$. The pairs of spaces $(\mathcal{B}, \mathcal{D})$ and $(\mathcal{C}, \mathcal{D})$ are compatible with $(T, \mathbb{K})$, and in particular the invariant hyperbolic splitting for $\mathbb{K}$ in $\mathcal{A}$ extends to an invariant hyperbolic splitting for $\mathbb{K}$ in $\mathcal{B}$. Viewed as a map from $\mathcal{B}$ into $\mathcal{C}$, a robust operator is C^1. It also satisfies a uniform Gateaux differentiability condition in $\mathcal{C}$ for points and directions in $\mathcal{B}$. Finally, as an operator in $\mathcal{B}$, it is reasonably well-approximated by the extension of its derivative at a point of $\mathbb{K}$ in $\mathcal{A}$ to a bounded linear operator in $\mathcal{B}$. It will take us considerable effort (see §8) to verify that the renormalization operator indeed satisfies all these conditions.

Let $T : \mathcal{O} \to \mathcal{A}$ be a C^2 operator having a compact hyperbolic basic set $\mathbb{K}$ with the property that the unstable subspace of the DT-invariant splitting of the tangent space at each point of $\mathbb{K}$ is one-dimensional. By standard invariant manifold theory (see [14]), we know that for all $g \in \mathbb{K}$ the local unstable manifold $W^u(g)$ of T at g exists and is C^2. In particular, we can find a C^2 parametrization

$$t \mapsto u_g(t) \in W^u(g) \subseteq \mathcal{A}$$

varying continuously with g such that $\mathbf{u}_g = u_g'(0)$ is a unit vector. We define a C^2 function $t \mapsto \hat{\delta}_g(t)$ by

$$T(u_g(t)) \;=\; u_{T(g)}(\hat{\delta}_g(t)) \;.$$

This function also varies continuously with g and $\hat{\delta}_g(t) = \delta_g t + O(t^2)$ for some $\delta_g > 0$. Recall that by hyperbolicity of $\mathbb{K}$, there exist $C_0 > 0$ and $\lambda > 1$ such that for every $g \in \mathbb{K}$ and every $m \geq 1$,

$$\delta_{T^{m-1}(g)} \cdots \delta_g > C_0 \lambda^m \;. \qquad (6.1.1)$$

Definition 6.1. Let $\mathcal{A} \subseteq \mathcal{B} \subseteq \mathcal{C} \subseteq \mathcal{D}$ be Banach spaces, where each inclusion is a compact linear operator. Let $\mathcal{O}_\mathcal{A} \subseteq \mathcal{A}$ and $\mathcal{O}_\mathcal{B} \subseteq \mathcal{B}$ be open sets in their respective spaces such that $\mathcal{O}_\mathcal{A} \subseteq \mathcal{O}_\mathcal{B}$. Let $\mathbb{K} \subset \mathcal{O}_\mathcal{A}$ be a hyperbolic basic set of a C^2 operator $T : \mathcal{O}_\mathcal{A} \to \mathcal{A}$. We say that T is *robust* with respect to $(\mathcal{B}, \mathcal{C}, \mathcal{D})$ if it has an extension to an operator $T : \mathcal{O}_\mathcal{B} \to \mathcal{B}$ that satisfies the following conditions.

B1. The pair $(\mathcal{B}, \mathcal{D})$ is 1-*compatible* with T, while the pair $(\mathcal{C}, \mathcal{D})$ is $\rho_\mathcal{C}$-*compatible* with $(T, \mathbb{K})$ for some $\rho_\mathcal{C} < \lambda$ (where λ is as in (6.1.1)).

B2. For each $m > 0$, the interior $\mathcal{O}_\mathcal{B}^{(m)}$ of the set $\{f \in \mathcal{O}_\mathcal{B} : T^i(f) \in \mathcal{O}_\mathcal{B},\ \forall i < m\}$ contains $\mathbb{K}$, and $T^m : \mathcal{O}_\mathcal{B}^{(m)} \to \mathcal{C}$ is C^1 and its derivative is uniformly continuous in some neighbourhood of $\mathbb{K}$. Furthermore, for all $f \in \mathcal{A} \cap \mathcal{O}_{\mathcal{B}}^{(m)}$ the linear map

$$DT^m(f) : \mathcal{B} \to \mathcal{C}$$

extends to a continuous linear operator $L_m : \mathcal{D} \to \mathcal{D}$ that satisfies $L_m(\mathcal{X}) \subseteq \mathcal{X}$, for $\mathcal{X} = \mathcal{B}, \mathcal{C}$.

B3. For every m there exists $C_{m,1} > 1$ with the property that for each $g \in \mathbb{K}$ there is an open set $\mathcal{V}_g \subseteq \mathcal{O}_{\mathcal{B}}$ containing g such that for all $f \in \mathcal{V}_g$,

$$\|DT^m(f)\mathbf{u}_g - DT^m(g)\mathbf{u}_g\|_{\mathcal{C}} \leq C_{m,1}\|f - g\|_{\mathcal{B}} \ .$$

B4. There exist $C_1 > 1$ and $\rho > 1$ with the property that for each $g \in \mathbb{K}$ there is an open set $\mathcal{V}_g \subseteq \mathcal{O}_{\mathcal{B}}$ containing g such that for all $f_1, f_2 \in \mathcal{V}_g$,

$$\|T(f_1) - T(f_2) - DT(f_2)(f_1 - f_2)\|_{\mathcal{C}} \leq C_1\|f_1 - f_2\|_{\mathcal{C}}^{\rho} \ .$$

B5. For all $m > 0$, there exists $C_{m,2} > 0$, and there exists $\nu_m > 0$ such that for all $g \in \mathbb{K}$ and for all $f \in \mathcal{B}$ with $\|f - g\|_{\mathcal{B}} < \nu_m$,

$$\|DT^m(f) - DT^m(g)\|_{\mathcal{C}} \leq C_{m,2}\lambda^m \ .$$

Moreover, there exists $m_0 > 0$ such that for all $m > m_0$, $C_{m,2} < C_0/8$ (where C_0 and λ are as in (6.1.1)).

B6. For all $m > 0$, there exists $C_{m,3} > 0$ such that for all $g \in \mathbb{K}$, for all $f \in \mathcal{A}$ with $\|f - g\|_{\mathcal{A}} < \nu_m$ and for all $v \in \mathcal{B}$ with $\|v\|_{\mathcal{B}} < \nu_m$,

$$\|T^m(f + v) - T^m(f) - DT^m(g)v\|_{\mathcal{B}} \leq C_{m,3}\|v\|_{\mathcal{B}} \ .$$

Moreover, there exists $m_0 > 0$ such that for all $m > m_0$, $C_{m,3} < 1/4$.

Example 6.1. As one might expect, the main example of a robust operator is provided by renormalization. We know from Theorem 2.4 that the renormalization operator $T = R^N : \mathbb{O} \to \mathbb{A}$ is hyperbolic over $\mathbb{K}$. We also know that this renormalization operator is well-defined as a map from an open set of $\mathbb{U}^\gamma$ containing $\mathbb{K}$ into $\mathbb{U}^\gamma \subset 1 + \mathbb{A}^\gamma \cong \mathbb{A}^\gamma$ for all $\gamma \geq 2$. We will show in Section 8 that T is robust with respect to the spaces $\mathcal{A} = \mathbb{A}$, $\mathcal{B} = \mathbb{A}^r$, $\mathcal{C} = \mathbb{A}^s$ and $\mathcal{D} = \mathbb{A}^0$ whenever $s < 2$ is close to 2 and $r > s + 1$ is not an integer.

6.2. *Stable manifolds for robust operators.* We can now formulate a general local stable manifold theorem for robust operators.

THEOREM 6.1. *Let $T : \mathcal{O}_{\mathcal{A}} \to \mathcal{A}$ be a C^k with $k \geq 2$ (or real analytic) hyperbolic operator over $\mathbb{K} \subset \mathcal{O}_{\mathcal{A}}$, and be robust with respect to $(\mathcal{B}, \mathcal{C}, \mathcal{D})$. Then conditions* (i), (ii), (iii) *and* (iv) *of Theorem* 2.3 *hold true for the operator T acting on $\mathcal{B}$. The local unstable manifolds are C^k with $k \geq 2$ (or real analytic) curves, and the local stable manifolds are of class C^1 and form a C^0 lamination.*

The proof of this theorem will occupy the rest of Section 6. In the end, the theorem will follow by putting together Corollary 4.2, Proposition 6.13 and Theorem 6.15.

6.3. *Uniform bounds.* Before proceeding we prove the following simple bounds that we will use quite often.

LEMMA 6.2. *There exist $\mu_0 > 0$ and $1 < \lambda < M$ such that for all $g \in \mathbb{K}$ and all $t \in \mathbb{R}$ with $|t| < \mu_0$, $u_g(t)$ and $\hat{\delta}_g(t)$ are well-defined and*

(i) $M^{-1}\lambda^n < \delta_{T^{n-1}(g)} \cdots \delta_g < M^n$ *and* $\left|\hat{\delta}_g(t)\right| < M|t|$;

(ii) $M^{-1} < \|\mathbf{u}_g\|_{\mathcal{B}} < M$ *and* $M^{-1} < \|\mathbf{u}_g\|_{\mathcal{C}} < M$;

(iii) $M^{-1} < \|\sigma_g\|_{\mathcal{B}} < M$ *and* $M^{-1} < \|\sigma_g\|_{\mathcal{C}} < M$;

(iv) $M^{-1}|t| < \|u_g(t) - g\|_{\mathcal{C}} < M|t|$ *and* $M^{-1}|t| < \|u_g(t) - g\|_{\mathcal{B}} < M|t|$;

(v) $|\sigma_g(u_g(t) - g)| > \frac{1}{2}|t|$.

Proof. (i) By Definition 2.1 and (6.1.1), there exist $1 < \lambda < M_1$ such that for all $g \in \mathbb{K}$ and all $n \geq 1$ we have $M_1^{-1}\lambda^n < \delta_{T^{n-1}(g)} \cdots \delta_g < M_1^n$ and also $\left|\hat{\delta}_g(t)\right| < M_1|t|$ for all $|t| \leq \mu_1$ (where $\mu_1 > 0$ is a uniform constant).

(ii) For $\mathcal{X}$ equal to $\mathcal{B}$ and $\mathcal{C}$, we have that $g \mapsto \mathbf{u}_g$ as a map $\mathbb{K} \to \mathcal{X}$ is continuous and does not vanish. Hence, by compactness of $\mathbb{K}$ there is $M_2 > 1$ such that $M_2^{-1} < \|\mathbf{u}_g\|_{\mathcal{X}} < M_2$.

(iii) Since $\sigma_g(\mathbf{u}_g) = 1$ and by property **B1** in Definition 6.1, the functional σ_g extends continuously to $\mathcal{X}$ and there is $M_3 > 1$ such that $M_3^{-1} < \|\sigma_g\|_{\mathcal{X}} < M_3$.

(iv) Since $t \to u_g(t)$ as a map $\mathbb{R} \to \mathcal{X}$ is C^1 and varies continuously with $g \in \mathbb{K}$, there is $M_4 > 0$ and $\mu_2 > 0$ such that $\|u_g(t) - u_g(s)\|_{\mathcal{X}} \leq M_4|t - s|$ for all $g \in \mathbb{K}$ and all t, s with $|t| < \mu_2$ and $|s| < \mu_2$. Moreover, since

$$\left.\frac{d}{dt}u_g(t)\right|_{t=0} = \mathbf{u}_g \neq 0$$

there exists $M_5 > 0$ and $\mu_3 > 0$ such that $|t - s| \leq M_5\|u_g(t) - u_g(s)\|_{\mathcal{X}}$ for all $g \in \mathbb{K}$ and all $|t| < \mu_3$. Hence (iv) follows by taking $s = 0$ and noting that $u_g(0) = g$.

(v) This follows from (iv) and the fact that $\sigma_g(\mathbf{u}_g) = 1$. □

6.4. *Contraction towards the unstable manifolds.* The one-dimensional unstable manifolds of T in $\mathcal{A}$ are embedded in $\mathcal{B}$, and remain invariant. The first important estimate given by the following lemma shows that in $\mathcal{B}$ the operator T contracts towards such manifolds. Therefore, if T is to have unstable manifolds in $\mathcal{B}$, these have to coincide with unstable manifolds in $\mathcal{A}$. In what follows, we fix $g \in \mathbb{K}$ and for simplicity of notation we write

$$\sigma_i = \sigma_{T^i(g)} \ , \ u_i = u_{T^i(g)} \ , \ \mathbf{u}_i = \mathbf{u}_{T^i(g)} \ ,$$

and

$$\delta_i^m = \prod_{j=i}^{m-1} \delta_{T^j(g)}, \ \hat{\delta}_i^m = \hat{\delta}_{T^{m-1}(g)} \circ \cdots \circ \hat{\delta}_{T^i(g)} \ .$$

Set $\mu_0 > 0$ as in Lemma 6.2.

LEMMA 6.3. *For every $m > 0$ there exist $0 < \eta_m < \mu_0$ and $B_m > 0$ such that for every $g \in \mathbb{K}$ and every $v \in \mathcal{B}$ with $\|v\|_{\mathcal{B}} < \eta_m$ and $t \in \mathbb{R}$ with $|t| < \eta_m$, $\hat{\delta}_0^m(t + \sigma_0(v)) < \mu_0$, $u_0(t) + v \in \mathcal{O}_{\mathcal{B}}^{(m)}$ and*

$$\left\| T^m(u_0(t) + v) - u_m\left(\hat{\delta}_0^m(t + \sigma_0(v))\right)\right\|_{\mathcal{B}} < B_m \|v\|_{\mathcal{B}} \ .$$

Furthermore, there is $m_1 > m_0$ (where m_0 is as in **B6***) such that for all $m > m_1$, $B_m < 1/2$.*

Proof. We prove the second inequality only. The first is proven in the same way. By property **B6** in Definition 6.1, there is $m_0 > 0$ such that for all $m > m_0$, all v with $\|v\|_{\mathcal{B}} < \nu_m$, and all $t \in \mathbb{R}$ with $|t| < \nu_m$,

$$\|T^m(u_0(t) + v) - T^m(u_0(t)) - DT^m(g)v\|_{\mathcal{B}} \leq \frac{1}{4}\|v\|_{\mathcal{B}} \ . \tag{6.4.1}$$

By property **B1** in Definition 6.1, $(\mathcal{B}, \mathcal{D})$ is 1-compatible with $(T, \mathbb{K})$. Hence, by Corollary 4.2, there exists $m_1 > m_0$ such that for all $m > m_1$ we have

$$\|DT^m(g)v - \delta_0^m \sigma_0(v)\mathbf{u}_m\|_{\mathcal{B}} \leq \frac{1}{8}\|v\|_{\mathcal{B}} \ . \tag{6.4.2}$$

Putting (6.4.1) and (6.4.2) together we get

$$\|T^m(u_0(t) + v) - T^m(u_0(t)) - \delta_0^m \sigma_0(v)\mathbf{u}_m\|_{\mathcal{B}} \leq \frac{3}{8}\|v\|_{\mathcal{B}} \ . \tag{6.4.3}$$

Now, we know that $T^m(u_0(t)) = u_m(\hat{\delta}_0^m(t))$ and $t \to u_m \circ \hat{\delta}_0^m(t)$ is C^2. Hence,

$$\begin{aligned} \Big\| u_m \circ \hat{\delta}_0^m(t + \sigma_0(v)) - u_m \circ \hat{\delta}_0^m(t) &- \delta_0^m \sigma_0(v)\mathbf{u}_m\|_{\mathcal{B}} \\ &\leq c_1\left((\sigma_0(v))^2 + |t|\sigma_0(v)\right) \\ &\leq c_2\left(\|v\|_{\mathcal{B}} + |t|\right)\|v\|_{\mathcal{B}} \ . \end{aligned} \tag{6.4.4}$$

Therefore, choosing $\eta_m < \nu_m$ so small that $C_2\eta_m < 1/16$ and putting 6.4.3 and 6.4.4 together, we see that if $|t| < \eta_m$ and $\|v\|_{\mathcal{B}} < \eta_m$ then

$$\left\| T^m(u_0(t) + v) - u_m \circ \hat{\delta}_0^m(t + \sigma_0(v))\right\|_{\mathcal{B}} \leq \frac{1}{2}\|v\|_{\mathcal{B}}$$

as desired. □

Lemma 6.4. *Let $m_1 > 0$ be as in Lemma 6.3. For all $m > m_1$ there exist small constants $0 < \varepsilon_2 < \varepsilon_1 < \varepsilon_0$ such that the following holds for every $\varepsilon < \varepsilon_2$. For every $g \in \mathbb{K}$ and every $v \in \mathcal{B}$ with $\|v\|_{\mathcal{B}} < \varepsilon$, the recursive scheme given by $f_0 = g + v$, $t_0 = 0$, $v_0 = v$ and*

$$\begin{aligned} f_{k+1} &= T^m(f_k), \\ t_{k+1} &= \hat{\delta}_{km}^{(k+1)m}(t_k + \sigma_{km}(v_k)), \\ v_{k+1} &= f_{k+1} - u_{(k+1)m}(t_{k+1}) \end{aligned} \tag{6.4.5}$$

is well-defined for all $k = 0, \ldots, k_0 - 1$ where

$$k_0 = k_0(g, f_0) = \min\{j : |t_j| \geq \varepsilon_1\}.$$

For all $k \leq k_0$,

$$\begin{aligned} \left\| T^{km}(g+v) - u_{km}(t_k) \right\|_{\mathcal{B}} &< 2^{-k}\|v\|_{\mathcal{B}} \quad \textit{and} \\ \left\| T^{km}(g+v) - T^{km}(g) \right\|_{\mathcal{B}} &< \varepsilon_0/M_0 \ , \end{aligned} \tag{6.4.6}$$

where $M_0 = M^{m+2} + B_1 + \cdots + B_m$, M is as in Lemma 6.2 and $B_1, \ldots, B_m$ are as in Lemma 6.3. Furthermore,

(i) $\varepsilon_1 \leq |t_{k_0}| < \varepsilon_0$;

(ii) $\left\|T^{k_0 m}(g+v) - T^{k_0 m}(g)\right\|_{\mathcal{B}} > \varepsilon_2$;

(iii) $\left|\sigma_{k_0 m}\left(T^{k_0 m}(g+v) - T^{k_0 m}(g)\right)\right| > \varepsilon_2$;

(iv) $\left\|T^{km+i}(g+v) - T^{km+i}(g)\right\|_{\mathcal{B}} < M_0 \left\|T^{km}(g+v) - T^{km}(g)\right\|_{\mathcal{B}}$ *(which is less than ε_0) for all $k \leq k_0$ and all $i = 0, \ldots, m$.*

Proof. For every $g \in \mathbb{K}$, let $\mathcal{B}(g, \varepsilon)$ be the open ball in $\mathcal{B}$ of radius ε centered at g. Let us fix $m > m_1$ and choose $\varepsilon_0 < \min\{\mu_0, \eta_1, \ldots, \eta_m\}$ such that all the properties **B1** to **B6** of Definition 6.1 are satisfied in $\bigcup_{g\in\mathbb{K}} \mathcal{B}(g, \varepsilon_0) \subset \mathcal{O}_{\mathcal{B}}^{(m)}$, where μ_0 is as in Lemma 6.2 and $\eta_1, \ldots, \eta_m$ are as in Lemma 6.3. Since $m > m_1$, we have that $B_m < 1/2$ where B_m is as in Lemma 6.3. Let us take $M > 1$ as in Lemma 6.2. We choose $0 < \varepsilon_2 < \varepsilon_1 < \varepsilon_0 < \mu$ such that

$$\begin{aligned} \varepsilon_1 &< \varepsilon_0/\left(3M_0 M^{m+2}\right) \ , \\ \varepsilon_2 &< \varepsilon_1/(2+2M) \ . \end{aligned} \tag{6.4.7}$$

Now we work by induction on k. Let us assume that f_k, t_k, and v_k have been defined so that (6.4.6) holds. Hence $\left\|f_k - T^{mk}(g)\right\|_{\mathcal{B}} \leq \varepsilon_0 \leq \mu$, and so $f_k \in \mathcal{O}_{\mathcal{B}}^{(m)}$ and $f_{k+1} = T^m(f_k)$ is well-defined. Since $|t_k| \leq \varepsilon_1$ and $2M^{m+1}\varepsilon_1 \leq \varepsilon_0 \leq \mu$, by Lemma 6.2 and (6.2), and by (6.4.5) and (6.4.7), we have that t_{k+1} is well-defined and

$$\begin{aligned} |t_{k+1}| &= \left|\hat{\delta}_{km}^{(k+1)m}(t_k + \sigma_{km}(v_k))\right| \leq M^m(|t_k| + |\sigma_{km}(v_k)|) \\ &\leq M^m\left(\varepsilon_1 + M\frac{\varepsilon}{2^k}\right) < 2M^{m+1}\varepsilon_1 < \varepsilon_0 \ . \end{aligned} \tag{6.4.8}$$

Thus, by Lemma 6.2, $u_{(k+1)m}(t_{k+1})$ and $v_{k+1} = f_{k+1} - u_{(k+1)m}(t_{k+1})$ are also well-defined. By Lemma 6.3 and by (6.4.5), we get

$$\begin{aligned}
(6.4.9)\quad \|v_{k+1}\|_{\mathcal{B}} &= \left\|T^m(f_k) - u_{(k+1)m}(t_{k+1})\right\|_{\mathcal{B}} \\
&= \left\|T^m(v_k + u_{km}(t_k)) - u_{(k+1)m}\left(\hat{\delta}_{km}^{(k+1)m}(t_k + \sigma_{km}(v_k))\right)\right\|_{\mathcal{B}} \\
&\leq 2^{-1}\|v_k\|_{\mathcal{B}} \leq 2^{k+1}\|v_0\|_{\mathcal{B}} \ .
\end{aligned}$$

Now, let us estimate $\left\|f_{k+1} - T^{(k+1)m}(g)\right\|_{\mathcal{B}}$. From (6.4.5) and (6.4.9), we get

$$(6.4.10)\qquad \left\|f_{k+1} - u_{(k+1)m}(t_{k+1})\right\|_{\mathcal{B}} \leq \|v_{k+1}\|_{\mathcal{B}} \leq \frac{\varepsilon}{2^k} \ .$$

From Lemma 6.2 and by (6.4.8), we obtain

$$(6.4.11)\qquad \left\|u_{(k+1)m}(t_{k+1}) - T^{(k+1)m}(g)\right\|_{\mathcal{B}} \leq M\,|t_{k+1}| \leq 2M^{m+2}\varepsilon_1 \ .$$

Thus, by (6.4.7), (6.4.10) and (6.4.11),

$$\begin{aligned}
\left\|f_{k+1} - T^{(k+1)m}(g)\right\|_{\mathcal{B}} &\leq \left\|f_{k+1} - u_{(k+1)m}(t_{k+1})\right\|_{\mathcal{B}} \\
&\quad + \left\|u_{(k+1)m}(t_{k+1}) - T^{(k+1)m}(g)\right\|_{\mathcal{B}} \\
&\leq \frac{\varepsilon}{2^k} + 2M^{m+2}\varepsilon_1 \\
&\leq 3M^{m+2}\varepsilon_1 < \varepsilon_0/M_0 \ .
\end{aligned}$$

This completes the induction.

Now, we must prove (i), (ii), (iii) and (iv). Property (i) follows from (6.4.8). Let us prove (ii). By property (i) and Lemma 6.2,

$$\left\|u_{k_0m}(t_{k_0}) - T^{k_0m}(g)\right\|_{\mathcal{B}} \geq M^{-1}\varepsilon_1 \ .$$

By (6.4.9), we get $\|v_{k_0}\|_{\mathcal{B}} \leq \varepsilon/2^{k_0}$. Thus, by (6.4.7), we obtain

$$\begin{aligned}
\left\|T^{k_0m}(g+v) - T^{k_0m}(g)\right\|_{\mathcal{B}} &= \left\|u_{k_0m}(t_{k_0}) + v_{k_0} - T^{k_0m}(g)\right\|_{\mathcal{B}} \\
&\geq \left|\left\|u_{k_0m}(t_{k_0}) - T^{k_0m}(g)\right\|_{\mathcal{B}} - \|v_{k_0}\|_{\mathcal{B}}\right| \\
&\geq M^{-1}\varepsilon_1 - \frac{\varepsilon}{2^{k_0}} \\
&\geq \varepsilon_2 \ .
\end{aligned}$$

Let us prove (iii). By property (i) and Lemma 6.2,

$$\left|\sigma_{k_0m}(u_{k_0m}(t_{k_0} - T^{k_0m}g))\right| \geq \varepsilon_1/2 \ .$$

Using Lemma 6.2 yet again and (6.4.9), we have $|\sigma_{k_0m}(v_{k_0})| \le M\varepsilon/2^{k_0}$. Thus, by (6.4.7),

$$\begin{aligned}\left|\sigma_{k_0m}\left(T^{k_0m}(g+v)-T^{k_0m}g\right)\right| &\ge \big|\, |\sigma_{k_0m}(u_{k_0m}(t_{k_0}))| - |\sigma_{k_0m}(v_{k_0})| \,\big| \\ &\ge \frac{1}{2}\varepsilon_1 - M\frac{\varepsilon_2}{2^{k_0}} \\ &\ge \varepsilon_2 \ .\end{aligned}$$

Finally, we prove (iv). Fix $0 \le k \le k_0$ and $0 \le i \le m$. Setting $w_k = T^{km}(f) - T^{km}(g)$ we have by (6.4.6) that $\|w_k\|_{\mathcal{B}} < \varepsilon_0/M_0 < \eta_i$ where η_i is as in Lemma 6.3. Hence $T^{km}(g) + w_k \in \mathcal{O}_{\mathcal{B}}^{(i)}$ and by Lemma 6.3,

$$\left\|T^i(T^{km}(g)+w_k) - u_{km+i}\left(\hat{\delta}_{km}^{km+i}(\sigma_{km}(w_k))\right)\right\|_{\mathcal{B}} \le B_i\|w_k\|_{\mathcal{B}} \ .$$

On the other hand, by Lemma 6.2, we have

$$\left\|u_{km+i}\left(\hat{\delta}_{km}^{km+i}(\sigma_{km}(w_k))\right) - T^{km+i}(g)\right\|_{\mathcal{B}} \le M^{m+2}\|w_k\|_{\mathcal{B}} \ .$$

Therefore,

$$\begin{aligned}\|T^{km+i}(f) - T^{km+i}(g)\|_{\mathcal{B}} &\le \left\|T^i(T^{km}(g)+w_k) - u_{km+i}\left(\hat{\delta}_{km}^{km+i}(\sigma_{km}(w_k))\right)\right\|_{\mathcal{B}} \\ &\quad + \left\|u_j\left(\hat{\delta}_{km}^{km+i}(\sigma_{km}(w_k))\right) - T^{km+i}(g)\right\|_{\mathcal{B}} \\ &\le (B_i + M^{m+2})\|w_k\|_{\mathcal{B}} \le \varepsilon_0 \ ,\end{aligned}$$

which ends the proof. □

6.5. *Local stable sets.* Let us now consider the local stable set $W_\varepsilon^s(g)$ of T at g in $\mathcal{B}$ which consists of all points $f \in \mathcal{B}(g,\varepsilon)$ such that for all $n > 0$, we have $T^n(f) \in \mathcal{B}(T^n(g),\varepsilon)$ and

$$\|T^n(f) - T^n(g)\|_{\mathcal{B}} \to 0 \text{ when } n \to \infty \ .$$

Our aim in this section is to give a finite characterization of $W_\varepsilon^s(g)$ and prove that T contracts in the $\mathcal{B}$-norm exponentially along $W_\varepsilon^s(g)$. This is done in Lemma 6.5 below (see also Remark 6.1).

From now on in this section, we let m_1 and $\varepsilon_0 > \varepsilon_1 > \varepsilon_2 > \varepsilon$ be as in Lemma 6.4. For all sufficiently small $0 < \varepsilon < \varepsilon_2$ and for all $f \in \mathcal{B}(g,\varepsilon)$, we let $k_0 = k_0(g,f)$ and $t_k = t_k(g,f)$ for $k = 0,\ldots,k_0$ be as in Lemma 6.4. We write $\mathcal{B}(g,\varepsilon) = V_\varepsilon^-(g) \cup V_\varepsilon^0(g) \cup V_\varepsilon^+(g)$ where

$$\begin{aligned}V_\varepsilon^-(g) &= \{f \in \mathcal{B}(g,\varepsilon) : -\varepsilon_0 < t_{k_0}(g,f) < -\varepsilon_1\} \ , \\ V_\varepsilon^+(g) &= \{f \in \mathcal{B}(g,\varepsilon) : \varepsilon_1 < t_{k_0}(g,f) < \varepsilon_0\} \ , \\ V_\varepsilon^0(g) &= \mathcal{B}(g,\varepsilon) \setminus \left(V_\varepsilon^-(g) \cup V_\varepsilon^+(g)\right) \ .\end{aligned}$$

LEMMA 6.5. *There exist an integer m and a positive constant C_2 with the following properties. For all $\varepsilon > 0$ sufficiently small and for all $g \in \mathbb{K}$, the sets $V_\varepsilon^-(g)$ and $V_\varepsilon^+(g)$ are open subsets of $\mathcal{B}(g, \varepsilon)$ (and so $V_\varepsilon^0(g)$ is relatively closed in $\mathcal{B}(g, \varepsilon)$), and for all $f \in V_\varepsilon^0(g)$*

$$\left\|T^j(f) - T^j(g)\right\|_{\mathcal{B}} \leq \varepsilon C_2 2^{-j/m} \ . \tag{6.5.1}$$

Furthermore, the local stable set $W_\varepsilon^s(g)$ is a relatively open subset of $V_\varepsilon^0(g)$ and

(6.5.2)

$$W_\varepsilon^s(g) = \left\{f \in V_\varepsilon^0(g) : \left\|T^j(f) - T^j(g)\right\|_{\mathcal{B}} < \varepsilon, \text{ for all } 0 \leq j \leq m \log C_2 / \log 2\right\}$$

Proof. The first assertion is a consequence of the definitions of $V_\varepsilon^-(g)$ and $V_\varepsilon^+(g)$ and Lemma 6.4. It follows from property (i) of Lemma 6.4 that

$$V_\varepsilon^0(g) = \{f \in \mathcal{B}(g, \varepsilon) : |t_k(g, f)| < \varepsilon_1, \text{ for all } k \geq 0\} \ .$$

It also follows from property (ii) of Lemma 6.4 that if $f \in \mathcal{B}(g, \varepsilon)$ and $|t_{k_0}(g, f)| \geq \varepsilon_1$ then $\left\|T^{k_0 m} f - T^{k_0 m} g\right\|_{\mathcal{B}} > \varepsilon$ where $k_0 = k_0(g, f)$. This shows that $W_\varepsilon^s(g) \subset V_\varepsilon^0(g)$, and therefore (6.5.1) implies (6.5.2). Furthermore, $W_\varepsilon^s(g)$ is a relatively open subset of $V_\varepsilon^0(g)$.

It remains to show that if $f \in V_\varepsilon^0(g)$ then (6.5.1) holds. Set $1 < \lambda < M$ as in Lemma 6.2. Since $\beta > 2$, by Lemma 6.2, there is m large enough such that $\delta_{km}^{(k+1)m} \geq M^{-1}\lambda^m > \beta > 2$ for every $k \geq 0$. By Lemma 6.4, for all $k \geq 0$, we know that $t_k = t_k(g, f)$ and $v_k = v_k(g, f)$ are well-defined, and satisfy $|t_k| < \varepsilon_1$ and $\|v_k\|_{\mathcal{B}} \leq \varepsilon 2^{-k}$. Furthermore, $t_{k+1} = \hat{\delta}_{km}^{(k+1)m}(t_k + \sigma_{km}(v_k))$. Since $\delta_{km}^{(k+1)m}$ is C^2 and $\|\sigma_{km}\|_{\mathcal{B}} < M$ (see Lemma 6.2), there is $c_0 > 1$ so that

$$\begin{aligned}\left|t_{k+1} - \delta_{km}^{(k+1)m} t_k\right| &\leq \left|\hat{\delta}_{km}^{(k+1)m}(t_k + \sigma_{km}(v_k)) - \hat{\delta}_{km}^{(k+1)m}(t_k)\right| \\ &\quad + \left|\hat{\delta}_{km}^{(k+1)m}(t_k) - \delta_{km}^{(k+1)m} t_k\right| \\ &\leq c_0\left(\|v_k\|_{\mathcal{B}} + |t_k|^2\right) \\ &\leq c_0\left(\varepsilon 2^{-k} + |t_k|^2\right) \ .\end{aligned}$$

Hence (6.5.3) gives

$$|t_k| \leq \varepsilon c_0 \beta^{-1} 2^{-k} + \beta^{-1}|t_{k+1}| + c_0\beta^{-1}|t_k|^2 \ .$$

Taking ε (in Lemma 6.4) so small that $c_0\beta^{-1}\varepsilon < 1/2$, we get

$$|t_k| \leq 2\left(|t_k| - c_0\beta^{-1}|t_k|^2\right) \leq \varepsilon 2 c_0 \beta^{-1} 2^{-k} + 2\beta^{-1}|t_{k+1}| \ , \tag{6.5.3}$$

for all $k \geq 0$. Since $2\beta^{-1} < 1$, using induction in (6.5.3) and the fact that t_k is bounded, we get $|t_k| \leq \varepsilon c_1 2^{-k}$ with $c_1 = 2c_0\beta^{-1}/\left(1 - 2\beta^{-1}\right)$ for all $k \geq 0$. Now this estimate together with Lemma 6.2 gives

$$\|u_{km}(t_k) - T^{km}(g)\|_{\mathcal{B}} \leq M|t_k| \leq \varepsilon c_1 M 2^{-k} \ .$$

Hence, using Lemma 6.4 again, we get

$$\begin{aligned}\left\|T^{km}(f) - T^{km}(g)\right\|_{\mathcal{B}} &\leq \|v_k\|_{\mathcal{B}} + \left\|u_{km}(t_k) - T^{km}(g)\right\|_{\mathcal{B}} \\ &\leq \varepsilon 2^{-k} + \varepsilon c_1 M 2^{-k} = \varepsilon c_2 2^{-k} \ .\end{aligned}$$

Therefore, by (iv) in Lemma 6.4, for all $i \in \{1, \ldots, m-1\}$,

$$\left\|T^{km+i}(f) - T^{km+i}(g)\right\|_{\mathcal{B}} \leq M_0 \left\|T^{km}(f) - T^{km}(g)\right\|_{\mathcal{B}} \leq \varepsilon c_3 2^{-k} \ ,$$

which ends the proof. □

Remark 6.1. Note that since the constant C_2 is uniform (independent of ε) in the above lemma, inequality (6.5.1) can be improved to

$$\left\|T^j(f) - T^j(g)\right\|_{\mathcal{B}} \leq C' 2^{-j/m} \left\|f - g\right\|_{\mathcal{B}} \ ,$$

where $C' = 2C_2$. Therefore, we have exponential contraction in $\mathcal{B}$ (along the local stable sets) in the strong sense.

6.6. *Tangent spaces.* Our next goal is to show that $V_\varepsilon^0(g)$ is a C^1 manifold provided ε is sufficiently small. The first step towards this goal is to find the natural candidate for the tangent space at every point $f \in V_\varepsilon^0(g)$. This will be accomplished in Lemma 6.7 below. The proof will require the following elementary bootstrapping result.

LEMMA 6.6. *Let (a_n) be a sequence of real numbers such that, for some $c_0 > 0$ and all $n \geq 1$,*

$$|a_{n+1}| \leq \frac{1}{4}|a_n| + \frac{c_0}{2^n}\sum_{j=1}^{n-1}|a_j| \ . \tag{6.6.1}$$

Then $|a_n| \leq c_1 2^{-n}$ for some $c_1 > 0$ and all $n \geq 1$.

Proof. We may assume that $c_0 \geq 1$. Let $n_0 > 0$ be such that $c_0 n_0 / 2^{n_0} < 1/2$, and set $b = \max_{1 \leq j \leq n_0}\{|a_j|\}$. Then we see by induction from (6.6.1) that $|a_n| \leq b$ for all $n \geq 1$, and so

$$\begin{aligned}|a_{n+1}| &\leq \frac{1}{4}|a_n| + \frac{nbc_0}{2^n} \\ &\leq \frac{1}{4}|a_n| + bc_0\left(\frac{3}{4}\right)^n \ .\end{aligned}$$

By induction, this yields $|a_n| \leq (2bc_0)(\frac{3}{4})^n$ for all $n \geq 1$. Hence $\sum_{n=1}^{\infty}|a_n| \leq 6bc_0$. Using (6.6.1) once more, we deduce that

$$|a_{n+1}| \leq \frac{1}{4}|a_n| + \frac{6bc_0^2}{2^n} \ ,$$

for all $n \geq 1$. Again by induction, this gives $|a_n| \leq (24bc_0^2)2^{-n}$ for all $n \geq 1$, which is the desired result. □

Lemma 6.7. *There exist an integer m, constants $C_3, C_4 > 0$ and $\varepsilon > 0$ small enough with the following properties. For every $g \in \mathbb{K}$ and for every $f \in V_\varepsilon^0(g)$ there exists a linear functional $\theta_{f,g} \in \mathcal{C}^*$ with norm bounded from above by C_3 and with the property that*

$$\left\| DT^j(f)v - \delta_0^j \theta_{f,g}(v)\mathbf{u}_j \right\|_{\mathcal{C}} \le C_4 \delta_0^j 2^{-j/m} \|v\|_{\mathcal{C}} \ , \tag{6.6.2}$$

for all $v \in \mathcal{C}$ and all $j \ge 1$. If $g_0, g_1 \in \mathbb{K}$ and $f \in V_\varepsilon^0(g_1) \cap V_\varepsilon^0(g_2)$ then

$$\theta_{f,g_1}|\mathcal{B} = \theta_{f,g_2}|\mathcal{B}.$$

Furthermore, the map $\Psi : \bigcup_{g \in \mathbb{K}} V_\varepsilon^0(g) \to \mathcal{B}^*$ given by $\Psi(f) = \theta_f = \theta_{f,g}|\mathcal{B}$ (where g is any point of $\mathbb{K}$ such that $f \in V_\varepsilon^0(g)$) is well-defined and uniformly continuous.

Remark 6.2. Condition (6.6.2) entails that for every $g \in \mathbb{K}$, $\mathcal{B}$ is the direct sum of the one dimensional unstable subspace E_g^u with the kernel of θ_f, i.e. $\mathcal{B} = E_g^u \bigoplus \ker(\theta_f)$, provided f is sufficiently close to g. To see this, note that we can write

$$v = \theta_f(v)(\theta_f(\mathbf{u}_g))^{-1}\mathbf{u}_g + \left(v - \theta_f(v)(\theta_f(\mathbf{u}_g))^{-1}\mathbf{u}_g\right) \ .$$

Thus, from the continuity of $f \mapsto \theta_f$ plus the fact that $\theta_g(\mathbf{u}_g) \neq 0$ it follows that if f is close to g then $\mathbf{u}_g$ is transversal to $\ker(\theta_f)$. The hyperplane $\ker(\theta_f)$ is the natural candidate to be the tangent space of $V_\varepsilon^0(f)$ at f since it corresponds to all vectors which expand under $DT^j(f)$ by a factor less than δ_0^j.

Proof of Lemma 6.7. Let $\varepsilon > 0$ be small enough such that Lemma 6.5 is satisfied and $\varepsilon C_2 < \nu_m$ (where ν_m is as in property **B5** in Definition 6.1 and C_2 is as in Lemma 6.5). Let $R_k = R_{f,k} = \left(\delta_0^{km}\right)^{-1} DT^{mk}(f)$ and write $f_k = T^{km}(f)$, and $g_k = T^{km}(g)$ for all $k \ge 0$. Then

$$R_{k+1}(v) = \left(\delta_{km}^{(k+1)m}\right)^{-1} DT^m(f_k) R_k(v) \ . \tag{6.6.3}$$

Let us take $v \in \mathcal{C}$ with $\|v\|_{\mathcal{C}} = 1$. We can write $R_k(v) = \alpha_k u_{km} + w_k$, where $\alpha_k \in \mathbb{R}$ and $w_k \in \mathcal{C}$ are defined recursively by $\alpha_0 = 0$, $w_0 = v$ and

$$\begin{aligned} \alpha_{k+1} &= \alpha_k + \sigma_{km}(w_k) \\ w_{k+1} &= \alpha_k \left(\delta_{km}^{(k+1)m}\right)^{-1} (DT^m(f_k) - DT^m(g_k))\mathbf{u}_{km} \\ &\quad + \left(\delta_{km}^{(k+1)m}\right)^{-1} (DT^m(f_k) - DT^m(g_k)) w_k \\ &\quad + \left(\delta_{km}^{(k+1)m}\right)^{-1} Q_{km}^{(k+1)m-1}(w_k) \ . \end{aligned} \tag{6.6.4}$$

Now, by Lemma 6.5,

$$\|f_k - g_k\|_{\mathcal{B}} \le \varepsilon C_2 2^{-k} \ . \tag{6.6.5}$$

Since, by property **B3** of Definition 6.1, the map $f \to DT^m(f)\mathbf{u}_{km}$ is Lipschitz at $f = g_k$ (as a map from $\mathcal{B}$ to $\mathcal{C}$), for all k large enough

$$\|DT^m(f_k)\mathbf{u}_{km} - DT^m(g_k)\mathbf{u}_{km}\|_{\mathcal{C}} \le c_1\|f_k - g_k\|_{\mathcal{B}} \le c_2 2^{-k} . \tag{6.6.6}$$

By property **B5** in Definition 6.1 and (6.6.5), for all m large enough we also have

$$\|DT^m(f_k) - DT^m(g_k)\|_{\mathcal{C}} \le \frac{\delta_{km}^{(k+1)m}}{8} . \tag{6.6.7}$$

Since, by property **B1** of Definition 6.1, $(\mathcal{C}, \mathcal{D})$ is ρ-compatible with $(T, \mathbb{K})$, by Corollary 4.2, for all m large enough,

$$\left\|Q_{km}^{(k+1)m-1}\right\|_{\mathcal{C}} \le \frac{\delta_{km}^{(k+1)m}}{8} . \tag{6.6.8}$$

Using Lemma 6.2 and putting (6.6.6), (6.6.7) and (6.6.8) in (6.6.4) we get

$$\begin{aligned} \|w_{k+1}\|_{\mathcal{C}} &\le \frac{1}{4}\|w_k\|_{\mathcal{C}} + c_3 2^{-k}|\alpha_k| \\ &\le \frac{1}{4}\|w_k\|_{\mathcal{C}} + \frac{c_3 M}{2^k}\sum_{j=0}^{k-1}\|w_j\|_{\mathcal{C}} . \end{aligned} \tag{6.6.9}$$

From (6.6.9) and Lemma 6.6, we deduce that $\|w_k\|_{\mathcal{C}} < c_4 2^{-k}$. Thus, by (6.6.4) we obtain $|\alpha_{k+1} - \alpha_k| \le c_5 2^{-k}$ for all $k \ge 0$. Therefore, $\theta_{f,g}(v) = \lim \alpha_k$ exists and

$$\|R_k(v) - \theta_{f,g}(v)\mathbf{u}_{km}\|_{\mathcal{C}} \le c_6 2^{-k} , \tag{6.6.10}$$

for all $v \in \mathcal{C}$ with $\|v\|_{\mathcal{C}} = 1$. If $v \in \mathcal{C}$ and $\|v\|_{\mathcal{C}} \ne 1$ then $\theta_{f,g}(v) = \|v\|_{\mathcal{C}}\theta_{f,g}(v/\|v\|_{\mathcal{C}})$. By (6.6.10) and by Lemma 6.2, for all $v, w \in \mathcal{C}$,

$$\begin{aligned} &|\theta_{f,g}(v) + \theta_{f,g}(w) - \theta_{f,g}(v+w)| \\ &\quad \le M\,\|\theta_{f,g}(v)\mathbf{u}_{km} + \theta_{f,g}(w)\mathbf{u}_{km} - \theta_{f,g}(v+w)\mathbf{u}_{km}\|_{\mathcal{C}} \\ &\quad \le M\,\|\theta_{f,g}(v)\mathbf{u}_{km} - R_k(v)\|_{\mathcal{C}} + M\,\|\theta_{f,g}(w)\mathbf{u}_{km} - R_k(w)\|_{\mathcal{C}} \\ &\qquad + M\,\|\theta_{f,g}(v+w)\mathbf{u}_{km} - R_k(v+w)\|_{\mathcal{C}} \\ &\quad \le c_7 2^{-k}\left(\|v\|_{\mathcal{C}} + \|w\|_{\mathcal{C}}\right) . \end{aligned}$$

Hence, letting k go to infinity we deduce that $\theta_{f,g}$ is a linear functional in $\mathcal{C}^*$. Again by (6.6.10), $\|\theta_{f,g}\|_{\mathcal{C}}$ is uniformly bounded and inequality (6.6.2) is satisfied for $j = km$. By (6.6.10) and by property **B3** in Definition 6.1, for $j = km + i$ with $i \in \{1, \ldots, m-1\}$,

$$\begin{aligned} &\|R_j(v) - \theta_{f,g}(v)\mathbf{u}_j\|_{\mathcal{C}} \\ &\quad \le \left\|\left(\delta_{km}^{km+i}\right)^{-1} DT^i(T^{km}f)(R_k(v) - \theta_{f,g}(v)\mathbf{u}_{km})\right\|_{\mathcal{C}} \\ &\qquad + \left\|\left(\delta_{km}^{km+i}\right)^{-1}\Big(DT^i(T^{km}f) - DT^i(T^{km}g)\Big)\theta_{f,g}(v)\mathbf{u}_{km}\right\|_{\mathcal{C}} \\ &\quad \le c_8 2^{-k}\,\|v\|_{\mathcal{C}} , \end{aligned} \tag{6.6.11}$$

which proves (6.6.2). In particular, there is $M_0 > 0$ such that

$$\|R_{k,f}(v)\|_{\mathcal{C}} \le M_0 \ , \tag{6.6.12}$$

for all $g \in \mathbb{K}$, all $f \in V_\varepsilon^0(g)$ and all $v \in \mathcal{B}$ with $\|v\|_{\mathcal{B}} = 1$.

Let us prove that the map $f \mapsto \theta_{f,g}|\mathcal{B}$ is continuous from $V_\varepsilon^0(g)$ into $\mathcal{B}^*$ for every $g \in \mathbb{K}$. By property **B1** of Definition 6.1, for every $k \ge 1$ the functional σ_{km} is continuous on $\mathcal{C}$ and its norm is uniformly bounded. By property **B2** in Definition 6.1, the map $f \mapsto R_{k,f}$ is continuous from $\mathcal{B}$ into $\mathcal{C}$. Hence, the mapping $V_\varepsilon^0(g) \to \mathcal{B}^*$ given by $f \mapsto \sigma_{km} \circ R_{k,f}$ is also continuous. By (6.6.11),

$$\begin{aligned} |\sigma_{km} \circ R_{k,f}(v) - \theta_{f,g}(v)| &= |\sigma_{km}\left(R_k(v) - \theta_{f,g}(v)\mathbf{u}_{km}\right)| \\ &\le c_9 2^{-k} \|v\|_{\mathcal{B}} \ . \end{aligned} \tag{6.6.13}$$

Therefore, the continuous maps $f \mapsto \sigma_{km} \circ R_{k,f}$ converge uniformly to $f \mapsto \theta_{f,g}$, which implies that $f \to \theta_{f,g}$ is also a continuous map from $V_\varepsilon^0(g)$ to $\mathcal{B}^*$.

Let us prove that $\theta_{f,g}|\mathcal{B}$ for $f \in \bigcup_{g\in\mathbb{K}} V_\varepsilon^0(g)$ does not depend on $g \in \mathbb{K}$. We take $f \in \bigcup_{g\in\mathbb{K}} V_\varepsilon^0(g)$ and $g_0, g_1 \in \mathbb{K}$ such that $f \in V_\varepsilon^0(g_0)$ and $f \in V_\varepsilon^0(g_1)$. By Lemma 6.5, for every $k \ge 1$,

$$\begin{aligned} \left\|T^{km}(g_1) - T^{km}(g_0)\right\|_{\mathcal{B}} &\le \left\|T^{km}(g_1) - T^{km}(f)\right\|_{\mathcal{B}} + \left\|T^{km}(f) - T^{km}(g_0)\right\|_{\mathcal{B}} \\ &\le c_{10}\varepsilon 2^{-k} \ . \end{aligned}$$

By property **B1** of Definition 6.1, the map $g \mapsto \sigma_g$ from $\mathbb{K}$ into $\mathcal{C}^*$ is uniformly continuous. Hence, for every $\varepsilon > 0$ there is $k_0 > 0$ large enough such that for all $k > k_0$ and all $w \in \mathcal{C}$ with $\|w\|_{\mathcal{C}} \le M_0$,

$$\left|\sigma_{T^{km}(g_1)}(w) - \sigma_{T^{km}(g_0)}(w)\right| \le \varepsilon/2 \ . \tag{6.6.14}$$

By (6.6.12), (6.6.13) and (6.6.14) and taking k large enough, we get

$$\begin{aligned} |\theta_{f,g_1}(v) - \theta_{f,g_0}(v)| &\le \left|\theta_{f,g_1}(v) - \sigma_{T^{km}(g_1)} \circ R_{k,f}(v)\right| \\ &\quad + \left|\sigma_{T^{km}(g_1)} \circ R_{k,f}(v) - \sigma_{T^{km}(g_0)} \circ R_{k,f}(v)\right| \\ &\quad + \left|\sigma_{T^{km}(g_0)} \circ R_{k,f}(v) - \theta_{f,g_0}(v)\right| \\ &\le 2c_9 2^{-k} + \varepsilon/2 \le \varepsilon \end{aligned}$$

for all $v \in \mathcal{B}$ with $\|v\|_{\mathcal{B}} = 1$. Thus, $\theta_{f,g_1}(v) = \theta_{f,g_0}(v)$ and so the map Ψ is well-defined.

Let us prove that the map Ψ is uniformly continuous. For every $\alpha_0 > 0$, we choose $k_0 > 0$ large enough such that $2c_9 2^{-k_0} \le \alpha_0/3$. Since the map $g \to \sigma_g$ is uniformly continuous, there is $\alpha_1 > 0$ small enough such that for all $g_0, g_1 \in \mathbb{K}$ with $\|g_1 - g_0\|_{\mathcal{C}} < \alpha_1$ and all $w \in \mathcal{B}$ with $\|w\|_{\mathcal{C}} \le M_0$,

$$|\sigma_{g_1}(w) - \sigma_{g_0}(w)| \le \alpha_0/3 \ . \tag{6.6.15}$$

Choose $k_1 > k_0$ large enough such that $\varepsilon C_2 2^{-k_1} \le \alpha_1/3$ where $C_2 > 0$ is the constant of Lemma 6.5. Since $T : \mathcal{O}_B \to \mathcal{C}$ is a C^1 operator, by property **B2** of Definition 6.1, (and compactness of $\mathbb{K}$), there is $\alpha_2 > 0$ small enough such that for all $f_0 \in V_\varepsilon^0(g_0)$ and $f_1 \in V_\varepsilon^0(g_1)$ with $\|f_1 - f_0\|_{\mathcal{B}} < \alpha_2$ we obtain that $\|T^{k_1 m}(f_1) - T^{k_1 m}(f_0)\|_{\mathcal{C}} \leq \alpha_1/3$. Hence, by Lemma 6.5,

(6.6.16)

$$\begin{aligned}\left\|T^{k_1 m}(g_1)-T^{k_1 m}(g_0)\right\|_{\mathcal{C}} \leq & \left\|T^{k_1 m}(g_1)-T^{k_1 m}(f_1)\right\|_{\mathcal{C}}+ \left\|T^{k_1 m}(f_1)-T^{k_1 m}(f_0)\right\|_{\mathcal{C}} \\ & + \left\|T^{k_1 m}(f_0)-T^{k_1 m}(g_0)\right\|_{\mathcal{C}} \\ \leq & \, 2\varepsilon C_2 2^{-k_1} + \alpha_1/3 \leq \alpha_1 \ .\end{aligned}$$

By (6.6.15) and (6.6.16),

$$\left|\sigma_{T^{k_1 m}(g_1)} \circ R_{k_1,f_1}(v) - \sigma_{T^{k_1 m}(g_0)} \circ R_{k_1,f_1}(v)\right| \leq \alpha_0/3 \ . \tag{6.6.17}$$

Using property **B2** of Definition 6.1, choose $0 < \alpha_3 < \alpha_2$ small enough such that for all $f_1 \in \bigcup_{g\in\mathbb{K}} V_\varepsilon^0(g)$ with $\|f_1 - f_0\|_{\mathcal{B}} < \alpha_3$,

$$\|R_{k_1,f_1}(v) - R_{k_1,f_0}(v)\|_{\mathcal{C}} \leq \alpha_0/(3M) \ ,$$

where $v \in \mathcal{B}$ with $\|v\|_{\mathcal{B}} = 1$ and M is as in Lemma 6.2. Hence,

$$\left|\sigma_{T^{k_1 m}(g_0)} \circ R_{k_1,f_1}(v) - \sigma_{T^{k_1 m}(g_0)} \circ R_{k_1,f_0}(v)\right| \leq \alpha_0/3 \ . \tag{6.6.18}$$

By (6.6.13), (6.6.17) and (6.6.18), we obtain that

$$\begin{aligned}|\theta_{f_1}(v) - \theta_{f_0}(v)| \leq & \left|\theta_{f_1}(v) - \sigma_{T^{k_1 m}(g_1)} \circ R_{k_1,f_1}(v)\right| \\ & + \left|\sigma_{T^{k_1 m}(g_1)} \circ R_{k_1,f_1}(v) - \sigma_{T^{k_1 m}(g_0)} \circ R_{k_1,f_1}(v)\right| \\ & + \left|\sigma_{T^{k_1 m}(g_0)} \circ R_{k_1,f_1}(v) - \sigma_{T^{k_1 m}(g_0)} \circ R_{k_1,f_0}(v)\right| \\ & + \left|\sigma_{T^{k_1 m}(g_0)} \circ R_{k_1,f_0}(v) - \theta_{f_0}(v)\right| \\ \leq & \, 2c_9 2^{-k_1} + 2\alpha_0/3 \leq \alpha_0 \ .\end{aligned}$$

Therefore, the map Ψ is uniformly continuous and Lemma 6.7 is proved. □

6.7. *The main estimates.* Besides aiming at proving that the local stable set is a C^1 manifold, we want to show that the local hyperbolicity picture holds (in $\mathcal{B}$) near $\mathbb{K}$. In other words we want to show that if the iterates $T^{km}(f_1)$ of a point $f_1 \in \mathcal{B}(g,\varepsilon)$ remain in $\mathcal{B}(T^{km}(g),\varepsilon)$ for a long time, that is for $k = 0, 1, \ldots, N$ with N large, then f_1 has to be very close to a point f_0 on the stable set $W_\varepsilon^s(g)$ at the outset, and in the end $T^{Nm}(f_1)$ has to be very close to the unstable manifold $W_\varepsilon^u(T^{Nm}(g))$.

To prove these facts, we consider in this section (see Lemma 6.11) an intermediate time l for which we can find a good quantitative estimate for the point on the unstable manifold $W_\varepsilon^u(T^{lm}(g))$ that best approximates $T^{lm}(f_1)$. This estimate is provided by the value of $\theta_{f_0}(f_1 - f_0)$, and its most important consequence is obtained when f_1 also belongs to the local stable set $W_\varepsilon^s(g)$. In this case we prove an inequality of the form $|\theta_{f_0}(f_1 - f_0)| \leq C\|f_1 - f_0\|_{\mathcal{B}}^{1+\tau}$

(see Lemma 6.12). As we shall see in §6.8, this is precisely what is needed to show that the tangent space to the stable set at f_0 varies continuously with f_0.

In this section we fix m large enough and $\varepsilon_0 > \varepsilon_1 > \varepsilon_2$ small enough such that Lemmas 6.4, 6.5 and 6.7 are satisfied for all $\varepsilon < \varepsilon_2$ suficiently small.

LEMMA 6.8. *There exist constants $C_5, C_6, \varepsilon > 0$ with the following property. For all $g \in \mathbb{K}$, all $f_0 \in V_\varepsilon^0(g)$, and all $f_1 \in \mathcal{B}(g, \varepsilon)$ such that $\|f_1 - f_0\|_\mathcal{C} \leq C_5\,(\delta_0^n)^{-1}$,*

$$\left\| T^k(f_1) - T^k(f_0) - DT^k(f_0)\,(f_1 - f_0) \right\|_\mathcal{C} \leq C_6\,(\delta_k^n)^{-\rho}\ , \tag{6.7.1}$$

for all $0 \leq k \leq \hat{k}_0(g, f_1)$, where

$$\hat{k}_0(g, f_1) = \min\left\{ j \in \{0, \dots, n\} : \|T^j(f_1) - T^j(g)\|_\mathcal{B} \geq \varepsilon_0 \right\}$$

and $\rho > 1$ is as in property **B4** *of Definition* 6.1.

Proof. By Lemmas 6.2 and 6.7, there are $c_0, c_1 > 0$ and $\lambda > 1$ such that

$$\left\| DT^{k-i}\left(T^i f_0\right) \right\|_\mathcal{C} \leq c_0 \delta_i^k \text{ and } \delta_i^k > c_1 \lambda^{k-i} \tag{6.7.2}$$

for all $0 \leq i < k$. Define a sequence $v_i \in \mathcal{C}$ as follows: $v_0 = T(f_1) - T(f_0)$ and

$$v_i = T^i(f_1) - T^i(f_0) - DT\left(T^{i-1}(f_0)\right)\left(T^{i-1}(f_1) - T^{i-1}(f_0)\right)\ ,$$

for all $0 < i \leq k$. Hence,

$$T^i(f_1) - T^i(f_0) - DT^i(f_0)\,(f_1 - f_0) = \sum_{j=1}^{i} DT^{i-j}\left(T^j f_0\right) v_j\ . \tag{6.7.3}$$

Applying property **B4** of Definition 6.1, we get

$$\begin{aligned} \|v_{i+1}\|_\mathcal{C} &\leq c_2 \left\|T^i(f_1) - T^i(f_0)\right\|_\mathcal{C}^\rho \\ &\leq c_2 \left\| DT^i(f_0)\,(f_1 - f_0) + \sum_{j=1}^{i} DT^{i-j}(T^j f) v_j \right\|_\mathcal{C}^\rho\ . \end{aligned} \tag{6.7.4}$$

Let us first choose $C_6 > 0$ such that

$$C_6^{1-\rho} > c_2 \left(\frac{2 c_0 c_1^{1-\rho}}{(\delta_i)^\rho (1 - \lambda^{1-\rho})} \right)^\rho\ , \tag{6.7.5}$$

for all $0 < i \leq k$, and then choose $C_5 > 0$ such that

$$C_5^\rho < \frac{C_6 (\delta_0)^\rho}{c_2}\ , \qquad C_5 < \frac{c_1^{1-\rho} C_6 (\delta_i)^{1-\rho}}{1 - \lambda^{1-\rho}}\ . \tag{6.7.6}$$

Let us prove inductively for $i = 1, 2, \dots, k$ that $\|v_i\|_\mathcal{C} \leq C_6\,(\delta_i^n)^{-\rho}$. Using inequality (6.7.4) and (6.7.6), we get

$$\|v_1\|_\mathcal{C} \leq c_2\,\|f_1 - f_0\|_\mathcal{C}^\rho \leq \frac{c_2 C_5^\rho}{(\delta_0)^\rho} (\delta_1^n)^{-\rho} \leq C_6 (\delta_1^n)^{-\rho}\ .$$

Using the inequalities (6.7.2), (6.7.4), (6.7.5) and (6.7.6), we get

$$\begin{aligned}
(6.7.7)\qquad \|v_{i+1}\|_{\mathcal{C}} &\le c_2 \left(C_5 c_0 \delta_0^i \left(\delta_0^n\right)^{-1} + \sum_{j=1}^{i} c_0 \delta_j^i C_6 \left(\delta_j^n\right)^{-\rho} \right)^{\rho} \\
&\le c_2 c_0^{\rho} \left(\frac{C_5}{\delta_i} + \frac{C_6}{(\delta_i)^{\rho}} \sum_{j=1}^{i} \left(\delta_j^i\right)^{1-\rho} \right)^{\rho} \left(\delta_{i+1}^n\right)^{-\rho} \\
&\le c_2 c_0^{\rho} \left(\frac{C_5}{\delta_i} + \frac{c_1^{1-\rho} C_6}{(\delta_i)^{\rho}\left(1-\lambda^{1-\rho}\right)} \right)^{\rho} \left(\delta_{i+1}^n\right)^{-\rho} \\
&\le C_6 \left(\delta_{i+1}^n\right)^{-\rho} ,
\end{aligned}$$

which ends the induction. Thus, using (6.7.2) and (6.7.7) in (6.7.3), we get

$$\begin{aligned}
\left\| T^k(f_1) - T^k(f) - DT^k(f)\,(f_1 - f_0) \right\|_{\mathcal{C}} &\le \sum_{i=1}^{k} c_0 \delta_i^k C_6 \left(\delta_i^n\right)^{-\rho} \\
&\le c_0 C_6 \left(\delta_k^n\right)^{-\rho} \sum_{i=1}^{k} \left(\delta_i^k\right)^{1-\rho} \le \frac{c_0 c_1^{1-\rho} C_6}{1-\lambda^{1-\rho}} \left(\delta_i^n\right)^{-\rho} .
\end{aligned}$$

This proves the lemma. □

LEMMA 6.9. *Let $C_5, C_6, \varepsilon > 0$, $\rho > 1$ and $\hat{k}_0(g, f_1)$ be as in Lemma 6.8. There exist $C_7, C_8 > 0$ such that for all $g \in \mathbb{K}$, all $f_0 \in V_\varepsilon^0(g)$ and all $f_1 \in \mathcal{B}(g,\varepsilon)$ such that $\|f_1 - f_0\|_{\mathcal{C}} \le C_5 \left(\delta_0^n\right)^{-1}$,*

$$\begin{aligned}
(6.7.8)\qquad & \left\| T^k(f_1) - T^k(g) - \delta_0^k \theta_{f_0}\,(f_1 - f_0)\,\mathbf{u}_k \right\|_{\mathcal{C}} \\
&\qquad \le C_6 \left(\delta_k^n\right)^{-\rho} + \varepsilon C_7 2^{-k/m} + C_8 2^{-k/m} \left(\delta_k^n\right)^{-1} ,
\end{aligned}$$

for all $k \le \hat{k}_0(g, f_1)$.

Proof. By Lemma 6.7,

$$(6.7.9)\qquad \left\| DT^k(f_0)\,(f_1 - f_0) - \delta_0^k \theta_{f_0}\,(f_1 - f_0)\,\mathbf{u}_k \right\|_{\mathcal{C}} \le C_4 2^{-k/m} \left(\delta_k^n\right)^{-1} .$$

By Lemma 6.5, we obtain that

$$(6.7.10)\qquad \left\| T^k(f_0) - T^k(g) \right\|_{\mathcal{C}} \le c_0 \left\| T^k(f_0) - T^k(g) \right\|_{\mathcal{B}} \le \varepsilon c_0 C_2 2^{-k/m} .$$

Combining (6.7.1), (6.7.9) and (6.7.10), we get (6.7.8). □

Definition 6.2. Given $g \in \mathbb{K}$ and $p \ge 1$ we denote by $l = l(g,p)$ the *smallest integer* such that

$$(6.7.11)\qquad \left(\delta_{lm}^{pm}\right)^{\rho} \le 2^l ,$$

where $\rho > 1$ is as in property **B4** of Definition 6.1.

Lemma 6.10. (i) *There exist* $0 < \mu_0 < \mu_1 < 1$ *with the property that* $\mu_0 p \le l = l(g,p) \le \mu_1 p$ *for all* $g \in \mathbb{K}$ *and all* $p \ge 1$.

(ii) *There exists* $0 < \tau_1 < 1$ *such that for all* $g \in \mathbb{K}$ *and all* $f_0, f_1 \in \mathcal{B}(g,\varepsilon)$, *if* $\|f_0 - f_1\|_C \ge C_5 \left(\delta_0^{(p+1)m}\right)^{-1}$ *then* $\left(\delta_{lm}^{pm}\right)^{-1} \le C_9 \|f_0 - f_1\|_C^{\tau_1}$, *where* C_9 *depends only upon* $C_5 > 0$.

Proof. Let us prove part (i). Set $1 < \lambda < M$ as in Lemma 6.2. Then, by (6.7.11),

$$\rho \log M^{-1} + (pm - lm)\rho \log \lambda \ < \ \rho \log \delta_{lm}^{pm} \ < \ l \log 2 \ < \ lm \log 2 \ .$$

Hence,

$$\left(1 + \frac{\log 2}{\rho \log \lambda}\right) lm \ \ge \ \frac{\log M^{-1}}{\log \lambda} + pm \ > \ \frac{pm}{2} \ ,$$

for all p such that $pm > N_0 = \max\{2m, |2\log M^{-1}/\log\lambda|\}$. Thus, taking

$$\mu_0' = 2^{-1}\left(1 + \frac{\log 2}{\rho \log \lambda}\right)^{-1} > 0 \ ,$$

we get $\mu_0' p \le l$ for all such values of p. By (6.7.11) and by Lemma 6.2 there exists a uniform constant $0 < c_0 \le 1$ such that $c_0 2^l \le \left(\delta_{lm}^{pm}\right)^\rho$ and so

$$\log c_0 + l \log 2 < \rho \log \delta_{lm}^{pm} \le \rho(pm - lm) \log M \ .$$

Letting $\alpha = \log 2/(\rho \log M) > 0$ and $\beta = \log c_0/(\rho \log M)$ we get

$$lm\left(1 + \frac{\alpha}{m}\right) \le pm - \beta \le pm\left(1 + \frac{\alpha}{2m}\right)$$

for all $p > -2\beta/\alpha$. Thus, taking

$$0 < \mu_1' = \frac{2m + \alpha}{2(m+\alpha)} < 1$$

we obtain that $lm \le \mu_1' pm$ for all such values of p. Since δ_g varies continuously with g in the compact set $\mathbb{K}$, we can extend the previous results to all $p \ge 0$ for some $\mu_0 \le \mu_0'$ and $\mu_1 \ge \mu_1'$.

Let us prove part (ii). Take $0 < \tau_1 = (1 - \mu_1)\log M/\log\lambda < 1$. Then, by Lemma 6.2,

$$\begin{aligned}\left(\delta_{lm}^{pm}\right)^{-1} &\le c_0 \lambda^{-(p-l)m} \le c_0 \lambda^{-(1-\mu_1)pm} \\ &\le c_0 M^{-\tau_1 pm} \le c_1 \omega \left(\delta_0^{(p+1)m}\right)^{-\tau_1} \\ &\le c_1 \|f_0 - f_1\|_C^{\tau_1} \ .\end{aligned}$$

□

LEMMA 6.11. *There exist $\varepsilon > 0$ sufficiently small and $C_{10} > 0$ such that the following holds for $g \in \mathbb{K}$, $f_0 \in V_\varepsilon^0(g)$ and $f_1 \in \mathcal{B}(g,\varepsilon)$. If p is the largest integer such that*

$$\|f_1 - f_0\|_C \le C_5 \left(\delta_0^{pm}\right)^{-1}$$

then $l = l(g,p) \le k_0 = k_0(g,f_1)$ and so $t_l = t_l(g,f_1)$ is well-defined (where l is as in Lemma 6.10, k_0 and t_l are as in Lemma 6.4, and $C_5 > 0$ is as in Lemma 6.9). *Furthermore,*

$$\left|t_l - \delta_0^{lm}\theta_{f_0}(f_1 - f_0)\right| \le C_{10}\left(\delta_{lm}^{pm}\right)^{-\rho} \tag{6.7.12}$$

where $\rho > 1$ is as in Lemma 6.9.

Proof. Let $\hat{k}_0 = \hat{k}_0(g,f_1)$ be as in Lemma 6.9. Let us prove that $l \le k_0$. By (iv) in Lemma 6.4, $mk_0 < \hat{k}_0$. Hence it is enough to prove that $\min\left\{lm, \hat{k}_0\right\} \le mk_0$. Let $\varepsilon > 0$ be small enough such that Lemmas 6.8 and 6.9 are satisfied. Let us show that $lm \le \hat{k}_0$. By inequality (6.7.8), for all k such that $mk \le \hat{k}_0$ we have

$$\begin{aligned}\left|\sigma_{km}\left(T^{km}(f_1) - T^{km}(g)\right)\right| \\ \le \|\sigma_{km}\|_C \left(\delta_0^{km}\left|\theta_f\left(f_1 - f_0\right)\right| + c_0\left(\delta_{km}^{pm}\right)^{-1} + \varepsilon c_1 2^{-k}\right).\end{aligned} \tag{6.7.13}$$

By Lemma 6.2 and Remark 6.2, there is $M_1 > 1$ such that $M_1^{-1} \le \|\sigma_{km}\|_C \le M_1$ and $M_1^{-1} \le \|\theta_{f_0}\|_C \le M_1$. Since by Lemma 6.2, we have $\left(\delta_{km}^{pm}\right)^{-1} \le M\lambda^{-(p-k)m}$ we deduce that

$$\delta_0^{km}\left|\theta_f\left(f_1 - f_0\right)\right| \le \left(\delta_{km}^{pm}\right)^{-1}\|\theta_f\|_C \le MM_1\lambda^{-(p-k)pm}. \tag{6.7.14}$$

By Lemma 6.10, there is $0 < \mu_1 < 1$ such that for all $p > 0$ and all $k \le l$ we have $p - k \ge p - l \ge (1-\mu_1)p$. Now, we make $\varepsilon > 0$ small enough (and so p large enough) such that the following inequalities are satisfied

$$\begin{aligned}(c_0 + M_1)M\lambda^{-(1-\mu_1)pm} &< \frac{\varepsilon_2}{2\|\sigma_{km}\|_C}, \\ \varepsilon c_1 2^{-k} &< \frac{\varepsilon_2}{4\|\sigma_{km}\|_C},\end{aligned}$$

for all k such that $km \le \min\{lm, \hat{k}_0\}$. Therefore, for all such k, combining (6.7.13) and (6.7.14) we deduce that

$$\left|\sigma_{km}\left(T^{km}(f_1) - T^{km}(g)\right)\right| < \varepsilon_2. \tag{6.7.15}$$

Since $f_1 \in \mathcal{B}(g,\varepsilon)$ and (6.7.15) reverses the inequality (iii) in Lemma 6.4, we obtain that $\min\{lm, \hat{k}_0\} \le mk_0$, and so $l \le k_0$.

Now, let us prove (6.7.12). Since $l \le k_0$, by (6.4.5) and (6.4.6) in Lemma 6.4, there is $t_l = t_l(g,f_1)$ such that

$$\|T^{lm}(f_1) - u_{lm}(t_l)\|_{\mathcal{B}} \le \varepsilon 2^{-l} \le \varepsilon\left(\delta_{lm}^{pm}\right)^{-\rho}. \tag{6.7.16}$$

Since $lm < \hat{k}_0$, by lemmas 6.9 and 6.10 we get

$$\|T^{lm}(f_1) - T^{lm}(g) - s_l \mathbf{u}_{lm}\|_{\mathcal{C}} \leq c_2 \left(\delta_{lm}^{pm}\right)^{-\rho} ,$$

where $s_l = \delta_0^{lm} \theta_{f_0}(f_1 - f_0)$. Thus, using (6.7.16), we obtain that

$$\|u_{lm}(t_l) - T^{lm}(g) - s_l \mathbf{u}_{lm}\|_{\mathcal{C}} \leq c_3 \left(\delta_{lm}^{pm}\right)^{-\rho} .$$

Since $t \to u_{lm}(t)$ is C^2 as a map $\mathbb{R} \to \mathcal{C}$,

$$\begin{aligned} \|u_{lm}(s_l) - T^{lm}(g) - s_l \mathbf{u}_{lm}\|_{\mathcal{C}} &\leq c_4 s_l^2 \\ &= c_4 \left|\delta_0^{lm} \theta_{f_0}(f_1 - f_0)\right|^2 \\ &\leq c_5 \left(\delta_{lm}^{pm}\right)^{-2} . \end{aligned}$$

Therefore,

$$\begin{aligned} &\|u_{lm}(t_l) - u_{lm}(s_l)\|_{\mathcal{C}} \\ &\quad \leq \|u_{lm}(t_l) - T^{lm}(g) - s_l \mathbf{u}_{lm}\|_{\mathcal{C}} + \|u_{lm}(s_l) - T^{lm}(g) - s_l \mathbf{u}_{lm}\|_{\mathcal{C}} \\ &\quad \leq c_3 \left(\delta_{lm}^{pm}\right)^{-\rho} + c_5 \left(\delta_{lm}^{pm}\right)^{-2} \\ &\quad \leq c_6 \left(\delta_{lm}^{pm}\right)^{-\rho} , \end{aligned}$$

because $1 < \rho < 2$. Hence, applying Lemma 6.2, we get

$$|t_l - s_l| \leq M^{-1} \|u_{lm}(t_l) - u_{lm}(s_l)\|_{\mathcal{C}} \leq c_7 \left(\delta_{lm}^{pm}\right)^{-\rho} . \qquad \square$$

LEMMA 6.12. *There exist constants $\tau, \varepsilon, C > 0$ with the following properties: for all $g \in \mathbb{K}$ and all $f_0, f_1 \in V_\varepsilon^0(g)$,*

$$|\theta_{f_0}(f_1 - f_0)| \leq C \|f_1 - f_0\|_{\mathcal{B}}^{1+\tau} .$$

Proof. We shall in fact prove a stronger inequality, with the $\mathcal{C}$-norm replacing the $\mathcal{B}$-norm. Let $\varepsilon > 0$ be so small that Lemmas 6.8 to 6.11 are satisfied ($\varepsilon > 0$ will be made even smaller in the course of the argument). Let p be such that $C_5 \delta_0^{-(p+1)m} < \|f_0 - f_1\|_{\mathcal{C}} \leq C_5 \delta_0^{-pm}$ where $C_5 > 0$ is as in Lemma 6.8. As in Lemma 6.4, set $k_0 = k_0(g, f_1)$, $t_j = t_j(g, f_1)$ and $v_j = v_j(g, f_1)$ for all $0 \leq j \leq k_0$. Also, let $l = l(g, p)$ be as in Lemma 6.10. By Lemma 6.11, we have $l \leq k_0$ and so t_l is well-defined. Thus, applying Lemmas 6.4, 6.5 and 6.10, we get

$$\begin{aligned} \|u_{lm}(t_l) - u_{lm}(0)\|_{\mathcal{B}} &\leq \left\|u_{lm}(t_l) - T^{lm}(f_1)\right\|_{\mathcal{B}} + \left\|T^{lm}(f_1) - T^{lm}(g)\right\|_{\mathcal{B}} \\ &\leq \varepsilon c_0 2^{-l} \leq \varepsilon c_0 \left(\delta_{lm}^{pm}\right)^{-\rho} . \end{aligned}$$

Hence, by Lemma 6.2 we see that $|t_l| \leq c_1 \left(\delta_{lm}^{pm}\right)^{-\rho}$. Let us write $t_j = \alpha_j \left(\delta_{jm}^{pm}\right)^{-1}$ for $l \leq j \leq k_0(g, f_1)$. Recalling that $\delta_{jm}^{(j+1)m} > \beta > 2$ for all j

and using Lemma 6.10, we have

$$(6.7.17)\qquad \left(\delta_{lm}^{pm}\right)^{-1} \le \beta^{-(1-\mu_1)p} \text{ and } \left(\delta_{lm}^{pm}\right)^{-(\rho-1)/2} \le \beta^{-\tau_2 p}$$

where $\tau_2 = (1-\mu_1)(\rho-1)/2$. Hence, making $\varepsilon > 0$ smaller if necessary (and so p large enough), we get

$$(6.7.18)\qquad \alpha_l < 4^{-1}\left(\delta_{lm}^{pm}\right)^{-(\rho-1)/2} < 4^{-1}\beta^{-\tau_2 p} < \varepsilon_1/2\ .$$

By Lemma 6.4, we have $\|v_j\|_{\mathcal{B}} \le \varepsilon 2^{-j}$ and $t_{j+1} = \hat{\delta}_{jm}^{(j+1)m}(t_j + \sigma_{jm}(v_j))$. Since $t \mapsto \hat{\delta}_{jm}^{(j+1)m}(t)$ is C^2 as a map $\mathbb{R} \to \mathcal{C}$, we deduce that

$$\begin{aligned}\left|t_{j+1} - \delta_{jm}^{(j+1)m} t_j\right| &\le c_2\left(|t_j|^2 + \|v_j\|_{\mathcal{B}}\right)\\ &\le c_2\left(|t_j|^2 + \varepsilon 2^{-j}\right)\\ &\le c_2\left(|t_j|^2 + \varepsilon 2^{-(j-l)}\left(\delta_{lm}^{pm}\right)^{-\rho}\right)\ .\end{aligned}$$

Therefore,

$$(6.7.19)\qquad |\alpha_{j+1} - \alpha_j| \le c_2\left(\alpha_j^2\beta^{-(p-j+1)} + \varepsilon(2\beta)^{-(j-l)}\left(\delta_{lm}^{pm}\right)^{-(\rho-1)}\right)\ .$$

We prove that $k_0(g, f_1) > p$. To do this, we need to show that $|t_j| < \varepsilon_1$ for all $j \le p$. We prove by induction a slightly stronger statement, namely, that $\alpha_j \le 2^{-1}\left(\delta_{lm}^{pm}\right)^{-(\rho-1)/2} < \varepsilon_1$ for all $j = l, \dots, p$. This is certainly satisfied for $j = l$, as can be seen from (6.7.18). Suppose it is satisfied for α_i for all $i = l, \dots, j$. Using (6.7.17) and (6.7.19), and making $\varepsilon > 0$ even smaller (and thus p large enough), we get

$$\begin{aligned}(6.7.20)\quad |\alpha_{j+1} - \alpha_l| &\le \sum_{i=l}^{j} |\alpha_{i+1} - \alpha_i|\\ &\le \frac{1}{4}\left(\sum_{i=l}^{j}\left(c_2\beta^{-(p-i+1)} + 4c_2\varepsilon(2\beta)^{-(i-l)}\right)\right)\left(\delta_{lm}^{pm}\right)^{-(\rho-1)}\\ &\le \frac{1}{4}\left(\frac{c_2\beta^{-1}}{1-\beta^{-1}} + \frac{4c_2\varepsilon}{1-(2\beta)^{-1}}\right)\beta^{-\tau_2 p}\left(\delta_{lm}^{pm}\right)^{-(\rho-1)/2}\\ &\le \frac{1}{4}\left(\delta_{lm}^{pm}\right)^{-(\rho-1)/2}\ .\end{aligned}$$

Since $\alpha_l \le 4^{-1}\left(\delta_{lm}^{pm}\right)^{-(\rho-1)/2}$, we deduce that $\alpha_{j+1} \le 2^{-1}\left(\delta_{lm}^{pm}\right)^{-(\rho-1)/2} < \varepsilon_1$ (in particular $j+1 \le k_0(g, f_1)$) which ends the induction.

Now set $s_j = s_j(g, f_0, f_1) = \delta_0^{jm}\theta_{f_0}(f_1 - f_0)$ for all j. Let us estimate $|t_p - s_p|$. By Lemma 6.11 and the above estimates on α_j's, we have

$$(6.7.21)\qquad |t_p - s_p| \le |\alpha_p - \alpha_l| + \delta_{lm}^{pm}|t_l - s_l| \le c_3\left(\delta_{lm}^{pm}\right)^{-(\rho-1)/2}\ .$$

On the other hand, from Lemmas 6.4, 6.5 and 6.10, we also know that

$$\begin{aligned}\|u_{pm}(t_p) - u_{pm}(0)\|_{\mathcal{B}} &\leq \|u_{pm}(t_p) - T^{pm}(f_1)\|_{\mathcal{B}} + \|T^{pm}(f_1) - T^{pm}(g)\|_{\mathcal{B}} \\ &\leq \varepsilon c_4 2^{-p} .\end{aligned}$$

Hence, again by Lemma 6.2, we have $|t_p| \leq \varepsilon c_5 2^{-p}$. Since $p \geq l$, we deduce from Lemma 6.10 that

$$|t_p| \leq \varepsilon c_5 2^{-l} \leq \varepsilon c_5 \left(\delta_{lm}^{pm}\right)^{-\rho} . \tag{6.7.22}$$

But Lemma 6.10, also tells us that there exists $\tau_1 > 0$ such that $\left(\delta_{lm}^{pm}\right)^{-1} \leq c_6 \|f_0 - f_1\|_{\mathcal{C}}^{\tau_1}$. Moreover, $(\delta_0^{pm})^{-1} \leq c_7 \|f_0 - f_1\|_{\mathcal{C}}$ by hypothesis. Therefore, combining these facts with (6.7.21) and (6.7.22), we get at last

$$\begin{aligned}|\theta_{f_0}(f_1 - f_0)| &\leq (\delta_0^{pm})^{-1} |s_p| \\ &\leq (\delta_0^{pm})^{-1} (|t_p| + |t_p - s_p|) \\ &\leq (\delta_0^{pm})^{-1} \left(\varepsilon c_5 \left(\delta_{lm}^{pm}\right)^{-\rho} + c_3 \left(\delta_{lm}^{pm}\right)^{-(\rho-1)/2}\right) \\ &\leq c_8 \|f_0 - f_1\|_{\mathcal{C}}^{1+\tau_1(\rho-1)/2} ,\end{aligned}$$

which finishes the proof. □

6.8. *The local stable sets are graphs.* We shall prove now that the local stable set of every $g_0 \in \mathbb{K}$ in a sufficiently small neighborhood of g is the graph of a function defined over $\mathrm{Ker}\theta_g \cap \mathcal{B}$ (and with values on the one-dimensional subspace $\mathbb{R}\mathbf{u}_g \subset \mathcal{B}$). The idea is to show that every "vertical line" of the form $f + \mathbb{R}\mathbf{u}_g$ with f close to g cuts the local stable set $W_\varepsilon^s(g_0)$ exactly at one point. All other points in the same vertical line escape exponentially fast away from $W_\varepsilon^s(g_0)$ under iteration by T and the time $k_0 m$ each such point f takes to escape is logarithmic on the reciprocal of its distance to $W_\varepsilon^s(g_0)$. Moreover, $T^{k_0 m}(f)$ will be exponentially close (in k_0) to $W_\varepsilon^u(T^{k_0 m}(g_0))$.

Proposition 6.13. *There exist* $0 < \alpha_0, \alpha_1, \alpha_2 < \varepsilon$, $0 < \mu_0 < \mu_1$ *and* $M_0 > 1$ *with the following properties. If* $g_0 \in \mathbb{K}$ *and* $g \in \mathbb{K}$ *is such that* $\|g - g_0\|_{\mathcal{B}} < \alpha_0$, *then for every* $v \in \mathrm{Ker}\ \theta_{g_0} \cap \mathcal{B}$ *with* $\|v\|_{\mathcal{B}} < \alpha_1$, *there exists* $-\alpha_2/2 < \tau(g, v) < \alpha_2/2$ *such that*

(i) $f_{\tau(g,v)} = g_0 + v + \tau(g, v)\mathbf{u}_{g_0} \in W_\varepsilon^s(g) \subset V_\varepsilon^0(g)$;

(ii) $f_t = g_0 + v + t\mathbf{u}_{g_0} \in V_\varepsilon^+(g)$ *for all* $\tau(g, v) < t < \alpha_2$;

(iii) $f_t = g_0 + v + t\mathbf{u}_{g_0} \in V_\varepsilon^-(g)$ *for all* $-\alpha_2 < t < \tau(g, v)$;

(iv) $-\mu_0 \log(|t - \tau(g, v)|) \leq k_0(g, f_t) \leq -\mu_1 \log(|t - \tau(g, v)|)$, *where* $k_0(g, \phi_t)$ *is as in Lemma* 6.4.

Proof. Let $\varepsilon > 0$ be sufficiently small such that Lemmas 6.8 to 6.11 are satisfied and $0 < \varepsilon' < \varepsilon$ such that Lemma 6.5 is satisfied. Let $M > 0$ be as in Lemma 6.2 and take positive numbers α_1 and α_2 such that

$$0 < 8\alpha_1 M < \alpha_2 \text{ and } \alpha_1 + 2\alpha_2 M < \varepsilon'/2 \ . \tag{6.8.1}$$

Take $g \in \mathbb{K}$ and $f \in V_\varepsilon(g_0)$ with $\|f - g\|_{\mathcal{B}} < \varepsilon'/2$. Let $v \in \operatorname{Ker} \theta_{g_0} \cap \mathcal{B}$ with $\|v\|_{\mathcal{B}} < \alpha_1$, and $t \in \mathbb{R}$ with $2M\|v\|_{\mathcal{B}} < |t| < 2\alpha_2$. By the second inequality in (6.8.1), we have $\phi_t = f + v + t\mathbf{u}_{g_0} \in \mathcal{B}(g, \varepsilon)$ and $\|\phi_t - g\|_{\mathcal{B}} < \varepsilon'$ for all $|t| < 2\alpha_2$. Now, we have the following claim.

CLAIM. *The family ϕ_t satisfies the following property:*

$$\begin{cases} \phi_t \in V_\varepsilon^+(g), & \text{if} \quad 2M\|v\|_{\mathcal{B}} < t < 2\alpha_2 \\ \phi_t \in V_\varepsilon^-(g), & \text{if} \quad -2\alpha_2 < t < -2M\|v\|_{\mathcal{B}} \ . \end{cases} \tag{6.8.2}$$

To prove this claim, let $C_5 > 0$ be as in Lemma 6.8 and let p be such that $C_5\delta_0^{(p+1)m} < \|\phi_t - f\|_{\mathcal{C}} \le C_5\delta_0^{pm}$. Set $k_0 = k_0(g, \phi_t)$, $t_j = t_j(g, \phi_t)$ and $v_j = v_j(g, \phi_t)$ for all $0 \le j \le k_0$ as in Lemma 6.4. Set $s_j = s_j(g, f, \phi_t) = \delta_0^{jm}\theta_f(\phi_t - f)$. Set $l = l(g, p)$ as in Lemma 6.10. By Lemma 6.2 and Remark 6.2, there exist $c_0 > 1$ and $\alpha_0 > 0$ sufficiently small such that if $\|g - g_0\|_{\mathcal{B}} < \alpha_0$ then

$$c_0^{-1}|t| \le |\theta_f(\phi_t - f)| \le c_0|t| \ , \tag{6.8.3}$$

(when $\|f - g\|_{\mathcal{B}} < \varepsilon'/2$ and $\varepsilon' > 0$ made smaller if necessary). Since $t\mathbf{u}_{g_0} = \phi_t - f + v$ and $2M\|v\|_{\mathcal{B}} < |t|$, by Lemma 6.2 there is $c_1 > 1$ such that

$$c_1^{-1}\left(\delta_0^{pm}\right)^{-1} \le |t| \le c_1\left(\delta_0^{pm}\right)^{-1} \ . \tag{6.8.4}$$

Hence, by (6.8.3), we obtain that

$$c_2^{-1}\left(\delta_0^{pm}\right)^{-1} \le |\theta_f(\phi_t - f)| \le c_2\left(\delta_0^{pm}\right)^{-1} \ .$$

Thus,

$$c_3^{-1}\left(\delta_{lm}^{pm}\right)^{-1} \le |s_l| \le c_3\left(\delta_{lm}^{pm}\right)^{-1} \ . \tag{6.8.5}$$

Recall that $\delta_{jm}^{(j+1)m} > \beta > 2$. By Lemma 6.10 we get $\left(\delta_{lm}^{pm}\right)^{-1} \le \beta^{-(1-\mu_1)pm}$. Let us suppose from now on that $\theta_f(\phi_t - f)$ is positive and so $s_l > 0$. Hence, by Lemma 6.11, by (6.8.5) and making α_1 and α_2 smaller if necessary (and so p large enough), we obtain that

$$\begin{aligned} t_l &\ge s_l - |t_l - s_l| \ge s_l\left(1 - c_4\left(\delta_{lm}^{pm}\right)^{-(\rho-1)}\right) \\ &\ge s_l\left(1 - c_4\beta^{-(\rho-1)(1-\mu_1)pm}\right) \\ &> s_l/2 > 0 \ . \end{aligned} \tag{6.8.6}$$

Thus, t_l is positive and so it has the same sign as $\theta_f(\phi_t - f)$. By induction on $j = l, \ldots, k_0(g_0, \phi_t)$ let us show that $t_{j+1} \geq t_j$ and so that each t_j is positive as well. By Lemma 6.4, $\|v_j\|_{\mathcal{B}} \leq \varepsilon 2^{-j}$ and $t_{j+1} = \hat{\delta}_{jm}^{(j+1)m}(t_j + \sigma_{jm}(v_j))$. Since $t \mapsto \hat{\delta}_{jm}^{(j+1)m}(t)$ is C^2 as a map $\mathbb{R} \to \mathcal{C}$, we obtain that $\left|t_{j+1} - \delta_{jm}^{(j+1)m} t_j\right| \leq c_5\left(|t_j|^2 + \|v_j\|_{\mathcal{B}}\right)$. Thus,

$$
\begin{aligned}
t_{j+1} &\geq \beta t_j - c_5|t_j|^2 - c_6\varepsilon 2^{-j} \\
&\geq t_j(\beta - c_5|t_j|) - c_6 \varepsilon 2^{-j} .
\end{aligned}
\tag{6.8.7}
$$

Let $\varepsilon_1 > 0$ be as given by Lemma 6.4 and recall that $|t_j| < \varepsilon_1$ and $\varepsilon < \varepsilon_1$. Since $\beta > 1$ and by taking $\varepsilon_1 > 0$ sufficiently small, there is $\tau' > 0$ with the property that $\beta - c_5|t_j| > \beta - c_5\varepsilon_1 > 1 + 2\tau'$. By (6.8.5) and (6.8.6), we get

$$
t_j(\beta - c_5|t_j| - 1 - \tau') > t_j\tau' \geq t_l\tau' > s_l\tau'/2 > c_7\left(\delta_{lm}^{pm}\right)^{-1} . \tag{6.8.8}
$$

By Lemma 6.10,

$$
2^{-j} \leq 2^{-l} \leq \left(\delta_{lm}^{pm}\right)^{-\rho} . \tag{6.8.9}
$$

Putting together (6.8.7), (6.8.8) and (6.8.9), we deduce that

$$
t_{j+1} \geq (1 + \tau')t_j + c_7\left(\delta_{lm}^{pm}\right)^{-1} - c_6\varepsilon\left(\delta_{lm}^{pm}\right)^{-\rho} . \tag{6.8.10}
$$

Making α_2 sufficiently small (and so p large enough) and recalling from Lemma 6.10 that l is a fraction of p, we obtain that $c_7\left(\delta_{lm}^{pm}\right)^{-1} - c_6\varepsilon\left(\delta_{lm}^{pm}\right)^{-\rho} \geq 0$. Thus, by (6.8.10),

$$
t_{j+1} > (1 + \tau')t_j \tag{6.8.11}
$$

which implies that t_{j+1} has the same sign as t_j and that $\phi_t \in V_\varepsilon^+(g)$. If we suppose that $\theta_{f_0}(\phi_t - f)$ is negative, the proof that t_l is negative and that $t_{j+1} < (1 + \tau')t_j$ follows in the same way for all $j = l, \ldots, k_0(g_0, \phi_t)$ and so $\phi_t \in V_\varepsilon^-(g)$. Therefore (6.8.2) is satisfied and the claim is proved.

Let us now prove the assertions of the lemma. Take $f = g_0$ and consider the family $\phi_t = g_0 + v + t\mathbf{u}_{g_0}$. Since $2M\|v\|_{\mathcal{B}} < 2M\alpha_1 < \alpha_2/4$, the claim tell us that

$$
\begin{cases}
\phi_t \in V_\varepsilon^+(g), & \text{if } \quad \alpha_2/4 \leq t < 2\alpha_2 \\
\phi_t \in V_\varepsilon^-(g), & \text{if } \quad -\alpha_2 < t \leq -\alpha_2/4 .
\end{cases}
$$

Thus, by Lemma 6.5, there is at least one value $-\alpha_2/4 < \tau(g, v) < \alpha_2/4$ such that $\phi_{\tau(g,v)} = g_0 + v + \tau(g, v)\mathbf{u}_{g_0} \in V_\varepsilon^0(g)$.

Next, take $f = \phi_{\tau(g,v)}$, and define a new family $\psi_t = \phi_{\tau(g,v)} + t\mathbf{u}_{g_0}$. Using the claim again, this time to the family ψ_t (for which $v = 0$), we obtain that

$$
\begin{cases}
\psi_t \in V_\varepsilon^+(g), & \text{if } \quad 0 < t < \alpha_2/2 \\
\psi_t \in V_\varepsilon^-(g), & \text{if } \quad -\alpha_2/2 < t < 0 .
\end{cases}
$$

Therefore, $\tau(g,v)$ is the only value of $t \in \mathbb{R}$ between $-2\alpha_2$ and $2\alpha_2$ such that $\phi_t \in V_\varepsilon^0(g)$. Since $\|\phi_{\tau(g,v)} - g\|_{\mathcal{B}} < \varepsilon'$ we deduce from Lemma 6.5 that $\phi_{\tau(g,v)} \in W_\varepsilon^s(g)$. This proves assertions (i), (ii) and (iii).

Let us now prove assertion (iv). Set $k_0 = k_0(g,\psi_t)$. Using (6.8.11) and then (6.8.6), we have

$$|t_{k_0}| \geq (1+\tau')^{k_0-l}|t_l| \geq (1+\tau')^{k_0-l}|s_l|/2 \ . \tag{6.8.12}$$

Combining (6.8.4) and (6.8.5), we see that

$$|s_l| \geq c_3^{-1}\delta_0^{lm}\left(\delta_0^{pm}\right)^{-1} \geq c_8\beta^l|t| \ . \tag{6.8.13}$$

Taking $\tau'' = \min\{1+\tau',\beta\}$ and putting (6.8.13) back into 6.8.12 we get

$$|t_{k_0}| \geq c_9(\tau'')^{k_0}|t| \ . \tag{6.8.14}$$

By Lemma 6.4, there are $0 < \varepsilon_1 < \varepsilon_0$ such that $\varepsilon_1 \leq |t_{k_0}| \leq \varepsilon_0$. Thus, by (6.8.14), there is $\mu_0 > 0$ such that $k_0 \geq -\mu_0 \log(t)$. By Lemma 6.4, $|t_{k_0}| \leq c_9\beta^{k_0}|t|$ and so there is $\mu_1 \geq \mu_0$ such that $k_0 \leq -\mu_1\log(t)$, which proves assertion (iv). □

6.9. *Proof of the local stable manifold theorem.* It will follow from Theorem 6.15 in this section that for every $g \in \mathbb{K}$ the local stable manifold at g is a C^1 submanifold varying continuously with g. The proof of this theorem will use the following basic fact of calculus.

LEMMA 6.14. *Let X, Y be Banach spaces, let $x_0 \in X$ and consider a map $\xi : B_X(x_0,\varepsilon) \to Y$ whose image in Y falls within $B_Y(\xi(x_0),\varepsilon)$. Suppose there is a bounded linear operator $L : X \to Y$ such that for all $v \in X$ with $\|v\|_X \leq \varepsilon$,*

$$\|\xi(x_0+v) - \xi(x_0) - L(v)\|_Y \leq c_0\left(\|v\|_X + \|\xi(x_0+v)-\xi(x_0)\|_Y\right)^{1+\tau} \tag{6.9.1}$$

where $c_0 > 0$ and $\tau > 0$. If $c_0(2\varepsilon)^\tau < 1$ then ξ is differentiable at x_0 and $D\xi(x_0) = L$.

Proof. Say $\|L(v)\|_Y \leq a\|v\|_X$ for some $a > 0$. Noting that

$$\|v\|_X + \|\xi(x_0+v)-\xi(x_0)\|_Y < 2\varepsilon \ ,$$

we have from (6.9.1) that

$$\|\xi(x_0+v)-\xi(x_0)\|_Y \leq \left(a + c_0(2\varepsilon)^\tau\right)\|v\|_X + c_0(2\varepsilon)^\tau\|\xi(x_0+v)-\xi(x_0)\|_Y$$

whence

$$\|\xi(x_0+v)-\xi(x_0)\|_Y \leq \frac{a+c_0(2\varepsilon)^\tau}{1-c_0(2\varepsilon)^\tau}\|v\|_X = c_1\|v\|_X \ .$$

Putting this back into the right-hand side of (6.9.1) we get

$$\|\xi(x_0+v)-\xi(x_0)-L(v)\|_Y \leq c_2\left(\|v\|_X\right)^{1+\tau}$$

and therefore $D\xi(x_0)$ exists and equals L. □

For every $g \in \mathbb{K}$ and $\alpha_1 > 0$, let us consider the following sets

$$E_{g,\alpha_1} = \{v \in \ker \theta_g : \|v\|_{\mathcal{B}} < \alpha_1\} ,$$
$$F_g = \{g + t\mathbf{u}_g : t \in \mathbb{R}\} ,$$
$$G_{g,\alpha_1} = \{g + v + t\mathbf{u}_g : v \in E_{g,\alpha_1} \quad \text{and} \quad t \in \mathbb{R}\} .$$

THEOREM 6.15. *Set* $0 < \alpha_0 < \alpha_1 < \varepsilon$ *and* $\tau(g, v)$ *as in Proposition* 6.13. *For every* $g_0 \in \mathbb{K}$ *and every* $g \in \mathbb{K}$ *with* $\|g - g_0\|_{\mathcal{B}} < \alpha_0$, *the map* $\xi_g : E_{g_0,\alpha_1} \to F_{g_0}$ *given by* $\xi_g(v) = g_0 + \tau(g, v)\mathbf{u}_{g_0}$ *is well-defined and has the following properties:*

(i) *The graph of* ξ_g *is equal to* $W^s_\varepsilon(g) \cap G_{g_0,\alpha_1}$;

(ii) ξ_g *is* C^1 *and varies continuously with* g.

Proof. Set $\alpha_1 < \alpha_2 < \varepsilon$ and $0 < \mu_0 < \mu_1$ as in Proposition 6.13. By Proposition 6.13, the map $\xi_g : E_{g_0,\alpha_1} \to F_{g_0}$ is well-defined and assertion (i) is satisfied. Let $\hat{\xi}_g : E_{g_0,\alpha_1} \to \mathbb{R}$ be given by $\hat{\xi}_g(v) = \tau(g, v)$ where $\tau(g, v)$ is given by (i) in Proposition 6.13. To prove assertion (ii), it is enough to show that $\hat{\xi}_g : E_{g_0,\alpha_1} \to \mathbb{R}$ is C^1 and varies continuously with g. Let $v_1, v_2 \in E_{g_0,\alpha_1}$ and set

$$f_1 = g_0 + v_1 + \hat{\xi}_g(v_1)\mathbf{u}_{g_0} \tag{6.9.2}$$

and $f_2 = g_0 + v_2 + \hat{\xi}_g(v_2)\mathbf{u}_{g_0}$. By Lemma 6.12, we get

$$\begin{aligned}\Big|\theta_{f_1}(v_2 - v_1) + \theta_{f_1}(\mathbf{u}_{g_0})(\hat{\xi}_g(v_2) - \hat{\xi}_g(v_1))\Big| &= |\theta_{f_1}(f_2 - f_1)| \\ &\le c_0 \Big(\|v_2 - v_1\|_{\mathcal{B}} + |\hat{\xi}_g(v_2) - \hat{\xi}_g(v_1)|\Big)^{1+\tau} .\end{aligned}$$

By Lemma 6.7, and taking $\varepsilon > 0$ sufficiently small, we have that $\theta_{f_1}(\mathbf{u}_{g_0})$ is uniformly bounded away from 0. Therefore, by Lemma 6.14, we deduce that $\hat{\xi}_g$ is differentiable at every v_1 with derivative given by

$$D\hat{\xi}_g(v_1) = -(\theta_{f_1}(\mathbf{u}_{g_0}))^{-1}\theta_{f_1} . \tag{6.9.3}$$

From Lemma 6.7, θ_{f_1} varies continuously with f_1 and so $D\hat{\xi}_g(v_1)$ also varies continuously in a neighborhood of v_1. Hence, $\hat{\xi}_g$ is a C^1 map.

Let us check that $\hat{\xi}_g$ varies continuously with g in the C^1 sense and, more precisely, that the map $\mathbb{K} \cap \mathcal{B}(g_0, \alpha_0) \to C^1(E_{g_0,\alpha_1}, \mathbb{R})$ given by $g \mapsto \hat{\xi}_g$ is continuous. Taking into account that $D\hat{\xi}_g$ is given by (6.9.3) and that f_1 is given by (6.9.2), and that by Lemma 6.7 the map $f_1 \mapsto \theta_{f_1}$ is uniformly continuous (as a map into $\mathcal{B}^*$), we see that it suffices to prove $g \mapsto \hat{\xi}_g$ is continuous as a map into $C^0(E_{g_0,\alpha_1}, \mathbb{R})$.

To do this, let $v \in E_{g_0,\alpha_1}$ be such that $g = g_0 + v + \hat{\xi}_g(v)\mathbf{u}_{g_0}$, take $g_1 \in \mathbb{K}$ with $\|g_1 - g_0\|_{\mathcal{B}} < \alpha_0$ and let $w \in E_{g_0,\alpha_1}$ be such that $g_1 = g_0 + w + \hat{\xi}_{g_1}(w)\mathbf{u}_{g_0}$. Now, we have the following claim.

Claim. *There exist $c_1 > 0$ and $0 < \gamma < 1$ such that*

$$c_1^{-1}|\hat{\xi}_{g_1}(w) - \hat{\xi}_g(w)|^{1/\gamma} \leq |\hat{\xi}_{g_1}(z) - \hat{\xi}_g(z)| \leq c_1|\hat{\xi}_{g_1}(w) - \hat{\xi}_g(w)|^{\gamma}, \tag{6.9.4}$$

for all $z \in E_{g_0,\alpha_1}$.

Let us assume this claim for a moment. Its geometric meaning is that the distances between corresponding points of the graphs of $\hat{\xi}_g$ and $\hat{\xi}_{g_1}$ along the vertical fibers $(\{z\} \times F_{g_0})$ are uniform. We want to control such distances in terms of $\|g_1 - g\|_{\mathcal{B}}$. The above claim tell us that it is enough to control $|\hat{\xi}_{g_1}(w) - \hat{\xi}_g(w)|$. Hence, write

$$\begin{aligned} g - g_1 &= v - w + \left(\hat{\xi}_g(v) - \hat{\xi}_{g_1}(w)\right)\mathbf{u}_{g_0} \\ &= v - w + (a+b)\,\mathbf{u}_{g_0} \end{aligned}$$

where $a = \hat{\xi}_g(w) - \hat{\xi}_{g_1}(w)$ and $b = \hat{\xi}_g(v) - \hat{\xi}_g(w)$. Since $\hat{\xi}_g$ is C^1, we have $|b| \leq c_2\|v - w\|_{\mathcal{B}}$. On the other hand, since $\mathcal{B} = \operatorname{Ker}\theta_{g_0} \bigoplus \mathbb{R}\mathbf{u}_{g_0}$ is a splitting into closed subspaces, there exists a constant $c_3 > 0$ such that

$$\max\{\|v - w\|_{\mathcal{B}}, |a+b|\} \leq c_3\|g - g_1\|_{\mathcal{B}}.$$

But then

$$\begin{aligned} |a| &\leq \|g - g_1\|_{\mathcal{B}} + \|v - w\|_{\mathcal{B}} + |b| \\ &\leq (1 + c_3 + c_2c_3)\|g - g_1\|_{\mathcal{B}}. \end{aligned}$$

Hence $\left|\hat{\xi}_g(w) - \hat{\xi}_{g_1}(w)\right| \leq c_4\|g - g_1\|_{\mathcal{B}}$, and given the claim this proves that $g \mapsto \hat{\xi}_g$ is indeed continuous.

Finally, let us prove the claim. For each $z \in E_{g_0,\alpha_1}$, let $h = g_0 + z + \hat{\xi}_{g_1}(z)\mathbf{u}_{g_0}$. Set

$$\begin{aligned} t'_k &= t_k(g, g_1), & t''_k &= t_k(g, h), \\ u'_k &= u_{T^{km}(g)}(t'_k), & u''_k &= u_{T^{km}(g)}(t''_k), \end{aligned}$$

as given by Lemma 6.4, and set (also as in that lemma)

$$\begin{aligned} k'_0 &= k_0(g, g_1) = \min\{j : |t'_j| \geq \varepsilon_1\} \\ k''_0 &= k_0(g, h) = \min\{j : |t''_j| \geq \varepsilon_1\}. \end{aligned} \tag{6.9.5}$$

Applying Lemma 6.4, we obtain, for all $k \leq \min\{k'_0, k''_0\}$, the estimates

$$\begin{aligned} \|T^{km}(g_1) - u'_k\|_{\mathcal{B}} &\leq 2^{-k}\|g_1 - g\|_{\mathcal{B}} \\ \|T^{km}(h) - u''_k\|_{\mathcal{B}} &\leq 2^{-k}\|h - g\|_{\mathcal{B}}. \end{aligned} \tag{6.9.6}$$

Since $h \in W^s_\varepsilon(g_1)$, we also have, by Lemma 6.5,

$$\|T^{km}(h) - T^{km}(g_1)\|_{\mathcal{B}} \leq \varepsilon c_5 2^{-k}. \tag{6.9.7}$$

Combining (6.9.6) and (6.9.7), we get

$$\|u'_k - u''_k\|_{\mathcal{B}} \le c_6 2^{-k} .$$

Hence, by Lemma 6.2, we get

$$|t_k(g, g_1) - t_k(g, h)| \le c_7 2^{-k} , \tag{6.9.8}$$

for all $k \le \min\{k'_0, k''_0\}$. Using (6.8.11) together with (6.9.8), we deduce that there exists a uniform constant $c_8 > 0$ such that

$$k'_0 - c_8 \le k''_0 \le k'_0 + c_8 . \tag{6.9.9}$$

On the other hand, applying (iv) in Proposition 6.13, we also have

$$\begin{aligned} -\mu_0 \log(|\tau(g_1, w) - \tau(g, w)|) &\le k'_0 \le -\mu_1 \log(|\tau(g_1, w) - \tau(g, w)|) \\ -\mu_0 \log(|\tau(g_1, z) - \tau(g, z)|) &\le k''_0 \le -\mu_1 \log(|\tau(g_1, z) - \tau(g, z)|) . \end{aligned} \tag{6.9.10}$$

Combining (6.9.9) and (6.9.10) and noting that

$$\tau(g_1, w) = \hat{\xi}_{g_1}(w) \ , \ \tau(g_1, z) = \hat{\xi}_{g_1}(z) \ , \tau(g, w) = \hat{\xi}_g(w) \text{ and } \tau(g, z) = \hat{\xi}_g(z) \ ,$$

we get at last

$$c_9^{-1} \left|\hat{\xi}_{g_1}(w) - \hat{\xi}_g(w)\right|^{\mu_0/\mu_1} \le \left|\hat{\xi}_{g_1}(z) - \hat{\xi}_g(z)\right| \le c_9 \left|\hat{\xi}_{g_1}(w) - \hat{\xi}_g(w)\right|^{\mu_1/\mu_0}$$

for some $c_9 > 1$. This proves the claim with $\gamma = \mu_0/\mu_1 < 1$ and $c_1 = c_9$. □

Remark 6.3. Note that by Proposition 6.13 there exists a uniform $0 < \tilde{\varepsilon} < \varepsilon$ such that $W^s_{\tilde{\varepsilon}}(g) \subset G_{g_0,\alpha_1}$ for all $g_0 \in \mathbb{K}$ with $\|g_0 - g\|_{\mathcal{B}} < \alpha_0$.

7. Smooth holonomies

In the previous section we proved that a robust operator T has C^1 local stable manifolds through each point of its hyperbolic basic set $\mathbb{K}$, and that such manifolds form a C^0 lamination (near each point of $\mathbb{K}$). A natural question that may be asked at this point is this: how smooth is the holonomy of this lamination? To answer this question we shall assume that there exists a homeomorphism $H : \Theta^{\mathbb{Z}} \to \mathbb{K}$ of a finite-type shift space onto $\mathbb{K}$ which conjugates the two-sided full shift $\sigma : \Theta^{\mathbb{Z}} \to \mathbb{K}$ to our robust operator T restricted to $\mathbb{K}$. Under this topological assumption, and an additional geometric assumption concerning the unstable manifolds of points in the attractor –both of which are satisfied by the renormalization operator– we shall prove below that the holonomies of the local stable laminations are $C^{1+\theta}$ for some $\theta > 0$.

7.1. *Smooth holonomies for robust operators.* Let $\mathbb{K} \subset \mathcal{O}_A$ be a hyperbolic basic set of a C^2 operator $T : \mathcal{O}_A \to \mathcal{A}$ which is topologically conjugated to a two-sided shift of finite type. For $\varepsilon_0 > 0$ small enough and for every $g \in \mathbb{K}$ let $t \to u_g(t)$ be a parametrization of the local unstable manifold $W^u_{\varepsilon_0}(g)$. Set

$$\mathbb{K}_g = \mathbb{K} \cap \overline{W^u_{\varepsilon_0}(g)} \text{ and } K_g = u_g^{-1}(\mathbb{K}_g) \ .$$

Let

$$\Sigma_{\ldots,\theta_{k-1},\theta_k} = \left\{ (\theta'_j) \in \Theta^{\mathbb{Z}} : \theta'_j = \theta_j \quad \text{for} \quad \text{all} j \leq k \right\} .$$

If $H(\Sigma_{\ldots,\theta_k}) \cap \mathbb{K}_g \neq \emptyset$ then denote by $\Delta_{\ldots,\theta_k}(g)$ the smallest interval in $\mathbb{R}$ such that $u_g(\Delta_{\ldots,\theta_k}(g)) \supset H(\Sigma_{\ldots,\theta_k}) \cap \mathbb{K}_g$. Let $\mathbf{C}_k(g)$ be the set of all these intervals $\Delta_{\ldots,\theta_k}(g)$.

Definition 7.1. We say that the local unstable manifolds $W^u_{\varepsilon_0}(g)$ have *geometry bounded* by $\alpha > 0$ if for every $g \in \mathbb{K}$, K_g has geometry bounded by $\alpha > 0$ with respect to the collection $(\mathbf{C}_k(g))_{k \geq 0}$ (in the sense of §3).

Now suppose in addition that the operator T is robust with respect to the Banach spaces $(\mathcal{B}, \mathcal{C}, \mathcal{D})$. By Theorem 6.1, the local stable manifolds of T in $\mathcal{B}$ form a C^0 lamination Let $F : [-\mu_0, \mu_0] \to \mathcal{B}(g, \varepsilon)$ be a C^2 curve transversal to the stable lamination. Let K_F be the set of all values $r \in [-\mu_0, \mu_0]$ such that

$$f_r = F(r) \in \bigcup_{g_0 \in \mathbb{K} \cap W^u_{\varepsilon_0}(g_0)} W^s_{\varepsilon_0}(g_0) \ .$$

The *holonomy map* $\phi_F : F(K_F) \to W^u_{\varepsilon_0}(g)$ associates to each f_r the point $\phi_F(f_r)$ such that $f_r \in W^s_{\varepsilon_0}(\phi_F(f_r))$. In local coordinates, the holonomy map ϕ_F is given by $\psi_F : K_F \to K_g$ where $\psi_F(t) = u_g \circ \phi_F \circ F^{-1}$ and $K_F, K_g \subset \mathbb{R}$. The C^2 curve $F : [-\mu_0, \mu_0] \to \mathcal{B}(g, \varepsilon)$ is an *ordered transversal to the stable foliation* if F is transversal to the stable foliation, $\phi_F : F(K_F) \to W^u_{\varepsilon_0}(g)$ extends to $F([-\mu_0, \mu_0])$ as a homeomorphism $\hat{\phi}_F$ over its image such that $\phi_F(F(K_F)) = \hat{\phi}_F(F(K_F)) \cap \mathbb{K}$.

We note that, by Remark 6.2 and by Theorem 6.15, there is $\varepsilon_1 < \varepsilon_0$ small enough such that a C^2 transversal to $W^s_{\varepsilon_1}(g)$ in a point f is an ordered transversal to the stable foliation in a small neighborhood of f.

THEOREM 7.1. *Let $\mathbb{K} \subset \mathcal{O}_A$ be a hyperbolic basic set of a C^2 operator $T : \mathcal{O}_A \to \mathcal{A}$ which is robust with respect to $(\mathcal{B}, \mathcal{C}, \mathcal{D})$. Suppose that there exits $\varepsilon_0 > 0$ such that the local unstable manifolds $W^u_{\varepsilon_0}(g)$ of $g \in \mathbb{K}$ have bounded geometry. There exists $0 < \varepsilon < \varepsilon_0$ with the property that for every C^2 ordered transversal $F : [-\mu_0, \mu_0] \to \mathcal{B}(g, \varepsilon)$ to the stable foliation in $\mathcal{B}$, the holonomy $\phi_F : F(K_F) \to W^u_{\varepsilon_0}(g)$ has a $C^{1+\theta}$ diffeomorphic extension to $F([-\mu_0, \mu_0])$ for some $\theta > 0$.*

Example 7.1. As we know from Theorem 2.4, the renormalization operator $T = R^N : \mathbb{O}_{\mathbb{A}} \to \mathbb{A}$ is hyperbolic over $\mathbb{K}$. As we shall see in Theorem

8.1, T is robust with respect to $(\mathbb{A}^r, \mathbb{A}^s, \mathbb{A}^0)$ provided $s > s_0$ with s_0 sufficiently close to 2 and $r > s+1$ is not an integer. By Theorem 2.1, there is a two-sided full shift topologically conjugated to $T|\mathbb{K}$. By Lemmas 9.3 and 9.6 respectively on pages 403 and 405 of Lyubich's paper [20], there is $\alpha > 0$ such that the local unstable manifolds $W^u_{\varepsilon_0}(g)$ have geometry bounded by α. Hence the renormalization operator T satisfies the hypotheses of Theorem 7.1.

In what follows, the notation $A = \mathcal{O}(B)$ means that $\mu_1^{-1}B \le A \le \mu_1 B$ and the notation $A = B(1 \pm \mathcal{O}(C))$ means that $B(1-\mu_2 C) \le A \le B(1+\mu_2 C)$ for some constants $\mu_1 > 1$ and $\mu_2 > 0$.

The proof of Theorem 7.1 will be a consequence of the following lemmas.

LEMMA 7.2. *For every C^2 curve $F : [-\mu_0, \mu_0] \to \mathcal{B}(g, \varepsilon)$ transversal to the stable foliation and for all $r, t \in [-\mu_0, \mu_0]$ such that $r < t$,*

$$\begin{aligned} \|f_t - f_r\|_{\mathcal{X}} &= \mathcal{O}(|t-r|) \ , \\ |\theta_{f_r}(f_t - f_r)| &= \mathcal{O}(|t-r|) \ , \end{aligned} \tag{7.1.1}$$

and for all $s, r, t \in [-\mu_0, \mu_0]$ such that $s < r < t$,

$$\begin{aligned} \frac{\|f_t - f_r\|_{\mathcal{X}}}{\|f_s - f_r\|_{\mathcal{X}}} &= \frac{|t-r|}{|s-r|}(1 \pm \mathcal{O}(|t-s|)) \ , \\ \frac{\|\theta_{f_r}(f_t - f_r)\|_{\mathcal{X}}}{\|\theta_{f_r}(f_s - f_r)\|_{\mathcal{X}}} &= \frac{|t-r|}{|s-r|}(1 \pm \mathcal{O}(|t-s|)) \ , \end{aligned} \tag{7.1.2}$$

where $\mathcal{X} \in \{B, C, D\}$.

Proof. By Lemma 6.2, there are $\nu_1, \nu_2 > 0$ such that for all $r \in K_f$, $\|\mathbf{u}_r\|_{\mathcal{X}} > \nu_1$ and $|\theta_{f_r}(\mathbf{u}_r)| > \nu_2$. Since F is C^2,

$$\begin{aligned} f_t - f_r &= (t-r)\mathbf{u}_r \pm \mathcal{O}(|t-r|^2), \\ \theta_{f_r}(f_t - f_r) &= (t-r)\theta_{f_r}(\mathbf{u}_r) \pm \mathcal{O}(|t-r|^2) \ , \end{aligned}$$

and so (7.1.1) follows. Taking $s < r < t$, we get

$$\begin{aligned} \frac{\|f_t - f_r\|_{\mathcal{X}}}{\|f_s - f_r\|_{\mathcal{X}}} &= \frac{\|\mathbf{u}_r\|_{\mathcal{X}}\,|t-r|(1 \pm \mathcal{O}(|t-r|))}{\|\mathbf{u}_r\|_{\mathcal{X}}\,|s-r|(1 \pm \mathcal{O}(|s-r|))} \\ &= \frac{|t-r|}{|s-r|}(1 \pm \mathcal{O}(|t-s|)) \ . \end{aligned}$$

The second estimate in (7.1.2) is obtained in similar fashion. □

In what follows, it will be more convenient to denote $\phi_F(f_r)$ by $g_{\psi_F(r)}$. We will also work with a fixed $0 < \varepsilon < \varepsilon_0$ for which Lemma 6.8 holds.

LEMMA 7.3. *Set $l = l(g_{\psi_F(r)}, p)$ as in Lemma 6.10. Let $F : [-\mu_0, \mu_0] \to \mathcal{B}(g, \varepsilon)$ be a C^2 curve transversal to the stable foliation. For all $p > 0$ sufficiently large and all $s, r, t \in K_F$ such that*

$$|t-r| = \mathcal{O}\left((\delta_0^{pm})^{-1}\right) \quad \textit{and} \quad |s-r| = \mathcal{O}\left((\delta_0^{pm})^{-1}\right) \ ,$$

we have

$$\left\|T^{lm}(f_t) - T^{lm}(f_r)\right\|_{\mathcal{C}} = \mathcal{O}\left(\left(\delta_{lm}^{pm}\right)^{-1}\right) , \tag{7.1.3}$$

$$\|T^{lm}(f_s) - T^{lm}(f_r)\|_{\mathcal{C}} = \mathcal{O}\left(\left(\delta_{lm}^{pm}\right)^{-1}\right) ,$$

$$\frac{\|T^{lm}(f_t) - T^{lm}(f_r)\|_{\mathcal{C}}}{\|T^{lm}(f_s) - T^{lm}(f_r)\|_{\mathcal{C}}} = \frac{|t-r|}{|s-r|}\left(1 \pm \mathcal{O}\left(\left(\delta_{lm}^{pm}\right)^{-(\rho-1)}\right)\right) .$$

Proof. Using Lemma 6.2 and (7.1.1), we get

$$|\theta_{f_r}(f_t - f_r)| = \mathcal{O}\left(\left(\delta_0^{pm}\right)^{-1}\right) \quad \text{and} \quad |\theta_{f_r}(f_s - f_r)| = \mathcal{O}\left(\left(\delta_0^{pm}\right)^{-1}\right) .$$

Thus, taking p sufficiently large and using Lemmas 6.9 and 6.10 we deduce that

$$\begin{aligned} \left\|T^{lm}(f_t) - T^{lm}(f_r)\right\|_{\mathcal{C}} &= \left|\delta_0^{lm}\theta_{f_r}(f_t - f_r)\right| \pm \mathcal{O}\left(\left(\delta_{lm}^{pm}\right)^{-\rho}\right) \\ &= \mathcal{O}\left(\left(\delta_{lm}^{pm}\right)^{-1}\right) \pm \mathcal{O}\left(\left(\delta_{lm}^{pm}\right)^{-\rho}\right) \\ &= \mathcal{O}\left(\left(\delta_{lm}^{pm}\right)^{-1}\right) . \end{aligned} \tag{7.1.4}$$

Similarly, $\left\|T^{lm}(f_s) - T^{lm}(f_r)\right\|_{\mathcal{C}} = \mathcal{O}\left(\left(\delta_{lm}^{pm}\right)^{-1}\right)$. This proves the first two inequalities in (7.1.3). Combining (7.1.1) with (7.1.4), we see that

$$\begin{aligned} \frac{\|T^{lm}(f_t) - T^{lm}(f_r)\|_{\mathcal{C}}}{\|T^{lm}(f_s) - T^{lm}(f_r)\|_{\mathcal{C}}} &= \frac{|\delta_0^{lm}\theta_{f_r}(f_t - f_r)| \pm \mathcal{O}\left(\left(\delta_{lm}^{pm}\right)^{-\rho}\right)}{|\delta_0^{lm}\theta_{f_r}(f_s - f_r)| \pm \mathcal{O}\left(\left(\delta_{lm}^{pm}\right)^{-\rho}\right)} \\ &= \frac{|\theta_{f_r}(f_t - f_r)|\left(1 \pm \mathcal{O}\left(\left(\delta_{lm}^{pm}\right)^{-(\rho-1)}\right)\right)}{|\theta_{f_r}(f_s - f_r)|\left(1 \pm \mathcal{O}\left(\left(\delta_{lm}^{pm}\right)^{-(\rho-1)}\right)\right)} . \end{aligned}$$

Therefore, by Lemma 7.2, we get

$$\frac{\|T^{lm}(f_t) - T^{lm}(f_r)\|_{\mathcal{C}}}{\|T^{lm}(f_s) - T^{lm}(f_r)\|_{\mathcal{C}}} = \frac{|t-r|}{|s-r|}\left(1 \pm \mathcal{O}\left(\left(\delta_{lm}^{pm}\right)^{-(\rho-1)}\right)\right) ,$$

and this proves the last inequality in (7.1.3). □

LEMMA 7.4. *Set $l = l\left(g_{\psi_F(r)}, p\right)$ as in Lemma 6.10. Let $F : [-\mu_0, \mu_0] \to \mathcal{B}(g, \varepsilon)$ be a C^2 curve transversal to the stable foliation. For every $s \in K_F$ and $s' = \psi_F(s) \in K_g$,*

$$\left\|T^{lm}(f_s) - T^{lm}(g_{s'})\right\|_{\mathcal{C}} \leq \mathcal{O}\left(\left(\delta_{lm}^{pm}\right)^{-\rho}\right) .$$

Furthermore, for all p large enough and all $s', r', t' \in K_g$ such that $s = \psi_F^{-1}(s')$, $r = \psi_F^{-1}(r')$, $t = \psi_F^{-1}(t') \in K_f$,

$$|t' - r'| = \mathcal{O}\left(\left(\delta_0^{pm}\right)^{-1}\right) \quad \textit{and} \quad |s' - r'| = \mathcal{O}\left(\left(\delta_0^{pm}\right)^{-1}\right) ,$$

we have

$$\frac{\|T^{lm}(f_t) - T^{lm}(f_r)\|_{\mathcal{C}}}{\|T^{lm}(f_s) - T^{lm}(f_r)\|_{\mathcal{C}}} = \frac{\|T^{lm}(g_{t'}) - T^{lm}(g_{r'})\|_{\mathcal{C}}}{\|T^{lm}(g_{s'}) - T^{lm}(g_{r'})\|_{\mathcal{C}}} \left(1 \pm \mathcal{O}\left(\left(\delta_{lm}^{pm}\right)^{-(\rho-1)}\right)\right) .$$

Proof. By Lemmas 6.5 and 6.10,

$$\left\|T^{lm}(f_s) - T^{lm}(g_{s'})\right\|_{\mathcal{C}} \le C_3 \varepsilon 2^{-l} \le \mathcal{O}\left(\left(\delta_{lm}^{pm}\right)^{-\rho}\right) .$$

Thus, applying Lemma 7.3 to the transversal given by the local unstable manifold $\{g_t\}$ we get

$$\frac{\|T^{lm}(f_t) - T^{lm}(f_r)\|_{\mathcal{C}}}{\|T^{lm}(f_s) - T^{lm}(f_r)\|_{\mathcal{C}}} = \frac{\|T^{lm}(g_{t'}) - T^{lm}(g_{r'})\|_{\mathcal{C}} \left(1 \pm \mathcal{O}\left(\left(\delta_{lm}^{pm}\right)^{-(\rho-1)}\right)\right)}{\|T^{lm}(g_{s'}) - T^{lm}(g_{r'})\|_{\mathcal{C}} \left(1 \pm \mathcal{O}\left(\left(\delta_{lm}^{pm}\right)^{-(\rho-1)}\right)\right)}$$

$$= \frac{\|T^{lm}(g_{t'}) - T^{lm}(g_{r'})\|_{\mathcal{C}}}{\|T^{lm}(g_{s'}) - T^{lm}(g_{r'})\|_{\mathcal{C}}} \left(1 \pm \mathcal{O}\left(\left(\delta_{lm}^{pm}\right)^{-(\rho-1)}\right)\right) . \square$$

Proof of Theorem 7.1. Let $p > 0$ be so large that Lemmas 7.3 and 7.4 are satisfied and let t, s, r, t', s', r' be as in Lemma 7.4. First, a claim.

CLAIM. *We have*

$$|t - r| = \mathcal{O}\left(\left(\delta_0^{pm}\right)^{-1}\right) \quad \text{and} \quad |s - r| = \mathcal{O}\left(\left(\delta_0^{pm}\right)^{-1}\right) .$$

Assuming this claim for a moment, we finish the proof of Theorem 7.1 as follows. Set $l = l(g_{r'}, p)$ as in Lemma 6.10. By Lemmas 6.10 and 7.2, there is $0 < \tau_1 < 1$ such that $\delta_{lm}^{pm} \le \mathcal{O}\left(|t' - r'|^{\tau_1}\right)$. Therefore, by Lemmas 7.3 and 7.4 we deduce that

$$\text{(7.1.5)} \qquad \begin{aligned} \frac{|t-r|}{|s-r|} &= \frac{\|T^{lm}(f_t) - T^{lm}(f_r)\|_{\mathcal{C}}}{\|T^{lm}(f_s) - T^{lm}(f_r)\|_{\mathcal{C}}} \left(1 \pm \mathcal{O}\left(\left(\delta_{lm}^{pm}\right)^{-(\rho-1)}\right)\right) \\ &= \frac{\|T^{lm}(g_{t'}) - T^{lm}(g_{r'})\|_{\mathcal{C}}}{\|T^{lm}(g_{s'}) - T^{lm}(g_{r'})\|_{\mathcal{C}}} \left(1 \pm \mathcal{O}\left(\left(\delta_{lm}^{pm}\right)^{-(\rho-1)}\right)\right) \\ &= \frac{|t'-r'|}{|s'-r'|} \left(1 \pm \mathcal{O}\left(|t'-r'|^{-(\rho-1)\tau_1}\right)\right) . \end{aligned}$$

Since $\mathcal{K}_g$ has bounded geometry and by Theorem 9.5 on page 549 of [26], the inequalities (7.1.5) imply that the map ψ_F has a $C^{1+\theta}$ diffeomorphic extension to $\mathbb{R}$ for some $0 < \theta < 1$.

Let us now prove the claim. Let $\hat{p}$ be such that $|t - r| = \mathcal{O}\left(\left(\delta_0^{\hat{p}m}\right)^{-1}\right)$. All that we have to show is that

$$\text{(7.1.6)} \qquad |\hat{p} - p| \le \mathcal{O}(1) .$$

Set $\hat{l} = \hat{l}(g_{r'}, \hat{p})$ as in Lemma 6.10. By Lemmas 6.9 and 6.11, and the estimates in (7.1.1), for $k \leq \min\{l, \hat{l}\}$,

$$\left\|T^{km}(g_{t'}) - T^{km}(g_{r'})\right\|_{\mathcal{C}} = \left|\delta_0^{km}\theta_{g_r}(g_{t'} - g_{r'})\right| \pm \mathcal{O}\left(\left(\delta_{km}^{pm}\right)^{-\rho} + 2^{-k}\right) , \tag{7.1.7}$$

$$\left\|T^{km}(f_t) - T^{km}(f_r)\right\|_{\mathcal{C}} = \left|\delta_0^{km}\theta_{f_r}(f_t - f_r)\right| \pm \mathcal{O}\left(\left(\delta_{km}^{\hat{p}m}\right)^{-\rho} + 2^{-k}\right) .$$

By Lemma 6.5, we get (for all $k \leq \min\left\{l, \hat{l}\right\}$)

$$\left\|T^{km}(f_t) - T^{km}(f_r)\right\|_{\mathcal{C}} = \left\|T^{km}(g_{t'}) - T^{km}(g_{r'})\right\|_{\mathcal{C}} \pm \mathcal{O}\left(2^{-k}\right) . \tag{7.1.8}$$

Let us consider separately the case (i) where $p \leq \hat{p}$ and the case (ii) where $p \geq \hat{p}$.

Case (i). Here $l \leq \hat{l}$ and by Lemma 6.10 we get

$$\left(\delta_{lm}^{\hat{p}m}\right)^{-1} \leq \left(\delta_{lm}^{pm}\right)^{-1} 2^{-l} \leq \mathcal{O}\left(\left(\delta_{lm}^{pm}\right)^{-\rho}\right) . \tag{7.1.9}$$

By Lemma 7.3 applied to the transversal given by the local unstable manifold $\{g_t\}$,

$$\left\|T^{lm}(g_{t'}) - T^{lm}(g_{r'})\right\|_{\mathcal{C}} = \mathcal{O}\left(\left(\delta_{lm}^{pm}\right)^{-1}\right) . \tag{7.1.10}$$

On the other hand, by (7.1.8),

$$\left\|T^{lm}(f_t) - T^{lm}(f_r)\right\|_{\mathcal{C}} = \left\|T^{lm}(g_{t'}) - T^{lm}(g_{r'})\right\|_{\mathcal{C}} \pm \mathcal{O}\left(2^{-l}\right) .$$

But 2^{-l} is much smaller than $(\delta_{lm}^{pm})^{-1}$. Hence by (7.1.10),

$$\left\|T^{lm}(f_t) - T^{lm}(f_r)\right\|_{\mathcal{C}} = \mathcal{O}\left(\left(\delta_{lm}^{pm}\right)^{-1}\right) . \tag{7.1.11}$$

Thus, by (7.1.7) and (7.1.9), we deduce that

$$\begin{aligned}\left|\delta_0^{lm}\theta_{f_r}(f_t - f_r)\right| &= \left\|T^{lm}(f_t) - T^{lm}(f_r)\right\|_{\mathcal{C}} \pm \mathcal{O}\left(\left(\delta_{lm}^{\hat{p}m}\right)^{-\rho}\right) \pm \mathcal{O}\left(2^{-l}\right) \\ &= \mathcal{O}\left(\left(\delta_{lm}^{pm}\right)^{-1}\right) \pm \mathcal{O}\left(\left(\delta_{lm}^{\hat{p}m}\right)^{-\rho}\right) .\end{aligned}$$

Since $\left(\delta_{lm}^{pm}\right)^{-1}$ is much larger than $\left(\delta_{lm}^{\hat{p}m}\right)^{-\rho}$, it follows that

$$\left|\delta_0^{lm}\theta_{f_r}(f_t - f_r)\right| = \mathcal{O}\left(\left(\delta_{lm}^{pm}\right)^{-1}\right) .$$

This shows that

$$\left|\theta_{f_r}(f_t - f_r)\right| = \mathcal{O}\left(\left(\delta_0^{pm}\right)^{-1}\right) .$$

Therefore, by Lemma 7.2, we get $\|f_t - f_r\|_{\mathcal{C}} = \mathcal{O}\left(\left(\delta_0^{pm}\right)^{-1}\right)$, which in turn implies (7.1.6).

Case (ii). Here $\hat{l} \leq l$, and Lemma 6.10 tells us that

$$\left(\delta_{\hat{l}m}^{pm}\right)^{-1} \leq \left(\delta_{\hat{l}m}^{\hat{p}m}\right)^{-1} 2^{-\hat{l}} \leq \mathcal{O}\left(\left(\delta_{\hat{l}m}^{\hat{p}m}\right)^{-\rho}\right) . \tag{7.1.12}$$

Applying Lemma 7.3, we get

$$\left\|T^{\hat{l}m}(f_t) - T^{\hat{l}m}(f_r)\right\|_{\mathcal{C}} = \mathcal{O}\left(\left(\delta_{\hat{l}m}^{\hat{p}m}\right)^{-1}\right) . \tag{7.1.13}$$

On the other hand, by (7.1.12) and (7.1.7),

$$\left\|T^{\hat{l}m}(f_t) - T^{\hat{l}m}(f_r)\right\|_{\mathcal{C}} = \mathcal{O}\left(\left|\delta_0^{\hat{l}m}\theta_{f_r}(f_t - f_r)\right|\right) \pm \mathcal{O}\left(\left(\delta_{\hat{l}m}^{\hat{p}m}\right)^{-\rho}\right) .$$

But $\left(\delta_{\hat{l}m}^{\hat{p}m}\right)^{-\rho}$ is much smaller than $\left(\delta_{\hat{l}m}^{\hat{p}m}\right)^{-1}$. Hence, by (7.1.13), we get

$$\left\|T^{\hat{l}m}(f_t) - T^{\hat{l}m}(f_r)\right\|_{\mathcal{C}} = \mathcal{O}\left(\left|\delta_0^{\hat{l}m}\theta_{f_r}(f_t - f_r)\right|\right) . \tag{7.1.14}$$

By (7.1.8), we have also

$$\left\|T^{\hat{l}m}(f_t) - T^{\hat{l}m}(f_r)\right\|_{\mathcal{C}} = \left\|T^{\hat{l}m}(g_{t'}) - T^{\hat{l}m}(g_{r'})\right\|_{\mathcal{C}} \pm \mathcal{O}\left(2^{-\hat{l}}\right) .$$

But $2^{-\hat{l}} \leq \mathcal{O}\left(\left(\delta_{\hat{l}m}^{\hat{p}m}\right)^{-\rho}\right)$. Hence, again by (7.1.13), we deduce that

$$\left\|T^{\hat{l}m}(f_t) - T^{\hat{l}m}(f_r)\right\|_{\mathcal{C}} = \mathcal{O}\left(\left\|T^{\hat{l}m}(g_{t'}) - T^{\hat{l}m}(g_{r'})\right\|_{\mathcal{C}}\right) . \tag{7.1.15}$$

By Lemma 7.2 and (7.1.7),

$$\left\|T^{\hat{l}m}(g_{t'}) - T^{\hat{l}m}(g_{r'})\right\|_{\mathcal{C}} = \mathcal{O}\left(\left(\delta_{\hat{l}m}^{pm}\right)^{-1}\right) + \mathcal{O}\left(\left(\delta_{\hat{l}m}^{\hat{p}m}\right)^{-\rho}\right) .$$

Therefore, by (7.1.13) and (7.1.15),

$$\left\|T^{\hat{l}m}(g_{t'}) - T^{\hat{l}m}(g_{r'})\right\|_{\mathcal{C}} = \mathcal{O}\left(\left(\delta_{\hat{l}m}^{pm}\right)^{-1}\right) .$$

But then, using (7.1.14) and (7.1.15) once more, we deduce that

$$|\theta_{f_r}(f_t - f_r)| = \mathcal{O}\left((\delta_0^{pm})^{-1}\right) .$$

Therefore, by Lemma 7.2, we get $\|f_t - f_r\|_{\mathcal{C}} = \mathcal{O}\left((\delta_0^{pm})^{-1}\right)$ which in turn implies (7.1.6). □

8. The renormalization operator is robust

From the very beginning, our main goal is to show that the renormalization operator is "hyperbolic" in $\mathbb{U}^r$, provided r is sufficiently large. More precisely, we want to establish Theorem 2.5 and Corollary 2.6. We have already at our disposal an abstract theorem (Theorem 6.1) showing that any robust operator is indeed "hyperbolic". Hence, our work has been essentially reduced to showing that the renormalization operator T, or any one of its powers, is robust (see Theorem 8.1 below). The proofs of Theorem 2.5 and Corollary 2.6 will be given in §8.6.

We emphasize the important role played by the geometric estimates of §5.2 in the verification of properties **B5** and **B6** of a robust operator (Definition 6.1) for an iterate of the renormalization operator (see §8.4). Properties **B2**, **B3**, and **B4** are relatively straightforward consequences of the properties of the composition operator studied in §8.1 and are proved in §8.2 and §8.3 .

In this section, we shall prove the following result (see §8.5).

THEOREM 8.1. *Let $T : \mathbb{O} \to \mathbb{A}$ be the renormalization operator given by Theorem 2.4. If $s > s_0$ with $s_0 < 2$ sufficiently close to 2 and $r > s+1$ not an integer, then T is a robust operator with respect to $(\mathbb{A}^r, \mathbb{A}^s, \mathbb{A}^0)$.*

We shall present in the sequel complete proofs of all the estimates that are necessary for establishing the above result, carefully checking all the properties of robustness along the way.

In our estimates we will often concern ourselves with a power T^m of T. For each $m \geq 1$, let $\mathbb{O}_m \subseteq \mathbb{O}$ be the (open) set of those f's which are mN times renormalizable. Then T^m is well-defined in $\mathbb{O}_m$ and we can write

$$T^m(f) = \frac{1}{\lambda_f} \cdot f^p \circ \Lambda_f \ ,$$

where $p = p(f, mN)$, $\lambda_f = f^p(0)$, and $\Lambda_f : x \mapsto \lambda_f x$ is the linear scaling. Note that p (and hence λ_f and Λ_f) depends on m, but if m is held fixed then p is a locally constant function of $f \in \mathbb{O}_m$. To keep track of the dependence of constants on m, we shall denote by K those constants that may depend on m, and by c those that are independent of m.

Likewise, we define $\mathbb{O}_m^r$ to be an open set in $\mathbb{U}^r$ containing $\mathbb{K}$, all of whose elements are mN times renormalizable, so that $T^m = R^{mN} : \mathbb{O}_m^r \to \mathbb{U}^r$ is well-defined.

8.1. *A closer look at composition.* From a differentiable viewpoint, composition is a notoriously ill-behaved operation. Such bad behavior is the source of most technical difficulties arising in this work. Fortunately, some positive results lie at hand. For example, it is well-known that if r is a positive integer then composition, viewed as an operator from $C^r \times C^{r-1}$ into C^{r-1}, is a C^1 map (see [11]). We shall need not only this result but also a less well-known generalization of it for Hölder spaces: if $r - 1 > s \geq 1$ are *real* numbers then composition, as an operator from $C^r \times C^s$ into C^s, is a C^1 map (we can say a little bit more – see Proposition 8.7 below). For related results on the smoothness of composition, see [19].

Before we can prove this fact, some auxiliary results are in order. In what follows, our definition of the C^r norm of $\varphi \in C^r(I)$ is this: for $r = k + \alpha$ with $k \in \mathbb{N}$ and $0 \leq \alpha < 1$ we write

$$\|\varphi\|_{C^r} = \max\{\|\varphi\|_0, \|\varphi'\|_0, \ldots, \|\varphi^{(k)}\|_0; \|\varphi^{(k)}\|_\alpha\} \ .$$

For $r = k + \mathrm{Lip}$ we define $\|\varphi\|_{C^r}$ as above with $\alpha = 1$. This norm is equivalent to the one introduced earlier in 5.1, and has the advantage that $\|\varphi\|_{C^r} = \max\{\|\varphi\|_{C^0}, \|\varphi'\|_{C^{r-1}}\}$ whenever $r \geq 1$. This allows us to prove certain estimates by induction on k, which will be very useful later.

LEMMA 8.2. *Given $0 \leq \alpha < 1$ and $0 \leq \varepsilon \leq 1 - \alpha$, let $w \in C^{\alpha+\varepsilon}(I)$, $\varphi, \psi \in C^1(I, I)$ and $\|\psi - \varphi\|_{C^1} \leq 1$.*

(i) *If $\varepsilon > 0$ then there exists $K = K(\|\psi\|_{C^1}) > 0$ such that*

$$\|w \circ \varphi - w \circ \psi\|_{C^\alpha} \leq K \|w\|_{C^{\alpha+\varepsilon}} \|\varphi - \psi\|^{\varepsilon}_{C^1} \ .$$

(ii) *If $\varepsilon = 0$ then there exist $c > 0$ and $K = K(\|\psi\|_{C^1}) > 0$ such that*

$$\begin{aligned}\|w \circ \varphi - w \circ \psi\|_{C^\alpha} \leq\, & c\|w\|_{C^\alpha} \|\psi'\|^{\alpha}_{C^0} \\ & + K\|w\|_{C^\alpha} \|\varphi - \psi\|^{\alpha}_{C^1} \ .\end{aligned}$$

Proof. Let us start proving part (i) of this lemma. By the mean value theorem, we obtain

$$\|w \circ \varphi - w \circ \psi\|_{C^0} \leq \|w\|_{C^{\alpha+\varepsilon}} \|\varphi - \psi\|^{\alpha+\varepsilon}_{C^0}. \tag{8.1.1}$$

If $|y - x| \leq \|\varphi - \psi\|_{C^\alpha}$ then

$$\begin{aligned}|w \circ \varphi(y) - w \circ \varphi(x)| &\leq c_0 \|w\|_{C^{\alpha+\varepsilon}} \|\varphi\|^{\alpha+\varepsilon}_{C^1} \|\varphi - \psi\|^{\varepsilon}_{C^\alpha} |y - x|^\alpha \ , \\ |w \circ \psi(y) - w \circ \psi(x)| &\leq c_1 \|w\|_{C^{\alpha+\varepsilon}} \|\psi\|^{\alpha+\varepsilon}_{C^1} \|\varphi - \psi\|^{\varepsilon}_{C^\alpha} |y - x|^\alpha \ .\end{aligned}$$

If $|y - x| > \|\varphi - \psi\|_{C^\alpha}$, by (8.1.1) then

$$\|w \circ \psi - w \circ \varphi\|_{C^0} \leq c_2 \|w\|_{C^{\alpha+\varepsilon}} \|\varphi - \psi\|^{\varepsilon}_{C^\alpha} |y - x|^\alpha \ ,$$

which ends the proof of part (i) of this lemma.

Let us prove part (ii) of this lemma. By the mean value theorem, we obtain

$$\|w \circ \varphi - w \circ \psi\|_{C^0} \leq \|w\|_{C^\alpha} \|\varphi - \psi\|^{\alpha}_{C^0}. \tag{8.1.2}$$

Furthermore,

$$|w \circ \varphi(y) - w \circ \varphi(x)| \leq c_3 \|w\|_{C^\alpha} \|\varphi'\|_{C^0}^\alpha |y-x|^\alpha \ ,$$
$$|w \circ \psi(y) - w \circ \psi(x)| \leq c_4 \|w\|_{C^\alpha} \|\psi'\|_{C^0}^\alpha |y-x|^\alpha \ ,$$

and so

$$\|w \circ \varphi - w \circ \psi\|_{C^\alpha} \leq c_5 \|w\|_{C^\alpha} (\|\varphi' - \psi'\|_{C^0} + \|\psi'\|_{C^0})^\alpha \tag{8.1.3}$$

which ends the proof of part (ii) of this lemma. □

We shall need also some estimates on polynomial operators coming from simple algebraic considerations. For every polynomial P of degree d in n variables $x_1, x_2, \ldots, x_n$ over $\mathbb{R}$, define $\nu(P)$ as the sum of the absolute values of the coefficients of P. This is a well-known valuation in the ring $\mathbb{R}[x_1, x_2, \ldots, x_n]$, but all that really matters to us is that $\nu(P+Q) \leq \nu(P) + \nu(Q)$ (sub-additivity), and that $\nu(\partial_{x_i} P) \leq d\nu(P)$.

LEMMA 8.3. *Let $P \in \mathbb{R}[x_1, x_2, \ldots, x_n]$ be a polynomial of degree d, and let $\phi_1, \phi_2, \ldots, \phi_n \in C^s(I)$. Then,*

$$\|P(\phi_1, \phi_2, \ldots, \phi_n)\|_{C^s} \leq \nu(P) 2^{sd} M^d \ ,$$

where $M = \max\{1, \|\phi_1\|_{C^s}, \|\phi_2\|_{C^s}, \ldots, \|\phi_n\|_{C^s}\}$. Moreover, if $\psi_1, \psi_2, \ldots, \psi_n \in C^s(I)$ also satisfy $\|\psi_i\|_{C^s} \leq M$ for all $1 \leq i \leq n$, then

$$\|P(\phi_1, \phi_2, \ldots, \phi_n) - P(\psi_1, \psi_2, \ldots, \psi_n)\|_{C^s} \leq d\nu(P) 2^{sd} M^{d-1} \sum_{i=1}^n \|\phi_i - \psi_i\|_{C^s} \ .$$

Proof. The first inequality is immediate from the definition of $\nu(P)$. To prove the second, note that $P : (C^s(I))^n \to C^s(I)$ is a C^1 map (norm of the sum in the domain of P). Using the mean value inequality and the first inequality, we see that

$$\|P(\phi_1, \phi_2, \ldots, \phi_n) - P(\psi_1, \psi_2, \ldots, \psi_n)\|_{C^s}$$
$$\leq 2^s \sup_{0 \leq t \leq 1} \max_i \|\partial_{x_i} P(t\phi_1 + (1-t)\psi_1, \ldots, t\phi_n + (1-t)\psi_n)\|_{C^s} \sum_{i=1}^n \|\phi_i - \psi_i\|_{C^s}$$
$$\leq 2^s (d\nu(P) 2^{s(d-1)} M^{d-1}) \sum_{i=1}^n \|\phi_i - \psi_i\|_{C^s} \ ,$$

which is the desired result. □

We can now use the estimate given in the above lemmas to prove the following general proposition. Let $r, s \geq 1$ be real numbers and for each $w \in C^r(I)$, let

$$\Theta_w : C^s(I, I) \to C^s(I)$$

be the operator given by $\Theta_w(\varphi) = w \circ \varphi$.

PROPOSITION 8.4. *Let $r, s > 1$ be real numbers both noninteger, and let $w \in C^r(I)$, $\varphi, \psi \in C^s(I, I)$ with $\|\varphi - \psi\|_{C^s} \leq 1$.*

(i) *If $r > s$ then there exists $K = K(\|\varphi\|_{C^s}) > 0$ such that*

$$\|w \circ \varphi - w \circ \psi\|_{C^s} \leq K \|w\|_{C^r} \|\varphi - \psi\|_{C^s}^{\varepsilon}$$

where $\varepsilon = \min\{1 - \{s\}, r - s\}$ ($\{s\}$ denotes the fractional part of s). In particular, $\Theta_w : C^s(I, I) \to C^s(I)$ is ε-Hölder continuous.

(ii) *If $r = s$ there exists $c > 0$ and $K = K(\|\varphi\|_{C^s}) > 0$ such that*

$$\|w \circ \varphi - w \circ \psi\|_{C^r} \leq c \|w\|_{C^r} \|\psi'\|_{C^0}^{r} + K \|w\|_{C^r} \|\varphi - \psi\|_{C^r}^{\alpha} ,$$

where $\alpha = \{s\}$ is the fractional part of s.

Part (ii) of the above proposition shows one of the main difficulties in this theory which is the fact that for $w \in C^s(I)$ the operator $\Theta_w : C^s(I, I) \to C^s(I)$ is not even C^0.

Proof. Let us write $s = k + \alpha$, with k an integer and $0 < \alpha = \{s\} < 1$, and let

$$A = w \circ \varphi - w \circ \psi .$$

Since $w, \varphi, \psi \in C^1$ and $\varepsilon \leq 1 - \alpha$, applying Lemma 8.2 we get

$$\|A\|_{C^\alpha} \leq K_1 \|w\|_{C^1} \|\varphi - \psi\|_{C^1}^{\varepsilon} .$$

By Faa-di-Bruno's Formula (see [13], p.42), for all $1 \leq l \leq k$ we can write

$$A^{(l)} = B_l(\varphi) - B_l(\psi)$$

where

$$B_l(\phi) = \sum_{j=1}^{l} w^{(j)} \circ \phi \cdot P_{l,j}(\phi', \phi'', \ldots, \phi^{(l-j)}) ,$$

each $P_{l,j}$ being a (universal, homogeneous) polynomial of degree j in $l - j$ variables (with integer coefficients explicitly computable from l and j; see [13, p. 42]). We only need the expression of $P_{l,j}$ for $j = l$; it is easy to check that $P_{l,l}(\phi') = (\phi')^l$. Then, we can decompose $A^{(l)} = C_l + D_l$, where

$$C_l = \sum_{j=1}^{l} w^{(j)} \circ \varphi \cdot \left(P_{l,j}\left(\varphi', \varphi'', \ldots, \varphi^{(l-j)}\right) - P_{l,j}\left(\psi', \psi'', \ldots, \psi^{(l-j)}\right) \right) ,$$

$$D_l = \sum_{j=1}^{l} \left(w^{(j)} \circ \varphi - w^{(j)} \circ \psi \right) \cdot P_{l,j}\left(\psi', \psi'', \ldots, \psi^{(l-j)}\right) .$$

By Lemma 8.3 applied to each $P_{l,j}$,

$$\left\| P_{l,j}\left(\varphi', \varphi'', \ldots, \varphi^{(l-j)}\right) - P_{l,j}\left(\psi', \psi'', \ldots, \psi^{(l-j)}\right) \right\|_{C^\alpha} \leq K_2 \|\varphi - \psi\|_{C^s} .$$

Therefore, for all $1 \le l \le k$ we get

$$\|C_l\|_{C^\alpha} \le K_3 \|w\|_{C^s} \|\varphi - \psi\|_{C^s} \ .$$

Let us now rewrite $D_l = E_l + F_l$ where,

$$E_l = \sum_{j=1}^{l-1} \left(w^{(j)} \circ \varphi - w^{(j)} \circ \psi \right) \cdot P_{l,j} \left(\psi', \psi'', \ldots, \psi^{(l-j)} \right) \ ,$$

$$F_l = \left(w^{(l)} \circ \varphi - w^{(l)} \circ \psi \right) \cdot \left(\psi' \right)^l \ .$$

In bounding the first summation in E_l, we apply Lemma 8.2. Since $w^{(j)}, \varphi, \psi$ is at least C^1 we get

$$\left\| w^{(j)} \circ \varphi - w^{(j)} \circ \psi \right\|_{C^\alpha} \le K_4 \left\| w^{(j)} \right\|_{C^1} \|\varphi - \psi\|_{C^1}^{1-\alpha}$$

for all, with $1 \le j \le l-1$. From this and Lemma 8.3, for all $1 \le l \le k$ we obtain that

$$\|E_l\|_{C^\alpha} \le K_5 \|w\|_{C^s} \|\varphi - \psi\|_{C^s}^{1-\alpha} \ .$$

Our task has been reduced to bounding the C^α norm of F_l. Here, we will do separately the proof of part (i) and part (ii) of this proposition.

Let us prove part (i) first. Here, for all $1 \le l \le k$ we have that φ, ψ are at least C^1 and that $w^{(l)}$ is at least $C^{\alpha+\varepsilon}$, and so by Lemma 8.2 we get

$$\left\| w^{(l)} \circ \varphi - w^{(l)} \circ \psi \right\|_{C^{\alpha+\varepsilon}} \le K_6 \|w\|_{C^{l+\alpha+\varepsilon}} \|\varphi - \psi\|_{C^1}^{\varepsilon} \ .$$

Thus, for all $1 \le l \le k$ we obtain

$$\|F_l\|_{C^\alpha} \le K_7 \|w\|_{C^r} \|\varphi - \psi\|_{C^s}^{\varepsilon} \ ,$$

which ends the proof of part (i).

Let us now prove part (ii). We know that $w^{(l)}, \varphi, \psi \in C^1$ for all $1 \le l \le k-1$, and so by Lemma 8.2 we have

$$\left\| w^{(l)} \circ \varphi - w^{(l)} \circ \psi \right\|_{C^{\alpha+\varepsilon}} \le K_8 \|w\|_{C^{l+\alpha+\varepsilon}} \|\varphi - \psi\|_{C^1}^{\alpha} \ .$$

Thus, for all $1 \le l \le k-1$ we obtain

$$\|F_l\|_{C^\alpha} \le K_9 \|w\|_{C^r} \|\varphi - \psi\|_{C^s}^{\alpha} \ .$$

Therefore, we just have to bound $\|F_k\|_{C^\alpha}$. Here, $w^{(k)}$ is only C^α. From the inequalities (8.1.2) and (8.1.3) in the proof of Lemma 8.2, we get

$$\left\| w^{(k)} \circ \varphi - w^{(k)} \circ \psi \right\|_{C^0} \le c_1 \|w\|_{C^r} \|\varphi - \psi\|_{C^0}^{\alpha} \ ,$$

$$\left\| w^{(k)} \circ \varphi - w^{(k)} \circ \psi \right\|_{C^\alpha} \le c_2 \|w\|_{C^r} \left(\left\|\psi'\right\|_{C^0} + \left\|\varphi' - \psi'\right\|_{C^0} \right)^\alpha \ ,$$

and so

$$\begin{aligned} \|F_k\|_{C^\alpha} \le\; & c_3 \|\psi\|_{C^{1+\alpha}}^k \|w\|_{C^r} \|\varphi - \psi\|_{C^0}^{\alpha} \\ & + c_4 \|\psi\|_{C^1}^k \|w\|_{C^r} \left(\left\|\psi'\right\|_{C^0} + \left\|\varphi' - \psi'\right\|_{C^0} \right)^\alpha \ , \end{aligned}$$

which ends the proof of part (ii). □

LEMMA 8.5. *Given $0 \leq \alpha < 1$ and $0 < \varepsilon \leq 1-\alpha$, let $f \in C^{1+\alpha+\varepsilon}(I)$, $g \in C^1(I,I)$ and $v \in C^1(I)$ with $\|v\|_{C^1} \leq 1$ and $g+v \in C^1(I,I)$. There exists $K = K(\|g\|_{C^1}) > 0$ such that*

$$\left\| f\circ(g+v) - f\circ g - f'\circ g\cdot v\right\|_{C^\alpha} \leq K\|f\|_{C^{1+\alpha+\varepsilon}}\|v\|_{C^1}^{1+\varepsilon} .$$

In particular, there exists $K = K(\|g\|_{C^1}) > 0$ such that

$$\|f\circ(g+v) - f\circ g\|_{C^\alpha} \leq K\|f\|_{C^{1+\alpha+\varepsilon}}\|v\|_{C^1} .$$

Proof. Note that we have the following identity:

$$(f\circ(g+v) - f\circ g - f'\circ g\cdot v)(x) = v(x)\int_0^1 \left[f'(g(x)+tv(x)) - f'(g(x))\right]\,dt .$$

Applying Lemma 8.2 with $w = f'$, $\varphi = g + tv$ and $\psi = g$, we can bound the C^α norm of the integrand by $Kt\|f'\|_{C^{\alpha+\varepsilon}}\|v\|_{C^1}^{\varepsilon}$. This proves the first stated estimate in slightly stronger form. The second estimate is an immediate consequence of the first. □

PROPOSITION 8.6. *Let $2 \leq s+1 < r$ be real numbers, and let $f \in C^r(I)$, $g \in C^s(I,I)$. There exists $K = K(\|g\|_{C^s}) > 0$ such that, for all $v \in C^s(I)$ with $\|v\|_{C^s} \leq 1$ and $g+v \in C^s(I,I)$,*

$$\left\| f\circ(g+v) - f\circ g - f'\circ g\cdot v\right\|_{C^s} \leq K\|f\|_{C^r}\|v\|_{C^s}^{1+\theta} , \tag{8.1.4}$$

where $\theta = \min\{1-\{s\}, r-s-1\}$. In particular, (a) *the operator $\Theta_f : C^s(I,I) \to C^s(I)$ is C^1 and its derivative is given by $D\Theta_f(g)v = f'\circ g\cdot v$, and* (b) *there exists $K = K(\|g\|_{C^s}) > 0$ such that for all v as above,*

$$\|f\circ(g+v) - f\circ g\|_{C^s} \leq K\|f\|_{C^r}\|v\|_{C^s} . \tag{8.1.5}$$

Proof. In this proof we use $K_1, K_2, \ldots$ to denote constants that depend only on $\|g\|_{C^s}$. Consider the remainder term

$$F = f\circ(g+v) - f\circ g - f'\circ g\cdot v ,$$

as well as its derivative $F' = A + B$, where

$$\begin{aligned} A &= \left(f'\circ(g+v) - f'\circ g - f''\circ g\cdot v\right)\cdot g' , \\ B &= \left(f'\circ(g+v) - f'\circ g\right)\cdot v' . \end{aligned}$$

We want to show that

$$\left\|F'\right\|_{C^s} \leq K_1\|f\|_{C^r}\|v\|_{C^s}^{1+\theta} .$$

The proof will be by induction on the integral part of s. Note however that the mean value theorem already gives us $\|F\|_{C^0} \leq K_2\|f''\|_{C^0}\|v\|_{C^0}^2$ independently of s.

First we deal with the base of induction, namely when $1 \leq s < 2$, say, $s = 1 + \alpha$. By Lemma 8.5, we have

$$\|A\|_{C^\alpha} \leq K_3 \|f'\|_{C^{1+\alpha+\theta}} \left(\|v\|_{C^1}\right)^{1+\theta} .$$

The same Lemma 8.5 yields

$$\|B\|_{C^\alpha} \leq K_4 \|f'\|_{C^{1+\alpha+\theta}} \left(\|v\|_{C^{1+\alpha}}\right)^2 .$$

This establishes the base of induction.

Now suppose that our lemma holds for $s > 1$. We will prove from this that it holds for $s + 1$. To do this, it suffices to show that

$$\left\|F'\right\|_{C^s} \leq K_5 \|f\|_{C^r} \|v\|_{C^{s+1}}^{1+\theta} . \tag{8.1.6}$$

The proof is more of the same. By the induction hypothesis applied to f', we have

$$\|A\|_{C^s} \leq K_6 \|f'\|_{C^{r-1}} \left(\|v\|_{C^s}\right)^{1+\theta} . \tag{8.1.7}$$

The same fact also gives

$$\|B\|_{C^s} \leq K_7 \|f'\|_{C^{r-1}} \left(\|v\|_{C^{1+s}}\right)^2 . \tag{8.1.8}$$

Putting (8.1.7) and (8.1.8) together we get (8.1.6), and so the induction is complete. □

Proposition 8.7. *Let $2 \leq s + 1 < r$ be real numbers. The composition operator $\Theta : C^r(I) \times C^s(I, I) \to C^s(I)$ given by $\Theta(f, g) = f \circ g$ is $C^{1+\theta}$ and its derivative is given by $D\Theta(f, g)(u, v) = u \circ g + f' \circ g \cdot v$. In particular, there exists $K = K(\|f\|_{C^r}, \|g\|_{C^s}) > 0$ such that, for all $\|u\|_{C^r} \leq 1$ and $\|v\|_{C^s} \leq 1$ with $g + v \in C^s(I, I)$,*

(8.1.9)
$$\|\Theta(f + u, g + v) - \Theta(f, g) - D\Theta(f, g)(u, v)\|_{C^s} \leq K(\|u\|_{C^r} + \|v\|_{C^s})^{1+\theta} ,$$

where $\theta = \min\{1 - \{s\}, r - s - 1\}$.

Proof. In this proof, we denote by $K_1, K_2, \ldots$ positive constants depending only on $\|g\|_{C^s}$. Let us take $u \in C^r(I)$ and $v \in C^s(I)$ such that $\|u\|_{C^r} \leq 1$ and $\|v\|_{C^s} \leq 1$, respectively. Now,

$$\begin{aligned} F &= \Theta(f + u, g + v) - \Theta(f, g) - u \circ g - f' \circ g \cdot v \\ &= f \circ (g + v) - f \circ g - f' \circ g \cdot v + u \circ (g + v) - u \circ g . \end{aligned}$$

Using Proposition 8.6, we see that

$$\|f \circ (g + v) - f \circ g - f' \circ g \cdot v\|_{C^s} \leq K_1 \|f\|_{C^r} (\|v\|_{C^s})^{1+\theta} .$$

The same Proposition 8.6 with u replacing f yields

$$\begin{aligned}\|u \circ (g+v) - u \circ g\|_{C^s} &\leq \|u' \circ g \cdot v\|_{C^s} + K_2 \|u\|_{C^r} (\|v\|_{C^s})^{1+\theta} \\ &\leq K_3 \|u\|_{C^r} \|v\|_{C^s} \ .\end{aligned}$$

Therefore we get

$$\|F\|_{C^s} \leq K_1 \|f\|_{C^r} (\|v\|_{C^s})^{1+\theta} + K_3 \|u\|_{C^r} \|v\|_{C^s} \ ,$$

which proves that Θ is C^1 and that (8.1.9) is satisfied. Now, we have that

$$D\Theta(f+\phi, g+\psi)(u,v) - D\Theta(f,g)(u,v) = A + B + C$$

where

$$\begin{aligned}A &= u \circ (g+\psi) - u \circ g \ ,\\ B &= \big(f' \circ (g+\psi) - f' \circ g\big) \cdot v \ ,\\ C &= \phi' \circ (g+\psi) \cdot v \ .\end{aligned}$$

By Proposition 8.4, we obtain that

$$\begin{aligned}\|A\|_{C^s} &\leq K_4 \|u\|_{C^{r-1}} \|\psi\|_{C^s}^{\theta} \ ,\\ \|B\|_{C^s} &\leq K_5 \|f\|_{C^r} \|\psi\|_{C^s}^{\theta} \cdot \|v\|_{C^s}.\end{aligned}$$

Letting k be the integral part of s and $\varphi = g + \psi$, and using Faa-di-Bruno's Formula, we have

$$(\phi' \circ \varphi)^{(k)} = \sum_{j=1}^{k} \phi^{(j+1)} \circ \varphi \cdot P_{k,j}(\varphi', \varphi'', \ldots, \varphi^{(k-j)}) \ ,$$

each $P_{k,j}$ being a (universal, homogeneous) polynomial of degree j in $k-j$ variables. Hence, using Lemma 8.3, we get that $\|C\|_{C^s} = K_6 \|\Delta f\|_{C^r} \|v\|_{C^s}$. Thus, Θ is a $C^{1+\theta}$ operator. □

COROLLARY 8.8. *Let $r, s > 0$ be real numbers with $r - 1 > s \geq 1$ and for each positive integer m, let $Q_m : C^r(I,I) \to C^s(I,I)$ be the operator given by $Q_m(f) = f^m$.*

(i) *Let $0 \leq t \leq r$ and let $U : C^t(I,I) \to C^s(I,I)$ be a $C^{1+\theta}$ operator for some $0 < \theta < 1$. Then the operator $U_m : C^r(I,I) \to C^s(I,I)$ given by $U_m(f) = Q_m \circ U(f)$ is $C^{1+\theta'}$ for some $0 < \theta' = \theta'(\theta, r, s) < 1$.*

(ii) *In particular, the operator $Q_m : C^r(I,I) \to C^s(I,I)$ is $C^{1+\theta''}$ for some $0 < \theta'' = \theta''(r,s) < 1$ and there exists $K = K(m, \|f\|_{C^r}) > 0$ such that*

$$\|Q_m(f+u) - Q_m(f) - DQ_m(f)u\|_{C^s} \leq K \|u\|_{C^r}^{1+\theta''} \ . \qquad (8.1.10)$$

Proof. First note that $U_{m+1}(f) = \Theta(f, U_m(f))$. The operator U_1 arises as the composition of the operator $C^r(I,I) \to C^r(I,I) \times C^s(I,I)$ given by $f \mapsto (f, U(f))$, which is $C^{1+\theta}$ because U is $C^{1+\theta}$ (and $C^r(I,I)$ embeds in C^t), with the composition operator $\Theta : C^r(I,I) \times C^s(I,I) \to C^s(I,I)$, which is $C^{1+\theta''}$ for some $0 < \theta'' = \theta''(r,s) < 1$ by Proposition 8.7. The desired result for part (i) then follows by induction. Part (ii) is a corollary of part (i), and by a computation (8.1.10) follows from (8.1.9). □

PROPOSITION 8.9. *Let r, s, t be real numbers with $2 \leq s+1 < r$ and $t \geq 0$. Let $U : C^t(I,I) \to C^s(I,I)$ be a C^1 operator. Then for each $\phi \in C^r(I)$ and each $\psi \in C^t(I,I)$ there exists a function $\sigma_\psi : \mathbb{R}^+ \to \mathbb{R}^+$ with $\sigma_\psi(h)/h \to 0$ as $h \to 0$, varying continuously with ψ, such that for all $v \in C^t(I)$ with $\psi + v \in C^t(I,I)$,*

$$\left\|\phi \circ U(\psi+v) - \phi \circ U(\psi) - \phi' \circ U(\psi) \cdot DU(\psi)v\right\|_{C^s} \leq \sigma_\psi(\|v\|_{C^t}) . \tag{8.1.11}$$

Proof. As before, we denote by $K_1, K_2, \ldots$ positive constants that depend only on $\|\psi\|_{C^t}$. We have that

$$\phi \circ U(\psi+v) - \phi \circ U(\psi) - \phi' \circ U(\psi) \cdot DU(\psi)v = A + B$$

where

$$\begin{aligned} A &= \phi \circ U(\psi+v) - \phi \circ U(\psi) - \phi' \circ U(\psi) \cdot (U(\psi+v) - U(\psi)) , \\ B &= \phi' \circ U(\psi) \cdot (U(\psi+v) - U(\psi) - DU(\psi)v) . \end{aligned}$$

Since U is C^1, there exists a continuous function $\nu_\psi : \mathbb{R}^+ \to \mathbb{R}^+$ with $\nu_\psi(h)/h \to 0$ as $h \to 0$, varying continuously with ψ, such that

$$\|U(\psi+v) - U(\psi) - DU(\psi)v\|_{C^s} \leq \nu_\psi(\|v\|_{C^t}) .$$

Hence, applying Proposition 8.6 with $f = \phi$ and $g = U(\psi)$ and v replaced by $U(\psi+v) - U(\psi)$, we get

$$\begin{aligned} \|A\|_{C^s} &\leq K_2 \|\phi\|_{C^r} \left(\|U(\psi+v) - U(\psi)\|_{C^s}\right)^{1+\theta} \\ &\leq K_3 \|\phi\|_{C^r} \left(\|DU(\psi)\|^{1+\theta} \|v\|_{C^t}^{1+\theta}\right) , \end{aligned}$$

and

$$\|B\|_{C^s} \leq K_4 \|\phi\|_{C^r} \nu_\psi(\|v\|_{C^t}) ,$$

where $K_3 = K_3(\|U(\psi)\|_{C^s}, \|DU(\psi)\|, \nu_\psi)$ and $K_4 = K_2(\|U(\psi)\|_{C^s})$. Therefore,

$$\|A + B\|_{C^s} \leq K_3 \|\phi\|_{C^r} \|v\|_{C^t}^{1+\theta} + K_4 \|\phi\|_{C^r} \nu_\psi(\|v\|_{C^t}) .$$

This completes the proof. □

COROLLARY 8.10. *Let r, s, t be real numbers with $r - 1 > s > 1$ and $0 \leq t \leq r$, and let $U : C^t(I, I) \to C^s(I, I)$ be a C^1 operator. For each positive integer n the operator $V_n : C^r(I, I) \to C^s(I)$ given by $V_n(f) = (f^m)' \circ U(f)$ is differentiable at every $g \in C^{r+1}(I, I) \subseteq C^r(I, I)$, and as a map from $C^{r+1}(I, I)$ into $\mathcal{L}(C^r(I), C^s(I))$, the derivative operator $g \mapsto DV_n(g)$ is continuous.*

Proof. First note that by the chain rule,

$$V_n(f) = \prod_{j=0}^{n-1} f' \circ \left(f^{(j)} \circ U(f)\right) = \prod_{j=0}^{n-1} f' \circ U_j(f) \ .$$

This reduces the problem to the case $n = 1$. We claim that the linear operator

$$L(v) = v' \circ U(g) + g'' \circ U(g) \cdot DU(g)v$$

is the derivative of V_1 at $g \in C^{r+1}(I, I)$. Indeed, we have

$$V_1(g + v) - V_1(g) - L(v) = A + B \ ,$$

where

$$\begin{aligned} A &= g' \circ U(g + v) - g' \circ U(g) - g'' \circ U(g) \cdot DU(g)v \ , \\ B &= v' \circ U(g + v) - v' \circ U(g) \ . \end{aligned}$$

By Proposition 8.9 applied to $\phi = g'$ and $\psi = g$, there exists $K_1 = K_1(\|g\|_{C^{r+1}})$ such that

$$\|A\|_{C^s} \leq K_1 \sigma_\psi(\|v\|_{C^r}) \ ,$$

where $\sigma_\psi : \mathbb{R}^+ \to \mathbb{R}^+$ is a continuous function varying continuously with ψ such that $\sigma_\psi(h)/h \to 0$ as $h \to 0$. On the other hand, by part (i) of Proposition 8.4 and since U is C^1, we have

$$\begin{aligned} \|B\|_{C^s} &\leq K_2 \|v\|_{C^r} \|U(g + v) - U(g)\|_{C^s}^{\varepsilon} \\ &\leq K_3 \left(\|v\|_{C^r}\right)^{1+\varepsilon} \ , \end{aligned}$$

where $0 < \varepsilon = \min\{1 - \{s\}, r - s\} < 1$, $K_2 = K_2(\|U(g)\|_{C^s})$ and $K_3 = K_3(\|U(g)\|_{C^s}, \|DU(g)\|, \sigma_\psi(\|v\|_{C^r}))$. Combining these inequalities, we deduce that V_1 is differentiable at g and $DV_1(g) = L$ as claimed. It is clear from the expression defining it that L varies continuously with $g \in C^{r+1}(I, I)$. □

8.2. *Checking properties* **B2** *and* **B3**. We now proceed to verify that the operator T satisfies properties **B2** and **B3** of robustness. They will follow respectively from Lemmas 8.12 and 8.13. First it is necessary to analyze the behavior of the linear scaling used in such operators. Let us fix a positive integer p and for each $f \in C^r(I, I)$ let Λ_f be the linear map $x \mapsto \lambda_f x$, where $\lambda_f = f^p(0)$.

LEMMA 8.11. *For $r > 2$, the maps $\Lambda : C^r(I,I) \to \mathcal{L}(\mathbb{R},\mathbb{R})$ given by $\Lambda(f) = \Lambda_f$ and $\lambda : C^r(I,I) \to \mathbb{R}$ given by $\lambda(f) = \lambda_f$ are both $C^{1+\theta}$ for some $0 < \theta = \theta(r,s) < 1$. In particular, there is $K = K(p, \|f\|_{C^r}) > 0$ such that for all $v \in C^r(I)$ with $\|v\|_{C^r} \leq 1$ and $f + v \in C^r(I,I)$,*

$$\|\lambda(f+v) - \lambda(f) - D\lambda(f)v\|_{C^r} \leq K\|v\|_{C^r}^{1+\theta} \ . \tag{8.2.1}$$

The above inequality also holds when λ is replaced by Λ.

Proof. Choosing $1 < s < r-1$, we see that $\lambda = E \circ Q_p$ where $Q_p : C^r(I,I) \to C^s(I,I)$ is the operator $Q_p(f) = f^p$, which is $C^{1+\theta}$ for some $0 < \theta = \theta(r,s) < 1$, and $E : C^s(I,I) \to \mathbb{R}$ is the evaluation map $E(g) = g(0)$, which is linear. Therefore, by Corollary 8.8, λ is $C^{1+\theta}$ and (8.2.1) follows from (8.1.10) and the linearity of E. The proof for Λ is entirely analogous. □

We will also need to use the operators $U_n : C^r(I,I) \to C^s(I)$ given by $U_n(f) = f^n \circ \Lambda_f$ for all $n \geq 0$.

Property **B2** for the operator T is a consequence of the following lemma (the first assertion in **B2** is actually a consequence of Lemma 8.14 below).

LEMMA 8.12. *For $2 < s+1 < r$ and for each $n \geq 0$, the operator $U_n : C^r(I,I) \to C^s(I)$ is $C^{1+\theta}$ for some $0 < \theta = \theta(r,s) < 1$. In particular, $T^m : \mathbb{O}^r_m \to \mathbb{U}^s$ is also a $C^{1+\theta}$ operator.*

Proof. This follows at once from Lemma 8.11 and Corollary 8.8 applied to $U = \Lambda$. □

The following lemma is all we need to verify property **B3** for the operator T. In this case g is a map in the limit set $\mathbb{K}$ of T, hence analytic, and $v = \mathbf{u}_g$ is a tangent vector to the unstable manifold of g, which is analytic as well.

LEMMA 8.13. *For $2 < s+1 < r$, the map $\mathbb{O}^r_m \to \mathbb{U}^s$ given by $f \mapsto DT^m(f)v$ is differentiable at $f = g \in \mathbb{K}$. Furthermore, for every $m \geq 1$ there exist $C_m > 1$ and $\nu_m > 0$ such that for each $g \in \mathbb{K}$ and $f \in \mathbb{O}^r_m$ with $\|f - g\|_{C^r} < \nu_m$ and all $v \in \mathbb{A}^r$ with $\|v\|_{C^r} = 1$,*

$$\|DT^m(f)v - DT^m(g)v\|_{C^s} \leq C_m\|f - g\|_{C^r} \ . \tag{8.2.2}$$

Proof. Let $E : C^s(I,I) \to \mathbb{R}$ be the evaluation map $E(g) = g(0)$, which is linear. Recall that the derivative of T^m is given by the expression

$$\begin{aligned}(8.2.3)\ DT^m(f)v = &\frac{1}{\lambda_f}\sum_{j=0}^{p-1}(f^j)' \circ U_{p-j}(f) \cdot v \circ U_{p-j-1}(f)\\ &+\frac{1}{\lambda_f}[\mathrm{id} \cdot (T^m f)' - T^m f]\sum_{j=0}^{p-1} E\left((f^j)' \circ U_{p-j}(f)\right) \cdot E\left(v \circ U_{p-j-1}(f)\right) \ ,\end{aligned}$$

where $\lambda_f = E \circ f^p$ and id : $\mathbb{R} \to \mathbb{R}$ is the identity map. Each term of the first summation in (8.2.3) is differentiable at $f = g$. To see this apply Lemma 8.12 and Corollary 8.10 to each of the operators $f \mapsto (f^j)' \circ U_{p-j}(f)$ as well as Proposition 8.6 to each of the operators $f \mapsto v \circ U_{p-j}(f)$. On the other hand, each term of the second summation in (8.2.3) equals the corresponding term in the first summation post-composed with the evaluation map E (which is linear), and is therefore differentiable at $f = g$. The analysis of the expression in square brackets in (8.2.3) is similar. By Lemma 8.11 and Corollary 8.10, the operator $f \mapsto T^m(f)' = (f^p)' \circ \Lambda_f$ is differentiable at $f = g$, and the operator $f \mapsto T^m(f) = \lambda_f \cdot f^p \circ \Lambda_f$ is also differentiable at $f = g$ by Lemma 8.11 and Corollary 8.8. From this fact and compactness of $\mathbb{K}$ the inequality (8.2.2) follows. □

8.3. *Checking property* **B4**. The fifth property is verified in Lemma 8.16 below. First we will need to prove the following two lemmas about the operators $U_i : C^{t+1+\varepsilon}(I, I) \to C^t(I)$ with $t \geq 1$. Recall that $U_i(f) = f^i \circ \Lambda_f$.

LEMMA 8.14. *For every $f \in C^{t+1+\varepsilon}(I, I)$ and all $v \in C^t(I)$ with small norm and such that $f + v \in C^t(I, I)$,*

$$\|U_i(f + v) - U_i(f)\|_{C^t} \leq K\|v\|_{C^t}$$

for all $0 \leq i \leq p$ where $K = K(p, \|f\|_{C^{t+1+\varepsilon}})$.

Proof. Note that

$$U_{i+1}(f + v) - U_{i+1}(f) = f \circ U_i(f + v) - f \circ U_i(f) + v \circ U_i(f + v) .$$

By Proposition 8.6, there is $K_1 = K_1(p, \|f\|_{C^{t+1+\varepsilon}})$ such that

$$\|U_{i+1}(f + v) - U_{i+1}(f)\|_{C^t} \leq K_1\|U_i(f + v) - U_i(f)\|_{C^t} + \|v \circ U_i(f + v)\|_{C^t}.$$

The required estimate now follows by induction, because U_0 is C^1. □

LEMMA 8.15. *For every $f \in C^{t+1+\varepsilon}(I, I)$ and all $v \in C^{t+1+\varepsilon}(I)$ with small norm and such that $f + v \in C^{t+1+\varepsilon}(I, I)$,*

$$\|U_i(f + v) - U_i(f) - DU_i(f)v\|_{C^t} \leq K\|v\|_{C^t}^{1+\theta} ,$$

for all $0 \leq i \leq p$, for some $0 < \theta = \theta(t, \varepsilon) < 1$ and $K = K(p, \|f\|_{C^t}) > 0$.

Proof. In this proof we denote by $K_1, K_2, \ldots$ the positive constants depending only on m and $\|U_i(f)\|_{C^{t+1+\varepsilon}}$. Again we use induction; the case $i = 0$ follows from the differentiability of the scaling $f \to \Lambda_f$ and inequality (8.2.1). We have

$$U_{i+1}(f + v) - U_{i+1}(f) - DU_{i+1}(f)v = A + B + C$$

where

$$\begin{aligned}A &= f \circ U_i(f+v) - f \circ U_i(f) - f' \circ U_i(f) \cdot (U_i(f+v) - U_i(f)) \ , \\ B &= f' \circ U_i(f) \cdot (U_i(f+v) - U_i(f) - DU_i(f)v) \ , \\ C &= v \circ U_i(f+v) - v \circ U_i(f) \ .\end{aligned}$$

By Proposition 8.6,

$$\|A\|_{C^t} \leq K_1 \|U_i(f+v) - U_i(f)\|_{C^t}^{1+\varepsilon}.$$

Using Lemma 8.14, we get

$$\|A\|_{C^{t-1}} \leq K_2 \|v\|_{C^t}^{1+\varepsilon} \ .$$

On the other hand, since v is $C^{t+1+\varepsilon/2}$, we know again from Proposition 8.6 that

$$\|C\|_{C^t} \leq K_3 \|v\|_{C^{t+1+\varepsilon/2}} \|U_i(f+v) - U_i(f)\|_{C^t} \leq K_4 \|v\|_{C^{r+1+\varepsilon/2}} \|v\|_{C^t} \ .$$

Since v has bounded $C^{t+1+\varepsilon}$ norm, by an interpolation of norms, we have $\|v\|_{C^{t+1+\varepsilon/2}} \leq K_5 \|v\|_{C^t}^{\theta_1}$ for some $\theta_1 > 0$. Therefore, taking $\theta = \min\{\varepsilon, \theta_1\}$ we get

$$\|C\|_{C^t} \leq K_6 \|v\|_{C^t}^{1+\theta} \ .$$

This allows the induction as desired. □

Property **B4** for the operator T is a direct consequence of the following lemma.

LEMMA 8.16. *For every $f \in \mathbb{O}^{s+1+\varepsilon}$ and all $v \in \mathbb{A}^{s+1+\varepsilon}$ with small norm such that $f + v \in \mathbb{O}^{s+1+\varepsilon}$,*

$$\|T(f+v) - T(f) - DT(f)(v)\|_{C^s} \leq K \|v\|_{C^s}^{1+\tau} \ ,$$

for some $0 < \tau = \tau(s, \varepsilon) < 1$ and $K(p, \|f\|_{C^{s+1+\varepsilon}}) > 0$.

Proof. In this proof we denote by $K_1, K_2, \ldots$ the positive constants depending only on m and $\|U_i(f)\|_{C^s}$. Start, observing that since $T(f) = \lambda_f^{-1} \cdot U_p(f)$,

$$T(f+v) - T(f) - DT(f)v = A + B + C$$

where

$$\begin{aligned}A &= \lambda_f^{-1} \cdot (U_p(f+v) - U_p(f) - DU_p(f)v) \ , \\ B &= \left(\lambda_{f+v}^{-1} - \lambda_f^{-1} - D\lambda_f^{-1}(v)\right) \cdot U_p(f+v) \ , \\ C &= D\lambda_f^{-1}(v) \cdot (U_p(f+v) - U_p(f)) \ .\end{aligned}$$

Applying Lemma 8.15 with $t = s$ we get

$$\|A\|_{C^s} = K_1 \|v\|_{C^s}^{1+\theta_1} \ ,$$

for some $0 < \theta_1 = \theta_1(s, \varepsilon) < 1$. By Lemma 8.11 there is $0 < \theta_2 = \{s\} < 1$ such that

$$\|B\|_{C^s} \leq K_2 \|v\|_{C^s}^{1+\theta_2} .$$

By Lemma 8.14, we have $\|U_p(f+v) - U_p(f)\|_{C^s} \leq K_3 \|v\|_{C^s}$ and so

$$\|C\|_{C^s} \leq K_4 \|v\|_{C^s}^2 .$$

Therefore, it is enough to take $\tau = \min\{\theta_1, \theta_2\}$. □

8.4. *Checking properties* **B5** *and* **B6**. We now move on to the task of proving that the operator $T = R^N$ of Theorem 2.4 satisfies properties **B5** and **B6** in the definition of robustness. Unlike the previous ones, the verification of these (last) two properties depends upon the geometry of the post-critical sets of maps near in $\mathbb{A}$ to the limit set $\mathbb{K}$ of T. The estimates performed here are the most delicate, and involve the results of §5.2.

Recall that T^m is well-defined on an open set $\mathbb{O}_m$ in the Banach space $\mathbb{A} = \mathbb{A}_{\Omega_a}$ (see §3), which contains $\mathbb{K}$. We shall denote the renormalization intervals $\Delta_{0,mN}, \Delta_{1,mN}, \ldots, \Delta_{p,mN}$ simply by $\Delta_i = \Delta_{i,mN}$ (this shortened notation should cause no harm, because N is fixed since Theorem 2.4, and m will be fixed in the particular estimates involving these intervals).

We can write the derivative of T^m in the following form

$$DT^m(f)v = A(f) \sum_{j=0}^{p-1} B_j(f) \cdot C_j(f) + A(f) \cdot D(f) \sum_{j=0}^{p-1} E \circ B_j(f) \cdot E \circ C_j(f)$$

where E is the evaluation map and

$$A(f) = (\lambda_f)^{-1} , \qquad B_j(f) = \left(f^j\right)' \circ U_{p-j}(f) ,$$
$$C_j(f) = v \circ U_{p-j-1}(f) , \qquad D(f) = \mathrm{id} \cdot (f^p)' \circ U_0(f) - \lambda_f \cdot U_p(f) .$$

To carry out our estimates for T^m, we shall use the operators $U_i : f \mapsto f^i \circ \Lambda_f$ $(i \geq 0)$. Note that $U_0(f) = \Lambda_f$, hence U_0 is C^1 in whichever space $C^r(I, I)$ we work in, because the scaling $f \mapsto \Lambda_f$ is C^1 by Lemma 8.11.

First we need some estimates for U_i. It is clear that $\|U_i(f)\|_{C^0} \leq 1$ always, but more is true.

LEMMA 8.17. *There exists $C > 0$ with the following property. For every $m > 0$, there exists an open neigbourhood $\mathcal{O}_m \subset \mathbb{O}_m$ of $\mathbb{K}$ such that for all $f \in \mathcal{O}_m$,*

$$\|B_j(f)\|_{C^0} \leq C \frac{|\Delta_0|}{|\Delta_{p-j}|} ,$$

for all $0 \leq j \leq p-1$. Furthermore, $\|U_i(f)'\|_{C^0} \leq C|\Delta_i|$, for all $0 \leq i \leq p$.

Proof. Use bounded distortion and the real bounds (see §5.2). □

LEMMA 8.18. *For all* $f \in \mathcal{O}_m$ *and all* $v \in \mathbb{A}^r$ *with small norm,*

$$\|U_i(f+v) - U_i(f)\|_{C^r} \leq K\|v\|_{C^r}$$

for all $0 \leq i \leq p$, *where* $K = K(m) > 0$.

Proof. This lemma follows from Lemma 8.14. □

Next, we show an essential result to prove that the renormalization operator satisfies properties **B5** and **B6**. Here, we use again in a crucial way the geometric properties of the postcritical set of $f \in \mathbb{O}_m$ proved in Section 5.

PROPOSITION 8.19. (i) *For every* $t > 2$ *which is not an integer there exist* $0 < \mu < 1$ *and* $C > 0$ *with the following property. For every* $g \in \mathbb{K}$ *and for every* $m > 0$, *there is an* $\eta > 0$ *such that for all* $f \in \mathbb{O}_m$ *with* $\|f - g\|_{\mathbb{A}} < \eta$ *and for all* $w \in \mathbb{A}^t$ *with* $\|w\|_{C^t} < \eta$,

$$\left\| A(f) \sum_{j=0}^{p-1} B_j(f) \left(C_j(f+w) - C_j(f)\right) \right\|_{C^t} \leq C\mu^m \|v\|_{C^t} . \tag{8.4.1}$$

(ii) *For every* $\mu > 1$ *close to one, there is* $s < 2$ *close to two and* $C > 0$ *with the following property: for every* $g \in \mathbb{K}$ *and for every* m, *there is an* $\eta > 0$ *such that for all* $f \in \mathbb{O}_m$ *with* $\|f - g\|_{\mathbb{A}} < \eta$ *and for all* $w \in \mathbb{A}^t$ *with* $\|w\|_{C^t} < \eta$, *the inequality* (8.4.1) *above is also satisfied.*

Proof. Below, the positive constants $c_1, c_2, \ldots$ depend only on t (and the real bounds), while the positive constants $K_0, K_1, K_2, \ldots$ may depend also on m.

Let k and $0 < \alpha < 1$ be respectively the integer and the fractional part of t (when $t = k + \mathrm{Lip}$, take $\alpha = 1$). We start observing that for each j,

$$\begin{aligned} \|B_j(f)\left(C_j(f+w) - C_j(f)\right)\|_{C^t} \leq{}& \|B_j(f)\|_{C^0}\|C_j(f+w) - C_j(f)\|_{C^t} \\ &+ K_0\|B_j(f)\|_{C^t}\|C_j(f+w) - C_j(f)\|_{C^k} . \end{aligned} \tag{8.4.2}$$

Note that on the right-hand side of (8.4.2) only the second term carries a constant K_0. By Lemma 8.17, there is $c_1 > 0$ such that for every integer m there is an open neighbourhood $\mathcal{O}_m$ of $\mathbb{K}$ with the property that for each $f \in \mathcal{O}_m$ we have

$$\|B_j(f)\|_{C^0} \leq c_1 \frac{|\Delta_0|}{|\Delta_{p-j}|} . \tag{8.4.3}$$

In that neighborhood, we also have $\|B_j(f)\|_{C^t} \leq K_1$. By Proposition 8.4 and Lemma 8.18, taking $0 < \varepsilon < 1$ such that $\alpha - \varepsilon > 0$, we obtain

$$\begin{aligned} \|C_j(f+w) - C_j(f)\|_{C^k} &\leq \|C_j(f+w) - C_j(f)\|_{C^{t-\varepsilon}} \\ &\leq K_2\|v\|_{C^t}\|U_{p-j-1}(f+w) - U_{p-j-1}(f)\|_{C^t}^{\varepsilon} \\ &\leq K_3\|v\|_{C^t}\|w\|_{C^t}^{\varepsilon} . \end{aligned} \tag{8.4.4}$$

On the other hand, putting together Proposition 8.4 with Lemma 8.17 and with Lemma 8.18, we get

$$\begin{aligned}(8.4.5)\quad \|C_j(f+w)-C_j(f)\|_{C^t} &\le c_2\|U_{p-j-1}(f)'\|_{C^0}^t\|v\|_{C^t} \\ &\quad + K_4\|U_{p-j-1}(f+w)-U_{p-j-1}(f)\|_{C^t}^{\alpha}\|v\|_{C^t} \\ &\le c_3|\Delta_{p-j-1}|^t\|v\|_{C^t} + K_5\|w\|_{C^t}^{\alpha}\|v\|_{C^t}\ .\end{aligned}$$

The first term on the last line of (8.4.5) looks a bit dangerous. What saves us here is the geometric control on the post-critical set of f (hence on the intervals Δ_i) that we have had at our disposal since §5.2. Substituting (8.4.3), (8.4.4) and (8.4.5) in (8.4.2) and adding up the terms with $j=0,\ldots,p-1$ we arrive at

$$\left\| A(f)\sum_{j=0}^{p-1} B_j(f)\,(C_j(f+w)-C_j(f))\right\|_{C^t} \le c_4\left\{\frac{1}{|\Delta_0|}\sum_{j=0}^{p-1}\frac{|\Delta_0|\cdot|\Delta_{p-j-1}|^t}{|\Delta_{p-j}|} + K_5\|w\|_{C^t}^{\varepsilon}\right\}\|v\|_{C^t}\ .$$

But as we have seen in §5.2:

(i) By Proposition 5.5 and Remark 5.1, if $t>2$ there exist $0<\gamma<1$ and $C>0$ with the following property. For every $g\in\mathbb{K}$ and every $m>0$, there exists an $\eta>0$ such that for all $f\in\mathbb{O}_m$ with $\|f-g\|_{\mathbb{A}}<\eta$ we have

$$(8.4.6)\qquad \sum_{j=0}^{p-1}\frac{|\Delta_j|^t}{|\Delta_{j+1}|}\le C\gamma^{mN}\ .$$

(ii) By Propostion 5.8 and Remark 5.1, for every $\gamma>1$ close to one, there exists $t<2$ close to two and $C>0$ with the following property. For every $g\in\mathbb{K}$ and every $m>0$, there exists $\eta>0$ such that for all $f\in\mathbb{O}_m$ with $\|f-g\|_{\mathbb{A}}<\eta$ we have that the inequality (8.4.6) above is also satisfied.

These last estimates end the proof of this proposition, provided $\mu=\gamma^N$ and $\eta<\mu^{m/\varepsilon}$. □

We arrive at last to the two main results of this section.

THEOREM 8.20. (i) *If $t>2$ is not an integer, there exist $0<\mu<1$ and $C>0$ with the following property. For every $g\in\mathbb{K}$ and for every $m>0$, there is an $\eta>0$ such that for all $f\in\mathbb{O}_m$ with $\|f-g\|_{\mathbb{A}}<\eta$ and for all $w\in\mathbb{A}^t$ with $\|w\|_{C^t}<\eta$,*

$$(8.4.7)\qquad \|DT^m(f+w)v-DT^m(f)v\|_{C^t}\le C\mu^m\|v\|_{C^t}\ .$$

(ii) *For every $\mu>1$ close to one, there exist $t<2$ close to 2 and $C>0$ with the following property. For every $g\in\mathbb{K}$ and every $m>0$, there exists $\eta>0$ such that for all $f\in\mathbb{O}_m$ with $\|f-g\|_{\mathbb{A}}<\eta$ and all $w\in\mathbb{A}^t$ with $\|w\|_{C^t}<\eta$, the inequality (8.4.7) above is also satisfied.*

Part (ii) of this theorem with $t = s$ implies property **B5** and part (i) is used later (for $t = r$) to prove property **B6**.

Proof. In this proof the positive constants $K_1, K_2, \ldots$ depend only on r and $\mathbb{O}_m$ and also on m. Let $E : C^t(I, I) \to \mathbb{R}$ be the evaluation map $E(f) = f(0)$, which is linear, and let $U_n : C^s(I, I) \to C^s(I)$ be as before. Let us write $DT^m(f + w)v - DT^m(f)(v) = E_1 + E_2 + E_3 + E_4 + E_5 + E_6 + E_7$, where

$$E_1 = (A(f + w) - A(f)) \sum_{j=0}^{p-1} B_j(f + w) \cdot C_j(f + w) ,$$

$$E_2 = A(f) \sum_{j=0}^{p-1} (B_j(f + w) - B_j(f)) \cdot C_j(f + w) ,$$

$$E_3 = A(f) \sum_{j=0}^{p-1} B_j(f) \cdot (C_j(f + w) - C_j(f)) ,$$

$$E_4 = (A(f + w) - A(f)) \cdot D(f + w) \sum_{j=0}^{p-1} E \circ B_j(f + w) \cdot E \circ C_j(f + w) ,$$

$$E_5 = A(f) \cdot (D(f + w) - D(f)) \sum_{j=0}^{p-1} E \circ B_j(f + w) \cdot E \circ C_j(f + w) ,$$

$$E_6 = A(f) \cdot D(f) \sum_{j=0}^{p-1} (E \circ B_j(f + w) - (E \circ B_j(f)) \cdot E \circ C_j(f + w) ,$$

$$E_7 = A(f) \cdot D(f) \sum_{j=0}^{p-1} E \circ B_j(f) \cdot (E \circ C_j(f + w) - E \circ C_j(f)) .$$

By Lemma 8.11,

$$|A(f + w) - A(f)| = \left|\lambda_{f+w}{}^{-1} - \lambda_f{}^{-1}\right|_{C^t} \leq K_1 \|w\|_{C^t} . \tag{8.4.8}$$

Hence,

$$\|E_1\|_{C^t} \leq K_2 \|w\|_{C^t} \|v\|_{C^t} \quad \text{and} \quad \|E_4\|_{C^t} \leq K_3 \|w\|_{C^t} \|v\|_{C^t} .$$

By Proposition 8.6 and Lemma 8.18, we obtain

$$\begin{aligned} \|B_j(f + w) - B_j(f)\|_{C^t} &\leq K_4 \|U_{p-j}(f + w) - U_{p-j}(f)\|_{C^t} \\ &\leq K_5 \|w\|_{C^t} . \end{aligned} \tag{8.4.9}$$

Since E is a bounded linear operator, from the last inequality, we obtain

$$\|E_2\|_{C^t} \leq K_6 \|w\|_{C^t} \|v\|_{C^t} \quad \text{and} \quad \|E_6\|_{C^t} \leq K_7 \|w\|_{C^t} \|v\|_{C^t} .$$

Taking $j = p$ in (8.4.9), we get

$$\|B_p(f + w) - B_p(f)\|_{C^t} \leq K_8 \|w\|_{C^t} .$$

By Lemma 8.18 and by (8.4.8), we get

$$\|\lambda_{f+w} \cdot U_p(f+w) - \lambda_f \cdot U_p(f)\|_{C^t} \leq K_9 \|w\|_{C^t} .$$

Combining the last two inequalities, we get $\|E_5\|_{C^t} \leq K_{10}\|w\|_{C^t}\|v\|_{C^t}$.

Let k and $0 < \alpha < 1$ be the integer and the fractional part of t, and let $0 < \varepsilon < 1$ be such that $\alpha - \varepsilon > 0$. From inequality (8.4.4), and since E is a bounded linear operator, we get

$$|E\,(C_j(f+w)) - E\,(C_j(f))\,| \leq K_{11}\|w\|_{C^t}^{\varepsilon}\|v\|_{C^t} .$$

Thus, $\|E_7\|_{C^t} \leq K_{12}\|w\|_{C^t}^{\varepsilon}\|v\|_{C^t}$. The only thing left to do is to bound $\|E_3\|_{C^t}$, and this follows at once from Proposition 8.19. □

Theorem 8.21. *If $r > 2$ is not an integer, there exist $0 < \mu < 1$ and $C > 0$ with the following property. For every $g \in \mathbb{K}$ and for every $m > 0$, there is an $\eta > 0$ such that for all $f \in \mathbb{O}_m$ with $\|f - g\|_{\mathbb{A}} < \eta$ and for all $v \in \mathbb{A}^r$ with $\|v\|_{C^r} < \eta$,*

$$\|T^m(f+v) - T^m(f) - DT^m(f)v\|_{C^r} \leq C\mu^m\|v\|_{C^r} . \tag{8.4.10}$$

This theorem together with Theorem 8.20 (i) for $t = r$ imply that the renormalization operator satisfies property **B6**.

Proof. In this proof the constants $\theta, \theta_1, \theta_2, \ldots$ are greater than zero and smaller than one and just depend upon r. The positive constants $c, c_1, c_2, \ldots$ depend only on r and $\mathbb{O}_m$, and the positive constants $K, K_1, K_2, \ldots$ depend also on m. Start by observing that since $T^m(f) = \lambda_f^{-1} \cdot U_p(f)$, we have $T^m(f+v) - T^m(f) - DT^m(f)v = A + B + C$, where

$$A = \lambda_f^{-1} \cdot (U_p(f+v) - U_p(f) - DU_p(f)v) ,$$

$$B = \left(\lambda_{f+v}^{-1} - \lambda_f^{-1} - D\lambda_f^{-1}(v)\right) \cdot U_p(f+v) ,$$

$$C = D\lambda_f^{-1}(v) \cdot (U_p(f+v) - U_p(f)) .$$

By Lemma 8.11, we have that $f \to \lambda_f^{-1}$ is C^1 and that there is θ_1 such that $\|B\|_{C^r} \leq K_1\|v\|_{C^r}^{1+\theta_1}$. Since $\|U_p(f+v) - U_p(f)\|_{C^r} \leq K_2\|v\|_{C^r}$, we have also $\|C\|_{C^r} \leq K_3\|v\|_{C^r}^2$. Hence inequality (8.4.10) will be established if we prove the following claim.

Claim. *If $r > 2$ there exist $0 < \mu < 1$ and $c_1 > 0$ with the following property: for every $g \in \mathbb{K}$ and for every m, there is an $\eta > 0$ such that for all $f \in \mathbb{O}_m$ with $\|f - g\|_{\mathbb{A}} < \eta$ and for all $v \in \mathbb{A}^r$ with $\|v\|_{C^r} < \eta$ we have*

$$\|U_p(f+v) - U_p(f) - DU_p(f)v\|_{C^r} \leq c_1\mu^m|\lambda_f|\|v\|_{C^r} . \tag{8.4.11}$$

To prove this claim, we will proceed recursively. Let us write for $i = 0, \ldots, p$,

$$R_i = U_i(f+v) - U_i(f) - DU_i(f)v .$$

Note that $R_{i+1} = E_i + F_i + f' \circ U_i(f) \cdot R_i$, where

$$E_i = f \circ U_i(f+v) - f \circ U_i(f) - f' \circ U_i(f) \cdot (U_i(f+v) - U_i(f)) \ ,$$
$$F_i = v \circ U_i(f+v) - v \circ U_i(f) \ .$$

Thus, working recursively from these expressions, we get

$$R_p = R_0 \cdot G_p + \sum_{i=0}^{p-1} (E_i \cdot G_{p-i-1} + F_i \cdot G_{p-i-1}) \ ,$$

where $G_{p-i-1} = (f^{p-i-1})' \circ U_{i+1}(f)$ and $R_0 = \Lambda_{f+v} - \Lambda_f - D\Lambda_f(v)$. Since $f \in \mathbb{O}_m$, by Proposition 8.6 and Lemma 8.18, we get

$$\|E_i\|_{C^r} \leq K_4 \|U_i(f+v) - U_i(f)\|_{C^r}^{1+\theta_2} \leq K_5 \|v\|_{C^r}^{1+\theta_2}$$

for $\theta_2 = 1 - \{r\}$. Therefore, $\left\| \sum_{i=0}^{p-1} E_i \cdot G_{p-i-1} \right\|_{C^r} \leq K_6 \|v\|_{C^r}^{1+\theta_2}$. By Lemma 8.11, there is θ_3 such that $\|R_0\|_{C^r} \leq K_7 \|v\|_{C^r}^{1+\theta_3}$. Hence, $\|R_0 \cdot G_p\|_{C^r} \leq K_8 \|v\|_{C^r}^{1+\theta_3}$. Finally, by Proposition 8.19, there exists $\theta_4 > 0$ such that

$$\left\| \sum_{i=0}^{p-1} F_i \cdot G_{p-i-1} \right\|_{C^r} \leq K_9 \|v\|_{C^r}^{1+\theta_4} + c_2 \mu^m |\lambda_f| \|v\|_{C^r} \ .$$

This proves our original claim. □

8.5. *Proof of Theorem* 8.1. All the pieces of the puzzle may now be put together. We want to check the robustness of T relative to the spaces $\mathcal{A} = \mathbb{A}$, $\mathcal{B} = \mathbb{A}^r$, $\mathcal{C} = \mathbb{A}^s$ and $\mathcal{D} = \mathbb{A}^0$. By Theorem 5.1, the pair $(\mathbb{A}^\gamma, \mathbb{A}^0)$ is ρ_γ-compatible with $(T, \mathbb{K})$ and $\rho_\gamma < \lambda$ for γ sufficiently close to 2 and is 1-compatible for $\gamma > 2$. Hence property **B1** is satisfied because $s > s_0$ with $s_0 < 2$ close to 2 and $r > s+1 > 2$. Since $r > s+1$, we know from Lemma 8.12 that T satisfies property **B2**. It also satisfies property **B3** by Lemma 8.13, and property **B4** by Lemma 8.16. Finally, T satisfies property **B5** by Theorem 8.20, and property **B6** by Theorem 8.21. Therefore the renormalization operator T is indeed robust with respect to $(\mathbb{A}^r, \mathbb{A}^s, \mathbb{A}^0)$.

8.6. *Proof of the hyperbolic picture.* Having established that the renormalization operator T is robust, we are now ready to show that the hyperbolic picture holds true for T acting on each of the spaces $\mathbb{U}^r$ and $\mathbb{V}^r$.

8.6.1. *Proof of Theorem* 2.5. We divide the proof of Theorem 2.5 into two cases: (i) r is not an integer (including the Lipschitz case $r = k + \mathrm{Lip}$), and (ii) $r = k$ is an integer.

Proof of case (i). Putting together Theorem 6.1 with Theorem 8.1 we deduce all the assertions of Theorem 2.5 except the fact that the holonomies are $C^{1+\beta}$ for some $\beta > 0$. This last fact follows if we combine Theorem 7.1 with Example 7.1.

Proof of case (ii). To prove this case, let us consider the Banach space $\mathbb{A}^{k-1+\mathrm{Lip}}$. Note that the natural inclusion $i : \mathbb{A}^k \to \mathbb{A}^{k-1+\mathrm{Lip}}$ is an isometric embedding. Indeed, for all $v \in \mathbb{A}^k$ we have $\|v\|_{C^k} = \|v\|_{\mathcal{B}}$, by the mean-value theorem. Applying case (i) to $\mathbb{A}^{k-1+\mathrm{Lip}}$, we see that for every $g \in \mathbb{K}$ the local stable set $W_\varepsilon^{s,k-1+\mathrm{Lip}}(g)$ is a codimension-one C^1 Banach submanifold of $\mathbb{A}^{k-1+\mathrm{Lip}}$. In fact, there exists a C^1 function $\Phi : \mathcal{O}_0 \to \mathbb{R}$, where $\mathcal{O}_0 \subseteq \mathbb{A}^{k-1+\mathrm{Lip}}$ is an open set containing g, such that $0 \in \mathbb{R}$ is a regular value for Φ, with

$$\Phi^{-1}(0) = \mathcal{O}_0 \cap W_\varepsilon^{s,k-1+\mathrm{Lip}}(g)$$

and such that $D\Phi(g)\mathbf{u}_g \neq 0$. Let $\mathcal{O}_1 = i^{-1}(\mathcal{O}_0) \subseteq \mathbb{A}^k$. Then $\mathcal{O}_1$ is open and $\Phi \circ i : \mathcal{O}_1 \to \mathbb{R}$ is C^1. Since $\mathbf{u}_g \in \mathbb{A}^k$ and $D(\Phi \circ i)(g)\mathbf{u}_g = D\Phi(g)\mathbf{u}_g \neq 0$, it follows that $0 \in \mathbb{R}$ is a regular value for $\Phi \circ i$ at g. Hence, by the implicit function theorem,

$$\mathcal{O}_1 \cap W_\varepsilon^{s,k}(g) = \mathcal{O}_1 \cap W_\varepsilon^{s,k-1+\mathrm{Lip}}(g) = \mathcal{O}_1 \cap (\Phi \circ i)^{-1}(0) ,$$

is a C^1, codimension-one Banach submanifold of $\mathbb{A}^k$. Since by case (i) the local stable manifolds in $\mathbb{A}^{k-1+\mathrm{Lip}}$ form a continuous lamination, we deduce that the same is true for the local stable manifolds in $\mathbb{A}^k$, because i is an isometric embedding. Finally, if F is a C^2 ordered transversal (in the sense of §7) to the stable lamination in $\mathbb{U}^k$, then $i \circ F$ is a C^2 ordered transversal to the stable lamination in $\mathbb{A}^{k-1+\mathrm{Lip}}$, and therefore by case (i) its holonomy in $\mathbb{U}^{k-1+\mathrm{Lip}}$ is $C^{1+\theta}$ for some $\theta > 0$. But then it follows that the holonomy of the transversal F in $\mathbb{U}^k$ is $C^{1+\theta}$ also.

8.6.2. *Proof of Corollary* 2.6. A similar argument to the one used in the proof of Theorem 2.5 can be used here. The map i is replaced throughout by the inclusion $j : \mathbb{B}^r \to \mathbb{A}^r$, which is a bounded linear operator (see §2.1). Hence the pre-images by j of the local stable leaves in $\mathbb{U}^r$ are C^1 manifolds and form a C^0 lamination in $\mathbb{V}^r$. Using [24] (see Remark 9.1 below), we see that the leaves of such a lamination contain the local stable sets of each $g \in \mathbb{K}$ in $\mathbb{V}^r$.

9. Global stable manifolds and one-parameter families

In this section, we prove Theorem 2.7. The first part will follow from Theorem 9.1 and the second part will follow from Theorem 9.2.

9.1. *The global stable manifolds of renormalization.* In this section we construct the global stable manifolds of the renormalization operator T in $\mathbb{V}^r$, for all r sufficiently large.

Let g be an element of the (bounded-type) invariant set $\mathbb{K}$ of T. Recall that the global stable set $W^{s,r}(g)$ of $g \in \mathbb{V}^r$ is given by

$$W^{s,r}(g) = \{f \in \mathbb{V}^r : \|T^n(f) - T^n(g)\|_{C^r} \to 0 \quad \text{when } n \to \infty\} .$$

From Corollary 2.6, we know that the convergence is exponential, and the exponential rate of convergence is independent of f and g, provided $r \geq 2+\alpha$ with $0 < \alpha < 1$ close to one.

THEOREM 9.1. *For every* $r \geq 3+\alpha$ *with* $\alpha < 1$ *sufficiently close to* 1, *and every* $g \in \mathbb{K}$, *the global stable set* $W^{s,r}(g)$ *is an immersed, codimension one* C^1 *Banach submanifold of* $\mathbb{V}^r$.

Remark 9.1. By [24], if the invariant set $\mathbb{K}$ of the renormalization operator is of bounded type then for every $r \geq 3$ and every $g \in \mathbb{K}$ we have that $W^{s,r}(g)$ coincides with the set of all maps $f \in \mathbb{V}^r$ with the same combinatorial type of g.

Proof. We already know that the local stable sets are C^1 submanifolds. The idea is to pull-back such a manifold structure by T using the implicit function theorem. More precisely, by Corollary 2.6 there exist $\varepsilon, \beta > 0$ so small that $W_\varepsilon^{s,r-1-\beta}(g)$ is a codimension one C^1 Banach submanifold of $\mathbb{V}^{r-1-\beta}$, for all $g \in \mathbb{K}$. We may assume that $\varepsilon > 0$ is so small that the vector $\mathbf{u}_g$ is transversal to the local stable set $W_\varepsilon^{s,r-1-\beta}(g)$ at each one of its points.

Now fix $g \in \mathbb{K}$ and let $f \in W^{s,r}(g)$. There exists $N = N(f) > 0$ so large that

$$T^N(f) \in W_\varepsilon^{s,r}(T^N(g)) \subset W_\varepsilon^{s,r-1-\beta}(T^N(g)) \ .$$

Since $v = \mathbf{u}_{T^N(g)}$ is transversal at $T^N(f)$ to $W_\varepsilon^{s,r-1-\beta}(T^N(g))$, There exist a small open set $\mathbb{O}_0 \subset \mathbb{V}^{r-1-\beta}$ containing $T^N(f)$ and a C^1 function $\Phi : \mathbb{O}_0 \to \mathbb{R}$ such that $\Phi^{-1}(0) = W_\varepsilon^{s,r-1-\beta}(T^N(g)) \subset \mathbb{O}_0$ for which $0 \in \mathbb{R}$ is a regular value and $D\Phi(T^N(f))v \neq 0$. The operator T^N is C^1 as a map from $\mathbb{V}^r$ into $\mathbb{V}^{r-1-\beta}$. Let $\mathbb{O}_1 \subset \mathbb{V}^r$ be an open set containing f such that $T^N(\mathbb{O}_1) \subset \mathbb{O}_0$. We want to show that $0 \in \mathbb{R}$ is a regular value for $\Phi \circ T^N : \mathbb{O}_1 \to \mathbb{R}$. Defining $F_t = T^N(f)+tv$ (for $|t|$ small), we get a C^1 family $\{F_t\}$ of maps in $\mathbb{V}^r$ which is transversal to $W_\varepsilon^{s,r-1-\beta}(T^N(g))$ at $F_0 = T^N(f)$. Now, we have the following claim.

Claim. There exists a C^1 *family* $\{f_t\}$ *with* $f_t \in \mathbb{V}^r$ *such that for all small* t *we have* $T^N(f_t) = F_t$.

Let us assume this claim for a moment. Setting

$$w = \left.\frac{d}{dt}\right|_{t=0} f_t \ ,$$

we obtain that

$$D(\Phi \circ T^N)(f)w = D\Phi(F_0)v \neq 0 \ .$$

Therefore, $\Phi \circ T^N$ is a C^1 local submersion at f. By the implicit function theorem $(\Phi \circ T^N)^{-1}(0)$ is a codimension one, C^1 Banach submanifold of $\mathbb{O}_1$

(or $\mathbb{V}^r$). Furthermore, if $h \in (\Phi \circ T^N)^{-1}(0)$ then $T^N(h) \in W_\varepsilon^{s,r-1-\beta}(T^N(g))$, and so h belongs to the global stable set $W^{s,r-1-\beta}(g)$. Using [24] (see Remark 9.1), we deduce that h belongs in fact to $W^{s,r}(g)$. This proves that $W^{s,r}(g)$ is an immersed C^1 manifold as asserted.

It remains to prove the claim. We first note that $F_t = h_t \circ F_0$ where each $h_t \in C^r(I,I)$ is a C^r diffeomorphism of $I = [-1,1]$. Since $T^N(f) = F_0$, there exist $p > 0$ and closed, pairwise disjoint intervals $0 \in \Delta_0, \Delta_1, \ldots, \Delta_{p-1} \subseteq I$ with $f(\Delta_i) \subseteq \Delta_{i+1}$ for $0 \le i < p-1$ and $f(\Delta_{p-1}) \subseteq \Delta_0$, such that

$$F_0 = T^N(f) = \Lambda_f^{-1} \circ f^p \circ \Lambda_f \ ,$$

where $\Lambda_f : I \to \Delta_0$ is the map $x \mapsto f^p(0)x$. Let $\overline{h}_t : \Delta_0 \to \Delta_0$ be the C^r diffeomorphism given by $\overline{h}_t = \Lambda_f \circ h_t \circ \Lambda_f^{-1}$. Consider a C^r extension of $\overline{h}_t$ to a diffeomorphism $H_t : I \to I$ with the property that $H_t|\Delta_i$ is the identity for all $i \neq 0$. Then let $f_t \in \mathbb{V}^r$ be the map $f_t = H_t \circ f$. Note that $f_t^i(0) = f^i(0)$ for all $0 \le i \le p$, that f_t is N-times renormalizable (under T) and that

$$\begin{aligned}
T^N(f_t) &= \Lambda_f^{-1} \circ f_t^p \circ \Lambda_f \\
&= \Lambda_f^{-1} \circ (H_t \circ f)\,|_{\Delta_{p-1}} \circ (H_t \circ f)\,|_{\Delta_{p-2}} \circ \cdots \circ (H_t \circ f)\,|_{\Delta_0} \circ \Lambda_f \\
&= \Lambda_f^{-1} \circ \overline{h}_t \circ f^p \circ \Lambda_f = \Lambda_f^{-1} \circ \Lambda_f \circ h_t \circ \left(\Lambda_f^{-1} \circ f^p \circ \Lambda_f\right) \\
&= h_t \circ F_0 = F_t \ ,
\end{aligned}$$

which proves the claim. □

9.2. *One-parameter families.* A one-parameter family of maps is a map $\psi : [0,1] \times I \to I$ (where $I = [-1,1]$ is the phase space) such that $\psi_t = \psi(t,\cdot)$ belongs to $\mathbb{V}^r$ for all $t \in [0,1]$. If ψ is a C^k map, then we say that ψ is a C^k family (of C^r unimodal maps). We often identify the family ψ with the curve $\{\psi_t\}_{0 \le t \le 1}$ of unimodal maps in $\mathbb{V}^r$. We shall denote by $\mathcal{UF}^k$ the space of all C^k families with the C^k topology ($\mathcal{UF}^k$ is a subset of $C^k([0,1] \times I)$).

We say that two families are C^{1+} equivalent if there exists a diffeomorphism from one into the other which sends each infinitely renormalizable map (with a fixed bounded combinatorial type) to a map with the same combinatorics. We are now in a position to state the result we have in mind.

THEOREM 9.2. *Let $r \ge 3 + \alpha$ with $\alpha > 0$ close to 1, and let $2 \le k \le r$. There exists an open and dense subset $\mathcal{O} \subseteq \mathcal{UF}^k$ of one-parameter C^k families of C^r unimodal maps having the following properties:*

(i) *Every family $\psi \in \mathcal{O}$ intersects the global stable lamination $\mathcal{L}^s$ of renormalization transversally.*

(ii) *For every $\psi \in \mathcal{O}$, there exist $0 = t_0 < t_1 < \cdots < t_n = 1$ such that for each $i = 0, 1, \ldots, n-1$ the sub-arc $\{\psi_t : t_i \le t \le t_{i+1}\}$ is $C^{1+\beta}$ diffeomorphic, via a holonomy-preserving diffeomorphism, to a corresponding sub-arc in the quadratic family. Here $\beta > 0$ is as given in Corollary 2.6.*

The proof will require a few lemmas. The first lemma says that every C^k family can be approximated (in the C^k sense) by a real analytic family.

LEMMA 9.3. *If $\psi \in \mathcal{UF}^k$, then for each $\varepsilon > 0$ there exists a real analytic family $f \in \mathcal{UF}^\omega$ such that $\|\psi - f\|_{C^k([0,1]\times I)} < \varepsilon$.*

Proof. Write each $\psi_t \in \mathbb{V}^r$ as $h_t \circ q$, where $q(x) = x^2$ and h_t is a diffeomorphism, and consider the C^k map $h : [0,1] \times I \to I$ given by $h(t,x) = h_t(x)$, a C^k family of C^r diffeomorphisms. To approximate h by a real analytic family of diffeomorphisms, consider the convolution of h with the heat kernel $k(t,x,\varepsilon) = e^{-(t^2+x^2)/4\varepsilon}$ for $\varepsilon > 0$ sufficiently small (see [1]). $\square$

Given this "denseness" result, the idea will be to show that arbitrarily close to an f as in Lemma 9.3 we can find a C^k family which is also transversal to the global stable lamination $\mathcal{L}^s$ of renormalization, by some kind of perturbation argument, to eliminate possible tangencies between $\{f_t\}$ and $\mathcal{L}^s$.

We will reduce our problem to the following general result about laminations with complex analytic leaves, whose elegant proof is due to Douady.

LEMMA 9.4. *Let $\mathcal{L} \subseteq \mathbb{C}^2$ be a C^0 lamination whose leaves are complex one-manifolds, and let $F : \mathbb{D} \to \mathbb{C}$ be a holomorphic function whose graph is tangent of finite order at $(0, F(0))$ to a leaf $L_0 \in \mathcal{L}$. Then the tangency is isolated: there exists a neighborhood of $(0, F(0))$ in $\mathbb{C}^2$ on which every other intersection of the graph of F with the leaves of $\mathcal{L}$ is a transversal intersection.*

Proof. Using a suitable chart, we may assume that the leaf L_0 is the horizontal plane $w = 0$ in $\mathbb{C}^2$, and that the other leaves of $\mathcal{L}$ in that chart are the graphs of holomorphic functions $\varphi_\mu : \mathbb{D} \to \mathbb{C}$ (with $\varphi_\mu(0) = \mu \in D$, where $D \subseteq \mathbb{C}$ is some open disk around zero, and $\varphi_0 \equiv 0$).

Since $\mathcal{L}$ is a C^0 lamination, φ_μ converges to 0 uniformly in $\mathbb{D}$ as μ tends to 0. Hence, for $|\mu|$ small enough, we have $\varphi_\mu(\mathbb{D}) \subset \mathbb{D}$. Moreover, $\varphi_\mu(z) \neq 0$ for all $z \in \mathbb{D}$ (leaves cannot intersect), so in fact $\varphi_\mu(\mathbb{D}) \subset \mathbb{D}^*$.

Now, we have $F(0) = F'(0) = \cdots = F^{(k-1)}(0) = 0 \neq F^{(k)}(0)$, for some $k \geq 2$. Composing the chart with a bi-holomorphic map if necessary, we may therefore assume that $F(z) = z^k$.

Let us fix $\mu \in D \setminus \{0\}$ and suppose that $z_0 \in \mathbb{D}$ is such that $\varphi_\mu(z_0) = F(z_0)$. We assume that $|z_0| < 1/2$ (taking $|\mu|$ small enough). To show that this intersection between φ_μ and F is transversal, it suffices to show that $\varphi'_\mu(z_0) \neq F'(z_0)$. But, by Schwarz's Lemma, the derivative $\varphi'_\mu(z_0)$ measured with respect to the Poincaré metrics of domain $\mathbb{D}$ and range $\mathbb{D}^*$ must be less than or equal to 1, that is to say

$$\left\|\varphi'_\mu(z_0)\right\|_P = \frac{|\varphi'_\mu(z_0)|\left(1 - |z_0|^2\right)}{|\varphi_\mu(z_0)|\log\left(|\varphi_\mu(z_0)|^{-1}\right)} \leq 1 .$$

Thus,

$$\left|\varphi'_\mu(z_0)\right| \le \frac{4}{3}k|z_0|^k \log\left(|z_0|^{-1}\right) .$$

On the other hand,

$$\left|F'(z_0)\right| = k|z_0|^{k-1} .$$

This shows that $\left|\varphi'_\mu(z_0)\right| / |F'(z_0)|$ converges to 0 as μ tends to 0, whence $\varphi'_\mu(z_0) \neq F'(z_0)$ for all sufficiently small $|\mu|$. Therefore $(0, F(0))$ is an isolated tangency as claimed. □

We may now state and prove the result on laminations with real analytic leaves which is needed for the proof of Theorem 9.2.

LEMMA 9.5. *Let $\mathcal{F} \subseteq [a,b] \times \mathbb{R}$ be a C^0 foliation whose leaves are the graphs of real analytic functions $\varphi_\mu : [a,b] \to \mathbb{R}$ with, say, $\varphi_\mu(a) = \mu \in [0,1]$. Let $\mathcal{L} \subseteq \mathcal{F}$ be a sub-lamination which is transversally totally disconnected (i.e. $K_0 = \{\mu \in [0,1] : gr(\varphi_\mu) \subseteq \mathcal{L}\}$ is a totally disconnected set). If $F : [a,b] \to \mathbb{R}$ is a real analytic function whose graph is not a leaf, then*

(i) $\mathrm{gr}(F)$ *is tangent to $\mathcal{F}$ at only finitely many points*;

(ii) *for all $\varepsilon > 0$ and all $k \geq 0$, there exists a real analytic $G : [a,b] \to \mathbb{R}$ such that $\|F - G\|_{C^k} < \varepsilon$ and all tangencies of* $\mathrm{gr}(G)$ *with $\mathcal{F}$ belong to $\mathcal{F} \setminus \mathcal{L}$; in particular,* $\mathrm{gr}(G)$ *is transversal to $\mathcal{L}$.*

Proof. (i) Complexifying $\mathcal{F}$ (i.e. the leaves φ_μ) as well as F, we put ourselves in the situation of Lemma 9.4. All tangencies are therefore isolated, and since $[a,b]$ is compact, there are only finitely many such, say at $x_i \in [a,b]$, $i = 1, 2, \ldots, n$.

(ii) Let d_i be the order of tangency of F with $\mathcal{F}$ at $(x_i, F(x_i))$. Then for every real analytic G sufficiently close to F in the C^k topology with k large ($k \geq \sum_{i=1}^n d_i$ will do), the number $n(G)$ of tangencies of $\mathrm{gr}(G)$ with $\mathcal{F}$ – not counting multiplicies – is bounded by $\sum_{i=1}^n d_i$. Hence we can find $G_0 : [a,b] \to \mathbb{R}$ real analytic with $\|F - G_0\|_{C^k} < \varepsilon/2$ such that $n(G_0)$ is maximal. All tangencies of G_0 with $\mathcal{F}$ must be first-order tangencies ($d_i = 1$). Indeed, if, say, $d_1 > 1$, then adding a suitable polynomial with small C^k norm to G_0, vanishing of very high order at $x_2, x_3, \ldots, x_{n(G_0)}$, we could unfold the tangency at x_1 to produce a new real analytic G with $n(G) > n(G_0)$. Now we may consider $G_t : [a,b] \to \mathbb{R}$ given by $G_t(x) = G_0(x) + t$ for $|t| < \varepsilon/2$. Since first-order tangencies are persistent, each tangency $(x_i, G_0(x_i))$ of G_0 with $\mathcal{F}$ generates a continuous, nonconstant path $(x_i(t), G_t(x_i(t))) \in \mathrm{gr}(\varphi_{\mu_i(t)})$ of (first-order) tangencies of G_t with $\mathcal{F}$. Each function $t \mapsto \mu_i(t)$, $i = 1, 2, \ldots, n(G_0)$, is continuous and nonconstant. Since K_0 is totally disconnected, there exists t (with $|t| < \varepsilon/2$) such that $\mu_i(t) \in [0,1] \setminus K_0$ for all i. Therefore, all tangencies of G_t with $\mathcal{F}$ fall in $\mathcal{F} \setminus \mathcal{L}$, whence G_t is transversal to $\mathcal{L}$. □

Proof of Theorem 9.2. Both properties (i) and (ii) are easily seen to be open, hence we concentrate on proving that they are dense. Let $\varepsilon > 0$.

Take any family $\psi \in \mathcal{UF}^k$. By Lemma 9.3, there exists a real analytic family $f \in \mathcal{UF}^\omega$ whose C^k distance from ψ is less than $\varepsilon/2$. The corresponding curve $\{f_t\}$ in $\mathbb{V}^r$ may fail to be transversal to the global stable lamination $\mathcal{L}^s$, so let us show how to perturb it locally to get a transversal family. Let $t_0 \in [0,1]$ be such that $f_{t_0} \in \mathcal{L}^s$ (and $\{f_t\}$ is tangent to $\mathcal{L}^s$ at f_{t_0}). Since f_{t_0} is infinitely renormalizable and real analytic, there exists $N > 0$ such that $R^N(f_{t_0}) \in \mathbb{A}_{\Omega_a}$ (where $a > 0$ is the constant in Theorem 2.4). Let $J \subseteq [0,1]$ be an interval containing t_0 such that $R^N(f_t)$ is well-defined and belongs to $\mathbb{A}_{\Omega_a}$ for all $t \in J$. We restrict our attention to the sub-family $\{f_t\}_{t\in J}$ from now on.

First we embed $\{f_t\}_{t\in J}$ in a two-parameter family in the following way. Note that each f_t belongs to $\mathbb{A}_{\Omega_\alpha}$ for some (fixed) $\alpha > 0$. As a map from (an open subset of) $\mathbb{A}_{\Omega_\alpha}$ into $\mathbb{A}_{\Omega_a}$, R^N is a real analytic operator.

CLAIM. *There exist analytic vectors $v \in \mathbb{A}_{\Omega_\alpha}$ and $w \in \mathbb{A}_{\Omega_a}$ with the property that $DR^N(f_{t_0})v = w$ and w is transversal to $\mathcal{L}_a^s = \mathcal{L}^s \cap \mathbb{A}_{\Omega_a}$ at $R^N(f_{t_0}) \in \mathcal{L}_a^s$.*

To see this, take any $w_0 \in \mathbb{A}_{\Omega_a}$ transversal to the (co-dimension one) lamination $\mathcal{L}_a^s$ at $R^N(f_{t_0})$. The same construction used in the proof of Theorem 9.1 yields a C^∞ vector v_0 at f_{t_0} such that $DR^N(f_{t_0})v_0 = w_0$. Now approximate v_0 by an analytic vector $v \in \mathbb{A}_{\Omega_\alpha}$ (in the C^m sense for $m \geq r$). Then $w = DR^N(f_0)v$ will still be transversal to $\mathcal{L}_a^s$. Shrinking J if necessary, we may in fact assume that $DR^N(f_t)v$ is transversal to $\mathcal{L}_a^s$ for all $t \in J$. Hence, we consider the two-parameter family of maps $f_{t,s} \in \mathbb{A}_{\Omega_\alpha}$ given by $f_{t,s} = f_t + sv$ with $t \in J$ and $|s| \leq \delta$ with δ small. Now,

$$W = \{f_{t,s} :\ t \in J,\ s \in [-\delta,\delta]\} \cong J \times [-\delta,\delta] \subseteq \mathbb{R}^2 \ ,$$

and $R^N|W : W \to A_{\Omega_a}$ is an injective, real analytic map. Recall now that in $\mathbb{A}_{\Omega_a}$ we have a C^0 foliation $\mathcal{F}$ with real analytic leaves (coming from hybrid classes, *cf.* §3) and that $\mathcal{L}_a^s \subseteq \mathcal{F}$ is the sub-lamination corresponding to the stable leaves of renormalization, which is transversally totally disconnected. Taking $\mathcal{F}_W = R^{-N}(\mathcal{F}) \subseteq W$ and $\mathcal{L}_W^s = R^{-N}(\mathcal{L}_a^s) \subseteq W$ and noting that $DR^N(f_{t,s})v = w$ is transversal to $\mathcal{L}_a^s$ for all $t \in J$, $s \in [-\delta,\delta]$ (making δ smaller if necessary) we deduce that $\mathcal{F}_W$ is a C^0 foliation (in W) by real analytic curves, and $\mathcal{L}_W^s \subseteq \mathcal{F}_W$ is a sub-lamination. Therefore we can apply Lemma 9.5 to this situation (with $\mathcal{F} = \mathcal{F}_W$ and $\mathcal{L} = \mathcal{L}_W^s$), obtaining a new analytic curve $\{g_t\}_{t\in J}$ with $\|f_t - g_t\|_{C^k} < \varepsilon/2$, transversal to $\mathcal{L}_W^s$ in W, and such that $\{R^N(g_t)\}$ is transversal to $\mathcal{L}_a^s$ at $R^N(g_{t_0})$. Since by Corollary 2.6 the holonomy of $\mathcal{L}_a^s$ is $C^{1+\beta}$ for some $\beta > 0$ (and the quadratic family is transversal to $\mathcal{L}_a^s$) we deduce that $\{g_t\}$ satisfies properties (a) and (b) of the statement. This completes the proof. □

10. A short list of symbols

For the reader's convenience, we present below a short list of symbols used in this paper.

p	Period of renormalization
λ_f	Scaling factor $\lambda_f = f^p(0)$
Λ_f	Linear scaling $\Lambda_f : x \to f^p(0) \cdot x$
R	Renormalization operator $R^N f = \Lambda_f^{-1} \circ f^p \circ \Lambda_f$
$\mathbb{K}$	Bounded type limit set of R
p_k	Number of renormalization intervals at level k
$\Delta_{j,k}(f)$	Renormalization intervals at level k $(0 \leq j \leq p_k - 1)$
$\mathcal{I}_f$	Post-critical set of f
$\mathbb{A}_V$	Real Banach space of continuous maps $\overline{V} \to \mathbb{C}$, holomorphic in V, symmetric about real axis
$T = R^N : \mathbb{O} \to \mathbb{A}_{\Omega_a}$	Real analytic operator for which $\mathbb{K} \subset \mathbb{O}$ is a hyperbolic basic set
$u_g(t)$	Parametrization of local unstable manifold $W^u_\varepsilon(g)$
$\mathbf{u}_g$	Unit vector tangent to $W^u_\varepsilon(g)$ at g
δ_g	Unique real number such that $DT(g)\mathbf{u}_g = \delta_g \mathbf{u}_{T(g)}$
$\delta_g^{(n)}$	The product $\delta_g \delta_{T(g)} \dots \delta_{T^{n-1}(g)}$
$L_f = DT(f)$	Derivative of T at f
$\mathbb{V}^r$	C^r unimodal maps with quadratic critical point at 0
$\mathbb{A}^r$	Tangent space to unimodal maps contained in $\mathbb{V}^r$

Instituto de Matemática e Estatística, Universidade de São Paulo, São Paulo SP, Brazil
E-mail address: edson@ime.usp.br

I.M.P.A., Rio de Janeiro, Brazil
E-mail address: demelo@impa.br

Universidade do Porto, Porto, Portugal
E-mail address: aapinto@fc.up.pt

References

[1] H. W. Broer and F. M. Tangerman, From a differentiable to a real analytic perturbation theory, applications to the Kupka Smale theorems, *Ergod. Theory & Dynam. Systems* **6** (1986), 345–362.

[2] M. Campanino and H. Epstein, On the existence of Feigenbaum's fixed point, *Comm. Math. Phys.* **79** (1981), 261–302.

[3] M. Campanino, H. Epstein, and D. Ruelle, On Feigenbaum's functional equation $g \circ g(\lambda x) + \lambda g(x) = 0$, *Topology* **21** (1982), 125–129.

[4] P. Coullet and C. Tresser, Itérations d'endomorphismes et groupe de renormalisation, *J. Phys. Colloque C* **539** (1978), C5-25 (1978).

[5] A. M. Davie, Period doubling for $C^{2+\varepsilon}$ mappings, *Comm. Math. Phys.* **176** (1996), 262–272.

[6] A. Douady and J.H. Hubbard, On the dynamics of polynomial-like mappings, *Ann. Sci. École Norm. Sup.* **18** (1985), 287–343.

[7] E. de Faria and W. de Melo, Rigidity of critical circle mappings I, *J. European Math. Soc.* **1** (1999), 339–392.

[8] ———, Rigidity of critical circle mappings II, *J. Amer. Math. Soc.* **13** (2000), 343–370.

[9] H. Epstein, New proofs of the existence of the Feigenbaum functions, *Comm. Math. Phys.* **106** (1986), 395–426.

[10] M. J. Feigenbaum, Qualitative universality for a class of nonlinear transformations, *J. Statist. Phys.* **19** (1978), 25–52.

[11] J. Franks, Manifolds of C^r mappings and applications to differentiable dynamics, *Stud. in Anal. Adv. Math. Suppl. Stud.* **4** (1979), 271–290.

[12] J. Guckenheimer, Sensitive dependence on initial conditions for one dimensional maps, *Comm. Math. Phys.* **70** (1979), 133–160.

[13] M. Herman, Sur la conjugaison différentiable des difféomorphismes du cercle à des rotations, *Publ. Math. IHES* **49** (1979), 5–234.

[14] M. Hirsch and C. Pugh, Stable manifolds and hyperbolic sets, in *Global Analysis*, *Proc. Sympos. Pure Math.* **XIV**, 133–164, A. M. S., Providence, RI (1970).

[15] L. Hörmander, The boundary problem of physical geodesy, *Arch. Rat. Mech. and Anal.* **62** (1976), 1–52.

[16] M. Irwin, On the stable manifold theorem, *Bull. London Math. Soc.* **2** (1970), 196–198.

[17] Y. Jiang, T. Morita, and D. Sullivan, Expanding direction of the period doubling operator, *Comm. Math. Phys.* **144** (1992), 509–520.

[18] O. E. Lanford III, A computer-assisted proof of the Feigenbaum conjectures, *Bull. Amer. Math. Soc.* **6** (1982), 427–434.

[19] R. de la Llave and R. Obaya, Regularity of the composition operator in spaces of Hölder functions, *Discrete Cont. Dynam. Systems* **5** (1999), 157–184.

[20] M. Lyubich, Feigenbaum-Coullet-Tresser universality and Milnor's hairiness conjecture, *Ann. of Math.* **149** (1999), 319–420.

[21] ———, Dynamics of quadratic polynomials I, II, *Acta Math.* **178** (1997), 185–297.

[22] C. McMullen, *Complex Dynamics and Renormalization*, *Ann. of Math. Studies* **135**, Princeton Univ. Press, Princeton, NJ, 1994.

[23] ———, *Renormalization and 3-manifolds which Fiber over the Circle*, *Ann. of Math. Studies* **142**, Princeton Univ. Press, Princeton, NJ, 1996.

[24] A. Avila, W. de Melo, and M. Martens, On the dynamics of the renormalization operator, in *Global Analysis on Dynamical Systems* (H. Broer, B. Krauskopf, and G. Vegter, eds.), Institute of Physics Publ., Philadelphia (2001), 449–460 .

[25] W. de Melo and A. A. Pinto, Rigidity of C^2 infinitely renormalizable unimodal maps, *Comm. Math. Phys.* **208** (1999), 91–105.

[26] W. de Melo and S. van Strien, *One-dimensional Dynamics*, Springer-Verlag, New York (1993).

[27] C. Robinson, *Dynamical Systems. Stability, Symbolic Dynamics, and Chaos*, *Studies in Advanced Math.*, CRC Press, Boca Raton, FL (1995).

[28] D. Sullivan, *Bounds, Quadratic Differentials and Renormalization Conjectures*, *AMS Centennial Publications* **2**, *Mathematics into the Twenty-first Century*, A.M.S., Providence, RI (1992).

[29] E. Vul, Ya. Sinai, and K. Khanin, Feigenbaum universality and the thermodynamic formalism, *Russ. Math. Surveys* **39** (1984), 1–40.

(Received January 15, 2001)
(Revised July 22, 2004)

Rigidity of Critical Circle Maps

Pablo Guarino, Marco Martens and Welington de Melo†

Duke Math. J. **167**(11), 2125–2188 (2018)

To our coauthor and friend Welington de Melo, in memorium

Abstract
We prove that any two C^4 critical circle maps with the same irrational rotation number and the same odd criticality are conjugate to each other by a C^1 circle diffeomorphism. The conjugacy is $C^{1+\alpha}$ for a full Lebesgue measure set of rotation numbers.

Contents

1. Introduction

By a *critical circle map* we mean an orientation-preserving C^4 circle homeomorphism, with exactly one nonflat critical point of odd criticality (see Definition 2.1 below). In 1984, Yoccoz [33] proved that if such a critical circle map has no periodic

DUKE MATHEMATICAL JOURNAL
Vol. 167, No. 11, © 2018 DOI 10.1215/00127094-2018-0017
Received 15 December 2016. Revision received 19 March 2018.
First published online 18 July 2018.
2010 *Mathematics Subject Classification*. Primary 37E10; Secondary 37E20.
†Welington de Melo: 1946–2016.

Originally published in *Duke Math. J.* **167**(11), 2125–2188 (2018).

points, all its orbits are dense. This implies the following topological rigidity result: if two critical circle maps have the same irrational rotation number, then there exists a unique conjugacy between them that sends the critical point to the critical point.

Numerical observations (see [9], [22], [26]) suggested in the early 1980s that this topological conjugacy could in fact be a smooth diffeomorphism, at least for bounded combinatorics. These observations led to the *rigidity conjecture*, posed in several works by Lanford [16], [17], Rand [23]–[25] (see also [22]), and Shenker [26] (see also [9]) among others. Our main result is the following.

THEOREM A (Rigidity)
Let f and g be two C^4 circle homeomorphisms with the same irrational rotation number and with a unique critical point of the same odd type. Let h be the unique topological conjugacy between f and g that maps the critical point of f to the critical point of g. Then
(1) *h is a C^1-diffeomorphism;*
(2) *h is $C^{1+\alpha}$ at the critical point of f for a universal $\alpha > 0$;*
(3) *for a full Lebesgue measure set of rotation numbers, h is a $C^{1+\alpha}$-diffeomorphism.*

See [3, Section 4.4] for the definition of the full measure set of rotation numbers considered in Theorem A(3). Let us point out that, by a result of Ávila [1], there exist two real analytic critical circle maps with the same irrational rotation number and the same criticality that are not $C^{1+\beta}$-conjugate for any $\beta > 0$.

Much effort has been devoted to addressing the rigidity problem (see [1]–[4], [11], [14], [15], [28]–[31]). In particular, Theorem A was proved for real analytic critical circle maps in a series of papers by de Faria and de Melo [4], Khmelev and Yampolsky [15], and Khanin and Teplinsky [14]. In the C^3-category, moreover, rigidity holds for bounded combinatorics: any two C^3 critical circle maps with the same irrational rotation number of *bounded type* and the same odd criticality are conjugate to each other by a $C^{1+\alpha}$ circle diffeomorphism, for some universal $\alpha > 0$ (see [11]). We point out that this is the precise statement of the rigidity conjecture mentioned above.

Remark 1.1
We do not know whether Theorem A holds for C^3-dynamics with unbounded combinatorics, nor do we know if rigidity holds in less regularity, for instance, $C^{2+\alpha}$-smoothness, even for bounded combinatorics. Moreover, we do not know how to deal with critical points of noninteger criticality, not even with fractional criticality (of odd type; see Definition 2.1 below).

By the famous rigidity result of Herman [12], improved by Yoccoz [32], any real analytic circle diffeomorphism whose rotation number satisfies a Diophantine condition is analytically conjugate to a rotation. We believe that there exist two real analytic critical circle maps with the same rotation number of bounded type and the same criticality that are not C^2-conjugate. Theorem A follows from the following theorem, which is our main result on the dynamics of the renormalization operator acting on the C^4-class.

THEOREM B (Exponential convergence in the C^2-distance)
There exists a universal constant $\lambda \in (0,1)$ such that given two C^4 critical circle maps f and g with the same irrational rotation number and the same criticality, there exists $C = C(f,g) > 0$ such that for all $n \in \mathbb{N}$ we have

$$d_2\big(\mathcal{R}^n(f), \mathcal{R}^n(g)\big) \leq C\lambda^n,$$

where d_2 is the C^2-distance in the space of C^2 critical commuting pairs.

This article is devoted to the proof of Theorem B. The fact that Theorem B implies Theorem A follows from well-known results by de Faria and de Melo [3, First Main Theorem, p. 341] and Khanin and Teplinsky [14, Theorem 2, p. 198], and it will be explained in Section 13.

1.1. Strategy of the proof of Theorem B
A C^4 critical circle map f with irrational rotation number generates a sequence $\{\mathcal{R}^n(f)\}_{n\in\mathbb{N}}$ of commuting pairs of interval maps, each one being the renormalization of the previous one (see Definition 2.6). To prove Theorem B, we need to prove the exponential convergence in the C^2-distance of the orbits generated by two C^4 critical circle maps with the same irrational combinatorics and the same criticality. Roughly speaking, the proof has three ingredients:

(1) the existence of a sequence $\{f_n\}_{n\in\mathbb{N}}$ that belongs to a universal C^ω-compact set of real analytic critical commuting pairs, such that $\mathcal{R}^n(f)$ is C^3 exponentially close to f_n at a universal rate, and both have the same rotation number (see Theorem 11.1);

(2) the uniform exponential contraction of renormalization when restricted to topological conjugacy classes of real analytic critical commuting pairs (see Theorem 2.8);

(3) the *key lemma* (Lemma 4.1): a Lipschitz estimate for the renormalization operator, when restricted to suitable bounded pieces of topological conjugacy classes of infinitely renormalizable C^3 critical commuting pairs with negative Schwarzian derivative (we would like to note that the Lipschitz continuity

holds *along topological classes*: we cannot compare the action of the renormalization operator when restricted to different topological classes).

The fact that (1), (2), and (3) imply Theorem B will be proved in Section 12. Theorem 11.1 will be obtained in Section 11, based on a previous construction developed by the first and third authors in [11]. Theorem 2.8 was proved by de Faria and de Melo [4] for rotation numbers of bounded type, and extended by Khmelev and Yampolsky [15] to cover all irrational rotation numbers.

Our main task in this article is to prove Lemma 4.1, and its proof will occupy Sections 5–10. For bounded combinatorics, it is not difficult to prove the key lemma (see [11, Section 3.5]). Let us explain here the main difficulties in the unbounded case. Given an infinitely renormalizable critical commuting pair (see Definition 2.3), one obtains the domain of its first renormalization by iterating a boundary point of its original domain (see Definition 2.6). The rotation number of the map determines the number of iterates involved to obtain the boundary of the renormalization. The number, denoted by a, plays a delicate role. Denote the obtained boundary point by x_a.

Under the K-controlled and the negative Schwarzian assumptions (see Section 3 for definitions), the geometry of this piece of monotone orbit is precisely described by a result due to Yoccoz (see Lemma 4.2 in Section 4): for large values of a, most of the points are concentrated in a small area of accumulation, with size comparable to $1/\sqrt{a}$.

Now consider two C^3 critical commuting pairs which are renormalizable with the same period $a \in \mathbb{N}$. Suppose that their C^2-distance is $\varepsilon > 0$. Denote the corresponding new boundary points by x_a^0 and x_a^1. These new boundary points can indeed be arbitrarily separated. A consequence of this is that renormalization is definitely *not* even Hölder-continuous on the space of *all* commuting pairs. However, we will prove that when one renormalizes two pairs with the same irrational rotation number, the new boundary points become close. Namely, $|x_a^0 - x_a^1| = O(\varepsilon)$. The systems are said to be *synchronized* (see Definition 6.11 and the synchronization lemma in Section 9). If a separation were to occur, meaning that $|x_a^0 - x_a^1|$ is not small, then one can find a future iterate of renormalization where the a-values (the corresponding periods of renormalization) will not be the same. This contradicts the fact that the systems had the same irrational rotation number.

Observe that when the pairs are very close, the a-values of arbitrarily many future renormalizations will play a role. From this, it seems to be an impossible task to obtain an estimate for the distance between the new boundary points. A natural notion of order is introduced in Section 8. One can interpret this order as a cone field associated with the unstable direction of renormalization. The rotation number is monotone with respect to the order. Using the order, synchronization follows.

Synchronization implies that the pieces of orbit involved in defining the boundaries of the domains of the renormalizations of both pairs are everywhere close together (see Proposition 7.8). This allows us to control the branches of the renormalizations. Hence, renormalization is uniformly Lipschitz on classes of *controlled* pairs with the same irrational rotation number (see Section 10).

Remark 1.2
The renormalization theory for critical circle maps was developed during the late 1970s and the 1980s (see [9], [16], [17], [22]–[26]) in parallel with the renormalization theory of *unimodal maps* of the interval (see [8, Chapter VI] and the references therein; see also [6], [21]). The fact that the exponential convergence of renormalization for real analytic dynamics implies the exponential convergence for finitely smooth unimodal maps was obtained in the late 1990s in [7]. Note that the methods of [7] do not apply for critical circle maps, not even for bounded combinatorics.

One source for this difference is the fact that, for infinitely renormalizable unimodal maps, the sum of the lengths of the cycle of intervals related to each level of renormalization goes to zero exponentially fast. This gives a strong control of the nonlinearity, even for C^2-dynamics. In the circle case, however, the intervals involved at each step of renormalization cover the whole circle (see Section 2). In this case, C^4-smoothness is needed in order to have C^3-bounded orbits of renormalization (see Theorem 3.5) and then to be able to control the nonlinearity with the help of the Schwarzian derivative. Another crucial ingredient in [7] is the *hybrid* lamination and the fact that its holonomy is quasiconformal (see [18] and the references therein). To the best of our knowledge, no similar structure has been obtained for critical circle maps.

This article is organized as follows. In Section 2, we briefly review some preliminaries, and we introduce the renormalization operator acting on critical commuting pairs. In Section 3, we define *controlled* commuting pairs (see Definition 3.3) and state some related properties. In Section 4, we state the previously discussed Lemma 4.1 (the key lemma) and Lemma 4.2 (Yoccoz's lemma). In Section 5, we construct a suitable 1-parameter family around a critical commuting pair, the *standard family*. These families are *transversal* to topological classes, and they may be regarded as *unstable manifolds* for renormalization (see especially Proposition 5.8, where we estimate the expansion of renormalization along these families). In Section 6, we introduce the notion of *synchronization* (see Definition 6.11) and we collect several estimates that hold under synchronization (see especially Propositions 6.18 and 6.19). In Section 7, we prove that the key lemma (Lemma 4.1) holds under the synchronization assumption (see Proposition 7.8). In Section 8, we define the already discussed notion of *order*. In Section 9, we prove the *synchronization lemma*, and in Section 10, we finally prove Lemma 4.1. Section 11 hosts the proof of Theorem 11.1, and in Section 12 we finally prove that Theorem B follows from Theorem 11.1, Theorem 2.8, and Lemma 4.1. Section 13 provides the precise references of the fact that Theorem B implies Theorem A, and in the Appendix we prove Proposition 11.2, stated and used in Section 11.

Now let us fix some notation that we will use in this article: $\mathbb{N}$, $\mathbb{Z}$, $\mathbb{Q}$, $\mathbb{R}$, and $\mathbb{C}$ denote, respectively, the set of natural, integer, rational, real, and complex numbers. With S^1 we denote the multiplicative group of complex numbers of modulus 1, that is, the unit circle $S^1 = \{z \in \mathbb{C} : |z| = 1\}$. We will identify S^1 with $\mathbb{R}/\mathbb{Z}$ under the universal covering map $\pi : \mathbb{R} \to S^1$ given by $\pi(t) = \exp(2\pi i t)$. We denote by $\mathrm{Hom}_+(S^1)$ the group (under composition) of orientation-preserving circle homeomorphisms, and we denote by $\mathrm{Diff}^r_+(S^1)$ its subgroup of C^r-diffeomorphisms for any $r \geq 1$. The function $\rho : \mathrm{Hom}_+(S^1) \to [0, 1)$ will denote the rotation number function. The Euclidean length of an interval I will be denoted by $|I|$. Uniform constants will always be denoted by K. Their value might change during estimates.

2. Preliminaries

Definition 2.1
By a *critical circle map* we mean an orientation-preserving C^4 circle homeomorphism f with exactly one critical point c such that, in a neighborhood of c, the map f can be written as $f(t) = f(c) + (\phi(t))^{2d+1}$, where ϕ is a C^4 local diffeomorphism with $\phi(c) = 0$, and $d \in \mathbb{N}$ with $d \geq 1$. The *criticality* (or *order*, or *type*, or *exponent*) of the critical point c is the odd integer $2d + 1$. We also say that the critical point c is *nonflat*.

As an example, consider the so-called *Arnold's family*, which is the 1-parameter family $\widetilde{f_\omega} : \mathbb{R} \to \mathbb{R}$ given by

$$\widetilde{f_\omega}(t) = t + \omega - \left(\frac{1}{2\pi}\right) \sin(2\pi t) \quad \text{for } \omega \in [0, 1).$$

Since each $\widetilde{f_\omega}$ commutes with unitary translation, it is the lift, under the universal cover $t \mapsto e^{2\pi i t}$, of an orientation-preserving real analytic circle homeomorphism f_ω. Each f_ω is a critical circle map with a single critical point of cubic type (at the point 1, the projection of the integers). (See also [5, Section 6] for examples of rational maps.)

We will assume throughout this article that the rotation number $\rho(f) = \theta$ in $[0, 1)$ is irrational. Let $[a_0, a_1, \dots, a_n, a_{n+1}, \dots]$ be its continued fraction expansion:

$$\theta = \lim_{n \to +\infty} \cfrac{1}{a_0 + \cfrac{1}{a_1 + \cfrac{1}{a_2 + \cfrac{1}{\ddots \frac{1}{a_n}}}}}.$$

We define recursively the *return times* of θ by

$$q_0 = 1, \qquad q_1 = a_0, \qquad \text{and} \qquad q_{n+1} = a_n q_n + q_{n-1} \quad \text{for } n \geq 1.$$

Recall that the numbers q_n are also obtained as the denominators of the truncated expansion of order n of θ:

$$\frac{p_n}{q_n} = [a_0, a_1, a_2, \dots, a_{n-1}] = \cfrac{1}{a_0 + \cfrac{1}{a_1 + \cfrac{1}{a_2 + \cfrac{1}{\ddots \frac{1}{a_{n-1}}}}}}.$$

2.1. Real bounds

Denote by I_n the interval $[c, f^{q_n}(c)]$, and define $\mathcal{P}_n$ as

$$\mathcal{P}_n = \{I_n, f(I_n), \dots, f^{q_{n+1}-1}(I_n)\} \cup \{I_{n+1}, f(I_{n+1}), \dots, f^{q_n-1}(I_{n+1})\}.$$

A crucial combinatorial fact is that $\mathcal{P}_n$ is a partition (modulo boundary points) of the circle for every $n \in \mathbb{N}$. We call it the *nth dynamical partition* of f associated with the point c. Note that the partition $\mathcal{P}_n$ is determined by the piece of orbit

$$\{f^j(c) : 0 \leq j \leq q_n + q_{n+1} - 1\}.$$

The transition from $\mathcal{P}_n$ to $\mathcal{P}_{n+1}$ can be described in the following easy way. The interval $I_n = [c, f^{q_n}(c)]$ is subdivided by the points $f^{jq_{n+1}+q_n}(c)$ with $1 \leq j \leq a_{n+1}$ into $a_{n+1} + 1$ subintervals. This subpartition is spread by the iterates of f to all the $f^j(I_n) = f^j([c, f^{q_n}(c)])$ with $0 \leq j < q_{n+1}$. The other elements of the partition $\mathcal{P}_n$, which are $f^j(I_{n+1})$ with $0 \leq j < q_n$, remain unchanged. Note that, since we are working with critical circle maps with a single critical point, our partitions in this article are always determined by the critical orbit.

THEOREM 2.2 (Real bounds)

There exists $K > 1$ such that, given a C^3 critical circle map f with irrational rotation number, there exists $n_0 = n_0(f)$ such that, for all $n \geq n_0$ and for every pair I, J of adjacent atoms of $\mathcal{P}_n$, we have

$$K^{-1}|I| \leq |J| \leq K|I|.$$

Moreover, if Df denotes the first derivative of f, then we have

$$\frac{1}{K} \le \frac{|Df^{q_n-1}(x)|}{|Df^{q_n-1}(y)|} \le K \quad \textit{for all } x, y \in f(I_{n+1}) \textit{ and for all } n \ge n_0,$$

$$\frac{1}{K} \le \frac{|Df^{q_{n+1}-1}(x)|}{|Df^{q_{n+1}-1}(y)|} \le K \quad \textit{for all } x, y \in f(I_n) \textit{ and for all } n \ge n_0.$$

Theorem 2.2 was proved by Świątek [27] (see also [10]) and Herman [13] (see also [3]). The control on the distortion of the return maps follows from Koebe's distortion principle (see [3, Section 3]). Note that for a rigid rotation, we have $|I_n| = a_{n+1}|I_{n+1}| + |I_{n+2}|$. If a_{n+1} is big, then I_n is much larger than I_{n+1}. Thus, even for rigid rotations, real bounds do not hold in general.

2.2. *Critical commuting pairs*

Definition 2.3

A C^r *critical commuting pair* $\zeta = (\eta, \xi)$ consists of two C^r orientation-preserving homeomorphisms $\eta : I_\eta \to \eta(I_\eta)$ and $\xi : I_\xi \to \xi(I_\xi)$, where

(1) $I_\eta = [0, \xi(0)]$ and $I_\xi = [\eta(0), 0]$ are compact intervals in the real line;

(2) $(\eta \circ \xi)(0) = (\xi \circ \eta)(0) \neq 0$;

(3) $D\eta(x) > 0$ for all $x \in I_\eta \setminus \{0\}$ and $D\xi(x) > 0$ for all $x \in I_\xi \setminus \{0\}$;

(4) the origin is a nonflat critical point for both η and ξ with the same odd criticality (i.e., there exist a positive integer d, an open interval C around the origin, and two orientation-preserving C^r-diffeomorphisms $\phi : C \to \phi(C)$ and $\psi : C \to \psi(C)$ fixing the origin such that $\eta(x) = (\phi(x))^{2d+1} + \eta(0)$ for all $x \in C \cap I_\eta$ and $\xi(x) = (\psi(x))^{2d+1} + \xi(0)$ for all $x \in C \cap I_\xi$);

(5) the left derivatives of the composition $\eta \circ \xi$ at the origin coincide with the corresponding right derivatives of the composition $\xi \circ \eta$ at the origin; that is, for each $j \in \{1, 2, \dots, r\}$ we have $D_-^j(\eta \circ \xi)(0) = D_+^j(\xi \circ \eta)(0)$.

Any critical circle map f with irrational rotation number θ induces a sequence of critical commuting pairs in a natural way. Let $\widehat{f}$ be the lift of f to the real line (for the canonical covering $t \mapsto e^{2\pi i t}$) satisfying $D\widehat{f}(0) = 0$ and $0 < \widehat{f}(0) < 1$. For each $n \ge 1$, let $\widehat{I_n}$ be the closed interval in the real line, adjacent to the origin, that projects under $t \mapsto e^{2\pi i t}$ to I_n. Let $T : \mathbb{R} \to \mathbb{R}$ be the translation $x \mapsto x + 1$, and define $\eta : \widehat{I_n} \to \mathbb{R}$ and $\xi : \widehat{I}_{n+1} \to \mathbb{R}$ as

$$\eta = T^{-p_{n+1}} \circ \widehat{f}^{q_{n+1}} \qquad \text{and} \qquad \xi = T^{-p_n} \circ \widehat{f}^{q_n}.$$

It is not difficult to check that $(\eta|_{\widehat{I_n}}, \xi|_{\widehat{I}_{n+1}})$ is a critical commuting pair, which we denote by $(f^{q_{n+1}}|_{I_n}, f^{q_n}|_{I_{n+1}})$ to simplify notation.

As pointed out in [3, p. 344], the commuting condition (2) in Definition 2.3 actually holds on an open interval around the origin.

LEMMA 2.4
There exist open intervals $V_- \supseteq I_\xi$ and $V_+ \supseteq I_\eta$ and C^r-homeomorphic extensions $\widehat{\xi} : V_- \to \widehat{\xi}(V_-) \subset \mathbb{R}$ and $\widehat{\eta} : V_+ \to \widehat{\eta}(V_+) \subset \mathbb{R}$ of ξ and η, respectively, satisfying $(\widehat{\eta} \circ \widehat{\xi})(x) = (\widehat{\xi} \circ \widehat{\eta})(x)$ for all x in the open interval C around the origin given by $C = \{x \in V_- \cap V_+ : \widehat{\eta}(x) \in V_- \text{ and } \widehat{\xi}(x) \in V_+\}$.

Proof of Lemma 2.4
Since the origin is a nonflat critical point of odd criticality, there exists an open interval C around it on which we can extend both η and ξ to C^r-homeomorphisms $\widehat{\eta} : C \to A$ and $\widehat{\xi} : C \to B$, where A is an open interval around $\eta(0)$ and B is an open interval around $\xi(0)$ (we may suppose that A, B, and C are pairwise disjoint). Moreover, since the criticality of both $\widehat{\eta}$ and $\widehat{\xi}$ at the origin is the same odd integer, the composition $\widehat{\xi} \circ \widehat{\eta}^{-1} : A \to B$ is actually a C^r-diffeomorphism. Let $V_- = A \cup I_\xi \cup C$, which is an open interval where I_ξ is compactly contained, and in the same way let $V_+ = C \cup I_\eta \cup B$.

Since the composition $\eta \circ \xi$ is already defined at the left part of C, the extension of η defined above (given by the nonflatness of the critical point) allows us to extend ξ to the left part of A in the following way: for any $y \in A$ there exists a unique $x \in C$ such that $\widehat{\eta}(x) = y$ (since $A = \widehat{\eta}(C)$ and $\widehat{\eta} : C \to A$ is invertible), and then we define $\widehat{\xi} : A \to \mathbb{R}$ as $\widehat{\xi}(y) = \eta(\xi(x)) = (\eta \circ \xi \circ \widehat{\eta}^{-1})(y)$ if $y < \eta(0)$ and $\widehat{\xi}(y) = \xi(y)$ if $y \geq \eta(0)$.

By Definition 2.3(5), the left derivatives of the composition $\eta \circ \xi \circ \widehat{\eta}^{-1}$ at the point $\eta(0)$ coincide with the corresponding right derivatives of ξ at $\eta(0)$; that is, $\widehat{\xi}$ is of class C^r at the point $\eta(0)$ (and therefore on the whole domain V_-). Note also that $\widehat{\xi}$ has no critical points on $V_- \setminus \{0\}$ since $\widehat{\xi} \circ \widehat{\eta}^{-1} : A \to B$ is a C^r-diffeomorphism and η has no critical points in $B \cap I_\eta$ by condition (3). In the same way, since the composition $\xi \circ \eta$ is already defined at the right part of C and since ξ is also defined on C, we extend η to the right part of B by imposing the commuting condition $\widehat{\eta} \circ \widehat{\xi} = \widehat{\xi} \circ \widehat{\eta}$ on C as before. □

The Möbius metric
Given two critical commuting pairs $\zeta_1 = (\eta_1, \xi_1)$ and $\zeta_2 = (\eta_2, \xi_2)$, let A_1 and A_2 be the Möbius transformations such that for $i = 1, 2$,

$$A_i\big(\eta_i(0)\big) = -1, \qquad A_i(0) = 0, \qquad A_i\big(\xi_i(0)\big) = 1.$$

Definition 2.5
For any $0 \le r < \infty$, define the C^r-metric on the space of C^r critical commuting pairs in the following way:

$$d_r(\zeta_1, \zeta_2) = \max\left\{\left|\frac{\xi_1(0)}{\eta_1(0)} - \frac{\xi_2(0)}{\eta_2(0)}\right|, \|A_1 \circ \zeta_1 \circ A_1^{-1} - A_2 \circ \zeta_2 \circ A_2^{-1}\|_r\right\},$$

where $\|\cdot\|_r$ is the C^r-norm for maps in $[-1,1]$ with one discontinuity at the origin, and where ζ_i is the piecewise map defined by η_i and ξ_i:

$$\zeta_i : I_{\xi_i} \cup I_{\eta_i} \to I_{\xi_i} \cup I_{\eta_i} \quad \text{such that } \zeta_i|_{I_{\xi_i}} = \xi_i \text{ and } \zeta_i|_{I_{\eta_i}} = \eta_i.$$

When we are dealing with real analytic critical commuting pairs, we consider the C^ω-topology defined in the usual way. We say that $(\eta_n, \xi_n) \to (\eta, \xi)$ if there exist two open sets $U_\eta \supset I_\eta$ and $U_\xi \supset I_\xi$ in the complex plane and $n_0 \in \mathbb{N}$ such that η and η_n for $n \ge n_0$ extend continuously to $\overline{U_\eta}$, are holomorphic in U_η and we have $\|\eta_n - \eta\|_{C^0(\overline{U_\eta})} \to 0$, and such that ξ and ξ_n for $n \ge n_0$ extend continuously to $\overline{U_\xi}$, are holomorphic in U_ξ and we have $\|\xi_n - \xi\|_{C^0(\overline{U_\xi})} \to 0$. We say that a set $\mathcal{C}$ of real analytic critical commuting pairs is *closed* if, every time we have $\{\zeta_n\} \subset \mathcal{C}$ and $\{\zeta_n\} \to \zeta$, we have $\zeta \in \mathcal{C}$. This defines a Hausdorff topology, stronger than the C^r-topology for any $0 \le r \le \infty$ (in particular, any C^ω-compact set of real analytic critical commuting pairs is certainly C^r-compact also, for any $0 \le r \le \infty$).

The affine metric
Given two critical commuting pairs $\zeta_1 = (\eta_1, \xi_1)$ and $\zeta_2 = (\eta_2, \xi_2)$, let $L_{\eta_1} : [0,1] \to I_{\eta_1}$ be $L_{\eta_1}(t) = |I_{\eta_1}|t$, and let $L_{\xi_1} : [-1,0] \to I_{\xi_1}$ be $L_{\xi_1}(t) = |I_{\xi_1}|t$. Define in the same way L_{η_2} and L_{ξ_2}, and consider

$$\begin{aligned} d_{r,\mathrm{Aff}}&(\zeta_1, \zeta_2 \\ &= \max\left\{\left|\frac{\xi_1(0)}{\eta_1(0)} - \frac{\xi_2(0)}{\eta_2(0)}\right|, \|\eta_1 \circ L_{\eta_1} - \eta_2 \circ L_{\eta_2}\|_{C^r([0,1])}, \right. \\ &\left. \quad \|\xi_1 \circ L_{\xi_1} - \xi_2 \circ L_{\xi_2}\|_{C^r([-1,0])}\right\}. \end{aligned}$$

Both d_r and $d_{r,\mathrm{Aff}}$ are not metrics but *pseudometrics*, since they are invariant under conjugacy by homotheties: if α is a positive real number, $H_\alpha(t) = \alpha t$ and $\zeta_1 = H_\alpha \circ \zeta_2 \circ H_\alpha^{-1}$, then $d_r(\zeta_1, \zeta_2) = d_{r,\mathrm{Aff}}(\zeta_1, \zeta_2) = 0$. In order to have metrics, we need to restrict to *normalized* critical commuting pairs (recall that for a commuting pair $\zeta = (\eta, \xi)$ we denote by $\tilde{\zeta}$ the pair $(\tilde{\eta}|_{\tilde{I}_\eta}, \tilde{\xi}|_{\tilde{I}_\xi})$, where the tilde means rescaling by the linear factor $\lambda = 1/|I_\xi|$).

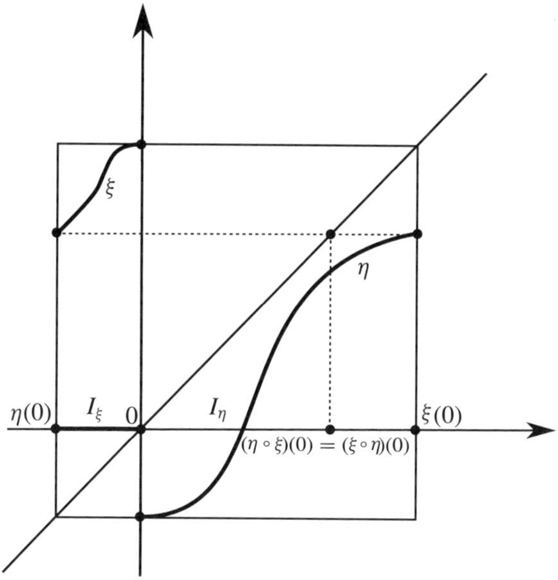

Figure 1. A commuting pair.

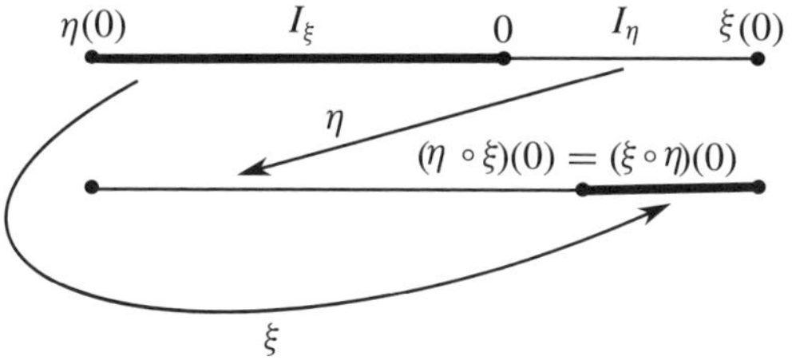

Figure 2. Scheme of a commuting pair.

2.3. The renormalization operator

As we just said, for a commuting pair $\zeta = (\eta, \xi)$, we denote by $\tilde{\zeta}$ the pair $(\tilde{\eta}|_{\tilde{I_\eta}}, \tilde{\xi}|_{\tilde{I_\xi}})$, where the tilde means rescaling by the linear factor $\lambda = 1/|I_\xi|$. Note that $|\tilde{I_\xi}| = 1$ and that $\tilde{I_\eta}$ has length equal to the ratio between the lengths of I_η and I_ξ. Equivalently, $\tilde{\eta}(0) = -1$ and $\tilde{\xi}(0) = |I_\eta|/|I_\xi| = \xi(0)/|\eta(0)|$. Let $\zeta = (\eta, \xi)$ be a C^r critical commuting pair according to Definition 2.3, and recall that $(\eta \circ \xi)(0) = (\xi \circ \eta)(0) \neq 0$. Let us suppose that $(\xi \circ \eta)(0) \in I_\eta$ (just as in both Figure 1 and Figure 2 above), and define the *period* $\chi(\zeta)$ of the commuting pair $\zeta = (\eta, \xi)$ as $a \in \mathbb{N}$ if

$$\eta^{a+1}\big(\xi(0)\big) < 0 \leq \eta^{a}\big(\xi(0)\big)$$

and $\chi(\zeta) = \infty$ if no such a exists (note that in this case, the map $\eta|_{I_\eta}$ has a fixed point, so when we are dealing with commuting pairs induced by critical circle maps with irrational rotation number we have finite period). Note also that the period of the pair $(f^{q_{n+1}}|_{I_n}, f^{q_n}|_{I_{n+1}})$ induced by a critical circle map f is exactly a_{n+1}, where $\rho(f) = [a_0, a_1, a_2, \ldots, a_n, a_{n+1}, \ldots]$ (because the combinatorics of f are the same as for the corresponding rigid rotation). For a pair $\zeta = (\eta, \xi)$ with $(\xi \circ \eta)(0) \in I_\eta$ and $\chi(\zeta) = a < \infty$, note that the pair

$$(\eta|_{[0,\eta^a(\xi(0))]}, \eta^a \circ \xi|_{I_\xi})$$

is again a commuting pair, and if $\zeta = (\eta, \xi)$ is induced by a critical circle map

$$\zeta = (\eta, \xi) = (f^{q_{n+1}}|_{I_n}, f^{q_n}|_{I_{n+1}}),$$

then we have

$$(\eta|_{[0,\eta^a(\xi(0))]}, \eta^a \circ \xi|_{I_\xi}) = (f^{q_{n+1}}|_{I_{n+2}}, f^{q_{n+2}}|_{I_{n+1}}).$$

This motivates the following definition (the definition in the case $(\xi \circ \eta)(0) \in I_\xi$ is analogous).

Definition 2.6
Let $\zeta = (\eta, \xi)$ be a critical commuting pair with $(\xi \circ \eta)(0) \in I_\eta$. We say that ζ is *renormalizable* if $\chi(\zeta) = a < \infty$. In this case, we define the *prerenormalization* of ζ as the critical commuting pair

$$p\mathcal{R}(\zeta) = (\eta|_{[0,\eta^a(\xi(0))]}, \eta^a \circ \xi|_{I_\xi}),$$

and we define the *renormalization* of ζ as the normalization of $p\mathcal{R}(\zeta)$; that is,

$$\mathcal{R}(\zeta) = \widetilde{p\mathcal{R}}(\zeta) = (\tilde{\eta}|_{[0,\widetilde{\eta^a(\xi(0))}]}, \widetilde{\eta^a \circ \xi}|_{\tilde{I}_\xi}).$$

Note in particular that $d_r(p\mathcal{R}(\zeta_0), p\mathcal{R}(\zeta_1)) = d_r(\mathcal{R}(\zeta_0), \mathcal{R}(\zeta_1))$ for any two critical commuting pairs ζ_0 and ζ_1 renormalizable with the same period.

A critical commuting pair is a special case of a *generalized interval exchange map* of two intervals, and the renormalization operator defined above is just the restriction of the *Zorich accelerated version* of the *Rauzy–Veech renormalization* for interval exchange maps. However, in this article we will keep the classical terminology for critical commuting pairs. (See Figure 3.)

Definition 2.7
Let ζ be a critical commuting pair. If $\chi(\mathcal{R}^j(\zeta)) < \infty$ for $j \in \{0, 1, \ldots, n-1\}$, then we say that ζ is *n-times renormalizable*, and if $\chi(\mathcal{R}^j(\zeta)) < \infty$ for all $j \in \mathbb{N}$, we say that ζ is *infinitely renormalizable*. In this case, the irrational number θ whose continued

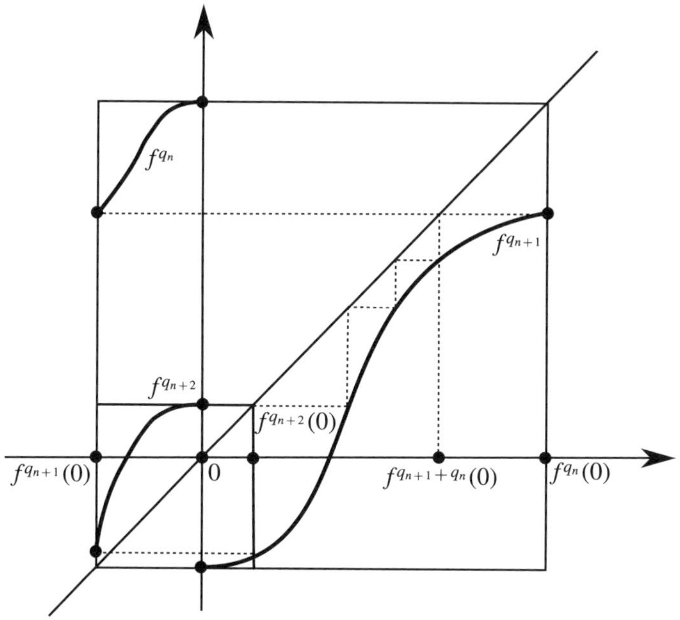

Figure 3. Two consecutive renormalizations of f, without rescaling (recall that f^{q_n} means $T^{-p_n} \circ \widehat{f}^{q_n}$; see Section 2.2). In this example, $a_{n+1} = 4$.

fraction expansion is equal to

$$\left[\chi(\zeta), \chi\big(\mathcal{R}(\zeta)\big), \ldots, \chi\big(\mathcal{R}^n(\zeta)\big), \chi\big(\mathcal{R}^{n+1}(\zeta)\big), \ldots\right]$$

is called the *rotation number* of the critical commuting pair ζ, and is denoted by $\rho(\zeta) = \theta$.

We immediately note that, when ζ is induced by a critical circle map with irrational rotation number, the pair ζ is automatically infinitely renormalizable and both definitions of rotation number coincide; any C^r critical circle map f with irrational rotation number gives rise to the orbit $\{\mathcal{R}^n(f)\}_{n \geq 1}$ of infinitely renormalizable C^r critical commuting pairs defined by

$$\mathcal{R}^n(f) = (\tilde{f}^{q_n}|_{\tilde{I}_{n-1}}, \tilde{f}^{q_{n-1}}|_{\tilde{I}_n}) \quad \text{for all } n \geq 1.$$

For any positive number θ, denote by $\lfloor \theta \rfloor$ the *integer part* of θ (i.e., $\lfloor \theta \rfloor \in \mathbb{N}$ and $\lfloor \theta \rfloor \leq \theta < \lfloor \theta \rfloor + 1$). Recall that the *Gauss map* $G : [0,1] \to [0,1]$ is defined by

$$G(\theta) = \frac{1}{\theta} - \left\lfloor \frac{1}{\theta} \right\rfloor \quad \text{for } \theta \neq 0 \qquad \text{and} \qquad G(0) = 0,$$

and note that ρ *semiconjugates* the renormalization operator with the Gauss map

$$\rho\big(\mathcal{R}^n(\zeta)\big) = G^n\big(\rho(f)\big)$$

for any ζ at least n-times renormalizable. In particular, the renormalization operator acts as a *left shift* on the continued fraction expansion of the rotation number: if $\rho(\zeta) = [a_0, a_1, \ldots]$, then $\rho(\mathcal{R}^n(\zeta)) = [a_n, a_{n+1}, \ldots]$.

2.4. Renormalization of real analytic critical commuting pairs

The main result on the dynamics of the renormalization operator acting on the real analytic category that we will need in this article is the following.

THEOREM 2.8 (see [4], [15])

There exists a universal constant λ in $(0,1)$ with the following property. Given two real analytic critical commuting pairs ζ_1 and ζ_2 with the same irrational rotation number and the same odd type at the critical point, there exists a constant $C > 0$ such that

$$d_r\big(\mathcal{R}^n(\zeta_1), \mathcal{R}^n(\zeta_2)\big) \le C\lambda^n$$

for all $n \in \mathbb{N}$ and for any $0 \le r < \infty$. Moreover, given a C^ω-compact set $\mathcal{K}$ of real analytic critical commuting pairs, the constant C can be chosen the same for any ζ_1 and ζ_2 in $\mathcal{K}$.

As we said in Section 1.1, Theorem 2.8 was proved by de Faria and de Melo [4] for rotation numbers of bounded type, and extended by Khmelev and Yampolsky [15] to cover all irrational rotation numbers.

2.5. Technical tools

We finish this preliminary section by stating some well-known distortion estimates that we will use throughout the text. For intervals $M \subset T$, we define the *space* of M inside T to be the smallest of the ratios $|L|/|M|$ and $|R|/|M|$, where L and R are the left and right components of $T \setminus M$.

THEOREM 2.9 (Koebe's distortion principle for real maps)

Let $f : S^1 \to S^1$ be a C^3-map, and let $S > 0$ and $\tau > 0$ be two positive constants. Then there exists a constant $K = K(S, \tau, f) > 1$ with the following property. If T is an interval in the unit circle such that $f^m|_T$ is a diffeomorphism and if $\sum_{j=0}^{m-1} |f^j(T)| \le S$, then for each interval $M \subset T$ for which the space of $f^m(M)$ inside $f^m(T)$ is at least τ and for all $x, y \in M$, we have

$$\frac{1}{K} \le \frac{|Df^m(x)|}{|Df^m(y)|} \le K.$$

A proof of Theorem 2.9 can be found in [8, Theorem 3.1, p. 295]. In Section 11 we will also need the following two classical results from complex analysis.

THEOREM 2.10 (Koebe's distortion principle for complex maps)
If K is a compact subset of a domain $\Omega \subset \mathbb{C}$, then there exists a constant M (depending on K) such that for every univalent function f on Ω and every pair of points $z, w \in K$ we have

$$\frac{1}{M} \leq \frac{|Df(z)|}{|Df(w)|} \leq M.$$

THEOREM 2.11 (Koebe's one-quarter theorem)
Let $f : \mathbb{D} \to \mathbb{C}$ be a univalent map from the open unit disk into the complex plane such that the origin is an indifferent fixed point (i.e., $f(0) = 0$ and $|Df(0)| = 1$). Then the image of f contains the open disk of radius $1/4$ around the origin.

3. Controlled pairs and real bounds

Definition 3.1
We say that a C^3 critical commuting pair $\zeta = (\eta, \xi)$ has *negative Schwarzian* if the Schwarzian derivatives of both η and ξ are negative in $I_\eta \setminus \{0\}$ and $I_\xi \setminus \{0\}$, respectively.

Recall that the *Schwarzian derivative* of a C^3-map f at a regular point x is defined as

$$Sf(x) = \frac{D^3 f(x)}{Df(x)} - \frac{3}{2}\left(\frac{D^2 f(x)}{Df(x)}\right)^2.$$

Remark 3.2
Let ζ be a C^3 critical commuting pair with negative Schwarzian. If ζ is renormalizable, then $\mathcal{R}(\zeta)$ has negative Schwarzian (this follows at once from the chain rule for the Schwarzian derivative). In particular, if ζ has negative Schwarzian and is infinitely renormalizable, then $\mathcal{R}^n(\zeta)$ has negative Schwarzian for all $n \in \mathbb{N}$.

Definition 3.3
Let $K > 1$, and let $\zeta = (\eta, \xi)$ be a normalized C^3 critical commuting pair which is renormalizable with some period $a \in \mathbb{N}$. We say that ζ is *K-controlled* if the following seven conditions are satisfied:

- $1/K \leq \xi(0) \leq K$;
- $\xi(0) - \eta(\xi(0)) \geq 1/K$;
- $\eta^{a-1}(\xi(0)) - \eta^a(\xi(0)) \geq 1/K$;

- $\eta^a(\xi(0)) \geq 1/K$;
- $\eta^{a+1}(\xi(0)) \leq -1/K$;
- $\|\xi\|_{C^3([-1,0])} \leq K$ and $\|\eta\|_{C^3([0,\xi(0)])} \leq K$;
- $D\eta(x) \geq 1/K$ for all $x \in [\eta^a(\xi(0)), \xi(0)]$.

Of course, if ζ is K_0-controlled and $K_1 \geq K_0$, then ζ is also K_1-controlled.

Definition 3.4
For $K > 1$, let $\mathcal{K} = \mathcal{K}(K)$ be the space of normalized C^3 critical commuting pairs which are K-controlled. For $K > 1$ and $a \in \mathbb{N}$, let $\mathcal{K}_a(K)$ be the space of normalized C^3 critical commuting pairs which are renormalizable with period a and are K-controlled.

From the real bounds (Theorem 2.2) we know that after a *finite* number of renormalizations, every C^4 critical circle map with arbitrary irrational rotation number gives rise to a *controlled* commuting pair. More precisely, we have the following.

THEOREM 3.5
There exists a universal constant $K_0 > 1$ with the following property. For any given C^4 critical circle map f with irrational rotation number there exists $n_0 = n_0(f) \in \mathbb{N}$ such that the critical commuting pair $\mathcal{R}^n(f)$ is K_0-controlled for any $n \geq n_0$.

The C^4-smoothness is needed so that the critical commuting pair $\mathcal{R}^n(f)$ is C^3-bounded for n big enough. (For a proof of Theorem 3.5, see [3, Section 3 and Appendix A]). Moreover, we have the following result.

THEOREM 3.6
For any given $K > 1$ there exists $n_0 = n_0(K) \in \mathbb{N}$ with the following property. If ζ is an infinitely renormalizable C^3 critical commuting pair with negative Schwarzian which is K-controlled, then $\mathcal{R}^n(\zeta)$ is K_0-controlled for all $n \geq n_0$, where the universal constant $K_0 > 1$ is given by Theorem 3.5.

We will also need the following fact.

LEMMA 3.7
Given $K > 1$, there exists $B = B(K) > K$ with the following property. Let $\zeta = (\eta, \xi)$ be a C^3 critical commuting pair which is K-controlled, renormalizable with some period $a \in \mathbb{N}$, and such that $\mathcal{R}(\zeta)$ is C^2-bounded by K. Then $\mathcal{R}(\zeta)$ is C^3-bounded by B.

Note that the constant B depends only on K and not on the period of renormalization a.

Proof of Lemma 3.7

Let $\zeta = (\eta, \xi)$ be a C^3 normalized critical commuting pair which is renormalizable with some period $a \in \mathbb{N}$. For $i \in \{0, \ldots, a\}$, let $x_i = \eta^i(\xi(0))$. Note that $x_i \in I_\eta = [0, \xi(0)]$ for all $i \in \{0, \ldots, a\}$. Denote by I_i, $i \in \{1, \ldots, a\}$, the fundamental domains of η given by $I_i = [\eta^i(\xi(0)), \eta^{i-1}(\xi(0))]$. By the commuting condition, $I_1 = \xi(I_\xi) = \xi([-1, 0])$. As before, the letter B will denote uniform constants whose value might change during estimates. We claim first that

$$\left|S\eta^a(x)\right| \leq B \quad \text{for all } x \in I_1, \tag{3.1}$$

where (as before) $S\eta^a$ denotes the Schwarzian derivative of η^a. Indeed, by the K-control we have $|S\eta(y)| \leq B$ for all $y \in [x_a, x_0] = [\eta^a(\xi(0)), \xi(0)]$ and then

$$\left|S\eta^a(x)\right| \leq \sum_{i=0}^{a-1} \left|S\eta\big(\eta^i(x)\big)\right| \times \left|D\eta^i(x)\right|^2 \leq B \sum_{i=0}^{a-1} \left|D\eta^i(x)\right|^2 \quad \text{for all } x \in I_1.$$

By bounded distortion (see Theorem 2.9) and the K-control, we have

$$\left|D\eta^i(x)\right| \leq \frac{B}{|I_1|} \times |I_{i+1}| \leq B|I_{i+1}| \quad \text{for all } x \in I_1,$$

and then we obtain for all $x \in I_1$ that

$$\begin{aligned}
\left|S\eta^a(x)\right| &\leq B \sum_{i=0}^{a-1} |I_{i+1}|^2 \\
&= B \sum_{i=1}^{i=a} |I_i|^2 \\
&\leq B \times \max_{i \in \{1, \ldots, a\}} |I_i| \times \sum_{i=1}^{i=a} |I_i| \\
&\leq B\left|\xi(0)\right|^2 \leq B,
\end{aligned}$$

as claimed. Note now that by hypothesis, we know that

$$\left|D\eta^a(x)\right| \leq K \qquad \text{and} \qquad \left|D^2\eta^a(x)\right| \leq K \quad \text{for all } x \in I_1.$$

Moreover, again by bounded distortion and the K-control, we also have $|D\eta^a(x)| \geq 1/B$ for all $x \in I_1$, since $|I_1|$ and $|I_a|$ are comparable. With this at hand

and claim (3.1), we obtain for all $x \in I_1$ that

$$\left|D^3\eta^a(x)\right| \le \left|D\eta^a(x)\right| \times \left|S\eta^a(x)\right| + \frac{3}{2}\frac{|D^2\eta^a(x)|^2}{|D\eta^a(x)|} \le BK + \frac{3}{2}BK^2.$$

Since $\|\eta^a\|_{C^2(I_1)} \le K$ by hypothesis, we obtain that $\|\eta^a\|_{C^3(I_1)} \le B$. Finally, from $p\mathcal{R}(\zeta) = (\eta|_{[0,\eta^a(\xi(0))]}, \eta^a \circ \xi|_{I_\xi})$ and the fact that ζ is K-controlled, we obtain that the critical commuting pair $\mathcal{R}(\zeta)$ is C^3-bounded by B. □

Remark 3.8
Let $\zeta = (\eta, \xi)$ be a C^3 critical commuting pair which is renormalizable with period a and is K_0-controlled, and denote by $N\eta$ the *nonlinearity* of η (see Section 7). Then $N\eta$ is Lipschitz-continuous in $[\eta^a(\xi(0)), \xi(0)]$ with some universal constant $K(K_0) > 1$. Indeed, this comes from the identity

$$D(N\eta) = \frac{D^3\eta}{D\eta} - \left(\frac{D^2\eta}{D\eta}\right)^2 = S\eta + \frac{1}{2}(N\eta)^2,$$

where (as before) $S\eta$ denotes the Schwarzian derivative of η. Note that actually $N\eta$ is C^1-bounded in $[\eta^a(\xi(0)), \xi(0)]$ whenever $\zeta \in \mathcal{K}_a$. Moreover, if $\zeta, \tilde{\zeta} \in \mathcal{K}_a$ and $x \in J = [\eta^a(\xi(0)), \xi(0)] \cap [\tilde{\eta}^a(\tilde{\xi}(0)), \tilde{\xi}(0)]$, we have $|N\eta(x) - N\tilde{\eta}(x)| \le K\|\eta - \tilde{\eta}\|_{C^2(J,\mathbb{R})}$.

Finally, it is not difficult to prove that given $K_0 > 1$ there exists $K = K(K_0) > 1$ such that both metrics d_2 and $d_{2,\mathrm{Aff}}$ are Lipschitz-equivalent, with constant K, when restricted to normalized K_0-controlled C^3 critical commuting pairs. This allows us to use both metrics in our estimates.

4. Lipschitz continuity

Sections 5–10 of this article are devoted to the proof of the following result.

LEMMA 4.1 (Key lemma)
For any given $K > 1$, there exist two constants $\varepsilon_0 = \varepsilon_0(K) \in (0,1)$ and $L = L(K) > 1$ with the following property. Let ζ_0 and ζ_1 be two infinitely renormalizable normalized C^3 critical commuting pairs which are K-controlled, let both ζ_0 and ζ_1 have negative Schwarzian, and let $\rho(\zeta_0) = \rho(\zeta_1) \in [0,1] \setminus \mathbb{Q}$ and $d_2(\zeta_0, \zeta_1) < \varepsilon_0$. Then we have

$$d_2\big(\mathcal{R}(\zeta_0), \mathcal{R}(\zeta_1)\big) \le L d_2(\zeta_0, \zeta_1),$$

where d_2 denotes the C^2-distance in the space of C^2 critical commuting pairs.

We remark that the Lipschitz constant L given by Lemma 4.1 depends only on K and does not depend on the common combinatorics of the critical pairs ζ_0 and ζ_1. As already explained in the Introduction, we note that the Lipschitz continuity holds *along topological classes* (see our comments in Section 1.1(3)).

4.1. Yoccoz's lemma

Let $\zeta = (\eta, \xi)$ be a C^3 K-controlled critical commuting pair which is renormalizable with some period $a \in \mathbb{N}$. As before (see the proof of Lemma 3.7 in Section 3), we will use the following notation: for $i \in \{0, \dots, a\}$, let $x_i = \eta^i(\xi(0))$. Note that $x_i \in I_\eta = [0, \xi(0)]$ for all $i \in \{0, \dots, a\}$. Denote by I_i, $i \in \{1, \dots, a\}$, the fundamental domains of η given by $I_i = [x_i, x_{i-1}] = [\eta^i(\xi(0)), \eta^{i-1}(\xi(0))]$. By the commuting condition, $I_1 = \xi(I_\xi) = \xi([-1, 0])$. The following result due to Yoccoz plays a fundamental role in our analysis.

LEMMA 4.2 (Yoccoz's Lemma)

Given $K > 1$ there exists $K_0 = K_0(K) > 1$ with the following property. Let ζ be a C^3 K-controlled critical commuting pair which is renormalizable with some period $a \in \mathbb{N}$. Assume that ζ has negative Schwarzian. Then for all $i \in \{1, \dots, a\}$, we have

$$\frac{1}{K_0} \frac{1}{\min\{i, a-i\}^2} \leq |I_i| \leq K_0 \frac{1}{\min\{i, a-i\}^2}.$$

(For a proof of Yoccoz's lemma, see [3, Section 4.1, p. 354 and Appendix B, p. 386].)

5. Standard families

Fix $K_0 > 1$, and let $\mathcal{K}$ be the space of normalized C^3 critical commuting pairs which are K_0-controlled (see Definition 3.3 in Section 3). We will consider in this section a C^3 critical commuting pair $\zeta = (\eta, \xi)$ with negative Schwarzian that belongs to $\mathcal{K}$ which is renormalizable with period $a \in \mathbb{N}$. For such a pair we will construct/define the corresponding *standard family*.

5.1. Gluing procedure and translations

In the notation of the proof of Lemma 2.4 in Section 2, we have the following.

LEMMA 5.1

There exists $s_0 = s_0(\mathcal{K}) > 0$ such that for any $\zeta = (\eta, \xi) \in \mathcal{K}$ both components of $A \setminus \{\eta(0)\}$ and both components of $B \setminus \{\xi(0)\}$ have Euclidean length greater than or equal to s_0.

Proof of Lemma 5.1
There exist positive constants δ and ρ (depending only on K_0) such that both components of $C \setminus \{0\}$ have Euclidean length greater than or equal to δ, $\inf_C\{D\phi\} > \rho$ and $\inf_C\{D\psi\} > \rho$. Then it is enough to take $0 < s_0 < (\delta\rho)^{2d+1}$, where the integer $2d+1$ is the criticality of η and ξ at the origin (see Definition 2.3(4)). □

Still in the notation of the proof of Lemma 2.4, let $M = V_- \cup V_+/\sim$, where $x \sim y$ if $x \in A$, $y \in B$ and $\widehat{\xi}(x) = \widehat{\eta}(y)$. Note that $\eta(0) \sim \xi(0)$ by the commuting condition (2) in Definition 2.3. Let $p : V_- \cup V_+ \to M$ be the canonical projection for the identification $\sim$, and note that M is a compact boundaryless 1-dimensional C^3-manifold since the map $\widehat{\eta}^{-1} \circ \widehat{\xi} : A \to B$ is a C^3-diffeomorphism (it can be proved that p is the restriction of a C^3 covering map from the real line to M, but this fact will not be needed in this article).

LEMMA 5.2
There exists a C^3-diffeomorphism $\psi : M \to S^1$ such that defining $P : V_- \cup V_+ \to S^1$ as $P = \psi \circ p$, we have for all $x, y \in A \cap I_\xi$, for all $x, y \in B \cap I_\eta$, and for all $x, y \in (I_\xi \cup I_\eta) \setminus (A \cup B)$,

$$\frac{|x-y|}{K} \le d\big(P(x), P(y)\big) \le K|x-y|$$

for some universal constant $K = K(\mathcal{K}) > 1$, where d denotes the Euclidean distance in the unit circle.

From now on, let $P : V_- \cup V_+ \to S^1$ be the C^3-map defined in Lemma 5.2. Given $t \in \mathbb{R}$, we define the *translation* by t on $I_\xi \cup I_\eta$ to be the C^3 map $T : I_\xi \cup I_\eta \times \mathbb{R} \to I_\xi \cup I_\eta$ given by

$$(P \circ T_t)(x) = e^{2\pi i t} P(x);$$

that is, $T(x,t) = T_t(x) = P^{-1}(e^{2\pi i t} P(x))$, whenever it is clear which preimage under P we choose for points in $P(A)$. In particular, T_0 is the identity on $I_\xi \cup I_\eta$. Note also that

$$\frac{\partial T}{\partial t}(x,t) = \frac{1}{DP(T_t(x))} \qquad \text{and} \qquad \frac{\partial T}{\partial x}(x,t) = \frac{DP(x)}{DP(T_t(x))},$$

and from Lemma 5.2 we get that $\frac{1}{K} \le \frac{\partial T}{\partial t}(x,t) \le K$ for all $x \in I_\xi \cup I_\eta$.

5.2. Standard families of commuting pairs (Figures 4 and 5)
By Definition 2.3(5), the discontinuous piecewise smooth map $\tilde{f}_\xi : I_\xi \cup I_\eta \to I_\xi \cup I_\eta$ given by

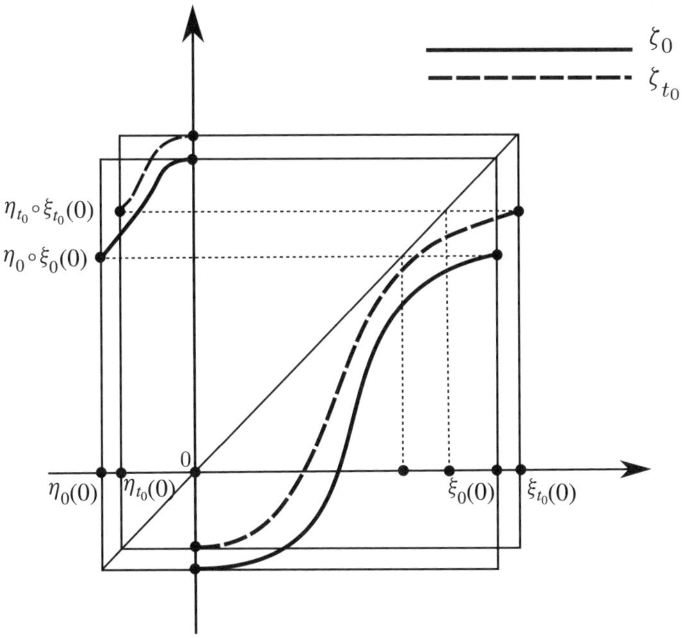

Figure 4. Standard families of critical commuting pairs (in this figure, the period of ζ_0 is equal to 3, while the period of ζ_{t_0} is 8).

$$\tilde{f}_\zeta(x) = \begin{cases} \xi(x) & \text{for } x \in I_\xi, \\ \eta(x) & \text{for } x \in I_\eta \end{cases}$$

projects under p to a C^3-homeomorphism of the quotient manifold M, and then it projects under P to a C^3 critical circle map f_ζ in S^1.

By Lemmas 5.1 and 5.2 above, the Euclidean length of both components of $P(A) \setminus \{f_\zeta(P(0))\}$ in S^1 is bounded from below by some positive constant l_0, universal in $\mathcal{K}$. For $t \in W = (-l_0, l_0)$, let $f_t : S^1 \to S^1$ be the C^3 critical circle map given by $f_t(z) = e^{2\pi i t} f_\zeta(z)$, and note that $f_0 = f_\zeta$. Since the critical value of f_t (which is $e^{2\pi i t} f_\zeta(P(0))$) belongs to $P(A)$, we can lift each f_t up to a C^3 critical commuting pair $\zeta_t = (\eta_t, \xi_t)$ with

$$\xi_t(x) = (T_t \circ \xi_0)(x) = T\big(\xi_0(x), t\big) \qquad \text{and}$$
$$\eta_t(x) = (T_t \circ \eta_0)(x) = T\big(\eta_0(x), t\big).$$

Note that

$$\frac{\partial \xi_t}{\partial t}(x) = \frac{1}{DP(\xi_t(x))} \qquad \text{and} \qquad \frac{\partial \eta_t}{\partial t}(x) = \frac{1}{DP(\eta_t(x))}.$$

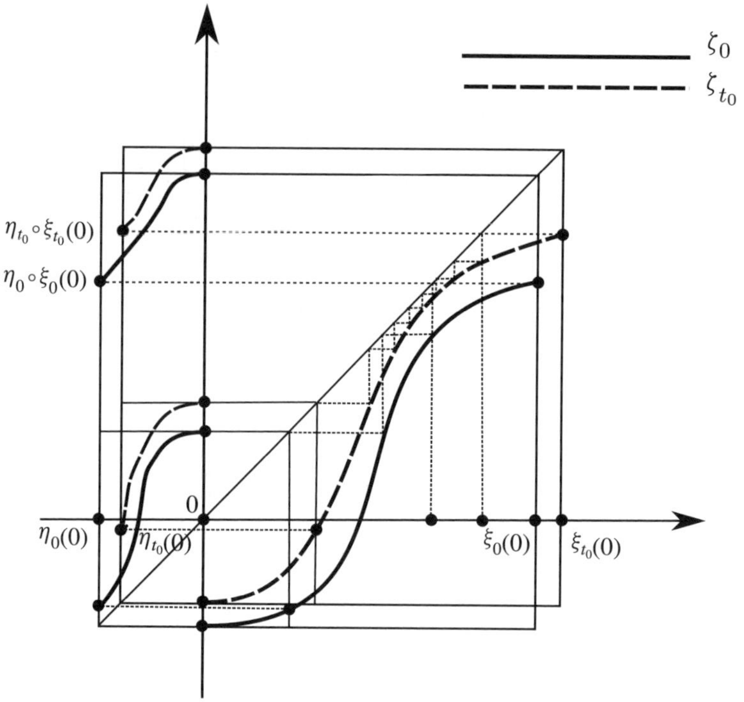

Figure 5. Both critical commuting pairs of Figure 4, and their renormalizations.

LEMMA 5.3
There exists $K = K(\mathcal{K}) > 1$ such that $|t|/K \leq d_2(\zeta_0, \zeta_t) \leq K|t|$ for all $t \in W$.

Now let $W_a \subset W$ be the set of all $t \in W$ such that ζ_t is renormalizable with period a, that is,

$$W_a = \left\{ t \in W : \left\lfloor \frac{1}{\rho(\zeta_t)} \right\rfloor = \left\lfloor \frac{1}{\rho(\zeta_0)} \right\rfloor = a \right\}.$$

LEMMA 5.4
There exists $a_0 = a_0(\mathcal{K}) \in \mathbb{N}$ such that if $a \geq a_0$, we have $\overline{W_a} \subset W$. If we denote the boundary points of W_a by $-w_-^a$ and w_+^a, that is, $W_a = [-w_-^a, w_+^a]$, then we have

$$\eta_{-w_-^a}^{a+1}\left(\xi_{-w_-^a}(0)\right) = 0 \qquad \text{and} \qquad \eta_{w_+^a}^{a}\left(\xi_{w_+^a}(0)\right) = 0.$$

Proof of Lemma 5.4
By Lemma 5.2, there exists a universal upper bound $K > 0$ for the first derivative of P in $V_- \cup V_+$. By Yoccoz's lemma (Lemma 4.2), it is enough to take $a_0 \gtrsim (K/|W|)^{1/2}$

in order to have $|W| \gtrsim K/a_0^2$. The assertion about the boundary of W_a follows by combinatorics. □

COROLLARY 5.5
Let $a_0 = a_0(\mathcal{K}) \in \mathbb{N}$ be given by Lemma 5.4. Let ζ be a normalized C^3 critical commuting pair that belongs to $\mathcal{K}$ which is renormalizable with period $a \geq a_0$. Given $x \in [0, \eta^a(\xi(0))]$, there exists $t_x \leq 0$ in $W_a(\zeta)$ such that $\eta_{t_x}^a(\xi_{t_x}(0)) = x$.

Finally, let $V = [-v_-, v_+] \subset W_a$ defined by

$$\eta_{-v_-}^{a+1}\big(\xi_{-v_-}(0)\big) = -1/K_0^2 \qquad \text{and} \qquad \eta_{v_+}^a\big(\xi_{v_+}(0)\big) = 1/K_0^2.$$

LEMMA 5.6
For any $t \in V$ and any $k \in \{1, \dots, a-1\}$ the C^3-diffeomorphism $\eta_t^{a-k} : I_k(t) \to I_a(t)$ has universally bounded distortion.

Here $I_i(t) = [x_i(t), x_{i-1}(t)]$ for all $i \in \{1, \dots, a\}$.

Proof of Lemma 5.6
Combine Koebe's distortion principle (Theorem 2.9) with the K-control. □

LEMMA 5.7
Let $a_0 = a_0(\mathcal{K}) \in \mathbb{N}$ be given by Lemma 5.4. Let $\zeta_0 = (\eta_0, \xi_0)$ and $\zeta_1 = (\eta_1, \xi_1)$ be two normalized C^3 critical commuting pairs that belong to $\mathcal{K}$ which are renormalizable with the same period $a \geq a_0$. Then there exists $t_0 \in V(\zeta_0) \subset W_a(\zeta_0)$ such that

$$\eta_{t_0}^a\big(\xi_{t_0}(0)\big) = \eta_1^a\big(\xi_1(0)\big) \qquad \textit{and} \qquad d_2(\zeta_0, \zeta_{t_0}) \leq K d_2(\zeta_0, \zeta_1),$$

where the constant $K = K(\mathcal{K}) > 1$ is given by Lemma 5.3.

Proof of Lemma 5.7
We may suppose that $\eta_0^a(\xi_0(0)) \geq \eta_1^a(\xi_1(0))$—that is, $\eta_1^a(\xi_1(0))$—belongs to the interval $[1/K_0, \eta_0^a(\xi_0(0))] \subset [1/K_0, K_0]$. By Corollary 5.5, there exists $t_0 < 0$ in $V(\zeta_0)$ such that $\eta_{t_0}^a(\xi_{t_0}(0)) = \eta_1^a(\xi_1(0))$. Note that $\eta_{t_0}^{a+1}(\xi_{t_0}(0)) \leq \eta_0^{a+1}(\xi_0(0)) \leq -1/K_0 < -1/K_0^2$. Now let $K = K(\mathcal{K}) > 1$ be given by Lemma 5.3. We claim that $|t_0| \leq K d_2(\zeta_0, \zeta_1)$. Indeed, if $|t_0| > K d_2(\zeta_0, \zeta_1)$, we would have $\xi_{t_0} < \xi_1$ and $\eta_{t_0} < \eta_1$ in the corresponding intersections of domains, but this implies that $\eta_{t_0}^a(\xi_{t_0}(0)) < \eta_1^a(\xi_1(0))$, which is a contradiction. Then $|t_0| \leq K d_2(\zeta_0, \zeta_1)$ and we are done. □

5.3. Renormalization of standard families

As before, fix $K_0 > 1$, and let $\mathcal{K}$ be the space of normalized C^3 critical commuting pairs which are K_0-controlled (see Definition 3.3 in Section 3). Again, we consider in this section a normalized C^3 critical commuting pair $\zeta = (\eta, \xi)$ in $\mathcal{K}$ with negative Schwarzian, which is renormalizable with some period $a \in \mathbb{N}$. Let $V(\zeta)$ be the parameter interval for the standard family around ζ constructed in Section 5.2, and consider the 1-parameter family of C^3 critical commuting pairs given by $G_t = p\mathcal{R}(\zeta_t)$ for each $t \in V$; that is, G_t is the prerenormalization of ζ_t (see Definition 2.6 in Section 2).

PROPOSITION 5.8

There exists $K = K(\mathcal{K}) > 1$ such that for all $t \in V$ and for all x in the domain of G_t, we have

$$\frac{\partial G_t}{\partial t}(x) \asymp a^3 \quad \text{if } x < 0, \qquad \text{and}$$

$$\frac{\partial G_t}{\partial t}(x) \asymp 1 \quad \text{if } x > 0.$$

Proof of Proposition 5.8

Note first that for $t \in V$ and $x \in I_{\xi_t}$, we have the identity

$$\frac{\partial G_t}{\partial t}(x) = \frac{\partial \xi_t}{\partial t}(x) D\eta_t^a\big(\xi_t(x)\big) + \sum_{k=1}^{k=a} \frac{\partial T}{\partial t}\big(\eta_0\big(\eta_t^{k-1}(\xi_t(x))\big), t\big) D\eta_t^{a-k}\big(\eta_t^k(\xi_t(x))\big). \tag{5.1}$$

Indeed, fix $x \in I_{\xi_t}$, and for each $j \in \{0, 1, \ldots, a\}$ let $y_j(t) = \eta_t^j(\xi_t(x))$. Note that $y_0(t) = \xi_t(x)$ and $y_a(t) = G_t(x)$ for $x < 0$. Since $y_{j+1}(t) = \eta_t(y_j(t)) = T(\eta_0(y_j(t)), t)$ for all $j \in \{0, 1, \ldots, a-1\}$, we see that

$$\begin{aligned} y'_{j+1}(t) &= y'_j(t) \frac{\partial T}{\partial x}\big(\eta_0(y_j(t)), t\big) D\eta_0\big(y_j(t)\big) + \frac{\partial T}{\partial t}\big(\eta_0(y_j(t)), t\big) \\ &= y'_j(t) D\eta_t\big(y_j(t)\big) + \frac{\partial T}{\partial t}\big(\eta_0(y_j(t)), t\big), \end{aligned} \tag{5.2}$$

since from $\eta_t(x) = T(\eta_0(x), t)$ we get $D\eta_t(x) = \frac{\partial T}{\partial x}(\eta_0(x), t) D\eta_0(x)$. By induction on (5.2), we obtain that for all $j \in \{1, \ldots, a\}$,

$$y'_j(t) = y'_0(t) \prod_{l=0}^{j-1} D\eta_t\big(y_l(t)\big) + \sum_{k=1}^{j-1} \frac{\partial T}{\partial t}\big(\eta_0(y_{k-1}(t)), t\big) \prod_{l=k}^{j-1} D\eta_t\big(y_l(t)\big)$$

$$
\begin{aligned}
&+\frac{\partial T}{\partial t}\big(\eta_0\big(y_{j-1}(t)\big),t\big)\\
&=y_0'(t)D\eta_t^j\big(y_0(t)\big)+\sum_{k=1}^{j-1}\frac{\partial T}{\partial t}\big(\eta_0\big(y_{k-1}(t)\big),t\big)D\eta_t^{j-k}\big(y_k(t)\big)\\
&+\frac{\partial T}{\partial t}\big(\eta_0\big(y_{j-1}(t)\big),t\big)\\
&=y_0'(t)D\eta_t^j\big(y_0(t)\big)+\sum_{k=1}^{k=j}\frac{\partial T}{\partial t}\big(\eta_0\big(y_{k-1}(t)\big),t\big)D\eta_t^{j-k}\big(y_k(t)\big).
\end{aligned}
$$

In particular,

$$
\frac{\partial G_t}{\partial t}(x)=y_a'(t)=y_0'(t)D\eta_t^a\big(y_0(t)\big)+\sum_{k=1}^{k=a}\frac{\partial T}{\partial t}\big(\eta_0\big(y_{k-1}(t)\big),t\big)D\eta_t^{a-k}\big(y_k(t)\big),
$$

and then we obtain for all $t\in V$ and all $x\in I_{\xi_t}$ the desired identity (5.1). Now by Lemma 5.2, the K_0-control, and Lemma 5.6, we have

$$
\begin{aligned}
0&\le\frac{\partial\xi_t}{\partial t}(x)D\eta_t^a\big(\xi_t(x)\big)\\
&=\Big(\frac{D\eta_0(\eta_t^{a-1}(\xi_t(x)))DP(\eta_0(\eta_t^{a-1}(\xi_t(x))))}{DP(\xi_t(x))DP(\eta_t^a(\xi_t(x)))}\Big)D\eta_t^{a-1}\big(\xi_t(x)\big)\\
&\le KD\eta_0\big(\eta_t^{a-1}\big(\xi_t(x)\big)\big)D\eta_t^{a-1}\big(\xi_t(x)\big)\\
&\le K\frac{|I_a(t)|}{|I_1(t)|}\le K.
\end{aligned}
$$

On the other hand, for all $k\in\{1,\dots,a\}$ we have

$$
\frac{\partial T}{\partial t}\big(\eta_0\big(\eta_t^{k-1}\big(\xi_t(x)\big)\big),t\big)=\frac{1}{DP(\eta_t^k(\xi_t(x)))}\in\Big[\frac{1}{K},K\Big]
$$

again by Lemma 5.2. Therefore, it follows from (5.1) that for any $x<0$ we have

$$
\frac{\partial G_t}{\partial t}(x)\asymp\sum_{k=1}^{a-1}D\eta_t^{a-k}\big(\eta_t^k\big(\xi_t(x)\big)\big)\quad\text{whenever } a>1.
$$

Again by Lemma 5.6 (bounded distortion) and the K_0-control, we have

$$
\frac{\partial G_t}{\partial t}(x)\asymp\sum_{k=1}^{a-1}\frac{|I_a(t)|}{|I_k(t)|}\asymp\sum_{k=1}^{a-1}\frac{1}{|I_k(t)|}.
$$

Therefore, by Yoccoz's lemma (Lemma 4.2), we obtain

$$\frac{\partial G_t}{\partial t}(x) \asymp \sum_{k=1}^{a-1} \min\{k, a-k\}^2 \asymp a^3 \quad \text{for any } x < 0.$$

Finally, recall that for $x \in [0, \eta_t^a(\xi_t(0))]$ we have $G_t(x) = \eta_t(x)$, and then

$$\frac{\partial G_t}{\partial t}(x) = \frac{\partial \eta_t}{\partial t}(x) = \frac{1}{DP(\eta_t(x))} \in \left[\frac{1}{K}, K\right]$$

by Lemma 5.2. □

With Proposition 5.8 at hand, we obtain the following.

COROLLARY 5.9

There exists $K = K(\mathcal{K}) > 1$ such that for all $t \in V$ and $x, y \in I_{\xi_t}$, we have

$$\frac{|\frac{\partial G_t}{\partial t}(x)|}{|\frac{\partial G_t}{\partial t}(y)|} \leq K.$$

In particular,

$$\frac{|G_t(x) - G_0(x)|}{|G_t(y) - G_0(y)|} = \frac{|\eta_t^a(\xi_t(x)) - \eta_0^a(\xi_0(x))|}{|\eta_t^a(\xi_t(y)) - \eta_0^a(\xi_0(y))|} \leq K$$

for all $t \in V \setminus \{0\}$ and $x, y \in I_{\xi_t} \cap I_{\xi_0} = [\max\{\eta_0(0), \eta_t(0)\}, 0]$.

6. Orbit deformations

We begin this section with the following fact.

LEMMA 6.1

Given $K > 1$, there exists $a_0 = a_0(K) \in \mathbb{N}$ with the following property. Let $\zeta = (\eta, \xi)$ be a normalized C^3 critical commuting pair with negative Schwarzian which is K-controlled and renormalizable with some period $a \geq a_0$. Then there exists a unique p in I_η such that $|\eta(p) - p| \leq |\eta(x) - x|$ for all $x \in I_\eta$. Moreover, the point p belongs to the interior of I_η, $D\eta(p) = 1$ and $D^2\eta(p) < 0$.

Proof of Lemma 6.1

Since ζ is renormalizable, we know that $x > \eta(x)$ for all $x \in I_\eta$. From the continuity of η and the compactness of its domain I_η, we obtain the existence of a point p such that $0 < |\eta(p) - p| \leq |\eta(x) - x|$ for all $x \in I_\eta$.

We claim first that, if $a_0 > K^2$ and $a \geq a_0$, then p belongs to the interior of I_η. Indeed, note first that the (positive) difference $\mathrm{Id} - \eta$ equals $|I_\xi|$ at the origin, and equals $|\xi(I_\xi)|$ at the point $\xi(0)$. In both cases it is greater than $1/K$, by the K-control hypothesis. If p is one of the boundary points of I_η, we would have $|\eta(x) - x| \geq 1/K$ for all $x \in I_\eta$, and since the period of ζ is a, we would have $a/K < |I_\eta|$. On the other hand, again by the K-control hypothesis, we have $a_0 > K^2 > K|I_\eta|$ and then $|I_\eta| < a_0/K$, which gives the desired contradiction.

With the claim at hand, we clearly have $D\eta(p) = 1$ and $D^2\eta(p) \leq 0$. Uniqueness of p follows at once from the minimum principle for maps with negative Schwarzian derivative (see [8, Section II.6, Lemma 6.1] for its statement). Now we claim that $D^2\eta(p)$ is strictly negative. Indeed, if $D^2\eta(p) = 0$ we would have $D^3\eta(p) = S\eta(p) < 0$, and then there would exist $\delta_0 > 0$ such that $D^2\eta(x) > 0$ for all $x \in (p - \delta_0, p)$. But then there would exist $0 < \delta_1 \leq \delta_0$ such that $|\eta(x) - x| < |\eta(p) - p|$ for all $x \in (p - \delta_1, p)$, which gives the desired contradiction. □

Remark 6.2

We can slightly improve the statement of Lemma 6.1: there exists $K_0 = K_0(K) > 1$ such that $D^2\eta(p) < -1/K_0$. Indeed, the fact that $D^2\eta(p)$ is uniformly bounded away from zero (by a constant depending only on K) follows from (the proof of) Yoccoz's lemma (Lemma 4.2) (see [3, Appendix B, pp. 386–389]).

Throughout this section fix $K > 1$, and let $\mathcal{K}$ be the space of normalized C^3 critical commuting pairs which are K-controlled (see Definition 3.3 in Section 3). Let $\zeta = (\eta, \xi)$ and $\tilde{\zeta} = (\tilde{\eta}, \tilde{\xi})$ be two C^3 critical commuting pairs with negative Schwarzian that belong to $\mathcal{K}$ which are renormalizable with the same period $a \geq a_0$, where $a_0 \in \mathbb{N}$ is given by Lemma 6.1. Denote by $\varepsilon > 0$ the C^2-distance between ζ and $\tilde{\zeta}$, that is, $\varepsilon = d_2(\zeta, \tilde{\zeta})$. We will assume that $\varepsilon < \varepsilon_0$, where $\varepsilon_0 > 0$ will be fixed later in this section (see the proof of Claim 6.7 below, during the proof of Lemma 6.6). Moreover, we will only consider in this section the special situation when

(1) $I_\eta = I_{\tilde{\eta}}$ and $I_\xi = I_{\tilde{\xi}} = [-1, 0]$,

(2) $p = \tilde{p}$, where $D\eta(p) = D\tilde{\eta}(\tilde{p}) = 1$ (see Lemma 6.1).

Let $H : I_\eta \to [-\varepsilon, \varepsilon] \subset \mathbb{R}$ be defined by $H(x) = \eta(x) - \tilde{\eta}(x)$, and let

$$h = H(p).$$

Observe that for every $x \in I_\eta$, we have

$$\big|H(x)\big| \leq |h| + \varepsilon(x - p)^2 \tag{6.1}$$

and

$$\big|DH(x)\big| \leq \varepsilon|x - p|. \tag{6.2}$$

Indeed, given $x \in I_\eta$ there exists $y \in I_\eta$ such that $DH(x) = D^2H(y)(x-p)$ and then $|DH(x)| = |D^2H(y)||x-p| \le \varepsilon|x-p|$, and there exists also $z \in [p,x] \subset I_\eta$ such that $H(x) = h + DH(z)(x-p)$ and then $|H(x)| \le |h| + |DH(z)||x-p| \le |h| + \varepsilon(x-p)^2$.

As before, we will use the following notation. For $i \in \{0,\dots,a\}$, let $x_i = \eta^i(\xi(0))$. Note that $x_i \in I_\eta = [0,\xi(0)]$ for all $i \in \{0,\dots,a\}$. Define $\tilde{x}_i = \tilde{\eta}^i(\tilde{\xi}(0))$ similarly. Denote by I_i, $i \in \{1,\dots,a\}$, the fundamental domains of η given by $I_i = [\eta^i(\xi(0)), \eta^{i-1}(\xi(0))]$. By the commuting condition, $I_1 = \xi(I_\xi) = \xi([-1,0])$. Define $\tilde{I}_i$ similarly. Let us state some consequences of Yoccoz's lemma (Lemma 4.2).

LEMMA 6.3
Let ζ be a C^3 K-controlled critical commuting pair which has negative Schwarzian and is renormalizable with some period $a \ge a_0$, where a_0 is given by Lemma 6.1. Let $N \in \{1,\dots,a\}$ be defined by $p \in I_{N+1}$, that is, $x_{N+1} \le p \le x_N$, where p is given by Lemma 6.1. Then we have

(1) *$N \asymp a$, that is, there exist two constants $\delta_0 = \delta_0(\mathcal{K})$ and $\delta_1 = \delta_1(\mathcal{K})$ with $0 < \delta_0 \le \delta_1 < 1$ such that $\delta_0 a \le N \le \delta_1 a$;*

(2) *$|I_i| \asymp \frac{1}{i^2}$ for $i \in \{1,\dots,N\}$ and $|I_i| \asymp \frac{1}{(a-i)^2}$ for $i \in \{N,\dots,a-1\}$;*

(3) *$|x_i - p| \asymp \frac{N-i}{iN} = \frac{1}{i} - \frac{1}{N} < \frac{1}{i}$ for all $i \le \min\{\lfloor a/2 \rfloor, N-1\}$.*

Proof of Lemma 6.3
To prove item (1), we claim that $|I_{N+1}| \asymp p - \eta(p)$. Indeed, note first that $p - \eta(p) \le |I_{N+1}| + |I_{N+2}|$. Being adjacent fundamental domains of η, we know from Yoccoz's lemma that $|I_{N+1}| \asymp |I_{N+2}|$, and then $p - \eta(p) \le (1+K_0)|I_{N+1}|$ for some $K_0 = K_0(K)$. On the other hand, we have

$$\begin{aligned}
\frac{|I_{N+2}|}{p-\eta(p)} &= \frac{x_{N+1} - x_{N+2}}{p-\eta(p)} \\
&= 1 + \frac{1}{p-\eta(p)} \int_{x_{N+1}}^{p} \big(D\eta(t) - 1\big)\,dt \\
&= 1 + \frac{1}{p-\eta(p)} \int_{x_{N+1}}^{p} (p-t)\big(-D^2\eta(p) + O(t-p)\big)\,dt \\
&\le 1 + \int_{x_{N+1}}^{p} \big(-D^2\eta(p) + O(t-p)\big)\,dt \\
&\le 1 + K_0(p - x_{N+1}) \\
&\le 1 + K_0|I_{N+1}|.
\end{aligned}$$

Again using the result that $|I_{N+1}| \asymp |I_{N+2}|$, we obtain the claim. Combining the comparability $|I_{N+1}| \asymp p - \eta(p)$ with Yoccoz's lemma (Lemma 4.2), we deduce that $|I_{N+1}| \asymp 1/a^2$ (recall that the positive number $p - \eta(p)$ is less than or equal to the length of any fundamental domain of η) and then $N \asymp a$, which implies item (1). Item (2) follows at once from item (1) and Yoccoz's lemma. To prove item (3), note first that by definition of N,

$$\sum_{j=i+1}^{j=N} |I_j| = |x_i - x_N| \leq |x_i - p| \leq |x_i - x_{N+1}| = \sum_{j=i+1}^{j=N} |I_j| + |I_{N+1}|$$

for all $i \in \{1, \dots, N-1\}$. By item (2) we have $\sum_{j=i+1}^{j=N} |I_j| \asymp \sum_{j=i+1}^{j=N} \frac{1}{j^2}$, that is,

$$\frac{1}{K_0} \sum_{j=i+1}^{j=N} \frac{1}{j^2} \leq |x_i - p| \leq K_0 \Big[\sum_{j=i+1}^{j=N} \frac{1}{j^2} + \frac{1}{(N+1)^2} \Big]$$

for all $i \leq \min\{\lfloor a/2 \rfloor, N-1\}$, where $K_0(K) > 1$ is given by Lemma 4.2. From the elementary estimates

$$\begin{aligned} \frac{N-i}{(i+1)(N+1)} &= \frac{1}{i+1} - \frac{1}{N+1} \\ &= \int_{i+1}^{N+1} \frac{dt}{t^2} \leq \sum_{j=i+1}^{j=N} \frac{1}{j^2} \leq \int_i^N \frac{dt}{t^2} = \frac{1}{i} - \frac{1}{N} = \frac{N-i}{iN} \end{aligned}$$

we obtain

$$\frac{1}{K_0} \frac{1}{(i+1)(N+1)} \leq \frac{|x_i - p|}{N-i} \leq K_0 \Big(\frac{1}{iN} + \frac{1}{(N+1)^2} \Big)$$

for all $i \leq \min\{\lfloor a/2 \rfloor, N-1\}$, which implies item (3). □

A similar application of Lemma 4.2 is given by the following.

LEMMA 6.4
There exists $K_0 = K_0(\mathcal{K}) > 1$ such that for any $\zeta \in \mathcal{K}$ renormalizable with period $a \in \mathbb{N}$ and for any $b < \lfloor a/2 \rfloor$, we have

$$|x_b - x_{a-b}| \leq \frac{K_0}{b}.$$

The constant K_0 does not depend on the period a.

Proof of Lemma 6.4
By Yoccoz's lemma (Lemma 4.2), we have

$$|x_b - x_{a-b}| = \sum_{i=b+1}^{i=a-b} |I_i| \leq K\Big(\sum_{i=b+1}^{i=a-b} \frac{1}{\min\{i, a-i\}^2}\Big).$$

To finish, note that

$$\sum_{i=b+1}^{i=a-b} \frac{1}{\min\{i, a-i\}^2} \leq \frac{2}{b}.$$

Indeed, by symmetry it is enough to prove that

$$\sum_{i=b+1}^{i=\lfloor a/2 \rfloor} \frac{1}{\min\{i, a-i\}^2} \leq \frac{1}{b},$$

and this follows again from elementary calculus:

$$\sum_{i=b+1}^{i=\lfloor a/2 \rfloor} \frac{1}{\min\{i, a-i\}^2} = \sum_{i=b+1}^{i=\lfloor a/2 \rfloor} \frac{1}{i^2}$$
$$\leq \int_b^{\lfloor a/2 \rfloor} \frac{dt}{t^2} \leq \int_b^{+\infty} \frac{dt}{t^2} = \frac{1}{b}. \qquad \square$$

Yet another consequence of Yoccoz's lemma is the following.

LEMMA 6.5
There exist $a_0 = a_0(\mathcal{K}) \in \mathbb{N}$ and $b = b(\mathcal{K}) \in \{1, \ldots, a_0\}$ with the following property. Given $\zeta, \tilde{\zeta} \in \mathcal{K}$ renormalizable with period $a \geq a_0$, we have

$$\tilde{x}_{\tilde{N}-b} \geq x_{N-1} \qquad \text{and} \qquad \tilde{x}_{\tilde{N}+b} \leq x_{N+2}.$$

Recall that we are assuming that $p = \tilde{p}$. The number b given by Lemma 6.5 will be used in the next lemma (Lemma 6.6) and also in Lemmas 6.12 and 6.17 below.

Proof of Lemma 6.5
Consider that

$$a_0 \gg \frac{1}{\delta_0} 2 \frac{K^2}{\delta^2} \qquad \text{and} \qquad 2\frac{K^2}{\delta^2} \leq b < \delta_0 a_0,$$

where K is given by Lemma 4.2, $\delta = \min\{\delta_0, 1-\delta_1\}$, and δ_0, δ_1, in turn, are given by Lemma 6.3(1). We claim that $|\tilde{x}_{\tilde{N}-b} - \tilde{x}_{\tilde{N}}| \geq |x_{N-1} - p|$. Indeed, on the one hand,

$$|\tilde{x}_{\tilde{N}-b} - \tilde{x}_{\tilde{N}}| = \sum_{i=\tilde{N}-b+1}^{i=\tilde{N}} |\tilde{I}_i| \geq \frac{1}{K} \sum_{i=\tilde{N}-b+1}^{i=\tilde{N}} \frac{1}{i^2} \geq \frac{1}{K} \frac{b}{(\tilde{N}-b+1)(\tilde{N}+1)}.$$

On the other hand,

$$\begin{aligned} |x_{N-1} - p| &\leq |I_N| + |I_{N+1}| \\ &\leq K\Big(\frac{1}{\min\{N, a-N\}^2} + \frac{1}{\min\{N+1, a-N-1\}^2}\Big) \leq \frac{2K}{\delta^2}\frac{1}{a^2}. \end{aligned}$$

Therefore, it is enough to have

$$b \geq \frac{2K^2}{\delta^2}\frac{1}{a^2}(\tilde{N}-b+1)(\tilde{N}+1),$$

which follows from our choice of b (recall that $\tilde{N} \leq \delta_1 a < a$ and then $\tilde{N}+1 \leq a$). Then $|\tilde{x}_{\tilde{N}-b} - \tilde{x}_{\tilde{N}}| \geq |x_{N-1} - p|$ as claimed, and this implies at once that $\tilde{x}_{\tilde{N}-b} \geq x_{N-1}$. The other estimate can be proved in the same way. □

The distance between corresponding critical iterates of ζ and $\tilde{\zeta}$ will be denoted by Δx_i, that is,

$$\Delta x_i = \tilde{x}_i - x_i = \tilde{\eta}^i\big(\tilde{\xi}(0)\big) - \eta^i\big(\xi(0)\big) \quad \text{for all } i \in \{0, 1, \ldots, a\}.$$

LEMMA 6.6
There exists $K = K(\mathcal{K}) > 0$ such that for $i \leq \min\{\lfloor a/2 \rfloor, N-b, \tilde{N}-b\}$, we have

$$|\Delta x_i| \leq K\Big(|h| \cdot i + \frac{\varepsilon}{i}\Big),$$

where b is given by Lemma 6.5.

Proof of Lemma 6.6
Let $x_0 = \xi(0) = \tilde{\xi}(0)$ be the common critical value of ξ and $\tilde{\xi}$, which is the right boundary point of $I_\eta = I_{\tilde{\eta}}$. Recall that, by definition, $x_i = \eta^i(x_0)$ and $\tilde{x}_i = \tilde{\eta}^i(x_0)$ for all $i \in \{1, \ldots, a\}$. We will consider the case $x_{\lfloor a/2 \rfloor} \leq \tilde{x}_{\lfloor a/2 \rfloor}$. Note that for any $i \in \{1, \ldots, \lfloor a/2 \rfloor\}$ and any $k \in \{0, \ldots, i-1\}$, we have by combinatorics

$$x_{a-i+k+1} \leq x_{\lfloor a/2 \rfloor+1} < x_{\lfloor a/2 \rfloor} \leq \tilde{x}_{\lfloor a/2 \rfloor} \leq \tilde{x}_{k+1} < \tilde{x}_k.$$

Therefore, $x_{\lfloor a/2 \rfloor+1} < \eta(\tilde{x}_k)$ and then $x_{a-i+k+1} < \eta(\tilde{x}_k)$, that is, both points $\eta(\tilde{\eta}^k(x_0))$ and $\tilde{\eta}^{k+1}(x_0)$ lie to the right of the point $x_{a-i+k+1}$. In particular, the iterate η^{i-k-1} is well defined in the interval with boundary points $\eta(\tilde{\eta}^k(x_0))$ and $\tilde{\eta}^{k+1}(x_0)$. This allows us to use a simple telescopic trick and the mean-value theorem in order to write for any $i \in \{1, \dots, \lfloor a/2 \rfloor\}$:

$$|\Delta x_i| = \Big|\sum_{k=0}^{i-1} \big(\eta^{i-k-1}\big(\eta(\tilde{\eta}^k(x_0))\big) - \eta^{i-k-1}\big(\tilde{\eta}^{k+1}(x_0)\big)\big)\Big|$$

$$\leq \sum_{k=0}^{i-1} \big|D\eta^{i-k-1}(y_k)\big|\big|H\big(\tilde{\eta}^k(x_0)\big)\big|, \tag{6.3}$$

where for each $k \in \{0, \dots, i-1\}$ the point y_k lies between $\eta(\tilde{\eta}^k(x_0))$ and $\tilde{\eta}^{k+1}(x_0)$ (the points $y_0, y_1, \dots, y_{i-1}$ depend also on each fixed i, but we will denote them just by y_k to simplify the notation). From (6.1) and Lemma 6.3, we get that

$$\big|H\big(\tilde{\eta}^k(x)\big)\big| \leq |h| + \frac{K\varepsilon}{(k+1)^2}. \tag{6.4}$$

For each $k \in \{0, \dots, i-1\}$, let us denote $D_k = |D\eta^{i-k-1}(y_k)|$. Our goal is, therefore, to estimate the sum

$$\big|\eta^i(x) - \tilde{\eta}^i(x)\big| \leq \sum_{k=0}^{i-1} D_k\Big(|h| + \frac{K\varepsilon}{(k+1)^2}\Big). \tag{6.5}$$

For each $k \in \{0, \dots, i-1\}$, let $m = m(k) \in \{1, \dots, a\}$ be such that $y_k \in I_m(\eta)$, where $I_m(\eta) = [\eta^m(x), \eta^{m-1}(x)]$ as before. Since we are assuming that $x_{\lfloor a/2 \rfloor} \leq \tilde{x}_{\lfloor a/2 \rfloor}$, we have $m \leq a/2 + 1$. We claim that $m(k) \asymp k$ for all $k \in \{0, \dots, i-1\}$. More precisely, we have the following.

CLAIM 6.7
There exists $C = C(\mathcal{K}) > 1$ such that $\frac{k}{C} < m < Ck$ for all $k \in \{0, \dots, i-1\}$ and for all $i \in \{1, \dots, \lfloor a/2 \rfloor\}$.

Proof of Claim 6.7
From Lemma 6.3 we know that $|y_k - p| \asymp \frac{1}{m}$, and then it is enough to prove that $|y_k - p| \asymp \frac{1}{k}$. Recall that $d_2(\zeta, \tilde{\zeta}) < \varepsilon_0$, where $\varepsilon_0 > 0$ will be fixed later in the proof. On the one hand, $|y_k - p| \leq |\tilde{\eta}^k(x_0) - p| \asymp \frac{1}{k}$. On the other hand, since $i \leq \min\{N - b, \tilde{N} - b\}$, the point p does not belong to the interval with boundary points $\eta(\tilde{\eta}^k(x_0))$ and $\tilde{\eta}^{k+1}(x_0)$, and then

$$|y_k - p| \geq \min\big\{\big|\tilde{\eta}^{k+1}(x_0) - p\big|, \big|\eta\big(\tilde{\eta}^k(x_0)\big) - p\big|\big\}$$

$$= \big|\tilde{\eta}^{k+1}(x_0) - p\big| - \big|\eta\big(\tilde{\eta}^k(x_0)\big) - \tilde{\eta}^{k+1}(x_0)\big|$$

$$= \big|\tilde{\eta}^{k+1}(x_0) - p\big| - \big|H\big(\tilde{\eta}^k(x_0)\big)\big|.$$

From (6.4) we get $|H(\tilde{\eta}^k(x_0))| \leq K(|h| + \frac{\varepsilon}{(k+1)^2}) \leq \frac{K}{(k+1)^2}$ since $|h| \leq K/a^2$ by Yoccoz's lemma (indeed, by Lemma 4.2, the length of the fundamental domain $(\eta(p), p)$ is bounded by $1/a^2$, up to a multiplicative constant, i.e., both $p - \eta(p)$ and $p - \tilde{\eta}(p)$ are bounded by $1/a^2$ up to a multiplicative constant, and then $|h| \leq K/a^2$). Therefore,

$$|y_k - p| \geq \frac{1}{K}\Big(\frac{1}{k+1} - \frac{K^2}{(k+1)^2}\Big) = \frac{1}{K}\Big(1 - \frac{K^2}{k+1}\Big)\frac{k}{k+1}\frac{1}{k} \geq \frac{1}{4k}$$

if $k \geq 2K^2 + 1$ and then $|y_k - p| \asymp \frac{1}{k}$ in this case. We choose $\varepsilon_0 > 0$ in order to have that if $k \leq 2K^2 + 1$, then both $\tilde{\eta}^{k+1}(x)$ and $\eta(\tilde{\eta}^k(x))$ belong to the interval $[\tilde{\eta}^{k+2}(x), \tilde{\eta}^k(x)]$, and again $|y_k - p| \asymp \frac{1}{k}$ as we wanted to prove. □

We have two claims regarding the values of D_k.

CLAIM 6.8
There exists $K = K(\mathcal{K}) > 0$ such that for all $k \in \{0, \ldots, i-1\}$ and $i \in \{1, \ldots, a/2\}$, we have $D_k \leq K$.

Proof of Claim 6.8
By bounded distortion and Yoccoz's lemma, we know that

$$\big|D\eta^{i-k-1}(y_k)\big| \asymp \frac{|I_{m+i-k-1}(\eta)|}{|I_m(\eta)|} \asymp m^2\big|I_{m+i-k-1}(\eta)\big|,$$

and then it is enough to prove that $|I_{m+i-k-1}(\eta)| \leq \frac{K}{m^2}$. To prove this, we have two cases to consider. If $\eta^{i-k-1}(y_k) \geq p$, then $|I_{m+i-k-1}(\eta)| \asymp \frac{1}{(m+i-k-1)^2}$ by Yoccoz's lemma, and since $i - k - 1 \geq 0$ we are done. If $\eta^{i-k-1}(y_k) < p$, then $|I_{m+i-k-1}(\eta)| \asymp \frac{1}{(a-m-i+k+1)^2}$, and since $a - m - i \geq 0$ we obtain $|I_{m+i-k-1}(\eta)| \leq \frac{K}{(k+1)^2}$. Since $m \asymp k$ by Claim 6.7, we obtain Claim 6.8. □

CLAIM 6.9
There exists $K = K(\mathcal{K}) > 0$ such that if $k < \frac{i}{4(C-1)}$, then $D_k \leq K\frac{k^2}{i^2}$.

Proof of Claim 6.9
Write $m = \lfloor \theta k \rfloor$ with $\frac{1}{C} < \theta < C$ (see Claim 6.7). If $m < k$, we have $\theta < 1$ and $i + m - k - 1 = \theta i + (1-\theta)i - (1-\theta)k - 1 = \theta i + (1-\theta)(i-k) - 1 \geq \theta i - 1 \geq \frac{1}{C}i$. Since $i + m - k - 1 \leq i \leq \frac{a}{2}$, we have $D_k \leq K\frac{C^2k^2}{(\frac{i}{C})^2} \leq K\frac{k^2}{i^2}$. On the other hand, if $m > k$ (i.e., $\theta > 1$), we have $m + i - k - 1 \leq i + (\theta - 1)k - 1 \leq i + (C-1)k - 1 \leq i + \frac{1}{4}i - 1 \leq \frac{3}{2}a$. Then $|I_{m+i-k-1}(\eta)| \asymp \frac{1}{(m+i-k-1)^2} \leq \frac{1}{(i-1)^2}$, and so we also have $D_k \leq K\frac{k^2}{i^2}$ in this case, since $\frac{1}{3}a < j < \frac{2}{3}a$ implies that $\frac{1}{a-j} > \frac{1}{2}\frac{1}{j} > \frac{1}{4}\frac{1}{a-j}$. □

With Claims 6.8 and 6.9 at hand, we are ready to estimate the sum (6.5) as follows:

$$\begin{aligned}\sum_{k=0}^{i-1} D_k\Big(|h| + \frac{K\varepsilon}{(k+1)^2}\Big) &= |h|\Big(\sum_{k=0}^{i-1} D_k\Big) + K\varepsilon\Big(\sum_{k=0}^{\lfloor \frac{i}{4(C-1)}\rfloor} \frac{D_k}{(k+1)^2}\Big)\\ &\quad + K\varepsilon\Big(\sum_{k=\lfloor \frac{i}{4(C-1)}\rfloor+1}^{i-1} \frac{D_k}{(k+1)^2}\Big)\\ &\le K|h|i + K\frac{\varepsilon}{i^2}\Big(\sum_{k=0}^{\lfloor \frac{i}{4(C-1)}\rfloor} \Big(\frac{k}{k+1}\Big)^2\Big)\\ &\quad + K\varepsilon\Big(\sum_{k=\lfloor \frac{i}{4(C-1)}\rfloor+1}^{i-1} \frac{1}{(k+1)^2}\Big)\\ &\le K|h|i + K\frac{\varepsilon}{i} + K\frac{\varepsilon}{i}.\end{aligned}$$

For the last inequality, we have used that both sequences

$$\frac{1}{i}\Big(\sum_{k=0}^{\lfloor \frac{i}{4(C-1)}\rfloor} \Big(\frac{k}{k+1}\Big)^2\Big) \quad \text{and} \quad i\Big(\sum_{k=\lfloor \frac{i}{4(C-1)}\rfloor+1}^{i-1} \frac{1}{(k+1)^2}\Big)$$

remain bounded when i goes to infinity, with constants depending only on C. We have proved Lemma 6.6. □

LEMMA 6.10
For every $a \ge 1$ there exists $K_a = K_a(a) > 0$ such that

$$|\Delta x_a| \le K_a \varepsilon.$$

Proof of Lemma 6.10
Observe that

$$|\Delta_{i+1}| = \big|\tilde{\eta}(\tilde{x}_i) - \eta(x_i)\big| = \big|\eta(\tilde{x}_i) - \eta(x_i) + H(\tilde{x}_i)\big| \le D|\Delta x_i| + \varepsilon,$$

where $D = \max\{D\eta\}$. So

$$|\Delta x_a| \le \varepsilon \cdot \sum_{k=0}^{a} D^{a-k}.$$

The lemma follows. □

The following definition is given for general commuting pairs which are contained in the previously discussed set $\mathcal{K}$ of K_0-controlled commuting pairs.

Definition 6.11
Given $L > 1$, we say that the commuting pairs $\zeta_0 = (\xi_0, \eta_0)$ and $\zeta_1 = (\xi_1, \eta_1)$, with $a_{\zeta_0} = a_{\zeta_1} = a$, are *L-synchronized* if

$$|\Delta x_a| \le L \cdot d_2(\zeta_0, \zeta_1).$$

By working just as in the proof of Lemma 6.6 but with backward iterations, we obtain the following.

LEMMA 6.12
Given $L > 0$, there exists $K = K(\mathcal{K}, L) > 0$ such that if $\zeta, \tilde{\zeta} \in \mathcal{K}$ are L-synchronized with $a_\zeta = a_{\tilde{\zeta}} = a$, then we have

$$|\Delta x_i| \le K\Big(|h|(a-i) + \frac{\varepsilon}{a-i}\Big)$$

for all $i \in \mathbb{N}$ such that $\max\{\lfloor a/2 \rfloor, N+b, \tilde{N}+b\} \le i \le a$.

PROPOSITION 6.13
For every $L > 0$ there exists $K = K(\mathcal{K}, L) > 1$ such that the following holds. If $\zeta, \tilde{\zeta} \in \mathcal{K}$ are L-synchronized with $a_\zeta = a_{\tilde{\zeta}} = a$, then we have

$$|h| \le K\frac{\varepsilon}{a^2}.$$

Proof of Proposition 6.13
Let us suppose that $\tilde{\eta}(p) = \eta(p) + h$ with $h > 0$. We want to prove that, under the synchronization assumption, the ratio $C = \frac{a^2 h}{\varepsilon}$ is uniformly bounded in $\mathcal{K}$. As before, let $N \in \{1, \dots, a\}$ be defined by $p \in [x_{N+1}, x_N]$. By Yoccoz's lemma (see in particular [3, Lemma B.1, p. 387]), there exists $K_0 = K_0(\mathcal{K}) > 1$ such that $N = \nu a$ with $1/K_0 \le \nu \le 1 - \frac{1}{K_0}$. In the same way, let $\tilde{N} = \tilde{\nu} a$ defined by $p \in [\tilde{x}_{\tilde{N}+1}, \tilde{x}_{\tilde{N}}]$ with $1/K_0 \le \tilde{\nu} \le 1 - \frac{1}{K_0}$.

By Lemma 6.4, there exists $K_1 = K_1(\mathcal{K}) > 1$ such that $(x_j, \tilde{x}_j) \subset (p - K_1/M, p)$ when $(1 - \frac{1}{K_0})a \le j \le a - M$, and $(x_j, \tilde{x}_j) \subset (p, p + K_1/M)$ when $M \le j \le a/K_0$ for any $M \in \{1, \dots, \lfloor a/K_0 \rfloor\}$. Let $K_2 = K_2(\mathcal{K}) > 1$ be the constant given by Lemma 6.6. By Lemma 6.12, we have

$$|\Delta x_{a-M}| \le K_3\Big(hM + \frac{\varepsilon}{M}\Big) \tag{6.6}$$

for some universal constant $K_3(L, \mathcal{K}) > 1$. Let $K = \max\{K_0, K_1, K_2, K_3\}$, and let us suppose that $a > K(4K+1)$ (otherwise we are done since $|h| \leq \varepsilon$). Fix $M \in \{1, \ldots, \lfloor a/2 \rfloor\}$ small enough in order to have

$$0 < \theta = \frac{M}{a} < \frac{1}{K(4K+1)} < 1.$$

Let $T = [p - K/M, p + K/M]$, and recall that $(x_j, \tilde{x}_j) \subset T$ for all $j \in \{M, \ldots, a - M\}$. The next three claims will show that if C is big enough, in terms of K and $\theta(K)$, the pairs ζ and $\tilde{\zeta}$ cannot be L-synchronized.

CLAIM 6.14
If $C \geq 2(\frac{K}{\theta})^2$, then $\tilde{\eta}(x) \geq \eta(x) + \frac{h}{2}$ for all $x \in T$.

Proof of Claim 6.14
As before,

$$\tilde{\eta}(x) - \eta(x) \geq h - \varepsilon(x-p)^2 \geq h - \varepsilon\Big(\frac{K}{M}\Big)^2 = h - \frac{\varepsilon}{a^2}\Big(\frac{K}{\theta}\Big)^2 \geq h - \frac{h}{2} = \frac{h}{2}.$$

In the last inequality, we have used that $\frac{\varepsilon}{a^2} \leq \frac{h}{2}(\frac{\theta}{K})^2$ since $\frac{a^2 h}{\varepsilon} \geq 2(\frac{K}{\theta})^2$. □

Note that $0 < \theta < \frac{1}{K(4K+1)}$ implies that $1 - 2\theta K^2 - \theta K \in (0,1)$.

CLAIM 6.15
If $C > \frac{1}{\theta}(\frac{2K^2}{1-2\theta K^2 - \theta K})$, then there exists $i_0 \in \{M, \ldots, a/K\}$ such that $x_{i_0} \leq \tilde{x}_{i_0}$.

Proof of Claim 6.15
We will first prove that

$$\Big(\frac{a}{K} - M\Big)\frac{h}{2} \geq K\Big(hM + \frac{\varepsilon}{M}\Big). \tag{6.7}$$

Indeed, since $1 - 2\theta K^2 - \theta K > \frac{2K^2}{C\theta}$, we have

$$\frac{1 - 2\theta K^2 - \theta K}{2\theta K} > \frac{K}{C\theta^2},$$

and then

$$hM\Big(\frac{1 - 2\theta K^2 - \theta K}{2\theta K}\Big) > K\frac{\varepsilon}{M}$$

since $\varepsilon/M = hM/C\theta^2$. From

$$\frac{1-2\theta K^2-\theta K}{2\theta K}=\frac{1}{2}\Big(\frac{1}{\theta K}-1-2K\Big)$$

we obtain

$$\frac{h}{2}\Big(\frac{a}{K}-M\Big)-KhM>K\frac{\varepsilon}{M},$$

which implies the desired estimate (6.7). Now estimate (6.7) combined with Lemma 6.6 gives us

$$|x_M-\tilde{x}_M|\leq\Big(\frac{a}{K}-M\Big)\frac{h}{2}. \tag{6.8}$$

With estimate (6.8) at hand, we are ready to prove Claim 6.15. Indeed, let $i\in\{M,\dots,a/K\}$ be such that $p\leq\tilde{x}_i<x_i\leq p+K/M$ (if no such i exists we are done). From Claim 6.14, we have

$$\begin{aligned}\tilde{x}_{i+1}-x_{i+1}&=\tilde{\eta}(\tilde{x}_i)-\eta(x_i)\\&\geq h/2+\eta(\tilde{x}_i)-\eta(x_i)\\&=h/2+D\eta(y_i)(\tilde{x}_i-x_i)\\&=h/2+\tilde{x}_i-x_i+D^2\eta(z_i)(y_i-p)(\tilde{x}_i-x_i),\end{aligned}$$

where $y_i\in[\tilde{x}_i,x_i]$ and $z_i\in[p,y_i]$ are given by the mean-value theorem. Since $D^2\eta(z_i)<0$, $y_i-p>0$, and $\tilde{x}_i-x_i<0$, we obtain

$$\tilde{x}_{i+1}-x_{i+1}\geq h/2+\tilde{x}_i-x_i,\quad\text{that is }\Delta x_{i+1}\geq h/2+\Delta x_i.$$

Therefore if the difference $\tilde{x}_{i+1}-x_{i+1}$ is still negative, it will be at least $h/2$ closer to zero than the previous difference $\tilde{x}_i-x_i$. What estimate (6.8) tells us is that we have enough time inside the interval $(p,p+K/M)$ in order to interchange the positions of the critical iterates. With this we have proved Claim 6.15. □

Claim 6.15 implies that $x_i\leq\tilde{x}_i$ for all $i\in\{i_0,\dots,a-M\}$, since $D\eta>0$ and $h>0$. Therefore, by Claim 6.14, we have

$$|\Delta x_{a-M}|\geq\frac{h}{2}\Big[a-M-\Big(1-\frac{1}{K}\Big)a\Big]. \tag{6.9}$$

Our third and last claim tells us that (6.9) contradicts the synchronization assumption. Note that $0<\theta<\frac{1}{K(4K+1)}$ implies that $1-\theta K(4K+1)\in(0,1)$.

CLAIM 6.16

$$\textit{If } C \geq \frac{1}{\theta}\Big[\frac{4K^2}{1-\theta K(4K+1)}\Big], \quad \textit{then } 2K\Big(hM + \frac{\varepsilon}{M}\Big) \leq \frac{h}{2}\Big[a - M - \Big(1-\frac{1}{K}\Big)a\Big].$$

Proof of Claim 6.16
Note first that

$$2K\Big(hM + \frac{\varepsilon}{M}\Big) = \frac{\varepsilon}{a}\Big[2K\Big(C\theta + \frac{1}{\theta}\Big)\Big]$$

and

$$\frac{h}{2}\Big[a - M - \Big(1-\frac{1}{K}\Big)a\Big] = \frac{\varepsilon}{a}\Big[\frac{C}{2}\Big(\frac{1}{K} - \theta\Big)\Big].$$

A straightforward computation shows that both conditions

$$C \geq \frac{1}{\theta}\Big[\frac{4K^2}{1-\theta K(4K+1)}\Big] \qquad \text{and} \qquad 2K\Big(C\theta + \frac{1}{\theta}\Big) \leq \frac{C}{2}\Big(\frac{1}{K} - \theta\Big)$$

are actually equivalent. □

We are ready to finish the proof of Proposition 6.13. Indeed, by combining estimates (6.6) and (6.9), we have

$$\begin{aligned}\frac{h}{2}\Big[a - M - \Big(1-\frac{1}{K}\Big)a\Big] &\leq |x_{a-M} - \tilde{x}_{a-M}| \\ &\leq K\Big(hM + \frac{\varepsilon}{M}\Big) < 2K\Big(hM + \frac{\varepsilon}{M}\Big),\end{aligned}$$

which contradicts Claim 6.16. Therefore,

$$C \leq \max\Big\{2\Big(\frac{K}{\theta}\Big)^2, \frac{1}{\theta}\Big(\frac{2K^2}{1-2\theta K^2 - \theta K}\Big), \frac{1}{\theta}\Big[\frac{4K^2}{1-\theta K(4K+1)}\Big]\Big\},$$

that is, the ratio $C = \frac{a^2 h}{\varepsilon}$ is bounded by a constant depending only on K and L. We have proved Proposition 6.13. □

With Proposition 6.13 at hand, we can improve Lemmas 6.6 and 6.12 under the synchronization assumption.

LEMMA 6.17
Given $L > 0$, there exists $K = K(\mathcal{K}, L) > 0$ such that if $\zeta, \tilde{\zeta} \in \mathcal{K}$ are L-synchronized with $a_\zeta = a_{\tilde{\zeta}} = a$, then we have

$$|\Delta x_i| \le \frac{K\varepsilon}{i} \quad \textit{for all } 1 \le i \le \min\{\lfloor a/2 \rfloor, N-b, \tilde{N}-b\},$$

$$|\Delta x_i| \le \frac{K\varepsilon}{a-i} \quad \textit{for all } a \ge i \ge \max\{\lfloor a/2 \rfloor, N+b, \tilde{N}+b\}.$$

Moreover, we have the following.

PROPOSITION 6.18
For every $L > 0$, there exists $K = K(\mathcal{K}, L) > 0$ such that the following holds. If ζ and $\tilde{\zeta}$ are L-synchronized, then

$$|\Delta x_i| \le K\varepsilon \cdot \frac{1}{i} \quad \textit{for all } i \in \{0, 1, \dots, a/2\}$$

and

$$|\Delta x_i| \le K\varepsilon \cdot \frac{1}{a-i} \quad \textit{for all } i \in \{a/2, \dots, a\}.$$

Proof of Proposition 6.18
By Lemma 6.17, we only need to estimate $|\Delta x_i|$ for the intermediate iterates $\min\{\lfloor a/2 \rfloor, N-b, \tilde{N}-b\} < i < \{\lfloor a/2 \rfloor, N+b, \tilde{N}+b\}$. We will prove only the first part of the statement (the other being the same), that is, we will prove that

$$|\Delta x_i| \le K\varepsilon \cdot \frac{1}{i} \quad \text{for all } i \in \{\min\{\lfloor a/2 \rfloor, N-b, \tilde{N}-b\}, \dots, a/2\}.$$

We use the same notation as in the proof of Proposition 6.13. By the choice of θ, we know that $M \le \min\{\lfloor a/2 \rfloor, N-b, \tilde{N}-b\}$ and $a - M \ge \max\{\lfloor a/2 \rfloor, N+b, \tilde{N}+b\}$.

Recall that $H : I_\eta \to [-\varepsilon, \varepsilon] \subset \mathbb{R}$ is defined as $H(x) = \eta(x) - \tilde{\eta}(x)$. By Proposition 6.13, we have $|H(x)| \le \varepsilon[\frac{K}{a^2} + (x-p)^2]$ and then $|H(x)| \le \frac{K\varepsilon}{a^2}$ whenever $x \in T$, since for $x \in T$ we have $|x - p| \le \frac{K}{M} \le \frac{K}{a}$. Therefore, by considering $\alpha = 1 + \frac{K}{a}$ and $\beta = \frac{K\varepsilon}{a^2}$, we obtain that $\Delta x_{i+1} \le \alpha \Delta x_i + \beta$, and then

$$\Delta x_{i+n} \le \alpha^n \Delta x_i + \beta\Big(\sum_{j=0}^{n-1} \alpha^j\Big) \quad \text{for all } 1 \le n \le (\delta_1 - \delta_0)a + 2b.$$

Note that $\sum_{j=0}^{n-1} \alpha^j = \frac{\alpha^n - 1}{\alpha - 1} = \frac{a}{K}(\alpha^n - 1)$. Moreover, since $n < a$ we have $\alpha^n = (\frac{K}{a} + 1)^n \le e^{\frac{Kn}{a}}$ is bounded. Therefore,

$$\Delta x_{i+n} \le \alpha^n \Delta x_i + \beta \frac{a}{K}(\alpha^n - 1) \le K\frac{\varepsilon}{i}\Big[\alpha^n + \frac{i}{a}(\alpha^n - 1)\Big] \le K\frac{\varepsilon}{i}\alpha^n \le K\frac{\varepsilon}{i}.$$

Finally, from $i \ge M = \theta a$ and $n \le (\delta_1 - \delta_0)a + 2b$, we get that $\frac{n}{i}$ is bounded and then $\Delta x_{i+n} \le K\frac{\varepsilon}{i+n}$, as we wanted to prove. □

For $i \in \{1, \dots, a\}$, let

$$\Delta_i = |\Delta x_i - \Delta x_{i-1}|.$$

PROPOSITION 6.19
For every $L > 0$, there exists $K = K(\mathcal{K}, L) > 0$ such that the following holds. If ζ and $\tilde{\zeta}$ are L-synchronized, then

$$\Delta_i \leq K\Big(\varepsilon \cdot \frac{\log i}{i^2} + \varepsilon^2 \cdot \frac{1}{i}\Big) \quad \textit{for all } i \leq a/2$$

and

$$\Delta_i \leq K\Big(\varepsilon \cdot \frac{\log(a-i)}{(a-i)^2} + \varepsilon^2 \cdot \frac{1}{a-i}\Big) \quad \textit{for all } i \geq a/2.$$

Proof of Proposition 6.19
The proof of the second part of this proposition can be obtained as in the first part by working backward. (See also the proof of Proposition 6.18.) We will only present the proof of the first part. Observe that for $i \geq 1$,

$$\begin{aligned}
\Delta_{i+1} &= \big|\big[\tilde{\eta}(x_i + \Delta x_i) - \eta(x_i)\big] - \big[\tilde{\eta}(x_{i-1} + \Delta x_{i-1}) - \eta(x_{i-1})\big]\big| \\
&= \big|\big[\eta(x_i + \Delta x_i) - \eta(x_i) + \tilde{\eta}(x_i + \Delta x_i) - \eta(x_i + \Delta x_i)\big] \\
&\quad - \big[\eta(x_{i-1} + \Delta x_{i-1}) - \eta(x_{i-1}) + \tilde{\eta}(x_{i-1} + \Delta x_{i-1}) - \eta(x_{i-1} + \Delta x_{i-1})\big]\big| \\
&= \big|\big[D\eta(\theta_i)\Delta x_i + H(x_i + \Delta x_i)\big] \\
&\quad - \big[D\eta(\theta_{i-1})\Delta x_{i-1} + H(x_{i-1} + \Delta x_{i-1})\big]\big| \\
&\leq \big|D\eta(\theta_i)\Delta x_i - D\eta(\theta_{i-1})\Delta x_{i-1}\big| + \big|DH(\theta)\tilde{I}_i\big|.
\end{aligned}$$

The intermediate point θ is in $\tilde{I}_i$. Hence, by using (6.2), Yoccoz's lemma (Lemma 4.2), and Lemma 6.3, we have

$$\big|DH(\theta)\tilde{I}_i\big| \leq K\varepsilon \cdot \frac{1}{i^3}. \tag{6.10}$$

The intermediate point θ_i is in $[x_i, x_i + \Delta x_i]$. Similarly, $\theta_{i-1} \in [x_{i-1}, x_{i-1} + \Delta x_{i-1}]$. This allows for the following estimate:

$$\begin{aligned}
&\big|D\eta(\theta_i)\Delta x_i - D\eta(\theta_{i-1})\Delta x_{i-1}\big| \\
&\quad \leq \frac{|I_{i+1}|}{|I_i|}\Delta_i + \Big|\Big(D\eta(\theta_i) - \frac{|I_{i+1}|}{|I_i|}\Big)\Delta x_i\Big| \\
&\qquad + \Big|\Big(D\eta(\theta_{i-1}) - \frac{|I_{i+1}|}{|I_i|}\Big)\Delta x_{i-1}\Big|
\end{aligned}$$

$$\leq \frac{|I_{i+1}|}{|I_i|}\Delta_i + K\big(|I_i| + |\Delta x_i|\big)|\Delta x_i| \\ + K\big(|I_i| + |\Delta x_{i-1}|\big)|\Delta x_{i-1}|.$$

Use Yoccoz's lemma (Lemma 4.2) and Proposition 6.18 to obtain

$$\big|D\eta(\theta_i)\Delta x_i - D\eta(\theta_{i-1})\Delta x_{i-1}\big| \leq \frac{|I_{i+1}|}{|I_i|}\Delta_i + K\Big(\varepsilon\frac{1}{i^3} + \varepsilon^2\frac{1}{i^2}\Big). \tag{6.11}$$

Combine (6.10) and (6.11) with the chain of estimates obtained at the beginning of the proof to deduce that

$$\Delta_{i+1} \leq \frac{|I_{i+1}|}{|I_i|}\Delta_i + K\Big(\varepsilon\frac{1}{i^3} + \varepsilon^2\frac{1}{i^2}\Big).$$

After iterating this recursive estimate and using Yoccoz's lemma (Lemma 4.2), one gets

$$\Delta_i \leq K\sum_{k=1}^{i-1}\Big(\varepsilon\frac{1}{k^3} + \varepsilon^2\frac{1}{k^2}\Big)\cdot\frac{|I_i|}{|I_{k+1}|} \\ \leq K\Big(\varepsilon\frac{1}{i^2}\sum_{k=1}^{i-1}\frac{1}{k} + \varepsilon^2\frac{1}{i^4}\sum_{k=1}^{i-1}k^2\Big) \\ \leq K\Big(\varepsilon\cdot\frac{\log i}{i^2} + \frac{\varepsilon^2}{i}\Big).$$

□

7. Composition

In this section we will discuss the composition of multiple diffeomorphisms. Let $I = [a,b]$ be a compact interval in the real line, and let $\mathcal{D} = \mathrm{Diff}^2_+([a,b])$ be the space of orientation-preserving C^2-diffeomorphisms of I, endowed with the C^2-metric. Let $X = C^0(I,\mathbb{R})$ be the vector space of continuous functions from $[a,b]$ to the real line, and recall that X is a Banach space when endowed with the sup norm. Consider the *nonlinearity* $N : \mathcal{D} \to X$ defined as

$$N\psi = \frac{D^2\psi}{D\psi} = D\log D\psi.$$

Note that N is a homeomorphism whose inverse is given by

$$(N^{-1}\phi)(x) = a + \Big(\frac{b-a}{\int_a^b \exp(\int_a^s \phi(t)\,dt)\,ds}\Big)\int_a^x \exp\Big(\int_a^s \phi(t)\,dt\Big)\,ds,$$

for any $x \in [a,b]$ and any $\phi \in X$. To prove that $N^{-1}\phi \in \mathcal{D}$, note that $DN^{-1}\phi > 0$, since $\frac{\partial}{\partial x}(\int_a^x \exp(\int_a^s \phi(t)\,dt)\,ds) = \exp(\int_a^x \phi(t)\,dt) > 0$.

In general, if $f : I \to \mathbb{R}$ is a C^2-map and x is a regular point of f, we define $Nf(x) = D^2 f(x)/Df(x)$. The chain rule for the nonlinearity is $N(f \circ g) = Nf \circ gDg + Ng$. The kernel of N is the group of affine transformations. In particular, $N(A \circ f) = Nf$ whenever A is affine. Note also that the nonlinearity goes to infinity around any nonflat critical point (elementary properties of nonlinearity can be found in [19]). On bounded sets it is bi-Lipschitz. In particular, we have the following.

LEMMA 7.1
Let B be a bounded set in $X = C^0(I, \mathbb{R})$. There exists $K = K(B) > 0$ such that for any pair ϕ, ψ in B, we have

$$d_2(N^{-1}\phi, N^{-1}\psi) \leq K d_{C^0}(\phi, \psi).$$

Proof of Lemma 7.1
Use the inverse of the nonlinearity to estimate the C^0-distance between $f = N^{-1}\phi$ and $g = N^{-1}\psi$, as in [19, Lemma 10.2, p. 579]. This gives $d_{C^0}(N^{-1}\phi, N^{-1}\psi) \leq K d_{C^0}(\phi, \psi)$. Since both $f = N^{-1}\phi$ and $g = N^{-1}\psi$ belong to $\mathrm{Diff}^2_+(I)$, there exists $t_0 \in I$ such that $Df(t_0) = Dg(t_0)$, and then $\log Df(t) - \log Dg(t) = \int_{t_0}^{t} (\phi - \psi)(s)\, ds$ for all $t \in I$. Therefore $d_{C^0}(\log Df, \log Dg) \leq |I| d_{C^0}(\phi, \psi)$, and since both f and g are C^1-bounded, we get $d_{C^0}(Df, Dg) \leq K d_{C^0}(\phi, \psi)$. Finally, note that for all $t \in I$ we have

$$\big|(D^2 f - D^2 g)(t)\big| \leq \big|(\phi - \psi)(t)\big|\big|Df(t)\big| + \big|(Df - Dg)(t)\big|\big|\psi(t)\big|. \qquad \square$$

As we said before, the nonlinearity allows us to identify the set $\mathfrak{D}$ of diffeomorphisms with the Banach space $X = C^0(I, \mathbb{R})$ of continuous functions. This defines the *nonlinearity* norm on $\mathfrak{D}$: $|f| = \|Nf\|_{C^0}$. The following lemma says that the composition of multiple diffeomorphisms on C^1-bounded sets is Lipschitz-continuous in the nonlinearity norm. This lemma is an adaptation of the "sandwich lemma" in [19, Lemma 10.5, p. 581].

LEMMA 7.2
Given $M > 0$, there exist $K = K(M) > 0$ such that for $f_1, \dots, f_n, g_1, \dots, g_n$ in $\mathrm{Diff}^3_+([0, 1])$ satisfying
- $\sum_{j=1}^{j=n} |Nf_j|_{C^0} \leq M$,
- $\sum_{j=1}^{j=n} |Ng_j|_{C^0} \leq M$,
- $\sum_{j=1}^{j=n} |DNf_j|_{C^0} \leq M$,
- $\sum_{j=1}^{j=n} |DNg_j|_{C^0} \leq M$,

we have

$$\left|N(\bigcirc_{j=1}^{j=n} f_j) - N(\bigcirc_{j=1}^{j=n} g_j)\right|_{C^0} \leq K \sum_{j=1}^{j=n} |Nf_j - Ng_j|_{C^0}.$$

In particular,

$$d_{C^2}(\bigcirc_{j=1}^{j=n} f_j, \bigcirc_{j=1}^{j=n} g_j) \leq K \sum_{j=1}^{j=n} |Nf_j - Ng_j|_{C^0}.$$

The branches of renormalizations are compositions of a homeomorphism and multiple diffeomorphisms. The composition of multiple diffeomorphisms can be controlled by Lemma 7.2. To control the effect of the first factor we need the following lemma, a basic property of composition.

LEMMA 7.3
For every $L > 0$, there exists $K = K(L) > 0$ such that the following holds. Let $q, \tilde{q} : [-1, 0] \to [0, 1]$ be C^3-homeomorphisms with one critical point, and let $Dq(0) = D\tilde{q}(0) = 0$. Let $f, \tilde{f} : [0, 1] \to [0, 1]$ be C^3-diffeomorphisms. If $|q|_{C^3}$, $|\tilde{q}|_{C^3}$, $|f|_{C^3}$, $|\tilde{f}|_{C^3} \leq L$, then

$$d_{C^2}(\tilde{f} \circ \tilde{q}, f \circ q) \leq K d_{C^2}(\tilde{f}, f) + d_{C^2}(\tilde{q}, q).$$

As before, fix $K_0 > 1$, and let $\mathcal{K}$ be the space of normalized C^3 critical commuting pairs which are K_0-controlled. Let $\zeta = (\eta, \xi)$ and $\tilde{\zeta} = (\tilde{\eta}, \tilde{\xi})$ be two C^3 critical commuting pairs with negative Schwarzian that belong to $\mathcal{K}$ which are renormalizable with the same period $a \in \mathbb{N}$. Denote by $\varepsilon > 0$ the C^2-distance between ζ and $\tilde{\zeta}$, that is, $\varepsilon = d_2(\zeta, \tilde{\zeta})$. We may assume in the computations that $\varepsilon \in (0, 1)$. We will only consider the special situation when

(1) $I_\eta = I_{\tilde{\eta}}$ and $I_\xi = I_{\tilde{\xi}}$,

(2) $p = \tilde{p}$ where $D\eta(p) = D\tilde{\eta}(\tilde{p}) = 1$ (see Lemma 6.1).

For each $i \in \{1, \ldots, a-1\}$, let $f_i \in \text{Diff}^3_+([0, 1])$ given by $f_i = A_{i+1}^{-1} \circ \eta \circ A_i$, where $A_i : [0, 1] \to I_i$ is the unique orientation-preserving affine diffeomorphism

$$\begin{aligned} A_i(x) &= |I_i| x + x_i \\ &= \left(\eta^{i-1}(\xi(0)) - \eta^i(\xi(0))\right)x + \eta^i(\xi(0)). \end{aligned}$$

Note that $\bigcirc_{i=1}^{a-1} f_i = A_a^{-1} \circ \eta^{a-1} \circ A_1$ in $\text{Diff}^3_+([0, 1])$.

LEMMA 7.4
There exists $K = K(\mathcal{K}) > 1$ such that for any ζ in $\mathcal{K}$ renormalizable with period $a \in \mathbb{N}$, we have

$$\sum_{i=1}^{a-1} |Nf_i(x)| \leq K \qquad \textit{and} \qquad \sum_{i=1}^{a-1} |D(Nf_i)(x)| \leq K \quad \textit{for all } x \in [0,1].$$

Proof of Lemma 7.4
Note that $Nf_i(x) = N(\eta \circ A_i)(x) = N\eta(A_i(x))|I_i|$ and that $D(Nf_i)(x) = D(N\eta)(A_i(x))|I_i|^2$ for all $x \in [0,1]$. Since $\zeta \in \mathcal{K}_a$, we know that $N\eta$ is C^1-bounded in $[\eta^a(\xi(0)), \xi(0)]$ (see Remark 3.8 at the end of Section 3). Then

$$\sum_{i=1}^{a-1} |Nf_i(x)| \leq K \sum_{i=1}^{a-1} |I_i| \leq K|I_\eta| \qquad \text{and}$$
$$\sum_{i=1}^{a-1} |D(Nf_i)(x)| \leq K \sum_{i=1}^{a-1} |I_i|^2 \leq K|I_\eta| \sum_{i=1}^{a-1} |I_i| \leq K|I_\eta|^2. \qquad \square$$

In the same way, let $\tilde{A}_i : [0,1] \to \tilde{I}_i$ be the unique orientation-preserving affine diffeomorphism, and define $g_i = \tilde{A}_{i+1}^{-1} \circ \tilde{\eta} \circ \tilde{A}_i \in \mathrm{Diff}_+^3([0,1])$. The first factors of the renormalizations are controlled by the following.

LEMMA 7.5
There exists $K > 0$ such that

$$|A_1^{-1} \circ \xi|_{C^3}, |\tilde{A}_1^{-1} \circ \tilde{\xi}|_{C^3} \leq K$$

and

$$d_{C^2}(A_1^{-1} \circ \xi, \tilde{A}_1^{-1} \circ \tilde{\xi}) \leq K\varepsilon.$$

Proof of Lemma 7.5
The four maps $\xi : [-1,0] \to I_1$, $\tilde{\xi} : [-1,0] \to \tilde{I}_1$, $A_1^{-1} : [0,K] \to \mathbb{R}$, and $\tilde{A}_1^{-1} : [0,K] \to \mathbb{R}$ are C^3-bounded by some constant $M > 1$ universal on $\mathcal{K}$. Similar to Lemma 7.3, we get

$$\begin{aligned} d_{C^2}(A_1^{-1} \circ \xi, \tilde{A}_1^{-1} \circ \tilde{\xi}) &\leq K\big(\|A_1^{-1} - \tilde{A}_1^{-1}\|_{C^2} + \|\xi - \tilde{\xi}\|_{C^2}\big) \\ &\leq K\big(\|A_1^{-1} - \tilde{A}_1^{-1}\|_{C^2} + \varepsilon\big). \end{aligned}$$

Observe that

$$
\begin{aligned}
\left|A_1^{-1}(x)-\tilde{A}_1^{-1}(x)\right| &= \left||I_1|^{-1}(x-x_1)-|\tilde{I}_1|^{-1}(x-\tilde{x}_1)\right| \\
&= \frac{|(x-x_1)(\tilde{x}_0-\tilde{x}_1)-(x-\tilde{x}_1)(x_0-x_1)|}{|I_1||\tilde{I}_1|} \\
&= \frac{|x(\tilde{x}_0-x_0)+x(x_1-\tilde{x}_1)+(x_0\tilde{x}_1-\tilde{x}_0x_1)|}{|I_1||\tilde{I}_1|} \\
&\le K\Big(\frac{\Delta x_0+\Delta x_1+|x_0\tilde{x}_1-\tilde{x}_0x_1|}{|I_1||\tilde{I}_1|}\Big) \\
&\le K\Big(\frac{\Delta x_0+\Delta x_1+|x_0||\tilde{x}_1-x_1|+|x_1||x_0-\tilde{x}_0|}{|I_1||\tilde{I}_1|}\Big) \\
&\le K(\Delta x_0+\Delta x_1)/|I_1||\tilde{I}_1| \\
&\le K\varepsilon,
\end{aligned}
$$

where we used Lemma 6.6. On the other hand,

$$
\begin{aligned}
\left|(A_1^{-1})'-(\tilde{A}_1^{-1})'\right| &= \big(|\tilde{I}_1|-|I_1|\big)/|I_1||\tilde{I}_1| \\
&\le (\Delta_0+\Delta_1)/|I_1||\tilde{I}_1|,
\end{aligned}
$$

and we finish in the same way as before. □

LEMMA 7.6
There exists $K>0$ such that for $i\le a$, we have

$$|Nf_i-Ng_i|_{C^0}\le K\big(\varepsilon|I_i|+\Delta_i+|\Delta x_i||I_i|\big).$$

Proof of Lemma 7.6
Observe that

$$|\tilde{A}_ix-A_ix|\le K\big(|\Delta x_i|+\Delta_i\big).$$

So,

$$
\begin{aligned}
\left|Nf_i(x)-Ng_i(x)\right| &= \left|Nf\big(A_i(x)\big)|I_i|-Ng\big(\tilde{A}_i(x)\big)|\tilde{I}_i|\right| \\
&\le \left|Nf(A_ix)|I_i|-Ng(A_ix)|\tilde{I}_i|\right| \\
&\quad+\left|DNg(\theta_i)\right|\cdot\big(|\Delta x_i|+\Delta_i\big)\cdot|\tilde{I}_i| \\
&\le K\big(\varepsilon|I_i|+\Delta_i+\big(|\Delta x_i|+\Delta_i\big)\big(|I_i|+\Delta_i\big)\big) \\
&\le K\big(\varepsilon|I_i|+\Delta_i+|\Delta x_i||I_i|\big).
\end{aligned}
$$

□

LEMMA 7.7
For every $L > 0$, there exists $K = K(\mathcal{K}, L) > 0$ such that the following holds. If ζ and $\tilde{\zeta}$ are L-synchronized, then

$$\sum_{i=1}^{a} |Nf_i - Ng_i|_{C^0} \leq K\varepsilon.$$

Proof of Lemma 7.7
Let $a_\varepsilon = \lfloor \frac{1}{\varepsilon} \rfloor$. Assume for a moment that $a \geq a_\varepsilon$. Then Lemma 6.4 implies that $|x_{a-a_\varepsilon} - x_{a_\varepsilon}|, |\tilde{x}_{a-a_\varepsilon} - \tilde{x}_{a_\varepsilon}| \leq K\varepsilon$. Hence,

$$\begin{aligned}\sum_{a_\varepsilon \leq i \leq a-a_\varepsilon} |Nf_i - Ng_i|_{C^0} &\leq \sum_{a_\varepsilon \leq i \leq a-a_\varepsilon} |Nf_i|_{C^0} + |Ng_i|_{C^0} \\ &\leq \sum_{a_\varepsilon \leq i \leq a-a_\varepsilon} |Nf|_{C^0} \cdot |I_i| + |Ng|_{C^0} \cdot |\tilde{I}_i| \\ &\leq K\big(|x_{a-a_\varepsilon} - x_{a_\varepsilon}| + |\tilde{x}_{a-a_\varepsilon} - \tilde{x}_{a_\varepsilon}|\big) \\ &\leq K\varepsilon. \end{aligned} \tag{7.1}$$

These estimates hold trivially when $a < a_\varepsilon$. Observe that

$$\begin{aligned}\sum_{i=1}^{a} |Nf_i - Ng_i|_{C^0} &= \sum_{i=1}^{a_\varepsilon} |Nf_i - Ng_i|_{C^0} + \sum_{i=a_0}^{a-a_\varepsilon} |Nf_i - Ng_i|_{C^0} \\ &\quad + \sum_{i=a-a_\varepsilon}^{a} |Nf_i - Ng_i|_{C^0}.\end{aligned}$$

The middle term is estimated by (7.1). The first (and third) term can be estimated by using Lemma 7.6, Yoccoz's lemma (Lemma 4.2), and Propositions 6.18 and 6.19. Namely,

$$\begin{aligned}\sum_{i=1}^{a_\varepsilon} |Nf_i - Ng_i|_{C^0} &\leq K \sum_{i=1}^{a_\varepsilon} \varepsilon|I_i| + \Delta_i + |\Delta x_i||I_i| \\ &\leq K \sum_{i=1}^{a_\varepsilon} \varepsilon\frac{1}{i^2} + \varepsilon \cdot \frac{\log i}{i^2} + \varepsilon^2 \cdot \frac{1}{i} + \varepsilon \cdot \frac{1}{i^3} \\ &\leq K\varepsilon + K \sum_{i=1}^{a_\varepsilon} \varepsilon^2 \cdot \frac{1}{i} \\ &\leq K\varepsilon + K\varepsilon^2 \log \frac{1}{\varepsilon} \leq K\varepsilon.\end{aligned}$$

The lemma follows. □

The following proposition holds for general critical commuting pairs with negative Schwarzian which are contained in the previously discussed set $\mathcal{K}$; that is, the set of normalized C^3 critical commuting pairs which are K-controlled.

PROPOSITION 7.8
For every $L > 0$, there exists $K = K(\mathcal{K}, L) > 0$ such that the following holds. If ζ_0 and ζ_1 are L-synchronized, then

$$d_2\big(p\mathcal{R}(\zeta_0), p\mathcal{R}(\zeta_1)\big) \le K d_2(\zeta_0, \zeta_1).$$

Proof of Proposition 7.8
There exists $K = K(\mathcal{K}) > 0$ such that the following holds. There exists a diffeomorphism $h : \mathrm{Dom}(\zeta_1) \to \mathrm{Dom}(\zeta_0)$ such that $\zeta = \zeta_0$ and $\tilde{\zeta} = h \circ \zeta_1 \circ h^{-1}$ satisfy the normalizations

(1) $I_\eta = I_{\tilde{\eta}}$ and $I_\xi = I_{\tilde{\xi}}$,
(2) $p = \tilde{p}$ where $D\eta(p) = D\tilde{\eta}(\tilde{p}) = 1$,

needed to apply the results from Sections 6 and 7. We may construct the conjugation such that

$$d_{C^3}(h, \mathrm{id}) \le K d_2(\zeta_0, \zeta_1)$$

and $h|\mathrm{Dom}(p\mathcal{R}(\zeta_1)) = \mathrm{id}$. This last condition implies that

$$p\mathcal{R}(\zeta_1) = p\mathcal{R}(\tilde{\zeta}).$$

In particular, it suffices to prove the proposition for the pairs ζ and $\tilde{\zeta}$.

Let $p\mathcal{R}(\zeta) = (\eta', \xi')$ and $p\mathcal{R}(\tilde{\zeta}) = (\tilde{\eta}', \tilde{\xi}')$. Because $\xi' = \xi$ and $\tilde{\xi}' = \tilde{\xi}$, it suffices to estimate the distance between η' and $\tilde{\eta}'$.

Let $I_{a+1} = [x_{a+1}, x_a]$, and let $A : [0, 1] \to I_{a+1}$ be the orientation-preserving affine diffeomorphism. Let

$$F = A^{-1} \circ \eta',$$

and similarly define $G = \tilde{A}^{-1} \circ \tilde{\eta}'$. Now apply Lemmas 7.2 and 7.7 to obtain

$$d_{C^2}(F, G) \le K\varepsilon,$$

where $\varepsilon = d_2(\zeta, \tilde{\zeta})$. By using a similar argument as in the proof of Lemma 7.5, one obtains $d_2(\eta', \tilde{\eta}') \le K\varepsilon$. This shows that prerenormalization is Lipschitz among synchronized pairs. □

8. Order

Commuting pairs might have different domains. Any natural definition of *order* between such systems has to include this difference of domains also. There are two cases:

$$\text{Case I:} \quad \eta \circ \xi(0) > 0, \qquad \text{Case II:} \quad \eta \circ \xi(0) < 0.$$

Definition 8.1

Let $\zeta_0 = (\xi_0, \eta_0)$ and $\zeta_1 = (\xi_1, \eta_1)$ be two commuting pairs, and let $t \geq 0$. If

(1) $\zeta_0(x) + t \leq \zeta_1(x)$ for $x \in \mathrm{Dom}(\zeta_0) \cap \mathrm{Dom}(\zeta_1)$ and

(2) $\eta_0(0) \leq \eta_1(0)$ and $\xi_0(0) \leq \xi_1(0)$,

we write

$$\zeta_0 \preceq_t \zeta_1.$$

LEMMA 8.2

Let $\zeta_0 = (\xi_0, \eta_0)$ and $\zeta_1 = (\xi_1, \eta_1)$ be two commuting pairs such that $\zeta_0 \preceq_t \zeta_1$.

If Case I holds, then

(1) $a_{\zeta_0} \leq a_{\zeta_1}$,

(2) *for* $x \in [\eta_1(0), 0]$ *and* $k = 0, 1, \ldots, a_{\zeta_0}$,

$$\eta_0^k \circ \xi_0(x) + t \leq \eta_1^k \circ \xi_1(x).$$

If Case II holds, then

(1) $a_{\zeta_0} \geq a_{\zeta_1}$,

(2) *for* $x \in [0, \xi_0(0)]$ *and* $k = 0, 1, \ldots, a_{\zeta_1}$,

$$\xi_0^k \circ \eta_0(x) + t \leq \xi_1^k \circ \eta_1(x).$$

The proof of Lemma 8.2 is different for Case I and Case II. We will only present the proof in Case I.

Proof of Lemma 8.2

Let $x \in [0, \xi_0(0)]$. The order condition of Definition 8.1(1) gives the statement of the lemma for $k = 0$, $\xi_0(x) + t \leq \xi_1(x)$. Property (2) follows inductively. Namely,

$$\begin{aligned}\eta_0^{k+1} \circ \xi_0(x) + t &= \eta_0\big(\eta_0^k \circ \xi_0(x)\big) + t \leq \eta_1\big(\eta_0^k \circ \xi_0(x)\big) \\ &\leq \eta_1\big(\eta_1^k \circ \xi_1(x)\big) \\ &= \eta_1^{k+1} \circ \xi_1(x).\end{aligned}$$

In particular, $\eta_0^{a_{\zeta_0}} \circ \xi_0(x) \leq \eta_1^{a_{\zeta_0}} \circ \xi_1(x)$. This implies that $a_{\zeta_0} \leq a_{\zeta_1}$. □

Prerenormalization preserves order. Namely, we have the following.

LEMMA 8.3
If $\zeta_0 \leq_t \zeta_1$ and $a_{\zeta_0} = a_{\zeta_1}$, then $p\mathcal{R}(\zeta_0) \leq_t p\mathcal{R}(\zeta_1)$.

Proof of Lemma 8.3
We will only present the proof in Case I. Let $a = a_{\zeta_0} = a_{\zeta_1}$. Observe that $\eta_{p\mathcal{R}(\zeta_0)}(0) = \eta_0(0) \leq \eta_1(0) = \eta_{p\mathcal{R}(\zeta_1)}(0)$. Hence, the left-hand side of the domains of the prerenormalizations satisfies the order condition of Definition 8.1(2). Consider the right-hand side of the domains of the prerenormalizations,

$$\xi_{p\mathcal{R}(\zeta_0)}(0) + t = \eta_0^a \circ \xi_0(0) + t \leq \eta_1^a \circ \xi_1(0) = \xi_{p\mathcal{R}(\zeta_1)}(0), \tag{8.1}$$

where we used Lemma 8.2(2). This means that the right-hand side of the domains of the prerenormalizations also satisfies the order condition of Definition 8.1(2). According to Lemma 8.2(2), the estimate (8.1) also holds for any $x \in [\eta_1(0), 0]$, instead of $x = 0$. This means that the prerenormalization also satisfies the order condition of Definition 8.1(1). □

The following proposition will play a key role in the proof of the synchronization lemma (see Section 9 below).

PROPOSITION 8.4
If $\zeta_0 \leq_t \zeta_1$ with $t > 0$, then

$$\rho_{\zeta_0} \neq \rho_{\zeta_1}.$$

Proof of Proposition 8.4
Assume that $a_{\zeta_0}(n) = a_{\zeta_1}(n)$ for $n \geq 0$. Applying Lemma 8.3, we have

$$(p\mathcal{R})^n(\zeta_0) \leq_t (p\mathcal{R})^n(\zeta_1).$$

Note that $\eta_{(p\mathcal{R})^n(\zeta_{0,1})}(0) \to 0$. Hence,

$$0 > \eta_{(p\mathcal{R})^n(\zeta_1)}(0) \geq \eta_{(p\mathcal{R})^n(\zeta_0)}(0) + t \geq \frac{1}{2}t > 0$$

for n large enough, which is a contradiction. □

9. Synchronization

In the next statement we refer to the constant $\varepsilon_0 > 0$ obtained in Section 6 (see in particular the proof of Claim 6.7, during the proof of Lemma 6.6).

SYNCHRONIZATION-LEMMA
For any given $K_0 > 1$, there exists $L = L(K_0) > 1$ such that the following holds. Let ζ_0 and ζ_1 be two C^3 critical commuting pairs which are K_0-controlled, let both ζ_0 and ζ_1 have negative Schwarzian, and let $\rho(\zeta_0) = \rho(\zeta_1) \in [0,1] \setminus \mathbb{Q}$ and $d_2(\zeta_0, \zeta_1) < \varepsilon_0$. Then ζ_0 and ζ_1 are L-synchronized.

We omit mention of the hypothesis $d_2(\zeta_0, \zeta_1) < \varepsilon_0$ in the proof presented below, but it is needed to allow us to apply the estimates obtained in Sections 6–8.

Proof of the synchronization lemma
We will only present the proof in Case I. Let $a = a_{\zeta_0} = a_{\zeta_1}$. Choose $a_0 \geq 1$ such that Lemma 5.7 applies. The synchronization lemma follows from Lemma 6.10 when $a \leq a_0$. We will assume that $a \geq a_0$.

We may assume that $x_a^1 \geq x_a^0$. There exists $K = K(K_0) > 0$ such that the following holds. There exists a diffeomorphism $h : \mathrm{Dom}(\zeta_1) \to \mathrm{Dom}(\zeta_0)$ such that $\zeta = \zeta_0$ and $\tilde{\zeta} = h \circ \zeta_1 \circ h^{-1}$ satisfy the normalizations

$$x_1(\zeta) = x_1(\tilde{\zeta}).$$

We may construct the conjugation such that

$$d_{C^3}(h, \mathrm{id}) \leq K d_2(\zeta_0, \zeta_1)$$

and $h | \mathrm{Dom}(p\mathcal{R}(\zeta_1)) = \mathrm{id}$. This last condition implies that

$$x_a(\zeta_1) = x_a(\tilde{\zeta}).$$

In particular, it suffices to prove synchronization for the pairs ζ and $\tilde{\zeta}$. Let $\varepsilon = d_2(\zeta, \tilde{\zeta}) \leq K d_2(\zeta_0, \zeta_1)$.

Apply Lemma 5.7 to obtain a commuting pair ζ_{t_0} in the standard family of ζ such that

$$\Delta x_a(\zeta_{t_0}, \tilde{\zeta}) = 0.$$

From Lemma 5.7, we get

$$0 \leq t_0 \leq K\varepsilon. \tag{9.1}$$

Note that if $t_0 > 0$ is much larger than $\varepsilon \geq d_{C^0}(\zeta, \tilde{\zeta})$, then $\xi_{t_0}(x) > \tilde{\xi}(x)$. This would imply that $x_a(\zeta_{t_0}) > x_a(\tilde{\zeta})$ because $x_1(\zeta) = x_1(\tilde{\zeta})$. Assume that

$$\tilde{x}_a = x_a + L\varepsilon, \tag{9.2}$$

where just as before $x_i = \eta^i(\xi(0))$ and $\tilde{x}_i = \tilde{\eta}^i(\tilde{\xi}(0))$ for $i \in \{0, \dots, a\}$. Note also that the assumption $x_a^1 \geq x_a^0$ implies that $\tilde{x}_a \geq x_a$.

We have to show that L is uniformly bounded. From (9.2) and Corollary 5.9 we get for every $x \in [\eta_{p\mathcal{R}(\zeta_{t_0})}(0), 0]$,

$$\begin{aligned} p\mathcal{R}(\zeta_{t_0})(x) - p\mathcal{R}(\zeta)(x) &\geq \frac{1}{K}\big(p\mathcal{R}(\zeta_{t_0})(0) - p\mathcal{R}(\zeta)(0)\big) \\ &= \frac{1}{K}\big(p\mathcal{R}(\tilde{\zeta})(0) - p\mathcal{R}(\zeta)(0)\big) \\ &= \frac{1}{K}(\tilde{x}_a - x_a) = \frac{1}{K}L\varepsilon. \end{aligned} \tag{9.3}$$

From Proposition 7.8 we get for every $x \in [\eta_{p\mathcal{R}(\zeta_{t_0})}(0), 0]$,

$$\begin{aligned} \big|p\mathcal{R}(\zeta_{t_0})(x) - p\mathcal{R}(\tilde{\zeta})(x)\big| &\leq K d_2(\zeta_{t_0}, \tilde{\zeta}) \\ &\leq K d_2(\zeta, \tilde{\zeta}) + K\varepsilon \\ &\leq K\varepsilon, \end{aligned} \tag{9.4}$$

where we also used (9.1). Combine (9.3) and (9.4) to get for every $x \in [\eta_{p\mathcal{R}(\zeta_{t_0})}(0), 0]$,

$$p\mathcal{R}(\tilde{\zeta})(x) \geq p\mathcal{R}(\zeta)(x) + \frac{1}{K}L\varepsilon - K\varepsilon. \tag{9.5}$$

As a matter of fact, (9.5) holds for $x \in [-1, 0]$. This follows from the following. Let $x \in [-1, \eta_{p\mathcal{R}(\zeta_{t_0})}(0)]$. Observe that, according to (9.1),

$$\big|[-1, \eta_{p\mathcal{R}(\zeta_{t_0})}(0)]\big| = t_0 \leq K\varepsilon.$$

This implies that

$$\begin{aligned} p\mathcal{R}(\tilde{\zeta})(x) &\geq p\mathcal{R}(\zeta)\big(\eta_{p\mathcal{R}(\zeta_{t_0})}(0)\big) + \frac{1}{K}L\varepsilon - K\varepsilon - \max\{Dp\mathcal{R}(\tilde{\zeta})\}t_0 \\ &\geq p\mathcal{R}(\zeta)(x) + \frac{1}{K}L\varepsilon - K\varepsilon. \end{aligned}$$

Hence, for $x \in [-1, 0]$ we have

$$p\mathcal{R}(\tilde{\zeta})(x) \geq p\mathcal{R}(\zeta)(x) + \frac{1}{K}L\varepsilon - K\varepsilon. \tag{9.6}$$

So, when $L \geq 2K^2$, then for the relevant $x < 0$,

$$(p\mathcal{R})^2(\tilde{\zeta})(x) > (p\mathcal{R})^2(\zeta)(x). \tag{9.7}$$

The last part of the proof will show that similar estimates hold for relevant positive points. The goal is to prove that $(p\mathcal{R})^2(\tilde{\zeta}) \geq_t (p\mathcal{R})^2(\zeta)$ for some positive t. The branches on the left-hand side of the second prerenormalizations, according to (9.7), satisfy the order condition of Definition 8.1(1). The right-hand side of the domains of the second prerenormalizations satisfy the order condition of Definition 8.1(2). Namely,

$$\mathrm{Dom}\big((p\mathcal{R})^2(\zeta)\big) \cap \{x \geq 0\} = [0, x_a] \subset [0, \tilde{x}_a] = \mathrm{Dom}\big((p\mathcal{R})^2(\tilde{\zeta})\big) \cap \{x \geq 0\}.$$

What remains is to describe the branches on the right-hand side and the domains on the left-hand side. Let $x \in \mathrm{Dom}((p\mathcal{R})^2(\zeta)) \cap \{x \geq 0\} = [0, x_a]$, and for $k \geq 1$ define

$$z_k(x) = \big(p\mathcal{R}(\zeta)\big)^k(x),$$

and similarly, $\tilde{z}_k(x) = (p\mathcal{R}(\tilde{\zeta}))^k(x)$. Observe that

$$\big|z_1(x) - \tilde{z}_1(x)\big| = \big|p\mathcal{R}(\zeta)(x) - p\mathcal{R}(\tilde{\zeta})(x)\big| = \big|\eta(x) - \tilde{\eta}(x)\big| \leq \varepsilon.$$

Hence, applying (9.6),

$$\begin{aligned}
\tilde{z}_2(x) &= p\mathcal{R}(\tilde{\zeta})(\tilde{z}_1) \\
&\geq p\mathcal{R}(\zeta)(\tilde{z}_1) + \frac{1}{K}L\varepsilon - K\varepsilon \\
&\geq z_2(x) - \max\big(Dp\mathcal{R}(\zeta)\big) \cdot \big|z_1(x) - \tilde{z}_1(x)\big| + \frac{1}{K}L\varepsilon - K\varepsilon \\
&\geq z_2(x) + \frac{1}{K}L\varepsilon - K\varepsilon > z_2(x),
\end{aligned}$$

when $L \geq 2K^2$. Let $b = a_{p\mathcal{R}(\zeta)} = a_{p\mathcal{R}(\tilde{\zeta})}$. By repeatedly applying (9.6) with $L \geq 2K^2$, we obtain

$$(p\mathcal{R})^2(\tilde{\zeta})(x) = \tilde{z}_b(x) > (p\mathcal{R})^2(\zeta)(x) = z_b(x). \tag{9.8}$$

In particular,

$$\begin{aligned}
\mathrm{Dom}\big((p\mathcal{R})^2(\tilde{\zeta})\big) \cap \{x \leq 0\} &= [\tilde{z}_b, 0] \subset [z_b, 0] \\
&= \mathrm{Dom}\big((p\mathcal{R})^2(\zeta)\big) \cap \{x \leq 0\}.
\end{aligned} \tag{9.9}$$

The estimates (9.8) and (9.9) finish the proof of

$$(p\mathcal{R})^2(\tilde{\zeta}) \geq_t (p\mathcal{R})^2(\zeta),$$

for some $t > 0$. However, this contradicts Proposition 8.4 because $(p\mathcal{R})^2(\tilde{\zeta})$ and $(p\mathcal{R})^2(\zeta)$ have the same rotation number. This contradiction establishes the synchronization with $L \leq 2K^2$. □

10. Lipschitz estimate

In this section we prove Lemma 4.1.

Proof of Lemma 4.1

According to the synchronization lemma from Section 9, we know that for $L = L(\mathcal{K})$, the pairs ζ_0 and ζ_1 are L-synchronized. Now the Lipschitz estimate for renormalization of synchronized pairs, Proposition 7.8, implies a Lipschitz estimate for prerenormalization along topological classes. The fact that the maps are synchronized implies that the domains of the prerenormalizations are also close. This means that the normalizations will not affect the Lipschitz property. □

11. The attractor of renormalization

This section is devoted to the following result.

THEOREM 11.1

There exists a C^ω-compact set $\mathcal{K}$ of real analytic critical commuting pairs with the following property. For any $r \geq 3$ there exists a constant $\lambda = \lambda(r) \in (0,1)$ such that, given a C^r critical circle map f with irrational rotation number, there exist $C > 0$ and a sequence $\{f_n\}_{n\in\mathbb{N}}$ contained in $\mathcal{K}$ such that

$$d_{r-1}\big(\mathcal{R}^n(f), f_n\big) \leq C\lambda^n \quad \textit{for all } n \in \mathbb{N},$$

and such that the pair f_n has the same rotation number as the pair $\mathcal{R}^n(f)$ for all $n \in \mathbb{N}$. Here d_{r-1} denotes the C^{r-1}-distance in the space of C^{r-1} critical commuting pairs.

We will apply Theorem 11.1 in the next section (Section 12, during the proof of Theorem B) with $r = 4$. The compact set $\mathcal{K}$ and the approximations $\{f_n\}_{n\in\mathbb{N}}$ given by Theorem 11.1 were constructed in [11], but the exponential convergence was only proved for the C^0-metric (see [11, Theorem D]). In this section we will show that the same estimate actually holds for the C^{r-1}-metric, whenever f is C^r. For this purpose, we will use the following fact from complex analysis.

PROPOSITION 11.2

Let I be a compact interval in the real line with nonempty interior, and let U be an open set in the complex plane containing I. Fix $M > 0$, and consider the family

$$\mathcal{F} = \big\{f : U \to \mathbb{C} \textit{ holomorphic: } \|f\|_{C^0(U)} \leq M\big\}.$$

Then for any $r \in \mathbb{N}$ and any $\alpha \in (0,1)$, there exists $L = L(r, \alpha, M) > 0$ such that

$$\|f\|_{C^r(I)} \leq L\big(\|f\|_{C^0(I)}\big)^{\alpha} \quad \textit{for all } f \in \mathcal{F},$$

where

$$\|f\|_{C^r(I)} = \sup_{\substack{z\in I\\ n\in\{0,1,\dots,r\}}} \big\{\big|f^{(n)}(z)\big|\big\}.$$

Remark 11.3
We mention that both components of each approximation f_n constructed in [11] have holomorphic extensions satisfying the conditions of Proposition 11.2 (uniformly bounded on a definite domain; see [11, Section 7] for the construction).

In this section we explain how Theorem 11.1 follows by combining [3, Appendix A] and [11, Theorem D] with Proposition 11.2. We postpone the proof of Proposition 11.2 until the Appendix. In the remainder of this section we assume, to simplify the exposition, that the criticality of the critical point is 3, that is, $d = 1$ in Definition 2.3(4).

Definition 11.4
Let $I = [0,a]$, and let $g : I \to g(I) = J$ be a real analytic orientation-preserving homeomorphism with a cubic critical point at zero. We say that g is an *Epstein* map if there exist a topological disk $U \supset I$, an open interval $L \supset J$, and a holomorphic three-fold branched covering map $G : U \to \mathbb{C}_L$ such that $G|_I = g$ (as usual, $\mathbb{C}_L$ denotes the open set $\mathbb{C} \setminus (\mathbb{R} \setminus L)$).

For any $\beta \in (0,1)$, denote by $\mathcal{E}_\beta$ the set of Epstein maps $g : I = [0,a] \to g(I) = J$ satisfying the following properties:
(1) $\beta \leq |I|/|J| \leq \beta^{-1}$;
(2) $\operatorname{dist}(I,J) \leq \beta^{-1}|J|$, where dist denotes the standard distance between compact sets in the real line;
(3) $g'(a) > \beta$;
(4) the length of each component of $L \setminus J$ is at least $\beta|J|$ and at most $\beta^{-1}|J|$.
In order to apply Proposition 11.2, we will need the following fact.

PROPOSITION 11.5
For any $\beta \in (0,1)$ there exist a Jordan domain $U_\beta \supset [0,1]$ and a positive constant M_β such that for any $g \in \mathcal{E}_\beta$, with normalization $I = [0,1]$, the holomorphic extension G given by Definition 11.4 is well defined in U_β and satisfies $|G(z)| \leq M_\beta$ for all $z \in U_\beta$.

Proof of Proposition 11.5
Note that it is enough to prove the result for any sequence in $\mathcal{E}_\beta$. Let $g_n : I = [0,1] \to J_n \subset L_n$ be a normalized sequence in the class $\mathcal{E}_\beta$. By Definition 11.4, these maps extend to triple branched coverings $g_n : U_n \to \mathbb{C}_{L_n}$, where U_n is a topological disk. Therefore, each g_n can be decomposed as $g_n = Q_{c_n} \circ h_n$, where $Q_{c_n}(z) = z^3 + c_n$ and $h_n : [0,1] \to [0, b_n]$ with $h_n([0,1]) = [0, b_n]$ is a univalent map $h_n : U_n \to \Omega(h_n)$ onto the complex plane with six slits, which triply covers $\mathbb{C}_{L_n}$ under Q_{c_n}. With this notation, $J_n = [c_n, c_n + b_n^3]$; in particular, $b_n^3 = |J_n| \le |I|/\beta$ and then the sequence of positive numbers $\{b_n = h_n(1)\}$ is bounded. Moreover, since the length of each component of $L_n \setminus J_n$ is at least $\beta|J_n| \ge \beta^2|I|$, there exists $\varepsilon > 0$ such that $\overline{B(w,\varepsilon)} \subset \Omega(h_n)$ for all $w \in [0, b_n]$ and for all $n \in \mathbb{N}$.

CLAIM 11.6
There exists $\lambda > 0$ such that $|(h_n^{-1})'(w)| > \lambda$ for all $w \in [0, b_n]$ and for all $n \in \mathbb{N}$.

Proof of Claim 11.6
We claim first that, by passing to a subsequence if necessary, the sequence of marked domains $(\Omega(h_n), 0)$ converges to some marked domain $(\Omega, 0)$ in the Carathéodory topology (for the definition of the Carathéodory topology, see McMullen [20, Chapter 5]).

Indeed, note that $\operatorname{dist}([0,1], J_n) \le |J_n|/\beta \le |I|/\beta^2$, and then the intervals J_n accumulate on an interval J. Moreover, since $|J_n| \ge \beta|I|$, the interval J has nonempty interior. Since the length of each component of $L_n \setminus J_n$ is at least $\beta|J_n| \ge \beta^2|I|$, the intervals L_n accumulate on an open interval L that contains J. Moreover, the length of L is finite since the length of each component of $L_n \setminus J_n$ is at most $\beta^{-1}|J_n| \le \beta^{-2}|I|$. Since c_n is the left boundary point of J_n, and $\operatorname{dist}([0,1], J_n) \le |I|/\beta^2$, the sequence $\{c_n\}$ is bounded, and then (by passing to a subsequence if necessary) the sequence $(\Omega(h_n), 0)$ converges to some $(\Omega, 0)$ in the Carathéodory topology.

Second, we claim that the sequence of biholomorphisms $\{h_n^{-1} : \Omega(h_n) \to U_n\}$ is normal in Ω (note that any compact subset of Ω is eventually contained in $\Omega(h_n)$, and then h_n is well defined on it; again see [20, Chapter 5] for more on the Carathéodory topology). Indeed, from $Q_{c_n} = g_n \circ h_n^{-1}$ we get

$$(h_n^{-1})'(b_n) = \frac{Q'_{c_n}(b_n)}{g'_n(1)} = \frac{3b_n^2}{g'_n(1)},$$

and then $|(h_n^{-1})'(b_n)| < (3/\beta)|b_n|^2$ is bounded (since b_n is bounded, as we already have seen). As we said before, since the length of each component of $L_n \setminus J_n$ is at least $\beta|J_n| \ge \beta^2|I|$, the points b_n stay away from the boundary of Ω, that is, $\inf_{n\in\mathbb{N}}\{d([0,b_n], \partial\Omega)\} \ge \varepsilon > 0$. Koebe's distortion theorem (Theorem 2.10) then implies that the family $\{h_n^{-1}\}$ is normal in Ω. Since b_n is bounded from above, any limit function is nonconstant and therefore univalent. In particular, it has no critical points, and this completes the proof of Claim 11.6. □

With Claim 11.6 at hand and Koebe's one-quarter theorem (Theorem 2.11), we obtain

$$B\big(h_n^{-1}(w), \lambda\varepsilon/4\big) \subset (h_n^{-1})\big(B(w,\varepsilon)\big)$$

for all $w \in [0, b_n]$ and for all $n \in \mathbb{N}$. Let $b = \sup_{n\in\mathbb{N}}\{b_n\}$, and consider the two bounded Jordan domains

$$U_\beta = \bigcup_{z\in[0,1]} B(z, \lambda\varepsilon/4) \qquad \text{and} \qquad V_\beta = \bigcup_{w\in[0,b]} B(w,\varepsilon).$$

For each $n \in \mathbb{N}$, we have seen that h_n is well defined in U_β and satisfies $h_n(U_\beta) \subset V_\beta$. Since the sequence $\{c_n\}$ is bounded (as we pointed out before in the proof of Claim 11.6), each g_n is (well defined and) uniformly bounded in U_β. This completes the proof of Proposition 11.5. □

11.1. Proof of Theorem 11.1

The proof of Theorem 11.1 given below relies on the following result of de Faria and de Melo [3, Theorem A.6, Appendix A, p. 382].

THEOREM 11.7

There exist $\beta \in (0,1)$ and $\lambda \in (0,1)$ with the following property. Let f be a C^r critical circle map with irrational rotation number, for some $r \geq 3$. For each $n \in \mathbb{N}$, denote by $\tilde{\xi}_n$ and $\tilde{\eta}_n$ the components of the normalized critical commuting pair $\mathcal{R}^n(f)$, as defined in Section 2.2. Then there exists $C > 0$ such that for each $n \in \mathbb{N}$ there exist η_n and ξ_n in $\mathcal{E}_\beta$ such that ξ_n has the same domain as $\tilde{\xi}_n$, η_n has the same domain as $\tilde{\eta}_n$, and moreover,

$$\|\xi_n - \tilde{\xi}_n\|_{C^{r-1}([-1,0])} \leq C\lambda^n \qquad \textit{and} \qquad \|\eta_n - \tilde{\eta}_n\|_{C^{r-1}([0,|I_n|/|I_{n+1}|])} \leq C\lambda^n.$$

Unfortunately, the pair (η_n, ξ_n) given by Theorem 11.7 is *not* a *commuting* pair in general. In particular, we have no information on the behavior of the renormalization operator acting on these pairs.

Proof of Theorem 11.1

By [11, Theorem D], there exists a C^ω-compact set $\mathcal{K}$ of real analytic critical commuting pairs and a constant $\lambda_0 \in (0,1)$ with the following property. Given a C^r critical circle map f, $r \geq 3$, with any irrational rotation number, there exist $C_0 > 0$ and a sequence $\{f_n\}_{n\in\mathbb{N}}$ contained in $\mathcal{K}$ such that

$$d_0\big(\mathcal{R}^n(f), f_n\big) \leq C_0\lambda_0^n \quad \text{for all } n \in \mathbb{N},$$

and such that the pair f_n has the same rotation number as the pair $\mathcal{R}^n(f)$ for all $n \in \mathbb{N}$. From Theorem 11.7, we obtain $C_1 > 0$, $\lambda_1 \in (0,1)$, and a sequence $\{(\eta_n, \xi_n)\}_{n\in\mathbb{N}}$ such that

$$d_{r-1}\big(\mathcal{R}^n(f), (\eta_n, \xi_n)\big) \leq C_1\lambda_1^n \quad \text{for all } n \in \mathbb{N}.$$

By considering $C_2 = 2\max\{C_0, C_1\} > 0$ and $\lambda_2 = \max\{\lambda_0, \lambda_1\} \in (0,1)$, we get that

$$d_0\big(f_n, (\eta_n, \xi_n)\big) \leq C_2\lambda_2^n \quad \text{for all } n \in \mathbb{N}.$$

By Proposition 11.2, Remark 11.3, and Proposition 11.5, the C^{r-1}-metric is Hölder-equivalent to the C^0-metric for the sequences $\{f_n\}$ and $\{(\eta_n, \xi_n)\}$. More precisely,

$$d_{r-1}\big(f_n, (\eta_n, \xi_n)\big) \leq C_3\lambda_3^n \quad \text{for all } n \in \mathbb{N},$$

where $C_3 = LC_2^\alpha > 0$ and $\lambda_3 = \lambda_2^\alpha \in (0,1)$, and the constants $L > 0$ and $\alpha \in (0,1)$ are given by Proposition 11.2. With this at hand, we finally obtain, by the triangle inequality, that

$$d_{r-1}\big(\mathcal{R}^n(f), f_n\big) \leq C_4\lambda_4^n \quad \text{for all } n \in \mathbb{N},$$

where $C_4 = 2\max\{C_1, C_3\} > 0$ and $\lambda_4 = \max\{\lambda_1, \lambda_3\} \in (0,1)$. □

12. Exponential convergence

In this section we prove the uniform exponential convergence of renormalization in the C^4-category (more precisely, we prove that Theorem 11.1 and Lemma 4.1 combined with Theorem 2.8 imply Theorem B).

Proof of Theorem B

Let f and g be two C^4 critical circle maps with the same irrational rotation number $\rho(f) = \rho(g) = [a_0, a_1, \ldots]$ and with the same order at their critical points. By Theorem 11.1, there exist a C^ω-compact set $\mathcal{K}_0$ of real analytic critical commuting pairs, two constants $\lambda_0 \in (0,1)$ and $C_0 > 1$, and two sequences $\{f_m\}_{m\in\mathbb{N}}$ and $\{g_m\}_{m\in\mathbb{N}}$ contained in $\mathcal{K}_0$ such that for all $m \in \mathbb{N}$ we have $\rho(f_m) = \rho(g_m) = [a_m, a_{m+1}, \ldots]$, and

$$d_3\big(\mathcal{R}^m(f), f_m\big) \leq C_0\lambda_0^m \qquad \text{and} \qquad d_3\big(\mathcal{R}^m(g), g_m\big) \leq C_0\lambda_0^m. \tag{12.1}$$

From [3, Theorem A.4, p. 379] we know that there exists $n_0 \in \mathbb{N}$ such that both critical commuting pairs $\mathcal{R}^n(f)$ and $\mathcal{R}^n(g)$ have negative Schwarzian bounded away from zero for all $n \geq n_0$. From Theorem 11.1 we also know that the closure for the C^3-metric of the orbit $\{\mathcal{R}^n(f)\}_{n \geq n_0}$ is a C^3-compact set that we denote by $\mathcal{K}_f$ (the ω-limit for the C^3-metric under renormalization of f is contained in $\mathcal{K}_0$, which is C^3-compact). Let $\mathcal{K}_g$ be the corresponding compact set for g, that is, $\mathcal{K}_g$ is the closure for the C^3-metric of the orbit $\{\mathcal{R}^n(g)\}_{n \geq n_0}$. By compactness, there exists $\beta \in (0,1)$ with the following property: any C^3 critical commuting pair ζ such that $d_{C^3}(\zeta, \mathcal{K}_f \cup \mathcal{K}_g) < \beta$ has negative Schwarzian.

From the real bounds (see Theorem 3.5) there exist a universal constant $K_0 > 1$ and $n_1 = n_1(f, g) \in \mathbb{N}$, with $n_1 > n_0$, such that the critical commuting pairs $\mathcal{R}^n(f)$ and $\mathcal{R}^n(g)$ are K_0-controlled for any $n \geq n_1$.

Let $n_2 = n_2(2K_0) \in \mathbb{N}$ be given by Theorem 3.6, and let $K = B^{n_2}(2K_0)$ be given by Lemma 3.7 (here, the power n_2 denotes iteration). Let $\varepsilon_0(K) \in (0,1)$ and $L(K) > 1$ be given by Lemma 4.1. Fix $\delta \in (0,1)$ such that $\delta > \frac{\log L}{\log L - \log \lambda_0}$. Let $\lambda_2 = L^{1-\delta}\lambda_0^{\delta}$, and note that $\lambda_2 \in (0,1)$ since $\delta \log \lambda_0 + (1-\delta)\log L < 0$. For each $n \in \mathbb{N}$, let $m = m(n) \in \mathbb{N}$ given by $m = \lfloor \delta n \rfloor$, and fix $m_0 \in \mathbb{N}$ such that

$$m_0 > \max\left\{\frac{n_2 \log L + \log C_0 + \log(1/\varepsilon_0)}{\log(1/\lambda_0)}, \frac{\log C_0 + \log(1/\beta)}{\log(1/\lambda_0)}, n_1\right\}.$$

From (12.1) we see that $d_2(\mathcal{R}^m(f), f_m) < \varepsilon_0$ and $d_3(\mathcal{R}^m(f), f_m) < \beta$ for all $m > m_0$, and also that $d_2(\mathcal{R}^m(g), g_m) < \varepsilon_0$ and $d_3(\mathcal{R}^m(g), g_m) < \beta$ for all $m > m_0$. In particular, both critical commuting pairs f_m and g_m are $2K_0$-controlled for all $m \geq m_0$, and moreover, both f_m and g_m have negative Schwarzian for all $m \geq m_0$ (and then the pairs $\mathcal{R}^j(f_m)$ and $\mathcal{R}^j(g_m)$ have negative Schwarzian for all $m \geq m_0$ and all $j \in \mathbb{N}$; see Remark 3.2).

By Theorem 3.6, the critical commuting pair $\mathcal{R}^j(f_m)$ is K_0-controlled for all $m > m_0$ and $j \geq n_2$. For $j \in \{0, 1, \dots, n_2\}$, we combine Lemma 4.1 with (12.1) to obtain that for all $m > m_0$,

$$d_2\big(\mathcal{R}^j(f_m), \mathcal{R}^{j+m}(f)\big) \leq L^j d_2\big(f_m, \mathcal{R}^m(f)\big) \leq L^j C_0 \lambda_0^m \leq L^{n_2} C_0 \lambda_0^m < \varepsilon_0.$$

In particular, $\mathcal{R}^j(f_m)$ is C^2-bounded by $2K_0$ for all $m > m_0$ and $j \in \{0, 1, \dots, n_2\}$, and therefore, $\mathcal{R}^j(f_m)$ is K-controlled for all $m > m_0$ and $j \in \{0, 1, \dots, n_2\}$ by Lemma 3.7. This allows us to combine Lemma 4.1 with (12.1) in order to obtain

$$\begin{aligned} d_2\big(\mathcal{R}^n(f), \mathcal{R}^{n-m}(f_m)\big) &\leq L^{n-m} \cdot d_2\big(\mathcal{R}^m(f), f_m\big) \\ &\leq C_0 L^{n-m} \lambda_0^m \\ &\leq \left(\frac{LC_0}{\lambda_0}\right)\lambda_2^n \end{aligned} \tag{12.2}$$

for all $n \in \mathbb{N}$ such that $m = \lfloor \delta n \rfloor > m_0$, since $L^{n-m-1}\lambda_0^{m+1} \leq (L^{1-\delta}\lambda_0^{\delta})^n = \lambda_2^n$. Let $C_3 = LC_0/\lambda_0$. Replacing f with g, we also get

$$d_2\big(\mathcal{R}^n(g), \mathcal{R}^{n-m}(g_m)\big) \leq C_3\lambda_2^n \quad \text{for all } n \in \mathbb{N} \text{ such that } m = \lfloor \delta n \rfloor > m_0. \tag{12.3}$$

Since f_m and g_m are real analytic and have the same combinatorics for each $m \in \mathbb{N}$, we know by Theorem 2.8 that there exist constants $C_1 > 1$ and $\lambda_1 \in (0,1)$ (both uniform in $\mathcal{K}_0$) such that

$$d_2\big(\mathcal{R}^{n-m}(f_m), \mathcal{R}^{n-m}(g_m)\big) \leq C_1\lambda_1^{n-m} \leq C_1(\lambda_1^{1-\delta})^n \tag{12.4}$$

for all $n \in \mathbb{N}$ such that $m = \lfloor \delta n \rfloor > m_0$. Finally, we define $\lambda = \max\{\lambda_1^{1-\delta}, \lambda_2\} = \max\{\lambda_1^{1-\delta}, L^{1-\delta}\lambda_0^{\delta}\} \in (0,1)$ and $C = C_1 + 2C_3 = C_1 + 2LC_0/\lambda_0 > 1$. Combining (12.2), (12.3), and (12.4), we get

$$d_2\big(\mathcal{R}^n(f), \mathcal{R}^n(g)\big) \leq C\lambda^n \quad \text{for all } n \in \mathbb{N} \text{ such that } m = \lfloor \delta n \rfloor > m_0. \qquad \square$$

13. Rigidity

As we said in the Introduction, the fact that Theorem B implies Theorem A follows from well-known results by de Faria and de Melo [3, First Main Theorem, p. 341] and Khanin and Teplinsky [14, Theorem 2, p. 198]. In this final section we just give more precise references.

Let f and g be two C^4 circle homeomorphisms with the same irrational rotation number and with a unique critical point of the same odd type. Let h be the unique topological conjugacy between f and g that maps the critical point of f to the critical point of g. Let $\{\mathcal{P}_n^f\}_{n\geq 1}$ and $\{\mathcal{P}_n^g\}_{n\geq 1}$ be the corresponding sequences of dynamical partitions (see Section 2.1), and note that the homeomorphism h identifies those partitions.

Khanin and Teplinsky [14, Section 3] proved that Theorem B implies the existence of two constants $\hat{C} > 0$ and $\hat{\lambda} \in (0,1)$ such that if I_f, J_f are adjacent atoms in $\mathcal{P}_n^f$, or they are contained in the same atom of $\mathcal{P}_{n-1}^f$, and if $I_g = h(I_f)$, $J_g = h(J_f)$ are the corresponding atoms in $\mathcal{P}_n^g$, then we have

$$\left|\log\frac{|I_g|}{|I_f|} - \log\frac{|J_g|}{|J_f|}\right| \leq \hat{C}\hat{\lambda}^n \quad \text{for all } n \geq 1. \tag{13.1}$$

Combining these estimates with the real bounds, it is not difficult to prove the first two conclusions of Theorem A (see [14, Proposition 1, p. 199] for Theorem A(1), and [14, Remark 5, p. 213] for Theorem A(2)). To prove Theorem A(3), however, it is not enough to have (13.1) for the dynamical partitions (indeed, note that (13.1) holds for Ávila's examples in [1] mentioned in the introduction).

De Faria and de Melo [3, Section 4] constructed suitable partitions $\{\mathcal{Q}_n^f\}_{n\geq 1}$ and $\{\mathcal{Q}_n^g\}_{n\geq 1}$ (the so-called *fine grids*; see [3, Sections 4.2, 4.3]) such that for a full

Lebesgue measure set of rotation numbers (see [3, Section 4.4] for its definition), Theorem B implies the existence of two constants $\tilde{C} > 0$ and $\tilde{\lambda} \in (0, 1)$ such that

$$\left| \frac{|I_f|}{|J_f|} - \frac{|I_g|}{|J_g|} \right| \leq \tilde{C} \tilde{\lambda}^n \quad \text{for all } n \geq 1, \tag{13.2}$$

for each pair of adjacent atoms I_f, J_f that belong to $\mathcal{Q}_n^f$. Estimate (13.2) is enough to prove that the derivative of the conjugacy h is Hölder-continuous on the whole circle (see [3, Proposition 4.3(b), p. 356]), which is precisely Theorem A(3).

Appendix. Proof of Proposition 11.2

In this appendix we prove Proposition 11.2. In the proof we follow the exposition of Lyubich [18, Lemma 11.5].

Proof of Proposition 11.2

Let V be a bounded Jordan domain containing the interval I, and compactly contained in U (as usual, a *Jordan domain* is an open, connected and simply connected set of the complex plane, whose boundary is a Jordan curve). Consider a continuous function $h : \overline{V} \to [0, 1]$ satisfying:

- h is harmonic and positive in the annulus $V \setminus I$,
- $h \equiv 0$ on ∂V and $h \equiv 1$ on I.

Recall that the existence of such a function h is a particular case of *Dirichlet's problem*.

To begin the proof, suppose first that $M = 1$, and let $f : U \to \mathbb{C}$ be a holomorphic function such that $|f(z)| \leq 1$ for all $z \in U$. Let $\varepsilon = \|f\|_{C^0(I)} \leq 1$, and note that

$$\log|f| \leq h \log \varepsilon \tag{A.1}$$

on $\partial(V \setminus I) = I \cup \partial V$. Since f is holomorphic, $\log|f|$ is harmonic where $f \neq 0$ and subharmonic in the whole domain V, and since h is harmonic in $V \setminus I$, we get from the maximum principle that inequality (A.1) also holds inside the annulus $V \setminus I$, that is, $|f(z)| \leq \varepsilon^{h(z)}$ for all $z \in V$. Given $\alpha \in (0, 1)$, let $W = \{z \in V : h(z) \in (\alpha, 1]\}$, and note that W is a Jordan domain containing I, compactly contained in V, and such that $h(z) = \alpha$ for all $z \in \partial W$. Since $\varepsilon \in [0, 1]$, we have $|f(z)| \leq \varepsilon^{h(z)} \leq \varepsilon^{\alpha}$ for all $z \in W$, that is,

$$\|f\|_{C^0(W)} \leq \left(\|f\|_{C^0(I)}\right)^{\alpha}.$$

The next step is just the standard application of Cauchy's integral formulas. Let $\rho \in (0, 1)$ be such that $\overline{B(z, \rho)} \subset W$ for all $z \in I$. Then for any $z \in I$ and any $n \in \{0, 1, \ldots, r\}$, we have

$$\begin{aligned}|f^{(n)}(z)| &= \left|\frac{n!}{2\pi i}\int_{\partial B(z,\rho)}\frac{f(w)}{(w-z)^{n+1}}\,dw\right| \\ &= \frac{n!}{2\pi}\left|\int_0^{2\pi}\frac{f(z+\rho e^{i\theta})}{(\rho e^{i\theta})^{n+1}}i\rho e^{i\theta}\,d\theta\right| \\ &\le \frac{n!}{2\pi}\frac{1}{\rho^n}\int_0^{2\pi}|f(z+\rho e^{i\theta})|\,d\theta \\ &\le \left(\frac{n!}{\rho^n}\right)\Big(\sup_{w\in\partial B(z,\rho)}\{|f(w)|\}\Big).\end{aligned}$$

Defining $L_1 = r!/\rho^r$, we obtain

$$\|f\|_{C^r(I)} \le L_1\|f\|_{C^0(W)} \le L_1\big(\|f\|_{C^0(I)}\big)^{\alpha}.$$

Therefore, Proposition 11.2 is true for the case $M = 1$. For the general case, note that for any $f \in \mathcal{F}$ we have $\|f\|_{C^r(I)} = M\|f/M\|_{C^r(I)} \le ML_1(\|f/M\|_{C^0(I)})^{\alpha} = M^{1-\alpha}L_1(\|f\|_{C^0(I)})^{\alpha}$, and therefore it is enough to consider $L = M^{1-\alpha}L_1$. □

Acknowledgments. We are grateful to Edson de Faria for several conversations about the subject. During the preparation of this article, the second author visited the Instituto de Matemática Pura e Aplicada and the first author visited Stony Brook University, and they thank both institutions for their hospitality.

Guarino's work was partially supported by Fundação de Amparo à Pesquisa do Estado de São Paulo grant 2012/06614-8, Fundação de Amparo à Pesquisa do Estado de Rio de Janeiro grant E-26/102.784/2012, and Coordenação de Aperfeicoamento de Pessoal de Nível Superior grant 23038.009189/2013-05. We also note that de Melo's work was partially supported by Conselho Nacional de Desenvolvimento Científico e Tecnológico grant 300813/2010-4.

References

[1] A. ÁVILA, *On rigidity of critical circle maps*, Bull. Braz. Math. Soc. (N.S.) **44** (2013), 611–619. MR 3167125. DOI 10.1007/s00574-013-0027-5. *(2126, 2183)*

[2] E. DE FARIA, *Asymptotic rigidity of scaling ratios for critical circle mappings*, Ergodic Theory Dynam. Systems **19** (1999), 995–1035. MR 1709428. DOI 10.1017/S0143385799133959. *(2126)*

[3] E. DE FARIA and W. DE MELO, *Rigidity of critical circle mappings, I*, J. Eur. Math. Soc. (JEMS) **1** (1999), 339–392. MR 1728375. DOI 10.1007/s100970050011. *(2126, 2127, 2132, 2133, 2140, 2143, 2151, 2159, 2178, 2180, 2181, 2183, 2184)*

[4] ———, *Rigidity of critical circle mappings, II*, J. Amer. Math. Soc. **13** (2000), 343–370. MR 1711394. DOI 10.1090/S0894-0347-99-00324-0. *(2126, 2128, 2138)*

[5] E. DE FARIA and P. GUARINO, *Real bounds and Lyapunov exponents*, Discrete Contin. Dyn. Syst. **36** (2016), 1957–1982. MR 3411549. DOI 10.3934/dcds.2016.36.1957. *(2130)*

[6] W. DE MELO, "Rigidity and renormalization in one-dimensional dynamical systems" in *Proceedings of the International Congress of Mathematicians, Vol. II (Berlin, 1998)*, Doc. Math. **1998**, 765–778. MR 1648124. *(2129)*

[7] W. DE MELO and A. A. PINTO, *Rigidity of C^2 infinitely renormalizable unimodal maps*, Comm. Math. Phys. **208** (1999), 91–105. MR 1729879. DOI 10.1007/s002200050749. *(2129)*

[8] W. DE MELO and S. VAN STRIEN, *One-Dimensional Dynamics*, Ergeb. Math. Grenzgeb. (3) **25**, Springer, Berlin, 1993. MR 1239171. DOI 10.1007/978-3-642-78043-1. *(2129, 2139, 2151)*

[9] M. J. FEIGENBAUM, L. P. KADANOFF, and S. S. SHENKER, *Quasiperiodicity in dissipative systems: A renormalization group analysis*, Phys. D **5** (1982), 370–386. MR 0680571. DOI 10.1016/0167-2789(82)90030-6. *(2126, 2129)*

[10] J. GRACZYK and G. ŚWIĄTEK, *Critical circle maps near bifurcation*, Comm. Math. Phys. **176** (1996), 227–260. MR 1374412. *(2132)*

[11] P. GUARINO and W. DE MELO, *Rigidity of smooth critical circle maps*, J. Eur. Math. Soc. (JEMS) **19** (2017), 1729–1783. MR 3646874. DOI 10.4171/JEMS/704. *(2126, 2128, 2177, 2178, 2180)*

[12] M.-R. HERMAN, *Sur la conjugaison différentiable des difféomorphismes du cercle à des rotations*, Publ. Math. Inst. Hautes Études Sci. **49** (1979), 5–233. MR 0538680. *(2127)*

[13] ———, *Conjugaison quasi-symétrique des homéomorphismes du cercle à des rotations*, preprint, http://www.math.kyoto-u.ac.jp/mitsu/Herman/qsconj1/index.html (accessed 6 June 2018). *(2132)*

[14] K. KHANIN and A. TEPLINSKY, *Robust rigidity for circle diffeomorphisms with singularities*, Invent. Math. **169** (2007), 193–218. MR 2308853. DOI 10.1007/s00222-007-0047-0. *(2126, 2127, 2183)*

[15] D. KHMELEV and M. YAMPOLSKY, *The rigidity problem for analytic critical circle maps*, Mosc. Math. J. **6** (2006), 317–351. MR 2270617. *(2126, 2128, 2138)*

[16] O. E. LANFORD, III, "Renormalization group methods for critical circle mappings with general rotation number" in *VIIIth International Congress on Mathematical Physics (Marseille, 1986)*, World Scientific, Singapore, 1987, 532–536. MR 0915597. *(2126, 2129)*

[17] ———, "Renormalization group methods for circle mappings" in *Nonlinear Evolution and Chaotic Phenomena (Noto, 1987)*, NATO Adv. Sci. Inst. Ser. B Phys. **176**, Plenum, New York, 1988, 25–36. MR 1109968. *(2126, 2129)*

[18] M. LYUBICH, *Feigenbaum-Coullet-Tresser universality and Milnor's hairiness conjecture*, Ann. of Math. (2) **149** (1999), 319–420. MR 1689333. DOI 10.2307/120968. *(2129, 2184)*

[19] M. MARTENS, *The periodic points of renormalization*, Ann. of Math. (2) **147** (1998), 543–584. MR 1637651. DOI 10.2307/120959. *(2166)*

[20] C. T. MCMULLEN, *Complex Dynamics and Renormalization*, Ann. of Math. Stud. **135**, Princeton Univ. Press, Princeton, 1994. MR 1312365. *(2179)*

[21] ———, "Rigidity and inflexibility in conformal dynamics" in *Proceedings of the International Congress of Mathematicians, Vol. II (Berlin, 1998)*, Doc. Math. **1998**, 841–855. MR 1648131. *(2129)*

[22] S. ÖSTLUND, D. RAND, J. SETHNA, and E. SIGGIA, *Universal properties of the transition from quasiperiodicity to chaos in dissipative systems*, Phys. D **8** (1983), 303–342. MR 0719630. DOI 10.1016/0167-2789(83)90229-4. *(2126, 2129)*

[23] D. RAND, *Global phase space universality, smooth conjugacies and renormalisation, I: The $C^{1+\alpha}$ case*, Nonlinearity **1** (1988), 181–202. MR 0928952. *(2126, 2129)*

[24] ———, "Universality and renormalisation in dynamical systems" in *New Directions in Dynamical Systems*, London Math. Soc. Lecture Note Ser. **127**, Cambridge Univ. Press, Cambridge, 1988, 1–56. MR 0953969. *(2126, 2129)*

[25] ———, *Existence, nonexistence and universal breakdown of dissipative golden invariant tori, I: Golden critical circle maps*, Nonlinearity **5** (1992), 639–662. MR 1166539. *(2126, 2129)*

[26] S. J. SHENKER, *Scaling behavior in a map of a circle onto itself: Empirical results*, Phys. D **5** (1982), 405–411. MR 0680574. DOI 10.1016/0167-2789(82)90033-1. *(2126, 2129)*

[27] G. ŚWIĄTEK, *Rational rotation numbers for maps of the circle*, Comm. Math. Phys. **119** (1988), 109–128. MR 0968483. *(2132)*

[28] M. YAMPOLSKY, *Complex bounds for renormalization of critical circle maps*, Ergodic Theory Dynam. Systems **19** (1999), 227–257. MR 1677153. DOI 10.1017/S0143385799120947. *(2126)*

[29] ———, *The attractor of renormalization and rigidity of towers of critical circle maps*, Comm. Math. Phys. **218** (2001), 537–568. MR 1828852. DOI 10.1007/PL00005561. *(2126)*

[30] ———, *Hyperbolicity of renormalization of critical circle maps*, Publ. Math. Inst. Hautes Études Sci. **96** (2002), 1–41. MR 1985030. DOI 10.1007/s10240-003-0007-1. *(2126)*

[31] ———, *Renormalization horseshoe for critical circle maps*, Comm. Math. Phys. **240** (2003), 75–96. MR 2004980. DOI 10.1007/s00220-003-0891-8. *(2126)*

[32] J.-C. YOCCOZ, *Conjugaison différentiable des difféomorphismes du cercle dont le nombre de rotation vérifie une condition diophantienne*, Ann. Sci. Éc. Norm. Supér. (4) **17** (1984), 333–359. MR 0777374. *(2127)*

[33] ———, *Il n'y a pas de contre-exemple de Denjoy analytique*, C. R. Math. Acad. Sci. Paris **298** (1984), 141–144. MR 0741080. *(2125)*

Guarino
Instituto de Matemática e Estatística, Universidade Federal Fluminense, Rio de Janeiro, Brazil;
pablo_guarino@id.uff.br

Martens
Department of Mathematics, Stony Brook University, Stony Brook, New York, USA;
marco@math.sunysb.edu

de Melo
Instituto de Matemática Pura e Aplicada, Rio de Janeiro, Brazil

List of Publications of Welington de Melo

Books

1. Zh. Palis and V. di Melu. Геометрическая теория динамических систем. Современная Математика: Вводные Курсы. [*Geometric theory of dynamical systems. Contemporary Mathematics: Introductory Courses*]. Translated from the English by V.N. Kolokoltsov, translation edited and with notes and an afterword by D.V. Anosov. "Mir", Moscow (1986).
2. Jacob Palis, Jr. and Welington de Melo. *Introdução aos sistemas dinâmicos*. Volume 6 of *Projeto Euclides* [Euclid Project], Instituto de Matemática Pura e Aplicada, Rio de Janeiro (1978).
3. Jacob Palis, Jr. and Welington de Melo. *Geometric theory of dynamical systems. An introduction*. Translated from the Portuguese by A.K. Manning. Springer-Verlag, New York-Berlin (1982).
4. Welington de Melo and Sebastian van Strien. *One-dimensional dynamics*. Volume 25 of Ergebnisse der Mathematik und ihrer Grenzgebiete (3) [Results in Mathematics and Related Areas (3)], Springer-Verlag, Berlin (1993).
5. Edson de Faria and Welington de Melo. *One dimensional dynamics*. Publicações Matemáticas do IMPA. [IMPA Mathematical Publications]. The mathematical tools, 23o Colóquio Brasileiro de Matemática [23rd Brazilian Mathematics Colloquium], Instituto de Matemática Pura e Aplicada (IMPA), Rio de Janeiro (2001).
6. Edson de Faria and Welington de Melo. *Mathematical tools for one-dimensional dynamics*. Volume 115 of Cambridge Studies in Advanced Mathematics, Cambridge University Press, Cambridge (2008).
7. Edson de Faria and Welington de Melo. *Mathematical aspects of quantum field theory*. With a foreword by Dennis Sullivan. 26o Colóquio Brasileiro de Matemática. [26th Brazilian Mathematics Colloquium]. Publicações Matemáticas do IMPA. [IMPA Mathematical Publications]. Instituto Nacional de Matemática Pura e Aplicada (IMPA), Rio de Janeiro (2007).

8. Edson de Faria and Welington de Melo. *Mathematical aspects of quantum field theory*. With a foreword by Dennis Sullivan. Volume 127 of Cambridge Studies in Advanced Mathematics, Cambridge University Press, Cambridge (2010).

Papers

1. Pablo Guarino, Marco Martens and Welington de Melo. Rigidity of critical circle maps. *Duke Math. J.* **167**(11), 2125–2188 (2018).
2. Pablo Guarino and Welington de Melo. Rigidity of smooth critical circle maps. *J. Eur. Math. Soc. (JEMS)* **19**(6), 1729–1783 (2017).
3. Welington de Melo, Bjorn Poonen, Jeremy Quastel and Anton Zorich. The work of the 2014 Fields medalists. *Notices Amer. Math. Soc.* **62**(11), 1334—1349 (2015).
4. Welington de Melo, Pedro A.S. Salomão and Edson Vargas. A full family of multimodal maps on the circle. *Ergodic Theory Dynam. Systems* **31**(5), 1325–1344 (2011).
5. W. De Melo. Renormalization in one-dimensional dynamics. *J. Difference Equ. Appl.* **17**(8), 1185–1197 (2011).
6. W. De Melo. Fine structure of hyperbolic diffeomorphisms [book review of MR2464147]. *Bull. Amer. Math. Soc. (N.S.)* **48**(1), 131–136 (2011).
7. W. de Melo. Dynamics on the circle. In: *Dynamics, games and science. I*, volume 1 of Springer Proc. Math., pp. 651–659, Springer, Heidelberg (2011).
8. V.V.M.S. Chandramouli, M. Martens, W. de Melo and C.P. Tresser. Chaotic period doubling. *Ergodic Theory Dynam. Systems* **29**(2), 381–418 (2009).
9. W. de Melo. Rigidity in dynamics. *Bull. Belg. Math. Soc. Simon Stevin* **15**(5: Dynamics in perturbations), 789–796 (2008).
10. Edson de Faria, Welington de Melo and Alberto Pinto. Global hyperbolicity of renormalization for C^r unimodal mappings. *Ann. of Math. (2)* **164**(3), 731–824 (2006).
11. Welington de Melo. Bifurcation of unimodal maps. *Qual. Theory Dyn. Syst.* **4**(2), 413–424 (2004).
12. Artur Avila, Mikhail Lyubich and Welington de Melo. Regular or stochastic dynamics in real analytic families of unimodal maps. *Invent. Math.* **154**(3), 451-–550 (2003).
13. Marco Martens and Welington de Melo. Universal models for Lorenz maps. *Ergodic Theory Dynam. Systems* **21**(3), 833–860 (2001).
14. Artur Avila, Welington de Melo and Marco Martens. On the dynamics of the renormalization operator. In: *Global analysis of dynamical systems*, pp. 449—460, Inst. Phys., Bristol (2001).
15. Edson de Faria and Welington de Melo. Rigidity of critical circle mappings. II. *J. Amer. Math. Soc.* **13**(2), 343—370 (2000).
16. Marco Martens and Welington de Melo. The multipliers of periodic points in one-dimensional dynamics. *Nonlinearity* **12**(2), 217-–227 (1999).
17. W. de Melo and A.A. Pinto. Rigidity of C^2 infinitely renormalizable unimodal maps. *Comm. Math. Phys.* **208**(1), 91-–105 (1999).

18. Edson de Faria and Welington de Melo. Rigidity of critical circle mappings. I. *J. Eur. Math. Soc. (JEMS)* **1**(4), 339—392 (1999).
19. W. de Melo. Rigidity and renormalization in one-dimensional dynamical systems. In: *Proceedings of the International Congress of Mathematicians, Vol. II (Berlin, 1998)*, number Extra Vol. II, pp. 765—778 (1998).
20. W. de Melo. On the cyclicity of recurrent flows on surfaces. *Nonlinearity* **10**(2), 311—319 (1997).
21. W. de Melo and B.F. Svaiter. The cost of computing integers. *Proc. Amer. Math. Soc.* **124**(5), 1377—1378 (1996).
22. W. de Melo. Full families of circle endomorphisms. In: *Dynamical systems and chaos, Vol. 1 (Hachioji, 1994)*, pp. 25--27, World Sci. Publ., River Edge, NJ (1995).
23. Welington de Melo and Charles Pugh. The C^1 Brunovský hypothesis. *J. Differential Equations* **113**(2), 300–337 (1994).
24. M. Martens, W. de Melo and S. van Strien. Julia–Fatou–Sullivan theory for real one-dimensional dynamics. *Acta Math.* **168**(3-4), 273–318 (1992).
25. Marco Martens, Sebastian van Strien, Welington de Melo and Pedro Mendes. On Cherry flows. *Ergodic Theory Dynam. Systems* **10**(3), 531–554 (1990).
26. W. de Melo and S. van Strien. A structure theorem in one-dimensional dynamics. *Ann. of Math. (2)*, **129**(3), 519--546 (1989).
27. W. de Melo and S. van Strien. One-dimensional dynamics: the Schwarzian derivative and beyond. *Bull. Amer. Math. Soc. (N.S.)* **18**(2), 159—162 (1988).
28. W. de Melo and F. Dumortier. A type of moduli for saddle connections of planar diffeomorphisms. *J. Differential Equations* **75**(1), 88–102 (1988).
29. W. de Melo and S.J. van Strien. Diffeomorphisms on surfaces with a finite number of moduli. *Ergodic Theory Dynam. Systems* **7**(3), 415–462 (1987).
30. W. de Melo. A finiteness problem for one-dimensional maps. *Proc. Amer. Math. Soc.* **101**(4), 721–727 (1987).
31. W. de Melo, G.L. dos Reis and P. Mendes. Equivariant diffeomorphisms with simple recurrences on two-manifolds. *Trans. Amer. Math. Soc.* **289**(2), 793--807 (1985).
32. W. de Melo, J. Palis, and S.J. van Strien. Characterising diffeomorphisms with modulus of stability one. In: *Dynamical systems and turbulence, Warwick 1980 (Coventry, 1979/1980)*, volume 898 of Lecture Notes in Math., pp. 266–285, Springer, Berlin-New York (1981).
33. W. de Melo and J. Palis. Moduli of stability for diffeomorphisms. In: *Global theory of dynamical systems (Proc. Internat. Conf., Northwestern Univ., Evanston, Ill., 1979)*, volume 819 of Lecture Notes in Math., pp. 318--339, Springer, Berlin (1980).
34. W. de Melo. Moduli of stability of two-dimensional diffeomorphisms. *Topology* **19**(1), 9—21 (1980).
35. A. Lins, W. de Melo and C.C. Pugh. On Liénard's equation. In: *Geometry and topology (Proc. III Latin Amer. School of Math., Inst. Mat. Pura Aplicada CNPq, Rio de Janeiro, 1976)*, pp. 335—357, Lecture Notes in Math., Vol. 597 (1977).

36. C. Gutiérrez and W. de Melo. The connected components of Morse–Smale vector fields on two manifolds. In: *Geometry and topology (Proc. III Latin Amer. School of Math., Inst. Mat. Pura Aplicada CNPq, Rio de Janeiro, 1976)*, pp. 230–251, Lecture Notes in Math., Vol. 597 (1977).
37. W. de Melo. Accessibility of an optimum. In: *Geometry and topology (Proc. III Latin Amer. School of Math., Inst. Mat. Pura Aplicada CNPq, Rio de Janeiro, 1976)*, pp. 429–440, Lecture Notes in Math., Vol. 597 (1977).
38. W. de Melo. Stability and optimization of several functions. *Topology* **15**(1), 1–12 (1976).
39. W. de Melo. On the structure of the Pareto set of generic mappings. *Bol. Soc. Brasil. Mat.* **7**(2), 121—126 (1976).
40. W. Melo. Structural stability on two-manifolds. In: *Dynamical systems (Proc. Sympos., Univ. Bahia, Salvador, 1971)*, pp. 255-–257 (1973).
41. W. de Melo. Structural stability of diffeomorphisms on two-manifolds. *Invent. Math.* **21**, 233–246 (1973).

List of Ph.D. Students of Welington de Melo at IMPA

Maria José Pacífico, 1980
Stability of vector fields on compact 3-dimensional manifolds with boundary.

Antonio Augusto Gaspar Ruas, 1982
Hyperbolic attractors of codimension 1 and isotopy classes on surfaces.

Edson Vargas, 1989
Frequency of bifurcations in families of unimodal maps.

Daniel Smania, 2001
Complex bounds and renormalization theory for multimodal maps.

Artur Ávila, 2001
Bifurcations of unimodal maps: the topological and metric picture.

Alejandro Kocsard, 2007
Toward the classification of cohomology-free vector fields.

Pablo Guarino, 2012
Rigidity conjecture for C^3 critical circle maps.

Credits

American Mathematical Society

1. Edson De Faria and Welington de Melo. Rigidity of critical circle mappings. II. *J. Amer. Math. Soc.* **13**(2), 343—370 (2000).
2. Welington De Melo and Benar F. Svaiter. The cost of computing integers. *Proc. Amer. Math. Soc.* **124**(5), 1377—1378 (1996).

Cambridge University Press

1. Welington De Melo and Sebastian J. van Strien. Diffeomorphisms on surfaces with a finite number of moduli. *Ergodic Theory Dynam. Systems* **7**(3), 415–462 (1987).
2. Marco Martens, Sebastian J. van Strien, Welington de Melo and Pedro Mendes. On Cherry flows. *Ergodic Theory Dynam. Systems* **10**(3), 531—554 (1990).

Duke University Press

1. Pablo Guarino, Marco Martens and Welington de Melo. Rigidity of critical circle maps. *Duke Math. J.* **167**(11), 2125—2188 (2018).

Elsevier

1. Welington De Melo. Moduli of stability of two-dimensional diffeomorphisms. *Topology* **19**(1), 9—21 (1980).

European Mathematical Society

1. Edson De Faria and Welington de Melo. Rigidity of critical circle mappings. I. *J. Eur. Math. Soc. (JEMS)* **1**(4), 339–392 (1999).

Princeton University Press

1. Welington De Melo and Sebastian J. van Strien. A structure theorem in one dimensional dynamics. *Ann. of Math.* (2) **129**(3), 519–546 (1989).
2. Edson De Faria, Welington de Melo and Alberto Pinto. Global hyperbolicity of renormalization for C^r unimodal mappings. *Ann. of Math.* (2) **164**(3), 731--824 (2006).

Springer Nature

1. Welington De Melo. Structural stability of diffeomorphisms on two-manifolds. *Invent. Math.* **21**, 233–246 (1973).
2. Marco Martens, Welington de Melo and Sebastian J. van Strien. Julia–Fatou–Sullivan theory for real one-dimensional dynamics. *Acta Math.* **168**(3-4), 273—318 (1992).
3. Artur Avila, Mikhail Lyubich and Welington de Melo. Regular or stochastic dynamics in real analytic families of unimodal maps. *Invent. Math.* **154**(3), 451—550 (2003).